THE PENGUIN DICTIONARY OF CHEMISTRY

David William Arthur Sharp was born in 1931 and educated at Ashford Grammar School, Harvey Grammar School, Folkestone, and Sidney Sussex College, Cambridge. From 1957 until 1961 he was a lecturer at the Imperial College of Science and Technology, London, and between 1965 and 1967 he was Professor of Chemistry at the University of Strathclyde, Glasgow. Since 1968 he has been Ramsay Professor of Chemistry at the University of Glasgow. He was a contributor to the third, fourth and fifth editions of *Miall's Dictionary* and editor of the fifth edition. He has written many papers for learned journals in his field.

THE PENGUIN DICTIONARY OF

CHEMISTRY

Edited by D. W. A. SHARP, M.A., PH.D.,

C.CHEM., F.R.S.C., F.R.S.E.

PENGUIN BOOKS

Penguin Books Ltd, Harmondsworth, Middlesex, England
Viking Penguin Inc., 40 West 23rd Street, New York, New York 10010, U.S.A.
Penguin Books Australia Ltd, Ringwood, Victoria, Australia
Penguin Books Canada Ltd, 2801 John Street, Markham, Ontario, Canada L3R 1B4
Penguin Books (N.Z.) Ltd, 182–190 Wairau Road, Auckland 10, New Zealand

An abridged edition of *Miall's Dictionary of Chemistry 5th Edition*, edited by Professor D. W. A. Sharp, first published by Longman Group Ltd, 1981

First published in Great Britain by Penguin Books 1983
Reprinted 1984, 1985

Made and printed in Great Britain by
Richard Clay (The Chaucer Press) Ltd, Bungay, Suffolk
Set in Monophoto Times

Preface

In this Dictionary I have aimed to provide an explanation of the terms used in the various branches of chemistry, together with brief accounts of important substances and chemical operations. The Dictionary is intended for use in schools, colleges and universities from the first study of chemistry up to about second year at university or college. We have necessarily had to be selective rather than comprehensive but most chemicals met with at this level should be included.

We have paid particular attention to industrial processes and have tried to give some indication of the use of particular chemicals. In order to give some idea of the relative magnitudes of production we have included data on annual production. In this, because of availability of data, we have not been entirely consistent in including common data or common units. (World, U.S. or European production figures are used as seem most appropriate or up-to-date – and it must be remembered that for most speciality chemicals U.S. production is dominant.)

The list of contributors is on page 6 and I am deeply indebted to them, for it is they who originally prepared *Miall's Dictionary of Chemistry* from which this Dictionary has been compiled. Errors and omissions are my responsibility and I would appreciate receiving notice of them.

Department of Chemistry
University of Glasgow
Glasgow G12 8QQ

D. W. A. Sharp

Contributors to previous editions

E. F. Armstrong
W. H. Bennett
J. O'M. Bockris
T. J. Bradley
C. J. W. Brooks
T. T. Cocking
L. R. Cox
F. G. Crosse
E. G. Cutbush
J. Dalziel
C. E. Dent
P. H. Elworthy
H. J. Emeléus
G. H. Ford
W. Francis
E. Fyleman
F. J. Garrick
H. Gibson
J. Harley-Mason
J. F. Herringshaw
S. Ignatiowicz
R. A. Jeffreys
A. Rees Jones
L. A. Jordan
F. Jordinson
A. King
J. A. Kitchener
G. R. Knox
H. Langwell
J. Maitland-Edwards
S. Mattana
P. Maynard
A. McAuley
L. M. Mcleod
L. M. Miall
S. Miall
R. H. Nuttall
K. G. A. Pankhurst
T. G. Pearson
A. Phillips
A. T. Picton
E. M. Pope
D. J. Poulton
A. R. Powell
J. E. Proctor
H. C. Rampton
C. A. Robb
E. R. Roberts
A. B. Searle
D. W. A. Sharp
K. J. Sinclair
A. G. Smith
K. Stewart
R. G. Stitchbury
R. B. Strathdee
A. G. K. Thomson
E. Tomalin
E. Vernon
E. S. Waight
V. J. Ward
G. Webb
A. J. E. Welch
A. F. Wells
T. W. Wight
L. A. Williams
W. S. Wilson

Abbreviations and nomenclature

Abbreviations and symbols commonly used in this book include:

Å	Ångström unit
at.no.	atomic number
at.wt.	atomic weight
b.p.	boiling point
Btu	British thermal unit
D-, DL-	*see* optical activity, chirality
E	energy
G	Gibbs function (*see* Gibbs free energy, free energy)
g	gram
h	Planck's constant
H	enthalpy
L-	*see* optical activity, chirality
L	Avogadro's number
lbs.	pounds
M	molar
M (in formulae)	any metal
ml	millilitre
mol.wt.	molecular weight
m.p.	melting point
nm	nanometres
p.a.	per annum
pH	*see* hydrogen ion concentration
R	gas constant
R	*see* optical activity, chirality
rH	*see* rH
S	entropy
S	*see* optical activity, chirality
X (in formulae)	any non-metal (usually a halogen)
μ	micron
ν	frequency
(+)- and (−)-	*see* optical activity, chirality

For simple compounds the main reference uses the systematic I.U.P.A.C. nomenclature, but other nomenclature is cross-referenced to the systematic name.

Note the symbol * is used for cross-referencing throughout.

A

AAS Atomic absorption spectroscopy*.

abherents, release agents, parting agents Liquid or solid films which reduce or prevent adhesion between surfaces; solid–solid, solid–paste, solid–liquid. Waxes, metallic soaps, glycerides (particularly stearates), polyvinyl alcohol, polyethene, silicones, and fluorocarbons are all used as abherents in metal, rubber, food, polymer, paper and glass processing.

abietic acid, $C_{20}H_{30}O_2$. A crystalline diterpenoid carboxylic acid m.p. 172–175°C, obtained from pine rosin by acid treatment. The commercial product may be glassy and of low m.p. Used extensively in the manufacture of plastics, paints, varnishes, paper sizes and detergents.

ab-initio calculations Quantum mechanical calculations using the Schrödinger wave equation, which do not rely upon any experimental data for solution.

ablation Erosion and disintegration due to heat. Ablation-resistant materials, particularly Nylon fibres in a phenolic resin, are used to protect space vehicles during re-entry into the atmosphere. Decomposition to gaseous products with a porous refractory residue is desirable.

abrasives Hard materials used to disintegrate other materials. Amongst those most widely used are SiC, Al_2O_3 (often containing Ti), diamond, tungsten carbides, BN, and metal abrasives, e.g. steel wool.

absolute temperature A temperature on the 'absolute' or Kelvin scale usually denoted by T; the zero of this scale is the temperature at which a perfect gas would occupy zero volume if it could be cooled indefinitely without liquefaction or solidification. The absolute zero (0 K) is −273·16°C, and one degree absolute is equivalent to one degree Centigrade;

$$T(K) = \text{temperature } (°C) + 273{\cdot}16.$$

absorptiometer An instrument used to measure the absorption of light by a liquid. Also the name of the apparatus used to determine the solubility of a gas in a liquid.

absorption *See* gas absorption.

absorption bands *See* absorption of light.

absorption coefficient of a gas The volume of gas measured at 0°C and 760 mm pressure which will dissolve in 1 ml of a liquid. The absorption coefficients in water at 0°C for several common gases are as follows: N_2, 0·024; O_2, 0·049; C_2H_4, 0·25; CO_2, 1·71; H_2S, 4·68; SO_2, 79·8; HCl, 506; NH_3, 1300.

absorption coefficient of light *See* Lambert's law and Beer's law.

absorption column, absorption tower The equipment normally used for absorption of gases consists of a column or tower, usually circular in cross-section, the absorbing liquid passing down counter-current* to the gas passing upwards.

When the main object of the absorption is to remove impurities these columns are often referred to as scrubbers*.

absorption of light When light falls on the surface of a transparent substance part of the light is reflected, the remainder is transmitted unchanged. If, however, the light falls upon the surface of a black substance, e.g. lampblack, it is neither reflected nor transmitted; all wavelengths are absorbed in a general absorption process. Many substances appear coloured because they have absorbed selectively all the wavelengths of white light except those corresponding to the particular colour which they appear. If the spectrum of the light transmitted by a coloured substance is examined it is found that certain wavelengths, called *absorption bands*, are missing. For gases in the atomic state the spectrum of the transmitted light shows dark lines rather than bands. These *absorption lines* correspond to the wavelengths of light absorbed by the atoms.

Relationships between the intensity of incident light, sample thickness, concentration and intensity of transmitted light are embodied in Beer's law* and Lambert's law*.

absorption spectroscopy Any spectroscopic technique, qualitative or quantitative, depending upon the measurement of an absorption spectrum.

absorption tower *See* absorption column.

ABS plastics A group of plastic materials based on blended copolymers of styrene-acrylonitrile (70:30) and butadiene-acrylonitrile (65:35) and on graft interpolymers of styrene and acrylonitrile with polybutadiene. Used in pipe (25%), appliances (20%), and automotive parts (15%). U.S. consumption 1979 530 000 tonnes.

Ac Actinium. Also used for acetate (ethanoate).

acac Abbreviation for acetylacetonato group.

accelerators Substances used to accelerate desired cross-linking reactions in polymers. In particular they assist the vulcanization* of rubber, in some cases conferring on it increased resistance to wear. Many types of organic compounds can be used, e.g. diphenylguanidine*, the thiazoles, e.g. mercaptobenzthiazole*, and the thiuram disulphides*. The term is also used for those substances which act as catalysts by increasing the rate at which thermosetting resins cure or harden.

acceptor A substance which, while normally not oxidized by oxygen or reduced by hydrogen, can be oxidized or reduced in presence of a substance which is itself undergoing oxidation or reduction.

Also an atom, molecule, or ion that is electron deficient and which can form a co-ordinate link* with an electron donor. Thus in the complex ion $[Co(NH_3)_6]^{3+}$ the cobalt(III) ion is an acceptor and the ammonia the electron donor. π-acceptors are molecules or atoms which accept electrons into π, p or d orbitals.

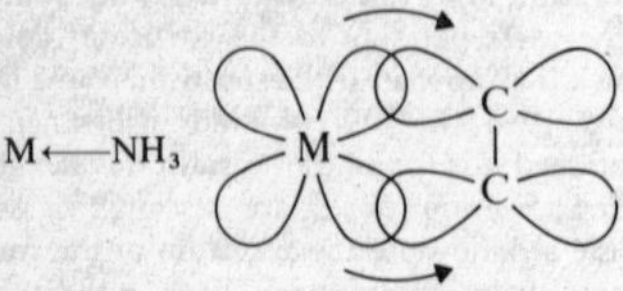

π bond (acceptor)

acenaphthene, $C_{12}H_{10}$. Colourless needles, m.p. 95°C, b.p. 278°C. It is important as a

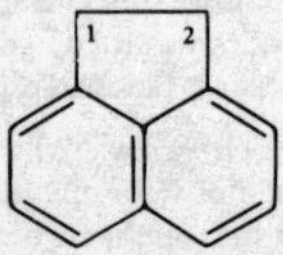

dyestuff intermediate, being used as a source of nitro- and amino-derivatives.

acenaphthenequinone, $C_{12}H_6O_2$. Yellow needles, m.p. 261°C. It is condensed with thioindoxyl and its derivatives to bright scarlet and red vat dyes.

acenaphthylene, $C_{12}H_8$. Yellow plates, m.p. 92–93°C (decomp.). It is derived from acenaphthene by catalytic dehydrogenation (1, 2) and polymerizes to give plastic products.

acetal, $C_6H_{14}O_2$, $CH_3{\cdot}CH(OEt)_2$. A pleasant-smelling liquid, b.p. 104–105°C. Prepared by mixing ethanal and ethanol in the presence of a catalyst, such as HCl, or by passing C_2H_2 into EtOH in the presence of a catalyst. Some acetals are useful solvents.

acetaldehyde *See* ethanal.

acetal resin *See* aldehyde polymers.

acetals Compounds of the general formula:

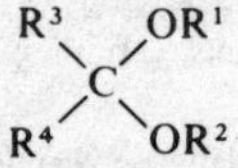

Derived from an aldehyde or ketone and an alcohol using an acid catalyst. Ethylene glycol or 1,3-dihydroxypropane are frequently used to give 5-or 6-member cyclic products.

acetamide *See* ethanamide.

acetanilide, C_8H_9NO, $PhNH{\cdot}CO{\cdot}CH_3$. White crystals, m.p. 114°C. Manufactured by reacting aniline with excess ethanoic acid or ethanoic anhydride. Chief use is in the manufacture of dye intermediates such as *p*-nitroacetanilide, *p*-nitroaniline and *p*-phenylenediamine, in the manufacture of rubber, and as a peroxide stabilizer.

Hydrolysed by dilute acids and alkalis to aniline. It chlorinates more slowly than aniline to *o*-and *p*-chloroacetanilides.

acetate fibres *See* cellulose acetate plastics.

acetates Salts or esters of ethanoic acid*.

acetic acid *See* ethanoic acid.

acetic anhydride *See* ethanoic anhydride.

acetic ester, acetic ether, $CH_3C(O)O{\cdot}C_2H_5$. *See* ethyl ethanoate.

acetins Acetates (ethanoates) of glycerol. There are five possible acetates, two mono-, two di-, and one tri-acetate. The commercial acetins are mixtures of the various acetates and form colourless or slightly brown syrupy liquids. Prepared from glycerol, ethanoic acid and sulphuric acid.

Monoacetin contains chiefly the 1-acetate $CH_2OH{\cdot}CHOH{\cdot}CH_2OOCCH_3$. Used as a solvent for the dyes employed in printing paper bags.

Diacetin is chiefly the 1:3-diacetate $CH_2OOCCH_3{\cdot}CHOH{\cdot}CH_2OOCCH_3$. Used as plasticizer for cellulose acetate lacquers and as a solvent for basic dyes.

Triacetin is about 90% glycerol triacetate and 10% diacetate. Used as a plasticizer for lacquers and as a solvent for certain gums and resins.

acetoacetic acid, acetonemonocarboxylic acid, butan-3-onoic acid, $C_4H_6O_3$, $CH_3CO{\cdot}CH_2COOH$. A colourless and strongly acid syrup. It is unstable, and decomposes into propanone and carbon dioxide below 100°C. Prepared from acetoacetic ester. It occurs in traces in normal urine, but is characteristically present in increased amount in the urine of diabetic patients.

acetoacetic ester *See* ethyl acetoacetate.

acetoin, acetyl methyl carbinol, 3-hydroxy-2-butanone, $C_4H_8O_2$, $CH_3CH(OH)COCH_3$. M.p. 15°C, b.p. 148°C. Produced from propene and butene glycols by the action of ethanoic acid bacteria, and from ethanal by yeast and by the reduction of diacetyl(2,3-butanedione). When distilled, it forms diacetyl.

acetol *See* hydroxypropanone.

acetolysis The process of removing acetyl groups from an organic compound. It is usually carried out by heating the acetyl compound with aqueous or alcoholic alkalis, whereby the acetyl groups are removed as ethanoic acid.

acetone *See* propanone.

acetone alcohol *See* hydroxypropanone.

acetone bodies *See* ketone bodies.

acetone dicarboxylic acid, β-ketoglutaric acid, $C_5H_6O_5$, $CO{\cdot}(CH_2COOH)_2$. Colourless needles, m.p. 135°C (decomp.). Prepared by the action of sulphuric acid on citric acid. Readily decomposed by boiling water, acids, or alkalis to acetone (propanone) and carbon dioxide. The acid or its diethyl ester reacts with sodium in a manner similar to acetoacetic ester. The ester is used in organic syntheses.

acetone monocarboxylic acid *See* acetoacetic acid.

acetonitrile, methyl cyanide, ethanenitrile, CH_3CN. Poisonous liquid, b.p. 82°C. Prepared from ethyne and ammonia or by dehydration of ethanamide. Widely used for dissolving inorganic and organic compounds, especially when a non-aqueous polar solvent of high dielectric constant is required, e.g. for ionic reactions.

acetonyl The group $CH_3COCH_2{-}$.

acetonylacetone *See* hexan-2,5-dione.

acetophenone, C_8H_8O, $PhCOCH_3$. Colourless plates, m.p. 20°C. Odour resembling bitter almonds. Prepared by the action of ethanoyl chloride upon benzene in the presence of aluminium chloride. Has typical ketonic properties. Oxidized by potassium permanganate to phenylglyoxylic acid. Used as a solvent for cellulose ethers.

acetoxy The group $CH_3COO{-}$.

acetylacetonates Metal derivatives of acetyl-

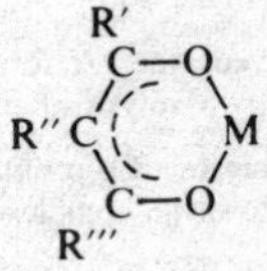

acetone generally containing the grouping shown (Macac) with some delocalization in the ring. The number of acetylacetonate groups bonded to the metal can vary from 1 to 4.

acetylacetone, $C_5H_8O_2$, $CH_3COCH_2COCH_3$. A diketone with enolic properties. The enolate ion acac forms stable complexes with many metals, e.g. tris(acetylacetonato)iron(III), $Feacac_3$. Metal derivatives are generally soluble in organic solvents and often appreciably volatile; they are used in solvent extraction and mass spectrometry. Derivatives of acetylacetone, e.g. $CF_3COCH_2COCH_3$, $CF_3COCH_2COCF_3$ and thienyl-$COCH_2COCF_3$ form particularly stable derivatives.

acetylation, ethanoylation A process for introducing acetyl groups into an organic compound containing $-OH$, $-NH_2$ or $-SH$ groups. It is carried out by heating the compound with ethanoic anhydride or ethanoic chloride usually in presence of an inert solvent such as benzene or ethanoic acid. In many cases, zinc chloride or pyridine is used to hasten the reaction.

acetyl chloride *See* ethanoyl chloride.

acetylcholine, $C_7H_{17}NO_3$, $(CH_3)_3N^+CH_2CH_2OOCCH_3OH^-$. Passage of a nerve impulse from a nerve end to another nerve or muscle cell causes release of acetylcholine to transmit the impulse across the intercellular space. The acetylcholine is quickly destroyed by the hydrolytic action of acetylcholine esterase. Injection of acetylcholine can cause dilatation of the arteries.

acetyl coenzyme A A reactive thioester of fundamental importance in metabolism and biosynthesis. Acetyl coenzyme A is the principal substrate of the citric acid cycle*. It also acts as an acetylating agent, e.g. in the formation of acetylcholine. It has a central role in three major biosynthetic pathways: (1) formation of fatty acids from acetyl coenzyme A and malonyl coenzyme A (itself arising by carboxylation of acetyl coenzyme A), (2) synthesis of the polyketides via acetoacetyl coenzyme A, (3) synthesis of the terpenoids and steroids via mevalonic acid.

acetylene *See* ethyne.

acetylene black, cuprene A form of carbon black prepared by pyrolysis of ethyne.

acetylene complexes Co-ordination compounds similar to olefin complexes*.

acetylene dicarboxylic acid, $HO_2CC{\equiv}CCO_2H$. *See* dimethylacetylenedicarboxylate.

acetylene dichloride *See* dichloroethenes.

acetylene tetrachloride *See sym*-tetrachloroethane.

acetylides Carbides* containing $C_2{}^{2-}$ or C_2R^- species. Formed by more electropositive elements (e.g. K, Ca, Al) and by some transition elements (e.g. Cu, Ag, Au). Hydrolysed to ethyne, C_2H_2. Most transition metal acetylides are explosive.

Metal derivatives of terminal alkynes, RC_2H. Transition metals form complex acetylides (e.g. $[M(C{\equiv}CR)_n]^{x-}$) often containing the metal in low oxidation states.

achiral The molecule of a compound which is not optically active* is achiral.

achromatic indicators Substances which give a grey end-point and generally find application in the titration of turbid liquids.

acid An acid on the aqueous system is defined as a substance which is capable of forming hydrogen ions when dissolved in water. Most inorganic acids may be regarded as a compound of an acidic oxide and water; where the oxide concerned is that of a metal, that oxide may exhibit amphoteric character, that is act sometimes as an acid and sometimes as a base. Aqueous solutions of acids have a sharp taste, turn litmus red, liberate carbon dioxide from a metallic carbonate and give reactions characteristic of the anion present.

Since free protons cannot exist, acidic properties can only be shown when the solvent can act as a proton acceptor, i.e. as a base. Thus aqueous solutions of acids contain the hydroxonium ion, H_3O^+.

Acids can also exist in non-aqueous solvents. Since ammonia can also solvate a proton to give the ammonium ion, $NH_4{}^+$, substances which dissolve in ammonia to give the ammonium ion, e.g. NH_4Cl, are acids in that system.

Liquid water is ionized

$$2H_2O \rightleftharpoons H_3O^+ + OH^-$$

this ionization being the reverse of the neutralization reaction in water; substances giving hydroxyl ions are bases in water. Liquid ammonia is ionized

$$2NH_3 \rightleftharpoons NH_4{}^+ + NH_2{}^-$$

and ionic amides are bases on this system.

The concept of acids and bases has been extended to solvents which are ionized and yet do not contain hydrogen: a substance giving the appropriate positive ion is an acid on that system. Thus bromine trifluoride ionizes

$$2BrF_3 \rightleftharpoons BrF_2{}^+ + BrF_4{}^-$$

and a substance giving the $BrF_2{}^+$ ion in solution, e.g. $BrF_3{\cdot}SbF_5$, is an acid in the system.

Typical organic acids contain the –C(O)OH group, but many other acid groupings, e.g. the sulphonic $-S(O)_2OH$ give acidic properties to organic compounds. Phenols have acidic properties and are classified with enols as pseudo-acids.

The term acid was extended by Lewis to include substances which are electron acceptors. Thus $AlCl_3$ can accept electrons from a chloride ion forming the $[AlCl_4]^-$ ion and is a Lewis acid.

The 'strength' of an acid is measured by the value of its dissociation constant, 'strong' acids, e.g. HCl, HNO_3, being substantially fully ionized in solution and 'weak' acids predominately unionized.

acid-base indicator A substance, a weak acid or weak base, which has a different colour in acid or base solution. The colour change is due to a marked difference in colour between the undissociated and ionic forms. For a good indicator the colour change must occur between narrow limits of pH, e.g. methyl orange is red at pH 3·1 and changes to yellow at pH 4·4.

acid dyes Dyestuffs containing an aromatic chromophoric group and a group conferring solubility in water, generally the SO_3H group as its sodium salt. They are relatively simple in application. The types of acid dyestuffs are:

Simple acid dyes contain no polyvalent metals and are not improved by after treatment with $Na_2Cr_2O_7$/dil. HCl.

Mordant acid dyes combine simultaneously with the mordanting agent (generally $Cr(OH)_3$) and the fibre; the dyestuff generally contains ortho OH–azo or OH–OH groups.

Premetallized acid dyes are similar to the simple acid dyes but are already complexed to a metal.

acid egg A non-mechanical pump for the handling of highly corrosive liquids. Liquid is admitted to a corrosion-resistant vessel and forced into the delivery line by compressed air.

acid exchange resins *See* ion exchange.

acid oil The alkali extract of phenol derivatives formed in the production of gasoline by cracking operations. Also referred to as cresylic acids* or 'phenols'.

acid sludge A complex acid residue produced when kerosines, lubricating oils or other petroleum products are refined with sulphuric acid or oleum. It consists of hydrocarbons, sulphonic acids and free sulphuric acid. Petroleum sulphonates can be recovered by extraction with alkali and used for the metal processing, textile and leather industries.

aconitic acid, $C_6H_6O_6$. Unsaturated acid prepared by dehydrating citric acid with 50% sulphuric acid.

aconitine Intensely poisonous alkaloid obtained from *Aconitum napellus*, monksbane, m.p. 197°C.

acridine, $C_{13}H_9N$. Colourless needles, m.p. 111°C.

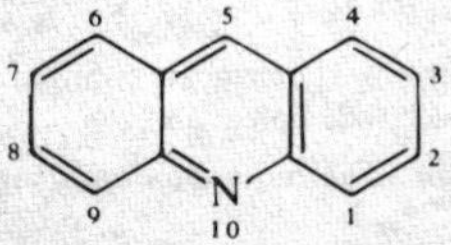

acriflavine A mixture of 2,8-diamino-10-methylacridinium chloride hydrochloride, and 2,8-diaminoacridine dihydrochloride. It is used for the same purposes as proflavine.

Acrilan A brand name for a synthetic fibre, based on a copolymer of acrylonitrile with minor proportions of other unspecified vinyl monomers. *See also* propenenitrile.

acrolein *See* propenal.

acrolein polymers, propenal polymers Polymers of $CH_2{=}CHCHO$ generally formed by free radical polymerization. Polymerization is generally through the vinyl group and the aldehyde groups are present as acetals. Polymerization with strong base causes polymerization through the carbonyl group. Acrolein polymers can be modified chemically and used as thickening agents and protective colloids. Also used in plastics and lacquers. Disacryl is an insoluble acrolein polymer.

acrylamide polymers *See* polyacrylamide.

acrylate resins and plastics Generally used to refer to polymeric methyl acrylate (methyl propenoate) and polymeric methyl methylacrylate (methacrylate) (methyl 2-methyl propenoate). Polymeric methyl acrylate is used, principally in an emulsion form, in textile and leather finishes, lacquers, paints, adhesives and safety glass interlayers. Polymeric methyl methylacrylate gives a clear solid material (Perspex) and is used in injection moulding and extrusion. U.S. production 1978 330 000 tonnes.

acrylic acid *See* propenoic acid.

acrylic acid polymers
Acrylic acid, $CH_2{=}CHCO_2H$, and methacrylic acid, $CH_2{=}CMeCO_2H$ undergo free radical polymerization to give polymers used as thickeners, in textile treatment, as drilling-mud additives, as flocculating agents, in paper making, and if co-polymerized with e.g. divinylbenzene as ion-exchange resins.

acrylonitrile *See* propenenitrile.

acrylonitrile polymers *See* polyacrylonitrile.

actinides The elements actinium, thorium, protactinium, uranium, neptunium, plutonium, americium, curium, berkelium, californium, einsteinium, fermium, mendelevium, nobelium, and lawrencium are collectively known as actinides. Those with atomic numbers 93 and above are artificial and are produced by irradiation of uranium or other artificial elements with neutrons, alpha particles, or carbon or nitrogen ions. In the actinide elements the 5f shell is being filled and they are thus analogous to the lanthanides or rare-earth elements. There is less shielding of the 5f electrons than of the 4f and oxidation states greater than three are common, particularly among lighter members of the series. Amongst the heavier elements the +2 state becomes stable.

All of the actinides are radioactive and present some health hazards; with the more radioactive elements all handling has to be by remote control and no contact with the operator is permissible.

actinium, Ac. At.no. 89, m.p. 1050°C, b.p. 3200 ± 300°C. Occurs naturally as a minor constituent in uranium ores but difficult to separate and best prepared by bombardment of radium with neutrons. Separated from other species by ion exchange or solvent extraction. The most stable isotope ^{227}Ac is very radioactive ($t_{\frac{1}{2}}$ 22 years), the metal glows blue and is oxidized in air. It has been prepared by reduction of AcF_3 with lithium vapour at 1200°C and is silvery white in colour.

actinium compounds Because of its intense radioactivity very few actinium compounds have been isolated. In its compounds the element is tripositive, forming very similar compounds to lanthanum*. Actinium salts are colourless.

activated adsorption Chemisorption often has an activation energy associated with it and is sometimes referred to as activated adsorption.

activated carbon *See* active carbon.

activated clay *See* bleaching earths.

activated molecule Molecules will only react on collision when they possess more than a certain minimum amount of energy. A molecule which has acquired more energy than the average amount possessed by other molecules, and is therefore in a more reactive condition, is said to be activated. Molecules are activated when they absorb a quantum of light, or after coming in contact with a hot surface.

activation analysis An analytical technique in which an artificial radioactive isotope is formed by irradiation (generally with neutrons) from the stable element to be determined and the amount of the artificial isotope is then estimated from its radioactivity.

activation energy The minimum energy which reacting species must possess in order to be able to form an 'activated complex' or transition state before proceeding to the products. The activation energy (E_a) may be derived from the temperature dependence of the reaction rate using the Arrhenius equation*.

active carbon Carbon (charcoal) treated at high temperature with steam, air or CO_2. Used as an adsorbent for removal of small traces of impurities from a gas or liquid. Used extensively in water and waste water treatment, air pollution control, as a catalyst, sugar refining, purification of chemicals and gases (gas masks), dry cleaning, rubber reclamation, cigarette filters. U.S. production 1980 60 000 tonnes. *See* carbon black.

active centres In heterogeneous catalysis adsorption is considered to occur at active centres on the surface of the catalyst. Active centres are the specific sites at which adsorption occurs. The concept of active centres has been extended to enzymes and bacterial action.

active earths *See* bleaching earths.

active mass The thermodynamic activity*. An old terminology.

active transport The biochemical transport of substances, usually against a concentration gradient, other than by osmosis or diffusion.

activity A thermodynamic quantity which measures the effective concentration or intensity of a particular substance in a given chemical system. The *absolute activity*, a°, of a substance is given by $\mu = RT \ln a^\circ$, where μ is the chemical potential* of the substance, R is the gas constant, T the absolute temperature. The *relative activity*, a, is given by $\mu = \mu^\circ + RT \ln a$; μ° being the chemical potential of the substance in its standard state*. For dilute, ideal solutions the activity is directly proportional to the concentration; for ideal gases, activity is proportional to the partial pressure of the gas.

activity coefficient (f or γ) A dimensionless factor by which the concentration (c) of a substance must be multiplied to give an exact measure of its thermodynamic activity (a) in a chemical system, i.e. $a = fc$. It is a measure of the deviation from ideal behaviour of the solution; f is unity for an ideal mixture and greater or less than unity for a non-ideal system. Activity coefficients of *electrolytes* are considered to be the geometrical mean of the single ion activities, the latter being hypothetical quantities which cannot be determined separately.

activity series The elements arranged in order of their electrode potentials*.

actomyosin The most important protein of muscle.

acyclic *See* aliphatic. Containing chains, possibly with branches but no rings.

acyl The general name for organic acid groups, which are the residues of carboxylic acids after removal of the –OH group, e.g. ethanoyl chloride, $CH_3CO{\cdot}Cl$, is the acyl chloride formed from ethanoic acid, $CH_3CO{\cdot}OH$. The names of the individual acyl groups are formed by replacing the *-ic* of the acid by *-yl*.

acylation A chemical transformation which substitutes the acyl (RCO–) group into a molecule, generally for an active hydrogen of e.g. an –OH group.

acyloin condensation The formation of an acyloin, very often cyclic by condensing two molecules of ester with sodium.

acyloins 1,2-Ketoalcohols of the type R·CO·CHR′·OH. *See* acetoin, benzoin.

adamantane, $C_{10}H_{16}$. Colourless hydro-

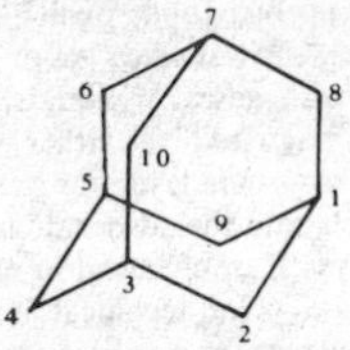

carbon, m.p. 269°C, but subliming readily at room temperature and atmospheric pressure. Occurs (up to 0·0004%) in some petroleum fractions, together with alkylated adamantanes. Has a rigid ring system composed of three fused chain cyclohexane rings, having the same configuration as the diamond lattice. Synthesized from tetrahydrodicyclopentadiene. 1-Adamantanamine hydrochloride has been found useful in viral infections, where it acts by preventing penetration of the host cell by the viral particle; it also finds application in treating Parkinson's disease. Adamantyl derivatives are used as lubricants and resins.

Adams' catalyst, platinum oxide, $PtO_2 \cdot H_2O$. Produced by fusion of H_2PtCl_6 with sodium nitrate at 500–550°C and leaching of the cooled melt with water. Stable in air, activated by hydrogen. Used as a hydrogenation catalyst for converting alkenes to alkanes at low pressure and temperature. Often used on SiO_2

adatom An adsorbed atom.

addition reactions Reactions in which an unsaturated system (C=C, C≡C, C=O, etc.) is saturated or part saturated by addition of a molecule across the multiple bond. Examples include the reaction of bromine with ethene to give 1,2-dibromoethane, hydrogen cyanide addition to an aldehyde giving a cyanhydrin, the Diels-Alder reaction, and addition polymerization. Also used in inorganic chemistry, e.g. the reaction of BF_3 with nucleophiles, e.g. ammonia, ether, to form complexes where the co-ordination number or the oxidation state of the metal atom is increased.

additives Compounds, usually added in small amounts, which will confer specific properties on the bulk material, e.g. anti-foaming additives for lubricating oils.

additive volumes, law of The volume occupied by a mixture of gases is equal to the sum of the volumes which would be occupied by the constituents under the same conditions of temperature and pressure.

adduct A phase (often a compound) formed by direct combination, generally in simple proportions, of two or more different compounds or elements.

adenine, 6-aminopurine, $C_5H_5N_5$. A constituent of the nucleic acid portion of nucleoproteins, and, combined, as adenosine pyrophosphate, it plays an important part in many metabolic processes.

adenosine *See* nucleosides.

adenosine diphosphate (ADP) Adenosine 5′-diphosphate (pyrophosphate). A precursor of ATP and is also formed from it during processes in which ATP is involved.

adenosine monophosphate (AMP) Normally adenosine 5′-phosphate (muscle adenylic acid), an important structural component of nucleic acids* and of several coenzymes. *See also* cyclic AMP.

adenosine triphosphate (ATP) The most

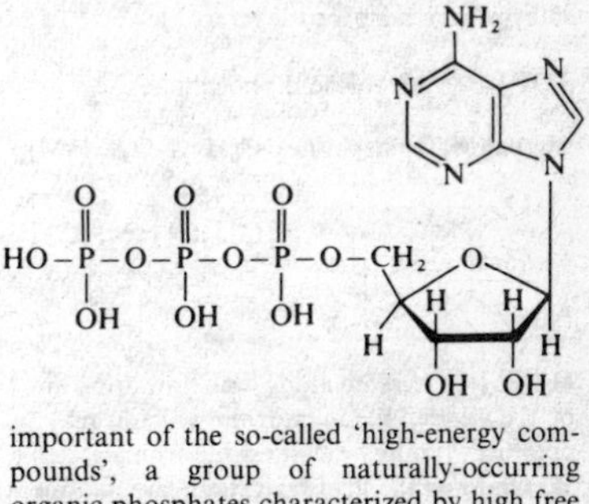

important of the so-called 'high-energy compounds', a group of naturally-occurring organic phosphates characterized by high free energies of hydrolysis, and playing a fundamental role in biosynthesis and active transport and muscle action. ATP is the primary source of energy in the metabolism of plant, animal and bacterial cells.

adenylic acid *See* adenosine monophosphate.

adhesion agents Any additive which will improve the adhesivity of a material. Surface-active agents used to prevent stripping of an adhesive from a surface by water. They are widely used in road bitumens, particularly cutback bitumens*, used in surface dressings and in coated macadam surfacings.

Soaps of heavy metals have been used but cationic surface-active agents have proved more suitable, notably organic amines of relatively high molecular weight.

adhesive A material which will wet two surfaces which are to be joined and which will subsequently solidify to form a join.

adhesives Materials used to bond solids together. Examples include glues (starches, proteins), thermoplastic resins, thermosetting resins, rubber, asphalt, sodium silicates. Used in construction, packaging, textiles, furniture, and electrical goods.

adiabatic change A change in which heat is neither allowed to escape from, nor is added to, the system during the change.

adipic acid, $C_6H_{10}O_4$, $HOOC{\cdot}[CH_2]_4{\cdot}COOH$. M.p. 153°C. Present in beet-juice; manufactured by the air or nitric acid oxidation of cyclohexanol, the latter being obtained by the air oxidation of cyclohexane. Distillation of its calcium salt gives cyclopentanone. It forms long chain polymers with diamines and is very largely used in the manufacture of Nylon. It is also used for the manufacture of plasticizers and certain vinyl and urethane plastics. U.S. production 1978 770 000 tonnes.

adlayer An adsorbed layer.

ADP *See* adenosine diphosphate.

adrenaline, epinephrine, $C_9H_{13}NO_3$. M.p.

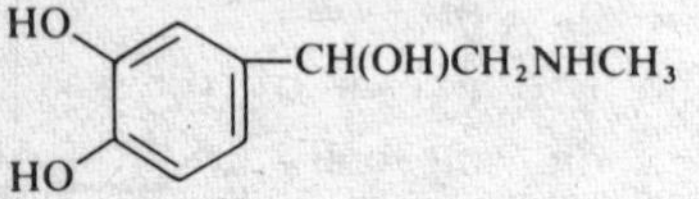

HO
HO
CH(OH)CH₂NHCH₃

212°C. It is secreted along with noradrenaline* by the adrenal medulla, from which it may be obtained. It may be synthesized from catechol. It is used as the acid tartrate in the treatment of allergic reactions and circulatory collapse. It is included in some local anaesthetic injections in order to constrict blood vessels locally and slow the disappearance of anaesthetic from the site of injection. Ultimately it induces cellular activation of phosphorylase which promotes catabolism of glycogen to glucose.

adsorbate The substance which is adsorbed onto the surface of the adsorbent.

adsorbent The substance upon whose surface the process of adsorption occurs.

adsorption Because of the abnormal condition of the atoms at the surface of a solid or a liquid compared with atoms in the bulk, the surface of a solid or liquid is a seat of free energy. The process of adsorption in which foreign atoms or molecules become attached to the surface lowers the free energy of the surface. Thus, e.g., adsorption of detergent molecules at liquid surfaces lowers the surface tension of the liquid.

Adsorption may in principle occur at all surfaces; its magnitude is particularly noticeable when porous solids, which have a high surface area, such as silica gel or charcoal are contacted with gases or liquids. Adsorption processes may involve either simple unimolecular adsorbate layers or multilayers; the forces which bind the adsorbate to the surface may be physical or chemical in nature.

Adsorption is of technical importance in processes such as the purification of materials, drying of gases, control of factory effluents, production of high vacua, etc. Adsorption phenomena are the basis of heterogeneous catalysis and colloidal and emulsification behaviour.

adsorption, chemical *See* chemisorption.

adsorption indicator An indicator which functions by adsorption on the surface of a precipitate. Thus in precipitating AgCl in the presence of excess Cl^- the surface has a layer of Cl^- and is negatively charged; in the presence of the indicator fluorescein the indicator is in solution. As soon as there is excess of Ag^+ the precipitate takes on a positive charge and the presence of the fluorescein anion as counter ion gives the precipitate a pinkish-red colour.

adsorption, industrial Adsorption is of considerable industrial importance, being employed, for example, in the refining of lubricating oils, the decolourizing of sugar solutions and the purifying and drying of gases. Industrial adsorption may employ *batch, fixed bed,* or *continuous countercurrent* operations.

adsorption, physical Many adsorption processes are not chemically specific and are readily reversible. The forces of attraction between the adsorbate and adsorbent are weak and similar in nature to those responsible for the cohesion of molecules in the liquid state, van der Waals' forces*. Physical adsorption often involves the formation of multiple layers at the adsorbent surface.

aerosol, aerogel A dispersion in which a finely divided solid is suspended in air and the particles are of colloidal dimensions, e.g.

smoke. Aerosols can often be formed by the rapid condensation of a vapour such as that of a metal oxide.

Aerosol sprays consist of a material dissolved or suspended in a liquid which when pressure is released volatilizes to produce a fine spray. The spray carries the active material. Used in hair lacquers, paints, etc.; the propellant should be inert and non-inflammable. Chlorofluorocarbons have been used extensively but are now being replaced.

Aerosol is also a trade name for certain wetting agents, most of which are esters of sodium sulphosuccinic acid.

aetiophyllin *See* chlorophyll.

affinity, chemical The term 'affinity of a chemical reaction' is used to denote the change of free energy in the system when the reaction takes place; this is a true measure of the tendency to react.

affinity chromatography A method of purifying natural macromolecules, especially proteins. Depends on covalently attaching a specific ligand to an insoluble inert support. The ligand must have a special and unique affinity for the macromolecule, so that on passage in solution down a column of the material the macromolecule is preferentially retarded and thus separated from contaminating molecules.

aflatoxins The toxic metabolites of the fun-

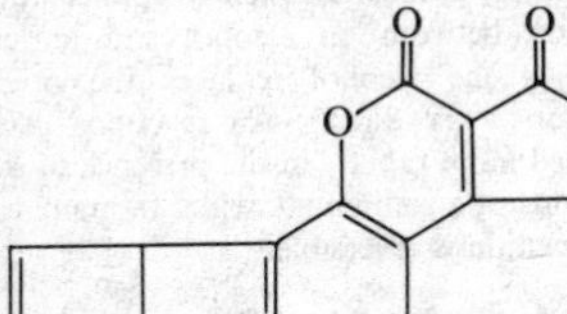

gus *Aspergillus flavus* Link *ex* Fries. The fungus is found in grain crops and peanuts and the aflatoxins are extremely toxic and carcinogenic. Four aflatoxin isomers [B_1, B_2, G_1 and G_2] of the structure shown are known.

Ag Silver.

agar, agar-agar A seaweed colloid*. A mixture of two polysaccharides, agarose and agaropectin. Agarose, the main constituent, is a well-defined neutral polysaccharide containing 3,6-anhydro-L-galactose and D-galactose as the repetitive unit. Agaropectin contains the carboxyl and sulphate groups of native agar.

Agar occurs as a cell-wall constituent of the red marine algae *Rhodophyceae*, from which it is extracted by hot water, and marketed as a dry powder, flakes, or strips. It dissolves in hot water and sets on cooling to a jelly at a concentration as low as 0·5%. Its chief uses are as a solid medium for cultivating micro-organisms, as a thickener, emulsion stabilizer in the food industry and as a laxative.

agate A form of SiO_2 similar to chalcedony with coloured bands of purple or brown. Used as a gemstone, for making pestles and mortars, and as a bearing surface in scientific instruments.

age hardening *See* precipitation hardening.

aggregate Pieces of material which are, or can be, united by a cementitious substance or bond, to form a solid mass. In concrete the aggregate consists of crushed stone, slag or clinker and sand, united by Portland or other cement. In bricks and other ceramic materials, particles of non-plastic materials form the aggregate; they are united by plastic clay when in the green or unfired state and later (in the burned state) by a crude glass which fuses and, on cooling, solidifies and forms the bond. The sizes, shapes and mechanical properties of the pieces forming an aggregate have a great influence on the properties of the bonded product.

aglycon(e) *See* glycosides.

air The volume composition of normal dry air is N_2 (78·08%), O_2 (20·94%), Ar (0·934%), CO_2 (0·03%), Ne (0·0018%), Kr (0·0001%), Xe (0·000009%), Rn (6×10^{-18}%). The proportion of CO_2 can vary slightly and the content of water vapour varies widely according to atmospheric conditions. The physical constants of dry air are *d* 1·2928 gl^{-1} at 273 K and 760 mm; specific heat capacity C_p 0·2396, C_v 0·1707 (ratio 1·403); gas constant R $2{\cdot}1529 \times 10^3$ (pressure mm; $cm^3 g^{-1}$). Air can be liquefied by cooling (under pressure) and by allowing rapid expan-

agar, agar-agar

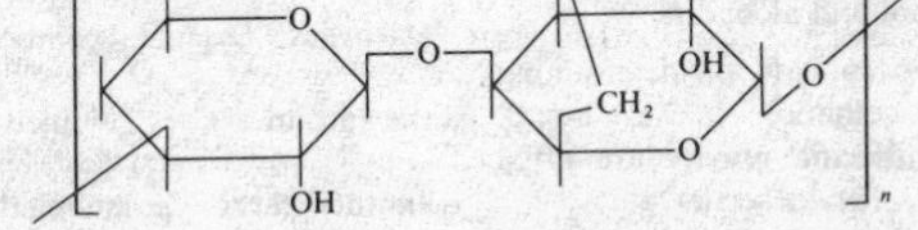

sion of gas; the liquid is pale blue because of the presence of O_2. The b.p. varies from $-192°C$ (air) to $-185°C$ (after N_2 evaporation), *d* approx. 0·9. Liquid air (and oxygen) must not be used in the preparation of low temperature baths with organic liquids or for cooling organic materials since extremely explosive mixtures are formed.

air-cooled heat exchangers The usual arrangement comprises a bank of externally-finned tubes, the fluid to be cooled passing through the tubes with air being forced or induced over the tubes by a fan.

air filters Filters designed to remove atmospheric dust from an air stream (or process dust from air or gas). In general they consist of a space packed with a filtering medium such as glass fibre or slag wool through which air is passed.

air–fuel ratio The normal way of expressing the composition of a mixture of fuel and combustion air, either on a volume or weight basis. The theoretical air–fuel ratio contains just sufficient oxygen to produce complete oxidation of the fuel components.

air hardening A method of hardening steel by allowing it to cool naturally in air, or an air blast (air quenching).

air-lift agitator A plant for the leaching of fine solids.

air-lift pump A device for pumping corrosive liquids in which a **U**-shaped tube is used in conjunction with injected air.

air quenching *See* air hardening.

Al Aluminium.

α-alanine, 2-aminopropanoic acid, $C_3H_7NO_2$, $CH_3{\cdot}CH(NH_2)COOH$. M.p. 297°C. One of the amino-acids obtained by the hydrolysis of proteins.

β-alanine, 3-aminopropanoic acid, $C_3H_7NO_2$, $H_2N{\cdot}CH_2{\cdot}CH_2{\cdot}COOH$. M.p. 200°C. Prepared by heating propenoic acid under pressure with ammonia. Part of the molecule of pantothenic acid and is a step in the manufacture of the vitamin both in the laboratory and the living organism. Not present in proteins.

albumins *See* proteins.

alcohol *See* ethanol and alcohols.

alcoholometry The process of determining the amount of ethanol in a liquid. Measurements of specific gravity are often used.

alcohols Organic compounds containing one or more hydroxyl groups attached directly to carbon atoms. Aromatic derivatives having hydroxyl groups joined directly to the carbon atoms of the ring are called phenols and have distinctive properties. Compounds with one –OH group are called monohydric alcohols; those with more than one are di-, tri-, or polyhydric alcohols. Aliphatic dihydric alcohols are known as glycols; the most important trihydric alcohol is glycerol; tetra-, penta-, and hexa-hydric alcohols are generally derived from sugars. Monohydric alcohols are given systematic names as derivatives of methanol, which, for this purpose, is called carbinol. It is usual to employ names ending in -ol derived from the names of the corresponding paraffin hydrocarbons, e.g. ethanol for ethyl alcohol, isopropanol for isopropyl alcohol. The monohydric alcohols form a series whose lower members are odorous liquids and higher members white, odourless solids. They react with metals such as sodium, calcium, and aluminium to give alkoxides*, and with acids to give esters. Sulphuric acid and other agents remove water to form alkenes. Primary alcohols on oxidation give aldehydes; secondary alcohols give ketones while tertiary alcohols break down to give compounds containing fewer carbon atoms than the original alcohol.

alcoholysis A reaction in which an alcohol plays a part similar to that of water in hydrolysis. It is most frequently applied to the reaction between an alcohol and an ester whereby the alcohol replaces the original alcohol of the ester. Such reactions usually proceed more rapidly in the presence of small amounts of a sodium alkoxide. In many cases the reaction is reversible.

aldehyde polymers Polymers with a $-\!\!(CHRO)\!\!-$ backbone formed by polymerization of the aldehydes RCHO. The commercially important aldehyde polymers are polyformaldehyde (polymethanal, Delrin) ($R=H$) formed by polymerization of CH_2O in polar solvents; polyacetaldehyde (polyethanal) ($R=CH_3$); polychloral ($R=CCl_3$). These materials are light but hard and rigid and fairly chemically resistant. They are used in machine parts, car parts, etc. Aldehydes telomerize to trimers, etc. (*see* under individual headings).

aldehydes Organic compounds containing the group $-C{\genfrac{}{}{0pt}{}{\nearrow O}{\searrow H}}$ joined directly on to another carbon atom: methanal, HCHO, is included in the class, but some of its properties are not those of a typical aldehyde. Aldehydes are usually colourless liquids (aliphatic) or solids (higher aromatic) with characteristic odours, and are oxidized to acids and reduced to primary alcohols. A number of preparations are available: oxidation of a primary alcohol or treatment of a Grignard reagent with ethyl orthoformate; the selective reduction of carboxylic acid derivatives (e.g. amides, nitriles, acid chlorides) by complex metal hydrides. Aromatic aldehydes may be prepared by heating aromatic hydrocarbons with carbon monoxide, hydrogen chloride, anhydrous aluminium chloride and copper(I) chloride (Gattermann–Koch synthesis). Alkaline solutions of phenols react with chloroform (Reimer-Tiemann reaction) to give phenolic aldehydes.

Most aliphatic and aromatic aldehydes give adducts with alkali hydrogen sulphites and form cyanhydrins, aldoximes*, semicarbazones and phenylhydrazones. Aldehydes, unlike ketones, restore the pink colour to Schiff's reagent, and reduce ammoniacal silver nitrate and Fehling's solution. They also form acetals with alcohols. Aliphatic aldehydes form aldehyde ammonias, and some undergo the aldol condensation. Aromatic aldehydes form unsaturated acids when heated with the sodium salt of a fatty acid and its anhydride (Perkin reaction). They form Schiff's bases with primary amines and react with phenols and dimethylaniline to give triphenylmethane derivatives. All aromatic aldehydes and some aliphatic aldehydes undergo the Cannizzaro reaction*.

aldicarb, $C_7H_{14}N_2O_2S$, $CH_3SC(CH_3)CH{=}NOC(O)NHMe$. A systemic carbamate-type insecticide and nematocide widely used on sugar beet and potatoes.

aldol, acetaldol, 3-hydroxybutanal, β-hydroxybutyraldehyde, $C_4H_8O_2$, $CH_3CHOHCH_2CHO$. A colourless oily liquid, b.p. 83°C/20 mm. Manufactured by treating ethanal with alkali carbonates, lime, borates and other alkaline condensing agents. Decomposes on heating to crotonaldehyde Reduction with hydrogen under pressure gives 1,3-butylene glycol.

aldol condensation A reaction between two molecules of an aliphatic aldehyde to give a β-hydroxyaldehyde. The simplest case is that of ethanal which gives aldol, $CH_3CHOHCH_2CHO$. The reaction is carried out at a low temperature in presence of a catalyst, usually potassium cyanide, sodium ethanoate or dilute sodium hydroxide. Only those aldehydes containing the $-CH_2{\cdot}CHO$ grouping can react in this way: other aldehydes undergo the Cannizzaro reaction. The β-hydroxyaldehydes readily lose water to give unsaturated aldehydes. Aliphatic ketones react in a similar manner with aliphatic ketones or aldehydes and with some aromatic aldehydes to give β-hydroxyketones, which also readily lose water, giving unsaturated ketones.

aldonic acid An acid derived from an aldose by oxidation of the aldehyde group to a carboxyl group, e.g. gluconic acid.

aldose A sugar containing a potential aldehyde (CHO) group. The group may be obscured by its inclusion in a ring system. Aldoses are called aldopentoses, aldohexoses, etc., according to the number of carbon atoms they contain.

aldosterone, $C_{21}H_{28}O_5$. The most active known steroid hormone secreted by the adrenal cortex.

aldoximes Compounds formed by the action of hydroxylamine on aldehydes containing the group $-CH{=}NOH$. They are also formed by the oxidation of primary amines by permonosulphuric acid. The aliphatic aldoximes are colourless liquids or low-melting solids, some of which are soluble in water. Aromatic aldoximes are crystalline solids, e.g. benzaldoximes. All aldoximes are capable of existing in two stereoisomeric forms known as the *syn*- and the *anti*-modifications, but in practice only the *anti*- form is isolated from

anti-aldoxime *syn*-aldoxime

most aliphatic aldehydes. Aromatic aldehydes yield *syn*-aldoximes which may frequently be isomerized by HCl to the *anti*-aldoximes. The *syn*-aldoximes yield acetyl derivatives with ethanoyl chloride but the *anti*- forms are dehydrated to the corresponding nitriles RCN. Dilute mineral acids may regenerate the aldehyde and hydroxylamine hydrochloride.

aldrin The insecticidal product containing

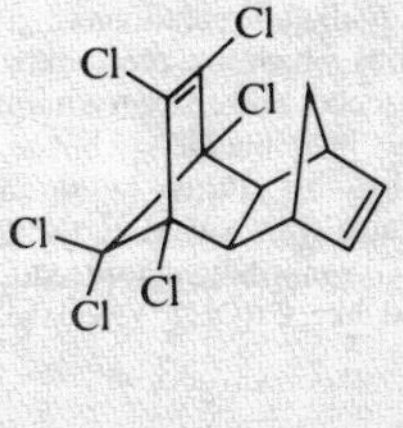

not less than 95% by weight of 1,2,3,4,10,10-hexachloro-1,4,4a,5,8,8a,hexahydro-*exo*-1,4-*endo*-5,8-dimethanonaphthalene.

Aldrin is obtained from the Diels-Alder addition product of cyclopentadiene and vinyl chloride by dehydrochlorination followed by condensation with hexachlorocyclopentadiene.

Aldrin is insecticidally active as a contact and stomach poison against a wide range of soil pests. It is non-phytotoxic and does not cause taint. Aldrin is toxic to humans and animals and is now less used.

algin A seaweed colloid* principally alginic acid which is D-mannuronic acid. Used mainly as the Na salt giving a very viscous solution with water. Used as a stabilizer for food products (ice cream, etc.), pharmaceuticals and as a dressing in the textile industry.

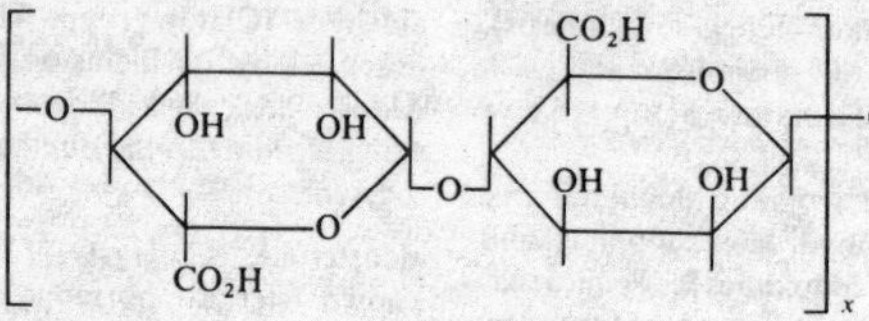

algin (*x* 200–1000)

alginic acid *See* algin.

alicyclic Cyclic carbon compounds which do not possess a benzene ring with its system of conjugated double bonds. They are cyclic compounds with aliphatic characteristics. Examples cyclopropane, cyclohexane.

aliphatic A compound having its carbon atoms in chains and not in closed rings. Originally used to describe the fats and fatty acids, which are typical of this structure.

alizarin, $C_{14}H_8O_4$. Orange-red dyestuff.

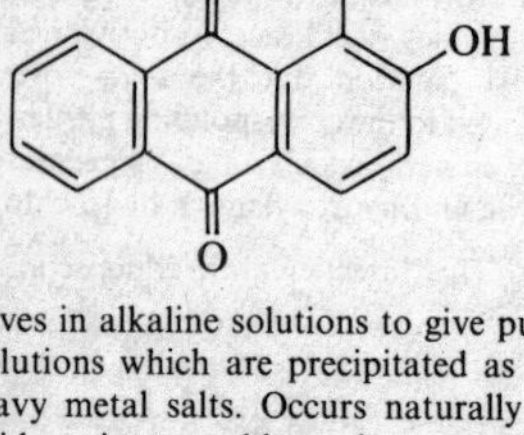

Dissolves in alkaline solutions to give purple-red solutions which are precipitated as lakes by heavy metal salts. Occurs naturally as a glucoside in madder but produced synthetically by fusing anthraquinone-2-sulphonic acid with NaOH and some $KClO_3$. Applied to the mordanted fibre. $Al(OH)_3$ gives a bright red lake, $Cr(OH)_3$ a red lake, $Fe(OH)_3$ a violet lake. Dyeing is carried out in boiling aqueous solution.

alkali A hydroxide of one of the alkali metals*. Also used for such substances as CaO, $Ca(OH)_2$, Na_2CO_3, which give an alkaline solution (pH > 7) in water. In aqueous solution the term alkali is virtually synonymous with the term base.

alkali metals The metals Li, Na, K, Rb, Cs and Fr. All are very electropositive with an s^1 outer electron shell and form M^+ ions.

alkaline An aqueous solution with a pH greater than 7.

alkaline earth metals The elements Mg, Ca, Sr, Ba and Ra which are very electropositive and form M^{2+} ions.

Historically an 'earth' was a non-metallic substance, nearly insoluble in water and unchanged on heating. The alkaline earth oxides, e.g. CaO, have an alkaline reaction in addition to being clearly 'earths'.

alkaloids Organic substances existing in combination with organic acids in great variety in many plants, and to which many drugs owe their medicinal properties. Many are very poisonous, and even in minute doses produce characteristic physiological effects. They vary considerably in their chemical properties and constitutions. All alkaloids are basic and combine with acids to form crystalline salts, which are usually water-soluble. The majority of alkaloids are only very sparingly soluble in water, but are usually readily soluble in organic solvents such as alcohol, chloroform and ether.

In the commercial extraction of alkaloids from the drugs in which they exist, the powdered drug, or an alcoholic extract of it, is treated with an alkali such as ammonia or lime to liberate the alkaloid and the alkaloid is then extracted by means of an organic solvent. The crude material thus obtained is purified and finally crystallized either as the base itself or as its water-soluble salts.

Alkaloids occur chiefly in flowering plants, especially in the *Ranunculaceae*, *Papaveraceae* and *Solanaceae*, but are widespread throughout the botanical kingdom. They are formed from amino-acids combining with methanoate or methionine, or ethanoate or mevalonate, or various higher transformation products of these including shikimic acid and terpenes. Many alkaloids are derivatives of pyridine, quinoline, isoquinoline, or pyrimidine. They are optically active and most dextrorotatory. They include a number of important drugs, e.g. morphine, caffeine, nicotine, atropine, cocaine, hyoscyamine, quinine, strychnine and pilocarpine. Their part in the physiology of plants is not known.

A minor chemical use for many of the commoner alkaloids is the resolution of racemic compounds (often acids) into their optically active enantiomorphs.

alkanals Aldehydes. Compounds containing terminal R–CHO groups.

alkanes, paraffins Aliphatic hydrocarbons of the general formula C_nH_{2n+2}. The first four members are gases, the higher members are liquids and those above $C_{16}H_{34}$ are waxy solids. They are insoluble in water but soluble in chloroform and benzene. They form the chief constituents of petroleum. They are remarkably resistant to chemical action and only chlorine and bromine will react with any ease. These give chloro- and bromo-substituted paraffins respectively. They are formed by the reduction of alkenes or by treating alkyl iodides with sodium in ethereal solution.

alkanolamines, alkylolamines Hydroxyamines produced by the reaction of an alkene oxide with aqueous ammonia at 50–60°C. They are hygroscopic solids of low m.p., and usually occur as rather viscous, colourless liquids of ammoniacal odour. They are used as accelerators in rubber manufacture, as catalysts in certain polymerization reactions, and as absorbents for acidic gases such as carbon dioxide and hydrogen sulphide. They form soaps with fatty acids; these are almost neutral in reaction and are used as detergents and emulsifying agents. *See* ethanolamines, isopropanolamines.

alkanolamine soaps *See* alkanolamines.

alkanols Alcohols. Compounds ROH where R is an alkyl group.

alkenes, olefins Aliphatic hydrocarbons of the general formula C_nH_{2n} containing one double bond; they are isomeric with the cycloparaffins. In physical properties they closely resemble the paraffins (alkanes); the lower members are gases, the intermediate liquids, while the higher members are waxy solids. They are insoluble in water but soluble in chloroform and benzene. They burn with a smoky, luminous flame. They occur in crude petroleum and in the gases from the cracking of petroleum. Alkenes are more reactive than alkanes, adding groups across the double bond. Reduced to alkanes by hydrogen in presence of catalysts. React with halogens to give dihalides and with halogen acids to give alkyl halides. Dilute chlorine or bromine solutions in water react to give chloro- or bromohydrins. They dissolve in concentrated sulphuric acid to give alkylsulphonic acids and dialkyl sulphates, and are oxidized by potassium permanganate or lead tetra-acetate to glycols. They polymerize with various catalysts or under pressure. May be prepared by heating aliphatic alcohols with sulphuric acid or by passing the alcohol vapour over heated alumina. They are used as fuels and are the starting materials for the preparation of alcohols, glycols and other substances.

alkoxides, alcoholates Metal derivatives of alcohol (e.g. $KOCH_3$, $Al(OC(CH_3)_3)_3$). Alkali metal derivatives can be prepared directly but many derivatives (e.g. transition metals) must be prepared indirectly (e.g. $Ti(OR)_4$ from $TiCl_4$, ROH and a tertiary amine). Although many alkoxides are polymeric many are volatile. Used as reagents (*see* aluminium) and as catalysts. Commercially $Al(OPr^i)_3$ is used for reductions, as a paint additive; NaOMe is used as a catalyst and reagent; $Ti(OPr^i)_4$ and other alkyl titanates are used in polymer technology and paints.

alkyd resins Products of the condensation of polyhydric alcohols and polybasic (or sometimes monobasic) acids, e.g. phthalic acid. Used widely in coatings, particularly in paints.

alkyl The residue left when a hydrogen atom is removed from an aliphatic hydrocarbon, e.g. $CH_3\cdot$, $C_2H_5\cdot$, etc. The free radicals are known and are intermediates in many reactions, e.g. the thermal degradation of alkanes, as in petrochemical processes.

alkylamides Metal and non-metal derivatives containing ENR_2 groups (e.g. $W(NMe_2)_6$). Prepared from the halide plus amine, silylamide, or lithium alkylamide, $LiNR_2$. Often volatile and less associated than alkoxides. Molecular species with low coordination numbers are known.

alkylation Formally the introduction of alkyl groups into hydrocarbon chains or aromatic rings. The production of branched-chain hydrocarbons by reacting alkenes with iso-alkanes or aromatics, e.g. the production of iso-octane as a high anti-knock gasoline by reacting iso-butane and 1-butene. While alkylation can be effected thermally, milder reaction conditions are possible if sulphuric acid or hydrofluoric acid catalysts are used.

alkylenes *See* alkenes.

alkylolamines *See* alkanolamines.

alkylphenols Made by alkylation of phenol or its homologues. Used to form thermosetting resins (with methanal) and, particularly for long chain derivatives, polyethers used as surfactants and detergents, with ethylene oxide. Hindered di-*t*-butyl derivatives are valuable anti-oxidants.

alkynes, acetylenes Compounds containing $C\equiv C$ groups. Thus methylacetylene, $CH_3C\equiv CH$ is propyne; dimethylacetylene, $MeC\equiv CMe$ is but-2-yne, etc.

allantoin, glyoxyldiureide, $C_4H_6N_4O_3$. M.p.

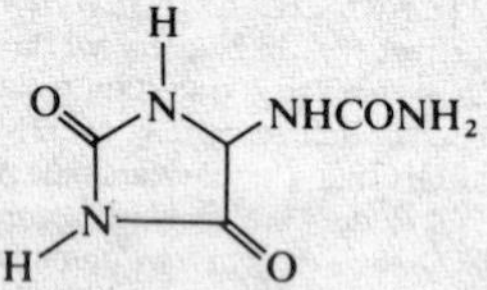

235–236°C. Allantoin is the end product of purine metabolism, and is excreted in the urine of most mammals, exceptions being man and the anthropoid apes.

allene *See* 1,2-propadiene.

allenes 1,2-Dienes having the $>C{=}C{=}C<$ skeleton and general formula C_nH_{2n-2}. They are derivatives of 1,2-propadiene (allene) and are isomeric with alkynes. They undergo typical reactions of alkenes with, e.g. hydrogen, hydrogen halides, halogens, etc. They are not as stable as alkenes which have conjugated, or isolated, double bonds. The action of base is usually sufficient to cause isomerization to an alkyne. Allenes are usually colourless liquids with garlic-like odours but the higher members are solids. They are prepared by removal of bromine and hydrogen bromide from 1,2,3-tribromopropane derivatives, usually by KOH followed by zinc and ethanol.

allethrin, $C_{19}H_{26}O_3$. A chrysanthemum carboxylate derivative. Technical allethrin is a viscous liquid. It is formulated in the same way as natural pyrethrin, and has similar toxicity to many insects, notably the house-fly. Piperonyl butoxide has been used as a synergist. Used mainly as the synthetic material.

allo- The prefix allo- before the name of an organic compound indicates that two compounds have the same empirical formula. With sterols and related substances the prefix allo- should indicate that rings A and B are in the *trans* position to each other, but allocholesterol is an exception.

allobarbitone, diallylbarbituric acid, $C_{10}H_{12}N_2O_3$. White powder, m.p. 172–

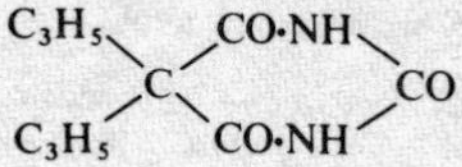

174°C. Has hypnotic properties similar to barbitone but its action is less prolonged. For manufacture, *see* barbiturates.

allomone A compound or mixture acting between different species in the environment which adaptively favours the emitting species, e.g. a defensive secretion, a floral scent attracting pollinating insects. *See also* kairomone and pheromone.

allophanic acid, carbamylcarbamic acid, $C_2H_4N_2O_3$, $NH_2CONHCOOH$. Unknown in the free state as it breaks down immediately to urea and CO_2. The NH_4, Ba, Ca, K and Na salts are known and are prepared by treating ethyl allophanate with the appropriate hydroxide. The esters with alcohols and phenols are crystalline solids, sparingly soluble in water and alcohol. They are formed by passing cyanic acid into alcohols or a solution of an alcohol or phenol in benzene. The amide of allophanic acid is biuret. Alcohols are sometimes isolated and identified by means of their allophanates.

allose *See* hexose.

allosteric effects The control of the functions of some enzymes by their interaction with small molecules at sites distant from the active sites. The binding of these other molecules (which may be products or substrates of the reactions) alters the conformation of the enzyme causing it to be either more or less effective. Allosteric effects enable the output from biosynthetic pathways to be tightly geared to requirements.

allotrope *See* allotropy.

allotropy The existence of more than one

physical form of the element. The difference may be in bonding, e.g. graphite and diamond, ozone and dioxygen, or in crystal form, e.g. most metallic elements. The various forms are termed allotropes (*see* polymorphism).

alloxantin, $C_8H_6N_4O_8$. Forms dihydrate,

$$OC \diagdown \begin{array}{c} NH-CO \\ NH-CO \end{array} \diagup C-C \diagup \begin{array}{c} CO-NH \\ CO-NH \end{array} \diagdown CO$$
$$\qquad\qquad \overset{|}{OH}\ \overset{|}{OH}$$

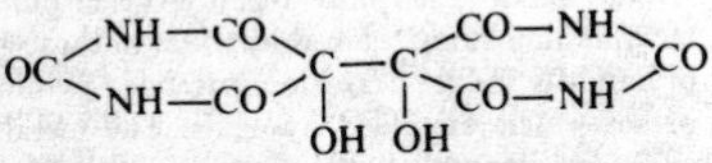

sparingly soluble in water. Found among the products of hydrolysis of the glycoside convincin from soya beans and in small amounts in beets. Prepared by oxidizing uric acid or reducing alloxan. Used in synthesis of riboflavin.

alloy An intimate association or compound of two or more metals or of a metal and one or more non-metallic elements, the alloy possessing in a marked degree all or most of those characteristics commonly described as metallic. Pure metals are too soft and possess insufficient strength for most engineering purposes. Alloying is one of the common methods of increasing useful properties in metals, e.g. strength and hardness.

alloy elements Elements added to give special properties to alloys, such as resistance to corrosion, heat, creep, etc.

allyl The trivial name given to the propenyl group $CH_2:CHCH_2-$.

allyl alcohol *See* propenol.

allyl derivatives of metals Two series are known. The h^1 derivatives, e.g. $CH_2=CH_2CH_2Mn(CO)_5$, and the h^3, e.g. $\langle\!\!\!\!\!\!\!-Mn(CO)_4$. The former are prepared using e.g., Grignard reagents; decarbonylation or frequently direct reaction gives the more important h^3 derivatives (also h^4-butadiene complex plus H^+). Allyl complexes are important catalysts and important as intermediates in many reactions involving metals. Allyl derivatives are the simplest members of a series of complexes of higher homologues.

allylene *See* propyne.

allylic rearrangement The migration of a double bond in a three carbon system from carbon atoms one and two to carbon atoms two and three. The substituent simultaneously migrates from carbon atom three to carbon atom one.

$$\overset{1}{C}=\overset{2}{C}-\overset{3}{C}-X \rightleftharpoons X-\overset{1}{C}-\overset{2}{C}=\overset{3}{C}$$

The simplest example involves the hydrolysis of crotyl chloride (2-butenyl chloride) to give crotyl alcohol (2-buten-1-ol) and α-methylallyl alcohol (1-buten-3-ol). An intermediate allyl (crotyl, etc.) carbonium ion is involved, which can react with a nucleophile at either end of the delocalized system.

$$CH_3 \cdot CH:CH \cdot CH_2Cl$$

$$\downarrow$$

$$CH_3 \cdot CH \cdot CH \cdot CH_2$$
$$\text{------ +}$$

$$CH_3 \cdot CH \cdot CH:CH_2 \qquad CH_3 \cdot CH:CH \cdot CH_2OH$$
$$\overset{|}{OH}$$

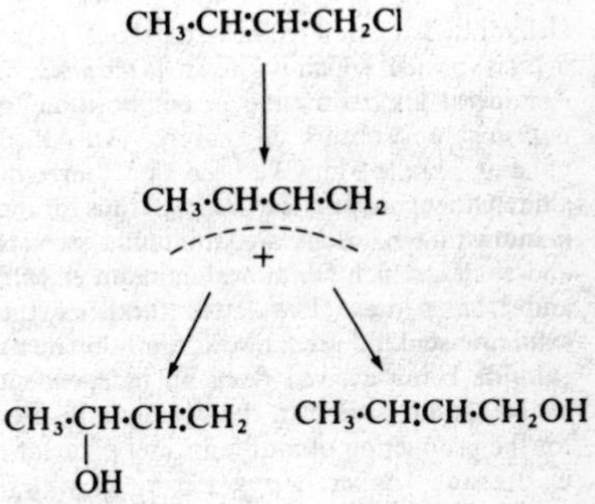

allyl isothiocyanate *See* propenyl isothiocyanate.

allyl polymers Polymers derived from monomers containing the $CH_2=CHCH_2-$, allyl, propenyl group. An important group of thermosetting polymers, polymerization is initiated thermally, photochemically, by radiation, or by catalysis. Triallyl cyanurate and isocyanurate are copolymerized with polyesters for use in glasscloth-reinforced laminates. Allyl ethers of carbohydrates and other polyols together with styrene and other vinyl monomers are used in automative parts and adhesives and as furniture finishes. Allyl esters, e.g. $RCOOCH_2CH=CH_2$, particularly diallyl phthalate and isophthalate readily polymerize to cross-linked thermoset materials with good abrasion and chemical resistance. Allylphosphates form polymers with good flame-resistant properties.

allylthiourea *See* propenylthiourea.

Alnico alloys Alloys containing Al, Ni, Co and Fe. Used for permanent magnets.

alpha particle, α-ray A He^{2+} ion, i.e. a helium nucleus with no outer electrons. α-Particles are emitted with high velocity ($c.\ 2 \times 10^9$ cm sec^{-1}) by Ra and other radioactive substances. For the detection and estimation of alpha particles, *see* counters.

α-Particles may be used, either with or without acceleration, as bombarding agents in nuclear disintegration reactions.

The energy, and thus the range, of α-particles, is characteristic of the source of emission.

altrose *See* hexose.

alum, potash alum, potassium alum, $KAl(SO_4)_2,12H_2O$. A substance which crystallizes in large colourless octahedra. Crystals become white on exposure to the air owing to the production of a basic salt. Dehydrated at a dull red heat, a porous friable mass is formed, which is known as *burnt alum*. A mineral approximating in composition to potash alum occurs in nature. An allied mineral, alunite* may be used as a source of alum. Other common starting materials for the manufacture of alum are aluminium schists and shales, which contain aluminium silicate and iron pyrites. The latter furnishes the sulphuric acid required, potassium sulphate or chloride being derived from an independent source. Alum is used in the dyeing industries for the production of mordants and pigments, in dressing leather, sizing paper, in waterproofing fabrics, in fireproofing, in gelatin hardening and medicinally as a styptic and astringent, and as a coagulant for water purification. *See* alums.

alumina Aluminium oxide*. *See also* aluminates.

alumina gel *See* aluminium hydroxide.

aluminates A group of compounds formally containing the Al^{3+} ion in anionic hydroxy- or oxo-complexes. Species containing $[Al(OH)_4]^-$ and $[(HO)_3Al\text{-}\mu\text{-}O\text{-}Al(OH)_3]^{2-}$ and higher polymers (cationic hydroxy species are also known – *see* aluminium hydroxide) are formed from Al^{3+}aq solutions at high pH. Solid salts are formed by crystallization or precipitation.

Aluminates formed by fusing together a metal oxide and Al_2O_3 are mixed metal oxides* and contain Al^{3+} in tetrahedral or octahedral co-ordination. E.g. spinel, $MgAl_2O_4$. $NaAlO_2$ (Al_2O_3 plus Na oxalate, prepared commercially by dissolving bauxite in NaOH solution) is used as a mordant, in brick manufacture, for paper sizing and in water softening. Barium aluminate is an efficient but costly water-softening agent. Calcium aluminates are constituents of Portland cements. So-called β-alumina is an aluminate of ideal composition $Na_2O{\cdot}11Al_2O_3$. It is an ionic conductor. There are variable amounts of Na^+ which can be replaced by other cations.

aluminium, Al. At.no. 13, at.wt. 26·98154, m.p. 660·37°C, b.p. 2467°C, *d* 2·702. Occurs naturally in many silicates (*see* aluminosilicates) including clays, and as the hydrated oxide (*see* bauxite). Manufactured by the electrolysis of Al_2O_3 dissolved in Na_3AlF_6 (cryolite) with added Li_2CO_3 using carbon electrodes; the product is 99% pure but contains some Fe and Si. Very pure Al is obtained by electrolysis of compounds such as $NaAlF(C_2H_5)_3$. Al reduces many oxides (Goldschmidt*, thermite* reactions). The pure metal is almost unattacked by water but traces of impurity induce corrosion: corroded by salt or sea-water; dissolved by acids. The metal, or more generally its alloys are used in applications requiring light weight and strength, e.g. in engineering, aircraft, kitchen utensils, overhead cables. Al film is used as a mirror coating and as a wrapping material. Aluminium salts are important in water purification (*see* sulphate), as catalysts (*see* alkyls and oxide); the oxide is important in refractories, abrasives, cement, ceramics, antiperspirants and gemstones (ruby – also lasers). World production Al 1976 17 megatonnes.

aluminium acetate *See* aluminium ethanoate.

aluminium alkoxides, $Al(OR_3)$. Formed by reaction between Al and the alcohol in the presence of $HgCl_2$ as catalyst or from $AlCl_3$ and RONa. The alkoxides are polymeric $[Al(OBu^t)_3]_2$, $[Al(OPr^i)_3]_4$, fairly low melting solids readily hydrolysed by water. Used as specific reducing agents for C=O groups, for formation of esters from aldehydes by the Cannizzaro reaction ($Al(OPr^i)_3$), oxidation of secondary alcohols to ketones in the presence of excess Me_2CO (Oppenauer oxidation – $Al(OBu^t)_3$), and with co-catalysts, e.g. $TiCl_4$, as olefine polymerization catalysts. The most important aluminium alkoxides are $Al(OBu^t)_3$ and $Al(OPr^i)_3$.

aluminium alkyls *See* aluminium organic derivatives.

aluminium alloys Important materials for their combination of light weight, strength and generally good corrosion resistance.

aluminium bromide, $AlBr_3$. Very similar to $AlCl_3$, formed from Al plus Br_2 or Br_2 over Al_2O_3 plus carbon. Forms hydrates.

aluminium bronze *See* bronze.

aluminium *t*-butoxide, aluminium *t*-butylate, $Al(OC(CH_3)_3)_3$, $Al(OBu^t)_3$. *See* aluminium alkoxides.

aluminium chemistry Al is an element of group III, electronic configuration $1s^22s^22p^63s^23p^1$. The chemistry at normal temperature is that of Al(III)

($E°$ $Al^{3+} \longrightarrow$ Al in acid solution − 1·66 volts) although gaseous Al(I) compounds (e.g. $AlCl_3$ plus Al gives AlCl, at high temperatures) are known. The normal co-ordination arrangements for Al(III) are octahedral ($AlCl_3$) and tetrahedral $[AlCl_4]^-$. Aluminium organic derivatives* are common and of industrial importance. Aluminium oxide* and aluminosilicates* (in which Al may be 6-co-ordinate or 4-co-ordinate replacing Si in $[SiO_4]^{4-}$ tetrahedra) are of great importance.

aluminium chloride
Aluminium trichloride, $AlCl_3$, is the stable chloride at room temperature. Colourless or yellowish solid *d* 2·44, sublimes 180°C. Soluble in water and many organic solvents. Prepared from Al and Cl_2 or HCl or Cl_2 over Al_2O_3 plus carbon. Forms complexes $AlCl_3{\cdot}L$ with e.g., H_2O, H_2S and amines, and $[AlCl_4]^-$ and $[Al_2Cl_7]^-$ complexes with chloride ions. Solid $AlCl_3$ contains octahedrally co-ordinated Al, dimeric, $[Cl_2Al(\mu\text{-}Cl)_2AlCl_2]$ in the gas phase. Forms volatile, chloride-bridged speci[illegible] with many halides. Used as a source of aluminium salts, and (also in the laboratory) in, e.g. Friedel-Crafts syntheses.

Aluminium monochloride, AlCl, is formed at very high temperatures from $AlCl_3$ and Al. The reaction is reversed on cooling.

aluminium ethanoate, aluminium acetate Prepared from $Al(OH)_3$ and ethanoic acid or $Pb(O_2CCH_3)_2$ or $Ba(O_2CCH_3)_2$ and $Al_2(SO_4)_3$. Normally occurs in the solid state as a partially hydrolysed material. The ethanoate is soluble in water to give extensively hydrolysed solutions. Used in dyeing as a mordant often with $Al_2(SO_4)_3$ ('red liquor') or $Al(NO_3)_3$. Also used in sizing paper, hardening cardboard, proofing paper and fabrics, and in pharmacy as an antiseptic and astringent.

aluminium ethoxide, aluminium ethylate, $(CH_3CH_2O)_3Al$. M.p. 139°C, b.p. 320°C. *See* aluminium alkoxides.

aluminium fluoride, AlF_3. A colourless substance, *d* 3·07, sublimation pt. 1257°C, prepared by the action of HF on $Al(OH)_3$ or by heating $(NH_4)_3AlF_6$ at 600°C. An inert material, relatively insoluble in most solvents unattacked by acid and alkali. Hydrates (3 or 9 H_2O) formed from aqueous solution. The $AlF_6{}^{3-}$ ion is formed in conc. HF solution and salts, e.g. Na_3AlF_6 (cryolite*), are known which contain 6-co-ordinate Al. Condensed octahedrally co-ordinated AlF_6 units are found in Tl_2AlF_5 and NH_4AlF_4. AlF_3 also contains octahedrally co-ordinated Al.

aluminium hydride, AlH_3. A white polymeric substance (H_2SO_4 plus $LiAlH_4$ in THF). A useful reducing agent (e.g. RCN gives RCH_2NH_2); reacts with alcohols to give alkoxy hydrides, $(RO)_nAlH_{3-n}$; forms complex hydrides including the important tetrahedral tetrahydroaluminate ion, $[AlH_4]^-$, $LiAlH_4$ (LiH plus $AlCl_3$ in ether) is a very important reducing agent (e.g. RCOOH gives RCH_2OH) and starting point for the formation of Al derivatives, e.g. RNH_2 gives $Li[Al(NHR)_4]$. Alkoxy derivatives, e.g. $Li[AlH(OR)_3]$ and $Na[AlH(OMe)_2(OEt)]$ are good and often specific reducing agents.

aluminium hydroxide A white or yellowish gelatinous mass precipitated from solutions of aluminium salts by ammonia and also existing as mineral forms. Formally hydrates of aluminium oxide (alumina). $Al(OH)_3$ exists in two forms, α, bayerite, and γ, gibbsite, hydrargillite; AlO(OH) also exists in two forms, α, diaspore, and γ, boehmite. On heating, diaspore gives $\alpha\text{-}Al_2O_3$ but other hydroxides and most Al salts initially give $\gamma\text{-}Al_2O_3$ and only give $\alpha\text{-}Al_2O_3$ on strong heating. Gels of aluminium hydroxide partially dried and converted to $\gamma\text{-}Al_2O_3$ valuable drying agents, catalysts and absorbents (alumina gel) used medicinally as an antacid. Sols of aluminium hydroxide are formed by dialysis of aqueous solutions of aluminium salts. Aluminium hydroxide is amphoteric, dissolving in acids and in alkalis.

aluminium iodide, AlI_3. Very similar to $AlCl_3$ formed from Al and I_2.

aluminium isopropoxide, aluminium isopropylate, $Al(OPr^i)_3$ M.p. 125°C. *See* aluminium alkoxides.

aluminium nitrate, $Al(NO_3)_3$, nH_2O (*n* commonly 9). *See* aluminium oxy-acid salts.

aluminium organic derivatives, aluminium alkyls Formed Al plus R_2Hg, $AlCl_3$ plus RMgX, or on a commercial scale =Al–H plus C_nH_{2n} (alkene) to give $=AlC_nH_{2n+1}$. The latter reaction can be carried out by the direct reaction Al plus H_2 plus AlR_3 to give AlR_2H which can react with alkenes. All aluminium organic derivatives are easily oxidized; the lower alkyls are spontaneously inflammable. The lower alkyls are polymeric liquids with bridging alkyl groups with multicentre bonds. Complexes, AlR_3L, are readily formed. Aluminium alkyls are used for introduction of alkyl groups and industrially for polymerization of alkenes

(Ziegler process) and telomerization of alkenes to medium chain derivatives for detergents and fats. Both processes operate by insertion of an alkene into AlR bonds.

aluminium oxide, alumina, Al_2O_3. The oxide of aluminium formed by heating aluminium hydroxides or most aluminium salts of oxy acids. Occurs naturally as the α-form, corundum, obtained by heating α-AlO(OH), m.p. 2045°C, b.p. 2980°C, almost insoluble in water. After strong ignition it becomes very inert. For structure *see* corundum. Moderate heat on other aluminium hydroxides produces γ-Al_2O_3 with a defect spinel structure (Al in both tetrahedral and octahedral co-ordination); γ-Al_2O_3 is a very important catalyst, because of its adsorptive power (activated alumina), and is used in the manufacture of artificial gemstones which are crystalline and contain traces of other metals (ruby and sapphire). α-Al_2O_3 is used as an abrasive (emery*) and as an inert material. β-alumina is an aluminate*. For hydrated aluminium oxide *see* aluminium hydroxide. For Al_2O *see* aluminium chemistry.

aluminium oxy-acid salts Aluminium oxy-acid salts are soluble in water (except the phosphate) and often form hydrolysed solutions. They are generally heavily hydrated. The most important salts are the ethanoate* (acetate), the nitrate (used as a mordant), and the sulphate* (used in paper sizing and water treatment. *See* alum).

aluminium silicates Many aluminosilicates* are known. Al_2SiO_5 (cyanite, sillimanite and andalusite*) contains discrete SiO_4^{4-} tetrahedra.

aluminium sulphate, $Al_2(SO_4)_3$, nH_2O ($n = 0$, 6, 10, 16, 18, 27). *See* aluminium oxy-acid salts. Occurs naturally as alunite*. *See* alums. U.S. production 1978 1 080 000 tonnes.

aluminoferric Impure $Al_2(SO_4)_3$ containing some Fe^{II} and Fe^{III}. Used in water and sewage purification.

aluminon, ammonium aurinetricarboxylate, $C_{22}H_{23}N_3O_9$. An organic reagent used for the detection and estimation of aluminium. It is a brownish-red powder, soluble in water which gives a red lake with aluminium which can be estimated colorimetrically. It can also be used for detecting scandium and indium.

aluminosilicates Silicates containing aluminium. The most important aluminosilicates have Al in tetrahedral positions replacing Si. An M^+ cation (or equivalent, e.g. $\frac{1}{2}(M^{2+})$ must be present to balance the charge for every Si replaced by Al. Silicates with framework structures based on SiO_2 must contain AlO_4 groups. Important aluminosilicates of this type are feldspars, micas, zeolites and some clay minerals. Al can also be present in octahedral co-ordination in silicates when the Al functions as a normal cation.

aluminous cement A hydraulic cement closely resembling Portland cement, but composed chiefly of calcium aluminate.

aluminum Aluminium*. Used in U.S. literature.

alums A group of crystalline double sulphates of general formula $M^IM^{III}(SO_4)_2{,}12H_2O$. The sulphate group may be replaced by SeO_4^{2-}, BeF_4^{2-}, or $ZnCl_4^{2-}$. The crystals are made up of $[M^I(H_2O)_6]^+$ and $[M^{III}(H_2O)_6]^{3+}$ cations and SO_4^{2-} anions (three different structures are known) and in solution the alums behave chemically as mixtures of the two constituent sulphates. $KAl(SO_4)_2{,}12H_2O$ is normally known as potash alum and the sodium and ammonium derivatives of aluminium sulphates are also commercial materials. These latter derivatives are prepared in the same manner as potash alum.

alumstone *See* alunite.

alundum An artificial form of corundum (α-Al_2O_3), made by fusing calcined bauxite in an electric furnace, and allowing the molten product to cool rather rapidly.

Alundum is used for highly refractory bricks (m.p. 2000–2100°C), crucibles, refractory cement and muffles; also for small laboratory apparatus used at high temperatures (combustion tubes, pyrometer tubes, etc.).

alunite, alumstone Mineral $KAl_3(SO_4)_2(OH)_6$; the potassium may be partially replaced by sodium. Used commercially as a source of potash alum and of potassium sulphate.

Am Americium.

amalgam Mercury–metal compounds which may be liquid or solid. *See* mercury amalgams.

amalgamation The process of forming an amalgam*.

amanitins A group of very toxic cyclic peptides found in the mushroom *Amanita phalloides*; they are more poisonous than the phalloidins* but slower in action – accounting for the delayed fatal effects of ingesting the fungus.

amaranth An important red azo dye used in foodstuffs.

amatol A mixture of NH_4NO_3 and TNT used as an explosive.

ambident anions Anions which are capable of reaction at two or more different sites, e.g. NO_2^- (from $AgNO_2$) can react with RX to give nitrites RONO and nitroalkanes, RNO_2.

ambident ligands Ligands which can use more than one co-ordinating site. E.g. NH_2CH_2COOH can co-ordinate through N or O.

amblygonite, $LiAl(F,OH)PO_4$. A lithium ore.

americium, Am. At.no. 95, m.p. 994°C, estimated b.p. 2600°C. The most stable isotopes are ^{243}Am (7650 years) and ^{241}Am (433 years) formed by multiple neutron irradiation of ^{239}Pu. Purified by ion exchange. Metallic Am is obtained by reduction of AmF_3 with barium at about 1200°C. The metal is electropositive, silvery in appearance and dissolves in acid. Hazardous to health because of its radioactivity. ^{243}Am is proposed as a neutron target for use in producing higher actinides.

americium compounds Americium is a typical actinide showing similar chemistry to uranium. Aqueous Am(III) is oxidized only with difficulty, Am^{4+} disproportionates to Am^{3+} and AmO_2^+. Americium(II) halides (Cl, Br, I) are prepared from the metal and HgX_2 at 300–400°C, all of the trihalides are known and tan AmF_4 is formed from fluorine and AmO_2. AmO_2 is the stable oxide in air, Am_2O_3 is also known and mixed oxides, e.g. $Li_6Am^{VI}O_6$ and $Li_3Am^{V}O_4$ are made by solid state reactions using oxides, peroxides and oxygen. Oxidation to aqueous AmO_2^{2+} is with peroxydisulphate.

amethocaine, $C_{15}H_{24}N_2O_2$. The hydrochloride is a white crystalline powder, m.p. 147–150°C. It is prepared from *β*-dimethylaminoethanol and *p*-*N*-butylaminobenzoic acid, and is used as a local anaesthetic.

ametryne, $C_9H_{17}N_5S$. A translocated and soil-acting herbicide. A triazine derivative. Forms colourless crystals, m.p. 84–86°C. It is used as a pre- and post-emergence selective herbicide for the control of broad-leaved and grass weeds in a variety of crops. It has a relatively low mammalian toxicity.

amides Organic compounds derived from ammonia by substitution of one or more of its hydrogen atoms by organic acid groups. $RCONH_2$ (primary); $(RCO)_2NH$ (secondary); $(RCO)_3N$ (tertiary). The amides are crystalline solids soluble in alcohol and ether; some are also soluble in water. Primary amides are prepared by the action of ammonia or amines on acid chlorides, anhydrides, or esters. Some amides are prepared by distillation of the ammonium salt of the appropriate acid. Secondary and tertiary amides are formed by treating nitriles or primary amides with organic acids or their anhydrides. Primary amides react with nitrous acid to give carboxylic acids: in many cases, heating with mineral acids or alkalis has the same effect. Primary amides are weakly basic. Compounds are formed with metals such as sodium, potassium, and mercury. HClO and HBrO react to give N-chlor- or bromamides; these give amines when treated with alkali. Alkylated amides, particularly dimethylacetamide and dimethylformamide are good solvents for conducting replacement reactions involving ionic reagents (e.g. replacement of a chlorine atom by fluorine using KF).

Inorganic amides contain the ion NH_2^-. They are formed by the action of ammonia on metals or by the ammonolysis of nitrides. The heavy metal amides are prepared by metathetical reactions in liquid ammonia, e.g.

$$Cd(SCN)_2 + 2KNH_2 \rightarrow Cd(NH_2)_2 + 2KSCN.$$

The alkali amides are stable, crystalline salts; the heavy metal amides are often explosive. The amides are the bases of the ammonia system.

amidines Strong monoacid bases forming

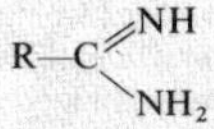

salts with strong acids. They are imido derivatives of amides to which they are converted by acid hydrolysis.

Amidines are best made in two stages: a nitrile reacts with dry HCl and anhydrous alcohols to give an imidic ester (imino-ether) which yields an amidine with NH_3.

amido *See* amino.

amidol *See* aminophenols.

amidone *See* methadone.

amine oxides, R_3NO. Prepared by alkylation of hydroxylamines or oxidation of tertiary amines with ozone or hydrogen peroxide. Long-chain aliphatic amine oxides are used in detergents and shampoos (alkylbis(2-hydroxyethyl)amine oxides and alkyldimethylamine oxides). Other amine oxides are physiologically active.

amines Organic compounds derived from ammonia by replacement of one or more of its hydrogen atoms by hydrocarbon groups. Replacement of one, two and three hydrogen atoms results in primary, secondary and tertiary amines respectively. Compounds in which the nitrogen forms part of a ring are usually known as heterocyclic bases. Aliphatic, aromatic and mixed aliphatic-aromatic amines are known. Primary amines are obtained by the action of ammonia on alcohols or chloro-derivatives of hydrocarbons. The ammonia may be replaced by primary and secondary amines to give secondary and tertiary amines. Some aldehydes react with ammonia and hydrogen under pressure in presence of nickel catalysts to give amines. Primary amines are also obtained by reduction of nitro-compounds, nitriles, ketoximes and amides, or by treating amides with NaOBr. Secondary amines are obtained by reducing Schiff's bases. A general preparation of amines is from potassium phthalimide and a halogen compound, with subsequent hydrolysis of the product. The aliphatic amines are strong bases; the lower members are soluble in water and are stronger bases than ammonia; they have ammoniacal or fishy odours. The higher members are odourless solids. Aromatic amines are not such strong bases. They are not generally soluble in water, and have characteristic odours. All types form crystalline salts with acids; adducts with $SnCl_4$ and $HgCl_2$ and weak complexes with other metal halides. The picrates and picrolonates are crystalline, rather insoluble salts used for the separation and identification of amines. With nitrous acid, primary aliphatic amines give alcohols. Aromatic amines give diazonium salts in the cold, but these decompose to phenols if the reaction is carried out at higher temperatures. Secondary amines of all types give nitrosamines. Schiff's bases, or azomethines, are formed by the action of aldehydes on primary or secondary amines. All types except tertiary amines give ethanoyl (acetyl) compounds with ethanoyl chloride or anhydride. These are usually crystalline, sparingly soluble solids. These amines also yield sulphonamides with aromatic sulphonyl chlorides. Primary amines react with chloroform and potassium hydroxide to give isocyanides, or carbylamines. Primary and secondary amines yield derivatives of thiocarbamic acid with alcoholic carbon disulphide, while aromatic amines give substituted thioureas. The aromatic amines are of great importance as dyestuff intermediates.

amino- A compound containing an amino group ($-NH_2$) joined directly to a carbon atom. Formerly the prefix amido was also used but this is now usually restricted to compounds containing the amide group ($-CO{\cdot}NH_2$).

aminoacetal, $C_6H_{15}NO_2$, $H_2N{\cdot}CH_2{\cdot}CH(OC_2H_5)_2$. Colourless oily liquid with an ammoniacal odour, b.p. 172–174°C. Prepared by the action of ammonia on chloroacetal. Hydrochloric acid converts it to aminoacetaldehyde. Condensation with aromatic aldehydes gives isoquinoline derivatives.

aminoacetic acid *See* glycine.

amino-acids A large class of organic compounds containing both the carboxyl, COOH, and the amino, NH_2, group, e.g. glycine, $H_2N{\cdot}CH_2{\cdot}COOH$. Their chief importance lies in the fact that many proteins are built up entirely of amino-acid groupings by condensation between the NH_2 and COOH groups, all the amino-acids in proteins being α-amino-acids, with the amino group attached to the same carbon atom as the carboxyl group and with the same L-configuration of asymmetric groups about the α-carbon atom. The following 23 amino-acids are found in very variable proportions as constituents of most proteins:

Ala	alanine
Arg	arginine
$Asp(NH_2)$ or Asn	asparagine
Asp	aspartic acid
CySH	cysteine
CyS	cystine (half)
Glu	glutamic acid
$Glu(NH_2)$ or Gln	glutamine
Gly	glycine
His	histidine
HyLys	hydroxylysine
Hypro	hydroxyproline
Ileu	isoleucine
Leu	leucine
Lys	lysine
Met	methionine
Phe	phenylalanine
Pro	proline
Ser	serine
Thr	threonine
Try	tryptophan
Tyr	tyrosine
Val	valine

In representations of peptides the shortened

form is used and the amino-acid first listed has the free amino group.

Certain other amino-acids occur in a few proteins, and others, not necessarily α- or L-amino-acids, are found naturally in the free state or as constituents of peptides.

The amino-acids are colourless, crystalline substances which melt with decomposition. They are mostly soluble in water and insoluble in alcohol.

As constituents of proteins the amino-acids are important constituents of the food of animals. Certain amino-acids can be made in the body from ammonia and non-nitrogenous sources; others can be made from other amino-acids, e.g. tyrosine from phenylalanine and cystine from methionine, but many are essential ingredients of the diet. The list of essential amino-acids depends partly on the species. *See also* peptides and proteins.

amino-acid analysis Peptides are hydrolysed by acids (e.g. 6MHCl) which destroy some amino-acids (e.g. tryptophane). Alkaline solution also hydrolyses peptides and destroys other amino-acids (e.g. arginine) and causes racemization. Enzymic hydrolysis is slow and incomplete. Use of successive methods allows degradation and hence determination of amino-acid sequence. Hydrolysed amino-acids are analysed by, e.g., ion exchange chromatography, paper chromatography, paper electrophoresis, n.m.r. Sequences of amino-acids can also be determined by specific reactions, e.g. reaction of the C-terminal with hydrazine; reduction of the terminal-COOH to CH_2OH with $LiBH_4$ or $LiAlH_4$; attack of carboxypeptidase on a free α-COOH. *See*, e.g., 1-fluoro-2,4-dinitrobenzene.

1-aminoanthraquinone, α-aminoanthraquinone, $C_{14}H_9NO_2$. Red prisms, m.p. 252°C. A typical aromatic amine. Best prepared by the prolonged action of concentrated ammonia solution at a high temperature upon anthraquinone-1-sulphonic acid in the presence of $BaCl_2$ and by reduction of the corresponding nitro compound or by amination of the chloroanthraquinone.

It is used in the dispersed form as a dye for acetate silk, though it has no affinity for other fibres. It is also used as a starting point for alkyl- or acyl-aminoanthraquinones which are used either as vat dyes or, after sulphonation, as acid wool dyes.

2-aminoanthraquinone, β-aminoanthraquinone, $C_{14}H_9NO_2$. Orange-red, m.p. 302°C. It is a typical aromatic amine. Best prepared by the prolonged action of ammonia at 200°C on anthraquinone-2-sulphonic acid in the presence of MnO_2 or a barium salt. It can also be prepared by amination of 2-chloroanthraquinone.

Its chief outlet is as an intermediate in the preparation of flavanthrene and indanthrene.

aminoazobenzene, $C_{12}H_{11}N_3$. Brownish-

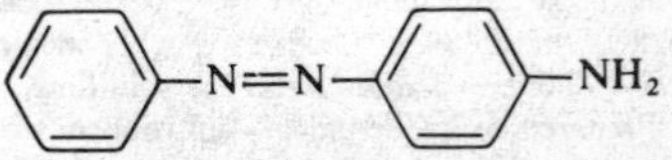

yellow needles, m.p. 127°C. The hydrochloride forms steel-blue needles.

It is prepared by an intramolecular transformation of diazoaminobenzene in the presence of aniline hydrochloride, or in one stage by diazotizing a solution of aniline and aniline hydrochloride with an insufficient amount of nitrous acid.

It is used as a first component in the preparation of azo-dyes.

aminoazo-dyes *See* azo-dyes.

***o*-aminobenzoic acid** *See* anthranilic acid.

4-aminobenzoic acid, P.A.B., $C_7H_7NO_2$. Yellowish red crystals, m.p. 186–187°C. 4-Aminobenzoic acid is an essential metabolite for certain bacterial cells and is regarded as a member of the vitamin B group.

aminocaproic acid, 6-aminohexanoic acid, $H_2N(CH_2)_5COOH$, $C_6H_{13}NO_2$. Prepared from ε-benzoylaminocapronitrile or from 1-hydroxycyclohexylhydroperoxide, m.p. 205°C. Aminocaproic acid is an antifibrinolytic agent, used to treat thrombosis in the deep veins.

aminoethyl alcohol. *See* ethanolamines.

5-aminolaevulinic acid, 5-amino-4-oxo-pentanoic acid, $C_5H_9NO_3$, $H_2N{\cdot}CH_2CO{\cdot}CH_2CH_2CO_2H$. The base unit in the biosynthesis of porphyrins* and so furnishing all the carbon and nitrogen atoms in the haem of haemoglobin, myoglobin, cytochromes, catalase and peroxidase as well as the dihydroporphyrin ring of chlorophyll. It is also incorporated into the corrin ring of vitamin B_{12}. 5-Aminolaevulinic acid is formed from succinyl coenzyme A and glycine. Condensation of two molecules yields porphobilinogen* (a pyrrole derivative) and four molecules of this condense to give the porphyrin skeleton.

aminomethylation *See* Mannich reaction.

aminonaphthols, $C_{10}H_9NO$. Usually prepared by reduction of the nitronaphthols. The sulphonated aminonaphthols are valuable dyestuffs' intermediates.

2-aminophenol, C_6H_7NO. Crystallizes in colourless scales, m.p. 174°C. The crystals darken on exposure to air. Forms a hydrochloride which is readily soluble in water and alcohol. Prepared by reduction of 2-nitrophenol with sodium sulphide, and used as a hair and fur dye.

4-aminophenol, rodinol, C_6H_7NO. Colourless leaflets, m.p. 184°C. Oxidizes quickly in air. Very soluble in acid and alkaline solutions.

Prepared by reduction of 4-nitrophenol or 4-nitrosophenol. Can be diazotized and used as a first component in azo-dyes. Chief outlet is for sulphur dyes in which it is fused with sodium polysulphides. Used as a photographic developer.

aminophenols Used as photographic developers* (reducing agents used in photography) including the important compounds rodinol (4-aminophenol), metol (4-methylamino phenol hemisulphate), glycin (*N*-4-hydroxyphenylglycine) and amidol (2,4-diaminophenoldihydrochloride). Also used extensively in dyestuff synthesis.

aminophylline Theophylline* (75–82%) with ethylenediamine (12–14%). Used as a diuretic and as a bronchodilator in the treatment of asthma.

amino resins and plastics An important group of organic nitrogen-rich polymers formed by condensation of amino derivatives (urea, melamine, thiourea, dicyanodiamide, acrylamide, aniline) with formaldehyde (methanal).

4-aminosalicylic acid, PAS, $C_7H_7NO_3$. M.p. 150°C. Prepared by heating *m*-aminophenol with ammonium carbonate or potassium bicarbonate in solution under pressure. One of the earlier drugs used, usually in combination with streptomycin*, to treat tuberculosis infections.

aminotriazole, 3-amino-1,2,4-triazole, amitrole, $C_2H_4N_4$. A translocated herbicide, m.p. 157–159°C, used as a non-selective herbicide on fallow land or in established orchards.

amitriptyline $C_{20}H_{23}N$. A psychotherapeutic

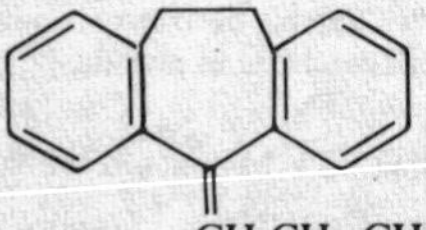

agent similar to imipramine used to treat pathological depression and anxiety.

amitrole *See* aminotriazole.

ammines Complexes containing ammonia co-ordinated to the acceptor atom, e.g. $[Co(NH_3)_6]Cl_3$, hexamminecobalt(III) chloride.

ammonia, NH_3. The most important of the nitrogen hydrides. Colourless gas with a characteristic smell. Readily liquefied by cooling or compression, b.p. −33·5°C, m.p. −77·7°C. NH_3 is very soluble in water, the saturated solution at room temperature has *d* 0·88 ('880 ammonia'); hydrates NH_3,H_2O and $2NH_3,H_2O$ are known. NH_3 occurs naturally in some gases from hot springs. It is manufactured (U.S. production 1983 15·5 megatonnes) by the Haber process (N_2 and H_2 at 500°C and 300 atm in presence of a catalyst, the H_2 is generally from the steam-reforming process) and used to produce HNO_3 (25%), urea (15%), NH_4NO_3 (15%), Nylon (10%) the ultimate products being fertilizers (80%) and explosives (5%) plastics, foams and films (10 %). Prepared in the laboratory from an ammonium salt and a strong alkali or by hydrolysis of a nitride. Just burns in oxygen; $NH_3:O_2$ mixture give NO over catalysts, oxidized to N_2 over heated metal oxides. NH_3 co-ordinates to many elements and forms ammines* reacts with Na or K to give amides MNH_2. Liquid ammonia is a good ionizing solvent (self dissociation $2NH_3 \rightleftharpoons NH_4^+ + NH_2^-$ pK = 30; NH_4^+ salts acids; NH_2^- salts bases). It is generally a better solvent for organic compounds than water but a poorer solvent for inorganic compounds (unless NH_3 complexes formed). Metals frequently dissolve to give blue reducing solutions containing solvated electrons. Detected by smell, blackening of $Hg_2(NO_3)_2$ paper and Nessler's reagent.

ammonia-soda process *See* sodium carbonate.

ammonium The unipositive cation $[NH_4]^+$ which forms a group of, generally, soluble salts similar to the alkali metal salts (formed NH_3 and acid). Most of the salts have strong hydrogen bonding between cation and anion. Substituted ammonium cations, e.g. $[Me_4N]^+$, $[Me_3NH]^+$ are formed from organic amines and acids.

ammonium bicarbonate *See* ammonium hydrogen carbonate.

ammonium bromide, NH_4Br. A colourless crystalline solid with a saline taste. Turns yellow in air. Readily soluble in water, sublimes on heating. Prepared HBr and NH_3 in the gaseous state or solution.

ammonium carbonate, $(NH_4)_2CO_3$. Pure $(NH_4)_2CO_3$ is formed from the commercial salt (see below) and NH_3. It is very soluble in water and decomposes to NH_3, CO_2 and H_2O on heating. Commercial ammonium carbonate is NH_4HCO_3, NH_2COONH_4 obtained from $CaCO_3$ and NH_4Cl or $(NH_4)_2SO_4$ by heating and sublimation. It smells of NH_3, decomposes in moist air to NH_4HCO_3. Used in baking powders, smelling salts (sal volatile), dyeing, in wool treatment and as an expectorant.

ammonium chloride, sal ammoniac, NH_4Cl. A white crystalline solid having a characteristic saline taste. Readily soluble in water, sublimes on heating. Obtained from NH_3 and HCl or commercially from $CaCl_2$ solution (from Solvay process) with NH_3 and CO_2 or by re-crystallization from solutions from the process, or by crystallization from $(NH_4)_2SO_4$ and NaCl (Na_2SO_4 crystallizes preferentially). Used in dry cells, as a mordant, and as a soldering and galvanizing flux. Not, generally, suitable as a fertilizer.

ammonium chromate, $(NH_4)_2CrO_4$. A golden yellow solid (NH_3 plus CrO_3 plus H_2O). Soluble in water, decomposes on heating to the dichromate, NH_3, and H_2O.

ammonium dichromate, $(NH_4)_2Cr_2O_7$. A red crystalline solid (NH_3 plus excess CrO_3 plus H_2O). Very soluble in water. Decomposes spectacularly on heating to N_2,H_2O, and Cr_2O_3.

ammonium fluoride, NH_4F. A white deliquescent solid (NH_3 and gaseous HF). The solutions are extensively hydrolysed and the solid is hydrolysed on standing in moist air. Strongly hydrogen-bonded in the solid, sublimes on heating. Used as a disinfectant and preservative, soluble in ethanol. Adds excess HF to give the hydrogen difluoride NH_4HF_2.

ammonium fluoroborate, NH_4BF_4. Colourless crystals (HBF_4 plus aqueous NH_3) used as an electrolyte.

ammonium hexachlorostannate (IV), chlorostannate, pink salt, $(NH_4)_2SnCl_6$. Precipitated from concentrated H_2SnCl_6 solution with NH_4Cl; used as a mordant.

ammonium hydrogen carbonate, ammonium bicarbonate, NH_4HCO_3. A white crystalline solid (NH_3 plus CO_2 plus water vapour; CO_2 and steam on ammonium carbonate solution; decomposition of commercial ammonium carbonate). A constituent of commercial ammonium carbonate and formed by decay of nitrogenous materials and present in guano. Used in baking powders. More stable as a solid than the carbonate and hence frequently used medicinally in place of the carbonate.

ammonium hydrogen fluoride, NH_4HF_2. *See* ammonium fluoride.

ammonium hydroxide An aqueous solution of NH_3 (*see* ammonia), postulated to contain the weak base NH_4OH but probably only contains ammonia hydrates.

ammonium iodide, NH_4I. Colourless salt (aqueous KI plus $(NH_4)_2SO_4$ plus ethanol, K_2SO_4 precipitated). Very soluble in water, sublimes, with partial oxidation in air. Readily forms polyiodides. Used in the photographic industry.

ammonium iron(II) sulphate, $(NH_4)_2Fe(SO_4)_2,6H_2O$. *See* iron ammonium sulphate.

ammonium iron(III) sulphate, $(NH_4)Fe(SO_4)_2,12H_2O$. *See* iron alums.

ammonium molybdates A range of salts formed from NH_3 plus molybdic acid. The commercial salt is $(NH_4)_6Mo_7O_{24},4H_2O$.

ammonium nitrate, NH_4NO_3. Colourless crystals, sublimes (NH_3 plus HNO_3). On gentle heating decomposes to N_2O. U.S. production 1978 6·54 megatonnes. Used mainly as a fertilizer but also, with a detonator, as an explosive or constituent of explosives (e.g. amatol, slurry explosives). Has been known to detonate spontaneously.

ammonium nitrite, NH_4NO_2. Yellowish solid formed by part oxidation of NH_3 or from $Ba(NO_2)_2$ and $(NH_4)_2SO_4$. Decomposes on heating to N_2 and H_2O.

ammonium perchlorate, NH_4ClO_4. White crystalline solid (NH_3 plus $HClO_4$). Used as an oxidizer in solid propellants.

ammonium persulphate, ammonium peroxodisulphate, $(NH_4)_2S_2O_8$. Colourless solid, soluble in water prepared by electrolysis of cooled saturated $(NH_4)_2SO_4$ in dil. H_2SO_4. A powerful oxidizing agent. An ammoniacal solution of ammonium persulphate is used to strip brass plating from iron.

ammonium phosphates, $NH_4H_2PO_4$, $(NH_4)_2HPO_4$ and $(NH_4)_3PO_4$. Formed from NH_3 and phosphoric acid and used as important fertilizers. The two hydrogen phosphates are used medicinally as diuretics. All of the ammonium phosphates lose NH_3 on heating. Other ammonium phosphates include $NH_4NaHPO_4,4H_2O$ microcosmic salt*.

ammonium phosphomolybdate, $(NH_4)_3Mo_{12}PO_{40}.xH_2O$. The bright yellow precipitate formed from a phosphate, ammonium molybdate and HNO_3 in solution. Used as a test for phosphates.

ammonium sulphamate, $H_6N_2O_3S$. A translocated and soil-acting herbicide. Its strong phytotoxic properties make it useful as a non-selective herbicide for the control of woody plants in agriculture and forestry. It is relatively non-toxic to mammals.

ammonium sulphate, $(NH_4)_2SO_4$. Colourless crystalline solid produced from NH_3 and H_2SO_4 or NH_3 and CO_2 and a solution of $CaSO_4$ (gypsum or anhydrite) in water ($CaCO_3$ precipitates). Used as a fertilizer, although tending to be replaced by fertilizers with higher nitrogen contents. Decomposes on heating, firstly to NH_4HSO_4 and then to N_2, NH_3, SO_2 and H_2O. U.S. production 1978 1 800 000 tonnes.

ammonium sulphides, $(NH_4)_2S$ (NH_3 and H_2S at −18°C) and NH_4HS (NH_3 and H_2S in CH_3OOCCH_3 solution) are relatively unstable. Yellow polysulphides are formed by oxidizing the sulphides, by passing H_2S through a strong aqueous solution of ammonia, or from solutions of $[NH_4]_2S$ and sulphur. Used to dissolve metal sulphides, particularly in qualitative analysis ('yellow ammonium sulphide').

ammonium thiocyanate, NH_4NCS. Colourless solid, soluble in water, prepared from CS_2 and NH_3 in EtOH or HCN and yellow ammonium sulphide. Forms isomeric thiourea, $SC(NH_2)_2$, on heating. Used as a source of NCS^- ions, in the explosives industry and in photography.

amorphous Substances which do not have an extended ordered arrangement of the constituent molecules or ions and are thus without crystalline form. Amorphous substances show no cleavage planes and give an irregular or conchoidal fracture. Many apparently amorphous substances have a microcrystalline structure. Glasses and supercooled liquids are amorphous.

AMP *See* adenosine monophosphate.

amperometric titration A method of analysis in which current flowing through a cell is plotted against added titrant. There are sharp breaks in the curves at the end points.

amphetamine, β-aminopropylbenzene, $C_9H_{13}N$, $PhCH_2CH(CH_3)NH_2$. Colourless liquid, b.p. 200°C (decomp.). It is prepared by the reduction of phenylacetone oxime. It is a vasoconstrictor and central stimulant, now little used due to its addictive properties.

amphiboles Silicates containing double chains of stoicheiometry $[Si_4O_{11}]_n^{6n-}$, e.g. tremolite $(OH)_2Ca_2Mg_5(Si_4O_{11})_2$. *See* asbestos.

amphiprotic Solvents (e.g. H_2O, MeOH) which both produce protons (protogenic, acidic) and react with protons (protophilic, basic).

ampholyte An amphoteric electrolyte, a substance which can behave as both an acid and a base.

amphoteric electrolyte *See* ampholyte.

amphoteric oxide or hydroxide An oxide or hydroxide which can have acid or basic functions, i.e. can combine with either an acid or a base to form a salt. ZnO dissolves in acids to form zinc salts (basic function) and in alkalis to form zincates (e.g. $Na_2Zn(OH)_4$) (acidic function). $Al(OH)_3$ is an amphoteric hydroxide.

amu, atomic mass units Units of atomic weight in terms of an arbitrary standard (^{12}C).

amygdalin, $C_{20}H_{27}NO_{11}$. Cyanophoric glycoside m.p. 215°C found in the kernels of most fruits belonging to the *Rosaceae*, particularly in bitter almonds. A gentiobiose mandelonitrile, $C_6H_5CH(CN){\cdot}O{\cdot}C_6H_{10}O_4{\cdot}O{\cdot}C_6H_{11}O_5$. Emulsin hydrolyses it to benzaldehyde, hydrogen cyanide and glucose. It is widely used as a flavouring material in the form of essence of bitter almonds.

amyl, iso, 3-methylbutyl
The group $(CH_3)_2CHCH_2CH_2$–.

amyl, normal, *n*-amyl, pentyl
The group $CH_3{\cdot}CH_2{\cdot}CH_2{\cdot}CH_2{\cdot}CH_2$–.

amyl, secondary, 1-methylbutyl The group

$$\begin{array}{l} CH_3CH_2CH_2\diagdown \\ \qquad\qquad\qquad CH-. \\ \qquad\quad CH_3\diagup \end{array}$$

amyl, tertiary, 1,1-dimethylpropyl The group

$$\begin{array}{l} \qquad\quad CH_3\diagdown \\ CH_3CH_2-C-. \\ \qquad\quad CH_3\diagup \end{array}$$

amyl acetate, 3-methylbutyl ethanoate
Colourless, volatile liquid with a strong pear-like odour, b.p. 138·5°C. Manufactured by heating amyl alcohol (1-pentanol) with potassium ethanoate and sulphuric acid or by heating amyl alcohol with ethyl ethanoate in the presence of a little sulphuric acid. Commercial

amyl acetate is usually a mixture of isoamyl acetate, $(CH_3)_2CHCH_2CH_2OOCCH_3$, with varying amounts of secondary amyl acetate, $CH_3CH_2CH(CH_3)CH_2OOCCH_3$. It is used as a solvent for cellulose acetate lacquers and paints: an alcoholic solution is used for flavouring purposes as 'essence of Jargonelle pears'.

amyl alcohols, $C_5H_{12}O$. Eight alcohols of this formula are possible. The amyl alcohol of commerce is obtained by distillation of fusel oil and is a mixture of isoamyl alcohol with from 13–60% secondary amyl alcohol; it has b.p. 128–132°C. It is used in the preparation of amyl acetate, amyl nitrite and amylene. The most important of the amyl alcohols are:

(1) *isoamyl alcohol, isobutyl carbinol, 3-methylbutanol* $(CH_3)_2CHCH_2CH_2OH$, a colourless liquid with an unpleasant odour; inhalation of its vapour causes violent coughing, b.p. 131°C. Obtained from fusel oil or by treating $(CH_3)_2CHCH_2CH_2Cl$ with sodium hydroxide. Oxidized by chromic acid to 3-methylbutanoic acid;

(2) *active amyl alcohol, sec. butyl carbinol, 2-methylbutanol,* $CH_3CH_2CH(CH_3)CH_2OH$, b.p. 128°C;

(3) *normal amyl alcohol, 1-pentanol,* $CH_3CH_2CH_2CH_2CH_2OH$, b.p. 137°C, obtained by treating 1-chloropentane with sodium hydroxide;

(4) *amylene hydrate, tert. amyl alcohol, 2-methyl-2-butanol,* $CH_3CH_2C(CH_3)_2OH$, liquid with a penetrating camphor-like odour, b.p. 102°C. Prepared from β-isoamylene (2-methyl-2-butene) by passing into sulphuric acid and then diluting with water.

Complex mixtures of various amyl alcohols are manufactured from chloropentanes derived from petroleum.

amylases, diastase Amylases are enzymes which break down starches and glycogen to the maltose stage. They are widely distributed in animal and plant tissues.

amylene hydrate, $CH_3CH_2C(CH_3)_2OH$. *See* amyl alcohols.

amyl ether Colourless liquid with a pleasant odour, b.p. 172·5–173°C. Manufactured by heating amyl alcohol with sulphuric acid. The commercial amyl ether is a mixture of isoamyl ether, $[(CH_3)_2CHCH_2CH_2]_2O$, with varying amounts of secondary amyl ether, $(CH_3CH_2CH_2CHMe_2)O$. It is used as a solvent for fats and essential oils and as a high boiling ether type solvent.

amylobarbitone, 5-ethyl-5-iso-amylbarbituric acid, $C_{11}H_{18}N_2O_3$. A white crystalline powder, m.p. 155–158°C. It is made by condensing ethyl ethylisopentyl malonate with urea. It is an intermediate acting barbiturate, used as a hypnotic and sedative.

amylocaine hydrochloride, $C_{14}H_{21}NO_2{\cdot}HCl$, $(Me_2NCH_2)(C_2H_5)(Me)C{\cdot}OOC{\cdot}Ph{\cdot}HCl$. Colourless crystalline powder with a bitter taste, m.p. 177–179°C. Prepared by the action of ethyl magnesium bromide on dimethylaminoacetone. It is a local anaesthetic, mainly used to produce spinal anaesthesia.

amylopectin *See* starch.

amylose *See* starch.

***p-t*-amylphenol,** $C_{11}H_{16}O$, $CH_3CH_2CMe_2C_6H_4OH$. M.p. 95°C, b.p. 265°C. Made by alkylation of phenol. Used to form thermosetting resins (with methanal) and surfactants (with ethylene oxide).

amyrin Pentacyclic triterpenoids, $C_{30}H_{50}O$.

anabolic agents Drugs which promote storage of protein and stimulate tissue metabolism, often used in convalescence. Androgenic steroid hormones, e.g. testosterone*, have marked anabolic properties, and a number of synthetic analogues are in therapeutic use.

anabolism *See* metabolism.

anaesthetics Compounds used to abolish the perception of pain and other stimuli. General anaesthetics produce total anaesthesia, e.g. N_2O, ethylvinyl ether, halothane, cyclopropane. Local anaesthetics only act at the site of application by freezing, e.g. ethyl chloride, or by affecting the nerves, e.g. cocaine.

analgesics Compounds which reduce pain perception, e.g. morphine.

analysis The identification of a substance or of the components in a mixture of substances is termed qualitative analysis. Analysis of inorganic substances is carried out by various physical methods; e.g. metals are analysed by spark spectra, mixtures may be analysed by ultra-violet or infra-red spectroscopy, or by identification of X-ray powder patterns. Modern chemical analysis is carried out using various organic reagents which are more or less specific for a given metal (*see*, e.g., aluminon, cupferron, magneson). Organic qualitative analysis is carried out by identification of the type of compound by physical methods or by specific reactions followed by complete identification by preparation of derivatives or by physical methods. The physical methods most used are ultra-violet, infra-red, or nuclear magnetic resonance spectroscopy, mass spectrometry, or molecular weight determination. Chromatography is much used.

Quantitative analysis is the estimation of the amount of element or group present in a mixture or compound. This is done by various methods, in volumetric analysis a titration, in gravimetric analysis a precipitation followed by a weighing, in colorimetric analysis the estimation of a coloured species. Other quantitative methods include infra-red spectroscopy, estimation of the opalescence of a precipitate (turbidimetry, nephelometry and fluorimetry), estimation of optical rotation, electrolytic decomposition, potentiometric, conductometric and amperometric titrations, and polarography. Organic quantitative analysis is generally carried out by physical methods or by conversion to known derivatives which can be estimated by weighing or by titration.

anaplerotic sequences Ancillary routes to the catabolic cycles in organisms, which operate to maintain levels of intermediates in these cycles despite tapping off for anabolic purposes. An example is the glyoxylate cycle*. Anaplerotic sequences are most obviously seen in micro-organisms where rapid biosynthesis occurs from simple carbon compounds.

anatase, octahedrite, TiO_2. A steel blue or yellow form of TiO_2.

anation Replacement of an uncharged ligand, e.g. H_2O, in a complex by an anion, e.g. Cl^-.

andalusite, Al_2SiO_5. A form of aluminium silicate. Used as a refractory material in the manufacture of electrical porcelain (spark plugs), firebricks and for metallurgical furnace linings. World production 1976 270 000 tonnes.

Andrews titration An important titration for the estimation of reducing agents. The reducing agent is dissolved in concentrated hydrochloric acid and titrated with potassium iodate(V) solution. A drop of carbon tetrachloride is added to the solution and the end point is indicated by the disappearance of the iodine colour from this layer. The reducing agent is oxidized and the iodate reduced to ICl, i.e. a 4-electron change.

androsterone, 3α-hydroxy-5α-androstan-17-one, $C_{19}H_{30}O_2$. M.p. 183°C. A typical urinary 17-oxosteroid arising by metabolic reduction of testosterone. Androsterone has androgenic activity but is less potent than testosterone.

anemometer An instrument for measurement of gas velocity.

anethole, 1-methoxy-4-prop-1-enylbenzene, $C_{10}H_{12}O$. White leaflets, with a strong smell and sweet taste, m.p. 22°C, b.p. 235°C. The chief constituent of anise and fennel oils and other essential oils, from which it is manufactured. It can also be prepared from anisole (methoxybenzene). It is widely used for flavouring pharmaceuticals and dentifrices, and in perfumery.

aneurine *See* thiamine.

angelic acid, cis-2-methyl-2-butenoic acid,

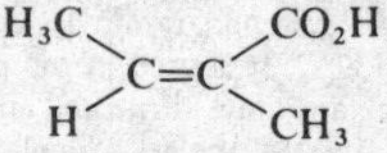

$C_5H_8O_2$. M.p. 45–46 °C, b.p. 185 °C. It occurs free in the roots of *Angelica archangelica.* When heated alone or with acids or alkalis it is converted to tiglic acid. (Z)-isomer.

Ångström unit, Å A unit of length equivalent to 10^{-8} cm, (10^{-10} m). Used to describe molecular and lattice dimensions and wavelengths of visible light.

anharmonicity Harmonic vibrations are characterized by a restoring force which is proportional to the displacement. While at very small amplitudes the vibrations of the atomic nuclei in molecules are harmonic, band spectra show that at larger amplitudes, especially those approaching the dissociation point of the bond, the vibrations are markedly anharmonic. The deviations from harmonic behaviour are expressed by the anharmonicity constant.

anhydride A substance formed by elimina-

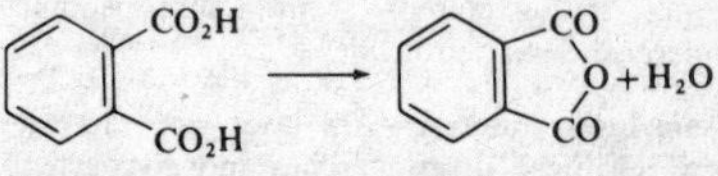

tion of one or more molecules of water from one or more molecules of an acid, or (less frequently) a base. Thus phthalic acid on heating gives phthalic anhydride. Similarly, sulphur trioxide, SO_3, is the anhydride of sulphuric acid, H_2SO_4.

anhydrite Mineral $CaSO_4$ generally associated with some gypsum, $CaSO_4,2H_2O$, and rock-salt, NaCl. Converted by water to gypsum with a 60% increase in volume. Used in the manufacture of sulphuric acid and ammonium sulphate.

anhydro A prefix for an organic compound indicating that one or more molecules of hydrogen have been removed.

anhydrone *See* magnesium perchlorate.

anilides A name applied to the acyl derivatives of aniline of which acetanilide is the commonest example.

aniline, C_6H_7N, $PhNH_2$. Colourless oily liquid, turning brown on oxidation, m.p. −6·2°C, b.p. 184°C. Manufactured from nitrobenzene by vapour phase hydrogenation in the presence of a copper catalyst, or by reduction with iron and water containing a trace of hydrochloric acid. Used in the manufacture of antioxidants and vulcanization accelerators for the rubber industry, but it is also employed for the manufacture of dyes and pharmaceuticals.

On acetylation it gives acetanilide. Nitrated with some decomposition to a mixture of 2- and 4-nitroanilines. It is basic and gives water-soluble salts with mineral acids. Heating aniline sulphate at 190°C gives sulphanilic acid. When heated with alkyl chlorides or aliphatic alcohols mono- and di-alkyl derivatives are obtained, e.g. dimethylaniline. Treatment with trichloroethylene gives phenylglycine. With glycerol and sulphuric acid (Skraup's reaction) quinoline is obtained, while quinaldine can be prepared by the reaction between aniline, paraldehyde and hydrochloric acid.

With sodium nitrite in dilute mineral acid solution at 0°C aniline gives benzene diazonium chloride, which loses nitrogen on warming and gives phenol. The diazonium group may also be replaced by halogeno- or cyano-groups by double decomposition with CuX or CuCN (Sandmeyer reaction); it can be coupled with phenols and amines to give coloured azo-compounds.

Aniline is readily halogenated, treatment with bromine water giving an instant precipitate of 2,4,6-tribromoaniline.

Oxidation with chromic acid gives 4-benzoquinone.

U.S. production 1978 275000 tonnes.

anils, *N*-phenylimides *See* Schiff's bases.

anion An atom which has gained one or more electrons or a negatively charged group of atoms. Anions are present in the solid state, e.g. Cl^- in NaCl, and in solutions and melts. In electrolysis anions travel to the anode where they are discharged.

anisaldehyde, 4-methoxybenzaldehyde, $C_8H_8O_2$. A colourless liquid, b.p. 248°C. It occurs in aniseed and is used in synthetic perfumery under the name 'aubepine' or artificial hawthorn.

***o*-anisidine, 2-methoxyaniline,** C_7H_9NO. A colourless oil, m.p. 2·5°C, b.p. 225°C. Prepared by the reduction of *o*-nitroanisole with iron and hydrochloric acid. It is used in making azo dyestuffs.

anisole, methoxybenzene, C_7H_8O, $PhOCH_3$. Colourless liquid, b.p. 155°C. Prepared by the action of dimethyl sulphate on a solution of phenol in excess of alkali.

anisotropic A substance is anisotropic when any of its physical properties (e.g. thermal or electrical conductivity, refractive index) are different in different principal directions. Crystals, other than those belonging to the cubic system, are anisotropic. Certain substances (e.g. *p*-azoxyanisole) melt to form a cloudy anisotropic liquid ('liquid crystals'), which again 'melts' at a higher temperature to a normal (isotropic) liquid.

annealing The heating and controlled cooling of a substance (generally a metal or glass) to relieve stresses. In general ordered structures are produced by annealing.

annulenes Simple conjugated cyclic polyalkenes: a prefix [*n*] is added to indicate the number of carbon atoms in the cycle. Thus [18]-annulene is a reasonably stable brown-

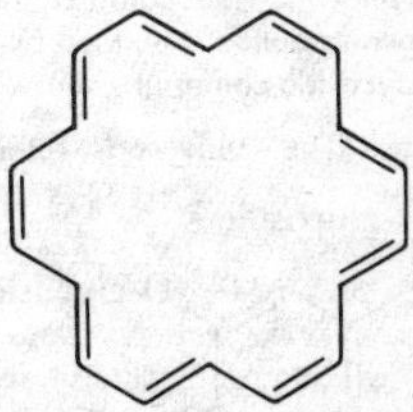

red crystalline solid which has *cis* and *trans* double bonds. Some annulenes are aromatic in character.

anode The electrode which carries the positive charge in an electrochemical cell. In electrolysis the anode is the electrode at which the negative ions are discharged.

anodic oxidation Process of oxidation in solution by means of an electric current.

anodizing Process of anodic oxidation of metals with the object of giving protection against corrosion. Commonly used for aluminium sheet which is made the anode of a sulphuric-chromate(VI) bath, and becomes coated with a layer of porous oxide. The pores are sealed by dipping in hot water; the layer may be coloured by adsorbed dyes.

anomers The specific term used to describe

carbohydrate stereoisomers differing only in configuration at the hemiacetal carbon atom (*), i.e. the carbonyl carbon atom in the open chain projection. These stereoisomers are referred to as the α and β anomers and for D-glucose, for example, both forms are known.

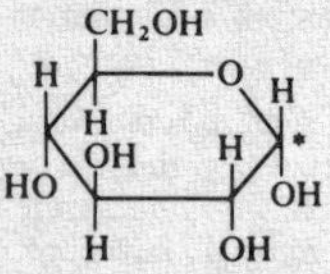

α-D-Glucose

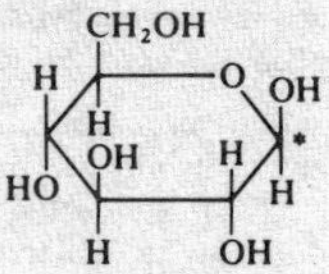

β-D-Glucose

anorthic system The triclinic* crystal system.

anserine, β-alanylmethylhistidine, $C_{10}H_{16}N_4O_3$. M.p. 238–239°C. A dipeptide constituent of the muscles of various species.

antabuse A drug used in the treatment of alcoholism. *See* disulphiram.

antacids Materials used to reduce the amount of acid in the stomach. There can be direct neutralization ($NaHCO_3$, MgO); buffering (sodium citrate, magnesium silicate); absorption of H^+ ($Al(OH)_3$); ion exchange (zeolites).

anthocyanidines The coloured aglycones (sugar-free compounds) obtained by hydrolysis of the glycoside-containing anthocyanines.

anthocyanins The universal soluble red,

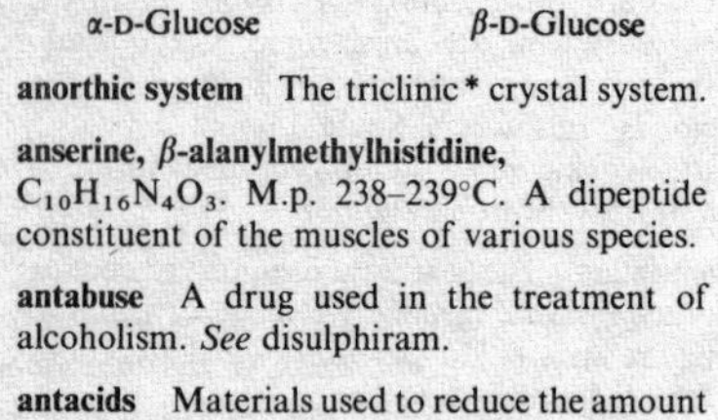

Cyanin

violet and blue pigments of flowers, fruits and leaves; glycosides which on hydrolysis yield one or more sugars and an anthocyanidin.

anthracene, $C_{14}H_{16}$. Colourless plates with

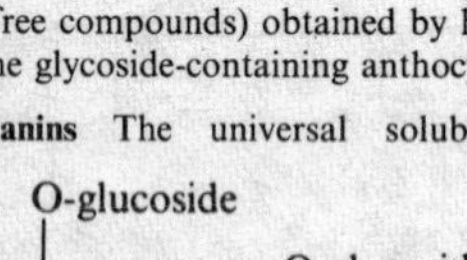

slight blue fluorescence. M.p. 217°C, b.p. 354–355°C.

Present in coal tar (½%), from which it is separated by fractional distillation. Can be chlorinated, sulphonated and nitrated by a mixture of nitric acid, ethanoic acid and ethanoic anhydride. Treatment with sulphuric and nitric acids gives anthraquinone derivatives.

Of little use commercially except as a route to anthraquinone. For this purpose it is oxidized with acid potassium dichromate solution, or better, by a catalytic air oxidation at 180–280°C, using vanadates or other metal oxide catalysts.

anthracite A bright, hard coal with a carbon content greater than 93%.

anthrahydroquinone *See* oxanthrol.

anthranilic acid, 2-aminobenzoic acid, $C_7H_7NO_2$. Colourless leaflets, m.p. 145°C. It is both acidic and basic.

It is prepared by the action of sodium hydroxide and sodium hypochlorite on phthalimide (Hofmann reaction). When heated with soda lime it gives aniline.

It is an important dyestuffs intermediate. It condenses with chloroethanoic acid to give phenylglycine-*o*-carboxylic acid for the synthesis of indigo. It can be diazotized and used as a first component in azo-dyes; it condenses also with chloroanthraquinones to give intermediates for anthraquinone dyes.

Decomposition of its internal salt (a zwitterion) provides a convenient source of benzyne for organic synthesis.

anthranol, 9-hydroxyanthracene, $C_{14}H_{10}O$. Crystallizes in golden brown needles which sinter at 120°C and melt to a clear liquid at 152°C. It couples easily with diazonium salts, and is rapidly oxidized by bromine or $FeCl_3$ to dihydrodianthrone.

It is prepared by acidifying an alkali solution of anthrone or by reduction of anthraquinone with aluminium powder and concentrated sulphuric acid.

Anthranol is isomeric with anthrone, and behaves in its reactions as a typical hydroxylic compound. The equilibrium mixture between the two compounds consists mainly of the keto form; anthranol is largely converted into anthrone on heating.

It is used in the preparation of benzanthrone by heating with glycerol and sulphuric acid (Skraup's reaction).

anthraquinone, $C_{14}H_8O_2$. Colourless needles or prisms which can be purified by sublimation or steam distillation. M.p. 285°C. It is manufactured by condensing benzene with phthalicanhydride in the presence of aluminium chloride to give *o*-benzoyl benzoic acid; this is then converted to anthraquinone by

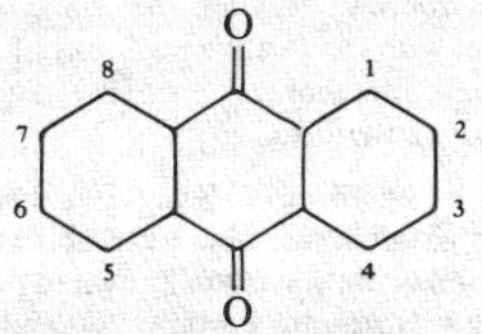

heating with concentrated sulphuric acid at 120–150°C. Another method is the condensation of 1,4-naphthoquinone with butadiene, the former being obtained by the vapour phase oxidation of naphthalene.

Anthraquinone is a very stable compound which more closely resembles a diketone than a quinone: it yields a monoxide with difficulty.

Reduction of anthraquinone gives dianthryl, anthrone* and finally anthracene.

It has been used as a bird repellant and is the parent compound of the anthraquinone vat dyes in which the dyeing is carried out by immersion in the reduced vat solution followed by air oxidation to the original insoluble compound.

Anthraquinone can be brominated, chlorinated directly to the tetrachloro (1, 4, 5, 8-) stage, nitrated easily in the 1-position, but gives the 1,5-and 1,8-dinitro-derivatives on prolonged nitration: the nitro groups in these compounds are easily displaced by neutral solutions of alkali sulphites yielding the corresponding sulphonic acids. Sulphonation with 20–30% oleum gives the 2-; 2,6-; and 2,7-derivatives; in the presence of Hg the 1-; 1,5-; and 1,8- derivatives are formed.

anthraquinone dyes A large group of dyestuffs based on anthraquinone. There are two main types; the alizarin and indanthrene dyestuffs. They are prepared via the sulphonic acids and/or amino derivatives.

anthraquinone glycosides Present in madder, used as dyestuffs (alizarin). The most important is ruberythric acid.

anthraquinone sulphonic acids For preparation *see* anthraquinone. The anthraquinone sulphonic acids are very important, as nearly all anthraquinone derivatives are obtained from them.

Anthraquinone-1-sulphonic acid. Colourless leaflets, m.p. 214°C. It is used in the preparation of 1-aminoanthraquinone.

Anthraquinone-2-sulphonic acid. The sodium salt, commonly called 'silver salt', is used for the preparation of alizarin and 2-aminoanthraquinone and to prepare Fieser's solution*. Aminoanthraquinone derivatives are the basis of many dyestuffs.

anthrone, $C_{14}H_{10}O$. Colourless needles.

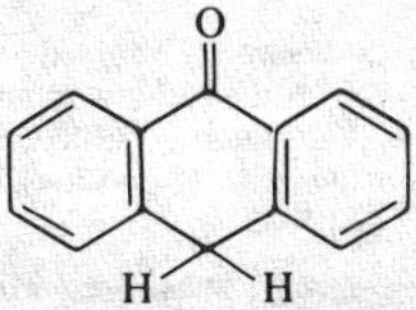

m.p. 154°C. Prepared by reduction of anthraquinone with tin and hydrochloric acid in glacial ethanoic acid solution or by prolonged treatment with sodium dithionite. On heating it is partly converted into the enol anthranol but does not couple with diazonium salts or show any of the reactions typical of a phenolic compound.

anti-aromatic Conjugate, cyclic systems which are thermodynamically less stable than the corresponding acyclic analogues; cyclobutadiene* and cyclo-octatetraene* are anti-aromatic since conjugation of the respective acyclic butadiene and octatetraene results in diminished stability.

antibiotic An organic substance which is produced by micro-organisms and is capable at low concentration of inhibiting the growth of, or destroying another micro-organism. Antibiotics have been isolated from numerous sources, but principally from bacteria (e.g. bacitracin*, polymyxin*, gramicidin*, actinomycetes (e.g. tetracycline*, streptomycin*, chloramphenicol) and fungi (e.g. penicillins*, cephalosporins*). Bacterial antibiotics are mostly polypeptides.

Many antibiotics are unsuitable for therapeutic use, frequently because of their general toxicity or as a result of other drawbacks such as instability, inadequate solubility or malabsorption.

antibodies Protein molecules present in the serum which are formed in the body in response to the presence of foreign substances called antigens*. For each antigen there is a specific antibody, i.e. a 'lock' and 'key' arrangement. The effect is to agglutinate or precipitate the antigens and if cellular, prepare them for the action of complement*. The antibody–antigen mechanism is the basis of the immune response.

antibonding orbitals *See* molecular orbitals.

anti-cathode *See* X-ray tube.

antiferromagnetism Positive magnetic behaviour in which the magnetic susceptibility drops with decreasing temperature. There is generally a characteristic temperature, the Néel point, above which magnetic behaviour is normal.

anti-fluorite structure Adopted by compounds M_2X with the cations occupying the F^- positions of the fluorite* structure and the anions occupying the Ca^{2+} positions. Adopted by e.g. K_2S and K_2O.

anti-foaming agents Substances which prevent the formation of foams. They are strongly adsorbed by the liquid medium but they do not have the electrical or mechanical properties required to form a foam. Examples of anti-foaming agents are polyamides which are used in boiler-feed water and octanol used in electroplating baths and in paper making. Low concentrations of silicones also find quite general use. For lubricating oils polysiloxanes are widely used.

anti-freeze additives This can be any additive which will lower the freezing point of the liquid concerned but it is generally applied to materials used to prevent freezing in the coolant system of internal combustion engines. Ethylene glycol* is the principal component of anti-freeze liquids although alcohols may also be employed.

anti-friction metal *See* white-metal alloys.

antigens Macromolecular proteins which, when injected into the blood of an animal, stimulate the production of antibodies*. Antigens may be bacteria, viruses (living or dead), foodstuffs, pollens, proteins, some polysaccharides, or nucleic acids.

antihistamines Substances which antagonize the physiological and pharmacological effects of histamine*. Examples: mepyramine*, dimenhydrinate*, cimetidine*.

antihormones Substances antagonizing natural hormones (e.g. anti-oestrogens and anti-androgens in contraceptive pills). Sometimes hormones occur in mutually antagonistic pairs (e.g. insulin and glucagon).

anti-isomer An isomer with particular,

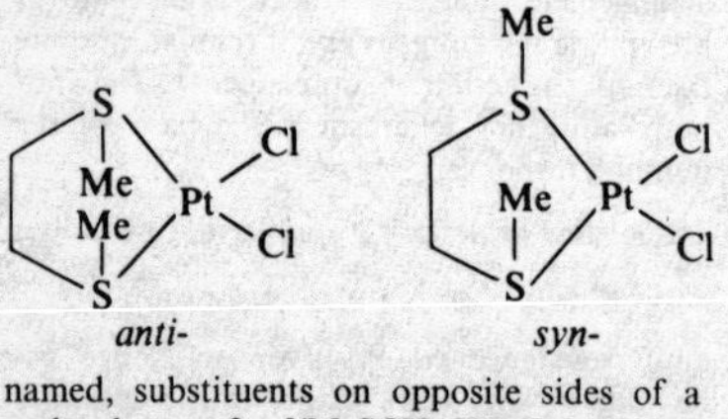

anti- *syn-*

named, substituents on opposite sides of a molecule, e.g. for $[(MeSCH_2CH_2SMe)PtCl_2]$. *See* isomerism.

anti-isomorphism Having the same crystal lattice but with the relative positions of anions and cations interchanged, e.g. ThO_2 and Li_2O (with CaF_2 structure).

anti-knock additives Several substances, added to gasoline, are able to inhibit the pre-combustion oxidation chain known to produce knocking*. Normally a mixture of lead alkyls with dibromoethane and dichloroethane (known as ethyl fluid) is added to gasoline, the lead being removed in the exhaust as volatile lead compounds. The bromine and chlorine additives are termed lead scavengers.

anti-knock value *See* octane number.

antimonates Antimonates(III), M^ISbO_2, are formed from alkalis and Sb_2O_3. The free acid is not known. Antimonates(V) are formed from hydrated Sb_2O_5 and contain the $[Sb(OH)_6]^-$ ion. Mixed oxides M^ISbO_3, $M^{III}SbO_4$ and $M_2{}^{II}Sb_2O_7$ are known and contain SbO_6 octahedra.

antimony, Sb. At.no. 51, at.wt. 121·75, m.p. 630·74°C, b.p. 1750°C, *d* 6·68. Main source is stibnite, Sb_2S_3. Extracted by reduction with Fe or C. The most stable form of the element has a metallic appearance, white or bluish white in colour. This form has a layer structure with three near neighbours and three at a greater distance. Sb will burn in air but is unattacked by water or dilute acids, it is attacked by oxidizing acids and by halogens. Sb is widely used in alloys (e.g. type-metal) its use in semi-conductors is important. Antimony compounds are used in flame-proofing, paints, ceramics, enamels, glass, pottery, rubber technology and dyestuffs. World Sb ore production 1976 69 000 tonnes. Antimony derivatives have considerable toxicity.

antimony bromides Antimony tribromide $SbBr_3$ (Br_2 plus Sb) m.p. 97°C, b.p. 280°C, similar to $SbCl_3$. Higher bromides are not known although complexes, M_2SbBr_6 (contains Sb(III) and Sb(V)) and $MSbBr_6$ are known.

antimony chemistry Antimony is a metalloid of Group V, electronic configuration $5s^25p^3$. Its chemistry is dominated by the +5 and +3 oxidation states which are generally covalent although some pseudocationic chemistry of the $(SbO)^+$ and Sb^{3+} species is known (cations are co-ordinated by oxygen). Antimonides are formed with metals and polymeric cations and anions, e.g. Sb_7^{3-}, are known. Alleged Sb(IV) compounds contain Sb(III) and Sb(V).

antimony chlorides
Antimony trichloride, $SbCl_3$, butter of anti-

mony. Colourless solid m.p. 73°C, b.p. 283°C (Sb compound in conc. HCl followed by distillation). Hydrolysed by water to basic chlorides. Forms many complexes including $(SbCl_4)^-$ (polymeric), $(SbCl_5)^{2-}$ (square pyramidal), and $(SbCl_6)^{3-}$, and complexes with, e.g. *N*-bonded ligands.

Antimony pentachloride, $SbCl_5$. M.p. 7°C, b.p. 79°C (Sb or $SbCl_3$ plus Cl_2). Readily hydrolysed by water, forms complexes, e.g. $[SbCl_6]^-$. Mixed antimony(III)/(V) complexes occur as salts M_2SbCl_6. $SbCl_5$ is used extensively as a chlorinating agent.

antimony fluorides
Antimony trifluoride, SbF_3, m.p. 292°C, sublimes 319°C (Sb_2O_3 in HF solution). Forms complex ions, e.g. $[SbF_4]^-$. Widely used as a fairly mild fluorinating agent.

Antimony pentafluoride, SbF_5, m.p. 7°C, b.p. 150°C is an associated liquid (Sb plus F_2 or $SbCl_5$ plus HF). Forms many complexes and complex ions including $[SbF_6]^-$, $[Sb_2F_{11}]^-$ and is a very powerful fluoride ion acceptor. Greatly enhances the dissociation of, e.g. HF and HSO_3F by forming anionic species (magic acid, super acid). Used as a fluorinating agent (sometimes in the form of its graphite intercalation compound).

antimony hydride, stibine, SbH_3.
An unstable colourless gas, m.p. −88°C, b.p. −17°C (Sb compound plus Zn plus HCl or HCl on Mg-Sb alloy) decomposes easily to Sb.

antimony iodide, SbI_3. Red or yellow solid m.p. 170°C, similar in properties to $SbCl_3$.

antimony, organic derivatives, stibines, SbR_3. A large range of such derivatives are prepared from, e.g., Grignard reagents and $SbCl_3$. Good donors, form tetra-organostibonium ions $[R_4Sb]^+$. SbR_5 derivatives are also known.

antimony oxide halides Compounds, e.g. SbOCl, $Sb_4O_5X_2$, formed by partial hydrolysis of antimony halides.

antimony oxides
Antimony trioxide, Sb_2O_3. The white pure oxide (Sb plus steam; Sb_2S_3 plus O_2) turns yellow on heating, m.p. 656°C, subl. 1550°C. Solid Sb_2O_3 contains Sb_4O_6 molecules, it forms SbO_2 on heating in air. Sb_2O_3 is reduced to Sb by H_2 at red heat, is insoluble in water, dissolves in acids (to give solutions of, e.g., $SbCl_3$, $Sb_2(SO_4)_3$) and in alkalis (to give antimonates(III)). Used as an opacifier in paints and as a flame retardant in plastics.

Antimony dioxide, SbO_2 (heat on Sb_2O_3 in air) is a yellowish compound containing Sb(III) and Sb(V).

Antimony pentoxide, Sb_2O_5, is yellow (Sb plus conc. HNO_3). Decomposes to SbO_2 on heating. Gives antimonates(V) with alkalis. Used as a flame retardant.

antimony potassium tartrate, tartar emetic, $KSbO(C_4H_4O_6).\frac{1}{2}H_2O$. Used medicinally as an emetic and is still injected in the treatment of kala-azar.

antimony sulphates The normal salt, $Sb_2(SO_4)_3$ (Sb_2O_3 plus conc. H_2SO_4) is readily available. It is hydrolysed by water. Complex species are present in H_2SO_4 solutions (e.g. $[SbO]^+$, $[Sb(OH)_2]^+$, $[Sb_3O_9]^{3-}$) of antimony (III) and (V).

antimony sulphides The trisulphide, Sb_2S_3, occurs in a black (stibnite) and a red (H_2S plus an Sb(III) compound in HCl) form. It is transparent to i.r. radiation. Forms thioantimonates(III), e.g. $[SbS_3]^{3-}$ with excess of sulphide ion. Antimony(V) sulphide is very unstable and readily loses sulphur on heating and may not have a finite existence. Thioantimonates(V), M_3SbS_4 (Sb_2S_3 plus S plus alkali) give Sb_2S_5 with acid.

antimonyl derivatives Compounds apparently containing the $[SbO]^+$ grouping although the actual species present are much more complex (*see* antimony sulphates).

antioxidants Substances which slow the oxidation rate in autoxidizable substances. They are added for the protection of foods, particularly fats; for stopping the ageing or deterioration of rubber and many plastics; for inhibiting gum formation in cracked petroleum; and for preserving many other products. Many antioxidants are highly substituted phenols, aromatic amines, or sulphur compounds. Certain sequestering agents act as antioxidants by inactivating metals that may catalyse oxidation. Many organic raw materials contain natural antioxidants. The u.v. rays of sunlight are responsible for activating rubbers etc. to oxidation by air. Hence u.v. absorbers are added if exposed to sunlight.

antiparticle An elementary particle with charge properties opposite to those normally found, e.g. a positive electron.

antipyretics Drugs used to reduce the temperature in the case of fever, e.g. salicylates.

antiseptics Materials which stop the growth of micro-organisms, generally applied to living tissue. I_2 is an example.

antistatic agents Compounds added to

materials (particularly fibres, plastics) to reduce charge accumulation with resultant dust and dirt build up. E.g. metals, water and hygroscopic salts.

antithyroid drugs Substances which interfere with production of thyroxine by the thyroid gland. E.g. perchlorate, thiocyanate, thiouracil, resorcinol. Antithyroid compounds are used in the treatment of hyperactivity of the gland.

antitoxins Antibodies* formed (usually over a period of months) in blood to which bacterial toxins have been introduced. A given antitoxin will only neutralize its own toxin, with which it apparently combines in a quantitative manner.

apatite, $Ca_5(PO_4)_3F$. A phosphate mineral containing fluoride and phosphate anions. Used for the production of phosphorus and its compounds including superphosphates. Hydroxyapatite, $Ca_5(OH)(PO_4)_3$ is isostructural with apatite. These compounds are of importance as components of teeth. The mineral contains useful quantities of uranium.

apo Prefix indicating a derived compound; thus apomorphine is derived from morphine.

apomorphine, $C_{17}H_{17}NO_2$. White crystals.

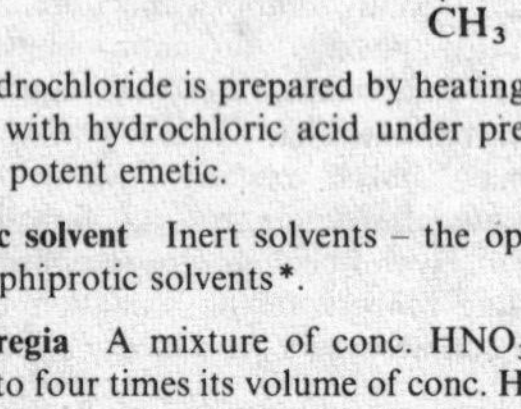

Its hydrochloride is prepared by heating morphine with hydrochloric acid under pressure. It is a potent emetic.

aprotic solvent Inert solvents – the opposite of amphiprotic solvents*.

aqua regia A mixture of conc. HNO_3 with three to four times its volume of conc. HCl, so called because it will dissolve 'noble' metals, such as Pt and Au. Its strong oxidizing action is due to nitrosyl chloride (NOCl) and chlorine produced by interaction of the two acids.

aquation The process of replacement of other ligands by water, e.g.

$$[Co(NH_3)_4(H_2O)(NO_3)]^{2+} \rightarrow (Co(NH_3)_4(H_2O)_2]^{3+}$$

The process of complexing by water molecules.

aquo ions Hydrated metal ions in aqueous solution; also applied to hydrated ions in complexes, e.g. $[Co(H_2O)_6]^{2+}$. Aquo ions can lose protons and act as acids, e.g.

$$[Fe(H_2O)_6]^{3+} \rightarrow [Fe(H_2O)_5(OH)]^{2+}$$

Ar Argon. Sometimes used as an abbreviation for an aromatic group.

L-arabinose, $C_5H_{10}O_5$. The pentose sugar

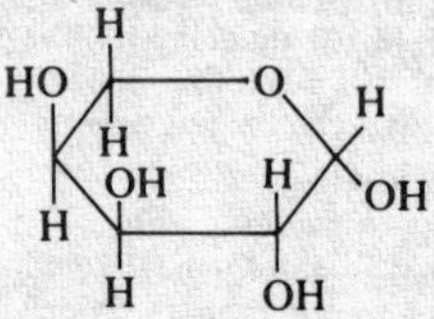

of the hemicelluloses, gums and mucilages, and of some glycosides, in particular the vicianosides. It crystallizes in the β-pyranose form, m.p. 160°C.

D-Arabinose is found in the glycoside barbaloin and in the polysaccharides of the tubercle bacillus.

arachidic acid, *n*-eicosanic acid, $C_{20}H_{40}O_2$, $CH_3{\cdot}[CH_2]_{18}{\cdot}COOH$. M.p. 75°C. A fatty acid occurring as glycerides in peanut and other vegetable oils.

arachidonic acid, $C_{20}H_{36}O_2$. *Cis,cis,cis,cis*-5,8,11,14-eicosatetraenoic acid. The principal

$$CH_3[CH_2]_4[CH{=}CHCH_2]_4CH_2CH_2CO_2H$$

fatty acid of the adrenal gland, is present also as an important constituent of brain and liver lipids. Arachidonic acid is an 'essential fatty acid'* in animal metabolism and is a biological precursor of prostaglandins E_2 and F_{2a}* and can be readily synthesized.

aragonite, $CaCO_3$. A form of calcium carbonate.

arbutin M.p. 195°C. A typical phenolic glycoside, hydroquinone-β-D-glucopyranoside, $C_6H_{11}O_5{\cdot}O{\cdot}C_6H_4OH$. A colourless, crystalline, bitter substance obtained from the leaves of the bearberry, and of most *Ericaceae*. Since hydroquinone is a powerful antiseptic, arbutin has pharmacological value; it also has diuretic action. The commercial product has 40% of methylarbutin.

arc spectrum The emission spectrum obtained when a substance has been excited by application of an electric arc rather than an electric spark which gives a *spark spectrum*. Because the spark causes a greater degree of excitation of the atoms than does the arc, the spark spectrum contains more lines and lines which are enhanced in intensity compared with the arc spectrum.

Arctons *See* Freons.

π-arene complexes Complexes in which an aromatic system is bonded to a metal through its π-electrons. Generally only applied to complexes of uncharged aromatic systems, e.g. $[(C_6H_6)_2Cr]$ but formally applied to any complex of an aromatic system, e.g. $[(C_5H_5)_2Fe]$ as a complex of $(C_5H_5)^-$.

argentates *See* silver oxides.

argentite, Ag_2S. An important silver ore.

argentous and argentic
Compounds of silver(I) and silver(II).

arginase The enzyme which decomposes arginine into ornithine* and urea*.

arginine, 2-amino-5-guanidino pentanoic acid, $C_6H_{14}N_4O_2$,
$H_2NC(=NH)NH[CH_2]_3CHNH_2CO_2H$.
M.p. 207°C. The naturally occurring substance is dextrorotatory. Arginine is one of the essential amino-acids and one of the most widely distributed products of protein hydrolysis. It is obtained in particularly high concentration from proteins belonging to the protamine and histone classes. It plays an important rôle in the production of urea as an excretory product.

argol Crude potassium hydrogen tartrate obtained from wine.

argon, Ar. At.no. 18, at.wt. 39·948, m.p. −189·2°C, b.p. −185·7°C. The most abundant of the noble gases (0·93% of dry air). Separated by fractionation from liquid air and used in light bulbs, as an inert shield in arc welding and for metal production (Ti, Si). World production 1976 440 000 tonnes. Argon has electronic configuration $3s^23p^6$ and although it is soluble in water and forms some clathrates forms no true compounds except excited state derivatives.

Arnd's alloy 60% Cu, 40% Mg. Used in estimating nitrates.

Arndt–Eistert synthesis A procedure for converting a carboxylic acid to its next higher homologue, or to a derivative of a homologous acid, e.g. ester or amide.

$$RCO_2H \longrightarrow RCOCl \xrightarrow[Et_2O]{CH_2N_2} RCOCH_2N_2 \xrightarrow{Ag^+} RCH_2CO_2H$$

aromatic The property (aromaticity) which allows benzene to undergo electrophilic substitution reactions, and to account for its stability. Compounds containing a benzene skeleton are said to be aromatic, whilst non-benzenoid compounds such as ferrocene and many heterocyclic compounds which also undergo electrophilic substitution are termed aromatic using this criterion only. Other definitions of aromaticity include a molecular orbital description which connects aromatic stability with the number of π electrons $(4n+2)$ associated with a planar cyclic conjugated system of double and single bonds, e.g. azulene 10 π electrons, $n=2$: tropylium cation 6 π electrons, $n=1$. Aromaticity has also been defined in terms of effect on the n.m.r. spectrum of the compound under consideration.

Arosolvan process A process for the extraction of benzene and toluene from a mixture of aromatic and saturated hydrocarbons using a mixture of water and N-methylpyrrolidone. The process is used when naphtha is cracked to produce alkenes. To prevent extraction of alkenes these are saturated by hydrogenation prior to extraction.

Arrhenius equation The variation in the rate of a chemical reaction with temperature can be represented quantitatively by the Arrhenius equation

$$k = A\exp(-E_a/RT)$$

where k is the velocity constant, A and E_a are constants, termed the pre-exponential factor and the activation energy*, of the particular reaction being considered (R is the gas constant). The pre-exponential factor A is a collision frequency factor, which is comparatively insensitive to temperature. $A = pZ$, where Z is the collision frequency and p is a 'steric factor', which is a measure of the efficiency of the collision for effective reaction. The Arrhenius equation applies to processes such as diffusion, electrolytic conduction, viscous flow, etc., as well as chemical reactions.

arsenates Arsenates(III) are formed from As_2O_3 and base but the free acid is unknown. Arsenates(III) containing the $(AsO_3)^{3-}$ and more complex species are known. Sodium and copper arsenates(III) (emerald green*, Scheele's green) are used as insecticides and were used as pigments. Sodium arsenate(III) is used as a standard reducing solution. Arsenates(V) are formed as the free acid $H_3AsO_4,\frac{1}{2}H_2O$ (As_2O_3 plus HNO_3), condensed arsenates are less stable to hydrolysis than condensed phosphates. $Na_2HAsO_4,7H_2O$ is used commercially in calico-printing.

arsenic, As. At.no. 33, at.wt. 74·9216, sublimes 613°C, *d* 5·7. The principal sources of As are as impurities in sulphide ores, orpiment As_2S_3, realgar As_4S_4, arsenolite As_2O_3, mispickel FeAsS, nickel glance NiAsS, cobalt bloom $Co_3(AsO_4)_2,8H_2O$. The ores are converted to As_2O_3 (white arsenic) by roasting in air and As_2O_3 is used as a source of As compounds, As_2O_3 is reduced to As with hydrogen or carbon. As is normally bright and metallic in appearance with a double-layer structure (3 contacts in plane, 3 next near neighbours). Yellow As (As_4 units) is formed by rapid condensation of the vapour; the vapour contains As_4 and As_2 units. Arsenic burns in air to As_2O_3 and reacts with halogens, conc. oxidizing acids and fused alkalis. Arsenic and its compounds are used in insecticides (*see* arsenates) as a dopant in solid-state devices (transistors, lasers), and in alloys where it has a hardening effect. Arsenic compounds in quantity are poisonous.

arsenic bromide, $AsBr_3$. *See* arsenic halides.

arsenic chemistry Arsenic is an element of Group V, electronic configuration $4s^24p^3$. It behaves as a metalloid and there is little evidence of cation chemistry although anions, e.g. $[AsO_4]^{3-}$, $[AsF_6]^-$ are well known. The stable oxidation states are +5 (co-ordination numbers 4, 5 and 6) and +3 (co-ordination numbers 3, 4). Arsenides are formed with many metals.

arsenic chlorides, $AsCl_3$ and $AsCl_5$. *See* arsenic halides.

arsenic fluorides, AsF_3 and AsF_5. *See* arsenic halides.

arsenic halides Arsenic forms AsF_3 (As_2O_3 plus HF or CaF_2 and H_2SO_4) used as a mild fluorinating agent (b.p. 57°C), AsF_5 (As plus F_2) (b.p. −53°C) both fluorides are readily hydrolysed by water and can act as acceptors with, e.g., F^- to give $[AsF_6]^-$ (hexafluoroarsenates). $AsCl_3$ is the only stable chloride (As or As_2O_3 plus Cl_2), b.p. 130°C. It forms many complexes in which it can act as acceptor and donor and is used as a starting point for organoarsenic compounds. A very unstable $AsCl_5$ ($AsCl_3$ plus Cl_2) is formed at low temperatures. $AsBr_3$ and AsI_3 are the only bromides and iodides and are very similar to $AsCl_3$.

arsenic hydrides, arsine, AsH_3.
M.p. −116·3°C, b.p. −55°C. An unstable poisonous gas (metal arsenide plus acid or arsenic compounds plus Zn plus dil. acid, $AsCl_3$ plus $LiAlH_4$) which decomposes to As on heating (Marsh test). Very poisonous, a strong reducing agent; gives arsenides with metal solutions.

arsenic iodide, AsI_3. *See* arsenic halides.

arsenic, organic derivatives Many organoarsenic compounds, e.g. R_3As, arsines, are known and are formed from Grignard reagents. They are good donors and extensive ranges of complexes are known. Compounds containing As–As bonds, e.g. the cacodyls (Me_2As derivatives) are formed.

arsenic oxides
Arsenic(III) oxide, As_2O_3 (white arsenic). Formed by burning As in air, contains As_4O_6 molecules or AsO_3 units joined by oxygen bridges. As_2O_3 is used in glass manufacture and as arsenates(III) in insecticides, weed-killers and defoliants. It is used as a standard reducing agent (with I_2 solution). It gives arsenates(III) with alkalis.

Arsenic(V) oxide, As_2O_5, is a rather indefinite compound (As_2O_3 plus conc. HNO_3) which loses O_2 and forms arsenates(V) with alkalis.

arsenic sulphides The sulphides As_4S_3, As_4S_4 (realgar*), As_2S_3 and As_2S_5 can be prepared by direct reaction of the elements or (As_2S_3 and As_2S_5) precipitated from acidified As(III) or As(V) solutions with H_2S. As_4S_3, As_4S_4 and As_2S_3 all have cage structures with As–As and As–S–As linkages. As_4S_4 is used in tanning and in pyrotechny, As_2S_3 is transparent to i.r. and has optical uses. Thioarsenates(III), e.g. K_3AsS_3, and (V), e.g. K_3AsS_4, are formed from the sulphides and alkalis or alkali sulphides.

arsenides Compounds between metals and arsenic.

arsenopyrite *See* mispickel*.

arsenurretted hydrogen, AsH_3. Arsine, *see* arsenic hydrides.

arsine, AsH_3. *See* arsenic hydrides.

arsines *See* arsenic, organic derivatives.

aryl When a hydrogen atom is removed from a hydrocarbon of the benzene series, the residue is called an aryl group.

arynes The transient intermediates, not capable of isolation, having two hydrogen atoms removed from adjacent non-ring junction carbon atoms of the corresponding aromatic compounds. Examples are known from benzene, naphthalene, pyridine and coumarone type compounds. The existence of arynes is inferred from the identity of products resulting from the trapping of such species in Diels–Alder reactions, and from the isomer orientation of elimination-addition reaction products of haloaromatic compounds. *See* benzyne.

As Arsenic.

asbestine A fibrous variety of talc mixed with tremolite asbestos. Used as a rubber filler and in paints.

asbestos The fibrous varieties of a number of silicate minerals containing SiO_4 groups linked into chains. The most important are the fibrous amphiboles (tremolites, crocidolites) and fibrous talcs and clay minerals (chrysotile). The fibres can be spun and woven into textiles and used in fabrics for heat and electrical insulation and fire-proof fabrics. Used as a binder in asbestos-cement building materials and in filtration. World production 1976 5·6 megatonnes. Inhalation of certain asbestos fibres can cause asbestosis, a lung disease which is frequently fatal.

ascorbic acid, vitamin C, $C_6H_8O_6$.

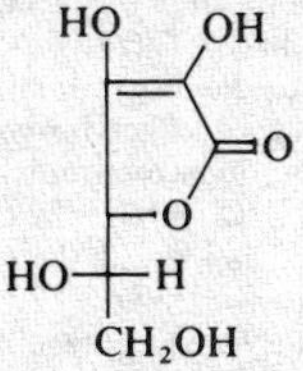

M.p. 190–192°C. The enolic form of 3-oxo-L-gulofuranolactone. It can be prepared by synthesis from glucose, or extracted from plant sources such as rose hips, blackcurrants or citrus fruits. Easily oxidized. It is essential for the formation of collagen and intercellular material, bone and teeth, and for the healing of wounds. It is used in the treatment of scurvy. Man is one of the few mammals unable to manufacture ascorbic acid in his liver. Used as a photographic developing agent in alkaline solution.

asparagine, aminosuccinamic acid, $C_4H_8N_2O_3$.

$$\begin{array}{c} CH_2{\cdot}CO{\cdot}NH_2 \\ | \\ H_2N{\cdot}CH{\cdot}COOH \end{array}$$

M.p. 234–235°C. Hydrolyses to aspartic acid. L-asparagine can be prepared from lupin seedlings, and DL-asparagine is synthesised from ammonia and maleic anhydride. L-asparagine is very widely distributed in plants, being found in all the *Leguminosae* and *Gramineae,* and in many other seeds, roots and buds.

aspartic acid, aminosuccinic acid, $C_4H_7NO_4$, $HO_2C{\cdot}CH_2{\cdot}CH(NH_2){\cdot}CO_2H$ M.p. 271°C. The naturally occurring substance is L-aspartic acid. One of the acidic-amino acids obtained by the hydrolysis of proteins.

asphalt In the U.K. the term asphalt generally refers to a natural mixture of bitumen* with a proportion of mineral matter, e.g. rock asphalt, or to a mechanical mixture of bitumen, aggregate, sand and filler, i.e. hot rolled asphalt.

In the U.S. and elsewhere the term asphalt is commonly applied to the material known in Britain as bitumen or asphaltic bitumen.

asphalt emulsions *See* bitumen emulsions.

asphaltenes Compounds of high molecular weight consisting largely of highly-condensed aromatic ring structures. Precipitated from asphaltic bitumen* by saturated hydrocarbon solvents or aromatic-free petroleum spirit.

asphaltic bitumen A black or brown viscous liquid or solid, consisting of hydrocarbons and their derivatives, obtained from crude oil as a distillation residue or from natural sources or combined with mineral matter in natural asphalt. It possesses good waterproofing and adhesion properties and is used largely for road construction, waterproof or protective coating and electrical insulation.

asphaltites Naturally occurring, hard bitumens of high purity containing only 1–2% of mineral matter.

aspirin, o–O-acetylsalicylic acid, $C_9H_8O_4$. Colourless crystals, m.p. 135–138°C. Manufactured by the action of ethanoic anhydride on salicylic acid. It is widely used in tablet form as the acid or a salt as an analgesic and antipyretic. U.S. production 1981 18 000 tonnes.

associated liquids Liquids which are composed wholly or partly of loosely combined aggregates of two or more molecules, rather than, as in the case of normal unassociated liquids, single molecules. Water and ethanol are examples of associated liquids, which show anomalous properties such as relatively high boiling points. Association in liquids arises from bonding between adjacent molecules, particularly hydrogen bonds in water and other hydroxylic compounds.

association A term applied to the combination of molecules of one substance with another to form more complex molecules. Substances showing this effect are said to be associated. The phenomenon is encountered in pure liquids, in vapours and in solutions and may be determined by ordinary methods of molecular weight determination, e.g. depression of freezing point; the associated molecules behave as single particles of increased molecular weight.

astatine, At At.no. 85. A radioactive halogen formed in natural radioactive decay series but studied on ^{211}At formed by α-bombardment of ^{209}Bi. Rather little is known of the chemistry of the element, electronic configuration $6s^26p^5$. Oxidation states established include −1 (At^- formed by reduction with SO_2 or Zn, AgAt insoluble), +1 ($[AtO]^-$ formed from At^-, water and Br_2 or Fe^{3+} also $[Atpy_2]^+$), and +5 ($[AtO_3]^-$ from $[AtO]^-$ and hypochlorite or $[S_2O_8]^{2-}$).

asulam, methyl (4-aminobenzenesulphonyl)-carbamate, $C_8H_{10}N_2O_4S$. A translocated herbicide, m.p. 143–144°C. It is used for post emergence control of *Rumex spp.* (e.g. docks) in grassland and orchards, and for the control of bracken on hill land. It is relatively non-toxic to wild-life.

asymmetric induction A term applied to the selective synthesis of one diastereomeric form of a compound, resulting from the influence of an existing chiral centre adjacent to the developing asymmetric carbon atom. This usually arises because, for steric reasons, the incoming atom or group does not have equal access to both sides of the molecule.

asymmetry *See* chirality.

At Astatine.

atactic polymer If in crystalline polymers it is considered that the carbon atoms in any chain lie in the same plane with a regular arrangement of the substituents the structure may be either isotactic* or syntactic*. If there is no stereospecificity of the substituents with respect to the carbon atoms then the polymer is said to be atactic.

Athabasca tar sands Naturally occurring mixtures of bitumen and sand found in Northern Alberta.

atmosphere *See* air.

atom The smallest part of an element which can exist as a stable entity.

atomic absorption spectroscopy An analytical technique in which characteristic radiation is absorbed by non-excited atoms in the vapour state. The sample is volatilized with a burner.

atomic emission spectroscopy An analytical technique used for the determination of trace metals. The sample is vaporized and excited in an arc or flame and the emission spectra observed.

atomic energy Mass and energy are equivalent, $E = mc^2$ (E, energy in ergs: m, mass in grams; c, velocity of light $= 3 \times 10^{10}$ cm/s^{-1}). In the building up of nuclei, part of the mass of the constituent particles appears as binding energy; the ratio (mass defect) : (atomic number) is known as the packing fraction. The elements of low atomic number and those of very high atomic number have a negative packing fraction; those of intermediate atomic number have a positive packing fraction and are more stable. If heavy nuclei are split to form nuclei of the more stable elements there is a loss of mass; there is a similar loss of mass if light nuclei are fused to give intermediate elements. This loss in mass is evolved as energy.

The effect of a slow or 'thermal' neutron on a nucleus of ^{235}U is to split it into one or more neutrons and into large fragments of approximately equal mass. There is a liberation of energy equal to the loss in total mass. If the neutrons produced effect further fissions, a 'chain-reaction' of successive fissions may be set up. ^{232}U, ^{233}U, ^{235}U, ^{239}Pu, ^{241}Am and ^{242}Am undergo fission with thermal neutrons; of these isotopes ^{235}U and ^{239}Pu are the most important as they are most readily obtainable. Other heavy nuclei require fast neutrons to induce fission; such neutrons are much more difficult to control into a self-sustaining chain-reaction.

The rapid fission of a mass of ^{235}U or another heavy nucleus is the principle of the atomic bomb, the energy liberated being the destructive power. For useful energy the reaction has to be moderated; this is done in a reactor where moderators such as water, heavy water, graphite, beryllium, etc., reduce the number of neutrons and slow those present to the most useful energies. The heat produced in a reactor is removed by normal heat-exchange methods. The neutrons in a reactor may be used for the formation of new isotopes, e.g. the transuranic elements, further fissile materials (^{239}Pu from ^{238}U), or of the radioactive isotopes used as tracers. The engineering difficulties in building a reactor are high. owing to the destructive effect of neutrons on normal structure materials and to the necessity to use materials which will absorb neutrons selectively. Further difficulties arise from the separation and disposal of the often intensely radioactive fission products.

If light nuclei can be induced to fuse together to produce heavier nuclei, energy will again be liberated. Such fusion can take place only when the nuclei have high thermal energies (temperatures 10^8 °C). Temperatures of this order occur in the centre of stars, where atomic fusions are the major source of energy, and in an atomic bomb. The high temperatures produced in an atomic explosion are used to effect the fusion of very light nuclei (^1H, ^2H, ^3H) in a hydrogen bomb. The practical difficulties in creating a fusion process in a controlled manner are even greater than those of deriving energy from an atomic fission and it is not yet possible to get useful energy from a fusion process.

atomic heat The atomic heat of an element is the product of its atomic weight and specific heat. It is the heat capacity, water being taken as unity, of one gram-atom of the element. *See* Dulong and Petit's Law.

atomic mass *See* atomic weight.

atomic number The atomic number of an element is the number of unit positive charges (protons) contained in the nucleus of an atom of the element. It is the property of an element which determines the position of the element in the periodic table*. The atomic number can be determined directly from the X-ray spectrum of the element.

atomic orbital The energy levels in an atom may be described by the four *quantum numbers*. In wave mechanics, the energy of a particular system may be described by the Schrödinger equation* and the wave function Ψ_{nlm} may be used to represent a solution of the wave equation in terms of these three numbers. Wave functions may be used to describe electron distribution and are thus sometimes referred to as atomic orbitals.

The wave function Ψ_{100} ($n=1, l=0, m=0$) corresponds to a spherical electronic distribution around the nucleus and is an example of an s orbital. Solutions of other wave functions may be described in terms of p and d orbitals.

atomic radii Half the closest distance of approach of atoms in the structure of the elements. This is easily defined for regular structures, e.g. close-packed metals, but is less easy to define in elements with irregular structures, e.g. As. The values may differ between allotropes (e.g. C–C 1·54 Å in diamond and 1·42 Å in planes of graphite). Atomic radii are very different from ionic* and covalent* radii.

atomic spectrum *See* line spectra.

atomic weights, at. wt. Defined relative to ^{12}C as 12·0000, atomic weights are ratios of masses and are not to be confused with the actual mass of an atom. Also termed relative atomic mass*.

Atomic weights

	Symbol	*At.no.*	*At.wt.*
actinium	Ac	89	(227)
aluminium	Al	13	26·982
americium	Am	95	(243)
antimony	Sb	51	121·75
argon	Ar	18	39·948
arsenic	As	33	74·922
astatine	At	85	(210)
barium	Ba	56	137·33
berkelium	Bk	97	(247)
beryllium	Be	4	9·012
bismuth	Bi	83	208·9804
boron	B	5	10·81
bromine	Br	35	79·904
cadmium	Cd	48	112·41
caesium	Cs	55	132·905
calcium	Ca	20	40·08
californium	Cf	98	(251)
carbon	C	6	12·01
cerium	Ce	58	140·12
chlorine	Cl	17	35·453
chromium	Cr	24	51·996
cobalt	Co	27	58·933
copper	Cu	29	63·546
curium	Cm	96	(247)
dysprosium	Dy	66	162·50
einsteinium	Es	99	(254)
erbium	Er	68	167·26
europium	Eu	63	151·96
fermium	Fm	100	(257)
fluorine	F	9	18·998
francium	Fr	87	(223)
gadolinium	Gd	64	157·25
gallium	Ga	31	69·72
germanium	Ge	32	72·59
gold	Au	79	196·966
hafnium	Hf	72	178·49
helium	He	2	4·003
holmium	Ho	67	164·930
hydrogen	H	1	1.00794
indium	In	49	114·82
iodine	I	53	126·904
iridium	Ir	77	192·22
iron	Fe	26	55·847
krypton	Kr	36	83·80
lanthanum	La	57	138·906
lawrencium	Lr	103	(260)
lead	Pb	82	207·2
lithium	Li	3	6·941
lutetium	Lu	71	174·967
magnesium	Mg	12	24·305
manganese	Mn	25	54·938
mendelevium	Md	101	(258)
mercury	Hg	80	200·59
molybdenum	Mo	42	95·94
neodymium	Nd	60	144·24
neon	Ne	10	20·179

	Symbol	At.no.	At.wt.
neptunium	Np	93	237·048
nickel	Ni	28	58·69
niobium	Nb	41	92·906
nitrogen	N	7	14·007
nobelium	No	102	(259)
osmium	Os	76	190·2
oxygen	O	8	15·999
palladium	Pd	46	106·42
phosphorus	P	15	30·974
platinum	Pt	78	195·08
plutonium	Pu	94	(244)
polonium	Po	84	(209)
potassium	K	19	39·0983
praseodymium	Pr	59	140·908
promethium	Pm	61	(145)
protactinium	Pa	91	231·036
radium	Ra	88	226·025
radon	Rn	86	(222)
rhenium	Re	75	186·2
rhodium	Rh	45	102·906
rubidium	Rb	37	85·468
ruthenium	Ru	44	101·07
samarium	Sm	62	150·36
scandium	Sc	21	44·956
selenium	Se	34	78·96
silicon	Si	14	28·086
silver	Ag	47	107.8682
sodium	Na	11	22·98977
strontium	Sr	38	87·62
sulphur	S	16	32·06
tantalum	Ta	73	180·9479
technetium	Tc	43	(97)
tellurium	Te	52	127·60
terbium	Tb	65	158·925
thallium	Tl	81	204·383
thorium	Th	90	232·038
thulium	Tm	69	168·934
tin	Sn	50	118·69
titanium	Ti	22	47·88
tungsten	W	74	183·85
uranium	U	92	238·029
vanadium	V	23	50·9415
xenon	Xe	54	131·29
ytterbium	Yb	70	173·04
yttrium	Y	39	88·906
zinc	Zn	30	65·38
zirconium	Zr	40	91·22

Bracketed figures are for the most commonly available isotope in the case of artificial elements.

The atomic weight of a particular sample depends on the origin of the sample.

Atomic weights were formerly defined relative to one atom of H and then to ^{16}O.

ATP *See* adenosine triphosphate.

atrazine, 2-chloro-4-ethylamino-6-isopropylamino-1,3,5-triazine, $C_8H_{14}ClN_5$. A soil-acting herbicide with some contact action. Colourless, m.p. 173–175°C. It is widely used as a selective pre- and post-emergence herbicide on many crops including maize, sorghum and sugar cane. It is relatively non-toxic to animal life.

atrolactic acid, 1-hydroxy-1phenylpropionic acid, $C_9H_{10}O_3$, $PhC(OH)(Me)CO_2H$. Colourless racemate, m.p. 94·5°C, soluble in water and polar organic solvents. Prepared from acetophenone cyanhydrin.

atropine, (±)-hyoscyamine, $C_{17}H_{23}NO_3$.

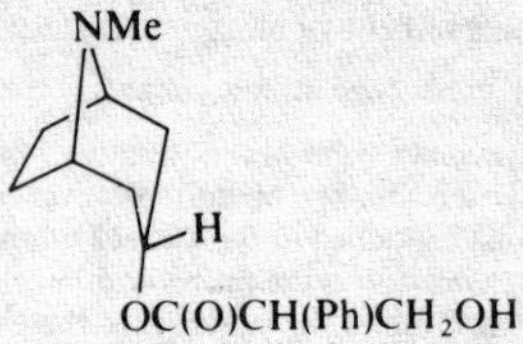

M.p. 114–116°C. Prepared by racemization of hyoscyamine. It and its salts are used to dilate the pupil of the eye. Given internally they reduce the secretion of saliva and relieve spasmodic pains.

attapulgite *See* fuller's earth.

attrition mill *See* size reduction equipment.

Au Gold.

Aufbau principle In building up the electronic configuration of an atom or a molecule in its ground state, the electrons are placed in the orbitals in order of increasing energy.

auger spectroscopy The photoinduced ejection of an electron from an atom causes the creation of an electron hole or vacancy (*see* photo-electron spectroscopy). This process is followed by an internal electronic reorganization in which an electron from a higher energy level falls into the vacancy. The excess energy associated with this process is released either as X-ray fluorescence, or by the ejection of a second electron (the auger electron). The kinetic energy of the auger electron (E_A) is characteristic of the element and its environment. Auger spectroscopy is particularly useful with the lighter elements and may be used to identify elements in the surface layers of solid samples.

auramine, Basic Yellow 2 Important yellow triarylmethane basic dyestuff used for wool or cotton.

$$[H_2NC(C_6H_4NMe_2\text{-}p)_2]^+Cl^-$$

aurates Gold-containing oxo compounds.

auric Gold(III) compounds.

aurine, rosolic acid, $(p\text{-}C_6H_4OH)_2\text{-}(p\text{-}C_6H_4O)_2C$. Yellow dyestuff. It is no longer used for dyeing textiles, but its lakes are used for staining paper, and it is also used as an indicator in alkalimetry.

aurous Gold(I) compounds.

austenite *See* iron, steel.

auto-catalysis The process whereby the reaction products catalyse the further reaction of the reactants.

autoclave Metal equipment, sometimes lined with glass or inert plastic, used for reaction between gases under pressure and liquid or solid reactants.

autofining process A process for the catalytic desulphurization* of oil fractions, all the hydrogen for the reaction being obtained from the dehydrogenation of naphthenes in the oil being treated.

autolysis The self-destruction which occurs in living systems after death. It is mainly due to the release of enzymes released from the lysosomes*.

autoxidation An oxidation reaction which proceeds only when another oxidation reaction is occurring simultaneously in the same system. For example, during the oxidation of oil of turpentine by atmospheric oxygen it is possible to simultaneously oxidize potassium iodide to iodine in the same system. The turpentine is termed the *autoxidator*; the iodide the acceptor.

autunite, $Ca(UO_2)_2(PO_4)_2,nH_2O$.
A uranium mineral.

auxins Plant hormones which promote the apical growth of plant shoots and have a number of other effects on the growth of plants, including yeasts and fungi. Synthetic auxins widely used in weed control include 2,4-dichlorophenoxyacetic acid (2,4-D), 2-methyl-4-chlorophenoxyacetic acid (MCPA) and 2,4,5-trichlorophenoxyacetic acid (2,4,5-T).

auxochrome A group containing lone pairs of electrons and thus modifying the colour of a chromophore.

aviation turbo-fuels Fuels used for aircraft jet engines, consisting either of kerosine* or 'wide-cut' mixtures of naphtha* and kerosine. While the main requirements are good combustion characteristics and a high energy content there are rigorous specifications for various physical and chemical properties.

Avogadro's number, *L* The number of particles (atoms or molecules) in one mole of any pure substance. $L = 6{\cdot}023 \times 10^{23}$. It has been determined by many methods including measurements of Brownian movement, electronic charge and the counting of α-particles.

axes of symmetry *See* symmetry elements.

axial *See* conformation.

axial ratios The ratio of the cell dimensions (a, b and c) in a crystal; b being taken as unity. In triclinic, monoclinic and orthorhombic systems

$$a \neq b \neq c$$

the axial ratios are, therefore, quoted as

$$a/b : 1 : c/b$$

In hexagonal and tetragonal systems

$$a = b \neq c$$

only $c{:}a$ is required. In rhombohedral and cubic systems $a = b = c$. *See* crystal structure and crystal systems.

axis of symmetry *See* crystal symmetry and symmetry elements.

azelaic acid, lepargylic acid, $C_9H_{16}O_4$, $HOOC{\cdot}(CH_2)_7COOH$. Colourless plates, m.p. 106°C. Made by the oxidation of oleic acid with ozones.

azeotropic distillation If two substances have similar boiling points or form an azeotrope, separation by fractional distillation is either very difficult or impossible. The method of azeotropic distillation consists of adding a third substance which forms azeotropic mixtures* with the other two; these azeotropes are then further treated to obtain pure compounds. Used in separating ethanol and water using benzene.

azeotropic mixtures Mixtures of liquids when distilled may reach a stage at which the composition of the liquid is the same as that of the vapour.

Many liquid mixtures exhibit a minimum boiling point (e.g. methanol and chloroform; *n*-propanol and water) whilst others show a maximum b.p. (e.g. hydrochloric acid and water) when vapour pressures are plotted as a function of change in composition of the mixture. An example of a system (A + B) where a minimum is seen is shown in the figure. The upper curve represents the composition of the vapour phase whilst the lower that of the liquid. When the liquid of composition x boils at temperature t it is seen that the composition of the vapour (v) differs in that it is richer in constituent A than the liquid, and an increase in the b.p. would eventually give pure B.

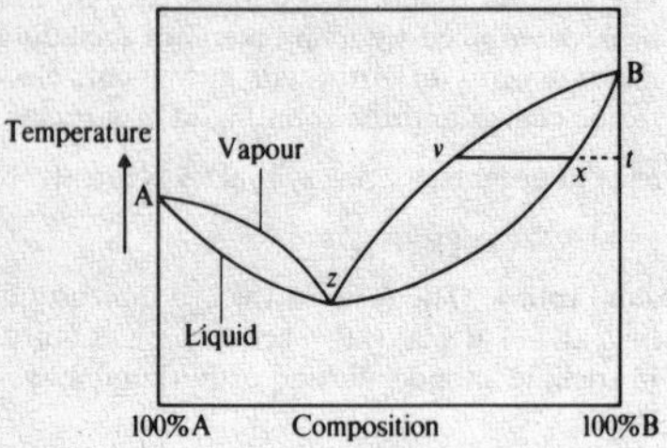

At z in the curve, however (the minimum of vapour pressure), the solution and vapour are in equilibrium and the liquid at this point will distil without any change in composition. The mixture at z is said to be azeotropic or a constant boiling mixture. The composition of the azeotropic mixture does vary with pressure.

azetidine   Colourless liquid, b.p. 61°C, smell-

$$CH_2 - NH$$
$$CH_2 - CH_2$$

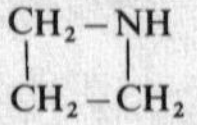

ing of ammonia and soluble in polar solvents. Azetidine-2-carboxylic acid has been isolated from leaves of lily-of-the-valley, and arises biosynthetically from methionine*. The fused azetidine ring is present in the penicillin* structure.

azides   Salts of hydrazoic acid, HN_3. Heavy metal azides (often insoluble) are extremely explosive and are used in percussion caps. Alkali azides, e.g. NaN_3 ($NaNH_2$ plus N_2O) are stable but decompose to, e.g., Na and N_2 on heating. The azide group is a good complexing agent. The azide ion is linear. Covalent azides are generally relatively stable.

azidocarbondisulphide, $(SCSN_3)_2$. A pseudo-halogen (oxidation of $MSCSN_3$ or MN_3 plus CS_2). White, shock-sensitive solid, slightly soluble in water.

azidodithiocarbonates   Compounds containing the $-NCSN_3$ group.

azimuthal quantum number, l   Also called the

angular momentum quantum number. *See* electronic configuration.

azine dyes   Series of related dyestuffs with the

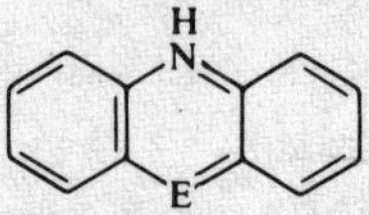

$$E = NR \text{ azines}$$
$$E = O^+ \text{ oxazines}$$
$$E = S^+ \text{ thiazines}$$

basic skeleton shown. Have very brilliant hues but are not very fast.

azines   Compounds containing $C=N-N=C$ groups prepared from $C=O$ compounds and hydrazine. Also used for compounds containing multiple N_2 groups and azine dyes.

azinphos-methyl, $C_{10}H_{12}N_3O_3PS_2$. A non-systemic insecticide and acaricide of long persistence. M.p. 73–74°C. It is used in horticulture and agriculture. It is hazardous to handle and toxic to wildlife, especially bees. A phosphorothiolothionate.

aziridine   *See* dihydroazirine.

azobenzene, $C_{12}H_{10}N_2$, PhN=NPh.
Orange-red crystals, m.p. 68°C. Reduces to aniline and also hydrazobenzene which can revert to benzidine under the reaction conditions. Prepared by the partial reduction of nitrobenzene.

It is the simplest compound containing the $-N=N-$ chromophore, but is of no practical importance since it lacks affinity for fibres. Normally it exists in the most stable *trans* form, but may be converted to a less stable, bright red *cis* form, m.p. 71·5°C by ultra-violet irradiation. *Cis* to *trans* isomerization occurs rapidly on heating.

azobisisobutyronitrile   White solid, m.p.

$$CH_3-\underset{\underset{CH_3}{|}}{\overset{\overset{CN}{|}}{C}}-N=N-\underset{\underset{CH_3}{|}}{\overset{\overset{CN}{|}}{C}}-CH_3$$

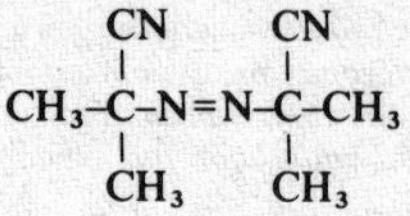

102°C, decomposes slowly into cyanopropyl radicals and is useful for initiation of radical reactions.

azo-compounds   A large group of compounds which contain the $-N=N-$ chromophore attached to two aromatic nuclei and which absorb light in the visible range of the spectrum. If the aromatic nuclei contain amino, hydroxyl, sulphonic acid or other salt-forming groups the compound will have some affinity for animal, vegetable or artificial fibres and will thus be a dyestuff.

Azo-compounds can be obtained by reduction of nitro-compounds, or by oxidation of hydrazo-compounds. They are usually prepared, however, by reacting a phenol or amine with a diazonium salt. The coupling usually takes place in the position para to the hydroxyl or amino group, but if this position is occupied it goes to the ortho position, e.g.

$$PhN_2Cl + PhOH \longrightarrow$$

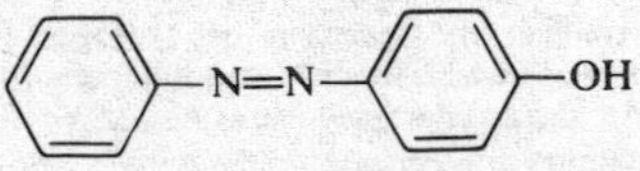

benzeneazophenol

The azo-compounds are usually very stable, and can be directly chlorinated, nitrated and sulphonated. On vigorous reduction the molecule splits at the azo group to give two molecules of primary amines, e.g. benzeneazophenol gives $PhNH_2$ and *p*-$HOC_6H_4NH_2$.

azo dyes Probably the most important dyestuffs. Contain one or more –N=N– chromophores. Prepared by coupling of a diazonium salt with a phenol or aromatic amine or by coupling heterocyclic hydrazines with aromatic amines. Many azo dyestuffs are sulphonic acid derivatives and are thus soluble in water. They are used as acid dyes for wool and affinity for cotton is increased by the number of auxochromes or azo groups. Alkylation of phenolic groups increases fastness. Azo dyes are used as pigments, in colour photography and are used as dyestuffs for most materials. They can be used as mordant or pre-metallized dyestuffs.

azoic dyes Water-insoluble azo dyes also known as ice colours and ingrain colours. In use they are formed within the fibre.

azoimides, azimino compounds Compounds

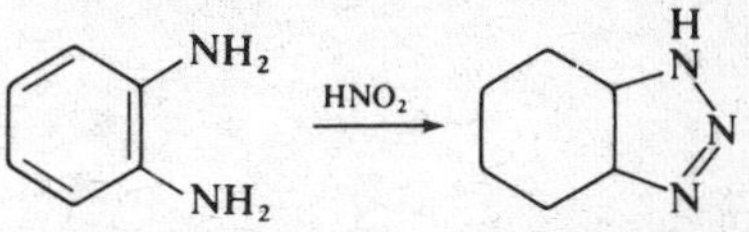

obtained on treating an *o*-arylene-diamine with nitrous acid when normal diazo compounds are not formed. Thus *o*-phenylenediamine (1,2-diaminobenzene) gives azoiminobenzene.

azomethines *See* Schiff's bases.

azophenols Compounds obtained by coupling a diazonium salt with a phenol.

azulene, $C_{10}H_8$. A blue-violet crystalline

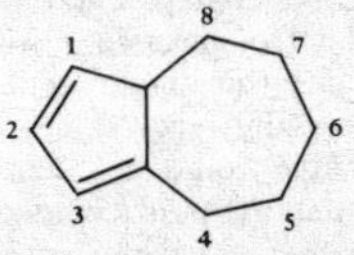

solid with a naphthalene-like smell, m.p. 99°C.

The generic term 'azulene' was first applied to the blue oils obtained by distillation, oxidation, or acid-treatment of many essential oils. These blue colours are usually due to the presence of either guaiazulene or vetivazulene. The parent hydrocarbon is synthesized by dehydrogenation of a cyclopentanocycloheptanol or the condensation of cyclopentadiene with glutacondialdehyde anil.

Azulene is an aromatic compound and undergoes substitution reactions in the 1-position. At 270°C it is transformed into naphthalene.

azurite, $Cu_3(OH)_2(CO_3)_2$. Blue basic carbonate of copper. An artificial form is known as Blue Verditer or Blue Ashes and is used as an artists' colour.

B

B Boron.

Ba Barium.

bacitracin Polypeptide antibiotic produced by a strain of *B. subtilis*. Bacitracin is chiefly active against gram-positive organisms, e.g. streptococci, staphylococci, gonococci and meningococci. It is used against organisms resistant to penicillin, especially for skin and eye infections. Prolonged internal administration may cause damage to the kidneys.

back bonding Overlap of a filled orbital, generally a d orbital, of an acceptor with a vacant orbital d, p, or π (bonding *or* antibonding) of a ligand. Formally relieves the build-up of negative charge on the central atom. Particularly important in complexes in low oxidation states, e.g. metal carbonyls. *See* acceptor.

baddeleyite One form of ZrO_2 and a principal mineral source of Zr.

bag filter A type of air filter*.

bakelite Phenol-formaldehyde (phenol-methanal) plastics and resins; a very widely used range of plastic materials.

baking powders Substitutes for yeast used in baking depending for their action on the slow evolution of CO_2. They consist of various acid substances (potassium hydrogen tartrate, calcium hydrogen phosphate, sodium hydrogen phosphates) with $NaHCO_3$ or NH_4HCO_3 together with added inert materials such as flour or starch.

BAL (British Anti-Lewisite), dimercaprol, dithioglycerol, $C_3H_8OS_2$, $CH_2SH{\cdot}CHSH{\cdot}CH_2OH$. Usually obtained as an oil, m.p. 77°C. Developed as an antidote to poisoning by organic arsenicals by external application, it is of use in poisoning by Hg, Cu, Zn, Cd but not Pb. It acts by forming a chelate with the metal and so removing it from the system.

balance An instrument used for weighing. Sensitive balances operate on a lever principle, less sensitive balances can use springs. The most sensitive balances – electrobalances – can weigh to a sensitivity of 0·1 μg.

ball mill *See* size reduction equipment.

Balmer series Frequencies of certain lines in the spectrum of hydrogen are simply related to each other, and can be expressed by a general formula. One group of lines is termed the Balmer series. Other series were later discovered in the spectrum of hydrogen by Lyman, Paschen, Brackett and Pfund.

All these series may be expressed by the equation

$$v = 109\,678{\cdot}8/n^2 - 109\,678{\cdot}8/m^2,$$

where v is the wave-number, and n and m are integers which can only have certain definite values for each series. Thus, in the Lyman series, $n=1$, $m=2$, 3, 4, etc. In the Balmer series, $n=2$; in the Paschen series, $n=3$; in the Brackett series, $n=4$; and in the Pfund series, $n=5$. 109678 is the Rydberg constant.

band spectrum The absorption of energy by a molecule may lead to the transition of an electron from an orbital of lower to one of higher energy (*see* line spectrum), but in addition there may be an increase in the energy of rotation of the molecule as a whole, and in the energy of vibration of the constituent atoms relative to each other. The simultaneous occurrence of these changes gives rise to many possible energy changes. The resulting spectrum is correspondingly complicated, and consists of numerous very closely spaced lines. The frequency of a line is determined by the algebraic sum of the energy changes involved. An electronic transition is associated with so much more energy than vibrational and rotational changes, that it determines the general position of the line in the spectrum. The spectrum thus consists of well-defined groups or *bands*, of closely packed lines. Such a spectrum, characteristic of a molecule, is termed a *band spectrum*.

band theory of solids In molecules, molecular orbitals arise by combination of atomic orbitals. The ideas of delocalized electrons can also be applied to large assemblies of atoms in solids. The quantum mechanical treatment of these electrons gives rise to the band theory of solids, in which there are bands of electron levels, sometimes overlapping, e.g. metals, sometimes with a forbidden energy gap, e.g. insulator oxides, between them.

barban, $C_{11}H_9Cl_2NO_2$. A translocated

herbicide, m.p. 75–76°C. Used to kill wild oats without serious damage to wheat, barley and various legumes. It is relatively non-toxic but an allergic reaction has been noted in man. A carbamate derivative.

Barbier-Wieland degradation A method for the stepwise degradation of aliphatic acids

Barbier-Wieland degradation

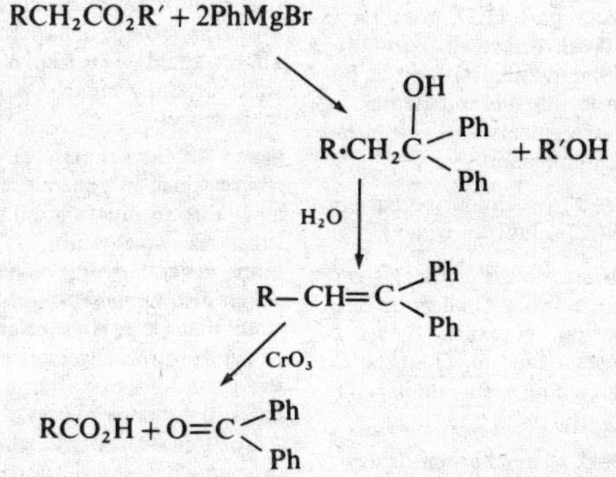

when one carbon atom is eliminated in each step. Opposite in objective to the Arndt–Eistert synthesis; especially useful for degrading side chains of steroid and other terpenoid molecules.

barbitone, diethylmalonylurea, $C_8H_{12}N_2O_3$. White crystals, m.p. 191°C. A barbituric acid derivative. The sodium salt is administered orally as a sedative.

barbiturates, barbituric acids Important group of central nervous system depressant drugs derived from barbituric acid by replacement of the hydrogen atoms on C_5 by alkyl or aryl groups or by substitution of S for O on C_2. They are colourless, crystalline solids, having a slightly bitter taste. Manufactured by heating the appropriate dialkylmalonic ester with urea and sodium ethoxide. They are administered orally as sedatives or intravenously as anaesthetics. Some are also effective in treating epilepsy.

barbituric acid, malonylurea, $C_4H_4N_2O_3$.

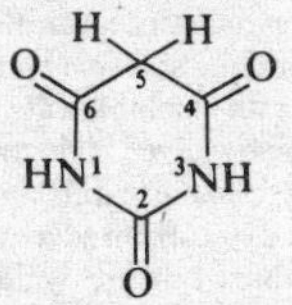

Crystallizes as the dihydrate in colourless prisms, m.p. 253°C. Prepared by heating malonic ester with urea and sodium ethoxide, or by heating alloxantin with sulphuric acid. It is a dibasic acid.

barium, Ba. At.no. 56, at.wt. 137·33, m.p. 725°C, b.p. 1640°C, *d* 3·51. An alkaline earth occurring principally as $BaSO_4$ (barytes) and as $BaCO_3$ (witherite, alstonite). The metal (bcc) is prepared by electrolysis of fused $BaCl_2$ or $BaCl_2$ plus Na. It is silver white in colour, oxidizes easily in air. Used as a getter for oxygen removal. Ba salts find extensive use, $BaSO_4$ used in paints (lithopone), X-ray diagnosis, glass, oil well drilling fluids; $BaCO_3$ as a rat poison. World production Ba minerals *c.* 4 megatonnes p.a. Ba compounds are poisonous.

barium bromide, $BaBr_2$. *See* barium halides.

barium carbonate, $BaCO_3$. A white crystalline solid occurring native (witherite) or prepared from a soluble Ba salt and Na_2CO_3.

barium chemistry Barium is an element of Group II, electronic configuration $6s^2$. It forms a single series of colourless M^{2+} salts

$$(E^\circ \text{ for } Ba^{2+} \longrightarrow Ba \ -\ 2{\cdot}90 \text{ volts in acid solution})$$

The co-ordination number of Ba^{2+} is generally high.

barium chloride, $BaCl_2$. *See* barium halides.

barium chromate, $BaCrO_4$. A yellow pigment (Ba salt plus Na_2CrO_4) rather insoluble in water.

barium diphenylamine-*p*-sulphonate A redox indicator, used particularly in CrO_4^{2-}/Fe^{2+} titrations. In the oxidized form the indicator is deep reddish-violet.

barium fluoride, BaF_2. *See* barium halides.

barium halides, BaX_2. Barium forms a single series of halides. BaF_2 (Ba salt plus fluoride) is only sparingly soluble in water, m.p. 1285°C, b.p. 2137°C. It has some optical uses. $BaCl_2$ forms $BaCl_2,2H_2O$ and is used for the electrolytic production of Ba. $BaBr_2$ and BaI_2 are similar to $BaCl_2$.

barium hydroxide, baryta, $Ba(OH)_2$. A white true hydroxide (BaO plus H_2O) forming a sparingly soluble octahydrate, also pentahydrate. Used in volumetric analysis and in the manufacture of lubricating oils and greases, as a plastic stabilizer, in pigments, as a vulcanization accelerator, gasoline additive.

barium nitrate, $Ba(NO_3)_2$. Colourless barium salt (BaO plus HNO_3) soluble in water.

barium oxides Barium monoxide, BaO, (ignition of $BaCO_3$) white powder. On heating BaO in dry oxygen barium peroxide, BaO_2, is formed; it decomposes to BaO and O_2 at 800°C. BaO_2 is used as a bleaching agent and in H_2O_2 production.

barium peroxide, BaO_2. *See* barium oxides.

barium sulphate, $BaSO_4$. Occurs native (heavy spar, barytes) and a principal source of Ba compounds after reduction to BaS with carbon. Also used as a pigment, etc. (*see* barium, lithopone). $BaSO_4$ is insoluble in water and is used for the gravimetric estimation of Ba^{2+} and SO_4^{2-}. Administered orally to allow X-ray study of the intestines.

barium sulphide, black ash, BaS. Important intermediate in Ba chemistry ($BaSO_4$ plus C). Strongly phosphorescent as normally prepared because of the impurities.

barium titanate, $BaTiO_3$. An important solid state compound. Piezoelectric, used in transducers, ultrasonics, gramophone pick-ups and for polishing. Prepared from $BaCO_3$ and TiO_2 at high temperatures.

barn In a nuclear reaction, the probability of reaction may be expressed in terms of the nuclear cross section, σ. If N target atoms per cm^3 are bombarded with I_0 particles, the number of particles transmitted is given by the equation

$$I = I_0 e^{-N\sigma x}$$

where x is the thickness of the target sample. The cross section of the target atoms σ is expressed in barn units; 1 barn $= 10^{-24}\ cm^2$.

barometric condenser A plant for condensing steam at pressures below atmospheric.

barrel A measure of volume widely used in the petroleum industry. A barrel is approximately 0·159 m^3 (5·61 ft^3) and contains 35 U.K. gallons.

baryta, $Ba(OH)_2$. Barium hydroxide*.

barytes, heavy spar, $BaSO_4$. A common mineral, it often forms well-developed orthorhombic crystals. Pure specimens are colourless or white, but brownish or bluish tints may be introduced by impurities. The mineral is used as a source of barium compounds; when finely ground it is used as a pigment and in oil-well drilling muds. World use 1976 5 megatonnes.

base On the simplest view a base is a substance which in aqueous solution reacts with an acid to form a salt and water only, i.e. is a substance which furnishes hydroxyl ions. A more general definition (Lowry, Brønsted), which also applies to non-aqueous solutions, states that a base is a substance with a tendency to gain protons. By this definition, OH^- and the anions of weak acids, e.g. CH_3COO^-, are bases in aqueous solution.

For ionized solvents which do not contain protons a base is a substance which reacts with the acid of that system to give a salt and the solvent. Thus the base $KBrF_4$ reacts with the acid BrF_2SbF_6 to give the salt $KSbF_6$ and BrF_3 in bromine trifluoride.

A Lewis base is a molecule with available electrons, e.g. ammonia.

As an adjective applied to metals base represents the opposite of noble, i.e. a base metal would be attacked by mineral acids.

base exchange An old term used to describe the capacity of soils, zeolites, clays, etc. to exchange their cations (Na^+, K^+, Ca^{2+}) for an equivalent of other cations without undergoing structural change. An example of the general process of ion exchange*.

base-pairing A term used with reference to a nucleic acid double helical structure indicating the specific hydrogen-bonded association of pairs of complementary purine and pyrimidine bases, viz. adenine with thymine or uracil, and guanine with cytosine.

basic or cationic dyes Dyestuffs containing ionic species. Chlorides are generally soluble in water and organic salts, stearates or oleates, are soluble in organic solvents and are used in printing inks. Form insoluble salts with hetero-polyanions which have high fastness and brilliant shades. Used in paints and wallpaper pigments. Basic dyestuffs need a mordant with most natural fibres but are widely used for acrylates and polyester fibres.

basic salts Compounds intermediate between

simple salts and oxides or hydroxides. Formed by hydrolysis of metal salts and often precipitated by sodium carbonate solution.

basic slag *See* slag.

bastnaesite A mineral, $CeFCO_3$, used as a source of the lighter lanthanides.

bathochromic Shifts absorption to longer wavelength, lower frequency, lower energy. Compare hypsochromic*.

batteries, electric cells Systems in which chemical change is used to produce electricity – the systems often being portable. Many batteries are effectively non-rechargeable. The most important is of the Leclanché type. $MnO_2 \longrightarrow Mn^{3+} \longrightarrow Mn^{2+}$,|| $Zn \longrightarrow Zn^{2+}$ in a NH_4Cl–MnO_2–Zn system. Other important cells are
$HgO \longrightarrow Hg$,||$Zn \longrightarrow Zn^{2+}$ – the alkaline HgO-zinc system;
$CuO \longrightarrow Cu$,||$Zn \longrightarrow Zn(OH)_4{}^{2-}$ Lalande cell; Ag_2O/Zn cell, Li/F_2/graphite.
Fuel cells involve use of gaseous reactants to produce electricity – most often H_2–O_2 within a porous electrode. Secondary cells are rechargeable. The most important systems are lead-acid $PbO_2 \longrightarrow PbSO_4$; nickel-cadmium (NiO(OH) + $Cd \longrightarrow Ni(OH)_2 + Cd(OH)_2$); Ag–Zn; Ni–Fe; Ag–Cd; MnO_2–Zn. *See* lead accumulator.

bauxite The principal ore of aluminium, a mixture of minerals formed by the weathering of aluminium-containing rocks and containing various forms of aluminium hydroxides*. Readily attacked by acids but forms Al_2O_3 after heating to 900°C and then becomes almost insoluble; used in the manufacture of aluminous refractories, as a source of Al and Al compounds.

bayerite, α-$Al(OH)_3$. *See* aluminium hydroxide.

bcc Body-centred cubic. *See* close-packed structures.

B.C.F., bromochlorodifluoromethane, $CBrClF_2$. A gas used in fire extinguishers.

Be Beryllium.

Beckmann rearrangement The conversion of

$$\begin{array}{c} C_6H_5\text{–}C\text{–}C_6H_5 \\ \| \\ N{\cdot}OH \end{array} \longrightarrow \begin{array}{c} O{=}C{\cdot}C_6H_5 \\ | \\ HN{\cdot}C_6H_5 \end{array}$$

ketoximes to amides by intramolecular rearrangement with e.g., PCl_5, ethanoyl chloride, $SbCl_3$, sulphuric acid, or hydrochloric acid in ethanoic acid. With the oximes of cyclic ketones, the rearrangement involves an enlargement of the ring by the inclusion of the nitrogen atom. This process is employed in the conversion of cyclohexanone oxime into caprolactam*, the precursor of Nylon-6.

Beckmann thermometer A very sensitive mercury thermometer with a small temperature range which can be changed by transferring mercury between the capillary and a bulb reservoir. Used for accurate temperature measurements in the determination of molecular weights by freezing point depression or boiling point elevation.

beclomethasone, 9-chloro-11β,17,21-trihydroxy-16β-methylpregna-1,4-diene-3,20-dione, $C_{22}H_{29}ClO_5$. The dipropionate ester, $C_{28}H_{37}ClO_7$, melts with decomposition at 212°C and is used in ointments and creams for inflammatory conditions of the skin and in an aerosol to treat severe asthma.

Beer's law This states that the proportion of light absorbed depends on the thickness (d) of the absorbing layer, and on the *molecular concentration* (c) of the absorbing substance in the layer. It is an extension of Lambert's law*, and may be written in the form

$$I = I_0\, e^{-Kcd},$$

where I_0 is the intensity of the incident radiation, and I that of the transmitted radiation; and K is a constant, the *molecular absorption coefficient*, which is characteristic of the absorbing substance for light of a given frequency. The *molecular extinction coefficient*, α_m, is the thickness in cm of a layer of a molar solution which reduces the intensity of light passing through it to one-tenth of its original value. $\alpha_m = 0{\cdot}4343\ K$. The law does not hold for substances whose absorption spectra and the species present are affected by concentration.

beeswax A mixture of mercyl palmitate, cerotic acid and esters, and some paraffins produced by bees. Used in floor waxes, shoe polishes, leather treatment.

behenic acid, n-docosanic acid, $C_{22}H_{44}O_2$, $CH_3{\cdot}(CH_2)_{20}{\cdot}COOH$. M.p. 80°C. A fatty acid occurring in oil of ben.

Beilstein, Handbuch der Organischen Chemie The definitive reference series on organic chemistry.

Beilstein's test A method for detecting the presence of halogen in an organic compound. A piece of copper gauze is heated in an oxidizing flame until the flame is no longer tinged green; the compound is placed on the gauze, which is reheated. If Cl, Br, or I is present the flame is tinged bright green.

belladonna, deadly nightshade *Belladonna* owes its medicinal activity largely to the alkaloid hyoscyamine* which is present in all parts of the plant. It is effective internally as an antispasmodic and in checking excessive secretion. Externally, it is used to relieve the pain of lumbago and neuralgia.

bemegride, β-ethyl- β-methylglutarimide,

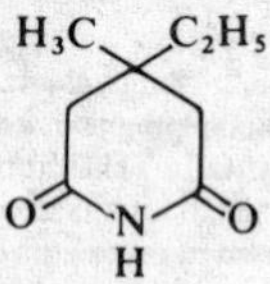

$C_8H_{13}NO_2$. White flaky or crystalline powder, m.p. 126–128°C. Used in the treatment of barbiturate poisoning, by intravenous injection.

benazolin, 4-chloro-2-oxobenzothiazolin-3-ylethanoic acid, $C_9H_6ClNO_3S$. Crystalline solid, m.p. 193°C. It is a post-emergence herbicide effective against chickweed and cleavers. It is relatively non-toxic.

bendrofluazide White crystalline powder,

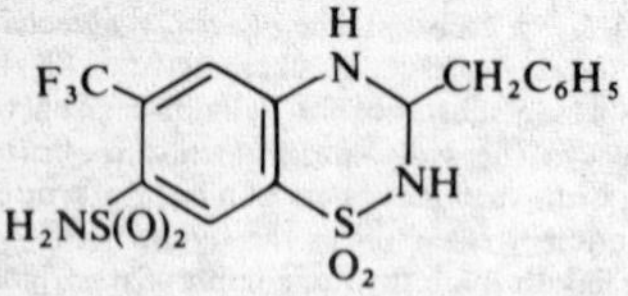

m.p. about 225°C with decomposition. A diuretic used against oedema and hypertension.

Benedict solution Aqueous solution of Na_2CO_3, $CuSO_4$, and sodium citrate used for testing for reducing agents, particularly sugars, which give red-yellow colours or precipitates.

Benfield process Removal of carbon dioxide from fuel gases, such as those obtained by gasifying coal in the Lurgi process*, by countercurrent scrubbing of the gases by hot potassium carbonate solution.

benomyl, methyl N-(1-(butylcarbamoyl)-2-benzimidazole carbamate, $C_{14}H_{18}N_4O_3$. A systemic fungicide; white crystalline solid which decomposes before melting. It is active against a wide variety of fungal pathogens especially the powdery mildews. It is non-toxic

to animal life and so is widely used by amateur gardeners.

bentonite A clay-like mineral consisting largely of montmorillonite. Forms gelatinous suspensions at low concentrations and used as a binder for foundry sand and in oil-drilling needs. One form is used for its absorptive properties.

benzalacetophenone, chalcone, 1,3-diphenyl-2-propen-1-one, $C_{15}H_{12}O$, Ph·CH=CHC(O)·Ph. Crystallizes in pale yellow prisms, m.p. 58°C. It can be prepared by condensing acetophenone and benzaldehyde in the presence of sodium ethoxide.

benzal chloride, benzylidene chloride, α,α-dichlorotoluene, $C_7H_6Cl_2$, PhCHCl_2.
Colourless strongly refracting lachrymatory liquid, b.p. 206°C. Prepared by the direct chlorination of toluene in the absence of metallic catalysts.

It is used for the preparation of benzaldehyde. For this purpose it is usually heated with slaked lime and water under pressure.

benzaldehyde, C_7H_6O, PhCHO. Colourless refractive liquid with almond-like odour. B.p. 180°C.

It occurs in nature as part of the glucoside, amygdalin. Manufactured from toluene, either by vapour phase air oxidation over a catalyst, or by chlorination to benzal chloride followed by hydrolysis with boiling water.

It is readily oxidized by air to benzoic acid. With aqueous KOH gives benzyl alcohol and benzoic acid. Gives addition products with hydrogen cyanide and sodium hydrogen sulphite.

Its chief importance is as a source of cinnamic acid by condensation with sodium ethanoate and ethanoic anhydride and as a source of triphenylmethane dyestuffs by condensation with pyrogallol, dimethylaniline, etc. It is also used in the manufacture of perfumes.

benzaldoxime, C_7H_7NO. There are two

$$C_6H_5·CH \qquad C_6H_5·CH$$
$$\alpha \quad \overset{\|}{N·OH} \qquad \beta \quad \overset{\|}{HO·N}$$

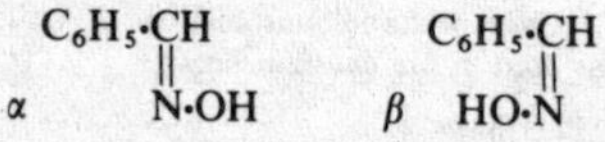

stereo isomers. The β-isomer cyclizes to a benz-*iso*oxazole and has the *anti*-configuration shown. The α-isomer is stable and is prepared by the action of hydroxylamine on benzaldehyde, m.p. 34°C. The β-isomer is obtained by irradiating a benzene solution of α-isomer, or by saturating an ethereal solution with hydrogen chloride, m.p. 127°C.

benzalkonium chloride Mixture of alkylbenzyldimethylammonium chlorides, generally in aqueous solution. A clear, colourless or pale yellow syrupy liquid. Used as an antiseptic detergent and preservative in some pharmaceutical preparations, e.g. eyedrops. The corresponding hydroxide in aqueous solution is a strong base used in phase-transfer chemistry.

benzamide, C_7H_7NO, $PhC(O)NH_2$. Colourless, m.p. 130°C, b.p. 288°C. Dehydrated by phosphorus pentoxide to benzonitrile, hydrolysed to benzoic acid by dilute acids or alkalis. Forms metallic salts, e.g. silver benzamide, $C_6H_5{\cdot}CONHAg$.

Prepared by the action of ammonia on benzoyl chloride or benzoic esters, or by partial hydrolysis of benzonitrile.

benzanthrone, $C_{17}H_{10}O$. Pale yellow needles,

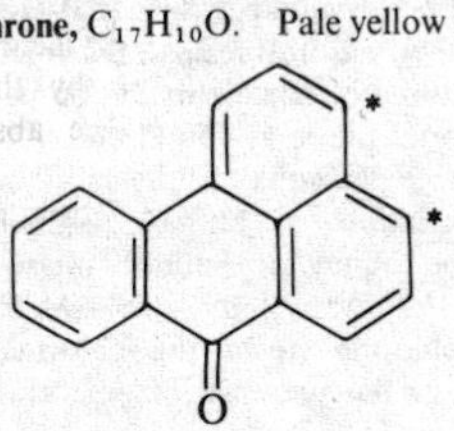

m.p. 170°C. It is prepared by heating anthrone or anthranol with glycerol and sulphuric acid for a few hours at 120°C.

It is a valuable dyestuffs intermediate. When fused with KOH, two molecules of benzanthrone combine at the positions marked * above to give a dibenzanthrone which is in itself a vat dye. Also used as an intermediate for the preparation of the valuable Caledon Jade Green.

Benzedrine A trade name for a brand of amphetamine.

benzene, C_6H_6. The benzene molecule con-

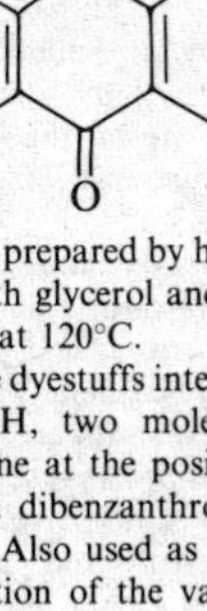

tains six C atoms arranged in the form of a regular hexagon; the C–C bond length is 1·39 Å. The six π-electrons, or aromatic sextet, represented by the double bonds, are not localized but are distributed uniformly around the ring. A thin, colourless, highly refractive liquid. Characteristic smell. Highly inflammable, burning with a smoky yellow flame. A good solvent for fats and lower m.w. aromatic compounds. Miscible with alcohol, ether, propanone and ethanoic acid. M.p. 5·49°C, b.p. 80·2°C, d^{15} 0·885. Shows strong absorption bands in the u.v. The vapour is toxic when inhaled over long periods.

Benzene was first isolated by Faraday in 1825 from the liquid condensed by compressing oil gas. It is the lightest fraction obtained from the distillation of the coal-tar hydrocarbons, but most benzene is now manufactured from suitable petroleum fractions by dehydrogenation (54%) and dealkylation processes. Its principal industrial use is as a starting point for other chemicals, particularly ethylbenzene, cumene, cyclohexane, styrene (45%), phenol (20%), and Nylon (17%) precursors. U.S. production 1979 2·6 B gals.

Structurally benzene is the simplest stable compound having aromatic character*, but a satisfactory graphical representation of its formula proved to be a perplexing problem for chemists. Kekulé is usually credited with description of two resonating structures which,

together with other structures, contribute to the structure of benzene. A full account of the historical controversies is given in most organic chemistry textbooks. Structures proposed by Dewar have since been recognized as different, distinct compounds.

The Hückel description of aromaticity was based in part on benzene, a cyclic fully conjugated hydrocarbon having $(4n+2)$ π-electrons $(n=1)$ in the closed shell (ring).

Benzene can undergo addition reactions which successively saturate the three formal double bonds, e.g. up to 6 chlorine atoms can be added under radical reaction conditions whilst catalytic hydrogenation gives cyclohexane.

The most widely used reactions are those of electrophilic substitution, and under controlled conditions a maximum of three substituting groups, e.g. $-NO_2$ (in the 1,3,5 positions) can be introduced by a nitric acid/sulphuric acid mixture. Hot conc. sulphuric acid gives sulphonation whilst halogens and a Lewis acid catalyst allow, e.g., chlorination or bromination. Other methods are required for introducing fluorine and iodine atoms. Benzene undergoes the Friedel–Crafts reaction*.

Nucleophilic substitution of benzene itself is not possible but the halogeno derivatives undergo nucleophilic displacement or elimination reactions (*see* arynes). Substituents located in the 1,2 positions are called *ortho-*; 1,3 *meta-* and 1,4 *para-*.

Benzene forms a wide range of organometallic complexes with the transition metals, using π-electrons for bonding to the metal, e.g. dibenzene chromium.

benzene-1,3-disulphonic acid, $m\text{-}C_6H_4(SO_3H)_2, C_6H_6O_6S_2$. Very deliquescent crystals, $+2\frac{1}{2}H_2O$. It is prepared by sulphonation of benzene at 225°C. It gives resorcinol on fusion with KOH.

benzene hexachloride *See* BHC.

benzenesulphonic acid, $C_6H_6O_3S$, $PhSO_3H$. Colourless deliquescent plates with $1\frac{1}{2}H_2O$ (from water), m.p. 43–44°C. Anhydrous acid, m.p. 65–66°C.

Prepared by the sulphonation of benzene in the liquid state or by passing benzene vapour into concentrated sulphuric acid at 150–180°C.

Forms water-soluble alkali and alkaline earth metal salts. Heating with KCN gives benzonitrile and phenol is formed by fusion with NaOH or KOH. Further sulphonation at 250°C gives benzene-1,3-disulphonic acid.

benzfuran (coumarone) ring system The

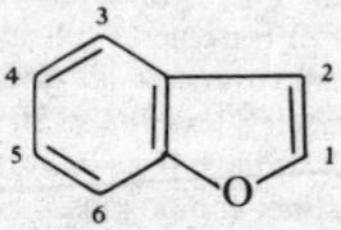

system numbered as shown.

benzhexol, trihexyphenidyl, 3-(1-piperidyl)-1-cyclohexyl-1-phenyl-1-propanol, $C_{20}H_{31}NO$. M.p. 115°C. An atropine*-like drug which acts mainly on the brain to give some relief from the rigidity of Parkinson's disease.

benzidine, 4,4′-diaminobiphenyl, $C_{12}H_{12}N_2$. M.p. 127·5°C. Dibasic, readily diazotizes. Prepared by treating hydrazobenzene with hydrochloric acid, intramolecular rearrangement taking place.

A potent carcinogen, it is now little used for dyestuffs for cotton. Still used for blood detection.

benzidine conversion The intramolecular rearrangement which occurs when the hydrazobenzenes are heated in acid solution. If both *para* positions in the hydrazobenzene are free a 4,4′ benzidine is produced; if one or both para positions are occupied by other groups then either a 2-benzidine or 2- or 4-semidines ($ArNHAr'NH_2$) are formed.

benzil, $C_{14}H_{10}O_2$, PhC(O)C(O)Ph. M.p. 95°C, b.p. 346–348°C (decomp.). Prepared by the oxidation of benzoin with nitric acid. It is thought to be carcinogenic.

benzine A term still used generically to describe special boiling point spirit* (SBP). This name is not recommended for petroleum solvent since it may be confused with the more toxic pure aromatic benzene.

benzoates Salts or esters of benzoic acid.

benzocaine, ethyl *p*-aminobenzoate, $C_9H_{11}NO_2$. White crystals, m.p. 90–91°C. Prepared from *p*-nitrotoluene by way of *p*-aminobenzoic acid. It is used as a local anaesthetic on mucous surfaces internally and by injection, and is taken internally to relieve gastric pain.

benzoic acid, $C_7H_6O_2$, PhCOOH. Colourless lustrous leaflets, m.p. 122°C, b.p. 249°C.

It was first described in 1608 when it was sublimed out of gum benzoin. It also occurs in many other natural resins. Benzoic acid is manufactured by the air oxidation of toluene in the liquid phase at 150°C and 4–6 atm. in the presence of a cobalt catalyst; by the partial decarboxylation of phthalic anhydride in either the liquid or vapour phase in the presence of water; by the hydrolysis of benzotrichloride (from the chlorination of toluene) in the presence of zinc chloride at 100°C.

It gives benzene when heated with soda lime. It is very stable towards oxidizing agents.

Much of the benzoic acid produced is converted to sodium benzoate, which is used as a food preservative (as is the acid) and a corrosion inhibitor. Other important uses of the acid are in the manufacture of alkyd resins, plasticizers, caprolactam, dyestuffs and pharmaceuticals.

benzoin, $C_{14}H_{12}O_2$, PhC(O)CH(OH)Ph. The (±)-compound has m.p. 137°C.

Usually prepared by the action of NaCN on benzaldehyde in dilute alcohol. It is oxidized by nitric acid to benzil, and reduced by sodium amalgam to hydrobenzoin PhCHOHCHOHPh by tin amalgam and hydrochloric acid to desoxybenzoin, $PhCH_2COPh$; and by zinc amalgam to stilbene PhCH=CHPh. It gives an oxime, phenylhydrazone and ethanoyl derivative. The α-oxime is used under the name 'cupron' for the estimation of copper and molybdenum.

The name is also given to a balsamic resin obtained from *Styrax benzoin*, which is carminative and mildly expectorant.

benzole A mixture of predominantly aromatic hydrocarbons produced by the carbonization of coal, obtained from coal gas by adsorption or from coal tar by distillation.

benzonitrile, C_7H_5N, PhCN. Colourless refractive liquid, b.p. 191°C. Slightly soluble in water, miscible with alcohol and ether in all proportions.

Prepared by the dehydration of benzamide. Hydrolysed by dilute acids and alkalis to benzoic acid. Good solvent.

benzophenone, $C_{13}H_{10}O$, PhC(O)Ph. Colourless rhombic prisms, m.p. 49°C, b.p. 306°C. Characteristic smell. It is prepared by the action of benzoyl chloride upon benzene in the presence of aluminium chloride (Friedel–Crafts reaction) or by the oxidation of diphenylmethane. It is much used in perfumery. Forms a ketyl* with sodium.

benzo(a)pyrene, 1,2-benzpyrene, $C_{20}H_{12}$.

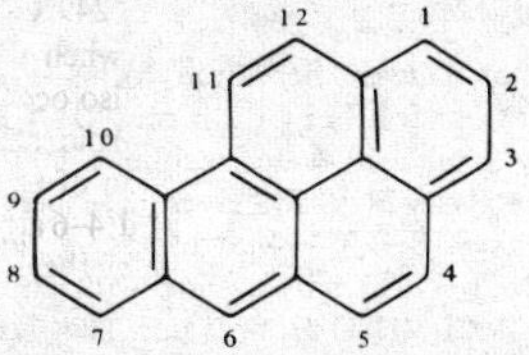

Pale yellow crystals, m.p. 179°C. A constituent of coal tar with strong carcinogenic properties.

benzoquinone, $C_6H_4O_2$. Yellow monoclinic prisms, m.p. 115·7°C, readily sublimes, volatile

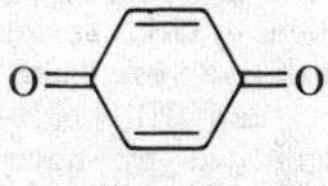

in steam, penetrating odour. Electrolytic reduction gives quinhydrone, reduction with hydrogen sulphide gives hydroquinone. Gives mono- and di-oximes. It is prepared by the oxidation of aniline with chromic acid. Used as a source of hydroquinone and of some sulphur dyes.

benzotrichloride, α,α,α-trichlorotoluene, $C_7H_5Cl_3$, $PhCCl_3$. Colourless liquid, b.p. 213–214°C. Insoluble in water, miscible with organic solvents.

It is prepared by fully chlorinating toluene. When heated with water at 100°C, or with lime, benzoic acid is obtained.

benzoyl The group PhC(O)-.

benzoyl chloride, C_7H_5ClO, PhC(O)Cl. Colourless lachrymatory liquid, b.p. 198°C. It has a pungent smell, and is slowly hydrolysed by cold water to give benzoic acid. With alcohols benzoic esters are obtained. It is prepared by heating benzoic acid with PCl_5 or thionyl chloride. It is much used as a benzoylating agent (i.e. to add the benzoyl group in place of H) in organic chemistry.

benzyl The group $PhCH_2$-.

benzyl alcohol, α-hydroxytoluene, C_7H_8O, $PhCH_2OH$. Colourless liquid, b.p. 205°C. Oxidizes to benzaldehyde and benzoic acid. Prepared by the hydrolysis of benzyl chloride. It is used in perfumery in the form of its esters.

benzylamine, phenylmethylamine, α-aminotoluene, C_7H_9N, $PhCH_2NH_2$. A colourless liquid, b.p. 185°C, which, like many amines, becomes dark coloured on exposure to air. It behaves as a typical primary amine.

benzyl benzoate, $C_{14}H_{12}O_2$, $PhCH_2OC(O)Ph$. White crystals, m.p. 20°C. It is used as an external application for the treatment of scabies.

benzyl chloride, C_7H_7Cl, $PhCH_2Cl$. Colourless liquid with a characteristic odour, b.p. 179°C. It is slowly hydrolysed by boiling water, yielding benzyl alcohol.

It is prepared by the direct chlorination of toluene in the presence of PCl_5. It is purified by fractionation from the unchanged toluene and the higher chlorinated products. It is used for benzylating amines and for preparing benzyl alcohol.

benzyl ether, $(PhCH_2)_2O$. Colourless liquid, b.p. 298°C. Prepared by dehydration of benzyl alcohol. Used as a plasticizer for cellulose acetate and a solubilizer for gums, resins and rubbers.

benzylidene chloride *See* benzal chloride.

benzyne, C_6H_4. The simplest aryne, or de-

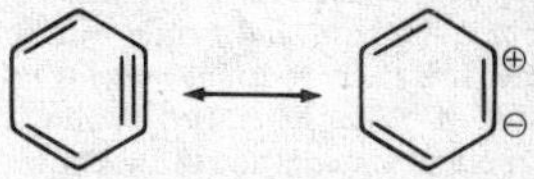

hydroaromatic compound. It is a highly reactive intermediate with an estimated lifetime of 10^{-5}–10^{-4} seconds in the vapour state, and cannot be isolated. A variety of methods are available for its production, including the decomposition of benzenediazonium-2-carboxylate and the base induced elimination of hydrogen halide from halobenzenes, e.g. phenyllithium and fluorobenzene. The reactivity of benzyne results from the formal incorporation of a triple bond in a six-membered ring: other evidence suggests that benzyne is better represented by a dipolar structure. Complexes to metals. *See* arynes.

berkelium, Bk. At.no. 97, m.p. 986°C, ^{249}Bk (314 days) is formed by the action of neutrons on ^{243}Am; ^{247}Bk (10^4 years) is much more stable but can only be produced in an accelerator. Bk is separated by ion exchange. The metal has been prepared by Li reduction of BkF_3, it has a double hexagonal close-packed structure and is a typical electropositive actinide.

berkelium chemistry Generally tripositive. BkF_4 (F_2 on BkF_3) and Bk(IV) formed BrO_3^- on Bk^{3+} ($Bk^{4+} \longrightarrow Bk^{3+}$ +1·6V). Cs_2BkCl_6 precipitated from HCl. All trihalides, BkOCl, Bk_2O_3 and BkO_2 are known. Bk is the last actinide to show the +4 state.

Berlin green, $FeFe(CN)_6$. *See* cyanoferrates.

Berry mechanism The mechanism postulated for the interchange of substituents in trigonal-bipyramidal 5-co-ordinate complexes, e.g. PF_5 and its substituted derivatives.

berthollide compound Solid phases showing a range of composition.

beryl, $Be_3Al_2Si_6O_{18}$. A beryllium mineral containing hexagonal rings of six linked SiO_4 tetrahedra. Transparent forms of beryl are rare; when coloured green by Cr^{3+} they are known as emerald and when blue-green as aquamarine. Used for the extraction of Be and the manufacture of Be refractories.

beryllates Anionic species formed by addition of hydroxyl ions to Be^{2+}.

beryllia, BeO. Beryllium oxide*.

beryllium, Be. At.no. 4, at.wt. 9·01218, m.p. 1278°C, b.p. 2970°C (5 mm), *d* 1·85. The lightest alkaline earth. The main source is beryl* although Be is present in many other minerals. The metal is obtained by electrolysis, generally of $BeCl_2$ with added NaCl. The metal is grey in colour quite hard and brittle with an hcp structure. It is fairly resistant to acids because of a layer of oxide. Be is used in alloys, as a light structural material and in nuclear reactors as a reflector or moderator. BeO is used in ceramics and also in reactors. World production 1976 16 tonnes. Be compounds are toxic and can cause serious respiratory conditions and dermatitis.

beryllium bromide, $BeBr_2$. *See* beryllium halides.

beryllium carbonate Various oxide and hydroxide carbonates are precipitated by addition of carbonate to a Be^{2+} solution.

beryllium chemistry Beryllium is an element of group II, electronic configuration $1s^22s^2$. The only stable oxidation state is +2

($E°$ for $Be^{2+} \longrightarrow$ Be(acid solution) – 1·85 volts)

The chemistry of Be is largely that of covalent species although ionic oxygen-co-ordinated species, e.g. $[Be(H_2O)_4]^{2+}$, $[Be_3(OH)_3]^{3+}$ are known. Be compounds with 2-, 3- and 4-co-ordination are known, the latter is the maximum. Many Be compounds achieve co-ordination number 4 by polymerization, e.g. $[BeCl_2]_\infty$.

beryllium chloride, $BeCl_2$. *See* beryllium halides.

beryllium ethanoate, beryllium acetate, $Be_4O(O_2CCH_3)_6$. The acetate is typical of the basic beryllium carboxylates ($Be(OH)_2$ plus ethanoic acid). The structures have O at the centre of a tetrahedron of Be with carboxylate spanning each edge of the tetrahedron. $Be(O_2CCH_3)_2$ is formed from $BeCl_2$ and glacial ethanoic acid.

beryllium fluoride, BeF_2. *See* beryllium halides.

beryllium halides Beryllium forms a single series of halides. BeF_2 (heat on $(NH_4)_2BeF_4$) is very soluble in water, sublimes 800°C. It forms a glassy mass and gives fluoroberyllates containing $(BeF_4)^{2-}$ with excess of fluoride ion. $BeCl_2$ (BeO plus carbon with chlorine) forms $BeCl_2,4H_2O$ from water. $(BeCl_4)^{2-}$ is formed from melts, $BeCl_2$ has m.p. 405°C and has an ordered polymeric structure. It is a catalyst similar to $AlCl_3$. $BeBr_2$ and BeI_2 are very similar to the chloride.

beryllium hydroxide, $Be(OH)_2$. Precipitated from Be solutions by OH^-. Dissolves in excess of OH^- to give beryllates containing $[Be(OH)_4]^{2-}$. $[Be_2(OH)]^{3+}$, $[Be_3(OH)_3]^{3+}$, further co-ordinated by water, are also formed.

beryllium iodide, BeI_2. *See* beryllium halides.

beryllium nitrate, $Be(NO_3)_2,3H_2O$. A readily available beryllium salt (beryllium carbonate plus nitric acid). Extensively hydrolysed in solution, the salt loses HNO_3 on exposure to air. Added in small quantities to solutions used in impregnating incandescent gas mantles.

beryllium oxide, beryllia, BeO. Obtained on ignition of $Be(OH)_2$, or the carbonate, etc. It is a very hard material used in ceramics and reactors and used as an additive in gas mantles.

beryllium sulphate, $BeSO_4,4H_2O$. A very soluble Be salt (BeO plus H_2SO_4).

Bessemer process A process for converting pig iron to steel by oxidation of the impurities (C, Si, P, Mn) by blowing air through the molten metal.

betaine, trimethylglycine, $C_5H_{11}NO_2$, $Me_3\overset{+}{N}\cdot CH_2CO_2^-$. Crystallizes 1 H_2O, which it loses when heated to 100°C, m.p. 293°C.

Soluble in water and alcohol. It is a very feeble base. It occurs in beets and mangolds and many other plants, and can conveniently be prepared from beet molasses.

betaines A group of feebly basic substances, resembling betaine, which occur chiefly in plants. They are intramolecular salts (zwitterions*) of e.g. quaternary ammonium compounds and include stachydrine, trigonelline and carnitine. Also the name given to the dipolar intermediates, $X^+-C-C-O^-$ assumed to be formed between ylids and a ketone or an aldehyde ($X = R_2S$, R_3P, R_3N).

betamethasone, 9α-fluoro-16β-methylprednisolone. Prepared by partial synthesis. M.p. about 246°C (decomp.). Used systemically as the disodium phosphate salt and topically as the 17-valerate in treating conditions such as inflammation, where its potency is greater than the natural steroid hormones*.

beta particle, beta ray An electron* which is emitted during certain radioactive disintegrations.

BHC, benzene hexachloride, HCH, hexachlorocyclohexane, $C_6H_6Cl_6$. Chlorinated hydrocarbon insecticide which exists in several isomeric forms. Crude BHC contains at least 5 isomers, which have widely differing melting points and other physical properties. It is active against several insects, the most active isomer being gamma-BHC, which is also known as lindane.

BHC is manufactured by chlorination of benzene in the presence of ultra-violet light. The gamma-isomer is obtained from the crude mixture by selective crystallization, and forms colourless crystals, m.p. 113°C. U.S. production 1980 400 tonnes.

BHT 2,6-Di-tert-butyl-4-methylphenol.

Bi Bismuth.

biacetyl *See* diacetyl.

bicarbonates Hydrogen carbonates. *See* carbonic acid.

Bi-Gas process A high-pressure operation for the conversion of solid fuel into substitute natural gas* (SNG) using two stages of gasification.

bile acids *See* cholic acid.

bile pigments Breakdown products of haemoglobin which are formed during the disintegration of erythrocytes.

bile salts The sodium salts of glycocholic and taurocholic acids, which are present in the bile of animals. They have a very low surface tension, and so act as stabilizers for the emulsion formed by the fat particles in the intestine, thus rendering further chemical action on the fats easier. They also serve to hold in solution the otherwise insoluble fatty acids.

bimolecular reaction A chemical reaction in which two molecules or other species react together, e.g. $H_2 + I_2 \longrightarrow 2HI$. The majority of reactions are bimolecular or proceed through a number of bimolecular steps.

binary compound A compound containing two elements, e.g. NiAs.

bioassay Quantitative and sometimes qualitative analysis by determining effect on a test organism. Used particularly for substances having a large physiological effect.

biogenesis *See* biosynthesis.

biogenic amines An important group of naturally occurring amines derived by enzymic decarboxylation of the natural amino-acids.

bioluminescence Light emission by plant or animal organisms, often due to the reaction of oxygen with an oxidizable substrate (luciferin*) catalysed by an enzyme (a luciferinase).

biose A carbohydrate with two carbon atoms. The only biose is glycollic aldehyde, $CHO \cdot CH_2OH$.

biosynthesis A term describing the processes by which living organisms build up the compounds and structures required for their further growth and reproduction. Ultimately the build-up is from CO_2, H_2O, light energy and inorganic compounds; however, the complete sequence only takes place in photosynthetic organisms. The term biogenesis is not quite synonymous, as it may also be used to refer to nonsynthetic biological transformations.

biotin One of the vitamin-B factors. Biotin is present in yeast, egg yolk, liver and other tissues. The biotin from liver is 2-keto-3,4-imidazolido-2-tetrahydrothiophen-*n*-valeric acid, $C_{10}H_{16}N_2O_3S$. Biotin serves as the prosthetic group of a series of enzymes, each of which catalyses the fixation of CO_2 into an organic linkage.

biphenyl *See* diphenyl.

bipy Generally 2,2′-dipyridyl.

bipyridyl *See* dipyridyl.

bisabolene, $C_{15}H_{24}$. One of the most widely distributed sesquiterpenes.

bisacodyl, $C_{22}H_{19}NO_4$, **bis(4-methyl benzoato)-2-pyridylmethane.** M.p. 133–135°C. Extensively used as a laxative.

2,4-bis(p-amino-benzyl)aniline Used as an isocyanate curing agent and in adhesives.

bis-2-chloroethyl formal, $CH_2(OCH_2CH_2Cl)_2$, $C_5H_{10}Cl_2O_2$. Formed from 2-chloroethyl alcohol* and methanal. B.p. 105°C/14 mm. Used extensively in the formation of polysulphide polymers*.

1,8-bis-(dimethylamino)-naphthalene, N,N,N′,N′-tetramethyl-1,8,-naphthalenediamine M.P. 51°C. A remarkably strong mono-acidic base (pK_a 12·3) which is almost completely non-nucleophilic and valuable for promoting organic elimination reactions (e.g. of alkyl halides to alkenes) without substitution.

4,4′-bis-(isocyanatophenyl)methane, methylene-bis(4-phenylisocyanate) A bifunctional isocyanate used for cross-linking polyurethanes.

Bismarck brown, Basic Brown 1 Basic azo-dyestuff, dyes wool (reddish brown), used for cotton with tannin as mordant. Used as hair dye.

bismuth, Bi. At.no. 83, at.wt. 208·9808, m.p. 271·3°C, b.p. 1560°C, *d*9·8. A Group V element occurring native in sulphide ores, and as Bi_2S_3 (bismuth glance) and in Cu, Sn and Pb ores. The ores are roasted to Bi_2O_3 and reduced to the metal with H_2 or C. The metal is brittle, reddish-white in colour with the As and Sb structure (double layers with 3 near neighbours and 3 far). Bi burns in air to Bi_2O_3 and is only slowly attacked by acids. It is used extensively in low melting alloys and for making castings. BiOCl is used in cosmetics and some Bi compounds are used medically. Bi is used in alloys used in electronics. World production 1976 4000 tonnes.

bismuth alloys Alloys of bismuth with Pb, Sn and Cd of low m.p. Used for fire protective devices, fuses, solders, etc. The alloys of bismuth with lead and antimony are used for stereotype plates.

bismuthates(V) Formed from, e.g. Na_2O_2 and Bi_2O_3 (impure $NaBiO_3$). Never formed pure, they are very strong oxidizing agents (e.g. Mn^{2+} to MnO_4^- in acid solution).

bismuth bromide, $BiBr_3$. *See* bismuth halides.

bismuth carbonates $Bi_2O_2CO_3,\frac{1}{2}H_2O$ (Bi^{3+} plus soluble carbonate) is precipitated from solution. Loses H_2O at 100°C. Used for the relief of indigestion.

bismuth chemistry Bismuth is an element of Group V, electronic configuration $6s^26p^3$. There are two stable oxidation states, +5 (strongly oxidizing in, e.g., $NaBiO_3$, BiF_5) and +3. There are also cluster ions found in e.g. BiCl. There is a considerable chemistry of complex cationic species, e.g. $[Bi_6(OH)_{12}]^{6+}$, $[Bi_6O_6(OH)_3]^{3+}$; the free Bi^{3+} ion does not seem to exist. There is a limited organic chemistry of Bi.

bismuth chlorides, $BiCl_3$, BiCl. *See* bismuth halides.

bismuth fluorides, BiF_5, BiF_3. *See* bismuth halides.

bismuth halides Bismuth forms fluorides. BiF_5 (Bi plus F_2) is a very powerful fluorinating agent, forms $[BiF_6]^-$ (MF plus Bi_2O_3 plus BrF_3) and BiF_3, a white rather insoluble solid (Bi^{3+} plus a fluoride). Bismuth trichloride, $BiCl_3$ (Bi^{3+} and conc. HCl) or Bi plus Cl_2. Forms $[BiCl_4]^-$ and $[Bi_2Cl_7]^-$ complex ions and is hydrolysed to BiOCl. BiCl ($BiCl_3$ melt plus Bi) consists of $[BiCl_5]^{2-}$, $[Bi_2Cl_8]^{2-}$, $[Bi_9]^{5+}$ species the latter being a cluster species. $BiBr_3$ and BiI_3 are similar to $BiCl_3$.

bismuth hydride, BiH_3. A very unstable toxic gas (Mg–Bi plus acid).

bismuth iodide, BiI_3. *See* bismuth halides.

bismuth nitrates The most important bismuth compound. $Bi(NO_3)_3,5H_2O$ (Bi plus HNO_3) forms other hydrates and an oxide nitrate, $BiONO_3,H_2O$. The nitrate is probably co-ordinated to Bi^{3+}.

bismuth oxides Bi_2O_3 (Bi^{3+} plus alkali to give $Bi(OH)_3$ then dehydration) is the only definite oxide. It is yellow and dissolves in acids to salts (e.g. nitrate, sulphate). Used in porcelain glazes and stained glass. Bi_2O_5 (oxidizing agents on Bi_2O_3) is unstable and has never been obtained pure.

bismuth sulphates
$Bi_2(SO_4)_3$ (white), $Bi_2(OH)_2SO_4$ (white), $Bi_2O_2SO_4$ (yellow), are formed from sulphuric acid solution.

bismuth sulphides Bi_2S_3 (Bi^{3+} solution plus H_2S – dark brown) and BiS_2 (Bi plus S) are known.

bisphenol A, 2,2-bis(*p*-hydroxyphenyl)propene, 4,4′-isopropylidenediphenol, $C_{15}H_{16}O_2$. White flaky solid, m.p. 152–153°C, b.p. 220°C/4 mm. Prepared by condensation of phenol and propanone under acid conditions. Important material for production of epoxy resins (e.g. with epichlorhydrin) and modified phenolic resins. U.S. use 1978, 210000 tonnes.

bisulphan, 1,4-dimethanesulphonyloxybutane, $C_6H_{14}O_6S_2$, $CH_3SO_2O[CH_2]_4OSO_2CH_3$. A cytotoxic alkylating agent used clinically to treat leukaemia.

bitumen *See* asphaltic bitumen.

bitumen emulsions Oil-in-water type emulsions prepared by passing asphaltic bitumen of suitable viscosity, along with an aqueous phase containing an emulsifier, through a colloid mill or other homogenizer. Bitumen emulsions flow easily at ambient temperature and are used without heating. They are used extensively for surface dressing and screed coating of roads and for soil stabilization.

bituminous coals The term applied to a wide variety of coals containing about 75–91% carbon on a dry, ash-free basis.

bituminous mastic A mixture of asphaltic bitumen with inert fillers.

bituminous plastics Compositions based on asphaltic bitumen, coal tar pitches and petroleum still residues which are used as thermoplastic moulding materials and as flooring compositions.

biuret, ureidoformamide, $C_2H_5N_3O_2$, $NH_2{\cdot}CO{\cdot}NH{\cdot}CO{\cdot}NH_2$. Crystallizes with 1 H_2O, m.p. 193°C (decomp.). Formed by the action of heat on urea.

biuret reaction Substances containing two C(O)NH– groups attached to one another, or to the same N or C atom, give a violet or pink colour when treated with sodium hydroxide and copper sulphate. The reaction therefore serves as a test for biuret, oxamide, peptides and proteins.

bixin, $C_{25}H_{30}O_4$. Violet-red needles, m.p. 198°C. The principal pigment of annatto, the colouring matter of the shrub, *Bixa orellana*, used for colouring foodstuffs.

Bk Berkelium.

blackash *See* barium sulphide.

black lead Powdered graphite* used for blackening and polishing.

blanc fixe Precipitated $BaSO_4$ used as a fine white pigment in coating paper. It does not absorb ink.

Blanc's rule All dicarboxylic acids, with the exception of oxalic and malonic acids, up to and including 1:5 acids, give anhydrides when treated with ethanoic anhydride and subsequently distilled; 1:6 and higher acids lose CO_2 to give ketones. Thus glutaric acid gives glutaric anhydride, adipic acid gives cyclopentanone. Some exceptions to this rule, presumably due to steric hindrance, and to ring strain, have been found.

bleach bath A bath used in photography to convert developed silver back to silver halide prior to removal (colour) or sulphide toning. $K_3Fe(CN)_6$ and KBr is a commonly used bleach bath.

bleaching agents Chemicals, generally oxidants, which remove colours, e.g. bleaching powder, chlorine, chlorates (I), perborates. Brightening agents, optical bleaches, act differently in that they convert u.v. radiation to blue light and thus mask colours. Typical brighteners are diaminostilbenedisulphonic acid derivatives. *See* fluorescent brightening agents.

bleaching earths There are two types: (1) *fuller's earth** possessing absorptive properties in the natural state, (2) *activated clay* in which the absorptive powers which are low in the raw state are enhanced by heat or chemical treatment (usually a mild acid leach). Used for purification and bleaching.

bleaching powder, chloride of lime Manufactured from Cl_2 and $Ca(OH)_2$ (slaked lime). The overall composition approximates to CaCl(OCl). Yields Cl_2 with dilute acid and sold on the basis of 'available chlorine' (about 35%). Used in bleaching.

blende, sphalerite, ZnS. Generally contains some Fe which gives a dark colour. *See* zinc blende.

block copolymerization Copolymerization in which individual units are joined together in relatively long sequences. Branched copolymers of this type are graft copolymers.

blood group substances Mucopolysaccharides present in red blood cells (erythrocytes) which react specifically with isoagglutins causing haemolysis when blood of an incompatible blood group is transfused.

blooming The white cloudy appearance sometimes seen on the surface of vulcanized rubber and caused by the separation of sulphur crystals.

blowing agents Gas-forming compounds used for making foam and sponge rubber and plastics. Agents used include carbonates (e.g. Na_2CO_3) and azo and nitroso compounds.

blown bitumen, oxidized bitumen A bitumen produced by air blowing a selected soft bitumen at about 300° C, often in the presence of catalysts. The product is more rubber-like than petroleum bitumen.

blow-off *See* lift-off.

blue-john Blue or violet fluorite*.

blue print paper Light sensitive paper (turns blue) impregnated with $[Fe(CN)_6]^{3-}$ and iron(III) citrate or oxalate. The paper is fixed by washing with water. Brown print paper contains a metal (Pt, Pd, Au, Hg, Cu) salt and Fe(III). Diazo (ozalid) papers are impregnated with azo-compounds which couple under the influence of light.

blue vitriol, $CuSO_4,5H_2O$. *See* copper sulphate.

blue water gas A mixture of CO and H_2, is produced by passing steam through a bed of incandescent coke followed by an air blast to counteract the temperature drop due to the endothermic water gas reaction. Typical blue water gas contains about 50% H_2 and 41% CO, the remainder being equally N_2 and CO_2, along with a trace of methane. Formerly used as a fuel and to enrich coal gas.

BMC A very powerful perchlorinating agent (to replace all H by Cl) prepared from S_2Cl_2, SO_2Cl_2 and $AlCl_3$.

boat form *See* conformation.

body centred lattice A lattice having a point (practically atom or molecule) at the corners of the cell and at the body centre. Body centred cubic lattices (bcc) are adopted by many metals*.

boehmite, γ-AlO(OH). *See* aluminium hydroxide.

bohr magneton The unit of magnetic moment. The theoretical values of magnetic moments for unpaired electrons, taking spin only into account are: 1 = 1·73 B.M., 2 = 2·83, 3 = 3·87, 4 = 4·90, 5 = 5·92, 6 = 6·93, 7 = 7·94. The observed values may be altered by metal–metal bonding, ferromagnetism*, antiferromagnetism*, spin-orbit coupling and other effects.

boiling point The temperature at which the vapour pressure of a liquid is equal to that of the atmosphere.

boiling-point diagram A graph showing equilibrium compositions of liquid and vapour plotted against temperature for a two-component system at any given pressure.

boiling point, elevation of The increase in boiling point of a solution, compared with that of the pure solvent, due to a dissolved substance. The elevation of boiling point is proportional to the weight of a particular solute. Molecular proportions of different solutes produce the same elevation. The following values represent the elevation of boiling point produced in different solvents by dissolving 1 mol of any solute in 100 g of solvent. This is known as the molecular elevation of boiling point.

Solvent	*Molecular elevation*
water	5·2°
chloroform	38·8°
ether	21·1°
acetone (propanone)	17·2°
benzene	25·7°
alcohol (ethanol)	11·5°

Boltzmann constant, k A fundamental constant. It is the gas constant per molecule, equal to R divided by Avogadro's number L.
$k = 1{\cdot}381 \times 10^{-16}$ erg degree^{-1} molecule^{-1}.

bomb calorimeter An instrument for measuring the heat of combustion of a substance in which the heat liberated, when the sample is ignited in an atmosphere, e.g. of oxygen under pressure, is measured. Used to obtain precise thermochemical data.

bond The linkage between atoms in molecules and between molecules and ions in crystals.

bond energy The energy required to break a particular bond and, precisely, to produce the products in particular electronic states. Bond energies of multiple bonds are generally greater than those of single bonds.

Average bond energies at 298 K

Bond	$\Delta H (kJ mol^{-1})$
C–C	348
C=C	612
C⩦C (benzene)	518
C≡C	837
C–H	412
C–Cl	338
Cl–Cl	242
H–H	436
I–Cl	211
N≡N	944
N–N	388
O=O	496
O–H	463

bonding orbitals *See* molecular orbitals.

Bone and Wheeler apparatus An apparatus

for the analysis of fuel and flue gases by absorption.

bone ash A white or creamy powder obtained by calcining bones. Essentially tricalcium phosphate with some calcium carbonate. It is a characteristic constituent of English china ware, mostly as bone china*.

bone black, animal charcoal The residue remaining after the destructive distillation of degreased bones. The black residue, bone black, contains about 10% amorphous carbon, disseminated through a very porous substrate of calcium phosphate (80%), and carbonate, etc. Treatment with mineral acid dissolves away the salts and leaves a charcoal known as Ivory Black, used in sugar refining as a decolorizing agent.

bone china The chief English china ware made of a mixture of china clay, ball clay, flint, Cornish stone and bone ash in various proportions. It resembles true porcelain, but is less refractory and not quite so hard. It permits special styles of decoration which are inapplicable to hard porcelain.

boost fluids Fluids added to both aviation gasoline and aviation turbo-fuel to obtain extra power or thrust for short periods, normally during aircraft take-off. Normally a combination of water and methanol is used and its effect is to cool an engine operating under enriched conditions or to cool the gases entering a turbine. Sufficient boost fluid is added to absorb the extra heat generated, as latent heat of vaporization.

borane Formally BH_3, i.e. borane pyridine, is $H_3BNC_5H_5$. BH_3 is only stable when complexed.

borane anions The anions $B_nH_n^{x-}$ formally formed from boron hydrides and having structures containing bonding similar to that of the boron hydrides. Prepared by the action of bases on boron hydrides or by hydride on, e.g., a borate ester or diborane. Specific salts are obtained by metathesis. Mixed anions, e.g. $[BH_3CN]^-$ are also known. The simplest borane anion is tetrahedral BH_4^-, the borohydride ion (NaH plus $B(OMe)_3$ gives $NaBH_4$; LiH plus B_2H_6 gives $LiBH_4$; $AlCl_3$ plus $NaBH_4$ gives $Al(BH_4)_3$). $NaBH_4$ is white crystalline stable in air, soluble in water and is widely used as a reducing agent, e.g. –COOH gives $-CH_2OH$. Covalent borohydrides are generally inflammable. Complex borohydrides are known for, e.g., Zr and Cu. The $[B_3H_8]^-$ ion ($NaBH_4$ plus B_2H_6 [diglyme 100°C] gives NaB_3H_8) is a further simple species. Borane anions with polyhedral (closo) structures include $[B_9H_9]^{2-}$, $[B_{10}H_{10}]^{2-}$, $[B_{12}H_{12}]^{2-}$ (NaB_3H_8 at 200–230°C under vacuum).

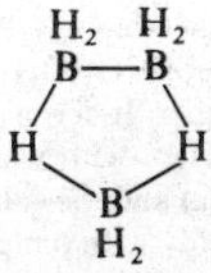

$[B_3H_8]^-$

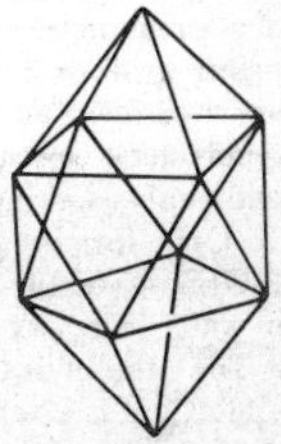

$[B_{10}H_{10}]^{2-}$

boranes Boron hydrides*. The simplest borane is BH_3 but this has only transitory existence and is normally dimerized.

borates Salts of boric acid H_3BO_3. Contain planar trigonal BO_3 and tetrahedral BO_4 groups either of which may be discrete or linked. Hydroxyl groups may be co-ordinated to boron. Pyroborates, e.g. $Co_2B_2O_5$, contain two BO_3 linked through one oxygen. Polyborates contain infinite chains or rings, e.g. B_3O_3 rings in metaborates $(B_3O_6)^{3-}$ or linear $[BO_2]_n^{n-}$ (in $LiBO_2$), two rings linked through a tetrahedral boron in $K(B_5O_6(OH)_4,2H_2O$. Borates readily form glasses. H_2O_2 or Na_2O_2 gives peroxyborates used in washing powders. Borate esters ($B(OH)_3$ plus ROH plus H_2SO_4) are known for most alcohols; anionic derivatives, e.g. $Na(BH(OR)_3)$ are formed and acyl borates $B(O_2CR)_3$ are also known. $(HO)_2B{\cdot}B(OH)_2$ forms esters.

borax, $Na_2[B_4O_5(OH)_4],8H_2O$. Occurs

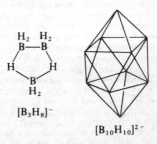

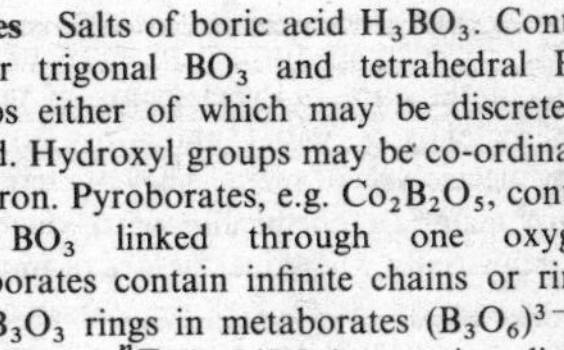

naturally as kernite ($Na_2B_4O_5(OH)_4,2H_2O$) and tincal ($Na_2B_4O_5(OH)_4.8H_2O$) and also present in the brines of Searles Lake, U.S.A. The anion is strongly hydrogen bonded to water. Used as a source of boron compounds. Also used as a standard in titrimetry and as a mild antiseptic.

Bordeaux mixture A mixture of $CuSO_4$ and $Ca(OH)_2$ made up in water and used as a fungicide.

boric acid, $B(OH)_3$. A white crystalline solid with a characteristic greasy feel. The crystal contains planar $B(OH)_3$ units linked in layers by hydrogen bonding. Obtained by treating borax, colemanite, or other natural borate with a mineral acid or by hydrolysis of boron halides, hydrides, etc. On heating, H_3BO_3 gives polymeric species, e.g. $[B_3O_3(OH)_4]^-$ predominantly as a ring polymer. It acts as an acid to OH^- forming $[B(OH)_4]^-$ with water. The $B(OH)_3$ is volatile in steam and the solubility in water increases markedly with temperature. The salts of $B(OH)_3$ and its polymers are borates*. It is a very weak acid but forms complexes with polyols (e.g. glycerol) which can be titrated using normal indicators. Boric acid is mildly bacteriostatic, inhibiting the growth of some bacteria. It is used as an ointment for wounds, and in aqueous solution as an eye lotion, mouthwash and skin lotion.

borides Borides are formed by most metals, by direct interaction of the elements, by reduction of metal oxides with C and boron carbide, and by electrolysis of melts. They are hard refractory materials, chemically inert, which are used industrially where inertness is of importance.

borine derivatives Derivatives of borane BH_3.

borneol, $C_{10}H_{18}O$. The secondary alcohol related to the ketone camphor.

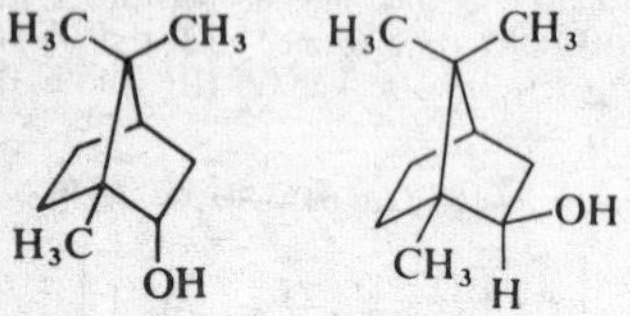

isoborneol

Borneol and isoborneol are respectively the *endo* and *exo* forms of the alcohol. Borneol can be prepared by reduction of camphor; inactive borneol is also obtained by the acid hydration of pinene or camphene. Borneol has a smell like camphor. The m.p. of the optically active forms is 208·5°C but the racemic form has m.p. 210·5°C. Oxidized to camphor, dehydrated to camphene.

Born–Haber cycle A thermodynamic cycle derived by application of Hess's law*. Commonly used to calculate lattice energies of ionic solids and average bond energies of covalent compounds. E.g. NaCl:

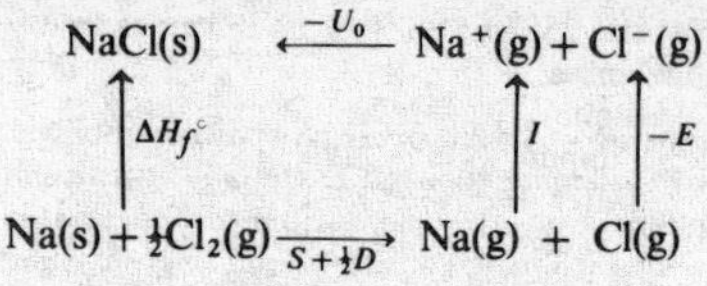

S = Heat of sublimation of sodium
D = Dissociation energy of chlorine
I = Ionization energy of sodium
E = Electron affinity of chlorine
U_0 = Lattice energy of sodium chloride
ΔH_f° = Heat of formation of sodium chloride.

From the Born–Haber cycle it follows that

$$\Delta H_f^\circ = S + \tfrac{1}{2}D + I - E - U_0$$

All terms in the equation can be determined experimentally except U_0, which can thus be calculated.

Similar cycles may be drawn for covalent compounds. E.g. PCl_5:

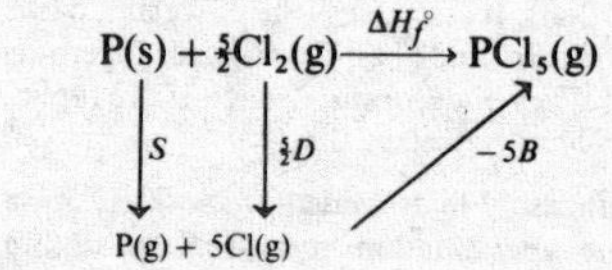

B = Average bond energy of P–Cl bond.

From the cycle it follows that:

$$\Delta H_f^\circ = S + \tfrac{5}{2}D - 5B$$

bornite, Cu_5FeS_4. An important Cu ore, dark bronze in colour. Often occurs mixed with chalcopyrite, $CuFeS_2$.

bornyl and isobornyl chlorides, $C_{10}H_{17}Cl$. Bornyl chloride, 'artificial camphor', is formed by the action of HCl on α-pinene, and is also formed together with its stereoisomer, isobornyl chloride, by the action of PCl_5 on borneol. Bornyl chloride is easily tautomerized to isobornyl chloride. By the elimination of HCl it forms camphene.

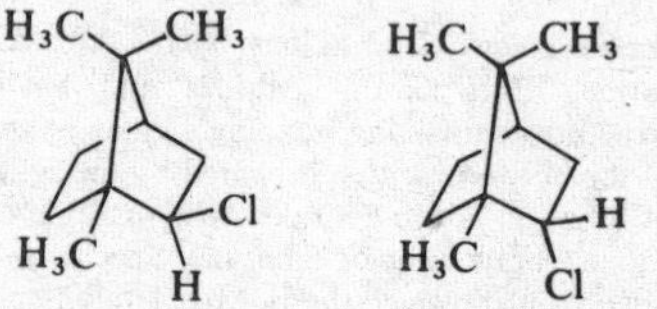

isobornyl chloride (*exo*) m.p. 147°C

bornyl chloride (*endo*) m.p. 132°C

borohydrides, $M(BH_4)_n$. *See* borane anions.

boron, B. At.no. 5, at.wt. 10·81, m.p. 2300°C,

sublimes 2550°C, *d* 2·34–2·37. Occurs in available forms as borates (rasorite, borax and colemanite). The element is obtained in a rather impure form by reduction of B_2O_3 with Mg. Purer forms result from reduction or pyrolysis of the halides. The various forms all contain icosahedral B_{12} units. The chief use of boron is as borosilicates in enamels and glasses; ^{10}B is used in nuclear reactors. Boron filaments and boron-containing materials find use as light-weight components. Boron itself is very inert and is only slowly attacked by oxidizing agents. World use 1976 340 000 tonnes.

boron bromides *Boron tribromide*, BBr_3, is prepared by passing bromine over boron. M.p. −46°C, b.p. 91°C, and has very similar properties to boron chloride but is a stronger Lewis acid. *Diboron tetrabromide*, B_2Br_4, is also known.

boron chlorides
Boron trichloride, BCl_3. Colourless mobile liquid, m.p. −107°C, b.p. 12·5°C. Obtained directly from the elements or by heating B_2O_3 with PCl_5 in a sealed tube. The product may be purified by distillation *in vacuo*. It is extremely readily hydrolysed by water to boric acid. Tetrachloroborates containing the BCl_4^- ion are prepared by addition of BCl_3 to metal chlorides.

Diboron tetrachloride, B_2Cl_4, m.p. −93°C, b.p. 55°C, is obtained by passing BCl_3 vapour through a glow discharge or by interaction of boron monoxide with BCl_3. Decomposes above 0° to *tetraboron tetrachloride*, B_4Cl_4, and other involatile chlorides including $B_{12}Cl_{11}$, B_9Cl_9 and B_8Cl_8.

boron chemistry The chemistry of boron is entirely that of covalent B−X bonds except for some complexed cationic species. Boron is commonly in the +3 oxidation state, 3- or 4-co-ordinate. B−B bonds are readily formed and boron clusters (e.g. B_{12} in the element, boron hydrides) are well established. BX_3 derivatives are strong Lewis acids. Organoboranes and boron hydrides are important synthetic reagents.

boron fluorides The normal fluoride is BF_3 (CaF_2 plus B_2O_3 plus H_2SO_4; heat on a diazonium fluoroborate) a colourless gas b.p. −100°C which fumes in moist air and gives boric acid and fluoroboric acid, HBF_4, with water. BF_3 is a strong Lewis acid and forms adducts with many donors (e.g. $F_3B{\cdot}NMe_3$). It is a very powerful catalyst in Friedel–Crafts and other reactions and is used as a starting point for the preparation of organoboron derivatives. Reacts with B_2O_3 to $(FBO)_3$. Lower fluorides are also known. B_2F_4 (BO plus SF_4) contains a single B−B bond. B plus BF_3 at very high temperatures gives unstable BF which may be used to prepare B_3F_5, B_8F_{12}, etc.

boron halides Boron forms BX_3, B_2X_4 and BX_4^- with each halogen and also some lower halides. The order of Lewis acidity is $BBr_3 > BCl_3 > BF_3$. Each of the halides is rapidly hydrolysed by water. *See* under individual entries.

boron hydrides A group of compounds con-

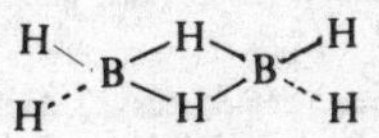

B_2H_6

taining B−H, B−B and B−H−B linkages. The simplest members are B_2H_6 (diborane(6)) m.p. −164·8°C, b.p. −92·6°C; B_4H_{10} (tetraborane (10)) m.p. −120°C, b.p. 18°C; B_5H_9 (pentaborane (9)) m.p. −46·6°C, b.p. 48°C; B_5H_{11} (pentaborane(11)) m.p. −123°C, b.p. 63°C; B_6H_{10} (hexaborane(10)) m.p. −62·3°C, b.p. 108°C; B_6H_{12} (hexaborane (12)) m.p. −82·3°C. Higher boron hydrides up to $B_{20}H_{16}$ are known. Diborane is prepared from MH, $LiAlH_4$, or $NaBH_4$ and BF_3. The higher hydrides are obtained by pyrolysis and fractionation. All of the boranes are electron deficient and there are insufficient electrons to allow conventional 2-electron bonds between atoms and the structures must be rationalized in terms of multicentre bonds. Diborane reacts with NaH to give sodium borohydride, $NaBH_4$, with anions and ligands to give complexes, e.g. $(BH_3CN)^-$, (BH_3CO). Olefines give insertion, e.g. B_2H_6 plus C_2H_4 gives $B(C_2H_5)_3$, in the hydroboration reactions*. Boron

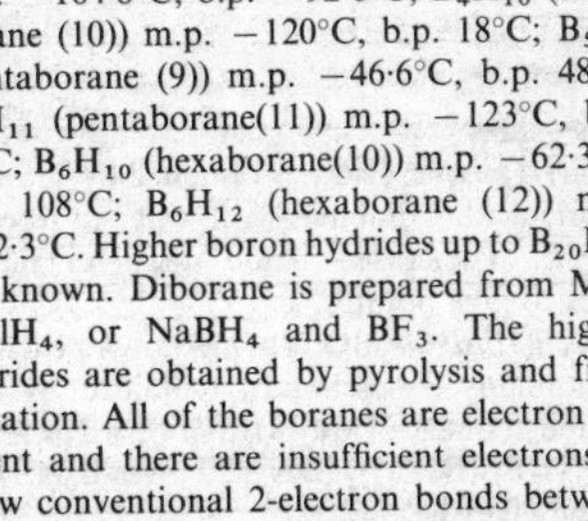

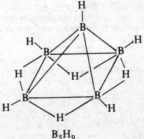

B_5H_9

trialkyls react with B_2H_6 to give alkyl boranes. Acetylenes give carboranes. Boron hydrides are potential high energy fuels and sources of pure boron.

boron iodides
Boron tri-iodide, BI_3 (BCl_3 plus HI at red heat or I_2 plus $NaBH_4$), m.p. 43°C, b.p. 210°C. It has very similar properties to boron trichloride.

Boron sub-iodide, B_2I_4, is prepared by the action of an electrical discharge on BI_3 vapour.

boron nitride, BN. Prepared by the action of nitrogen or ammonia on boron at high temperatures (thus part of the amide series $B(NH_2)_3$, $B_2(NH)_3$, BN). Has graphite-like and diamond-like modifications. It is almost as hard as diamond.

boron-nitrogen compounds Aminoborane derivatives (e.g. $B(NMe_2)_3$) are formed by reaction between boron halides and amines. Amino derivates of borane clusters and also cyclic $(RBNR^1)_n$ polymers (borazines) are also known.

boron organic derivatives *See* organoboranes.

boron oxides
Boron(III) oxide, B_2O_3, is obtained by ignition of boric acid. Combines with water to reform $B(OH)_3$. The fused oxide dissolves metal oxides to give borates.

Boron(II) oxide, BO, is of unknown structure but contains B–B bonds. Obtained by the reaction sequence $(Me_2N)_2BCl$ plus Na gives $(Me_2N)_2BB(NMe_2)_2$ plus acid gives $B_2(OH)_4$, heating gives BO.

borosilicates Boron-containing silicates found naturally, e.g. danburite, $CaB_2Si_2O_8$, and prepared as glasses by fusing together B_2O_3, SiO_2 and a metal oxide (the latter is not a necessary constituent). Pyrex glasses are borosilicate glasses.

boundary layer When a fluid flows over a surface, due to the frictional resistance there is a layer of fluid in the immediate vicinity of the surface which has a velocity smaller than that in the bulk of the fluid. This is known as the boundary layer. In the part of the layer nearest the surface flow is laminar, that is, the layers of fluid move parallel to each other and there is no mixing between them. Transfer of heat or material across the boundary layer can only occur by the relatively slow process of molecular diffusion*, and it is therefore of the greatest technical importance in that it determines the heat and mass transfer properties of the system.

Bouveault-Blanc reduction The reduction of esters to alcohols by nascent hydrogen generated from sodium, and most usually, ethanol.

$$RCO_2R' + 4[H] \longrightarrow R{\cdot}CH_2OH + R'OH$$

Now superseded by $LiAlH_4$.

Boyle's law At constant temperature the volume of a given mass of gas is inversely proportional to the pressure. Although exact at low pressures, the law is not accurately obeyed at high pressures because of the finite size of molecules and the existence of intermolecular forces. *See* van der Waals' equation.

Br Bromine.

bradykinin A nonapeptide occurring in normal blood plasma as the precursor bradykininogen, but released by the action of trypsin. Bradykinin is a smooth-muscle hypotensive agent. Together with histamine it is associated with shock and inflammatory reactions.

Arg·Pro·Pro·Gly·Phe·Ser·Pro·Phe·Arg

Brady's reagent A solution of 2,4-dinitrophenylhydrazine sulphate in methanol. Gives characteristic crystalline yellow to deep red 2,4-dinitrophenylhydrazone products with aldehydes and ketones.

Bragg equation When a beam of monochromatic X-rays of wavelength λ impinges on a crystal, strong scattering occurs in certain directions only: this is the phenomenon of X-ray diffraction*. For diffraction the path difference between waves scattered from successive planes of atoms in the crystal must equal an integral number of wavelengths, n ($n = 1,2,3,\ldots$). This condition is expressed by the equation

$$n\lambda = 2d \sin\theta$$

where d is the distance separating successive planes in the crystal and θ is the angle which the incident beam of X-rays makes with the same planes.

Bragg scattering Coherent elastic scattering of monochromatic neutrons by a set of crystal planes.

brass Cu–Zn alloy containing up to 40% Zn. There are two main groups of alloys: (a) the α-brasses contain up to 30% Zn; the structure is single-phase α-solid solution which is very ductile and so is suitable for severe cold forming, e.g. thin sheet, wire, tubes, cartridge cases, etc., (b) the $\alpha\beta$-brasses contain about 40% Zn; the structure is two-phased ($\alpha + \beta$) since the solubility of Zn in Cu has been exceeded. The β phase is an ordered or disordered bcc structure and is stronger than α but is lacking in ductility making this alloy unsuitable for cold shaping. It is widely used for castings of all sizes and also for hot forming, particularly extrusion. Additions of small quantities of other metals give improved properties.

The γ-brass structure is adopted by a series of alloys with an electron: atom ratio of 21 : 13, e.g. Cu_5Zn_8, Cu_9Al_4, Cu_3Sn_8.

braunite Brown Mn_2O_3 containing some SiO_2. Used as a source of manganese and as a colourizing agent for bricks and pottery.

brazing metal *See* solder.

Bredt's rule A double bond cannot be introduced at the bridgehead carbon of a bridged bicyclic or polocyclic system with small- or medium-sized rings.

breunnerite Impure magnesite ($MgCO_3$) containing 5–30% iron carbonate. Also known as giobertite and mesitite. Used as a source of MgO and for forming refractories when the Fe produces stronger (although less refractory) bricks.

brighteners, optical *See* bleaching agents.

bright stock Viscous lubricating oil obtained from the residue of secondary distillation of crude oil (short residue) or the residue from steam distillation (steam refined stock). Refining of these residues may involve de-asphalting*, dewaxing*, acid treatment and clay treatment*.

brilliant green A basic dyestuff, bis-(*p*-diethylaminophenyl)phenylmethyl hydrogen sulphate, also used as a bactericide.

Brillouin zones Electronic theory of metals divides the electronic states of a metal into a series of broad energy levels known as Brillouin zones.

brine Concentrated NaCl in water. NaCl is generally extracted as brine.

British Standards National standards involving test methods, specifications for quality, safety, performance, etc., and codes of practice. Standards are prepared by the British Standards Institution. *Ref.*: British Standards Year Book.

British thermal unit (Btu) The most commonly used industrial heat unit; the amount of heat required to raise 1 lb of water through 1°F under specified conditions. Since the specific heat of water varies, particularly with temperature, the actual value of Btu is dependent on the conditions chosen as standard. The value corresponding to the international table calorie is 1055·06 joules, and this is the value used by the British Standards Institution.

bromacil, 5-bromo-6-methyl-3-*sec*-butyluracil, $C_9H_{13}BrN_2O_2$. A soil-acting herbicide. White crystalline solid, m.p. 158–159°C. It is a non-selective inhibitor of photosynthesis used for weed control in citrus and cane fruit plantations. It is relatively non-toxic to animal life.

bromal *See* tribromoethanal.

bromates Salts containing bromine oxy-anions; hypobromates, $[BrO]^-$, bromates(I); bromates (the usual meaning), $[BrO_3]^-$, bromates(V); perbromates, $[BrO_4]^-$, bromates(VII). Bromates(I) are formed from Br_2 and base, but BrO^- disproportionates to Br^- and $[BrO_3]^-$ above 0°C. Bromates(V) are stable at elevated temperatures ($Ba(BrO_3)_2$ plus H_2SO_4 gives $HBrO_3$). Bromates (VII) are formed by electrolytic, XeF_2, or F_2 oxidation of $[BrO_3]^-$; the free acid $HBrO_4$, $2H_2O$ can be crystallized. $[BrO_4]^-$ is a strong but sluggish oxidizing agent

E° (acid solution) $[BrO_4]^- \rightarrow [BrO_3]^-$ + 1·76 volts

E° (acid solution) $[BrO_3]^- \rightarrow [BrO]^-$ + 0·56 volts

$$[BrO]^- \rightarrow Br_2 + 0{\cdot}45 \text{ volts}$$

bromelin, bromelain A potent proteolytic enzyme obtained from the stems of *Ananas comosus* (pineapple). It depolymerizes fibrin matrix and is also used for its anti-inflammatory activity.

bromic acid, $HBrO_3$. *See* bromates.

bromides Derivatives of HBr prepared by solution of the metals in aq. HBr or metal plus Br_2 (direct or in methanol) or HBr plus alkali carbonate (Na, K, etc.). Some bromides (NH_4, K, Na) are used medicinally as sedatives.

bromine, Br. At.no. 35, at.wt. 79·904, m.p. −7·2°C, b.p. 58·78°C, *d* 3·11. Bromine is a dark red liquid, the vapour is red and poisonous, containing Br_2 molecules. Bromine occurs naturally as bromides in sea-water and in some natural brines and salt deposits. It is obtained by conversion to Br_2 with Cl_2 followed by sweeping out with air. Br_2 is very reactive acting as an oxidizing agent and forming bromides with many elements. Its chief use is for making ethylene dibromide used in the manufacture of anti-knock agents but is also used in the manufacture of fumigants, flame-proofing agents, disinfectants and water-purification materials, photographic materials. U.S. production 1978 205 000 tonnes.

bromine chemistry Bromine is a typical halogen, a non-metal, electronic configuration $4s^2 4p^5$. The most stable oxidation state is -1 occurring as the Br^- ion

$$E^\circ \; Br_2 \longrightarrow Br^- \; +1{\cdot}07 \text{ volts in acid solution}$$

and in many covalent bromides. The bromide group is generally bonded to only one other atom but can bridge two or three metals. There is no simple cationic chemistry but unstable species, e.g. $[Brpy_2]^+$ ($AgClO_4$ plus Br_2 plus pyridine) and $[BrF_2]^+$ (in some BrF_3 adducts) are known. Bromine forms covalent compounds in oxidation states $+1$ (BrF), $+3$ (BrF_3, $[BrF_4]^-$), $+5$ (BrF_5, $[BrO_3]^-$) and $+7$ ($[BrO_4]^-$) and as indicated most of these also form anionic species. *See* bromates.

bromine chlorides *See* bromine halides.

bromine fluorides *See* bromine halides.

bromine halides Fluorides, BrF, m.p. -33°C, b.p. 20°C, readily decomposes to Br_2 and BrF_3; BrF_3, m.p. 9°C, b.p. 126°C, T structure, a yellow liquid which has found great use as a combined fluorinating agent and non-aqueous solvent. BrF_5, m.p. -60°C, b.p. 41°C square-pyramidal; Br chloride BrCl red brown; iodide BrI.

bromine oxides Bromine oxides are relatively unstable. Br_2O, m.p. -17°C, (Br_2 on HgO); BrO_2, Br_3O_8 are even less stable (Br_2 plus O_2 in discharge tube). The oxides are formally acidic and a range of oxyanions is known – *see* bromates.

bromoacetic acid *See* bromoethanoic acid.

bromoacetone, bromopropanone, $CH_3C(O)CH_2Br$. Colourless liquid which rapidly becomes violet in colour; it is a powerful lachrymator; b.p. 136°/725 mm. Manufactured by treating aqueous propanone with bromine at 30–40°C; it is usual to add sodium chlorate(V) to convert the hydrobromic acid formed by the reaction back to bromine. It is not very stable and decomposes on standing.

bromoacids, complex Many elements, e.g. Pt, form complex anions, e.g. $[PtBr_6]^{2-}$, in concentrated HBr solution. The free acids are generally not stable.

bromobenzenes
Bromobenzene, C_6H_5Br. B.p. 155°C. Prepared from benzene by direct bromination in the presence of a carrier (I_2, Fe, $AlCl_3$) or by treatment of the diazonium salt with CuBr. Used for making diphenyl and diphenyl ether and derivatives, and for introduction of a phenyl group into a molecule via a Grignard reagent; the halogen atom can be removed by treatment with magnesium, sodium or copper.

The disubstituted derivatives $C_6H_4Br_2$ *o-dibromobenzene*, m.p. 7·8°C, b.p. 224°C, and *p-dibromobenzene*, m.p. 89°C, b.p. 219°C, also yield substituted diphenyl derivatives by removal of halogens.

bromoethanoic acid, bromoacetic acid, $CH_2Br{\cdot}COOH$. White crystalline solid, m.p. 50°C, b.p. 208°C. Soluble in water and alcohol. Prepared by the action of dry bromine on dry ethanoic acid in presence of small amounts of red phosphorus. Produces sores upon the skin; used in chemical syntheses. *See* Reformatski reaction.

bromoform, tribromomethane, $CHBr_3$. Colourless liquid; m.p. 8°C; b.p. 151°C (slight decomp.). Sparingly soluble in water; miscible with alcohol and ether. Prepared by the action of bromine and sodium hydroxide on ethanol or propanone or by warming bromal with alkalis. Decomposed by light and air more readily than chloroform; stabilized by 4% alcohol. Converted to carbon monoxide and potassium bromide by potassium hydroxide.

bromomethane, methyl bromide, CH_3Br. B.p. 5°C. Generated from sodium bromide, methanol and conc. sulphuric acid. Used for methylation and commercially, very widely, as a fumigant. It is very toxic to man. The current TLV is 15 p.p.m. (= 60 mg/m³) but it cannot be detected by smell below about 10 000 p.p.m.

bromonaphthalenes, $C_{10}H_7Br$.

1-Bromonaphthalene, m.p. 5°C, b.p. 279°C.

2-Bromonaphthalene, m.p. 59°C, b.p. 282°C.

The 1-isomer can be prepared by the direct action of bromine on naphthalene. The 2-isomer is obtained by treating diazotized 2-naphthylamine with CuBr. 1-Bromonaphthalene is used as a standard for refractive index measurements; also as an immersion fluid in the determination of the refractive index of crystals.

***N*-bromosuccinimide, NBS,** $C_4H_4BrNO_2$.

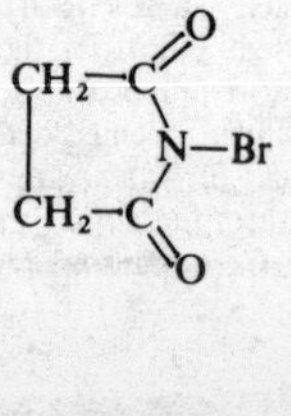

A white solid, m.p. 178°C. Primarily of interest as a brominating agent which will replace activated hydrogen atoms in benzylic or allylic positions, and also those on a carbon atom α to a carbonyl group. Activating influences can produce nuclear substitution in a benzene ring and certain heterocyclic compounds; also used in the oxidation of secondary alcohols to ketones.

bromothymol blue Indicator pH range 6·0 (yellow) to 7·6 (blue).

bromotrifluoromethane *See* BTM.

bronze An alloy of copper and tin. The Sn hardens and strengthens the Cu. *Gun metal, phosphor bronze, aluminium bronze,* and *beryllium bronze* all have particular uses.

bronzes Compounds with metallic appearance and properties formed by reduction of heavy transition metal oxides with non-stoicheiometric amounts of, e.g. alkali metals. Formed particularly by Nb, Ta and W. E.g. $Sr_{0·8}NbO_3$ and Na_xWO_3.

brosyl The trivial abbreviation for the

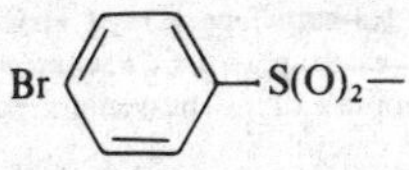

p-bromobenzenesulphonyl group: hence brosylate, etc.

Brownian movement The rapid and random movement of particles of a colloidal sol, observed brightly lit against a dark ground. First observed with a pollen suspension. The Brownian movement is due to the impact on the dispersed particles of the molecules of the dispersion medium. As the particles increase in size, the probability of unequal bombardment from different sides decreases, and eventually collisions from all sides cancel out and the Brownian movement becomes imperceptible at a particle size of about 3–4μ. From the characteristics of the movement, Perrin calculated Avogadro's number *L*.

brown print paper *See* blue print paper.

brown ring test A test for nitrates*.

brucine, $C_{23}H_{26}N_2O_4$. An alkaloid found with strychnine in *Nux vomica* and other plants. Crystallizes with $4H_2O$, m.p. 105°C (178°C, when anhydrous). The dimethoxy derivative of strychnine.

brunswick black A black paint.

brunswick green A green pigment obtained from lead ethanoate, $FeSO_4$, $K_4Fe(CN)_6$ and $Na_2Cr_2O_7$ solutions together with $BaSO_4$ as diluent.

BTM, bromotrifluoromethane, CF_3Br. A colourless gas, b.p. −59°C/740 mm. Used as a relatively non-toxic propellant gas in fire fighting apparatus, e.g. dry-powder extinguishers. Made by the bromination of fluoroform, CHF_3.

Btu *See* British thermal unit.

BTX A mixture of low boiling point aromatics, i.e. benzene, toluene and xylenes.

bubble-cap plate, bubble-cap tray A vapour-liquid contacting device used in distillation and absorption columns.

Bucherer reaction Bucherer discovered that the interconversion of 2-naphthol and 2-naphthylamine through the action of alkali and ammonia could be facilitated if the reaction was carried out in the presence of $[HSO_3]^-$ at about 150°C. This reaction is exceptional for the ease with which an aromatic C−OH bond is broken. It is not of general application. It is probable that the reaction depends upon the addition of $[HSO_3]^-$ to the normally unstable keto-form of 2-naphthol, and subsequent displacement of −OH by $-NH_2$.

buffer solutions It is often desirable to prepare a solution of definite pH, made up in such a way that this pH alters only gradually with the addition of alkali or acid. Such a solution is called a buffer solution, and generally consists of a solution of a salt of a weak acid in the presence of the free acid itself, e.g. sodium ethanoate and ethanoic acid. The pH of the solution is determined by the dissociation equilibrium of the free acid:

$$\text{e.g.} \quad \frac{[H^+][CH_3COO^-]}{[CH_3COOH]} = K$$

The sodium ethanoate which is largely dissociated, serves as a source of ethanoate ions, which combine with any hydrogen ions which may be added to the solution to yield more of the acid. The addition of hydrogen ions has therefore much less effect on such a solution than it would have on water. In a similar manner, the solution of the salt of a strong acid and a weak base, in the presence of a weak base, has a pH that is insensitive to additions of alkali.

bufotenin, $C_{12}H_{16}N_2O$ 5-Hydroxy-3-dimethylaminoethylindole, m.p. 146–147°C, found in the venoms of several species of toads, together with the corresponding betaine, bufotenidin. It has vasoconstrictor action (causing a rise in blood pressure) and also evokes hallucinations.

bufotoxin The suberylarginine ester of a steroid isolated from toad venom.

buna rubbers Synthetic elastomers based on butadiene copolymers. Buna-N, nitrile rubber, NBR, is a butadiene (70)-acrylonitrile (30)-copolymer. Buna-S, SAR, is a butadiene (70)-styrene (30)-copolymer. These synthetics have very similar properties to natural rubber.

burette Apparatus used for measuring the quantity of a liquid or gas in a chemical operation. In volumetric analysis the burette generally consists of a vertical tube, graduated in fractions of a millilitre, provided with a tap at the lower end, by means of which the amount of liquid which is allowed to flow from the graduated tube may be controlled. In more precise work, a weight-burette is employed. It consists of a flat-bottomed flask provided with a ground stopper, and a narrow side limb provided with a ground glass cap. The burette is weighed before and after releasing the requisite amount of liquid from the narrow side-limb.

In gas analysis the burettes are generally vertical graduated tubes provided with a tap at the upper end. The lower end is connected by means of tubing to a reservoir containing mercury or water, by means of which the pressure on the gas enclosed between the tap and the liquid surface may be adjusted and ascertained.

Burettes can be automated for continuous use.

Burger's vector A measure of the crystal lattice displacement resulting from the passage of a dislocation.

burners For the combustion of solid, gaseous or liquid fuels on an industrial scale special burners are usually necessary.

Solid fuels. Large coal-fired equipment normally uses pulverized fuel blown into the combustion chamber by a blast of air, similar to oil droplets.

Liquid fuels. Industrial burners for liquid fuels usually atomize the fuels in hot air so that droplets will evaporate during combustion. For more volatile fuels such as kerosine, *vaporizing burners* of various types are employed, usually for domestic purposes.

Gaseous fuels. Gas burners can be *diffusion flame burners* or *pre-aerated burners*. Diffusion flame burners may be relatively simple, with fuel gas burning at an orifice in the presence of surrounding air which diffuses into the gas. These flames are normally luminous because of the incandescence of carbon particles produced by cracking of the fuel. Pre-aerated burners are designed so that gas and air are pre-mixed in the burner and emerge as a flammable mixture. The laboratory bunsen burner is a low temperature aerated burner.

burning oils Kerosines for domestic use. *See* fuel oils.

burning velocity The speed at which the flame front in a burning gas mixture enters the mixture of unburned gases, i.e. the flame speed. A *stationary flame* in a burner is maintained due to the balance of a steady flow of reactants by an equal and opposite burning velocity. If burning velocity of the fuel/air mixture is increased for a particular appliance there is *flashback**, while if the gas flow rate is increased there is *lift-off* of the flame.

butacaine, $C_{18}H_{30}N_2O_2$. Its sulphate, a white

$p\text{-}H_2NC_6H_4CO_2[CH_2]_3{\cdot}N([CH_2]_3{\cdot}CH_3)_2$

crystalline powder, m.p. 100–103°C, is used as a local anaesthetic.

butadiene, 1,3-butadiene, $CH_2{:}CH{\cdot}CH{:}CH_2$. Colourless gas; b.p. −5°C. Prepared by passing the vapours of 1,3-butylene glycol, butylene oxide or cyclobutanol over heated catalysts. Manufactured by the catalytic cracking of *n*-butenes from oil or the catalytic dehydrogenation of butane. Also made by the dehydrogenation of ethanol to ethanal, followed by the condensation of ethanol and ethanal. When heated, e.g. with sodium, it polymerizes to a rubber-like material. Used for the preparation of artificial rubbers (styrene/butadiene 50%, polybutadiene 20%, nitrile rubbers 10%). Also oligomerized to 4-vinylcyclohexene-1 by heat and to cyclo-octa-1,5-diene and isomeric forms of 1,5,9-cyclododecatriene and higher oligomers by transition metal catalysts. Forms complexes, e.g. butadiene–$Fe(CO)_3$, with metal compounds. U.S. production 1982 1·4 M tonnes.

butadiene polymers Polymeric elastomers derived from butadiene and its derivatives. Polybutadiene and particularly copolymers between butadiene and acrylonitrile or styrene are used as rubbers. U.S. production 1980 ABS 400 000 tonnes, others 300 000 tonnes. *See* buna rubbers.

butadienes Unsaturated hydrocarbons containing the grouping $>C=C-C=C<$. The simplest member, C_4H_6, is a gas, the substituted butadienes are liquids or solids which with alkyl substituents, readily polymerize to rubber-like solids. They are formed by removal of water or halogen from suitable glycols or chloroparaffins, or in certain cases from alkynes. They are very reactive and combine with either one or two molecules of halogens, halogen acids, or HOCl and also with N_2O_4 and SO_2. Give Diels–Alder adducts with maleic anhydride (derivatives of tetrahydrophthalic anhydride). *See* butadiene, chloroprene, isoprene.

Unconjugated butadienes, i.e. buta-1,2-diene, $CH_2 = C = CH - CH_3$, also exist; *see* allenes.

butaldehydes *See* butanals.

butamer process A process for the isomerization of *n*-butane on a platinum-based catalyst to isobutane.

butanals, butaldehydes, butyraldehydes
n-butanal, $CH_3CH_2CH_2CHO$. Colourless liquid with a pungent odour, b.p. 75°C. Manufactured by reduction of crotonaldehyde with hydrogen and a metallic catalyst, or by passing the vapour of *n*-butanol over heated copper oxide. Polymerizes in the presence of $ZnCl_2$ to parabutaldehyde. Used in the preparation of rubber accelerators.

Isobutanal, Me_2CHCHO. Also known. B.p. 64°C.

butane, C_4H_{10}. The lowest member of the paraffin series to exhibit isomerism, there being two possible isomers.

n-Butane, $CH_3CH_2CH_2CH_3$. Colourless gas with a faint odour, b.p. −0·3°C. It occurs in natural gas, and is obtained in large amounts during the cracking of petroleum. For chemical properties *see* paraffins. Used in refrigeration plant; it is also compressed in cylinders for use as an illuminant, or for heating purposes, e.g. 'camping gas'. It has a high calorific value.

Iso-butane, $CH_3CH(CH_5)CH_3$. Colourless gas, b.p. −10·3°C. It also occurs in natural gas and petroleum gas and may be separated from *n*-butane by fractional distillation under pressure.

butanediols *See* butylene glycols.

butanoic acids, butyric acids, $C_4H_8O_2$. Two acids are known.

Butanoic acid, $CH_3CH_2CH_2COOH$, colourless syrupy liquid with a strong odour of rancid butter; b.p. 162°C. Occurs in butter as the glycerol ester. Prepared by oxidation of 1-butanol or by the fermentation of sugary or starchy materials by *B. subtilis* etc. Oxidized by nitric acid to succinic acid. Generally forms soluble salts with metals. Cellulose derivatives are used in lacquers and as moulding plastics. Butanoates are used in flavouring and as plasticizers.

Isobutyric acid, dimethylacetic acid, 2-methylpropanoic acid, $(CH_3)_2CH{\cdot}COOH$, colourless syrupy liquid with an unpleasant odour; b.p. 154°C. Prepared by oxidation of 2-methylpropanol with $K_2Cr_2O_7$ and H_2SO_4. Salts soluble in water. Used in alkaline solution for sweetening* gasoline.

butanols, butyl alcohols, $C_4H_{10}O$. There are 4 butanols. The liquid alcohols are important solvents for resins and lacquers; they are used to prepare butyl ethanoates, also important lacquer solvents: other butyl esters are used in artificial flavouring essences and perfumes.

Normal butyl alcohol, propyl carbinol, n-butanol, 1-butanol, $CH_3CH_2CH_2CH_2OH$. B.p. 117°C. Manufactured by reduction of crotonaldehyde (2-butenal) with H_2 and a metallic catalyst. Forms esters with acids and is oxidized first to butanal and then to butanoic acid. U.S. production 1978 300 000 tonnes.

Isobutyl alcohol, isobutanol, 2-methylpropanol, isopropyl carbinol, Me_2CHCH_2OH. B.p. 108°C. Occurs in fusel-oil. Oxidized by potassium permanganate to 2-methylpropanoic acid; dehydrated by strong sulphuric acid to 2-methylpropene.

Secondary butyl alcohol, methylethyl carbinol, 2-butanol, $CH_3CH_2CH(Me)OH$. B.p. 100°C. Manufactured from the butane–butene fraction of the gas from the cracking of petroleum. Used to prepare butanone.

Tertiary butyl alcohol, trimethyl carbinol, tertiary butanol, 2-methyl-2-propanol, Me_3COH. Colourless prisms, m.p. 25°C, b.p. 83°C. Prepared by absorbing isobutene (2-methylpropene) in sulphuric acid, neutralizing and steam distilling the liquor. Converted to isobutene by heating with oxalic acid. Potassium-*t*-butoxide is a very strong base.

butanone, methyl ethyl ketone, M.E.K., C_4H_8O, $CH_3COCH_2CH_3$. Colourless liquid with a pleasant odour, b.p. 80°C. It occurs with propanone in the products of the destructive distillation of wood. Manufactured by the liquid or vapour phase dehydrogenation of 2-butanol over a catalyst. Used as a solvent, particularly for vinyl and acrylic resins, and for nitrocellulose and cellulose acetate, also for the dewaxing of lubricating oils. U.S. production 1978 300 000 tonnes.

butanoyl Groups formed from the butanoic acids by removal of the OH. Thus butanoyl, *n*-butyryl, $CH_3CH_2CH_2C(O)-$, 2-methylpropanoyl, isobutyryl, $(CH_3)_2CHC(O)-$.

2-butene-1,4-diol, butene diol, $HOCH_2CH = CHCH_2OH$. Colourless stable liquid used in the manufacture of insecticides, resins and pharmaceuticals. Both *cis*- and *trans*-isomers are known. Prepared by hydrogenation of butynediol over a catalyst.

butene polymers Isobutene $Me_2C = CH_2$ readily undergoes ionic polymerization and incorporation of some isoprene or butadiene gives cross-linking and elastomers – butyl rubbers*. Halogenation of butyl rubbers extends the usefulness of these elastomers. 1-butene, $CH_3CH_2CH = CH_2$ can be polymerized with Ziegler–Natta catalysts to a useful material.

butenes, C_4H_8. Colourless gases with unpleasant odours. There are three isomers obtained from the appropriate butanols. All three are present in the gas from the cracking of petroleum.

1-Butene, α-butylene, $CH_3CH_2CH:CH_2$. Prepared by passing the vapour of 1-butanol over heated alumina.

2-Butene, β-butylene, $CH_3CH:CHCH_3$. Prepared by heating 2-butanol with sulphuric acid. It occurs as two geometrical isomers.

2-Methylpropene, isobutylene, isobutene, $Me_2C:CH_2$. Prepared by heating *t*-butanol with oxalic acid.

For general reactions *see* olefins. The butylenes are used to prepare 2-butanol. 1-Butene and isobutene are formed into widely used polymers.

butobarbitone, 5-butyl-5-ethylbarbituric acid, $C_{10}H_{16}N_2O_3$. White crystalline powder, m.p. 122–124°C. Prepared by condensing ethyl butylethylmalonate with urea. It is used as a sedative and hypnotic.

butter of antimony, $SbCl_3$. *See* antimony chlorides.

butyl, iso The 2-methylpropyl group Me_2CHCH_2-.

butyl, normal, *n*-butyl The group $CH_3CH_2CH_2CH_2-$.

butyl, secondary, sec.-butyl The 1-methylpropyl group, CH_3CH_2CHMe-.

butyl, tertiary, tert.-butyl, *t*-butyl The group Me_3C-.

butyl acetates *See* butyl ethanoates.

butyl alcohols *See* butanols.

butyl Cellosolve A trade name for ethylene glycol monobutyl ether.

butylene glycols, dihydroxybutanes, $C_4H_{10}O_2$. There are five glycols of this formula, three chiral. They are colourless, rather viscous liquids. The important isomers are:

1,3-Dihydroxybutane, β-butylene glycol, $CH_3CH(OH)CH_2CH_2OH$, b.p. 204°C. Manufactured by reduction of aldol or by the action of yeast on aldol. Used to prepare butadiene. Used in brake fluids, in gelling agents and as an intermediate in plasticizers.

2,3-Dihydroxybutane, ψ-butylene glycol, $CH_3CH(OH)CH(OH){\cdot}CH_3$. The glycol produced by fermentation consists largely of the optically inactive meso-form, while the synthetic glycol is mainly the optically inactive (±)-form; b.p. 177–180°C. Manufactured by the fermentation of potato mash or molasses by organisms of the *Aerobacter* and *Aerobacillus* groups.

1, 4- Dihydroxybutane, $HOCH_2CH_2CH_2CH_2OH$. B.p. 228°C. Prepared ethyne plus methanal, hydrogenated to butanediol. Used in production of γ-butyrolactone and 2-pyrrolidone. Widely used in polyurethane products.

butylenes *See* butenes.

butyl ethanoates, butyl acetates There are four esters corresponding to the four butyl alcohols. They are colourless liquids with fruity odours: the normal, iso- and secondary butyl ethanoates are important solvents for cellulose lacquers.

Normal butyl ethanoate, $CH_3CH_2CH_2CH_2O{\cdot}OCCH_3$; b.p. 126°C.

Isobutyl ethanoate, $(CH_3)_2CHCH_2O{\cdot}OCCH_3$; b.p. 118°C.

Secondary butyl ethanoate, $CH_3CH_2CH(Me)O{\cdot}OCCH_3$, b.p. 112–113°C. Manufactured by heating 2-butene with ethanoic and sulphuric acids in water.

***t*-butyl hypochlorite,** $(CH_3)_3COCl$. A yellow liquid with a lachrymatory action, which is prepared by passing chlorine into an aqueous alkaline solution of *t*-butanol. It may be distilled, b.p. 77–78°C, but is liable to decompose violently. It is used for *N*- and *C*-chlorinations (reaction similar to *N*-bromosuccinimide) and for dehydrating alcohols.

butyl lithium *See* lithium organic derivatives. Used in organic syntheses, as polymerization catalyst (ethanal, isoprene, butadiene) and for the preparation of LiH and lithio derivatives.

***p*-*t*-butylphenol,** $Me_3CC_6H_4OH$, $C_{10}H_{14}O$. M.p. 98°C, b.p. 239°C. Insoluble in water, soluble in alkaline solutions. Made with phenol plus 2-methylpropene. Used in phenolic coating materials (condensation with methanal).

butyl rubber A copolymer of $Me_2C{:}CH_2$ (isobutene) and $H_2C{:}C(Me)CH{:}CH_2$ (isoprene) (1–3%) that can be vulcanized in the normal way with sulphur. Less permeable than natural rubber and used in tyre inner tubes. It is resistant to atmospheric ozone and so can be used out-of-doors in black sheet form for roofing and as a water impermeable membrane to prevent seepage from reservoirs. U.S. production 1978 15·4 megatonnes.

2-butyne-1,4-diol, butyne diol, $HOCH_2C \equiv CCH_2OH$. White solid, m.p. 58°C, b.p. 238°C prepared by the high pressure reaction between ethyne and methanol and also from BrMgCCMgBr and methanal. Used in electroplating (Ni), as a corrosion inhibitor, and in paint and varnish removal.

butyric acids *See* butanoic acids.

γ-butyrolactone, 4-hydroxybutanoic acid lactone, $C_4H_6O_2$. Colourless liquid with a

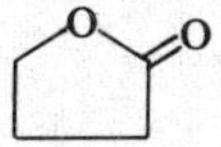

pleasant odour; b.p. 206°C. It is used as a solvent for various polymers and as an intermediate in the preparation of, e.g. polyvinylpyrrolidone and piperidine.

butyrone, di-*n*-propylketone, 4-heptanone, C_7H_4O, $(CH_3CH_2CH_2)_2CO$. A colourless odorous liquid; b.p. 144°C. Used as a solvent for resins, particularly the glyptal and vinyloid resins, and lacquers.

butyryl *See* butanoyl.

C

C Carbon.

Ca Calcium.

cacodyl derivatives Organoarsenic compounds containing the $AsMe_2$ grouping.

cadaverine, pentamethylenediamine, $H_2N{\cdot}[CH_2]_5{\cdot}NH_2$. A syrupy fuming liquid, b.p. 178–180°C. Soluble in water and alcohol. Cadaverine is one of the ptomaines and is found, associated with putrescine, in putrefying tissues, being formed by bacterial action from the amino-acid lysine. It is found in the urine in some cases of the congenital disease cystinuria. The free base is poisonous, but its salts are not.

cadinene, $C_{15}H_{24}$. A sesquiterpene of very wide distribution.

cadmium, Cd. At.no. 48, at.wt. 112·41, m.p. 320·9°C, b.p. 765°C, *d* 8·64. Cadmium occurs in Zn and also Pb and Cu ores but also as greenockite, CdS. Cd is more volatile than Zn and is collected in the dust in flues of furnaces for producing zinc. CdO is reduced with C or Cd^{2+} is reduced in solution with zinc dust. Cd is a white lustrous metal with a deformed hcp structure with 6 near (in plane) and 6 far neighbours. The metal reacts with oxygen on heating and with acids. Cd is used extensively for electroplating (35%) and in alloys (Cu for tramway wires, Al for casting and with Sn and Bi in fusible alloys), batteries, atomic reactors; cadmium compounds are used as pigments, plasticizers and as phosphors in colour TV tubes. Cadmium telluride has many electronic applications. World production 1978 17 200 tonnes. Cadmium compounds are very toxic.

cadmium chemistry Cadmium is an electropositive element with a filled d shell, electronic configuration $4d^{10}5s^2$. The only stable oxidation state is +2 frequently in a mainly ionic state.

$E°$ for $Cd^{2+} \longrightarrow$ Cd in acid solution −0·402 volts

An unstable lower oxidation state is found in Cd − $CdCl_2$ and Cd − $CdCl_2$ − $AlCl_3$ melts and may contain Cd − Cd bonds. The common co-ordination number for Cd^{2+} is 6 (octahedral) although 4 and 5 co-ordination are known.

cadmium chloride, $CdCl_2$. *See* cadmium halides. The $CdCl_2$ structure is an important lattice type, being a layer structure based on ccp of Cl.

cadmium halides All dihalides, CdX_2, are known. CdF_2 is ionic with the rutile structure. $CdCl_2$, m.p. 868°C (structure *see* cadmium chloride) forms hydrates and chloro-complexes. The bromide and iodide are similar to the chloride.

cadmium hydroxide *See* cadmium oxide.

cadmium iodide, CdI_2. *See* cadmium halides. The CdI_2 is an important lattice type being a layer structure based on hcp of I^-.

cadmium, organic derivatives Derivatives R_2Cd and RCdX are prepared from RLi and CdX_2 or Cd plus RI in a polar solvent. They are fairly unstable thermally. The reactions are similar to those of Grignard reagents but are used specifically to prepare ketones RCOR′ from acyl chlorides RCOCl.

cadmium oxide, CdO; **cadmium hydroxide** $Cd(OH)_2$. The hydroxide is precipitated from aqueous solution by OH^-, it does not dissolve in excess OH^-. Ignition of $Cd(OH)_2$ or $CdCO_3$ gives CdO which varies in colour from red-brown to black because of lattice defects.

cadmium oxy-acid salts Colourless salts prepared by solution of the oxide or carbonate in the appropriate acid [$Cd(O_2CCH_3)_2{\cdot}2H_2O$, $Cd(NO_3)_2{\cdot}4H_2O$, $Cd(ClO_4)_2{\cdot}6H_2O$, $3CdSO_4{\cdot}8H_2O$]. The carbonate, $CdCO_3$, is precipitated from an aqueous solution of Cd^{2+} by $CO_3{}^{2-}$, the other salts are soluble in water although the solutions are hydrolysed.

cadmium red, orange, scarlet Pigments consisting of CdS(Se) prepared from a solution of $CdSO_4$ with added BaS and Se.

cadmium sulphate, $3CdSO_4{\cdot}8H_2O$. The most important Cd salt. *See* cadmium oxy-acid salts.

cadmium sulphide, CdS. Occurs naturally as greenockite and obtained as a yellow precipitate from Cd^{2+}(aq) and H_2S. Used as a yellow pigment.

cadmium yellow Natural cadmium sulphide. Used as a pigment.

caesium, Cs. At.no. 55, at.wt. 132·9054, m.p. 28·40°C, b.p. 678°C, *d* 1·785. An alkali metal occurring in traces in salt deposits and as pollucite $Cs(AlSi_2O_6),xH_2O$. Separated from the other alkali metals by ion-exchange and fractional crystallization, the metal (bcc) is obtained by electrolysis or better by $CsAlO_2$ (from caesium alum) and Mg. The metal is extremely reactive and reacts violently with water, oxygen, halogens, etc. The metal is used as an oxygen-getter, in photoelectric cells, as a hydrogenation catalyst and in a number of specialized applications. World production 1976 13 tonnes. ^{137}Cs is an important fission product used in deep-ray therapy.

caesium chemistry Caesium is an element of group I, electronic configuration $6s^1$. It forms a single series of components in the +1 oxidation state; most compounds are predominantly ionic. It also forms a series of lower oxides, e.g. Cs_7O, $Cs_{11}O_3$ containing cluster cations.

caesium chloride, CsCl. A typical alkali

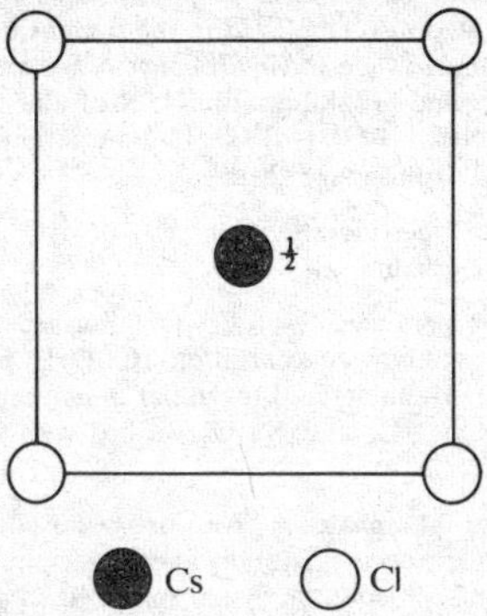

halide formed from Cs_2CO_3 and HCl. The structure is that adopted by many compounds AX with a large radius ratio of A:X. Each ion has a co-ordination number eight. Adopted by CsCl, CsI, AgLi, HgTl, etc.

caffeine, 1,3,7-trimethylxanthine, theine, $C_8H_{10}N_4O_2$. An alkaloid occurring in tea, coffee and guarana, from which it may be prepared by extraction. It is also manufactured by the methylation of theobromine and by the condensation of cyanoacetic acid with urea. Crystallizes with H_2O or anhydrous from organic solvents. M.p. (anhydrous) 235°C, sublimes at 176°C. Odourless, and with a very bitter taste. Caffeine acts as a stimulant and diuretic, and is a constituent of cola drinks, tea and coffee.

cage compounds Clathrate compounds*.

calamine, $ZnCO_3$ (British), $ZnSiO_4,H_2O$ (U.S.). Medicinal calamine is a basic zinc carbonate coloured with Fe_2O_3. Formerly obtained from the carbonate now obtained by precipitation. Used either in lotions or as a powder for sunburn, sore skin and dermatitis.

calciferol *See* vitamin D.

calcined bauxite A very hard, almost insoluble form of alumina produced by calcining bauxite* at temperatures above 1175K. It has a high resistance to abrasion and polishing and is used, often in conjunction with epoxy resin/bitumen mixes, as aggregate for highly skid-resistant road surfacings in specially dangerous locations.

calcite A form of $CaCO_3$. In the massive condition it forms limestone, marble, chalk. The crystals have many local names, including calcspar, Iceland spar, dogtooth spar. Calcite is the most stable form of calcium carbonate. *See* aragonite.

calcium, Ca. At.no. 20, at.wt. 40·08, m.p. 839°C, b.p. 1484°C, *d* 1·54. Calcium compounds are widely distributed in nature occurring as $CaCO_3$ (marble, limestone, chalk, calcite, aragonite, dolomite), $CaSO_4$ (gypsum), $Ca_3(PO_4)_2$ halides and silicates (various feldspars, anorthite). The metal is obtained by electrolysis of the fused chloride, or by $CaCO_3$ plus Al at high temperature and pressure. It is a soft silvery white metal ccp at room temperature hcp above 450°C. It tarnishes rapidly in air, reacts violently with water and combines with oxygen and the halogens. Ca is used as a reducing agent in the preparation of metals such as Th, V, Zr, as a deoxidizer and as an alloying agent. Calcium oxide is the most important base of the chemical industry. Ca^{2+} ions are important in biological systems.

calcium acetylide, CaC_2. *See* calcium carbide.

calcium aluminates Present in cements, particularly in aluminous cements*.

calcium bromide, $CaBr_2$. Very similar to the chloride.

calcium carbide, calcium acetylide, CaC_2. Prepared commercially by the action of an electric arc on coke and CaO in an electric furnace. Gives ethyne, H_2C_2, with water. The structure is ionic containing C_2^{2-} ions. Used for producing ethyne and calcium cyanamide.

calcium carbonate, $CaCO_3$. Occurs naturally (*see* calcium). Precipitated from $Ca(OH)_2$ solution by CO_2 but dissolves in excess CO_2 to form the hydrogen carbonate $Ca(HCO_3)_2$.

calcium chemistry Calcium is a Group II element electronic configuration $4s^2$. It shows a single oxidation state of +2 (E° for Ca^{2+} to Ca in acid solution −2·76 volts). The Ca^{2+} ion shows normal co-ordination arrangements of octahedral and cubic. An extensive range of salts is known and are used. Complexes are formed with oxygen-ligands particularly chelating ligands such as EDTA and polyphosphates (remove hardness of water).

calcium chloride, $CaCl_2$. Occurs naturally as tachydrite, $CaCl_2{\cdot}2MgCl_2,12H_2O$ and in sea and mineral waters. Prepared pure from $CaCO_3$ and HCl and gives $CaCl_2,6H_2O$, m.p. 30°C, $CaCl_2,4H_2O$, $CaCl_2,2H_2O$ and the anhydrous salt. Gives adducts with alcohols and ammonia. Obtained commercially from natural brines and from the ammonia-soda process for Na_2CO_3. Used for de-icing roads, dust control, as an antifreeze in concrete mixes and in solution in refrigeration plants. U.S. production 1978 1 100 000 tonnes.

calcium cyanamide, CaNCN. Prepared from N_2 and CaC_2 at 1100°C. Used as a fertilizer; on hydrolysis gives NH_3 and urea in the soil. Gives cyanides on fusion with C and Na_2CO_3 at 1200°C.

calcium ethanoate, $Ca(O_2CCH_3)_2$,2 or $1H_2O$. $CaCO_3$ plus ethanoic acid. Used in the production of other ethanoates and in calico printing.

calcium fluoride, fluorite, fluorspar, CaF_2. Occurs naturally as fluorospar. Relatively insoluble in water; m.p. 133°C. Used in optical systems for u.v. and near i.r. radiation and also to lower the melting point of molten salt baths used in Ca and Al production. The structure is ionic (*see* fluorite).

calcium glycerophosphate, $CaC_3H_5(OH)_2PO_4,H_2O$. A constituent of many proprietary nerve tonics.

calcium hydride, CaH_2. White solid (Ca (solid in NH_3) and hydrogen). Releases hydrogen with water (hydrolith), used as a drying agent and a reducing agent in metallurgy.

calcium hydrogen carbonate, $Ca(HCO_3)_2$. *See* calcium carbonate.

calcium hydroxide, slaked lime, $Ca(OH)_2$. The powder from the action of an equivalent quantity of water on CaO is slaked lime; a suspension with excess water is milk of lime, a clear solution of $Ca(OH)_2$ in water is lime-water. Used as an industrial alkali and in the preparation of mortar (slaked lime plus sand) which sets to a solid by reconversion of the $Ca(OH)_2$ to $CaCO_3$.

calcium iodide, CaI_2. Forms a hexahydrate. Similar to calcium chloride*.

calcium nitrate, $Ca(NO_3)_2,4H_2O$. A deliquescent solid.

calcium oxalate, CaC_2O_4,H_2O. One of the least soluble calcium salts and used (by oxidation of the oxalate) in the estimation of Ca^{2+}. Loses H_2O at 180°C, gives $CaCO_3$ and eventually CaO on heating in air. Occurs in rhubarb and other plants.

calcium oxide, lime, quicklime, CaO. A white solid with the NaCl structure produced by heating $CaCO_3$ to high temperatures. Commercially, lime is produced by heating $CaCO_3$ to a temperature between 900°C and 1200°C.

Although much lime is used in building and agriculture, the main use is in the chemical process industries. Thus large quantities are used in the manufacture of Na_2CO_3 and NaOH, SO_2 absorption, steel (45%), refractories, CaC_2, glass, pulp and paper, and sugar; it is also employed in the treatment of water and sewage and in ore concentration and refining and soil stabilization. U.S. production CaO plus $Ca(OH)_2$ 1982 15.5 megatonnes. Second top tonnage chemical.

calcium perchlorate, $Ca(ClO_4)_2,6H_2O$. A soluble calcium salt.

calcium peroxide, $CaO_2,8H_2O$. Formed anhydrous at high concentrations ($Ca(OH)_2$ plus H_2O_2). Manufactured for use as an antiseptic from $Ca(OH)_2$ and Na_2O_2 washed with ice-water.

calcium phosphates The orthophosphate, $Ca_3(PO_4)_2$ occurs naturally and is precipitated from a Ca^{2+} solution by phosphate (used as a polymerization aid for styrene and nutritional supplement). $CaHPO_4,2H_2O$ ($CaCl_2$ solution plus Na_2HPO_4) and $Ca(H_2PO_4)_2,H_2O$ (other phosphates plus H_3PO_4) are also known. The latter compound is the principal active ingredient of superphosphates*. Calcium phosphates are used in foodstuffs.

calcium silicates Important minerals also present in slag and cements.

calcium sulphate, $CaSO_4$. Occurs naturally as anhydrite, and as $CaSO_4,2H_2O$ as gypsum, selenite, satin-spar, alabaster. Natural anhydrite and $CaSO_4$ prepared by dehydrating the dihydrate above 650°C is only slowly soluble in water and is used as a filler for paper and other materials (*see also* satin white). $CaSO_4$ prepared by dehydration of the dihydrate at 60–90°C is used as a drying agent.

$CaSO_4 \cdot 0{\cdot}5H_2O$, prepared by heating gypsum at 130°C, is used as plaster of Paris*. Gypsum and anhydrite are used for H_2SO_4 production and gypsum is used as a soil additive and as an inert additive to pharmaceuticals and insecticides.

calcium sulphide, CaS. Formed from H_2S and limewater or $CaSO_4$ and charcoal at 1000°C. The impure material is luminescent. Used as a depilatory.

calcium sulphite, $CaSO_3$. Formed from SO_2 and limewater or $CaCl_2$ plus a sulphite as a precipitate, dissolves in excess SO_2. Used for sterilization and for dissolving the lignin in wood pulp in paper manufacture.

calcium superphosphate *See* superphosphates.

calcium titanate, perovskite, $CaTiO_3$. A mixed metal oxide which has given its name to the perovskite* structure.

calgon Sodium metaphosphate used as a washing powder and cleansing agent. *See* phosphates.

caliche Impure sodium nitrate*.

californium, Cf. At.no. 98. ^{252}Cf (961 days) is formed by neutron bombardment of ^{243}Am and ^{244}Cm and ^{249}Cf (360 years) by decay of ^{249}Bk. Cf is purified by ion exchange chromatography. The silver-grey metal has been obtained by reduction of Cf_2O_3 with lanthanum metal, double-hexagonal close-packed and ccp structures seem established. ^{242}Cf is a potential source of ^{248}Cm.

californium compounds Californium is the first actinide to show the +2 state found in $CfBr_2$ formed by reduction of $CfBr_3$ with hydrogen. The normal oxidation state is +3 found in the halides ($CfBr_3$ light green), and the oxide Cf_2O_3. CfO_2 and CfF_4 are formed by strong oxidation.

calomel, Hg_2Cl_2. Mercury(I) chloride.

calomel electrode The hydrogen electrode is not easy to use as a general standard for laboratory use, and it is often replaced by a calomel electrode which has a potential known in terms of that of the hydrogen electrode. It consists of a pool of mercury covered with a paste of Hg_2Cl_2 (calomel) and Hg, in contact with a standard solution of KCl saturated with Hg_2Cl_2. The potential of the cell is developed on the Hg and a suitable contact is made with it by means of a platinum wire. Three types of calomel electrode have been used, viz. with 0·1 M, 1·0 M and saturated KCl. Their respective potentials in volts on the hydrogen scale at 298 K are: −0·3338, −0·2800, −0·2415.

Calor gas, Calor propane The trade name for a liquified petroleum gas* (LPG) sold in cylinders for domestic and industrial heating. Calor gas is the name used for commercial butane while commercial propane is sold as Calor propane.

calorie A heat unit widely used in scientific and industrial work. Defined in terms of the joule and referred to as the international table calorie (cal_{IT}) which is equivalent to 4·1868 joules.

For fuels it is quite usual to employ kilocalories per kilogram as a measure of calorific value while for foods the unit used is kilocalories per gram. This, however, is often abbreviated to 'Calories', so that a value for carbohydrates of 4·1 'Calories' per gram is 4100 calories per gram.

calorific value The amount of heat given out by complete combustion of unit weight of a solid or liquid or unit volume of a gas in a suitable calorimeter*.

calorimeter A device for the measurement of calorific values of fuels and foods or for the measurement of heats of chemical reactions. For solid and liquid fuels the bomb calorimeter* is used, while for gases a suitable gas calorimeter* is used.

campesterol, (24*R*)-24-methyl-5-cholesten-3*β*-ol, $C_{28}H_{48}O$. M.p. 158°C. Occurs with sitosterol* in many plants, e.g. rape seed.

camphane, bornylane, $C_{10}H_{18}$. The parent

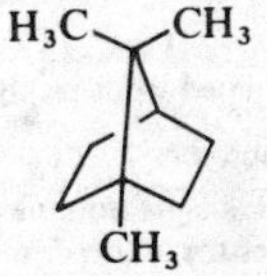

hydrocarbon of the camphor group. It is obtained by reduction of bornyl chloride or iodide or isobornyl chloride. M.p. 153–154°C, b.p. 160–161°C. It is very volatile and smells like borneol. It is optically inactive, and is chemically very inert.

camphene, $C_{10}H_{16}$. M.p. 51°C, b.p. 159°C.

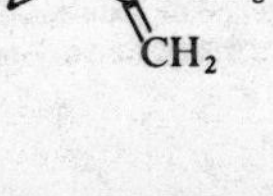

Present in citronella and valerian oils, turpentine, ginger, rosemary and spike oils. It is produced artificially by the elimination of hydrogen chloride from bornyl chloride (artificial camphor) or from isobornyl chloride, by the dehydrogenation of borneol and isoborneol and by the action of ethanoic anhydride on bornylamine. Chiral.

camphor, $C_{10}H_{16}O$. M.p. 179°C, b.p. 209°C.

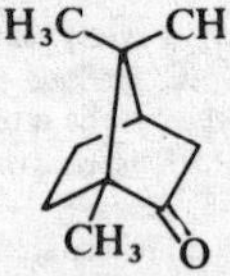

Ordinary commercial camphor is (+)-camphor, from the wood of the camphor tree. *Cinnamonum camphora.* Camphor is of great technical importance, being used in the manufacture of celluloid and explosives, and for medical purposes. It is manufactured from pinene through bornyl chloride to camphene, which is either directly oxidized to camphor or is hydrated to isoborneol, which is then oxidized to camphor. A large number of camphor derivatives have been prepared, including halogen, nitro and hydroxy derivatives and sulphonic acids.

Used medicinally as a carminative and stimulant. It is injected in olive oil solution in cases of circulatory collapse. It is a popular remedy for colds and is a constituent of many linaments. Camphorated oil is 20% solution in olive oil.

canavanine, $C_5H_{12}N_4O_3$. An amino-acid,

$$(HN)(H_2N)C{\cdot}NH{\cdot}O{\cdot}CH_2{\cdot}CH_2{\cdot}CH(NH_2){\cdot}COOH$$

m.p. 184°C, isolated from jackbean meal.

cane sugar Sucrose*.

Cannizzaro reaction Two molecules of many aldehydes, under the influence of dilute alkalis, will interact, so that one is reduced to the corresponding alcohol, while the other is oxidized to the acid. Benzaldehyde gives benzyl alcohol and benzoic acid. Compare the aldol condensation.

canonical form A single structure based on

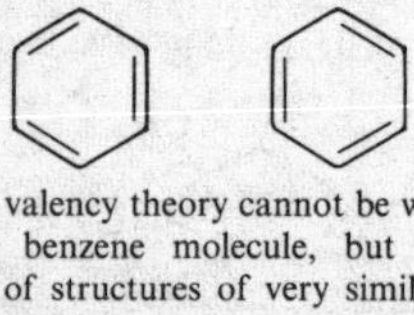

classical valency theory cannot be written for, e.g. the benzene molecule, but instead a number of structures of very similar energy, e.g. the Kekulé structures, canonical forms can be envisaged.

capillary condensation When a substance wets the wall of a capillary tube in which it is contained, and possesses a concave meniscus, the vapour pressure above the meniscus is less than that in contact with a plane surface of the liquid at the same temperature. This is expressed by the equation

$$\log_\varepsilon \frac{p}{p_s} = 2\frac{\sigma v}{rRT}$$

(p is the pressure at the concave surface, p_s the normal vapour pressure at the same absolute temperature T, σ the surface tension, r the radius of the capillary tube, R the gas constant and v the mol. volume of the liquid). There is thus a considerable tendency for a vapour to condense in the fine capillaries of a porous material such as charcoal. Capillary condensation is certainly the mechanism of the sorption of water by silica gel, and probably also at high humidities by charcoal, but it is unsatisfactory as a general theory of physical adsorption.

capric acid, *n*-decanoic acid, $C_{10}H_{20}O_2$, $CH_3{\cdot}[CH_2]_8{\cdot}COOH$. M.p. 31·5°C, b.p. 268–270°C. A fatty acid, occurring in wool as the potassium salt, as esters in fusel oil, and as glycerides in cows' and goats' milk and coconut and palm oils.

caproic acid, *n*-hexanoic acid, $C_6H_{12}O_2$, $CH_3{\cdot}[CH_2]_4{\cdot}CO_2H$. An oil, m.p. −3·4°C, b.p. 205°C. It occurs as glycerides in the milk of cows and goats, in coconut oil and palm oil.

caprolactam, $C_6H_{11}NO$. Prepared by

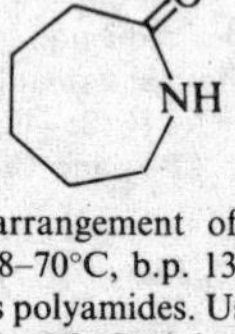

Beckmann rearrangement of *cyclo*hexanone oxime. M.p. 68–70°C, b.p. 139°C/12 mm. On heating it gives polyamides. Used in the manufacture of Nylon[6]. Cyclohexanone oxime is formed from cyclohexane and nitrosyl chloride. U.S. production 1978 410 000 tonnes.

capryl alcohol *See* 2-octanol.

caprylic acid *See* octanoic acid.

capsicum Dried ripe fruit of various species of *Capsicum*. It contains the intensely pungent principle capsaicin. Capsicum is used as a linament or ointment for external application as a counter-irritant particularly in lumbago and rheumatism.

captafol, *N*-(1,1,2,2-tetrachloroethylthio)-3*a*,4,7,7*a*-tetrahydrophthalimide, $C_{10}H_9Cl_4NO_2S$. A protective non-systemic fungicide used principally for the control of potato blight. M.p. 160–161°C. It is relatively non-toxic to most animal life although it is harmful to fish.

captan, *N*-(trichloromethylthio)-3*a*,4,7,7*a*-tetrahydrophthalimide, $C_9H_8Cl_3NO_2S$. A fungicide produced by the reaction of trichloromethyl-sulphenyl chloride with tetrahydrophthalimide. It forms white crystals m.p. 178°C. The technical product is an amorphous yellow solid with a pungent odour. It is relatively non-toxic and may be used by amateur gardeners.

caramel The brown substance obtained by heating cane sugar or other carbohydrate materials. Its chemical nature is unknown, and its reactions vary with its method of preparation. It is soluble in water and is used as a colouring agent for foodstuffs and drinks.

carbaboranes *See* carboranes.

carbachol, carbamylcholine chloride,

$$[H_2N{\cdot}COCH_2CH_2NMe_3]^+Cl^-$$

$C_6H_{15}ClN_2O_2$. Colourless, hygroscopic, m.p. 210–212°C (decomp.). Prepared from *β*-chloroethyl carbamate and trimethylamine. It has a physiological action similar to that of acetylcholine*, but more prolonged, as it is less readily hydrolysed. It is used for intestinal atony following operations, and can be given orally.

carbamates Salts and esters of carbamic acid.

carbamic acid, $H_2N{\cdot}COOH$. The acid is not known in the free state, but salts and esters exist. Ammonium carbamate is formed by the action of dry carbon dioxide on dry ammonia. Commercial ammonium carbonate* contains appreciable quantities of ammonium carbamate. Urethanes, widely used in plastics technology, are esters of carbamic acid.

carbamido- The group $-NHCONH_2$.

carbamyl- The group $-CONH_2$.

carbanilide, *sym*-diphenylurea, $C_{13}H_{12}N_2O$, $(PhNH)_2CO$. Silky needles, m.p. 235°C. Prepared by the action of phosgene or phenylisocyanate upon aniline.

carbanions, R_3C^- Produced by cleavage of certain C–H; C–halogen, C–metal and C–C bonds. They have high reactivity with air, water, etc. In many cases the existence of these negatively charged species may only be inferred from mechanistic studies. The delocalization of charge (unshared electron pair) over multiple bond systems increases their stability: e.g. triphenylmethyl sodium and cyclopentadienyl sodium are relatively stable.

carbazole, $C_{12}H_9N$. Occurs with anthracene

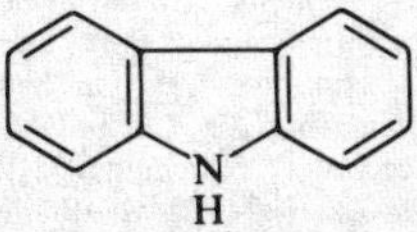

in the solid which separates from anthracene oil. It is used in dyestuff preparation. M.p. 238°C, b.p. 335°C. Insoluble in water and sparingly soluble in most organic solvents. It sublimes easily.

carbene, R_2C:. Reactive species, containing a formally divalent carbon atom. Produced, e.g., by photolysis of diazoalkanes or by an *α*-elimination of HX from 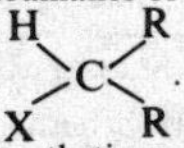 . Dihalocarbenes have greatest synthetic value, e.g. Cl_2C: may be generated by the thermal decomposition of $Cl_3C{\cdot}CO_2Na$ and used *in situ*. The reaction of base with haloforms under Riemer–Tiemann conditions, and the Arndt–Eistert synthesis are two reactions which proceed via carbene intermediates. Can act as ligands to transition metals when stable complexes are often formed.

carbenium ions, carbonium ions Positively-charged species containing a trivalent carbon atom R_3C^+. They exhibit a broad spectrum of reactivity and stability, the latter decreasing in the order tertiary > secondary > primary. They are postulated as intermediates in many reactions, e.g. the dissolution of alkenes in strong acids, the dehydrohalogenation of alkyl halides and certain solvolytic reactions. Carbenium ions may be detected by spectroscopic (n.m.r.) methods; tropylium and triphenylmethyl carbenium ions are stable and may easily be isolated; alkyl carbenium ions (as SbF_6^- salts) can be prepared *in situ* from the corresponding alkyl fluoride and SbF_5 at low temperature in SO_2 or $ClSO_2F$, or from alcohols or alkenes in HF/SbF_5 or FSO_3H/SbF_5 solutions and are the most powerful cationic type alkylating agents. All carbenium ions have a strong affinity for nucleophiles, e.g. water.

carbethoxy- The group $-C(O){\cdot}OCH_2CH_3$.

carbide formers Alloying elements added to steel (e.g. Mn, Cr, Mo, W, Ti, V, Nb) which stabilize carbides and confer wear resistance when present in a coarse form or creep resistance when present in a fine coherent form.

carbides Derivatives of carbon with elements of lower electronegativity. Carbides are formed by direct interaction of the elements or metal plus suitable hydrocarbon. The most electropositive elements form salt-like carbides which are essentially ionic. Be_2C and Al_4C_3 contain C^{4-} and give CH_4 on hydrolysis. Group I, II, Cu group, Zn, Cd, and Al, La, Pr, Th, form acetylides* containing $[C_2]^{2-}$ ions (HC_2H with water). Other lanthanide and actinide carbides contain C_2 units but give hydrocarbons in addition to ethyne on hydrolysis. Interstitial carbides, e.g. Fe_3C, W_2C, have carbon in holes in approximately cp metal lattices, there are related examples of carbon atoms in metal clusters, e.g. $[Ru_6C(CO)_{14}(\text{mesitylene})]$. Carbides formed with elements close to carbon in the periodic table, e.g. SiC, B_4C, are essentially covalent and are generally extremely hard and infusible. WC, TiC, TaC and NbC are of major importance for use in cutting tools and wear resistant materials.

carbinol A name for methanol, introduced by Kolbe in 1860 as a basis for the systematic naming of monohydric alcohols, e.g. triphenylcarbinol or triphenylmethanol Ph_3COH.

carbitols The trivial name given to the monoalkyl ethers of diethylene glycol, e.g. butylcarbitol, $C_4H_9OCH_2CH_2OCH_2CH_2OH$. *See also* Cellosolve and diethyleneglycol monoethylether.

carbocyclic A generic term relating to organic compounds which are derived from one or more rings of carbon atoms, e.g. benzene, naphthalene, diphenyl.

carbodiimides Derivatives $RC{=}N{=}CR^1$ (*see* dicyclohexylcarbodiimide). MeNCNMe is formed from MeNHC(S)NHMe and HgO; it is stable at low temperatures. Carbodiimides are formed by heating isocyanates and can be polymerized in the presence of catalysts. The dimer of cyanamide is a carbodiimide, $H_2N{\cdot}C({=}NH){\cdot}N{=}C{=}NH$.

carbohydrases Enzymes which act on carbohydrates including the amylases, glucosidases, cellulase, invertase and maltase.

carbohydrates One of the principal classes of naturally occurring organic compounds. An approximate formula is $(CH_2O)_x$ and for various values of x include sugars, starches and cellulose, each essential to plant and animal life. They are produced in plants as the result of photosynthesis, the chlorophyll catalysed combination of carbon dioxide and water under the influence of light. The simple carbohydrates are mono-, di- or poly-saccharides, having repeating units usually containing 5 or 6 carbon atoms joined through oxygen linkages. For example, sucrose and lactose are disaccharides which are simple carbohydrates; cellulose, a polysaccharide, is a polymer containing approximately 2000–3000 glucose units per molecule. The basic sugar skeleton of carbohydrates, involving hydroxyl groups, gives them their properties such as water solubility and sweet taste. As many carbon atoms are asymmetric the carbohydrates can exist in many stereochemical and structural modifications.

carbolan dyes A range of azo or anthraquinone dyes containing long-chain substituents and having high light fastness.

carbolic acid An old name for phenol*.

carbomethoxy- The group $-CO{\cdot}OCH_3$.

carbon, C. At.no. 6, at.wt. 12·011, m.p. 3550°C, b.p. 4830°C. Occurs in both free and combined (CO_2, $(CO_3)^{2-}$, etc.) states. The crystalline forms are diamond (metastable) and graphite – for details of structure *see* under individual headings. Various forms of amorphous carbon are described: most contain microcrystalline material. Active carbon* and carbon black* are widely used in industry. Steel is an iron-carbon system. Carbon fibres, produced by pyrolysis, confer great strength on plastic materials. ^{12}C is the standard of atomic mass, ^{13}C has nuclear spin of $\frac{1}{2}$ and can be used for n.m.r. measurements, ^{14}C is radioactive and is used for carbon dating since it is formed in the atmosphere and absorbed only by living materials. Carbon burns in oxygen and reacts with halogens.

carbonates Salts of the weak acid H_2CO_3. The normal carbonates include many minerals and compounds of great technical importance (e.g. $CaCO_3$, $MgCO_3$, Na_2CO_3, $ZnCO_3$). The hydrogen carbonates contain the $[HCO_3]^-$ ion and important salts include $NaHCO_3$ and $Ca(HCO_3)_2$. The carbonates of the alkali and alkaline earth metals are stable (the latter insoluble). Most other metal carbonates are precipitated as basic carbonates, e.g. $2PbCO_3{\cdot}Pb(OH)_2$. All carbonates except the alkali metal carbonates lose CO_2 on strong heating. Carbonate esters are important solvents.

carbonation Addition (frequently insertion) of CO_2 into derivatives such as RMgX to give RC(O)OMgX.

carbon black A range of technologically useful materials obtained by the incomplete combustion of natural gas or liquid hydrocarbons. The different varieties vary in particle size and colour value. Used in rubber manufacture, as a filler (tyres 65%, other rubber 25%), as a pigment for paint, ink, polish, etc. and for decolourizing. U.S. production 1981 1·2 megatonnes.

carbon chalcogenides Carbon forms CS_2 (*see* carbon disulphide), CSSe, CSe_2, CSTe. These are all fairly similar to CS_2.

carbon chemistry Carbon is an element of Group IV, electronic configuration $1s^22s^22p^2$. It shows a formal oxidation state of +2 in CO and in carbenes but this state is relatively reactive. The majority of carbon compounds are in the +4 oxidation state and the chemistry is dominated by the tendency to catenation (C−C bond formation) and to the formation of multiple bonds (C=C in alkenes, C≡C in alkynes, conjugated bonds and aromatic systems) and the formation of stable bonds to most other elements (*see* element organic derivatives) but particularly to O, N, S, halogens and H. Anionic derivatives of carbon are found in carbanions, carbides and carbonates; cations in carbenium ions.

carbon dating ^{14}C is continuously formed in the atmosphere by the reaction

$$^{14}N + {}^{1}n \longrightarrow {}^{14}C + {}^{1}H$$

The ^{14}C exchanges with ^{12}C in living organisms, but exchange ceases on death. The radioactive ^{14}C content decays with a half-life of 5730 years. Hence the age of a once living material may be established by determining the amount of ^{14}C.

carbon dioxide, CO_2. *See* carbon oxides.

carbon disulphide, CS_2. M.p. −112°C, b.p. 46°C. An important solvent and reagent (C plus S). The technical grade has a disagreeable odour owing to the presence of impurities but pure CS_2 is pleasant smelling. It is toxic and exceedingly inflammable. Used in the manufacture of viscose silk and CCl_4. Adds SH^- to give trithiocarbonates*, $(CS_3)^{2-}$, OR^- to give xanthates*, $(ROCS_2)^-$, amines to give dithiocarbamates*, $(RHNCS_2)^-$. U.S. production 1978 220 000 tonnes.

carbon electrodes Rods used in arcs and in electrolysis (e.g. F_2 production).

carbon fibres Graphite fibres used in the strengthening of composites.

carbon fluorides *See* fluorocarbons.

carbonic acid, H_2CO_3. A very weak dibasic acid formed from CO_2 and H_2O, the anhydrous acid can be prepared in ether at −30°C. The acid forms carbonates* as salts.

carbonic anhydrase A very important enzyme which reversibly catalyses the reaction

$$H_2O + CO_2 \rightleftharpoons H_2CO_3$$

in an organism. In respiration, it catalyses the formation of carbon dioxide from hydrogen carbonate in the blood vessels of the lung, whereas in kidney function it promotes the formation of carbonic acid. At the pH of blood most of the CO_2 is present as H_2CO_3. The enzyme is widely distributed and can be isolated from red blood cells. The enzyme contains Zn as a prosthetic group.

carbonium ions *See* carbenium ions.

carbonization The process of destructive distillation of coal, that is decomposition by heat in the absence of air. This produces a residue of coke along with coal gas, coal tar and other liquors. In high temperature carbonization (HTC) coal is heated in refractory retorts to about 1100°C and almost all the volatile matter is driven off leaving a non-reactive coke. HTC has been used to produce town gas* in continuous vertical retorts and for hard, metallurgical coke in coke ovens. Low temperature carbonization (LTC) is used to produce 'semi-coke' or 'smokeless fuel' by heating coal to about 900 K.

carbon monoxide, CO. *See* carbon oxides.

carbon oxides
Carbon monoxide, CO. M.p. −205°C, b.p. −191°C. Colourless, odourless, toxic gas formed by incomplete combustion of C or carbon compounds, C plus CO_2 at 900–1000°C, C plus H_2O (with H_2-blue water gas*), methanoic acid plus conc. H_2SO_4. Burns in air with pale blue flame and can react with oxygen explosively. Acts as a reducing agent (e.g. PbO or Fe_2O_3 to Pb or Fe at high temperatures), $[IO_3]^-$ to I_2 at 90°C (used in analysis) gives carbonyls with many transition metals. Cl_2 gives phosgene. Used industrially in synthesis (e.g. carbonylation of alkynes to carboxylic acids), and in metallurgical reduction processes and the purification of nickel.

Carbon dioxide, CO_2. Sublimes −78·5°C. A colourless gas at room temperature, occurs naturally and plays an important part in animal and plant respiration. Produced by the complete combustion of carbon-containing materials (industrially from flue gases and from synthesis gas used in ammonia production) and by heating metal carbonates or by the action of acid on carbonates. Soluble in water to give mainly a hydrate but there is some reaction to the weak acid, H_2CO_3. Detected by the formation of a white precipitate with a solution of $Ca(OH)_2$. Used in mineral waters under pressure, as a refrigerant (the solid 'dry-ice' or as the liquid), as a chemical reagent (RMgX gives RCO_2MgX), as an inert gas, as a fire extinguisher. Has a linear molecule, O=C=O. U.S. production 1982 4·1 megatonnes.

Carbon suboxide, C_3O_2, OCCCO. M.p. −107°C, b.p. 6·8°C. A toxic gas (malonic acid plus P_2O_5) which polymerizes at room temperature. Reforms malonic acid with water.

carbon residue The tendency for diesel fuel, fuel oil or lubricating oil to form carbon on heating or burning in the absence of excess air can be measured by evaluating the carbon residue.

carbon sulphides *See* carbon chalcogenides.

carbon tetrabromide, CBr_4. M.p. 94°C, b.p. 190°C. White crystalline solid (Br_2 plus CS_2 plus I_2).

carbon tetrachloride, tetrachloromethane, CCl_4. M.p. −23°C, b.p. 76·5°C. Manufactured from CS_2 plus Cl_2 or by chlorination of CH_4 and other hydrocarbons. In presence of H_2O gives HCl and phosgene Cl_2CO. Relatively toxic to the liver. The simplest member of the series of perchlorocarbons. Used as a solvent (use tends to be replaced by higher homologues and chloroethane derivatives) as a fumigant and in fire extinguishers. U.S. production 1978 330 000 tonnes.

carbon tetrafluoride, CF_4. M.p. −150°C, b.p. −128°C. An inert gas (C plus F_2).

carbon tetraiodide, CI_4. Red crystalline solid (CCl_4 plus CH_3I in presence of $AlCl_3$). Decomposes on heating.

carbon value The measure of the tendency of an oil to form carbon when used as lubricant.

carbonylate ions Ions containing metal carbonyls. Anions are prepared by the action of a base or an alkali metal on the carbonyl (e.g. $Mn_2(CO)_{10}$ plus pyridine or Na gives $(Mn(CO)_5)^-$). Cations are prepared from a Lewis acid and a co-ordinating group on the carbonyl halide (e.g. $Mn(CO)_5Cl$ plus $AlCl_3$ plus CO gives $[Mn(CO)_6]^+$). These ions generally have a noble gas electronic structure (both $[Mn(CO_5]^-$ and $[Mn(CO)_6]^+$ have the Kr electronic structure for Mn).

carbonylation The reaction of an organic or intermediate organometallic compound, with CO. Alkynes will react with carbon monoxide in the presence of a metal carbonyl (e.g. $Ni(CO)_4$) and water to give propenoic acids ($R{\cdot}CH{=}CH{\cdot}CO_2H$), with alcohols (R′OH) to give propenoic esters, $RCH{:}CHCO_2R'$ and with amines ($R'NH_2$) to give propenoic amides RCH:CHCONHR. Using alternative catalysts, e.g. $Fe(CO)_5$, alkynes and carbon monoxide will produce cyclopentadienones or hydroquinols. A commercially important variation of this reaction is hydroformylation (the oxo reaction*).

carbonyl chloride, Cl_2CO. *See* carbonyl derivatives.

carbonyl derivatives Derivatives containing the C=O group. The category includes aldehydes*, RCHO; ketones*, RR′CO; metal carbonyls*; and carbonyl halides. The halides, all slowly hydrolysable, X_2CO; carbonyl fluoride, F_2CO, m.p. −114°C, b.p. −83°C (Cl_2CO plus KF); carbonyl chloride, Cl_2CO, phosgene, m.p. −118°C, b.p. 8°C, a colourless poisonous gas (CO plus Cl_2 plus light or over a catalyst – often charcoal). Gives urea, $OC(NH_2)_2$, and NH_4Cl with ammonia, used to form toluene diisocyanate used in polyurethane plastics, 1-naphthyl-*N*-methyl carbamate (insecticide) and polycarbonates. Carbonyl chalcogenides include SCO, m.p. −138°C, b.p. −50°C (CO plus S). An explosive, inflammable gas.

carbonyl group The $>C{=}O$ group in such compounds as aldehydes, ketones, metallic carbonyls and phosgene. For the general reactions of this group *see* aldehydes and ketones.

carbonyl halides Applied to either the derivatives X_2CO (*see* carbonyl derivatives) or to metal carbonyl derivatives, e.g. $Mn(CO)_5Cl$.

carbonyl hydrides The metal carbonyl hydrides, e.g. $HMn(CO)_5$. These compounds have a weak acidic character and form carbonylate anions, e.g. $[Mn(O)_5]^-$. The H−M bond is very short.

carbonyloxime A misnomer for fulminic acid*.

carbonyls *See* metal carbonyls.

carboranes Derivatives of the borane anions in which a CH group has formally replaced a BH^-. Prepared by the action of acetylenes and borane derivatives. The series contain closo (polyhedral) carboranes, e.g. 1,2- and 1,7-$B_{10}C_2H_{12}$ (based on an icosahedron), $B_7C_2H_9$ and nido (open) carboranes, e.g. $B_7C_2H_{13}$. Nido carborane anions, e.g. $[B_9C_2H_{11}]^{2-}$ form metal complexes, e.g. $[(B_9C_2H_{11})_2Fe]^{2-}$

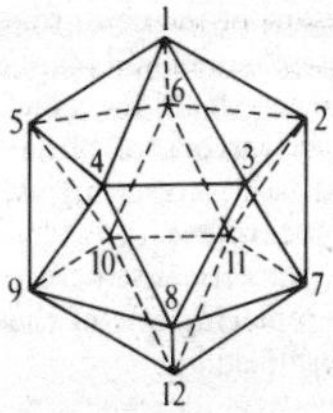

closo - $B_{10}C_2H_{13}$
(C at 1,2 - or 1,7-)

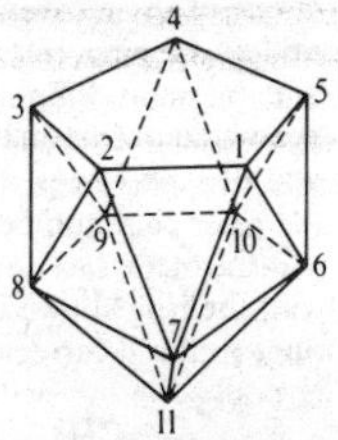

nido - $[B_9C_2H_{11}]^{2-}$
(C at 1,2 - or 3,5-)

through the open pentagonal faces of the ions (dicarbollide derivatives).

carborundum *See* silicon carbide.

Carbowaxes Trade name for a wide range of waxy polyethylene glocols of the general formula $HOCH_2(CH_2OCH_2)_xCH_2OH$. Carbowaxes of high molecular weight are used in rubber mixes, as dispersing agents in the textile industry and for pharmaceutical purposes. Also used as the stationary phase in GLC columns.

carboxy- A prefix denoting that the sub stance contains a carboxyl group, $-CO_2H$.

carboxyhaemoglobin The combination of haemoglobin with CO instead of O_2, by co-ordination, gives carboxyhaemoglobin. The affinity of haemoglobin for carbon monoxide is, in humans, 500 times as great as for oxygen; this accounts for the high toxicity of carbon monoxide as it prevents haemoglobin from fulfilling its natural function. (Does not contain a $-CO_2H$ group, *see* carboxy-.)

carboxylase An enzyme present in yeasts, bacteria and plants which catalyses the decarboxylation of α-ketoacids, in particular pyruvic acid, which is converted to ethanal.

Decarboxylases, which are highly specific for individual amino-acids, decarboxylate these to amines.

carboxyl group The $-CO_2H$ group.

carboxylic acids Organic compounds containing one or more carboxyl (−COOH) groups; the number of these groups is indicated by the prefixes mono-, di-, tri-, etc. These acids are much weaker than the mineral acids such as hydrochloric, but they are partially ionized in aqueous solution and all liberate carbon dioxide from metallic carbonates and hydrogen carbonates. The strength of the acid is markedly affected by the nature of the remainder of the molecule, e.g. $Cl_3C{\cdot}CO_2H$, trichloroethanoic acid is a very strong acid. They form salts with metals and with organic bases, and form esters with alcohols. Anhydrides containing −C(O)·O·C(O)− groups are formed from most acids by elimination of a molecule of water between two carboxyl groups. This elimination may be either inter- or intra-molecular. The hydroxyl (−OH) group is replaceable by halogens (with, e.g. PCl_5) to give acyl halides. The acids are formed by oxidation of the corresponding alcohol or aldehyde; by heating nitriles with alkalis; or by treating Grignard reagents with carbon dioxide and decomposing the product with sulphuric acid. Many of these acids occur naturally in plants and animals, either free or as salts or esters.

carboxypeptidase A pancreatic enzyme which detaches terminal amino groups of peptides.

carbylamine reaction A qualitative test for primary amines carried out by warming the suspected amine with chloroform and an alcoholic solution of potassium hydroxide. Under these conditions primary amines produce the intensely nauseating odour characteristic of the carbylamines (isonitriles). Sometimes used as a test for chloroform, but it is not specific, as bromoform, iodoform and chloral also react in the same manner.

carbylamines *See* isonitriles.

carbyloxime, CNOH. *See* fulminic acid.

carbyne derivatives Metal derivatives containing a M≡CR grouping.

carcinogens Substances capable of causing the development of malignant cells in animals. Many compounds have this effect in test animals. Benzpyrene which occurs in tar and soot, is one example of a carcinogen. Many amino aromatic, nitro and other derivatives including methyl (chloromethyl) ether, ethyleneimine, propiolactone are carcinogenic and workers must not be exposed to such materials.

cardiac aglucones *See* genins.

carenes Dicyclic terpenes. Colourless oils found naturally.

Carius method The quantitative determination of S and halogens in covalent (organic) compounds by complete oxidation of the compound with conc. nitric acid and subsequent estimation of precipitated AgX or $BaSO_4$.

Carius tube A thick-walled glass tube originally used in the Carius method* but used in any reaction involving volatile materials.

carmine A red lake pigment once obtained by extracting cochineal with boiling water and precipitating the extract with alumina. The modern equivalent is known as alizarin carmine, a lake prepared from 1,2-dihydroxyanthraquinone.

carminic acid, $C_{22}H_{20}O_{13}$. The red colouring matter of cochineal obtained from the *Coccus* insects. Used as a stain.

carnallite, KCl, $MgCl_2,6H_2O$. A principal component of many salt deposits. Used as a source of potassium salts.

carnauba wax The dried exudation from the leaves of the South American copaiba palm. The crude wax is yellow or dirty green, brittle and very hard. The principal constituent is melissyl cerotate. It is used in many polishes and varnishes, as when mixed with other waxes it makes them harder and gives more lustre.

carnitine, α-hydroxy-γ-butyro-betaine, $C_7H_{15}NO_3$,$(CH_3)_3N^+{\cdot}CH_2{\cdot}CHOH{\cdot}CH_2COO^-$. Isolated from skeletal muscle. It acts as a carrier for ethanoyl groups and fatty acyl groups across the mitochondrial membrane during the biosynthesis or oxidation of fatty acids.

carnosine, β-alanylhistidine, $C_9H_{14}N_4O_3$. M.p. 246–250°C (decomp.). A dipeptide present in mammalian muscle. Like anserine it contains the amino-acid β-alanine which is not found in proteins.

carnotite, $KUO_2VO_4,1{\cdot}5H_2O$. An important uranium ore.

Carnot's cycle A hypothetical scheme for an ideal heat machine. Shows that the maximum efficiency for the conversion of heat into work depends only on the two temperatures between which the heat engine works, and not at all on the nature of the substance employed.

carone, $C_{10}H_{16}O$. A saturated bicyclic ketone.

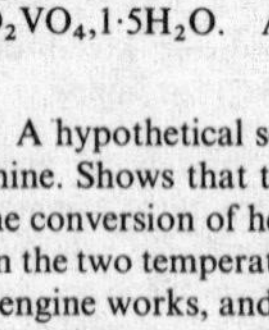

Caro's acid, H_2SO_5. *See* persulphuric acid.

carotene, $C_{40}H_{56}$. M.p. 181–182°C. Carotene is one of the most important of the colouring matters of green leaves. It is found in all plants and in many animal tissues. It is the chief yellow colouring matter of carrots, of butter and of yolk of egg. It exists in several forms, the most common being β-carotene.

α-Carotene has an unsymmetrical structure with the ring shown here replacing one of the rings in the β-carotene formula,

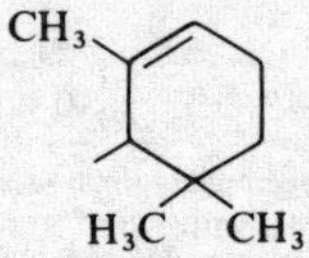

Carotene is found in plant chloroplasts. When ingested into animals it serves as a precursor of vitamin A (the transformation occurs in the liver).

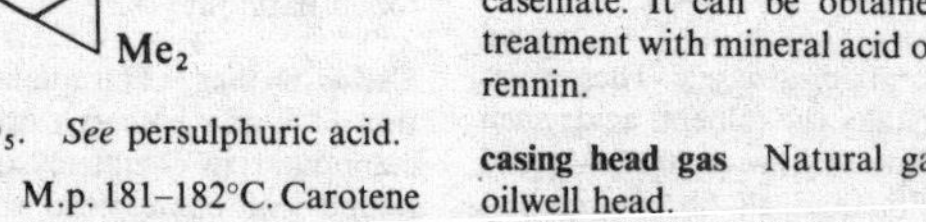

β-carotene

carotenoids Pigments related to carotene which occur in plants, particularly in the leaves, and in certain animal tissues. They include the hydrocarbons carotene and lycopene and their related hydroxyl compounds, the xanthophylls.

carragheenan An important seaweed hydrocolloid used in foodstuffs.

carvacrol, 2-hydroxycymene, $C_{10}H_{14}O$. A colourless liquid, b.p. 237–238°C; a constituent of many essential oils.

carveol, $C_{10}H_{16}O$. A monocyclic terpene alcohol.

carvestrene The racemic form of sylvestrene*.

carvone, $C_{10}H_{14}O$. A ketone derived from the terpene limonene. B.p. 230°C.

caryophyllene, $C_{15}H_{24}$. The sesquiterpene hydrocarbon which is the main hydrocarbon constituent of oil of cloves. It is a colourless oil, b.p. 123°–125°C/10 mm.

casein A phosphoprotein which occurs in the milk of mammals as a suspension of calcium caseinate. It can be obtained from milk by treatment with mineral acid or rennet. *See also* rennin.

casing head gas Natural gas evolved at an oilwell head.

cassel yellow *See* lead chlorides.

cassiterite, tinstone, SnO_2. The principal ore of tin. *See* tin oxides.

cast iron, pig iron *See* iron. Iron containing 2·5–4% C either in the form of cementite, Fe_3C (white cast iron, brittle but wear resistant) or graphite (grey cast iron, low strength). Alloying metals are added to improve properties.

Alloy cast irons. Alloy additions are made to cast irons to improve the properties for particular purposes. Alloy cast irons can be used in engineering applications where plain cast iron is unsuitable and may even replace steel for some components such as crankshafts.

Castner–Kellner cell An electrolytic cell for the production of sodium hydroxide*.

castor oil Fixed oil expressed from the seeds of *Ricinus communis*. It consists almost entirely of the glyceryl esters of ricinoleic acid. It is miscible with alcohol. It is a mild purgative. Used industrially for the manufacture of dehydrated castor oil and also in the production of polyurethanes.

catabolism *See* metabolism.

Catacarb process An extraction process used to remove carbon dioxide from process gases by scrubbing the hot gases with potassium carbonate solution containing additives which increase the hydration rate of the gas in the solution. The Vetrocoke process is similar. *See* Benfield process.

catalase A collective name for enzymes which destroy hydrogen peroxide and so prevent the accumulation of other toxic peroxides that might be formed. They contain haematin or related compounds as the prosthetic group. Their activity is related to the oxidation and reduction of the iron in the molecule. (*See also* peroxidases.)

catalysis The use of a catalyst* to affect the rate of a reaction.

catalyst A substance which when added to a reaction mixture changes the rate of attainment of equilibrium in the system without itself undergoing a permanent chemical change. Catalysts may be homogeneous or heterogeneous, but all comply with the following criteria:

(1) Although the catalyst affects the rate of reaction, it cannot affect the position of equilibrium in a reversible reaction.

(2) In theory, the catalyst can be recovered chemically unchanged at the end of the reaction, although it may be changed physically.

catalytic cracking The most widely used method of producing gasoline from heavy petroleum distillates is to subject the heavy fractions to catalytic cracking. Nowadays fluidized bed catalytic cracking is almost universally used, where the powdered catalyst is moved through the system in a fluidized state.

Many catalysts have been used but the standard catalysts are generally mixtures of silica and alumina or natural or synthetic aluminium silicate zeolites.

The major product is gasoline but gases and light and heavy gas oils are produced in proportions depending on feedstocks and operating conditions. The gasoline obtained is usually of a fairly high octane number* due to the high proportion of unsaturated hydrocarbons. *See* cracked gasoline.

catalytic reforming Catalytic reforming processes are employed to improve the octane number* of straight-run gasoline by increasing the proportion of aromatics and branched-chain alkanes. Feedstocks consist mainly of alkanes and naphthenes* and the main reactions are dehydrogenation of napthenes to produce aromatics, isomerization of alkanes and naphthenes, dehydrocyclization of *n*-alkanes yielding aromatics and hydrocracking.

Earlier processes (e.g. hydroforming) used $MoO_3-Al_2O_3$ catalysts but platinum-based catalysts are now extensively used, enabling longer on-stream times before catalyst replacement.

catechol *See* 1,2-dihydroxybenzene.

catecholamines The collective name given to amines of biomedical interest found in urine and plasma; dopamine, adrenaline and their metabolites being most common.

catenation The formation of element–element bonds as in hydrocarbons and higher silanes. Carbon is the element showing the greatest tendency to catenation.

catforming A process of catalytic reforming.

cathode The electrode which carries the negative charge in an electrochemical cell. In electrolysis the positively charged ions are discharged at the cathode.

cathodic protection A method of metal protection based upon the fact that there is no dissolution of metal ions from the cathode of an electrochemical cell, since the cathode is more negative than the electrode potential of the particular metal. Cathodic protection is used to prevent corrosion of submerged or underground metal work such as ships, piers, boiler installations or underground pipe-lines. For example, a moored ship may be made the cathode of a cell by suspending magnesium anodes some distance below the ship and short-circuiting them to the hull. Alternatively, in fresh water, which is a poor conductor, a direct current supply may be used instead of galvanic electrodes. Thus a negative lead may be connected to an underground pipe-line while a positive lead is connected to a graphite or iron anode some distance away, moist soil acting as the electrolyte.

cation An atom which has lost one or more electrons or a positively charged group of atoms, e.g. $[NH_4]^+$. Cations are present in the solid state, e.g. Na^+ in NaCl, and in solutions and melts. In electrolysis cations travel to the cathode.

cationic detergents Detergents which, in aqueous solution, form positively charged surface-active ions. They are often based on quaternary ammonium compounds, derivatives of pyridine or fatty acid amides. Although not used extensively as detergents they are used in the textile industry while those with bactericidal properties are used in disinfecting detergents.

caustic potash, KOH. Potassium hydroxide*.

caustic soda, NaOH. Sodium hydroxide*.

ccp Cubic close-packed. *See* close-packed structures.

Cd Cadmium.

C.D. Circular dichroism.

Ce Cerium.

celestine, $SrSO_4$. Native strontium sulphate, a major source of Sr compounds.

cell A device used as a source of electrical energy. Electrochemical cells consist of two electrodes dipping into one or more electrolyte solutions. If the electrodes are in separate solutions, electrical contact between solutions is achieved by means of a salt-bridge, porous disc or ionic conductor. The electrodes may be of the same metal as the electrolyte or of another inert material, e.g. platinum.

cell dimensions The dimensions of the unit cell in a crystal.

cellobiose, $C_{12}H_{22}O_{11}$. 4-[β-D-gluco-

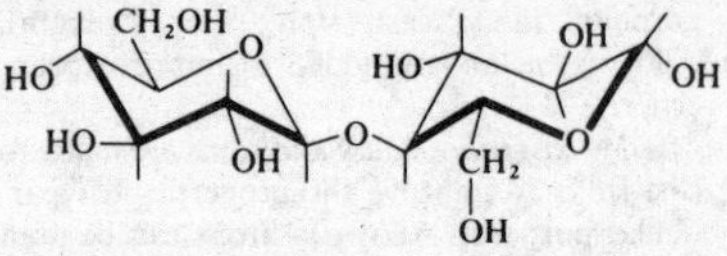

pyranosido]-D-glucopyranose. It is the disaccharide that forms the repeating unit of cellulose.

cellophane A transparent sheet cellulose prepared by extruding cellulose xanthate solution into a bath of acid. Sensitive to humidity although can be rendered waterproof by varnishing. Highly inflammable. Used as a wrapping material for foodstuffs and cigarettes but being replaced by polypropene.

Cellosolve A trade name for ethylene glycol monoethyl ether. Also known as Oxitol.

Cellosolve acetate A trade name for ethylene glycol monoethyl ether acetate.

cellular plastics The various forms of foamed and expanded plastics which have a form containing cells formed by gases (blowing agents). Have light weight and frequently favourable thermal properties. Used in packaging, thermal insulation, light-weight partitions and roofing. Plastics used include polystyrene, polyurethanes, PVC, polyethylene, urea and phenolic resins.

celluloid A little used inflammable thermoplastic based on cellulose nitrate plasticized with camphor.

cellulose, $(C_6H_{10}O_5)_x$. The chief constituent of the cell wall of all plants, and the most abundant organic substance found in nature. It is a polymer of glucose, with over 3500 repeat units in a chain. The glucoside linkage is β, whereas in starch it is α.

Strong acids completely hydrolyse cellulose to glucose; very mild hydrolysis gives hydrocelluloses with shorter chains and lower viscosity and tensile strength. Under special conditions a large yield of cellobiose is obtained.

Cellulose dissolves in strong mineral acids, in NaOH and in cuprammonium solution. It forms a triacetate (tri-ethanoate), a trinitrate

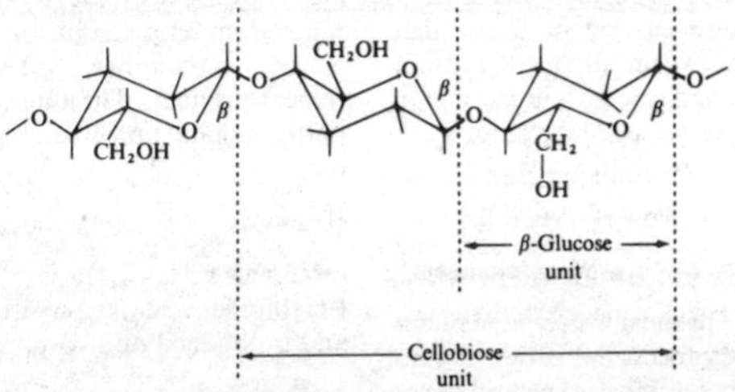

cellulose

and ethers with methyl, ethyl and benzyl alcohol; all of these are of great industrial importance. Its largest use is in the rayon* industry. Wood pulp and cotton linters are the most important sources from which it is prepared commercially.

cellulose acetate plastics Formed by treatment of cellulose (cotton, purified wood pulp) with ethanoic anhydride, ethanoic acid and sulphuric acid. Complete acetylation gives the triacetate used as a plastic material. Part acetylation also gives very useful materials. Processed by injection moulding. Used to form fibres and for use in textiles. Has relatively low strength and is moisture-sensitive. U.S. production 1978 140 000 tonnes.

cellulose ethers Formed by alkyl- and aryl halides on cellulose in alkali solution. Used as plastic materials. Methyl cellulose is water-soluble and is used as an emulsifying, sizing and priming material.

cellulose nitrate, nitrocellulose Prepared by treating cellulose (cotton waste or wood pulp) with HNO_3 or HNO_3 plus sulphuric or phosphoric acid. If the N content is greater than 13% it is known as gun cotton or smokeless powder; colloidon cotton is a rather less nitrated species. Very inflammable, shipped as the moist material. Used as a propellant and in blasting gelatine. Mainly used as a lacquer and for making some solid objects, although these are very inflammable. Formerly used for film.

cement A substance used to make other substances cohere. There are many kinds of cement, but they may be divided into two main groups: (1) cements used to bind together numerous particles so as to form a coherent mass of considerable strength and (2) those used to unite two or more separate masses (e.g. portions of a broken article); this second class includes glues and various organic cements. Cements in the first group include Portland cement, high alumina cement and many others. Cements which form a plastic paste with water and set hard on standing are known as hydraulic cements. Cements are also known by their chief constituent, e.g. calcareous cements, aluminous cements, siliceous cements; by their characteristic property, e.g. acid-resisting cements, quick-setting cements, rapid-hardening cements; by their reputed origin, e.g. Roman cement, and by their fancied resemblance to some other material, e.g. Portland cement which was, at one time, supposed to resemble Portland stone. Cements which can be used at high temperatures, e.g. for repairing furnaces, are known as refractory cements. The term 'cement' is also used for any argillaceous limestone capable of producing cement when sintered.

Cements are commonly made by heating a mixture of limestone and clay to about 1700°C. The product is ground with gypsum. Chemically cements consist of a mixture of calcium silicates and aluminates with some sulphate present. World production 1976 730 megatonnes.

cement fondu A hydraulic cement composed chiefly of calcium aluminate. *See* cement, high alumina.

cement, high alumina A rapid-hardening dark-coloured cement produced by heating a mixture of bauxite and limestone to about 1850 K. Varying amounts of iron oxides and silica are also present.

This cement is very resistant to attack by sulphates, sea water and acid waters. It is also used with crushed firebrick to produce refractory concrete.

cement, quick-setting Portland* or other which has a shorter setting time than ordinary Portland cement.

cement, rapid-hardening Portland* or other cements which harden rapidly so that concrete may be used in three to four days.

For concrete work in cold weather a specially rapid-hardening cement may be made by addition of calcium chloride to the cement.

cement, supersulphated A cement made from granulated blast furnace slag of fairly high alumina content, anhydrite or burnt gypsum and about 5% Portland cement*. It has a high sulphate resistance and is used largely for work underground and for work in sea water.

cementite, Fe_3C. *See* iron.

centre of symmetry *See* symmetry elements.

centrifugal pump The most widely used pump in the chemical and process industries. It consists of a rotor with a number of curved vanes rotating inside a flat, cylindrical casing. Liquid enters the centre of the casing along the rotor axis and is swept along the vanes into the volute which is a ring-shaped chamber.

centrifugal separators *See* centrifuges.

centrifugation Separation of substances by means of a centrifuge*.

centrifuges Machines which employ centrifugal force to obtain high rates of sedimentation or filtration in order to separate a solid and a liquid or two immiscible liquids. *See also* ultracentrifuge.

cephalins, kephalins Cephalins are phosphatides and are similar to lecithins except that ethanolamine or serine replaces choline in the molecule. They are found in all animal and vegetable tissues. They are easily hydrolysed to fatty acid, etc.

cephalosporin C Antibiotic* produced by *Cephalosporium acremonium*. Converted to other semi-synthetic antibiotics.

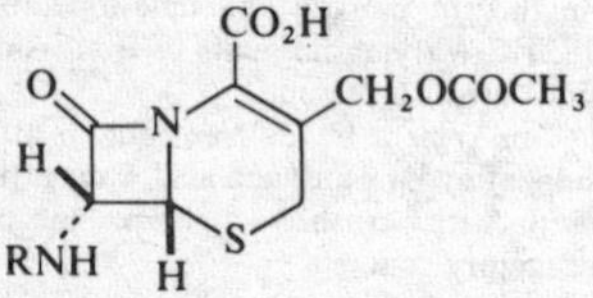

Cephalosporin C:

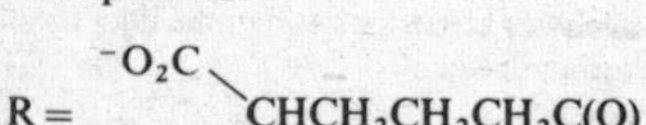

ceramics Engineering materials and products made from inorganic chemicals (excluding metals and alloys) by high temperature processing. They are resistant to chemical attack, can range from semi-conductors to insulators and include refractories, glasses, cement and cement products, vitreous enamels, abrasives, pottery, china, porcelain, clay wares, alumina, etc.

ceramides *See* cerebrosides.

cerebronic acid, $C_{24}H_{48}O_3$. M.p. 100–101°C. Cerebronic acid is 2-hydroxylignoceric acid and is obtained on hydrolysis of phrenosin*.

cerebrosides A group of glycosidic substances which occur most abundantly in the myelin sheaths of nerves together with sphingomyelins. The sphingosine-fatty acid portion of cerebrosides is usually known as a ceramide.

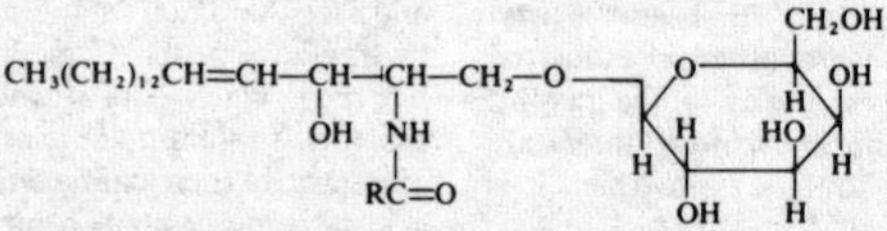

cerebrosides

ceresin The name formerly given to a hard, white, odourless wax obtained from fully refined ozokerite*. The term now often refers to certain forms of hard, brittle waxes obtained from petroleum.

cerium, Ce. At.no. 58, at.wt. 140·12, m.p. 798°C, b.p. 3257°C, *d* 6·66. The commonest of the lanthanides,* obtained in large quantities after extraction of thorium from monazite. The metal is used in alloys for improving the malleability of cast iron and the mechanical properties of Mg alloys; Ce alloys are pyrophoric; CeO_2 is used as a glass polish, ceramic coatings, and in incandescent gas mantles. Cerium sulphate is a catalyst in 'self-cleaning ovens'.

cerium compounds Cerium shows oxidation states +4 and +3. Ce(III) compounds are typical lanthanide compounds*

Ce^{3+} (f^1 colourless) $\longrightarrow$ Ce (−2·48 volts in acid) CeF_4($CeO_2 + F_2$) is the only cerium(IV) halide although chlorocerates(IV) containing the $CeCl_6^{2-}$ ion are stable. Most Ce compounds give CeO_2 on heating in oxygen, intermediate phases are present in the $Ce_2O_3 - CeO_2$ system. Various cerium(IV) sulphates, e.g. $Ce(SO_4)_2 \cdot 2H_2SO_4$ are obtained from CeO_2 and H_2SO_4 and are used in analysis; complex nitrates, $M_2Ce(NO_3)_6$, are also stable. The Ce(IV) state is the most stable +4 state of the lanthanides and is the only +4 state stable in aqueous solution

Ce^{4+} (yellow-orange) $\longrightarrow$ Ce^{3+} (+1·28 volts in HCl)

CeI_2 (CeI_3 + Ce) is probably Ce^{3+}, $2I^-$, e^-.

cermets Pressed or sintered ceramic–metal mixtures, e.g. $Cr-Al_2O_3$, used in high temperature applications, e.g. jet engines.

cerotic acid, hexacosanoic acid, $CH_3{\cdot}[CH_2]_{24}{\cdot}COOH$. M.p. 88°C. Occurs free in many waxes.

cerussite, $PbCO_3$. A native lead carbonate used as a Pb ore.

ceryl alcohol, 1-hexacosanol, $C_{26}H_{54}O$, $CH_3{\cdot}[CH_2]_{24}{\cdot}CH_2OH$. Colourless crystals, m.p. 79°C. Occurs as esters in various waxes, ceryl palmitate being the chief component of opium wax and ceryl cerotate of Chinese wax.

cesium The North American spelling for caesium*.

cetane *See* hexadecane, $C_{16}H_{34}$.

cetane number *See* knock rating.

cetrimide, cetyltrimethylammonium bromide, CTAB, $[C_{16}H_{33}N(CH_3)_3]^+Br^-$. The commercial product is made by the condensation of cetyl bromide with trimethylamine, and contains other alkyl ammonium bromides. It is a creamy-white powder, soluble in water to give a readily foaming solution. It is a cationic detergent and wetting agent.

cetyl alcohol, hexadecanol, $C_{16}H_{34}O$, $CH_3{\cdot}[CH_2]_{14}{\cdot}CH_2OH$. Colourless crystals, m.p. 49°C. It occurs as esters in various waxes, the most important being cetyl palmitate, which is the chief component of spermaceti. Used extensively in the pharmaceutical and cosmetic industries and in gel stabilizers for greases.

cetyl trimethylammonium bromide *See* cetrimide.

Cf Californium.

chain reactions Reactions which proceed by means of a chain reaction mechanism, e.g.

(a) $Br_2 \longrightarrow 2Br\cdot$
(b) $Br\cdot + H_2 \longrightarrow HBr + H\cdot$
(c) $H\cdot + Br_2 \longrightarrow HBr + Br\cdot$
(d) $H\cdot + HBr \longrightarrow H_2 + Br\cdot$
(e) $2Br\cdot \longrightarrow Br_2$

The first stage (a) of the reaction represents the dissociation of bromine into bromine atoms. Both steps (b) and (c) lead to production of HBr, and since bromine atoms are regenerated the process can be repeated. Reaction (d) leads to an inhibition in the chain reaction and step (e) represents chain termination.

The reaction between hydrogen and chlorine is probably also of this type and many organic free radical reactions (e.g. the decomposition of ethanal) proceed via chain mechanisms.

In nuclear chemistry, a fission reaction (*see* atomic energy) may be initiated by a neutron and may also result in the production of one or more neutrons, which if they reacted in like manner could start a chain reaction. Normally, moderators such as cadmium rods which absorb neutrons are placed in the reactor to control the rate of fission.

chair form *See* conformation.

chalcogens The elements oxygen, sulphur, selenium, tellurium, polonium of Group VI. Chalcogenides contain the X^{2-} species.

chalcones The trivial name given to the α,β-unsaturated ketones $ArCH=CHCOAr'$ obtained by condensing an aromatic aldehyde with an aryl methyl ketone in the presence of base. Benzalacetophenone* is the simplest example.

chalconides The chalcogenides. *See* chalcogens.

chalcopyrite, copper pyrites, $CuFeS_2$. The most important ore of copper, brassy yellow in colour with a metallic lustre.

chalk A naturally occurring form of $CaCO_3$ of marine origin.

charcoal A form of carbon produced by slow combustion of wood in a deficiency of oxygen. Formerly used as a fuel and reducing agent. Used to absorb gases. Has a low-order graphitic structure.

charge transfer band or spectrum An absorption band which corresponds to transfer of an electron from one atom or group to another.

charge transfer complexes Complexes in which there is weak interaction, charge transfer, between the donor and the acceptor properties which are generally induced in the acceptor. Examples of charge transfer complexes include the complexes formed between aromatic derivatives and halogens. The class is practically indistinguishable from true complexes. Charge (electron) transfer takes place in all complexes.

Charles's law At constant pressure the volume of a given mass of gas is directly proportional to the absolute temperature.

chelate compound A compound in which co-

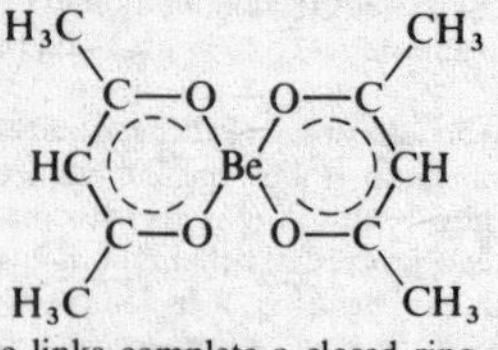

ordinate links complete a closed ring, e.g. in beryllium acetylacetonate.

chelate effect The enhanced stability of a complex containing chelate rings as compared with the complex containing monodentate ligands. E.g. ethylenediamine complexes, are

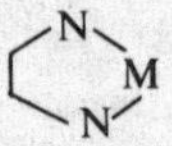

more stable than ethylamine complexes.

In complexes of chelates there are a number of types of isomerism which may occur. In a tris(ethylenediamine) octahedral complex two optically active isomers occur (often denoted Δ and Λ).

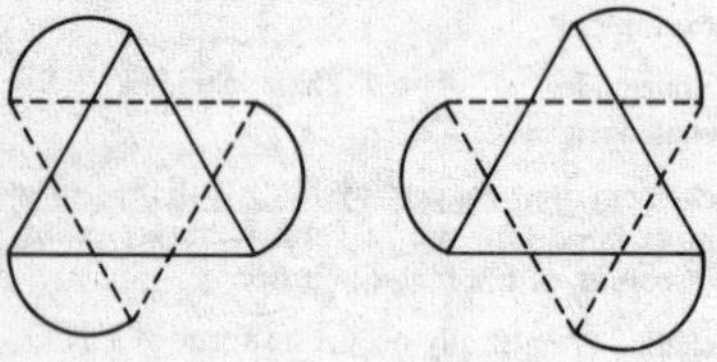

Λ Δ

A ligand such as ethylenediamine is not planar and has a spiral form (gauche) giving rise to further forms. When the direction of one C–C bond in one ethylenediamine is parallel to the 3-fold axis the isomer is termed the *lel* form, when it is inclined to the axis it is termed the *ob* form.

chemical development The photographic process of reducing the exposed Ag^+ in the halide grains of a photographic emulsion to Ag.

chemical equivalent The weight of substance which will combine with or displace 8 parts by weight of oxygen. May be related to weights of other elements (e.g. 35·5 g Cl, 1 g H). In redox reactions, expressed as the number of gram equivalent to a Faraday (1 mole electrons).

chemical potential (μ) Loosely described as a measure of the 'intensity' of a chemical substance in a given system (e.g. solution or mixture). Mathematically, if the total Gibbs free energy G of a system increases by an increment dG when an infinitesimal quantity, dn_A mole, of component A is added to the system under conditions of constant temperature and pressure, the chemical potential of A in the system denoted μ_A, is given by

$$\mu_A = (dG/dn_A)_{P,T}$$

μ is therefore sometimes called the partial molar free energy. Chemical potential is a valuable function in the theory of heterogeneous equilibria, since it can readily be proved that, if such a system is not initially at equilibrium, substances will tend to pass from one phase to another (e.g. solid dissolving or crystallizing) until a uniform chemical potential is attained for each substance in every phase.

chemical shift *See* nuclear magnetic resonance.

chemical warfare Materials used, or usable in, chemical warfare include toxic chemicals, flame, incendiaries, smokes and defoliating agents.

chemiluminescence The emission of light during a chemical reaction, e.g. the oxidation of yellow phosphorus. The light emitted by the fire-fly or glow-worm, and luminous combustion, are examples of this very common phenomenon.

chemiosmotic hypothesis The hypothesis that the function of the electron transport chain* is to translocate protons and thereby create a pH difference and a membrane potential. This is called the proton motive force and is used by the proton pump of the mitochondrial ATP-ase operating in reverse to generate ATP.

chemisorption The adsorption of a substance at a surface involving the formation of chemical bonds between the adsorbate, the species undergoing adsorption and the adsorbent, the surface at which adsorption is occurring. It is of importance in heterogeneous catalysis.

chemotaxonomy The study of the chemical constituents of plants, animals and other organisms as a guide to their phylogeny and classification.

chemotherapy The selective destruction of pathogenic organisms or control of e.g. cancer within a host by the use of chemicals.

chicle A resin obtained from the Mexican tree *Achras sapita*; softens at about 50°C, used in chewing gum manufacture.

Chile saltpetre, $NaNO_3$. *See* sodium nitrate.

china clay, kaolin A white powdery material

arising from the decomposition of feldspars in granites. Used in paper and pottery manufacture and as a filler in textiles and paint. *See* kaolinite.

Chinese blue *See* cyanoferrates, Prussian blue.

Chinese white, ZnO. *See* zinc oxide.

chirality A term which may be applied to any dissymmetric object or molecule. The property of non-identity of an object with its mirror image is termed chirality and implies 'handedness', i.e. the object and mirror image relationship of a right hand is a left hand. Traditionally this phenomenon has been discussed in older textbooks in terms of asymmetry.

A molecule is chiral if it cannot be superimposed on its mirror image (or if it does not possess an alternating axis of symmetry) and would exhibit optical activity*, i.e. lead to the rotation of the plane of polarization of polarized light. Lactic acid, which has the structure (2 mirror images) shown exhibits molecular chirality. In this the central carbon atom is said to be chiral but strictly it is the environment which is chiral.

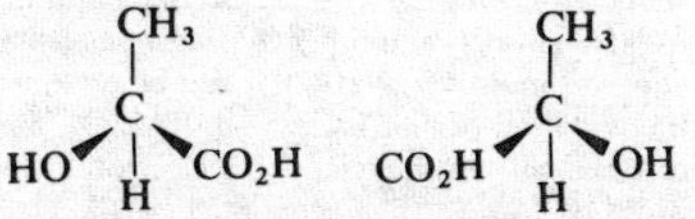

In certain crystals, e.g. in quartz, there is chirality in the crystal structure. Molecular chirality is possible in compounds which have no chiral carbon atoms and yet possess non-superimposable mirror image structures. Restricted rotation about the C=C=C bonds in an allene *ab*C=C=C*ba* causes chirality and the existence of two optically active forms (i) and (ii).

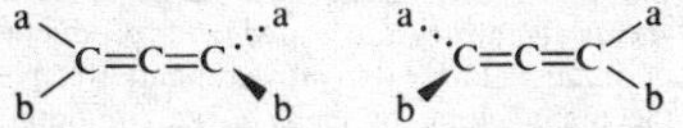

(i) (ii)

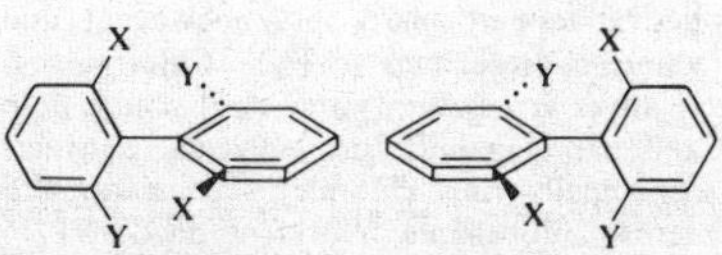

(iii) (iv)

Restricted rotation about single bonds caused by bulky substituents X and Y allows the existence of enantiomers* of the chiral diphenyl derivatives (iii) and (iv). E.g. $X=NO_2$, $Y=CO_2H$.

chitin A structural polysaccharide in which

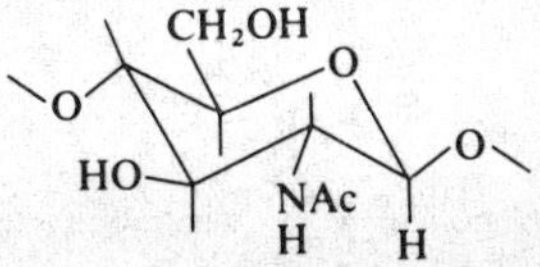

N-acetyl-D-glucosamine units are linked in the *β*-(1 ⟶ 4)-positions. Forms the integument of crustaceans and insects and is also found in some fungi.

chitinase An enzyme of the digestive juices of snail and mould fungi which can hydrolyse chitin and related compounds of crustaceans and fungi. N-acetyl glucosamine.

chloral, trichloroacetaldehyde, trichloroethanal, $CCl_3\cdot CHO$. A colourless oily liquid with a pungent odour; b.p. 98°C. Manufactured by the action of chlorine on ethanol; it is also made by the chlorination of ethanal. When allowed to stand, it changes slowly to a white solid. Addition compounds are formed with water (*see* chloral hydrate), ammonia, sodium hydrogen sulphite, alcohols, and some amines and amides. Oxidized by nitric acid to trichloroethanoic acid. Decomposed by alkalis to chloroform and a methanoate; a convenient method of obtaining pure $CHCl_3$. It is used for the manufacture of DDT. It is also used as a hypnotic.

chloral hydrate, trichloroethylidene glycol, $CCl_3CH(OH)_2$, $C_2H_3Cl_3O_2$. Crystallizes in large, colourless prisms having a peculiar odour; m.p. 57°C, b.p. 97·5°C. Manufactured by adding the calculated amount of water to chloral. For other properties *see* chloral. Its chief use is as a hypnotic.

chloramine, $ClNH_2$. *See* haloamines. Also used to denote organic compounds containing N–Cl bonds.

chloramine T, sodium *p*-toluenesulphonchloroamide, $C_7H_7ClNNaO_2S,3H_2O$, $[p\text{-}CH_3C_6H_4S(O)_2NCl]Na$. White crystals which lose water at 100°C and decompose suddenly at 175–180°C. Soluble in water and alcohol. Prepared from *p*-toluenesulphonamide and NaOCl. It is a powerful antiseptic and is used in dilute aqueous solution for washing wounds.

chloranil, tetrachloro-1,4-benzoquinone, $C_6Cl_4O_2$. A fungicide mainly used as a seed protectant. Yellow crystals, m.p. 290°C.

Prepared by oxidizing phenol with an acid and

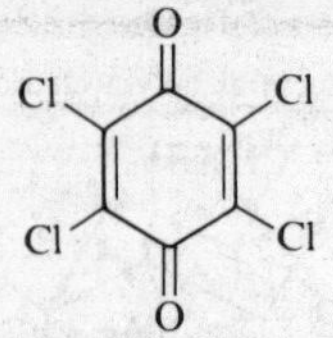

potassium chlorate(V). Chemically it is used as a dehydrogenating agent.

chlorates Salts of the various chlorine oxy-acids. Chlorates(I), hypochlorites, contain the $[OCl]^-$ ion and are formed from Cl_2 and H_2O, or HgO and Cl_2 and cold NaOH solution. The free acid is formed from H_2O and Cl_2 in the gas phase. Chlorates(III), chlorites, contain the bent $(ClO_2)^-$ ion (ClO_2 plus base). The free acid is a weak acid and cannot be isolated. Both chlorates(I) and (III) are used as bleaching agents. Chlorates(V), chlorates, contain the pyramidal $(ClO_3)^-$ ion and are obtained from Cl_2 plus OH^- in hot solution. The free acid is a strong acid. Chlorates(VII), perchlorates, contain the tetrahedral $(ClO_4)^-$ ion and are obtained by electrolytic oxidation of $(ClO_3)^-$. The free acid is a very strong acid and is widely used as an oxidizing agent, it may explode on contact with organic materials. Chlorates(VII) are used as non-complexing inert electrolytes. All of the chlorates are oxidizing species

$$(ClO_4)^- \longrightarrow Cl^- \quad +0{\cdot}56 \text{ volts}$$
$$(ClO_3)^- \longrightarrow Cl^- \quad +0{\cdot}63 \text{ volts}$$
$$(ClO_2)^- \longrightarrow Cl^- \quad +0{\cdot}78 \text{ volts}$$
$$(ClO)^- \longrightarrow Cl^- \quad +0{\cdot}89 \text{ volts}$$

(all in acid solution). In all cases the sodium salt is readily available; $KClO_4$ is sparingly soluble.

chlordane, $C_{10}H_6Cl_8$. A technical product containing 1,2,4,5,6,7,10,10-octachloro-4,7,8,9-tetrahydro-4,7 -endomethyleneindane, and made by condensation of hexachlorocyclopentadiene with cyclopentadiene followed by further chlorination of the product. Non-phytotoxic and of low toxicity to mammals.

chlorhexidine, $C_{22}H_{30}Cl_2N_{10}$, $[p\text{-}ClC_6H_4NHC(NH)NHC(NH)(CH_2)_3]_2$. The dihydrochloride is a white crystalline powder, m.p. about 225°C (decomp.). Used as an antiseptic and skin sterilizing agent, and as a bacteriostat in some pharmaceutical formulations.

chloric acid, $HClO_3$. *See* chlorates(V).

chloride of lime Bleaching powder*.

chlorides Formally compounds of Cl(-1) formed by all elements except the noble gases ($XeCl_2$ has some stability). Chlorides of the more electropositive elements are ionic and the more electronegative elements form covalent chlorides. High oxidation state chlorides are covalent. Chlorides are formed by use of Cl_2 or HCl on the elements or by metathesis (e.g. $AgNO_3$ plus NaCl gives AgCl), anhydrous chlorides are obtained by use of ethanoyl chloride or thionyl chloride. Chloride is a good complexing agent and can act as a monodentate or bridging ligand, e.g. in $Cl_2Al(\mu\text{-}Cl)_2AlCl_2$.

chlorin The dihydroporphin ring system present in chlorophylls.

chlorinated rubbers Materials containing up to 70% Cl formed by reacting rubber (solid, solution or latex) with Cl_2. The reaction is rapid to $C_{10}H_{14}Cl_2$ followed by cyclization and then formation of homogenous materials. Stable materials used as blending agents for alkyd resins and as film-formers in corrosion resistant paints. Also used as adhesives.

HCl reacts with rubber to form hydrochlorinated rubber – also used in paints.

chlorine, Cl. At.no. 17, at.wt. 35·453, m.p. −100·98°C, b.p. −34·6°C. Occurs naturally as NaCl, $MgCl_2$ and other chlorides; sea water contains up to 1% NaCl. Cl_2 can be obtained by oxidation of NaCl with, e.g. $KMnO_4$, and is obtained commercially by electrolysis of brine or $MgCl_2$. The economics of the electrolysis of brine depend upon the simultaneous formation of NaOH. Chlorine occurs as a greenish poisonous gas containing Cl_2 molecules. It is stored as liquid Cl_2. Chlorine is a very reactive element and combines directly with most other elements. It dissolves in water and from cold water gives chlorine hydrate, $Cl_2{\cdot}7{\cdot}27H_2O$ with a clathrate structure; a small proportion reacts to give Cl^- and $(OCl)^-$. Chlorine is one of the basic materials of the chemical industry and is used in the production of organochlorine derivatives and indirectly in the manufacture of many other compounds which are obtained via chlorine or chloride derivatives particularly anti-knock agents and chloro-hydrocarbons (55%). Chlorine derivatives are used in water sterilization, pulp and paper manufacture, solvents, polymers (principally vinyl chloride), refrigerants and aerosol propellants. World production 1976 25 megatonnes; U.S. production 1981 10·8 megatonnes.

chlorine chemistry Chlorine is one of the halogens, electronic configuration $3s^23p^5$. Its typical chemistry is that of a non-metallic element in the -1 oxidation state ($E°$ $Cl_2 \longrightarrow Cl^- + 1{\cdot}36$ volts in acid solution) forming ionic chlorides and covalent chloro derivatives. Chlorine also forms positive oxidation states – *see* chlorates – mainly with oxygen, and fluorine, and with other halogens.

chlorine fluorides *See* chlorine halides.

chlorine halides
Chlorine pentafluoride, ClF_5. M.p. $-103°C$, b.p. $-14°C$, square pyramidal.

Chlorine trifluoride, ClF_3. M.p. $-76°C$, b.p. 12°C, T-shaped, a powerful fluorinating agent.

Chlorine monofluoride, ClF. M.p. $-157°C$, b.p. $-100°C$, dissociates to the elements. Each of the chlorine fluorides are hydrolysed by water (ClF_3 explosively) and are prepared from Cl_2 and F_2.

Chlorine also forms BrCl (*see* bromine halides) and ICl and ICl_3 (*see* iodine chlorides). *See also* interhalogen compounds.

chlorine hydrate, $Cl_2{\cdot}7{\cdot}27H_2O$. *See* chlorine.

chlorine oxides
Chlorine monoxide, Cl_2O. M.p. $-116°C$, b.p. 4°C, yellow-red gas (Cl_2 plus HgO), dissolves in water to give some HOCl. Dissociates to Cl_2 plus O_2.

Chlorine dioxide, ClO_2. M.p. $-6°C$, b.p. 10°C, paramagnetic yellow, explosive gas ($NaClO_3$ plus H_2SO_4). Strong oxidizing agent, gives $HClO_2$ and $HClO_3$ with water. Used as a bleach for wood-pulp.

Dichlorine tetroxide, $ClOClO_3$, *chlorine perchlorate* ($CsClO_4$ plus chlorine fluorosulphate, $ClOSO_2F$); *dichlorine hexoxide*, Cl_2O_6, oily red liquid (ozone plus ClO_2). Both unstable.

Dichlorine heptoxide, Cl_2O_7. M.p. $-91{\cdot}5°C$, b.p. 82°C, the most stable chlorine oxide but still explosive ($HClO_4$ plus P_2O_5), gives $HClO_4$ with water.

chlorine oxide fluorides, ClO_2F and $ClOF_3$. Formed by hydrolysis of the fluorides, ClO_2F gives $(ClO_2)^+$ salts with e.g. BF_3.

Chlorine trioxide fluoride, ClO_3F, *perchloryl fluoride*. M.p. $-148°C$, b.p. $-47°C$ ($KClO_4$ plus HF plus SbF_5 or F_2 or HSO_3F on $KClO_4$) a toxic gas, thermally stable to 500°C. It is used as a reagent to replace H in C–H-containing compounds with F – it also gives organic perchlorates, $RClO_3$ (e.g. PhLi gives $PhClO_3$).

chlorites A type of aluminosilicate. Also the salts of $HClO_2$, chlorates(III)*.

chloroacetone, chloropropanone, $CH_2Cl{\cdot}CO{\cdot}CH_3$. Colourless lachrymatory liquid; b.p. 119°C. Manufactured by treating propanone with bleaching powder or chlorine. It is used as a tear gas and is usually mixed with the more potent bromoacetone.

chloro acids Complex chloroanions are formed by most elements of the periodic table by solution of oxides or chlorides in concentrated hydrochloric acid. Potassium salts are precipitated from solution when potassium chloride is added to a solution of the chloro acid, the free acids are generally unstable.

chloroalkylamines,
***N*-dialkylchloroalkylamines,** $R_2N(CH_2)_nCl$. Formed by treating the alkanolamine with thionyl chloride. Of importance in the synthesis of pharmaceuticals, particularly tranquillizers, e.g. amidons.

chloroanilines, aminochlorobenzenes Formed (mono-, di-, and tri-chloroanilines) by the action of SO_2Cl_2 on aniline. Less basic than aniline.

chlorobenzene, C_6H_5Cl, PhCl. Colourless liquid, b.p. 132°C. Nitrates in the *o*- and *p*-positions. Prepared by the direct chlorination of benzene in the presence of an iron catalyst. Largely used as an intermediate in the manufacture of other chemicals, particularly phenol, DDT and aniline.

2-chlorobutadiene polymers *See* neoprene.

chlorocarbonic ester *See* ethyl chloroformate.

chlorocarbons, chlorohydrocarbons An extremely important range of chemicals prepared by direct chlorination of hydrocarbons or by Cl_2 or HCl addition to olefines. HCl is frequently a waste product of chlorination. Subsequent treatment is by cracking. Used as cleaning fluids (tetrachloroethene, CH_3CCl_3), insecticides (DDT), polymers (vinyl chloride chloroprene), solvents (CCl_4), as chemical intermediates (CH_3Cl, C_2H_5Cl).

chlorochromate(VI) salts, $MCrO_3Cl$. Obtained by treating dichromates with concentrated hydrochloric acid or by adding CrO_2Cl_2 to a saturated solution of a metal chloride. The salts contain the tetrahedral anion CrO_3Cl^-; the free acid is unknown. The solutions decompose on boiling to chlorine and chromium(III) salts. The salts are powerful oxidizing agents, the pyridinium salt is used for oxidizing alcohols to aldehydes. Chlorochromates $M_2{}^1CrOCl_5$ result from the action of concentrated HCl and alkali metal chloride on CrO_3 at 0°C.

chloroethane *See* ethyl chloride.

chloroethanoic acids, chloroacetic acids *Monochloroethanoic acid*, $CH_2Cl{\cdot}COOH$. B.p. 189°C. Manufactured by heating trichloroethene with sulphuric acid, or by treating glacial ethanoic acid with chlorine in the presence of red phosphorus, sulphur or ethanoic anhydride. Reacts with ammonia to give glycine, and with many types of organic compounds. It is largely used as a chemical intermediate, particularly in the manufacture of the chlorophenoxyacetic acid weedkillers, thiocyanate insecticides and various pharmaceuticals.

Dichloroethanoic acid, $CHCl_2COOH$. Low-melting solid, m.p. 5–6°C, b.p. 194°C. Prepared by the action of copper powder on trichloroethanoic acid or by the action of sodium cyanide on chloral hydrate.

Trichloroethanoic acid, CCl_3COOH. A crystalline solid which rapidly absorbs water vapour; m.p. 58°C, b.p. 196·5°C. Manufactured by the action of chlorine on ethanoic acid at 160°C in the presence of red phosphorus, sulphur or iodine. It is decomposed into chloroform and carbon dioxide by boiling water. It is a much stronger acid than either the mono- or the dichloro-acids and has been used to extract alkaloids and ascorbic acid from plant and animal tissues. It is a precipitant for proteins and may be used to test for the presence of albumin in urine. The sodium salt is used as a selective weed-killer.

2-chloroethyl alcohol, ethylene chlorohydrin, $ClCH_2{\cdot}CH_2OH$. Colourless liquid with a faint ethereal odour; b.p. 129°C. Manufactured by passing ethene into dilute chlorine water or from ethylene oxide plus HCl. Reacts with solutions of sodium hydrogen carbonate to give ethylene glycol and with solid sodium hydroxide to give ethylene oxide; with concentrated sulphuric acid at 100°C to give $\beta\beta'$-dichloroethyl ether; with ammonia and amines to give aminoethyl alcohols and with sodium salts of organic acids to give glycol esters. It is used in the preparation of these compounds and in the synthesis of phenylethyl alcohol, procaine and indigo. Also used as an insecticide and for forcing the early sprouting of potatoes.

chloroform, trichloromethane, $CHCl_3$. Colourless liquid with a sweet, pleasant odour; b.p. 60–61°C. Very slightly soluble in water; the solution has a sweet taste. Manufactured by the chlorination of methane. Oxidized by air and sunlight to phosgene; the addition of a small amount of alcohol prevents this. It is an excellent solvent for oils, fats, waxes, rubber and many other organic substances. Characteristically forms phenyl isocyanide when it is warmed with aniline and potassium hydroxide. Bromoform, iodoform and chloral also give this carbylamine reaction*.

Chloroform is a potent volatile anaesthetic, but is little used due to its potential hepatotoxicity. It is used principally for the manufacture of chlorofluorohydrocarbon refrigerants ('Arctons' and 'Freons') and certain polymers.

chloroformic ester *See* ethyl chloroformate.

chlorohydrins Organic compounds containing the $C(OH){\cdot}CCl$ group. Formed by treating compounds containing a double bond with chlorine water, or by treating glycols or epoxides with hydrochloric acid. Converted to glycols when heated with weak alkalis such as sodium hydrogen carbonate.

chlorohydrocarbons *See* chlorocarbons.

chlorohydroxypropane *See* propylene chlorohydrins.

chloroisopropyl alcohol *See* propylene chlorohydrins.

chloromethane, methyl chloride, CH_3Cl. Colourless gas with a pleasant ethereal odour, b.p. −24°C. Manufactured principally by the reaction of methanol and hydrogen chloride in the presence of a catalyst, either in the vapour or liquid phase. The chlorination of methane is also used, but to a lesser extent. Reacts with chlorine to give methylene chloride, chloroform and carbon tetrachloride. Forms methanol when treated with $Ca(OH)_2/H_2O$ under pressure. The chief use of methyl chloride is in the production of silicones. It is also used in the manufacture of lead tetramethyl (anti-knock additive), butyl rubber, methyl cellulose and for quaternizing organic bases, e.g. Paraquat. U.S. production 1978 206000 tonnes.

chloromethylation The introduction of the $-CH_2Cl$ group into aromatic, especially activated aromatic compounds using hydrogen chloride, methanal and anhydrous zinc chloride. *See* Mannich reaction.

$$R-H+CH_2O+HCl \longrightarrow R-CH_2Cl+H_2O$$

chloromethyl methyl ether, $ClCH_2OCH_3$. Colourless liquid; b.p. 60°C. Made by passing hydrogen chloride into a mixture of formalin and methanol. Used for chloromethylation*, preparation of methoxymethyl ethers (used as protecting groups) and with anhydrous aluminium chloride as catalyst for introducing the formyl (−CHO) group into certain aromatic compounds. A potent carcinogen.

chloronaphthalenes
1-chloronaphthalene, α-chloronaphthalene. $C_{10}H_7Cl$. Colourless, b.p. 263°C. Prepared by direct chlorination of naphthalene in the presence of $FeCl_3$. Used as an industrial solvent.

Other chlorinated naphthalenes. The other monochloronaphthalene (2-), the ten theoretically possible dichloronaphthalenes, and the fourteen trichloronaphthalenes have all been prepared, generally from the corresponding amino-derivatives by diazotization and treatment with CuCl. They are of little industrial importance.

***m*-chloroperbenzoic acid,**
m-$ClC_6H_4C(O)OOH$. White solid; m.p. 92°C. Prepared by treating *m*-chlorobenzoyl chloride with hydrogen peroxide and sodium carbonate. This per-acid* is a reasonably stable, selective, active oxidizing agent. Used for the detection and estimation of alkenic bonds (Prileschaiev reaction) which give epoxides, the formation of N-oxides and the oxidation of primary amines to nitro-compounds. A superior reagent, for most purposes, to other percarboxylic acids.

chlorophenols Relatively acidic materials produced by direct chlorination of the aromatic nucleus of phenols. Used widely in phenol-resins (with methanol) the compounds are often effective antiseptics, disinfectants, germicides, insecticides, herbicides, wood preservatives. Also used in dyestuffs.

chlorophyll The green colouring matter of plants. Two chlorophylls are present in plants, *chlorophyll a*, $C_{55}H_{72}MgN_4O_5$, which is obtained as a blue-black powder, m.p. 150–153°C, and *chlorophyll b*, $C_{55}H_{70}MgN_4O_6$, dark green powder, m.p. 120–130°C. Both are soluble in alcohol and ether. The methyl group marked with an asterisk is replaced by an aldehyde group in chlorophyll *b*. The ratio of chlorophyll *a* to chlorophyll *b* in green leaves is about 3:1, and the two pigments can be separated by extracting a solution of them in petroleum ether with methanol in which chlorophyll *b* is preferentially soluble. Treatment of chlorophyll *a* with acids removes the Mg, replacing it by two H atoms, and gives an olive-brown solid, phaeophytin *a*. Hydrolysis of this splits off phytol, and gives phaeophorbide *a*. Similar compounds are obtained from chlorophyll *b*. Alkaline degradation of chlorophyll yields a series of phyllins – magnesium porphyrin compounds – the final member of the series being aetiophyllin, and treatment of the phyllins with acids gives porphyrins, resembling, but not identical with, those obtained from the animal blood pigments.

Chlorophyll is present in all green plants, and is essential to the life of the plant, as it acts as a catalyst in the photosynthesis of carbohydrates from carbon dioxide and water.

Crude chlorophyll is prepared commercially from alfalfa meal or nettles by extraction with alcohol and partition into benzene. It is used as a colouring matter, particularly for foods and pharmaceutical products.

Chlorophyll *a*

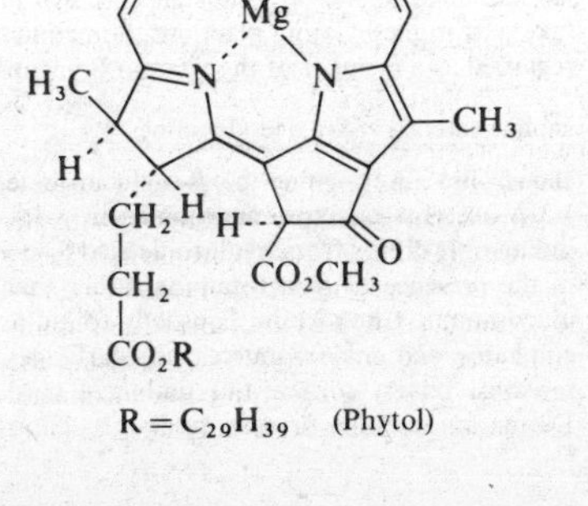

$R = C_{29}H_{39}$ (Phytol)

chloropicrin, nitrotrichloromethane, CCl_3NO_2. A colourless, lachrymatory, toxic liquid; b.p. 112°C. Manufactured by treating sodium picrate with chlorine, or calcium picrate with bleaching powder. Poison gas, it is used as an insecticide, parasiticide and for disinfecting cereals.

chloroprene, 2-chlorobutadiene, C_4H_5Cl, $CH_2{:}CCl{\cdot}CH{:}CH_2$. Colourless liquid; b.p. 59°C. Manufactured by treating vinylacetylene with hydrochloric acid at 30°C in the presence of a copper ammonium chloride catalyst. Used to prepare synthetic rubbers which are resistant to oil and are not affected by ozone, e.g. neoprene.

2-chloropropane *See* isopropyl chloride.

2-chloropropyl alcohol *See* propylene chlorohydrins.

3-chloropropylene glycol *See* glycerol monochlorohydrins.

chlorosulphonic acid *See* chlorosulphuric acid.

chlorosulphuric acid, chlorosulphonic acid, $HOS(Cl)O_2$. Colourless liquid; m.p. −80°C, b.p. 158°C (SO_3 plus HCl or H_2SO_4 plus PCl_5 or $POCl_3$). Hydrolysed by water. Used as a chlorinating and sulphonating agent.

chlorotoluenes, C_7H_7Cl. *2-Chlorotoluene*, b.p. 159°C; *4-chlorotoluene*, b.p. 162°C. Prepared by direct chlorination of toluene in the presence of a catalyst. 2-chlorotoluene can be oxidized to 2-chlorobenzaldehyde.

chlorous acid, $HClO_2$. Unknown in the free state but chlorates(III)*, chlorites, well known and used as oxidizing agents and as bleaches.

chloroxylenol, 4-chloro-2,6-dimethylphenol Anti-microbial, preservative, also used as a chemical intermediate.

chlorpromazine, 3-chloro-10-(2-dimethylamino-propyl) phenothiazine $C_{17}H_{19}ClN_2S{\cdot}HCl$. White or cream coloured powder; m.p. 194–197°C. It is a major tranquillizer, also an anti-emetic. The chlorine-free compound *promazine,* $C_{17}H_{20}N_2S{\cdot}HCl$, is widely used as a minor tranquillizer and as an adjunct to anaesthesia.

chlorthiamid 2,6-dichlorothiobenzamide, $C_7H_5Cl_2NS$, $PhC(S)NH_2$, a herbicide with a high level of biological activity.

cholane ring system The C_{24} skeleton of the

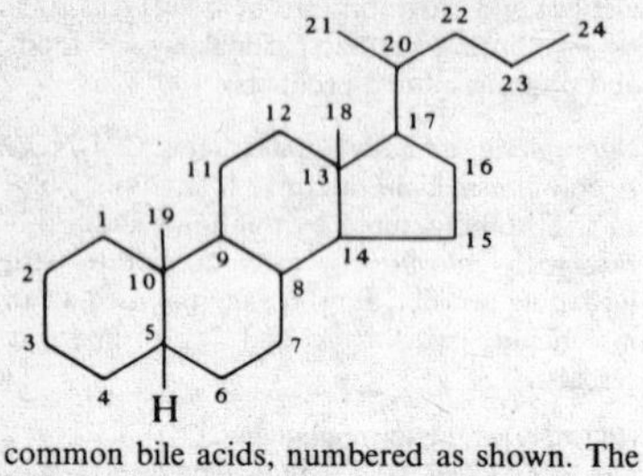

common bile acids, numbered as shown. The hydrocarbon depicted is 5*β*-cholane, m.p. 90°C, which has the *cis*-ring junction as found in the major natural bile acids.

cholesterol, 5-cholesten-3*β*-ol, $C_{27}H_{46}O$. M.p. 149°C. The principal sterol of animals, it

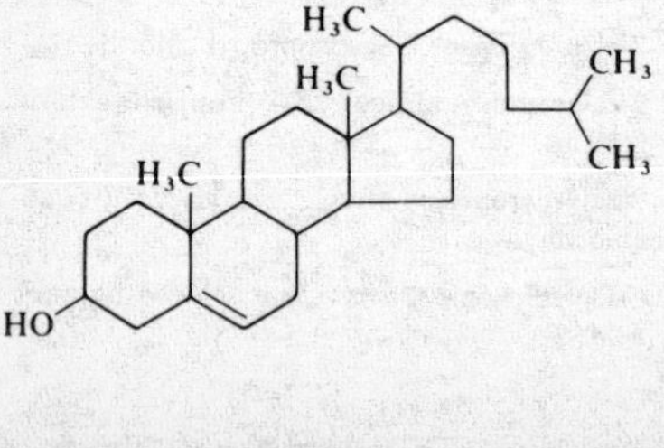

is found in free and esterified forms in all parts of the body. It was first isolated from gall-stones. Cholesterol is obtained commercially from beef spinal cords or from wool 'wax'. The sterol is synthesized by animals, from ethanoate units via mevalonic acid*, squalene* and lanosterol*. Besides its own biological rôles (e.g. as a constituent of membranes) it acts as a precursor of other steroids required in metabolism: notably the bile acids, sex hormones and adrenocortical hormones.

cholesterol oxidase An enzyme from various soil micro-organisms that oxidizes cholesterol to cholest-4-en-3-one using molecular oxygen, with the concomitant production of hydrogen peroxide. Many methods of assaying cholesterol using this enzyme are based on measuring H_2O_2 production.

cholic acid, 3*α*,7*α*,12*α*-trihydroxy-5*β*-cholanic acid, $C_{24}H_{40}O_5$. M.p. 195°C.

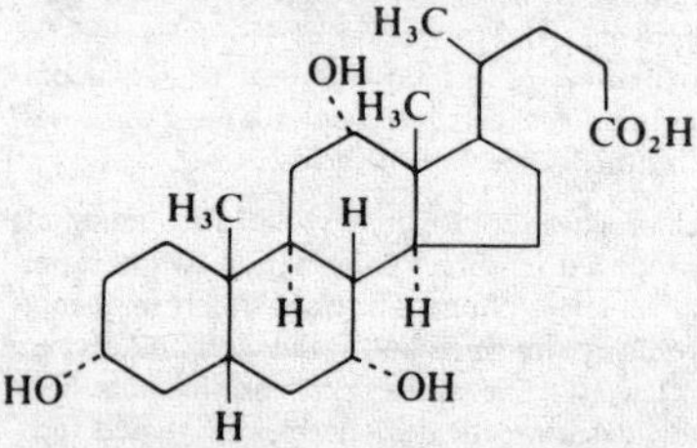

Mammalian bile contains sodium salts of 'conjugated' bile acids, e.g. glycocholic acid and taurocholic acid, in which cholic acid is combined (amide linkage) with glycine and taurine respectively.

choline, trimethyl-2-hydroxyethylammonium hydroxide, $[(CH_3)_3NCH_2{\cdot}CH_2OH]^+OH^-$. Colourless syrup crystallizing with difficulty to a hygroscopic mass. Strongly alkaline. It is present as a constituent of lecithin in all animal and vegetable tissues, and less commonly as the free base.

It is important in the body as, except for methionine, it is the only substance known to take part in methylating reactions. Sometimes regarded as a member of the vitamin B group.

choline esterase *See* acetylcholine.

chondroitin A polymer of *β*-D-glucuronide, 1,3-*N*-acetyl-D-galactosamine joined in *β*-1,4-linkages. It differs from hyaluronic acid* only in the presence of galactosamine rather than glucosamine. Chondroitin is usually found as sulphates and major sources are cartilages, tendons, bones, cornea and umbilical cord. Dermatan sulphate is very similar to chondroitin sulphate except that it contains L-iduronic acid instead of D-glucuronic acid.

D-chondrosamine, 2-amino-2-deoxy-D-galactose, $C_6H_{13}NO_5$. Obtained from chondroitin.

chorionic gonadotropin A glycoprotein hormone produced by the placenta and excreted in the urine during pregnancy. Its effect on the gonads of female rats and rabbits has been used as a method of detecting pregnancy.

chorismic acid, $C_{10}H_{10}O_6$. M.p. of mono-

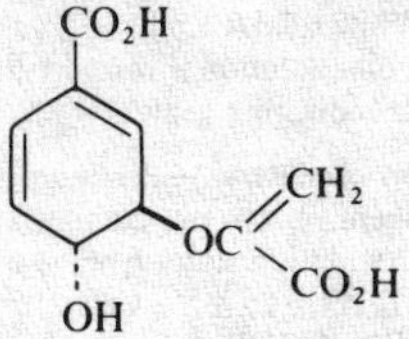

hydrate 148–149°C. An intermediate in aromatic biosynthesis via the 'shikimic acid pathway'. From shikimic acid* the sequence is through 5-phosphoshikimic acid, 3-enolpyruvylshikimic acid 5-phosphate, to chorismic acid. This appears to be the final common precursor of *p*-aminobenzoic acid, *p*-hydroxybenzoic acid, anthranilic acid (and tryptophan), and (via prephenic acid*) of phenylalanine and tyrosine.

chroman A strongly refractive liquid, b.p. 214–215°C which smells like peppermint.

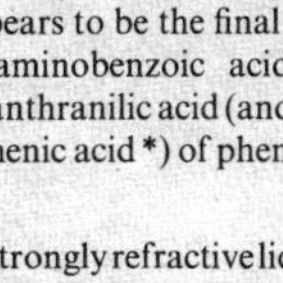

It may be regarded as the parent of a number of important classes of compounds derived from the γ-pyrone skeleton (e.g. flavone, xanthone) and the important chroman derivatives called the tocopherols (vitamin E).

chromates Formally any oxygen-containing chromium species (*see* below): Normally used to describe chromates(VI).

Chromates(VI). Derivatives of CrO_3 which forms the tetrahedral $(CrO_4)^{2-}$ ion and polymeric anions, e.g. $Cr_2O_7^{2-}$, $[O_3Cr(\mu\text{-}O)CrO_3]^{2-}$, dichromate, containing oxygen-bridges, in acid solution. The CrO_4^{2-} ion is yellow and the $Cr_2O_7^{2-}$ ion red. Free H_2CrO_4 is not known. Potassium salts are non-deliquescent and anhydrous; sodium salts are $Na_2CrO_4,10H_2O$ and $Na_2Cr_2O_7,2H_2O$. The chromates of the heavy metals are in general insoluble and are often used as pigments. Chromates(VI) are obtained from chromite* by heating with Na_2CO_3 in air at 1100°C followed by leaching and crystallization. Chromates(VI) are strong oxidizing agents ($E°$ $Cr_2O_7^{2-} \longrightarrow Cr^{3+}$ in acid + 1·33 volts) and are used as such and also in tanning, photographic processing printing and in pigments.

Chromates(V), $(CrO_4)^{3-}$. Dark green species which hydrolyse and disproportionate to Cr(III) and Cr(VI) (K_2CrO_4 plus KOH in melt).

Chromates(IV). Mixed oxides, e.g. $M_2^{II}CrO_4$ prepared by solid state reactions.

Chromates(III). Mixed oxides, e.g. $FeCr_2O_4$, having spinel structures and prepared by solid state reactions.

chromatography, chromatographic analysis A series of closely related techniques of separation of components of a mixture because of the distribution of the components between a liquid phase and a solid phase (stationary phase often silica or alumina) which usually has a large surface area. If the stationary phase is in a tube the technique is column chromatography, when paper or thin layers of stationary phase are used the technique is a sheet technique. The active solid stationary phase is the absorbent, an inert support is the support. The whole process of separation of the sample components into separate bonds or zones is the development of the chromatogram and is carried out by elution – the mobile phase being the eluting agent. The whole process gives a chromatogram.

The chromatogram can finally be used as the series of bands or zones of components or the components can be eluted successively and then detected by various means (e.g. thermal conductivity, flame ionization, electron capture detectors, or the bands can be examined chemically). If the detection is non-destructive, preparative scale chromatography can separate measurable and useful quantities of components. The final detection stage can be coupled to a mass spectrometer (GCMS) and to a computer for final identification.

Among the various chromatographic methods are adsorption column chromatography (use of a liquid phase in a solid column of adsorbent), partition column chromatography (distribution between two liquids in a column), thin-layer chromatography (partition on an open thin sheet), paper chromatography (use of a paper sheet as stationary phase), high pressure liquid chromatography, hplc, (partition column chromatography under high pressure), ion exchange chromatography (ion exchange), gas chromatography (distribution of a gaseous solute between a gas and a liquid or solid phase), zone electrophoresis (sheet chromatography in the presence of an electric field).

Gel permeation chromatography, exclusion chromatography, gel filtration chromatography. A technique for separating the components of a mixture according to molecular volume differences. A porous solid phase (a polymer, molecular sieve) is used which can physically entrap small molecules in the pores whilst large molecules pass down the column more rapidly. A solvent pressure up to 1000 psi may be used.

chrome Having a chromium plated surface.

chrome alum, $KCr(SO_4)_2,12H_2O$. A typical alum* used as a gelatin hardener*.

chrome orange *See* lead chromates.

chrome yellow *See* lead chromates.

chromic Outdated nomenclature for chromium(III) compounds.

chromic acid A name used incorrectly for CrO_3.

chromite, $FeCr_2O_4$. Brownish-black mineral with the spinel structure used as a source of Cr and its compounds.

chromium, Cr. At.no. 24, at.wt. 51·996, m.p. 1857°C, b.p. 2672°C, *d* 7·19. The most important chromium mineral is chromite, $FeCr_2O_4$, from which Cr is obtained by fusion with alkali carbonate in the presence of air, extraction with water as chromate(VI), reduction to Cr_2O_3 by carbon and reduction to metal with Al. Pure Cr can be obtained by electrolytic reduction of $[CrO_4]^{2-}$. The metal is hard silver-white in colour, bcc. The metal is very resistant to oxidation although it dissolves in non-oxidizing acids and reacts with O_2, halogens, S, etc. at high temperatures. It becomes passive in HNO_3. It is used widely in electroplating, as an additive for steel. Cr compounds are used as pigments (Cr_2O_3 and chromates(VI)) and for colouring glass, etc. in leather tanning, textiles, as catalysts, as oxidizing agents and in refractories. Chromium compounds are toxic in large quantities. World production Cr_2O_3 ore 1976 2 megatonnes.

chromium bromides Chromium(III) bromide, $CrBr_3$, and chromium(II) bromide, $CrBr_2$, are very similar to the corresponding chlorides.

chromium carbonyl, $Cr(CO)_6$. White solid; m.p. 149°C. Prepared from $CrCl_3$ plus reducing agent plus CO. Forms an extensive range of substituted derivatives.

chromium chlorides
Chromium(VI) dioxide dichloride, CrO_2Cl_2. *See* chromyl compounds; $[CrO_3Cl]^-$ *see* chlorochromates.

Chromium(IV) chloride, $CrCl_4$. Formed $CrCl_3$ and Cl_2 at 600–700°C followed by rapid cooling. Decomp. above −80°C.

Chromium(III) chloride, chromic chloride, $CrCl_3$. Violet solid (Cr plus Cl_2, hydrate plus $SOCl_2$) only soluble in water in presence of Cr^{2+}. Forms many complexes including the hydrates $[Cr(H_2O)_6]Cl_3$ – violet, $[Cr(H_2O)_5Cl]Cl_2,H_2O$ – green, $[Cr(H_2O)_4Cl_2]Cl,2H_2O$ – green.

Chromium(II) chloride, chromous chloride, $CrCl_2$. White solid (Cr plus HCl gas) dissolving to give a blue solution. Forms hydrates, widely used as a reducing agent.

chromium chemistry Chromium is a transition element, electronic configuration $3d^54s^1$ showing oxidation states from +6 to −2: +6 – CrO_3, $[CrO_4]^{2-}$; +5 – CrF_5, $[CrO_4]^{3-}$; +4 – $[CrF_6]^{2-}$, $Cr(OBu)_4$; +3 – $CrCl_3$; +2 – CrF_2, +1 – $[Cr(bipy)_3]^+$; 0 – $[Cr(CO)_6]$; −1 – $[Cr_2(CO)_{10}]^{2-}$; −2 – $[Cr(CO)_5]^{2-}$. In its higher oxidation states the Cr compounds are predominantly covalent; Cr(III) and Cr(II) compounds are largely ionic with Cr–Cr bonding in some derivatives. Chromium(II) compounds show Jahn–Teller distortion with 4 near and 2 far neighbours. The higher oxidation states are strongly oxidizing in acid solution

$[Cr_2O_7]^{2-} \longrightarrow Cr^{3+}$ $E°$ + 1·33 volts in acid
$[CrO_4]^{2-} \longrightarrow Cr(OH)_3$ − 0·13 volts in alkali

but Cr(III) and Cr(II) are stable in solution

$Cr^{3+} \longrightarrow Cr$ −0·74 volts
$Cr^{2+} \longrightarrow Cr$ −0·56 volts in acid

and Cr^{2+} is thus a fairly strong reducing agent. Chromium(III) forms a very extensive range of largely octahedral, inert complexes particularly with *O*- and *N*-ligands. There is an extensive aqueous chemistry of Cr(II) (blue) and Cr(III) (green) aquo-ions and of Cr(VI) anionic species.

chromium(II) ethanoate, chromium(II) acetate, $[Cr(O_2CCH_3)_2]_2,2H_2O$. Red insoluble compound formed from sodium ethanoate and $CrCl_2$ in aqueous solution. The most stable Cr(II) compound; contains a Cr–Cr bond.

chromium fluorides
Chromium hexafluoride, CrF_6. Yellow unstable compound (CrF_3 plus F_2 at high temperatures and pressure).

Chromyl fluoride, CrO_2F_2 and *fluorochromates(VI)*, containing the $[CrO_3F]^-$ ion are known.

Chromium pentafluoride, CrF_5. Red, m.p. 30°C (CrF_3 plus F_2 at 350–500°C).

Fluorochromates(V) containing the

$[CrOF_4]^-$ ion are formed from BrF_3 and $M_2Cr_2O_7$.

Chromium tetrafluoride, CrF_4. Green, sublimes 100°C (CrF_3 plus F_2).

Hexafluorochromates(IV), M_2CrF_6, are formed by the action of F_2 on 2MF plus CrF_3.

Chromium(III) fluoride, CrF_3. Green rather insoluble material (HF plus Cr_2O_3 or $CrCl_3$). Forms hydrates and complexes, e.g. M_3CrF_6.

Chromium(II) fluoride, CrF_2 plus HF at 600°C.

The higher chromium fluorides, CrF_6, CrF_5, CrF_4, are all extremely strong oxidizing agents, immediately hydrolysed by water.

chromium hydroxide It seems unlikely that a true chromium(III) hydroxide exists but hydroxy complexes, e.g. $[Cr(H_2O)_5OH]^{2+}$, $[(H_2O)_4Cr(\mu\text{-}OH)_2Cr(H_2O)_4]^{4+}$, are formed by addition of base to aquo-complexes and polymeric species and ultimately gels are formed by addition of further base.

chromium, organic derivatives Chromium forms a range of organic derivatives, e.g. Cr(O) $(h^6-C_6H_6)_2Cr$; Cr(I) $[(h^5-C_5H_5)Cr(CO)_3]_2$; Cr(II) $Li_4Cr_2Me_8$; Cr(III) $CrPh_3{\cdot}3THF$, Li_3CrMe_6. These are prepared by the action of LiR or RMgX on the appropriate chromium derivatives.

chromium oxides

Chromium trioxide, CrO_3. Red precipitate from $[CrO_4]^{2-}$ plus conc. H_2SO_4, m.p. 198°C, loses oxygen at 420°C. CrO_3 is a powerful oxidizing agent and is used as such. Acidic, gives $[CrO_4]^{2-}$ with water.

Chromium dioxide, CrO_2 (H_2O plus O_2 on CrO_3 at high temperature). Black solid with the rutile structure forming chromates(IV) in solid state reactions. Used in magnetic tapes.

Chromium(III) oxide, Cr_2O_3 (heat on $(NH_2)_2Cr_2O_7$, hydrolysis and ignition of chromium(III) compounds). Green compound with the corundum structure. Gives chromates(III) in solid state reactions, anionic species with hydroxyl ions and $[Cr(H_2O)_6]^{3+}$ with acids. Used as a green pigment (viridian) and to give green colours in glass and porcelain.

chromium oxy-salts Chromium(III) forms a sulphate which forms alums (chrome alum*), hydrates and sulphate complexes. The nitrate perchlorate and phosphate are also known; all are prepared from Cr_2O_3 and the acid, and form hydrates. Chromium(II) salts can be prepared by electrolytic reduction of the chromium(III) derivatives and include the ethanoate* and a blue sulphate. Chromium(IV) derivatives, $Cr(OBu^t)_4$ (oxidation of $Cr(NEt_2)_3$ followed by alcoholysis) is monomeric and there are many O-complexes including acetylacetonates.

chromium sulphates *See* chromium oxy-salts.

chromophore The group responsible for the colour of the compound, i.e. in practice for causing absorption in the visible and u.v. region of the spectrum, e.g. $-C{=}C-$, $-C{=}O$, $-N{=}N-$.

chromoproteins *See* proteins.

chromous Obsolete nomenclature for chromium(II).

chromyl chloride, CrO_2Cl_2. *See* chromyl compounds.

chromyl compounds Derivatives containing the CrO_2 unit. Chromyl chloride (NaCl plus $K_2Cr_2O_7$ plus H_2SO_4) is the best known derivative. Dark red liquid, m.p. −96°C, b.p. 117°C; violently hydrolysed by water. Powerful oxidizing agent for, e.g., P and S, used in Étard's reaction* (oxidation of alkyl aromatic to a ketone or aldehyde).

chrysanthemum carboxylic acids These are the acids, esters of which are the active constituents of pyrethrum*. In dicarboxylic acids, the side-chain −COOH group may be esterified: monocarboxylic acids are known. The ring −COOH group may be esterified. *See also* allethrin, pyrethrins and cinerins.

chrysene, 1,2-benzophenanthrene, $C_{18}H_{12}$. Crystallizes in colourless plates; m.p. 254°C, b.p. 448°C. Polycyclic aromatic hydrocarbon.

chymotrypsins An important group of proteolytic enzymes which are secreted into the intestine from the pancreas. It is in an inactive form as chymotrypsinogen but this is converted to chymotrypsin by the action of another intestinal enzyme, trypsin*. Chymotrypsins are the major proteases of the intestine. Chymotrypsins act specifically on peptide linkages next to phenylalanine or tyrosine.

ciment fondu An aluminous cement* which sets slowly but hardens rapidly. Resistant to sea water and to sulphates in solution. Also used as a refractory cement.

cinchonidine, $C_{19}H_{22}N_2O$. A stereoisomer (H, vinyl interchange) of cinchonine, and also obtained from cinchona. M.p. 210°C.

cinchonine, $C_{19}H_{22}N_2O$. Colourless needles;

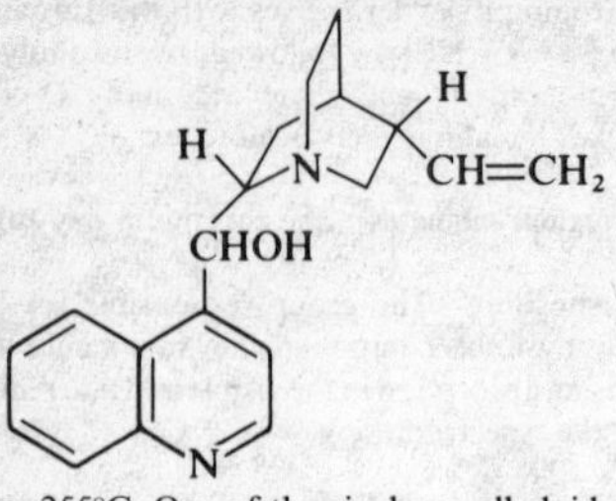

m.p. 255°C. One of the cinchona alkaloids.

1,8-cineole, eucalyptol, $C_{10}H_{18}O$. A very

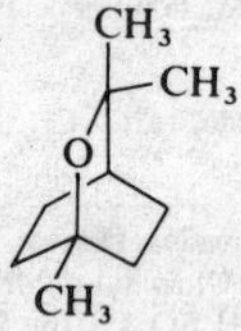

widely distributed constituent of essential oils; often considered as the active constituent of medicinal eucalyptus oil. Also present in wormseed oil, cajuput oil and in various eucalyptus oils. It is a colourless, viscous oil of characteristic camphor-like smell. M.p. −1°C, b.p. 174·4°C.

cinerins Constituents of pyrethrum*. Esters of cinerolone and chrysanthemum monocarboxylic acids.

cinerolone A ketonic alcohol, esters of which

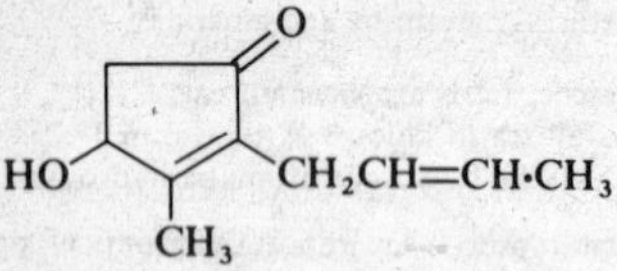

with chrysanthemum carboxylic acids are the cinerins*. *See* pyrethrins.

cinnabar, HgS. The principal ore of Hg. *See also* vermilion red.

cinnamic acid, 3-phenylpropenoic acid, $C_9H_8O_2$, $PhCH{=}CHCO_2H$. Colourless crystals. Decarboxylates on prolonged heating. Oxidized by nitric acid to benzoic acid. Ordinary cinnamic acid is the *trans*-isomer, m.p. 135–136°C; on irradiation with u.v. light it can be isomerized to the less stable *cis*-isomer, m.p. 42°C.

Prepared by heating benzaldehyde with sodium ethanoate and ethanoic anhydride (Perkin reaction) or with ethyl ethanoate and sodium ethoxide. Occurs in storax, or liquid amber, as the ester cinnamyl cinnamate made from cinnamyl alcohol, $PhCH{=}CH{\cdot}CH_2OH$. Cinnamic acid and its derivatives are used in flavours, perfumery, cosmetics and pharmaceuticals.

circular dichroism In the Cotton effect change in the polarization angle is seen as a function of wavelength. This is due to differences in the specific extinction coefficients for left- and right-handed polarized light. As a linearly-polarized light wave passes through a substance the differing absorptions cause an elliptically polarized wave to be produced. This phenomenon is known as circular dichroism. The magnitude of the effect is expressed by the equation

$$\varphi = \pi/\lambda(\eta_l - \eta_r)$$

where φ is the ellipticity (in radians) of the emerging beam and η_l and η_r are the absorption indices of the left- and right-handed circularly-polarized light respectively. When the ellipticity is plotted as a function of wavelength, a curve results with a maximum corresponding to the wavelength of zero angle in the optical rotatory dispersion curve. Optical isomers give circular dichroism curves which are identical except that in one case the effect is positive, i.e. φ is positive throughout whereas for the other isomer the effect is negative.

cis Term used in designating isomers to indi-

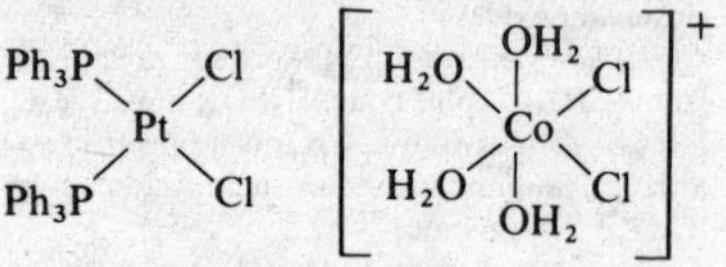

cate isomer with adjacent like ligands or groups. *cf. trans.*

citraconic acid, methylmaleic acid, $C_5H_6O_4$. A colourless solid crystallizing in fine needles; m.p. 91°C (decomp.). Prepared by the addition of

$$\begin{array}{c} CH_3{\cdot}C{\cdot}COOH \\ \| \\ H{\cdot}C{\cdot}COOH \end{array}$$

water to citraconic anhydride which is itself prepared by the rapid distillation of anhydrous citric acid. It loses water when heated and forms an anhydride (citraconic anhydride, methylmaleic anhydride). Reduced by hydrogen to pyrotartaric acid. Electrolysis of the acid yields propyne.

citral, $C_{10}H_{16}O$. A terpene aldehyde. A volatile oil of pleasant odour forming the main constituent of lemon-grass oil from *Cymbopogon flexuosus*, and also found in other essential oils. Lemon-grass oil is an important article of commerce. The citral found in natural products is a mixture of isomers.

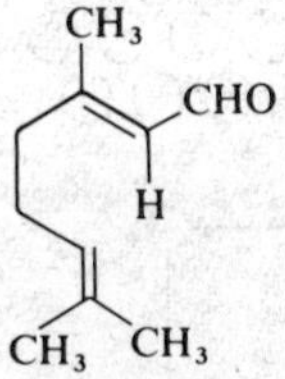

Citral-a

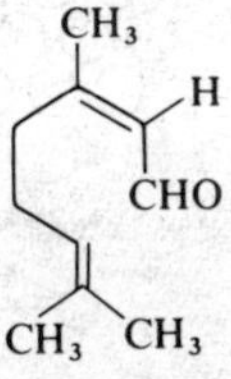

Citral-b

With dilute sulphuric acid citral forms *p*-cymene. Citral can be condensed with propanone to form a ketone, pseudoionone, $C_{13}H_{20}O$, which is technically important, as it is readily convertible into α and β-ionone.

citric acid, $C_6H_8O_7$. Crystallizes from water

$$\begin{array}{c} CH_2COOH \\ | \\ HOC\,COOH \\ | \\ CH_2COOH \end{array}$$

below 37°C as the monohydrate; the anhydrous acid has m.p. 153°C. It occurs in the juice of citrus fruits and in beets, cranberries and certain other acid fruits. Manufactured by the fermentation of sugar by moulds of the *Aspergillus niger* group. The acid is tribasic, and forms three series of salts; the citrates of the alkali metals are soluble, but the neutral citrates of calcium and barium are insoluble. When heated to 175°C it loses water to give aconitic acid: at higher temperatures aconitic and citraconic anhydrides are formed. Oxidation with potassium permanganate or heating with fuming sulphuric acid gives acetone dicarboxylic acid. It is used extensively in the soft drinks and food industries.

citric acid cycle, tricarboxylic acid cycle, Krebs's cycle A cyclic sequence of reactions in cell metabolism, by which is achieved the controlled oxidative breakdown of acetyl-coenzyme A, derived from carbohydrates (*see* glycolysis) and fatty acids (*see* β-oxidation). Other molecules, such as amino-acids, can be altered so that they too can be fed into the cycle. The products of the cycle are carbon dioxide together with reduced coenzymes (CoASH, NADH, NADPH, $FADH_2$) which serves as energy sources; under aerobic conditions this occurs through the respiratory chain. The citric acid cycle is of fundamental importance being at the centre of both the production of energy and the formation of complex cell constituents.

citrulline, α-amino-δ-ureidovaleric acid, $C_6H_{13}N_3O_3$, $H_2N{\cdot}CO{\cdot}NH(CH_2)_3{\cdot}CHNH_2{\cdot}CO_2H$. M.p. 222°C. Soluble in water, insoluble in alcohol. Citrulline is an intermediate in the urea cycle* in the excretion of excess nitrogen from the body.

Cl Chlorine.

Claisen condensation Condensation of an ester with another ester, a ketone or a nitrile in the presence of sodium ethoxide, sodium or sodamide, with the elimination of an alcohol. The result is the formation of a β-ketonic ester, ketone, or nitrile respectively, e.g.

$$CH_3COOC_2H_5 + HCH_2COOC_2H_5 \longrightarrow CH_3COCH_2COOC_2H_5 + C_2H_5OH.$$

ethyl acetoacetate

The reaction is of general application and of great importance, and a large number of syntheses have been effected by its use.

Claisen reaction Condensation of an aldehyde with another aldehyde or a ketone in the presence of sodium hydroxide with the elimination of water. Thus benzaldehyde and methanal give cinnamic aldehyde, PhCH:CH·CHO.

Clapeyron–Clausius equation A thermodynamic equation applying to any two-phase equilibrium for a pure substance. The equation states:

$$\frac{dP}{dT} = \frac{\Delta H_V}{T\Delta V}$$

where P is the pressure, T the absolute temperature, ΔH_V the molar latent heat of the vaporization and ΔV the corresponding change in molar volume. If one phase is the vapour, the equation reduces to

$$\ln P = C - \Delta H_V/RT$$

where C is a constant for the particular substance and R is the gas constant.

clarification The removal from a liquid of small amounts of suspended matter with the object of obtaining a clear product. Clarification may be achieved by filtration*, centrifugation*, or by the use of a clarifier*.

clarifier Large tanks with continuous feed and outflow in which the suspended matter is allowed to settle and is removed. Clarifiers are generally similar to thickeners* except that they handle smaller quantities of solids.

Clark electrode *See* oxygen cathode.

classification The process of separating a mixture of particles into two or more fractions according to size, shape, density, magnetic properties, etc.

clathrate compounds Molecular compounds formed by the inclusion of molecules of one type in holes in the lattice of another. Used in the separation of, e.g. gases.

Claude process A process similar to the Linde process* for the liquefaction of air, except that additional cooling is produced by allowing the expanding gas to do external work.

Clausius–Clapeyron equation *See* Clapeyron–Clausius equation.

Clausius–Mosotti law The molecular polarization (P) of a substance of molecular weight M, density d and dielectric constant D is:

$$P=\frac{(D-1)}{(D+2)}\cdot\frac{M}{d}$$

clays Natural aluminosilicates which occur as a plastic paste or can be converted to a paste by grinding or mixing with water. When dried and ground clay particles can be suspended in water almost indefinitely; suspension is aided if a small amount of Na_2CO_3 is added. Gels may be formed in concentrated suspension.

Clays have layers of linked (Al, Si)O_4 tetrahedra combined with layers of $Mg(OH)_2$ or $Al(OH)_3$. Clays are very important soil constituents and are used in pottery, ceramics, as rubber, paint, plastic and paper fillers, as adsorbents and in drilling muds.

clay treatment Removal of by-products and acid-sludge* from oils by adsorption on clay.

cleavage planes Directions in a crystal in which there is ready cleavage. Cleavage planes often correspond to layers of atoms or molecules in the lattice.

Clemmensen reduction Aldehydes and ketones may generally be reduced to the corresponding hydrocarbons by heating with amalgamated zinc and hydrochloric acid.

More selective than reduction with $LiAlH_4$ and, e.g., will not reduce the ester group in keto-esters.

close-packed structures The majority of

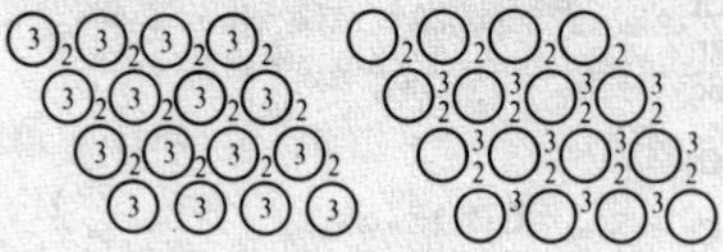

hcp ccp

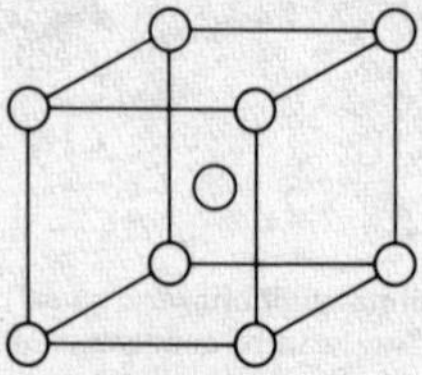

bcc

metals, alloys and ionic inorganic compounds, and also many compounds which are appreciably covalent, have structures which can be described in terms of close-packing of the largest species, atoms or ions. In each layer of a true close-packed structure each atom has six neighbours in an hexagonal array. A similar second layer lies with its atoms above the holes of the first layer. The atoms of the third layer can be directly above the atoms of the first layer (hexagonal close-packing, hcp) or above holes in both first and second layers (cubic close-packing, ccp). In the latter case the fourth layer lies over the first. More complex sequences of layers are known. The body-centred cubic structure, bcc, does not make quite such full use of space as hcc or ccp. Many structures are described in terms of smaller cations occupying appropriate holes (octahedral Δ or tetrahedral T) between. For each close-packed atom there is one octahedral hole and two tetrahedral holes (in the diagrams the number n designates an atom in the nth layer, the first layer is designated ○).

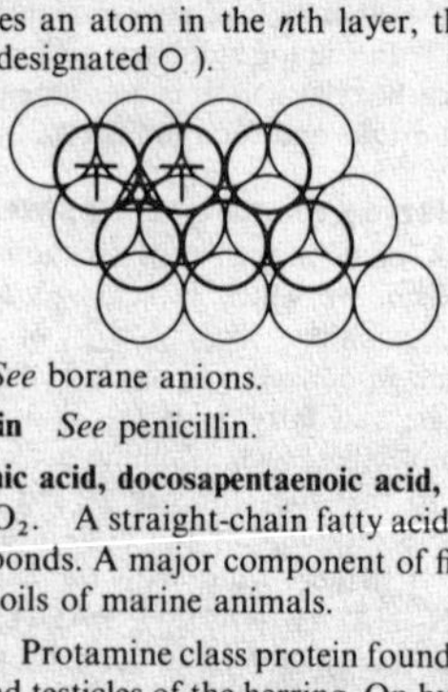

closo *See* borane anions.

cloxacillin *See* penicillin.

clupadonic acid, docosapentaenoic acid, $C_{22}H_{34}O_2$. A straight-chain fatty acid with 5 double bonds. A major component of fish oils and the oils of marine animals.

clupeine Protamine class protein found in the sperm and testicles of the herring. On hydrolysis it gives about 90% of argenine.

cluster compounds Compounds containing groups of metal atoms joined by metal–metal bonds, e.g. $[Rh_6(CO)_{16}]$, $[Mo_6Cl_8]^{4+}$.

Cm Curium.

Co Cobalt.

coacervation The process of separation of lyophilic (hydrated colloid) sols into two immiscible liquid phases, each of which has a different concentration of the dispersed phase.

coagel A gelatinous precipitate; e.g. rigid silica gel.

coagulation The stability of a colloidal particle is due to its surface electric charge, or its possession of a hydration sheath, or both. Any effect, chemical or physical, which removes these causes of stability produces coagulation. Thus the passage of an electric current through a colloidal sol makes the particles move to the electrodes. They are discharged and coagulated. The electrical precipitation of smokes and mists is due to a similar action.

The action of small quantities of electrolytes in coagulating hydrophobic (i.e. unhydrated) colloids is important. The colloid is coagulated by the adsorption of ions of charge opposite to that which it carries. Thus an arsenic sulphide sol (negative) can be coagulated by unipositive cations, but dipositive cations are more efficient and tripositive cations much more so, the latter producing coagulation even at great dilutions. The addition of a large excess of the coagulant may stabilize a sol of opposite charge to that of the initial colloid, by adsorption of a large excess of (in the above example, positive) ions. Other coagulating influences include ultrasonic vibrations, u.v. light and boiling (as in denaturization of proteins); the two latter effects are essentially chemical. Hydrated or hydrophilic colloids are also coagulated by electrolytes, but only in high concentrations. This effect is termed 'salting out'* in contrast to the simple adsorption effect outlined above. *See* colloid, electrophoresis, peptization, 'salting out'.

coal A naturally occurring solid fuel which exists in the form of seams at varying depths below the earth's surface. Formed by the arrested bacterial decay of vegetation that grew 40–300 million years ago, followed by chemical processes of condensation and polymerization under the influence of temperature, pressure and time. By these processes the differences between the original woody material and anthracite, the highest ranking coal, can be explained. As the rank increases the carbon content increases from about 50 to 95%, and the oxygen content decreases from about 40 to 3%. The calorific value* increases (average $3{\cdot}3 \times 10^7\,Jkg^{-1}$) through the range of coals: peat – lignites – bituminous – semi-bituminous and carbonaceous coals – to the anthracites.

The properties of the various coals in this ranking will vary considerably and many are used only for specific purposes.

coal briquettes Blocks of coalpowder mixed with a binder (coal-tar pitch or bitumen) and heated with pressing. Used as fuels, particularly domestically.

coal gas *See* town gas.

coalite process A process for the carbonization of coal at 600°C to give coke, coal-tar oil, hydrocarbons, carbon monoxide, ammonium compounds.

coal tar The by-product obtained from the high-temperature carbonization* of coal in coke ovens and gas retorts and from the low-temperature carbonization processes for smokeless fuel* production.

The composition of coal tar varies with the carbonization method but consists, largely, of mononuclear and polynuclear aromatic compounds and their derivatives. Coke oven tars are relatively low in aliphatic and phenolic content while low-temperature tars have much higher contents of both.

Crude tar is normally distilled in continuous plant into distillate fractions which can vary in boiling range and in name, leaving pitch as a residue.

Higher-boiling fractions supply the creosotes*, absorbing oils, anthracene, coal tar fuels*, road tar*, etc.

coal tar fuels (CTF) Industrial fuels produced by blending distillate and residual fractions from coal tar distillation*.

coal tar pigments Generally metallic lakes or insoluble metal salts of organic dyestuffs. Not now prepared from coal tar*.

coated paper To give a fine surface for printing and for preventing ink absorption paper is coated with a colloidal mix of e.g. kaolin, satin white, with an adhesive such as glue, casein, starch, plus a plasticizer to harden the protein and prevent frothing.

cobalamin *See* cobalt, organic derivatives.

cobalt, Co. At.no. 27, at.wt. 58·9332, m.p. 1495°C, b.p. 2870°C, *d* 8·92. Obtained commercially from silver ores (arsenides and sulphides) and Ni, Cu and Pb arsenide ores (speisses). Smaltite, $CoAs_2$, and cobaltite,

CoAsS, are also used as sources. The ore is roasted and Co is precipitated as the hydroxide and then reduced to Co with carbon (hcp below 417°C, ccp to m.p.). The metal is silvery white and readily polished. It dissolves in dilute acids and is slowly oxidized in air. Adsorbs hydrogen strongly. The main use of cobalt is in alloys. Cobalt compounds are used in paints and varnishes, catalysts. Cobalt is an essential element in the diet. World production 1976 32 000 tonnes metal.

cobalt alloys Up to 80% of cobalt is used in alloys, 20% of this for magnets. The most important cobalt alloys are stellite (up to 30% Cr, 18% W, 2½% C used in valves and cutting tools) and vitallium (20% Cr, high temperature uses). Co is also used in high-speed tools.

cobaltammines Ammonia complexes of cobalt(III), almost entirely with octahedral co-ordination.

cobalt bloom, $Co_3(AsO_4)_2,8H_2O$. An important source of arsenic.

cobalt bromide *See* cobalt halides.

cobalt carbonyls Cobalt in CO at 150°C and 300 atm. gives $Co_2(CO)_8$, $[(OC)_3Co(\mu\text{-}CO)_2Co(CO)_3]$ with a metal–metal bond; m.p. 51°C. $Co_2(CO)_8$ gives $Co_4(CO)_{12}$ on heating. With OH^- $Co_2(CO)_8$ gives $[Co(CO)_4]^-$ which gives $[HCo(CO)_4]$ with dilute acid. $Co_2(CO)_8$ and $HCo(CO)_4$ and their derivatives are important industrial catalysts in hydroformylation reactions. Many substituted derivatives of the cobalt carbonyls are known.

cobalt chemistry Cobalt is a late transition element, electronic configuration $3d^74s^2$. The common oxidation states in aqueous solution are +2

$E°$ for $Co^{2+} \longrightarrow Co$ −0·28 volts in acid

which exists as the hydrated ion and +3 which is very unstable as the hydrate but stable as ammine and other low spin complexes

$E°$ for $[Co(NH_3)_6]^{3+} \longrightarrow [Co(NH_3)_6]^{2+}$ +0·1 volts

There are many ionic Co(II) compounds, e.g. halides and a great range of largely octahedral Co(III) complexes. Other oxidation states for Co are −1, e.g. $[Co(CO)_4]^-$; 0, $Co_2(CO)_8$; +1, $CoBr(PR_3)_3$ and $[Co(NCR)_5]^+$; +4, $(CoF_6)^{2-}$. Cobalt complexes are oxygen-carriers. It can be present in enzymes and may be of great importance in their action.

cobalt fluorides *See* cobalt halides.

cobalt halides

Cobalt(II) fluoride, CoF_2. Forms 4,3,2,0 hydrates and gives perovskites, e.g. $KCoF_3$.

Cobalt(III) fluoride, CoF_3. Brown powder (CoF_2 plus F_2) also forms a green hydrate by electrolytic oxidation. CoF_3 is widely used in the fluorination of organic derivatives. Gives complexes e.g. M_3CoF_6.

Cobalt(IV) fluoride is not known but M_2CoF_6(Cs_2CoCl_4 plus F_2) is stable.

Cobalt chloride, $CoCl_2$. Obtained as red crystals of $CoCl_2{\cdot}6H_2O$ from aqueous solution, $CoCl_2,H_2O$ and $CoCl_2$ are blue as is the $(CoCl_4)^{2-}$ ion. No higher chloride is known although cobalt-haloammines, e.g. $(Co(NH_3)_5Cl)^{2+}$ are stable.

Cobalt(II) bromides and iodide are similar to the chloride.

cobalt(II) hydroxide, $Co(OH)_2$. Precipitated from cobalt(II) salts by strong base; pink or blue. Dissolves in excess base to give $[Co(OH)_4]^{2-}$ (deep blue) and $[Co(OH)_6]^{4-}$. Hydroxo complexes of cobalt(III), including hydroxy bridged species are known.

cobaltic Cobalt(III) derivatives.

cobaltite, CoAsS. A cobalt mineral.

cobalt, organic derivatives Low oxidation state derivatives are well established, e.g. $MeCo(CO)_4$ ($NaCo(CO)_4$ plus MeI). Vitamin B_{12} is a cobalt(III) organometallic and there is an extensive chemistry of alkyl, particularly methylcobalamin (vitamin B_{12} is cyanocobalamin) derivatives. Olefine and cyclopentadienyl complexes are known, olefine complexes are intermediates in hydroformylation reactions.

cobaltous Cobalt(II) derivatives.

cobalt oxides

Cobalt(II) oxide, CoO. Olive green solid (heat on $Co(OH)_2$ or cobalt(II) oxyacid salt in absence of air) with NaCl structure.

Tricobalt tetroxide, Co_3O_4. Black solid (ignition of CoO in air). Has spinel structure and other spinels MCo_2O_4 are also known.

Impure CoO_2 (oxidizing agent on alkaline Co(II)) and some mixed oxides of cobalt(IV) and (V), e.g. K_3CoO_4, are known.

cobalt oxyacid salts Cobalt(II) forms an extensive range of oxyacid salts, generally as hydrates, generally soluble in water. Examples are $Co(NO_3)_2,6H_2O$; $Co(ClO_4)_2,6H_2O$; $CoSO_4,7H_2O$. Few cobalt(III) oxy-acid salts are known. $Co_2(SO_4)_3,18H_2O$ is formed by electrolytic oxidation and forms alums; $Co(NO_3)_3$ contains co-ordinated nitrate (CoF_3 plus N_2O_5).

cocaine, benzoylmethylecgonine, $C_{17}H_{21}NO_4$. Colourless prisms, m.p. 98°C. Obtained from coca, either by direct purification, or by acid hydrolysis of the mixed alkaloids to ecgonine, which is then methylated and benzoylated. Coca consists of the dried leaves of *Erythroxylum coca* and *Erythroxylum truxillense*, shrubs growing in Bolivia and Peru.

Cocaine is the oldest of the local anaesthetics. It is a central nervous system stimulant and is habit-forming. *See* ecgonine.

cochineal The dried body of the female insect *Dactylopius coccus*. The deep red colour of the cochineal is due to carminic acid. It is used as a colouring matter for foodstuffs and drugs, and as an indicator.

co-current flow When two streams of material are brought into contact with each other with the object of transferring either heat or matter between them, if the streams move in the same direction flow is said to be co-current. *See also* counter-current flow.

codeine, *O*-methylmorphine, $C_{18}H_{21}NO_3$. Colourless crystals with one molecule of water, m.p. 155°C. Prepared by methylating morphine or directly from opium. It is useful in the treatment of coughs and as an analgesic.

cod-liver oil Oil expressed from the fresh liver of the cod, *Gadus callarias*. It is a pale yellow liquid, slightly soluble in alcohol, miscible with organic solvents. It contains about 1000 units of vitamin A* activity per g and about 100 units of antirachitic activity (vitamin D*) per g. It is a valuable source of the above vitamins and other food factors and is used in the prevention and cure of rickets in growing children.

codon *See* nucleic acids and genetic code.

coenzyme A The most prominent acyl group transfer coenzyme in living cells. It is concerned in a variety of reactions. Structurally coenzyme A is composed of adenosine-3′,5′-diphosphate and pantetheine phosphate moieties. Pantetheine is made up of pantoic acid, *β*-alanine and mercaptoethylamine, i.e. of pantothenic acid* and mercaptoethylamine. Acetyl coenzyme A* and malonyl coenzyme A are the most important forms in which coenzyme A is found biosynthetically. Acyl derivatives of coenzyme A are usually formed in ATP dependent reactions and are involved in many acyl group transfers.

coenzymes Defined as small thermostable molecules that are necessary for the action of enzymes. This term will include certain metals, but the word has come to be used only for substances which play a known part in the reaction catalysed by the enzyme. The difference between coenzymes and prosthetic groups (defined as the non-protein portions of enzymes) is one of degree rather than of kind, depending on the firmness of the combination of the protein, thus the tightly bound haematins are usually referred to as prosthetic groups and the phosphopyridine nucleotides as coenzymes, though both play essentially similar parts in biological oxidations. Coenzymes may be structurally altered in the course of the reaction but are usually regenerated in subsequent reactions.

Coenzymes may be classified into three main groups:

(1) *Coenzymes effecting transfer of hydrogen.* These include the pyridine nucleotides, nicotinamide-adenine dinucleotide and nicotinamide-adenine dinucleotide phosphate; the flavin nucleotides such as flavin-adenine dinucleotide; and lipoic acid.

(2) *Coenzymes effecting transfer of groups.* Examples of this class are adenosine triphosphate (ATP), biotin, coenzyme A and pyridoxal phosphate.

(3) *Coenzymes effecting isomerization.* Pyridoxal phosphate also falls into this class,

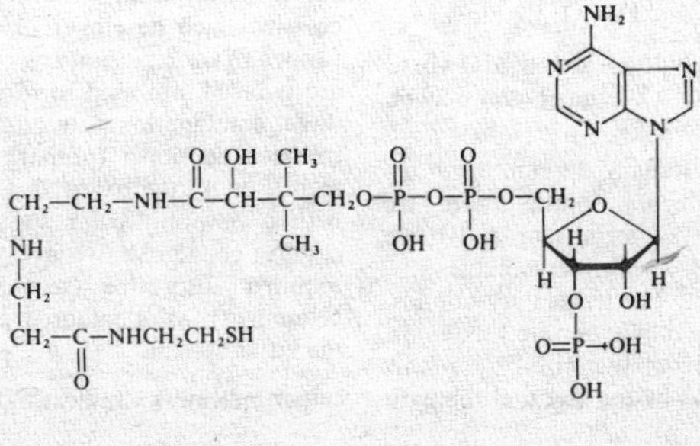

coenzyme A

in respect of its catalysis of the enzymic decarboxylation of amino-acids.

coherent precipitate A precipitate in a solid alloy system which is still bonded directly to the parent lattice and does not have a distinct grain boundary. Generally associated with high strength in the alloy.

coinage metals The elements Cu, Ag, Au.

coke The dense product of the carbonization of coal in an oven. Coke varies in colour from dark matter to a silvery lustrous material. Used as a fuel, as a reducing agent in blast furnaces, for producer gas formation.

coking Relatively severe cracking processes converting residual materials such as pitch or tar into a mixture of gases, naphtha, kerosine, gas oil and coke. The gas oil is used mainly as feedstocks for catalytic cracking.

coking coal Coal which, after carbonization, will give a marketable coke.

colemanite, $CaB_3O_4(OH)_3,H_2O$. An important source of borates.

collagen The most abundant protein in the body, containing mainly glycine, hydroxyproline and proline. Collagen is essentially insoluble but when boiled with water, the strands separate and undergo some hydrolysis, affording gelatin*. The conversion of collagen to leather involves the formation of cross linkages between collagen molecules. Collagen is found in all connective tissue, e.g. skin, cartilage, tendons, ligaments.

collagenase An enzyme for degrading collagen*.

s-collidine *See* 2,4,6-trimethylpyridine.

colligative properties Properties of solutions, e.g. vapour pressure, osmotic pressure, boiling point, freezing point, which depend only upon the number of particles (atoms, ions or molecules) present in solution, and are independent of the nature of the solute.

Collman's reagent $Na_2Fe(CO)_4$, $\frac{2}{3}$ dioxane.

collodion A nitrocellulose widely used as a base for lacquers, in which a ketone or ester may be used as the solvent.

colloid In 1860 Graham differentiated between substances which, in solution, were able to penetrate a dialysis membrane and those which were retained. The former (soluble salts, acids, cane sugar, etc.) he termed crystalloids, the latter (gelatin, albumin, etc.) colloids. Colloids are further characterized by a slow rate of diffusion and by the fact that the path of a beam of light is illuminated on passage through a colloidal solution. A substance is now defined as being in the colloidal state if its particles, of approximate size between 1 and 100 mm, are dispersed in a continuous medium. Colloids are thus intermediate between coarse suspensions on the one hand and molecular or ionic solutions on the other. The lines of demarcation are indefinite, fine suspensions showing some colloidal effects, whilst large molecules are essentially colloidal in behaviour.

As it is now possible by choice of suitable conditions to prepare most compounds in this form, the colloid state should be considered as a physical state in which all substances can be made to exist. Many materials such as proteins, vegetable fibres, rubber, etc. are most stable or occur naturally in the colloidal state. In the colloidal state the properties of surface are all-important.

colloidal electrolyte Although the sodium salts of the lower fatty acids behave as normal weak electrolytes, with higher members of the series, association begins to occur and colloidal character is assumed. The anion is no longer a discrete fatty acid anion, but consists of an aggregate of fatty acid anions together with undissociated salt molecules; it is a giant polyvalent ion, dissociating at the surface and compensated by an equal number of sodium ions in close proximity to the giant ion. Soaps and many dyestuffs may be classed as colloidal electrolytes.

colloid mills Devices for producing colloidal suspensions or emulsions where the particle and droplet sizes are less than one micron.

Colloid mills are used in the paint industry, and in the pharmaceutical and food-processing industries.

colour couplers Compounds used in photographic development of colour films in which the anion can react with developer to give appropriate dyestuffs.

colour development The photographic process in which developer oxidation products, formed in the emulsion layer where silver ions are reduced, are used to form dyes either by direct combination with another molecule of oxidized developer (primary colour development), or by the oxidized developer reacting with a new molecular species known as a colour coupler* to give a dye of the required absorption (secondary colour development). Multicoloured processes employ the latter system.

colour indicators Indicators* (acid-base, oxidation-reduction) which depend for their effect on colour change. Screened indicators have an additional dye added which acts as a colour filter and makes the colour change more visible. Thus the colour change of methyl orange with pH is made more obvious by addition of some methylene blue. A universal indicator is a mixture of indicators which shows a gradual but well marked series of colour changes over a very wide range of pH.

colour index, CI The definitive listing of dyestuffs and pigments which includes information on commercial names, method of application, dyefastness, etc.

colour photography The process of preparation of prints or transparencies in colour. White light is split into three components – red, green and blue – by filters – yellow, magenta, cyan. The particular colour activates grains of silver halide (generally in specific layers for each colour) which contain sensitizing dyes on the grain. Colour development* then gives rise to a coloured image.

columbite (Fe, Mn)(Nb, Ta)$_2$O$_6$ with excess Nb. The principal ore of niobium*.

columbium, Cb. A former name for niobium*.

combining volumes, law of *See* Gay-Lussac's law.

combustion The rapid, high-temperature oxidation of fuels, converting carbon to carbon dioxide (or carbon monoxide) and hydrogen to water vapour. Any sulphur in the fuel is oxidized to the dioxide or trioxide depending on combustion conditions, while nitrogen either remains unreacted or is converted to nitrogen oxides. Most combustion reactions occur in the gas phase except for the burning of the fixed carbon in solid fuels.

The release of chemical energy during combustion of gases produces a luminous, radiating zone which is seen as the flame or flame front*.

common-ion effect In a solution of a weak electrolyte, e.g. ethanoic acid, the concentration of ions is governed by the equilibrium

$$HAc \rightleftharpoons H^+ + Ac^-$$

Addition of excess H^+ ions to this solution will cause the equilibrium to move towards undissociated acid thereby decreasing the concentration of Ac^-. This effect is known as the common-ion effect and is of considerable practical importance. Thus, e.g. in the precipitation of metal ions as insoluble sulphides, the concentration of S^{2-} in aqueous solution is controlled by the equilibrium

$$H_2S(aq) \rightleftharpoons 2H^+(aq) + S^{2-}(aq)$$

Addition of acid will reduce the concentration of S^{2-}, whilst in alkaline solution the concentration of S^{2-} will increase. Since, in order for precipitation to occur, the solubility product of the sulphide must be exceeded, i.e.

$$K_{sp} = [M^{x+}][S^{2-}]^{x/2}$$

the actual precipitation can be controlled by varying the pH of the solution.

complement The combination of an antibody and an antigen* on the surface of a foreign cell leads to the 'fixation' of a group of proteins present in normal serum, collectively known as complement. The result is the activation of destructive enzymes which cause lysis of the cell.

complex Any compound in which the bonding is by interaction of the electrons of the donor with empty orbitals of the acceptor. In some complexes the electron flow may take place in both directions simultaneously – *see* back-bonding. The interaction may take place between charged or uncharged species.

Where the structure is known a complex species comprising the acceptor and its ligands is formulated within square brackets, e.g. $[Co(NH_3)_6]Cl_3$. Bridging ligands are designated μ-L, e.g. $Fe_2(CO)_9$ is $[(OC)_3Fe(\mu\text{-}CO)_3Fe(CO)_3]$. Hapto designates the number of ligand atoms actually bonded to the acceptor, e.g. $[(h^5-C_5H_5)Mn(CO)_3]$ has 5 carbon atoms (plus 3 carbonyls) bonded to manganese.

complex ion A complex ion is formed by the co-ordination of other ions or molecules to an ion to form a stable entity. Thus the Co^{3+} ion and ammonia give $[Co(NH_3)_6]^{3+}$ whilst the Fe^{3+} ion and cyanide ions give the $[Fe(CN)_6]^{3-}$ complex. Depending upon the stability of the complex a solution of a complex salt may or may not give the reactions of the individual components of the complex. It should be noted that hydrated salts generally contain aquo-complexes, e.g. $[Cu(H_2O)_6]^{2+}$ and that the anions of oxy-salts may be considered as being complexes formed by (mostly hypothetical) positive ions and oxide ions, e.g. the NO_3^- ion as N^{5+} and $3O^{2-}$.

complexometric indicator *See* metallochromic indicator.

complexometric titration A titration* involving formation or decomposition of a complex. Thus Fe^{3+} plus SCN^- is intense red; titration of the Fe^{3+} with EDTA causes decomposition of the red $Fe^{3+}-SCN^-$ complex and formation of the colourless $Fe^{3+}-EDTA$ complex.

complexone *See* sequestering agent.

component For the phase rule* the number of components in a chemical system is the smallest number of substances in terms of which the compositions of each of the phases in the system may be described separately. The number of components in a system need not necessarily equal the number of chemical species present. For example, if calcium carbonate is heated in a closed system:

$$CaCO_3 \rightleftharpoons CaO + CO_2$$

the concentration of $CaCO_3$ can be expressed in terms of those of CaO and CO_2, i.e. only two components are present.

composites High performance materials consisting of a plastic matrix reinforced by fibres (glass, graphite, boron, etc.). Used in spacecraft, aircraft, car components, sports goods, machinery.

compound The usage of chemists in respect of the word compound is not perfectly definite; all would agree that gunpowder is a mixture and common salt a compound. Not much difficulty is found with a compound with a fixed composition such as sodium chloride, cane sugar or water. It is not so easy to understand the orthodox usage in connection with substances the composition of which may vary, the properties of the substance probably also varying. Glass, steel, iron oxides and starch are examples of such substances; in them every atom is attracted by all the adjacent atoms so as to make a uniform or very nearly uniform solid; such bodies are clearly not mere mechanical mixtures; they are chemical compounds of varying or indefinite composition; the feldspars and many other silicates also have no definite composition; similarly many plastics and other polymers are completely chemically combined and must be considered as chemical compounds of varying compositions; alloys such as pewter and brass are compounds not mere mixtures. It is not useful to give a precise definition of compounds, and there are intermediate states in between the typical compound and the typical mixture.

compounding The mixing process used to produce a homogeneous mixture for use in, e.g., a plastic or rubber composition when for example rubber vulcanizing agents, fillers, anti-oxidants, pigments are added.

concentrated Containing a high proportion (e.g. of solute in a solution).

concentration cell The potential of a piece of metal immersed in a solution containing its ions (*see* electrode potential) depends on the concentration of the ions. Thus a cell may be set up which derives its electromotive force from the difference in concentration of solutions of the same electrolyte surrounding the two electrodes. Any cell which depends on this principle is called a concentration cell.

conchoidal fracture A surface resulting from the fracture of a solid which shows no regular crystal faces (cleavage planes). The fracture is generally made up of shell-like curved surfaces and is characteristic of amorphous glassy materials.

concrete The construction material resulting from the hardening of a mixture of cement, sand (or similar) and water. Can be made lighter by incorporation of perlite or polystyrene.

condensation reactions Addition of one molecule to another with the elimination of a simple molecule such as water, ammonia, an alcohol, etc., e.g. the Claisen condensation.

condensers Heat exchangers used for condensing vapours, the heat normally being removed by cooling water.

conductiometric titration A titration in which the end point is ascertained by observing the changes in conductivity of the solution.

conductivity The property by virtue of which the substance allows the passage of an electric current. The reciprocal of the resistance of a circuit is called the conductivity. The reciprocal of the specific resistance is the specific conductivity or conductance of a substance.

The conductivity of a solution containing 1 gram equivalent of solute when measured between two large parallel electrodes at a distance of 1 cm apart is called the equivalent conductivity Δ.

$$\Delta = \frac{1000\,K}{c}$$

where K = specific conductance and c = concentration in g equiv. per litre.

conductivity, solids Solids may conveniently be classified in terms of their electrical conductivity into electronic conductors, semiconductors and insulators. Metals are typical conducting solids, their conductivity increasing with decreasing temperature. Semiconductors may be classified as either intrinsic semiconductors, e.g. pure germanium or silicon and a number of transition metal oxides, or as impurity semiconductors, e.g. Al or P doped Si and Ge. Semiconductors may be differentiated from conducting solids in that the conductivity of the former increases markedly with increasing temperature. Refractory oxides, e.g. alumina, silica, magnesium oxide are typical insulators. Impurity doped semiconductors find extensive use in electronic microcircuits as transistors, etc. Ionic conductors conduct by movement of ions through the solid.

Condy's fluid A disinfectant solution of calcium and potassium permanganates.

configuration The spatial arrangement of atoms or groups in molecules. *See* isomerism and conformation.

conformation The term usually restricted to the potentially dynamic spatial arrangements of atoms or groups in a molecule which may be in equilibrium with other conformations. No single conformation constitutes a discrete

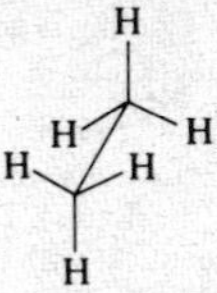

staggered

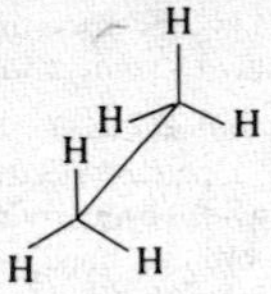

eclipsed

and isolable substance under usual conditions, in contrast to configurational isomers. Two extreme conformations of ethane are the *eclipsed* and *staggered* forms, which are easily interconvertible by rotation about the C–C bond. Cycloalkanes present other conformational differences. For example cyclohexane molecules exist as rapidly interconverting species with *boat* or *chair* conformations, and it is not possible to separate these, although at room temperature the chair conformation is considerably more stable and comprises more than 99% of the equilibrium mixture.

chair

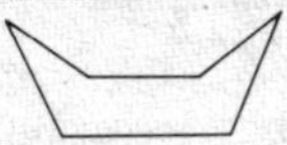

boat

A monosubstituted cyclohexane, e.g. methylcyclohexane, exists theoretically in two isomeric forms with a chair-form ring, and the methyl substituent either *axial* or *equatorial*. Since these rapidly interconvert through a process known as *ring inversion*, physical separation of the isomers is not possible. Physical evidence suggests that the equatorial form of methylcyclohexane predominates over the axial form. *Cis-trans*-isomers of cyclohexane derivatives have the additional possibility of conformational isomerism.

axial

equatorial

coniferin, $C_{16}H_{22}O_8$. The glucoside of coniferyl alcohol, present in fir trees. Used for the synthesis of vanillin which is formed by oxidation with chromic acid. M.p. 185°C.

coniferyl alcohol, $C_{10}H_{12}O_3$. $(3\text{-}CH_3O)\text{-}(4\text{-}HO)C_6H_3CH{=}CHCH_2OH$. Prisms; m.p. 73–74°C. Occurs to a small extent in wood and is a constituent of the glycoside coniferin*. Oxidized to vanillin. It resinifies with mineral acids.

coning and quartering A sampling technique involving the formation of a cone, flattening, rejection of opposite quarters, repetition to obtain a sample of appropriate size.

conjugate base In Brønsted–Lowry acid-base theory an acid is considered to dissociate into a proton and the conjugate base of the acid. Thus the conjugate base of an acid is its anion.

conjugate solutions Solutions of two substances in one another (e.g. phenol in water and water in phenol) which are in equilibrium at a particular temperature.

conjugation Alternating double (or triple) and single bonds as in –C=C–C=C–, –C=C–C=O, etc.

conproportionation The opposite of disproportionation*, e.g. Mo + MoF_6 to MoF_5.

conservation of energy, law of In systems of constant mass, energy can neither be created nor destroyed. One form of energy may disappear, but another takes its place. Thus the energy possessed by a hammer may be converted into heat on striking a surface. In an isolated system, the total energy of the system remains constant. Energy and mass are interconvertible.

conservation of matter, law of This law, which was understood by the Greeks, was first clearly formulated by Lavoisier in 1774 in the statement that matter can neither be destroyed nor created. However, emission of radiation must be accompanied by a loss of mass, equal to E/mc^2, where E is the energy of the radiation and c the velocity of light. E/mc^2 is usually small compared with the masses of material used in ordinary chemical manipulation. There is a conservation of mass and energy considered together. *See* atomic energy.

constantan A Ni (45%) Cu (55%) alloy with high resistance and low resistance-temperature coefficient used in resistances and thermocouples.

constant boiling mixture *See* azeotropic distillation and mixtures.

constant proportions, law of The statement that the composition of a pure chemical compound is independent of its method of preparation. In fact many compounds have a range of compositions – *see* compound, defect structures.

contact process *See* sulphuric acid.

continuity of state As the temperature is increased towards the critical point, the properties of liquids and vapour become increasingly similar, until at the critical temperature they are identical. Although the change from liquid to vapour, or vice-versa, is normally discontinuous, a gradual transition with continuity of state is possible.

continuous counter-current decantation A method for the continuous washing of finely-divided solids to free them from impurities. A series of continuous thickening* is used, the solids passing through them in series, counter-current to the solution.

continuous spectrum The occurrence of well defined lines (*see* line spectrum), or of bands (*see* band spectrum) consisting of well defined lines, arises because electronic, vibrational and rotational energy changes in a molecule can only occur in definite steps (quanta) corresponding to transitions between allowed energy levels. Certain changes, e.g. the dissociation of a molecule, are not quantized processes. Hence, if light is emitted during the occurrence of such a process, its frequency no longer has definite values, but may take any value over a continuous range of frequencies. The spectrum of such a system appears to be continuous. A continuous spectrum is characteristic of an unquantized process, such as dissociation.

contraceptive drugs Drugs used to prevent conception. The drugs are mainly hormonal steroid derivatives.

convergence frequency *See* convergence limit.

convergence limit The lines in a spectrum represented by a given series (e.g. the Balmer series) become closer and closer together as the wavelength of the lines becomes shorter, and eventually approach a limit, known as the convergence limit, which may be expressed as a frequency, or wave number, the convergence frequency.

conversion processes A general term used to describe an industrial reaction involving conversion of products, e.g. conversion of coal fuel to gaseous fuel by gasification.

converting A metallurgical oxidation process carried out by blowing air through the molten charge in a converter.

cool flames Quiescent, incomplete combustion exhibited by hydrocarbons, aldehydes and ethers. Under certain conditions the oxidation reaction can occur at low temperatures (580–800 K) emitting little heat and light and showing no flame front* or reaction zone. Also called homogeneous or slow combustion since oxidation apparently occurs simultaneously at different points in the mixture.

cooling towers Towers used to cool water for use in condensers and coolers. Cooling is generally by contact with atmospheric air.

co-ordinate bond The linkage of two atoms by a pair of electrons, both electrons being provided by one of the atoms (the donor). The co-ordinate bond is formally identical with the covalent bond. An atom capable of accepting the electrons is the acceptor, the molecule donating the electrons is the donor or ligand. Co-ordinate linkages occur widely in inorganic complexes. *See* valency, theory of.

co-ordination arrangements The arrangements of atoms or ligands around a central atom depends upon the sizes of the atoms and the electronic configuration of the central atom (lone-pairs of electrons occupy positions in space). For details of the most common co-ordination arrangements see under the individual headings. Common arrangements are:

co-ordination number		
2	linear	$BeCl_2$, CO_2
	bent	H_2O
3	trigonal planar	BF_3
	pyramidal	NH_3
4	tetrahedral*	CH_4, $[NiCl_4]^{2-}$
	square planar	$(PtCl_4]^{2-}$
5	trigonal bipyramidal*	PF_5
	square pyramidal	IF_5
6	octahedral*	NaCl, $[Co(NH_3)_6]^{3+}$
	trigonal prismatic	NiAs
7	pentagonal bipyramid	ReF_7
	face-centred octahedron	$[NbOF_6]^{3-}$
	face-centred trigonal prism	$[TaF_7]^{2-}$

8	cubic*	CsCl
	square antiprism*	$[Mo(CN)_8]^{4-}$
	dodecahedral	$[Mo(CN)_8]^{4-}$
12	icosahedral*	$[Ce(NO_3)_6]^{2-}$
	cubic close-packed*	perovskite

co-ordination compound A compound containing a co-ordinate bond* or bonds.

co-ordination isomerism Isomerism in co-ordination compounds in terms of the actual arrangement of the atoms.

co-ordination numbers The number of groups or ions surrounding a particular molecule or ion in a crystal or in solution; in complexes the total number of co-ordinated and covalently bonded groups surrounding the central atom. For elements Li–F the maximum co-ordination number for covalent compounds is 4, but 6 co-ordination can occur in ionic compounds. For elements Na–Cl the maximum co-ordination number is 6.

copolymer A complex polymer resulting from the polymerizing together of two or more monomers. A copolymer is a true compound and often has properties quite distinct from those of a physical mixture of the separately polymerized monomer components. Important industrial copolymers include those obtained from vinyl chloride–vinyl acetate and styrene–butadiene monomer mixtures. *See* block and graft copolymers.

copper, Cu. At.no. 29, at.wt. 63·546, m.p. 1083·4°C, b.p. 2567°C, *d* 8·92. One of the coinage metals; occurs native (contains Ag, Bi, Pb, etc.) and as cuprite Cu_2O, tenorite or melaconite CuO, copper pyrites or chalcopyrite $CuFeS_2$, erubescite or bornite Cu_3FeS_3; malachite $CuCO_3{\cdot}Cu(OH)_2$ (bright green) and azurite $2CuCO_3{\cdot}Cu(OH)_2$ (bright blue) are naturally occurring minerals used as pigments. The sulphide ores and native copper are used as sources of copper. Copper ores are concentrated by flotation and sulphur is oxidized in converters. The crude copper is refined electrolytically. The metal is ccp and reacts with oxygen at red heat. It is attacked by halogens, oxidizing acids, ammonia and KCN solutions (the latter two in the presence of oxygen). Approx. 55% of copper is used for electrical purposes and about 15% in pipes (plumbing). Much copper is used as alloys (brasses, bronzes, gold). Other copper uses include catalysis (CuCl, $CuCl_2$, CuO), fungicides ($CuCl_2$, CuO), printing, dyeing, pigments, paints. World production 1976 7·8 megatonnes. Copper is present in some enzymes (ascorbic acid oxidase, cytochrome oxidase, tyrosinases) and is part of the oxygen carrier in snails, crabs and some crustaceans.

copper bromides, $CuBr_2$ and CuBr. Very similar to copper chlorides.

copper chemistry Copper is a typical transition element, electronic configuration $3d^{10}4s^1$. Very low oxidation states are not stable although unstable carbonyls are formed by matrix isolation methods.

Copper (I) salts are very similar to Ag(I) salts with an insoluble chloride and iodide, they are colourless or pale yellow. Complexes are formed with phosphines and CO.

Copper(II) salts (blue in aqueous solution) are typical M(II) salts but generally have a distorted co-ordination (Jahn–Teller distortion, 4 near plus 2 far neighbours). Extensive ranges of complexes are known, particularly with *N*-ligands.

Copper(III) is known in complex oxides and fluorides and in amino-acid complexes.

Electrode potentials in acid are $Cu^+ \longrightarrow Cu + 0{\cdot}34$ volts, $Cu^{2+} \longrightarrow Cu^+ + 0{\cdot}158$ volts, Cu^+ disproportionates to Cu plus Cu^{2+} in aqueous solution but is stable in many non-aqueous solvents. Many cluster compounds, e.g. $[Cu_4I_4(PMe_3)_4]$, $[Cu_6H_6(PPh_3)_6]$ are known and Cu – Cu bonding is found in many copper(II) derivatives particularly those containing bridging ligands (e.g. $[Cu(O_2CCH_3)_2,H_2O]_2$.

copper chlorides
Copper(II) chloride, $CuCl_2$. Dark brown (Cu plus excess Cl_2) forms green $CuCl_2,2H_2O$ and many complexes, e.g. $CuCl_2$, pyridine, generally containing bridging chlorines. Complex species, e.g. $[Cu_2Cl_6)^{2-}$, $(CuCl_4)^{2-}$ yellow or green, are also known.

Copper(I) chloride, CuCl. White solid ($CuCl_2$ plus HCl plus excess copper or SO_2). Gives carbonyl and phosphine complexes.

copper chromite Mixed oxide, often with added Ba^{2+}, used as a catalyst for reduction of ketones and esters to alcohols.

copper cyanide, CuCN. White compound (Cu(II) plus KCN solution, $(CN)_2$ liberated) forms $K_3Cu(CN)_4$ with excess KCN.

copper ethanoate, $[Cu(O_2CCH_3)_2,H_2O]_2$. Dimeric compound with Cu – Cu interaction (copper carbonate plus ethanoic acid).

copper fluoride, $CuF_2,2H_2O$. Anhydrous CuF_2 and some complexes are known.

copper hydroxides Basic copper(II) salts are formed by precipitation with, e.g. NaOH. A rather indefinite copper(II) hydroxide is precipitated from copper(II) solutions. On heating converted to a hydrated oxide.

copper iodide Copper(II) iodide is not known. Addition of I^- to a solution of a copper(II) salt gives a precipitate of copper(I) iodide, CuI (I_2 liberated quantitatively).

copper naphthenates Copper(II) salts of carboxylic acids derived from crude petroleum oils and used as wood preservatives.

copper nitrate, $Cu(NO_3)_2$. Formed with $3H_2O$ (CuO, Cu, $CuCO_3$ in dil. HNO_3). Forms 9 and 6 hydrates and decomposes on heating to the oxide. Anhydrous $Cu(NO_3)_2$ is formed from Cu, liquid N_2O_4 and $C_2H_5O_2CCH_3$; it is appreciably volatile.

copper, organic compounds Copper(I) forms alkyls and aryls and also complexes, e.g. $(h^5-C_5H_5)CuPPh_3$, $(C_6H_5Cu)_2$. Copper derivatives formed *in situ* are important in coupling reactions.

copper oxides

Copper(I) oxide, Cu_2O. Red solid formed from Cu(II) salt and hydrazine or heat on CuO. Gives a cheap red glass and a cuprate, KCuO (K_2O plus Cu_2O).

Copper(II) oxide, CuO. Black solid formed by heating $Cu(OH)_2$, $Cu(NO_3)_2$, etc. Dissolves in acid to Cu(II) salts, decomposes to Cu_2O at 800°C. Forms cuprates in solid state reactions. A cuprate(III), $KCuO_2$, is also known.

copper perchlorates

Copper(II) perchlorate, $Cu(ClO_4)_2{,}6H_2O$. Formed from CuO and $HClO_4$.

Copper(I) perchlorate, $CuClO_4$. Formed in solutions of organic solvents (Et_2O, C_6H_6) by displacement of Ag by Cu.

copper sulphates

Copper(II) sulphate, $CuSO_4$. Blue crystals (blue-vitriol) of $CuSO_4{,}5H_2O$ from CuO and H_2SO_4; gives pale blue trihydrate, colourless monohydrate which loses most of its water at 250°C (used – blue colour – to detect water). Obtained industrially from copper ores and H_2SO_4 or Cu plus H_2SO_4 plus air. Used in agriculture, and water treatment, as a wood preservative.

Copper(I) sulphate, Cu_2SO_4. Grey solid, Cu_2O plus dimethyl sulphate.

copper sulphides

Copper(II) sulphide, CuS. Black solid, Cu plus excess S or copper(II) salt plus H_2S. Decomposes to copper(I) sulphide, Cu_2S, on heating.

coprecipitation A substance when precipitated from solution may be impure due to the presence of foreign substances adsorbed on the surface or occluded within the bulk of the precipitate particles. The contamination of the precipitate by substances which are normally soluble in the mother liquor is termed coprecipitation.

correlation diagram A diagram of calculated energies of a molecule against assumed molecular geometry including the energies of the various molecular orbitals. Used in approaches to molecular geometry.

corrin, vitamin B_{12} Composed of the nucleotide and the highly substituted, reduced, porphyrin-like corrin ring.

corrosion Most metals and alloys are attacked by oxygen, moisture and acids. Some are attacked by alkalis. The process of attack is called corrosion. Corrosion may result in a uniform attack which is generally not very serious; it may attack preferentially, particularly at grain boundaries, resulting in severe weakening of the metal without much visible deterioration; it may attack locally at areas where conditions are varied resulting in perforation; finally it may produce a passive oxide layer as on Cr and Al which gives protection from further corrosion. There are two main types of corrosion: atmospheric oxidation and tarnishing in dry air, which is not serious, and electrolytic corrosion under conditions of moisture or submersion. Any two areas on a metal or alloy surface not chemically and physically identical can act as anode and cathode. Thus two phases in an alloy such as steel act as microelectrodes. The anode reaction is solution of metal or corrosion $M - 2e \longrightarrow M^{++}$ for a dipositive metal. The cathode reaction is by hydrogen evolution in acid solution or oxygen reduction in neutral solution. Differential aeration is a common cause of corrosion. If one area of a metal has a higher oxygen potential than another, then it becomes cathodic

$$\tfrac{1}{2}O_2 + H_2O + 2e \longrightarrow 2OH^-$$

making the area of low oxygen potential anodic with consequent corrosion. Thus corrosion is more severe under deposits of silt or in general where an area is loosely covered. Underground corrosion is caused by

(1) electrolytic cells, the moist soil acting as electrolyte

(2) stray electric currents from earthing points, etc.

(3) anaerobic bacteria which in the presence of iron and traces of $CaSO_4$ depolarize the microcathodes with the production of H_2S.

The main methods of combating corrosion are

(1) treatment of water systems with inhibitors

(2) cathodic protection

(3) protection by paint, plating, phosphating, etc.

See also rusting, cathodic protection.

cortisol *See* hydrocortisone.

cortisone, $C_{21}H_{28}O_5$. M.p. 215°C. A steroid.

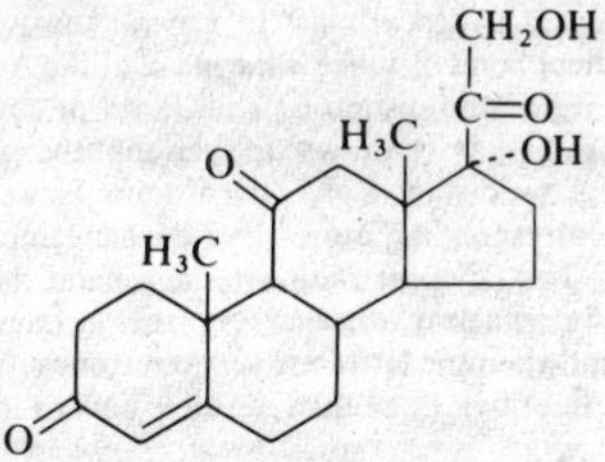

Cortisone is a hormone produced by the cortex of the adrenal glands. As with other adrenal corticoid steroids, administration of cortisone leads to an increased deposition of liver glycogen. It can remove features of rheumatoid arthritis, but does not check the underlying disease: it is used in various diseases of the eye, and is an antiallergic and antifibroplastic agent.

corundum, α-Al_2O_3. *See* aluminium oxide. The corundum structure has close-packed oxygen with Al in octahedral holes. Ruby, sapphire and emery are mineral varieties of corundum. Articles made from corundum are extremely inert and resistant to corrosion. Used for grinding. World use 1976 10 000 tonnes.

Cotton effect Within an absorption band,

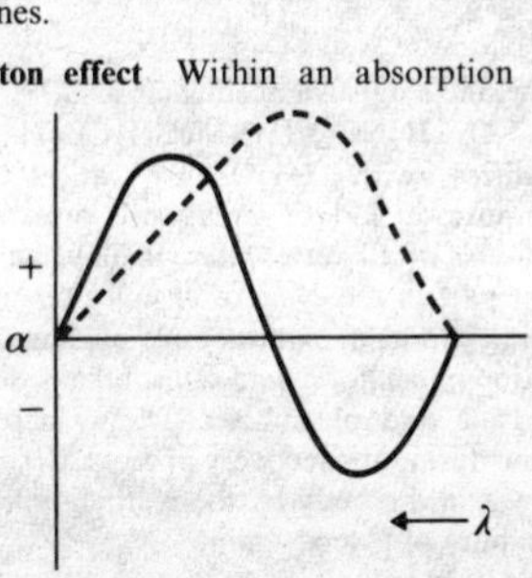

there is anomalous rotatory dispersion. In the figure, the broken line represents an absorption band and the full line the rotatory dispersion curve. As the wavelength decreases the rotation angle increases, passing through a maximum and then decreases, passing through zero angle at the wavelength corresponding to the maximum of absorption. On proceeding to lower wavelengths, the angle falls until it passes through a minimum and then rises again. This phenomenon is known as the Cotton effect. This effect is general for coloured compounds as well as for colourless substances with bands in the ultra-violet. *See* optical rotatory dispersion, circular dichroism.

coulomb The unit quantity of electricity; 96 494 coulombs, 1 Faraday, corresponds to 1 mole of electrons. *See* Faraday's laws of electrolysis.

coulometer An instrument or device for measuring the amount of charge (number of coulombs) passing in an electrical circuit. Thus a silver coulometer consists of 2 platinum electrodes dipping into a solution of silver nitrate. For each coulomb of charge passed $1{\cdot}118 \times 10^{-3}$ g of silver is deposited at the cathode. Thus, from the increase in weight of the cathode the charge passing can be calculated.

coulometry Analytical technique involving the measurement of quantities of electricity. Constant current coulometry can be used for coulometric titrations, e.g. Br_2 can be prepared electrically and used for oxidation (e.g. of olefines) *in situ* and the end point detected spectrophotometrically; Ag^+ can be generated and used to titrate Cl^-. In controlled (constant) potential coulometry one species may be oxidized or reduced and thus determined at a specific potential.

o-coumaric acid, *trans*-2-hydroxycinnamic acid,

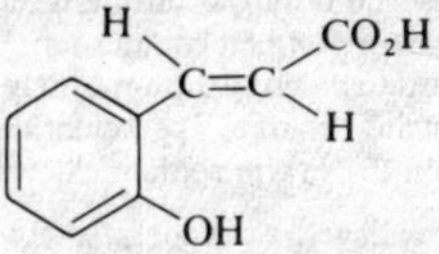

$C_9H_8O_3$. Colourless crystals; m.p. 108°C. Obtained by boiling coumarin with sodium ethoxide. Irradiation of *o*-coumaric acid produces coumarinic acid*. The stable form of the acid is the *trans* form.

coumarin, $C_9H_6O_2$. Colourless crystals; m.p.

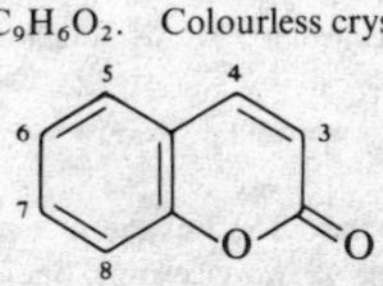

70°C, b.p. 290°C. Occurs in the Tonka bean, of which it is the odorous ingredient. Prepared synthetically by heating salicylaldehyde with ethanoic anhydride and sodium ethanoate. It is the δ-lactone of coumarinic acid. Hydrolysis with sodium hydroxide produces coumaric acid*. It is largely used in perfumery. Warfarin is a coumarin derivative.

coumarin glycosides Widely distributed in plants. The aglucones consist of hydroxycoumarins containing 1, 2 or 3 hydroxyls or their methyl ethers.

coumarinic acid, *cis*-2-hydroxycinnamic acid, $C_9H_8O_3$. An unstable acid which spontaneously forms a δ-lactone, coumarin, when the free acid is generated from its salts. *See* coumaric acid.

coumarone *See* benzfuran.

coumarone and indene resins Prepared from purified naphtha of boiling range 150–200°C which is treated with H_2SO_4 to produce a resin. Used particularly in floor tiles and hot melt compositions.

counter-current flow When two streams of material moving in opposite directions are brought into contact for the purpose of transferring heat or matter, flow is said to be counter-current. Counter-current operation is used in preference to co-current flow* whenever possible since it results in greater heat or mass transfer for a given set of operating conditions.

counter-ions The charge on the surface of a colloidal particle is compensated for by an equal and opposite ionic charge in the liquid in immediate contact with it. The system as a whole is electrically neutral. Those compensating ions, which occupy a diffuse layer around the particle, are termed counter-ions.

Also used to describe the oppositely charged ion balancing charge, e.g. chloride is the counter-ion to Na^+ in sodium chloride.

counters, radioactive Apparatus for the detection and quantitative estimation of radioactivity.

coupling The interaction between nuclei or electrons, particularly used in nuclear magnetic resonance*.

coupling constant, J *See* nuclear magnetic resonance.

covalency maximum *See* co-ordination numbers.

covalent bond The linkage of two atoms by the sharing of two electrons, one contributed by each of the atoms. The electrons are only shared equally when the atoms are identical, and in most covalent bonds the electrons are held to a greater extent by one atom than by the other (electronegativity*). Covalent links are formed where electrovalent linkages would involve the interaction of small, highly charged cations with large diffuse anions. *See* valency, theory of.

covalent radius The equilibrium distance between two atoms joined by a covalent bond of a given type (i.e. single, double or triple) is found to be, within close limits, constant. Thus the C–C bond distance of 1·54 Å found in diamond is the same as that found in cyclic saturated hydrocarbons. In general the A–B covalent bond distance is the mean of the A–A and B–B bond distances. Thus, a set of covalent radii may be drawn up such that the sum of any two distances gives the distance between the corresponding atoms in a covalent molecule. The covalent radii are, in general, half the internuclear distances in the elements, except when the latter are very electronegative (e.g. fluorine). For values *see* table under radii.

Cr Chromium.

cracked gasoline, cracked spirit Gasoline obtained from heavier petroleum distillates by a thermal* or, more usually, catalytic* cracking process. These gasolines have a fairly high octane number* being higher in alkenes than straight-run gasolines. Their unsaturated nature leads to formation of gum unless inhibitors are added.

cracking The thermal decomposition of a substance into fractions of lower molecular weight. Cracking processes, widely used in the petroleum industry, may be purely thermal or catalytic. *See* thermal cracking, catalytic cracking.

cream of tartar *See* potassium hydrogen tartrate.

creatine, methylguanidinoethanoic acid, $C_4H_9N_3O_2$, $H_2NC(NH)NMeCH_2CO_2H$. Crystallizes with $1H_2O$ lost at 100°C. Decomposes at 291°C. Creatine is present in the muscles of all vertebrates. It plays an important part in the cycle of chemical changes involving muscular contraction; phosphagen, a creatine phosphoric acid ester, breaks down to creatine and phosphoric acid, which recombine during the recovery process. Creatine is formed in the body by the methylation (by methionine) of glycocyamine.

creatine phosphate, phosphocreatine, phosphagen *See* creatine.

creatinine, 1-methylglycocyanidine, $C_4H_7N_3O$. Crystallizes with $2H_2O$; m.p. 260°C (decomp.). It is the internal anhydride of creatine and is

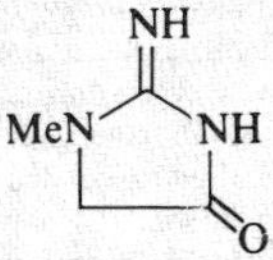

formed from creatine in alkaline solution. Found in human urine, 1–2 g daily being excreted formed as a breakdown product of the body tissues.

creep The slow yielding of metals and alloys (generally at high temperatures) under a load which does not cause fracture. The process is diffusion controlled.

creosote Fractions obtained in the distillation of coal tar. *Light*, *medium* and *heavy* fractions are classified by their boiling ranges (approx. 475–540; 540–570; 570–630 K respectively). Creosote fractions are also referred to as naphthalene oil, wash oil, light anthracene oil, etc. Creosotes have a characteristic odour and consist of a mixture of hydrocarbons, phenols and other aromatic derivatives. They are used for timber preservation, for fluxing pitch and bitumen, and as fuels.

Medicinal creosote is a mixture of phenols, chiefly guaiacol and creosol (4-methyl-2-methoxyphenol), obtained by distillation of wood tar. B.p. 480–500 K. It is almost colourless with a characteristic odour and is a strong antiseptic, less toxic than phenol.

cresols, hydroxytoluenes, C_7H_8O. The cresols are colourless liquids or crystalline solids. They are volatile in steam, and are reduced to toluene by Zn dust. They occur in coal tar and cracked naphtha.

o-cresol, 2-hydroxytoluene,
m.p. 31°C, b.p. 191°C
m-cresol, 3-hydroxytoluene,
m.p. 12°C, b.p. 203°C
p-cresol, 4-hydroxytoluene,
m.p. 35°C, b.p. 202°C

Usually prepared from the corresponding sulphonic acids by alkali fusion, methylation of phenol or from the aminotoluene by treatment with nitrous acid followed by boiling. Both *o*- and *p*-cresol are used as end components in azo dyes.

For uses *see* cresylic acids. U.S. production of cresylic acid and mixed cresols 1978 65 000 tonnes.

cresotic acid 2-hydroxy-3-methylbenzoic acid.

cresylic acids Methyl derivatives of phenol. E.g. $C_6H_4(Me)OH$ are cresols; $C_6H_3Me_2OH$ are xylenols. Also called tar acids as obtained from coal distillation or sweetening petroleum distillants. Used as starting materials for pesticides and herbicides and for phenolic resins, disinfectants, antioxidants and plasticizers (cresyl phosphate).

CRG process (catalytic rich gas process) A catalytic low-pressure process producing a natural gas substitute from naphtha feedstock and steam over a nickel catalyst at about 775 K. The product is a mixture of methane, carbon dioxide and some hydrogen. This may require methanation*, carbon dioxide removal and propane blending before satisfactory properties are obtained.

cristobalite, SiO_2. The high-temperature form of silica*, stable from 1470°C to the m.p. 1710°C.

critical humidity The humidity at which the vapour pressure at the surface of a solid or solution is equal to the partial pressure of water in the atmosphere. At humidities above the critical value water will tend to be absorbed; below the critical value moisture will be lost to the atmosphere.

critical phenomena Above the critical temperature it is impossible to liquefy a gas. The minimum pressure required to cause liquefaction at the critical temperature is the critical pressure. The critical volume is the volume occupied by one mole of the gas at its critical temperature and pressure.

critical solution temperature The temperature at which the mutual solubility of two phases becomes equal. For two liquids, e.g. phenol and water, the solubility of each in the other increases until at the critical solution temperature only one phase exists.

cross section A number denoting the effective area of a nucleus for a scattering or absorption process. *See* barn.

crotonaldehyde, 2-butenal, C_4H_6O, $CH_3{\cdot}CH{:}CH{\cdot}CHO$. Colourless lachrymatory liquid with a pungent odour. B.p. 104°C. Manufactured by the thermal dehydration of aldol. May be oxidized to crotonic acid* and reduced to crotonyl alcohol and 1-butanol; oxidized by oxygen in the presence of V_2O_5 to maleic anhydride. It is an intermediate in the production of 1-butanol from ethanol.

crotonic acids, 2-methylacrylic acids, 2-butenoic acids, $C_4H_6O_2$.

α-crotonic acid, trans-crotonic acid. Colourless needles or large plates; m.p. 72°C, b.p. 180°C. Prepared by the oxidation of crotonaldehyde or by heating the substance formed by the action of ethanol on malonic

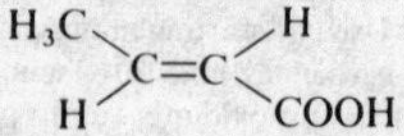

acid. Reduced by Zn and sulphuric acid to butanoic acid. Oxidized by potassium permanganate to dihydroxybutanoic acid. Reacts with chlorine to give 2,3-dichlorobutanoic acid, and with hydrogen bromide to give 3-bromobutanoic acid. Used in resins, surface coatings, plastic resins and pharmaceuticals.

Isocrotonic acid, β-crotonic acid, cis-crotonic acid. Colourless needles; m.p. 14°C, b.p. 169°C. Prepared by distilling β-hydroxyglutaric acid under reduced pressure. Converted to α-crotonic acid by heating at 180°C, or by the action of bromine and sunlight on an aqueous solution.

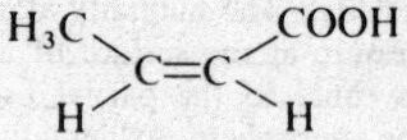

crotonyl The trivial name given to the 2-butenoyl group, $-COCH{:}CHCH_3$.

crotyl The trivial name given to the 2-butenyl group, $-CH_2CH{:}CHCH_3$.

crotyl alcohol, 2-buten-1-ol, C_4H_8O, $CH_3CH{:}CHCH_2OH$. Colourless liquid, b.p. 118°C. Prepared by the reduction of crotonaldehyde*.

crown ethers Macrocyclic polyethers containing repeating $(O-CH_2{\cdot}CH_2)_n$ units. The

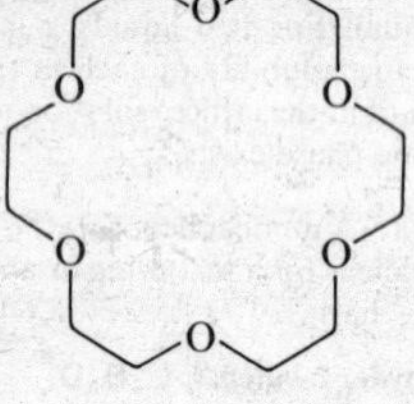

compound depicted would be called 18 – crown – 6 (18 – the total number of atoms in the ring, crown is the class name and 6 is the total number of hetero atoms in the ring), or 1,4,7,10,13,16-hexaoxacyclooctadecane. Simplest examples are prepared by the cyclic oligomerization of ethylene oxide. They act as complexing agents which solubilize alkali metal ions in non-polar solvents, complex alkaline earth cations, transition metal cations and ammonium cations, e.g. 12 – crown – 4 is specific for the lithium cation. Used in phase-transfer chemistry*.

crude oil A naturally occurring mixture consisting largely of hydrocarbons along with sulphur, nitrogen and oxygen derivatives. Crude oil is normally removed from reservoir rock in liquid form but is often associated with quantities of gas, water or solids. The properties of crude oils vary widely from light-coloured, mobile fluids to dark, viscous materials.

Crum Brown's rule A guide to substitution in benzene derivatives. This rule states that a substance C_6H_5A yields the *meta* disubstituted product if the compound HA can be oxidized directly to HOA; otherwise a mixture of the *o*- and *p*-compounds will be obtained. Not universally applicable. *See* Hammick and Illingworth's rules.

crushing and grinding Operations relating to particle size reduction. The term *crushing* is used where compressive forces are mainly employed, while *grinding* refers more to reduction by attrition and shearing actions.

cryohydric point An alternative name for the eutectic point*, usually used when one of the components is water.

cryolite, Na_3AlF_6. Mineral Na_3AlF_6, occurs native in Greenland. Made artifically from sodium aluminate, NaF, and $NaHCO_3$. Used as the electrolyte in Al manufacture and for white glass, glazes and enamels. The cryolite structure is adopted by many compounds of stoichiometry A_3BX_6. It is closely related to the ReO_3 and perovskite structures.

cryoscopic constant An alternative name for the freezing point depression constant.

cryoscopy The determination of molecular weights by the freezing-point depression method.

cryptates Polyheteroatom bicyclic macro-

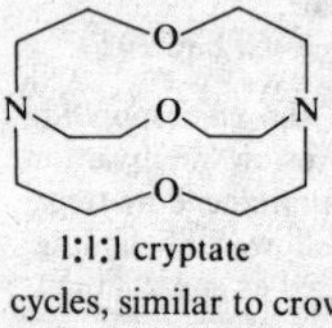

1:1:1 cryptate

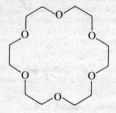

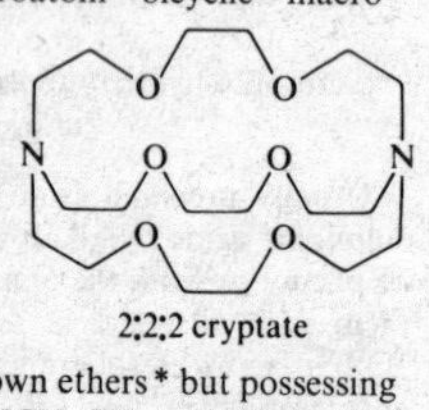

2:2:2 cryptate

cycles, similar to crown ethers* but possessing an additional $(-XCH_2CH_2)_n$ bridge. The chemical applications of these substances are closely related to crown ethers, the [2:2:2] compound specifically complexing K^+.

crystal A discrete solid particle bounded by definite faces intersecting at definite angles, and showing certain symmetry characteristics. The external form of a crystal is due to the ordered arrangement of atoms or molecules or ions within the bulk of the solid.

crystal drawing Crystals are drawn in a conventional projection and in a conventional position. For the cubic system the x axis does not point to the eye of the observer, but is rotated clockwise through an angle of 18° 26′, the tangent of which is $\frac{1}{3}$; the y axis is not drawn at right angles to the vertical c axis because the eye is elevated at an angle of 9° 28′, the tangent of which is $\frac{1}{6}$. The axes are therefore drawn as shown.

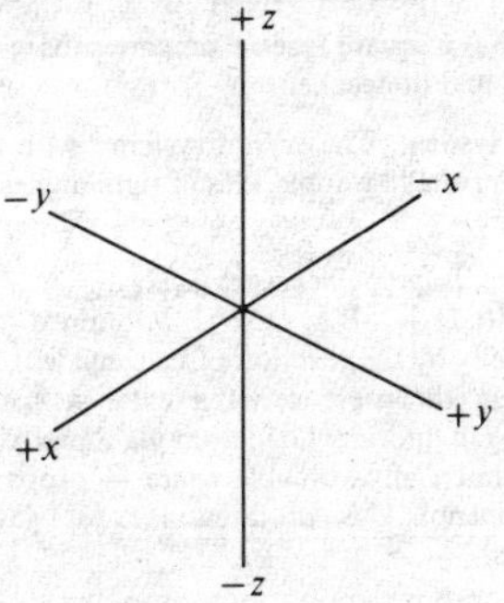

crystal field theory An approach to transition metal compounds which explains the u.v. and visible spectra (electronic) and magnetic properties in terms of a central ion surrounded by electron-rich ligands which split the d orbitals into a pattern determined by the geometry of the co-ordination, the distance of approach of the ligands and the tendency to covalent bonding.

crystal habit The description of the relative development of different faces and types of face on crystals which are of the same class.

crystalline state The form of the solid state with an ordered arrangement of atoms, etc. in the solid. The pattern is characteristic of the solid and generally gives rise to characteristic bounding faces of the crystal.

crystallization The removal of a solid from solution by increasing its concentration above the saturation point in such a manner that the excess solid separates out in the form of crystals. Successive crystallization or fractional crystallization (the successive removal of the least soluble component) may be used in separation and purification.

crystallizers Equipment for producing crystals from solutions of crystallizable materials.

crystallographic axes The (arbitrary) axes defining the repeat unit (unit cell) in a crystal. *See* crystal structure.

crystal nucleus A minute crystal which serves as a centre of formation for larger crystals which would not otherwise be formed. E.g. crystallization of a supersaturated solution may be induced by formation of a crystal nucleus through mechanical shock, inoculation by dust, etc.

crystals, liquid *See* liquid crystals.

crystal structure The arrangement of atoms and molecules in a crystal, as revealed by X-ray and electron diffraction methods.

A characteristic of a crystal is that the atoms, molecules or ions are arranged in a regular way in three dimensions (a lattice). It follows that if we start from an arbitrary origin in the crystal and proceed in any direction we shall arrive, after travelling a certain distance, at a point with the same environment as the point of origin. We may perform this process in two other directions and find the corresponding 'repeat distances'. These three axes of reference, x, y and z, and the three repeat distances, a, b and c, define a parallelopiped, defined by the symmetry elements but generally including the smallest volume, which contains a representative portion of the crystal structure, and which if indefinitely repeated in those directions will reproduce the crystal. A characteristic parallelopiped is termed the '*unit cell*', and the repeat distances or *cell dimensions* (the sides of the unit cell) are written a, b and c, corresponding to the x, y and z axes. The angles between the axes taken in pairs, are called α, β and γ (between yz, zx and xy respectively). The choice of axes in the crystal is not completely arbitrary, except in the triclinic system, but is determined by the positions of the symmetry elements. The shape of the unit cell, and hence the constants required to define it, depends on the symmetry of the crystal, and the following table summarizes the data required in the various systems:

Triclinic a, b, c, α, β, γ (no planes or axes of symmetry).
Monoclinic a, b, c, β, $(\alpha=\gamma=90°)$ – one 2-fold axis and/or one plane of symmetry.
Orthorhombic a, b, c, $(\alpha=\beta=\gamma=90°)$ – one 2-fold axis at the intersection of two planes or perpendicular to two other 2-fold axes.
Trigonal a $(=b)$ c, $(\gamma=120°$, $\alpha=\beta=90°)$ – one 3-fold axis perpendicular to other axes of symmetry.
Hexagonal a $(=b)$ c, $(\gamma=120°$, $\alpha=\beta=90°)$ – one 6-fold axis perpendicular to other axes of symmetry.

Tetragonal a (=b), c, ($\alpha=\beta=\gamma=90°$) – one 4-fold axis perpendicular to other axes of symmetry.

Cubic a (=b, =c), ($\alpha=\beta=\gamma=90°$) – four 3-fold axes at the tetrahedral angles to one another.

The crystal structure determines not only the arrangement of atoms in the lattice but also the external form of the crystal.

crystal symmetry A term describing the regularities in the position and arrangement of faces and edges of a crystal and also of the atoms within the crystal. Such regularities are defined in terms of planes of symmetry, axes of symmetry and centres of symmetry. A plane of symmetry divides the crystal into two identical halves, each of which is the mirror image of the other. An axis of symmetry is an imaginary line drawn through the crystal in such a way that, on rotating the crystal about this line as axis, the crystal presents an identical aspect to the observer more than once in a complete rotation. If the same aspect is presented four times (as would be the case for a cube rotated about a line through the centres of opposite faces) the axis is called an axis of four-fold symmetry. Similarly, there are axes of two-, three- and six-fold symmetry. A crystal has a centre of symmetry if like faces are arranged in matched pairs in corresponding positions on opposite sides of a point. There are a limited number of possible arrangements of planes, centres and axes of symmetry, which lead to a classification of crystals according to crystal systems*. In order to describe the external form of crystals three crystallographic axes are chosen. One face, making intercepts on all the axes, is taken as the standard or parametral face to which all other faces are referred. The description of crystal symmetry given above refers to the macroscopic crystal. The symmetry elements are also present in the unit cell, where they act on the atoms in the cell.

crystal systems Crystals are placed in one of the seven crystal systems: cubic, tetragonal, hexagonal, trigonal, orthorhombic, monoclinic and triclinic; according to their symmetry. For the characteristics of each of these systems *see* crystal structure and the separate headings (e.g. cubic system).

crystal violet *See* methyl violet.

Cs Caesium.

CS gas *See* malononitrile.

CTAB *See* cetyltrimethylammonium bromide.

Cu Copper.

cubic close-packing, ccp *See* close-packed structure.

cubic co-ordination Co-ordination by eight ligands at the corners of a cube as in CsCl. A

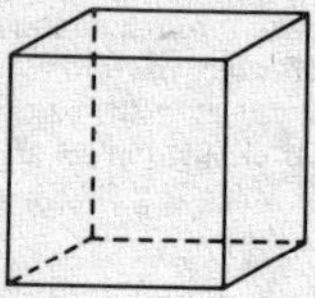

cube has 6 square faces. Compare square-antiprism and dodecahedron.

cubic system The crystal system* with three equal crystallographic axes at right angles, e.g. NaCl.

cumene, isopropylbenzene, C_9H_{12}, $PhCH(CH_3)_2$. B.p. 152°C. Obtained commercially by the reaction of benzene with propene in the presence of a catalyst such as hydrogen fluoride. Oxidation via a peroxide is a commercially valuable route to propanone and phenol. U.S. production 1978 1 480 000 tonnes.

cumyl-α-hydroperoxide, $C_9H_{12}O_2$, $PhC(CH_3)_2OOH$. A commercially important intermediate obtained by passing air through liquid cumene in the presence of a trace of a basic catalyst. On treatment with acids it gives phenol and propanone.

cupferron, ammonium N-nitrosophenylhydroxylamine, $[NH_4][ON(NO)C_6H_5]$. A reagent originally suggested for use in the detection of Cu but now used for the separation of Fe/Ti and Zr which it precipitates from acid solutions. Cupferron is a brownish-white crystalline substance, soluble in water.

cuprammonium, $[Cu(NH_3)_4]^{2+}$. The deep blue copper(II) ammine (excess aqueous ammonia to a copper(II) salt). The solution dissolves cellulose and is used in one process of rayon manufacture.

cuprates *See* copper oxides.

cuprene *See* acetylene black.

cupric Copper(II) species.

cuprite, red copper ore, Cu_2O. Bright red Cu ore. The cuprite structure has linear two-fold co-ordination by oxygen.

cupron *See* benzoin.

cuprous Copper(I) species.

curie The standard unit of radioactivity. 1 Ci = $3{\cdot}6 \times 10^{10}$ disintegrations per second; the same number of disintegrations as produced by 1 g of radium.

Curie–Weiss law The molar magnetic susceptibility of a substance may be represented as:

$$\chi = C/T,$$

where C is the Curie constant, and χ contains both diamagnetic and paramagnetic contributions. Not all substances conform to this equation; a more general relationship, the Curie–Weiss law, states:

$$\chi = \frac{C}{T-\theta}.$$

θ is the Weiss constant and is characteristic of the particular substance being considered.

curium, Cm. At.no. 96, m.p. 1340°C, b.p. ~3540°C, ^{242}Cm (162 days) and ^{244}Cm (18 years) are formed by multiple neutron irradiation of ^{239}Pu; ^{248}Cm ($4{\cdot}7 \times 10^5$ years) is formed by decay of ^{252}Cf, the compounds are purified by ion exchange chromatography. The metal is formed by reduction of CmF_3 with Ba, it has the double-hexagonal cp structure. The metal is silvery in colour and is rapidly oxidized. ^{244}Cm on neutron irradiation is a potential source of higher actinides.

curium compounds These are mainly in the +3 state and include all of the trihalides.

$$Cm^{4+} \xrightarrow{>2{\cdot}8\text{volt}} Cm^{3+}$$

The +4 state is stabilized in aqueous solution by fluoride ions. CmO_2 and CmF_4 are formed by strong oxidation or the action of fluorine.

Curtius transformation An alternative to the Hofmann transformation* for obtaining an amine from an ester via the hydrazide, azide and isocyanate. Thus ethyl ethanoate is converted into methylamine by the following series of reactions:

$$CH_3COOC_2H_5 + NH_2NH_2 \longrightarrow CH_3CONH{\cdot}NH_2$$
in hot alcohol

$$CH_3CONH{\cdot}NH_2 + HNO_2 \longrightarrow CH_3CO{\cdot}N_3$$
in aqueous solution at 0°C

$$CH_3CO{\cdot}N_3 \longrightarrow CH_3{\cdot}NCO + N_2$$
on warming

$$CH_3{\cdot}NCO + H_2O \longrightarrow CH_3NH_2 + CO_2$$
on boiling with dilute acid

Alternatively, the azide may be boiled with alcohol to convert it into the urethane, and the latter then hydrolysed with dilute acid. In general, the Hofmann method is rather more convenient, but the Curtius method must be employed when the compound in question is attacked by alkalis or easily brominated.

cutback bitumen A form of bituminous road binder produced by 'cutting-back' the viscosity of medium-hard bitumen with volatile solvents.

cutting fluids Fluids used as coolants and lubricants in metal cutting operations. They may be low-viscosity oils containing emulsifiers which form stable oil-in-water emulsions on dilution. Alternatively aqueous mixtures based on sodium nitrite, triethanolamine sebacate, polyglycols, etc. may be used. For heavier work straight lubricating oils are often used.

cyamelide *See* cyanic acid.

cyanamide, carbamic acid nitrile, carbodiimide, H_2NCN. Colourless, deliquescent solid; m.p. 41°C. Prepared HgO plus thiourea, CO_2 over heated $NaNH_2$, acid plus CaNCN. Normally obtained as the dimer, $H_2NC(=NH)NHCN$ or $H_2NC(=NH)N=C=NH$ (*see* carbodiimides). Acts as a weak acid and forms salts (*see* e.g., calcium cyanamide) and as a base with strong acids. Gives urea with acids, thiourea with H_2S, guanidine with NH_3. Polymerizes to melamine, $(H_2NCN)_3$.

cyanates Salts of cyanic acid, HNCO. KNCO formed from KCN plus PbO. NH_4NCO (KNCO plus NH_4Cl) isomerizes on heating to urea. Some cyanate complexes and covalent compounds, e.g. $P(NCO)_3$ (PCl_3 plus AgNCO), are mainly *N*-bonded.

cyanic acid, HNCO. Mobile volatile liquid formed by distillation of cyanuric acid*. A weak acid, polymerizes to a solid polymer, cyamelide, above 0°C. Hydrolysed to NH_3, H_2O, CO_2 in aqueous solution. Forms cyanates* and organic isocyanates, RNCO, and cyanates, NCOR.

cyanides The salts of hydrogen cyanide, HCN. Covalent derivatives RCN are generally termed cyanides or nitriles; derivatives RNC are isocyanides, isonitriles. Cyanides are formed from HCN, by metathesis (e.g. $AgNO_3$ plus KCN precipitates AgCN; AgCN plus ECl gives ECN), or by specific methods (*see* potassium cyanide, sodium cyanide). Cyanides are slowly hydrolysed (to CO_3^{2-} and NH_3), have solubilities similar to those of chlorides (pseudohalides) and are extremely poisonous. Many complex cyanides (cyanometallates) are known, *see*, e.g., cyanoferrates. Low-oxidation state compounds are particularly stable.

cyanine dyes Polymethine dyes containing $N-C(=C-C)_n=N$ and related systems. Not very fast to light and hence little used as dyestuffs but very powerful spectral sensitizers widely used in colour photography.

cyanoacetic ester *See* ethyl cyanoethanoate.

2-cyanoacrylates, $CH_2=C(CN)CO_2R$. Compounds formed by condensation of alkyl cyanoethanoates with methanal to give a polymer which can be depolymerized to the monomer. The methyl ester, b.p. 48–49°C at 2·5 mm is commercially available. Readily polymerized (through the vinyl group) by free radical or anionic polymerization and used as an adhesive by polymerization *in situ*.

2-cyanobenzamide, $C_8H_6N_2O$. Colourless needles; m.p. 173°C. On heating to its m.p. it is converted into imidophthalimide, m.p. 203°C. Readily hydrolysed to phthalic acid. Prepared by the partial dehydration of phthalamide with ethanoic anhydride.

It has a limited use in the preparation of the phthalocyanine pigments into which it is readily converted on heating with metallic salts.

cyanocobalamine *See* vitamin B_{12}.

cyanoethanoic acid, cyanoacetic acid, malonic acid mononitrile, $NC{\cdot}CH_2COOH$. Crystallizes in large prisms which liquefy on exposure to atmospheric moisture; m.p. 66°C. Prepared by sodium cyanide on sodium chloroethanoate followed by acidification with sulphuric acid. Decomposes at 160°C into carbon dioxide and acetonitrile. Concentrated acids or alkalis convert it to malonic acid. Condenses with aldehydes to give unsaturated α-cyano acids or unsaturated nitriles. The ethyl ester is used in organic syntheses.

cyanoethylation A particular example of a Michael reaction*. Some nucleophiles, especially carbanion species, readily add across the double bond of acrylonitrile, forming a cyanoethyl derivative. E.g. propanone, in the presence of a base, can undergo cyanoethylation to give mono-, di- or tricyanoethylation products.

$$CH_3COCH_3 \longrightarrow CH_3COCH_2CH_2CH_2CN \longrightarrow CH_3COCH(CH_2CH_2CN)_2 \longrightarrow CH_3COC(CH_2CH_2CN)_3$$

The base used is most often benzyltrimethylammonium hydroxide, but aqueous or ethanolic alkali is suitable. In addition to compounds with active methylene groups, primary and secondary amines, alcohols and phenols can undergo cyanoethylation. *See* propenitrile.

cyanoferrates
Hexacyanoferrates(III), $[Fe(CN)_6]^{3-}$, *hexacyanoferrates(II)*, $[Fe(CN)_6]^{4-}$. Formed by addition of alkali cyanide to a solution of an Fe(III) or Fe(II) salt. Hexacyanoferrates(III) are reddish in colour, $K_3Fe(CN)_6$ is generally obtained by oxidizing the $[Fe(CN)_6]^{4-}$ ion with Cl_2. With Fe^{2+} it gives a deep blue precipitate of Prussian blue. Heavy metal derivatives are generally insoluble. In alkaline solution $[Fe(CN)_6]^{3-}$ has oxidizing properties (e.g. can be used to titrate Ce^{3+} in the presence of K_2CO_3). Hexacyanoferrates(II) are yellow. The heavy metal salts are insoluble. Prussian blue and Turnbull's blue are $Fe_4^{III}[Fe^{II}(CN)_6]_3$, soluble Prussian blue is $KFe^{III}[Fe^{II}(CN)_6]$, Berlin green is $Fe^{III}[Fe(CN)_6]$, all containing octahedral $[M(CN)_6]^{n-}$ ions. Other complex cyanides have very similar structures. In the pigment industry these compounds are called iron blues or cyanide blues; with slightly different shades made under slightly different conditions. Substituted cyanoferrates containing anions such as $[Fe(CN)_5NO]^{2-}$ (nitroprusside), $[Fe(CN)_5NO_2]^{2-}$ are known.

cyanogen, C_2N_2, NCCN. Colourless, inflammable gas (m.p. −28°C, b.p. −21°C) with characteristic smell of bitter almonds. Very poisonous. Formed from HCN by oxidation with Cl_2 over a catalyst or NO_2, from Cu^{2+} plus CN^-, or $Hg(CN)_2$ plus $HgCl_2$. Paracyanogen, $(CN)_x$, is formed in some of these reactions. Behaves as a pseudohalogen and adds oxidatively to e.g. Pd(0) and gives CN^- and NCO^- with base.

cyanogen halides, XCN.
Cyanogen fluoride, FCN. Colourless gas (b.p. −46°C) prepared by pyrolysis of cyanuric fluoride. Polymerizes to $(FCN)_3$, cyanuric fluoride, at room temperature.

Cyanogen chloride, ClCN. Colourless liquid, m.p. −7°C, b.p. 13°C (aqueous CN^- plus Cl_2). Linear molecule, polymerizes to cyanuric chloride $(ClCN)_3$. Extremely poisonous.

cyanoguanidine, dicyandiamide, $C_2H_4N_4$, $(H_2N)_2C=N-CN$. Colourless crystalline solid; m.p. 209°C, soluble in water and alcohols. Prepared commercially by the dimerization of cyanamide in the presence of bases, and used in the manufacture of plastics, but mainly for the preparation of melamine, benzoguanine and diallylmelamine – all used in the formation of amino resins.

cyanohydrins Organic compounds containing the C(OH)CN group. They are formed by the action of hydrogen cyanide on aldehydes or ketones, or by the addition of sodium cyanide to a cooled suspension of the hydrogen sulphite compound of an aldehyde or ketone. Alkalis give a cyanide and the original compound; mineral acids convert them to α-hydroxycarboxylic acids.

cyanophoric glycosides Those glycosides containing hydrocyanic acid which is split off on hydrolysis; they are of frequent occurrence in plants. Amygdalin is the best known. Several glycosides of mandelonitrile are also known.

cyanuric acid, 2,4,6-triazinetriol, tricyanic acid, trihydroxycyanidine, $C_3H_3N_3O_3$. Stable

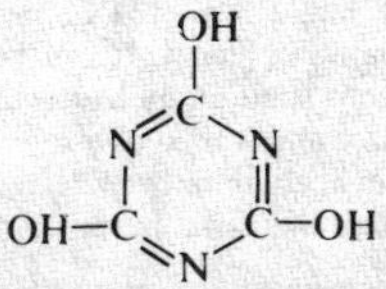

solid sparingly soluble in water. Prepared by melting urea or by hydrolysis of cyanuric chloride. *N*-esters are known.

cyanuric chloride, trichloro-*s*-triazene $C_3N_3Cl_3$. M.p. 154°C, b.p. 190°C. Cyclic derivative prepared by polymerization of ClCN

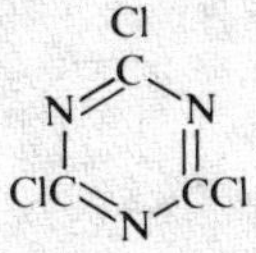

or from Cl_2, HCN and HCl. Used in the manufacture of dyestuffs, pharmaceuticals, triazene herbicides, plastics, explosives, bleaches and disinfectants. Gives cyanuric fluoride, $C_3N_3F_3$, with NaF.

cyclamate sodium, $C_6H_{11}NHSO_3Na$. Sodium salt of *N*-cyclohexylsulphamic acid, itself obtained by sulphonation of cyclohexylamine.

cyclenes Cyclic hydrocarbons containing one or more double bonds.

cycle oil, cycle stock In petroleum catalytic cracking processes the middle distillate fraction produced is recycled to the process. It may also be marketed as 'cycle oil' as a blending component of diesel oil or heating oil.

cyclic Compounds which contain a closed ring system, such as the cyclohexane ring, as opposed to the open chain aliphatic compounds. Cyclic compounds can be subdivided into homocyclic compounds, in which the ring is composed only of carbon atoms, and heterocyclic compounds, in which the ring is composed of atoms of more than one kind.

cyclic AMP, adenosine 3′,5′-monophosphate It now appears that many hormones (e.g. glucagon and adrenaline) in both animals and plants exert their effects by, as a first step, decreasing or increasing cyclic AMP within the cell. This may possibly occur by modification of the activity of the enzyme AMP cyclase which generates cyclic AMP from ATP.

cyclic process A system which has undergone a series of changes and has returned to its original state and thus has completed a cycle. The whole series of changes is a cyclic process. *See* e.g. Born–Haber cycle.

cyclic voltammetry Analytical method by which a species is successively oxidized and reduced by application of a varying potential. The current is plotted against voltage.

cyclized rubbers Modified rubbers containing cyclic systems produced by heating rubber with catalysts ($SnCl_4$, H_2SnCl_6). They are hard, thermoplastic and non-elastic, used in corrosion-resistant paints and adhesives, particularly for rubber–metal joins.

cyclitols The trivial name given to the mixed isomers of 1,2,3,4,5,6-hexahydroxycyclohexane.

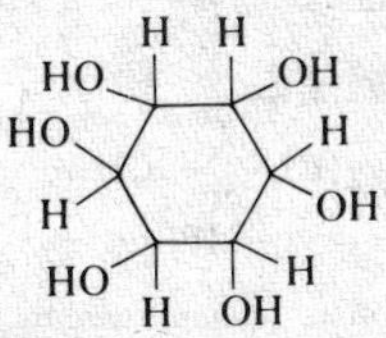

cyclobarbitone, $C_{12}H_{12}N_2O_3$. M.p. 173°C.

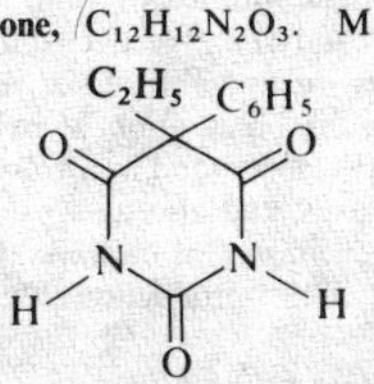

Prepared by condensing urea with ethyl ethyl cyanoethylcyclohex-1-enyl ethanoate and hydrolysing the product. It is a short-acting barbiturate*, used as a hypnotic.

cyclobutadiene, C_4H_4. An alkenic hydrocarbon of high reactivity and very short lifetime (<5 seconds) in the free state. It can be generated by degradation of a cyclobutadiene-metal complex, e.g. $C_4H_4Fe(CO)_3$ with Ce^{4+}, and its chemistry studied *in situ*, e.g. the addition of alkynes under Diels–Alder conditions

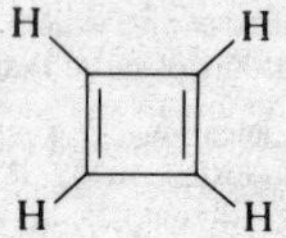

to give 'Dewar' benzene* derivatives. C_4H_4 behaves like a conjugated diene and an alkenic diradical.

cyclobutane, tetramethylene, C_4H_8. Colourless gas which burns with a luminous flame.

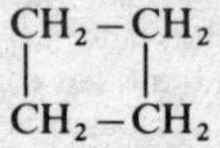

Insoluble in water, soluble in organic solvents; b.p. −15°C. Prepared by treating 1,4-dibromobutane with metallic sodium. Reduced to *n*-butane by hydrogen at 200°C in presence of nickel catalysts.

1,3,5-cycloheptatriene, tropilidene, C_7H_8. Prepared by heating norbornadiene to 450°C (also from atropine and cocaine). It is a liquid, b.p. 116–118°C, but yields solid salts of the tropylium cation by hydride abstraction, and tropolone by alkaline permanganate oxidation. Forms many metal complexes.

cyclohexane, C_6H_{12}. Colourless liquid, m.p.

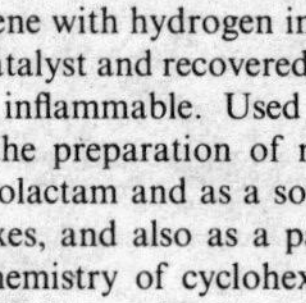

6·5°C, b.p. 81°C. Manufactured by the reduction of benzene with hydrogen in the presence of a nickel catalyst and recovered from natural gases. It is inflammable. Used as an intermediate in the preparation of nylon [6] and [66] via caprolactam and as a solvent for oils, fats and waxes, and also as a paint remover. For stereochemistry of cyclohexane *see* conformation. U.S. production 1980 1 megatonne.

cyclohexanol, hexalin, hexahydrophenol, $C_6H_{11}(OH)$, $C_6H_{12}O$. Colourless liquid; m.p. 24°C, b.p. 161°C. Manufactured by heating phenol with hydrogen under pressure in the presence of suitable catalysts. Oxidized to adipic acid (main use as intermediate for nylon production); dehydrogenated to cyclohexanone.

Dehydrated (e.g. $AlCl_3$) to cyclohexene. Used in the manufacture of celluloid, esters (plasticizers), detergents and printing inks.

cyclohexanone, 'pimelic ketone', $C_6H_{10}O$.

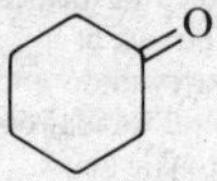

Colourless liquid with a strong peppermint-like odour; b.p. 155°C. Manufactured by passing cyclohexanol vapour over a heated copper catalyst. Volatile in steam. Oxidized to adipic acid. Used in the manufacture of caprolactam, Nylon, adipic acid, nitrocellulose lacquers, celluloid, artificial leather and printing inks.

cyclohexylamine, $C_6H_{11}NH_2$, $C_6H_{13}N$. Colourless liquid, b.p. 134°C, miscible with water, volatile in steam. It can be extracted from aqueous solution by hydrocarbons. It is a strong base and is used as a solvent for dyestuffs, as an acid inhibitor in degreasing baths and for other industrial purposes.

1,5-cyclo-octadiene, C_8H_{12}. The *cis,cis*-isomer is obtained by a catalytic dimerization of butadiene using e.g. derivatives of nickel carbonyl. It is a colourless, mobile liquid with terpene-like odour, b.p. 151°C, m.p. −69°C. The *trans,trans*-isomer is also known.

cyclo-octatetraene, COT, C_8H_8. A golden-yellow liquid, m.p. −7°C, b.p. 142–143°C.

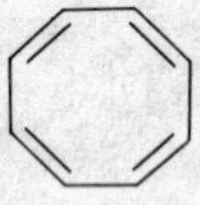

COT is prepared by the polymerization of ethyne at moderate temperature and pressure in the presence of nickel salts. The molecule is non-planar and behaves as a typical cyclic olefin, having no aromatic properties. It may be catalytically hydrogenated to cyclo-octene, but with Zn and dil. sulphuric acid gives 1,3,6-cyclooctatriene. It reacts with maleic anhydride to give an adduct, m.p. 166°C, derived from the isomeric structure bicyclo-4,2,0-octa-2,4,7-triene(I):

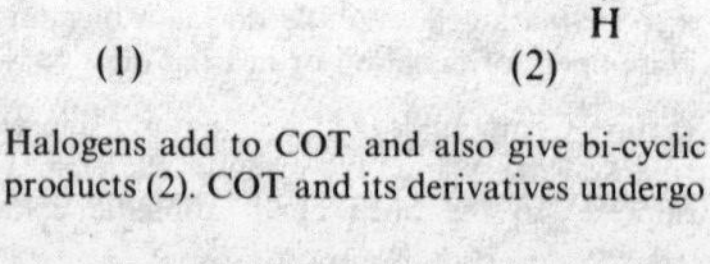

(1) (2)

Halogens add to COT and also give bi-cyclic products (2). COT and its derivatives undergo other interesting ring-contraction reactions. Forms metal complexes, e.g. $(C_8H_8)_2U$ derivatives, uranocene.

cyclo-octene, C_8H_{14}. Obtained in its most

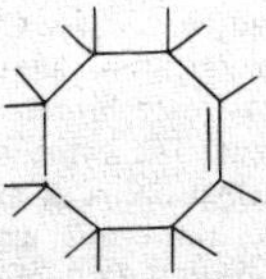

stable *cis* form by the partial hydrogenation of *cis,cis*-1,5-cyclo-octadiene. It is a colourless liquid with a terpene-like odour, b.p. 138°C, m.p. −124°C. Oxidation with potassium permanganate produces suberic acid.

The *trans* form, b.p. 143°C, m.p. −6°C is of stereochemical interest since it is the smallest carbocyclic ring capable of accommodating a *trans* double bond.

cyclopentadiene, C_5H_6. A colourless liquid

with a sweet, distinctive smell, b.p. 41·5–42·0°C. Insoluble in water, but soluble in all organic solvents. Obtained during the cracking of petroleum hydrocarbons, and from the distillates produced in the carbonization of coal. It polymerizes very readily on standing at room temperature to dicyclopentadiene (m.p. 32·5°C) and higher polymers, via an intermolecular Diels–Alder reaction. These are a more convenient form for handling cyclopentadiene which is easily regenerated by 'cracking' the oligomers. One hydrogen atom of the methylene group is acidic; sodium dissolves in ethereal cyclopentadiene solutions with the evolution of hydrogen, yielding $C_5H_5^-Na^+$ which is colourless to pink in the absence of air. The cyclopentadienyl anion $C_5H_5^-$ is an aromatic system. Fulvenes* are produced in basic media from ketones or aldehydes. Cyclopentadienylides* are produced with many metals. It undergoes typical Diels–Alder reactions; used in the preparation of plastics and insecticides. It is the parent hydrocarbon of the metallocenes and other metal cyclopentadienyls.

cyclopentadienylides Metal derivatives of cyclopentadiene containing C_5H_5 groups. Formed by use of Grignard reagents, NaC_5H_5, LiC_5H_5, TlC_5H_5 on metal halides, carbonyls, etc. The C_5H_5 group may be pentahapto $(h^5-C_5H_5)$ as in ferrocene $((h^5-C_5H_5)_2Fe$ or monohapto, e.g. $((h^5-C_5H_5)Fe(CO)_2$-$(h^1-C_5H_5))$. In pentahapto derivatives the C_5H_5 group is bonded symmetrically to the metal, monohapto derivatives show fluxional behaviour and each C−H may be equivalent on an n.m.r. time scale. Indene, fluorene and other cyclopentadiene derivatives form related complexes. The co-ordinated C_5H_5 group has considerable aromatic character and undergoes acylation under Friedel–Crafts conditions, etc. The C_5H_5 group can be considered to donate 5 electrons to the metal.

cyclophanes Benzene derivatives bridged between *p*- and/or *m*-positions with methylene groups; hence paracyclophane, etc. It is not usual to include the ortho-derivatives since alternative names based on conventional ring systems are possible. The value of the prefix [*m*] indicates the number of methylene groups in the ring whilst the number of prefixes [*m*, *n*,

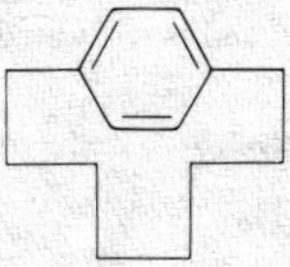

[8] paracyclophane

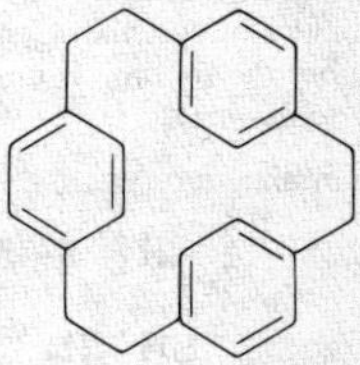

[2.2.2.]-paracyclophane

etc.] refers to the benzene rings in the structure, i.e. [8]-paracyclophane, [2,2,2]-paracyclophane. Systems based upon heterocyclic molecules are also known.

cyclophosphamide, $C_7H_{15},Cl_2N_2O_2P,H_2O$. Fine, white crystalline powder, m.p. 49·5–

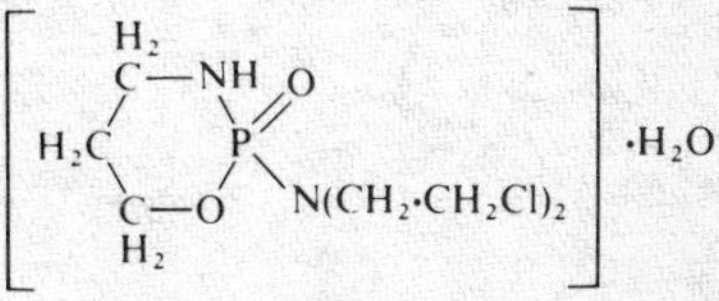

53°C. It is a cytotoxic agent used in the treatment of neoplastic disease, e.g. breast cancer.

cyclopropane, trimethylene, C_3H_6. Colourless gas with a sweetish odour, b.p. −34·5°C.

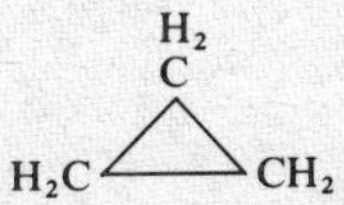

Prepared by treating 1,3-dibromopropane with zinc. It is a powerful gaseous anaesthetic, non-irritant and non-toxic to the liver and kidneys, but it is a respiratory depressant.

cymenes, isopropylmethylbenzenes, $C_{10}H_{14}$. p-Cymene is a colourless liquid, b.p. 177°C. Insoluble in water; miscible with organic solvents. Occurs in many essential oils, such as cumin, thyme and chenopodium. Obtained from the oil in the waste liquors of the sulphite wood pulp process. Prepared by heating camphor with $ZnCl_2$, or from oil of turpentine. Oxidized by chromic acid to terephthalic acid and by nitric acid to p-toluic acid. Used as a thinner for paints and in the manufacture of thymol. The other isomers are of little significance.

cysteine, α-**amino-**β-**thiolpropionic acid,** β-**mercaptoalanine,** $C_3H_7NO_2S$, $HSCH_2 \cdot CHNH_2 \cdot CO_2H$. Cysteine is a reduction product of cystine. It is the first step in the breakdown of cystine in the body, one molecule of cystine splitting to give two molecules of cysteine. Cysteine is soluble in water but the solution is unstable, and is reoxidized to cystine.

cystine, dicysteine, $C_6H_{12}N_2O_4S_2$. Reduced

$$
\begin{array}{ccc}
CH_2 - S - S - CH_2 \\
| & | \\
CH \cdot NH_2 & CH \cdot NH_2 \\
| & | \\
COOH & COOH
\end{array}
$$

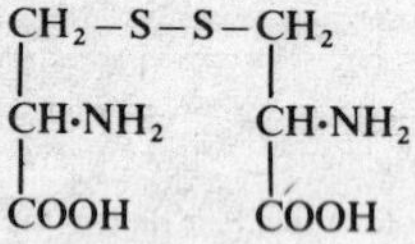

to cysteine. Cystine is abundant in the proteins of the skeletal and connective tissues of animals and in hair and wool, from which it is readily prepared.

cytidine See nucleosides.

cytochromes Widely distributed respiratory catalysts concerned in the electron transport chain of living cells. They are haemoproteins differing in their porphyrin groups. They do not combine with substrates but alternate between Fe^{2+} and Fe^{3+} states. Cytochrome C is present in the greatest amounts but cytochromes A_1, A_3 (cytochrome oxidase) and B are also present. The cytochrome system is reduced by dehydrogenases and cytochrome A_3, the last in the chain, uses atmospheric oxygen to convert Fe^{2+} to Fe^{3+}. The cytochrome system thus acts as an oxygen carrier system to the dehydrogenases. See also electron transport chain. Cytochrome P_{450} differs from other cytochromes in that it utilizes a reducing source and molecular oxygen to introduce a hydroxyl group into molecules, e.g. in steroid biosyntheses and drug metabolism.

cytosine, 2-oxy-4-aminopyrimidine, $C_4H_5N_3O$.

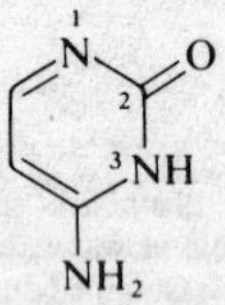

A hydrolysis product of nucleic acids.

cytotoxic agents Chemicals injurious to living cells. The term is applied particularly with respect to tumour formation. Many cytotoxic agents are bifunctional alkylating agents, e.g. mustard gas, nitrogen mustards.

D

D Deuterium.

2,4-D, 2,4-dichlorophenoxyacetic acid, $C_8H_6Cl_2O_3$. A selective growth regulator herbicide for use against broad-leaved plants. White solid; m.p. 138°C.

Dalton's law of partial pressures The total pressure (P) exerted by a mixture of gases is equal to the sum of the partial pressures (p) of the components of the gas mixture. The partial pressure is defined as the pressure the gas would exert if it was contained in the same volume as that occupied by the mixture.

Daniell cell A $Zn/Zn^{2+} || Cu^{2+}/Cu$ cell. The e.m.f. of the Daniell cell is 1·10 V and is virtually independent of temperature.

dansyl 5-Dimethylaminonaphthalene-1-sulphonyl derivatives used (e.g. chloride and fluoride) for active site investigations in serine proteinases.

2,4-DB, 4-(2,4-dichlorophenoxy)butyric acid, $C_{10}H_{10}Cl_2O_3$. Colourless crystals; m.p. 117–119°C. A herbicide of similar properties to 2,4-D, but with greater selectivity.

DCO Dehydrated castor oil*.

DDT, dichlorodiphenyltrichloroethane, 1,1-bis(4-chlorophenyl)-2,2,2-trichloroethane M.p. 108·5°C. Ordinary DDT contains about 15% of the 2,4′-isomer, and is prepared from chloral, chlorobenzene and sulphuric acid. It is non-phytotoxic to most plants. It is a powerful and persistent insecticide, used most effectively to control mosquitoes in countries where malaria is a problem. It is stored in the bodies of animals and birds.

d-d transition Electronic transition occurring between d orbitals which have been split by a crystal field. Generally occurs in the visible region but may occur in the u.v. or near i.r.

deactivating collision An intermolecular collision between an 'activated' and a normal molecule, which results in the deactivation of the former without it undergoing reaction. *See* deactivation.

deactivation The process whereby the chemical reactivity of a substance is diminished or entirely removed. For example, the catalytic activity of metals such as platinum or nickel for various hydrocarbon reactions may be considerably reduced or completely removed by the presence of trace amounts of catalyst poisons such as sulphur or mercury which cause deactivation of the catalyst.

Alternatively, in a photochemical reaction a molecule absorbs light. This results in it possessing more energy than an ordinary molecule and, consequently, more reactivity. However, collision of this activated molecule with another molecule may cause the former to lose its energy without reacting. The activated molecule is said to have undergone a deactivating collision and has itself become deactivated.

deactivators Additives used in gasoline to prevent the catalysis of gum formation by metals (e.g. Cu).

deaeration Removal of oxygen and other dissolved gases from solvents by physical or chemical means. Carried out on a large scale from boiler feed water with a view to reducing corrosion.

deasphalting A single-solvent precipitation process whereby petroleum residues are treated with liquid propane or similar light solvent in a counter-current column. The asphaltenes* are precipitated and the high quality lubricating oil stock (bright stock) dissolves in the propane and may be further refined by solvent extraction.

de Broglie equation In 1924 de Broglie suggested that the wavelength of electrons (λ) is given by the equation:

$$\lambda = (h/p) = (h/mv)$$

where h = Planck's constant, m = mass of electron, v = velocity and p = momentum of the electron. Wavelengths calculated from this equation are often called de Broglie wavelengths.

Debye–Hückel theory The activity coefficient of an electrolyte depends markedly upon concentration. In dilute solutions, due to the Coulombic forces of attraction and repulsion, the ions tend to surround themselves with an 'atmosphere' of oppositely charged ions. Debye and Hückel showed that it was possible to explain the abnormal activity coefficients at least for very dilute solutions of electrolytes.

For dilute solutions, the Debye–Hückel equation by calculations based on these Coulombic interactions is

$$\ln \gamma_i = \frac{-e^3 z_i^2}{(\varepsilon kT)^{\frac{3}{2}}} \sqrt{\frac{2\pi LI}{1000}}$$

where γ_i is activity coefficient of the ion i; z_i is the ionic charge; e is the electronic charge; ε, the dielectric constant of the solution; L, Avogadro's number and I is the ionic strength* of the solution.

decalin, decahydronaphthalene, $C_{10}H_{18}$. There are two stereoisomers of decalin. The *cis*

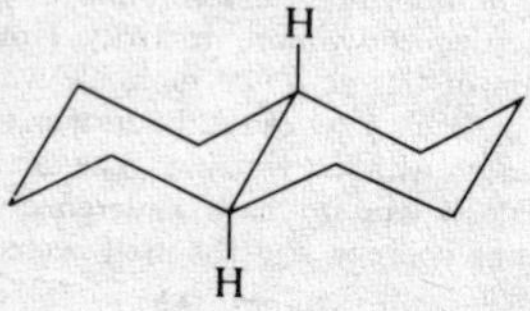

trans

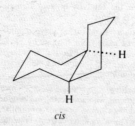

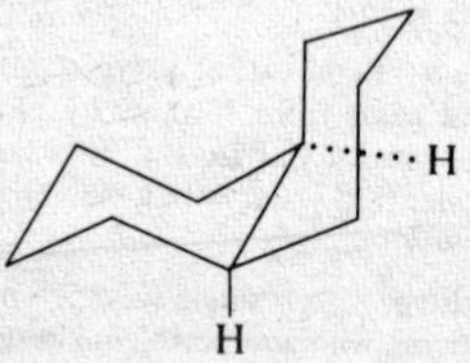

cis

form has m.p. −45°C, b.p. 198°C. The *trans* form has m.p. −32°C, b.p. 185°C. The *cis* form can be quantitatively converted into the *trans* form by $AlCl_3$. Commercial decalin is manufactured by the high-pressure hydrogenation of naphthalene at high temperatures, and contains 90% *cis* and 10% *trans* decalin. The two forms can be separated by fractional distillation. Decalin is used as a solvent.

decantation The removal of a liquid from a suspension or from an immiscible heavier liquid by pouring while leaving the latter in the original container.

decarboxylases *See* carboxylase.

decay constant Radioactive decay occurs according to the exponential rate law $N = N_0 e^{-\lambda t}$ where N_0 is the number of atoms present at time zero and N is the number of atoms present at time t. λ is the decay constant, which is equal to $0{\cdot}693/t_{\frac{1}{2}}$, $t_{\frac{1}{2}}$ being the half-life of the radioactive species, and has a characteristic value for a particular nuclide.

decoctions Pharmaceutical solutions made by boiling a drug with water and filtering.

decomposition point In a laboratory distillation the thermometer reading coinciding with the first indications of thermal decomposition of the liquid in the flask.

decomposition voltage The smallest voltage which will cause the electrolysis of an electrolyte. The decomposition voltage or potential depends both upon the nature of the electrolyte and of the electrodes.

decyl alcohols, decanols, $C_{10}H_{21}OH$. These are of considerable commercial importance, being used in the manufacture of plasticizers and detergents. They are manufactured by the oxo reaction *from nonenes, CO and H_2 with reduction of the resultant decanals to decyl alcohols. The composition of the product depends on the composition of the feed, the most common commercial grade being 'iso-decanol', a mixture of 10-carbon branched-chain primary alcohols with a narrow boiling range.

n-Decyl alcohol (1-decanol) is manufactured by the telomerization of ethene in the presence of aluminium triethyl to aluminium trialkyls which are oxidized to the alkoxides and hydrolysed by sulphuric acid to a mixture of straight-chain alcohols. From these *n*-decyl alcohol may be fractionated, though it is usual to market a mixture of alcohols. Pure *n*-decyl alcohol is a colourless liquid with a sweet fat-like odour, m.p. 7°C, b.p. 233°C. Insoluble in water, soluble in alcohol.

de-emulsification *See* demulsification, demulsibility.

defect structures Structures in which there are irregularities in the lattice. Defects may be of many types but include the non-alignment of crystallites, orientational disorder (e.g. of molecules or ions), vacant sites with the migrated atom at the surface (Schottky defect), vacant sites with an interstitial atom (Frenkel defects), non-stoicheiometry. Crystals exhibiting defects show anomalous physical properties, particularly density, colour, electrical conductivity. Applications of defect solids include catalysis, semi-conductors, luminescent materials.

degeneracy Having the same energy; particularly applied to wave-functions. In a transition element gaseous atom the five d orbitals are degenerate. In a crystal field the degeneracy is removed.

degenerate orbitals *See* degeneracy.

degree of hydrolysis The degree of hydrolysis

of a salt is defined as the fraction of the total salt hydrolysed by water. Thus, if in a solution of the salt AB, 90% of it is hydrolysed to yield the base AOH and the acid BH, the degree of hydrolysis is 0·90. It may be expressed also as a percentage, i.e. 90%.

degrees of freedom From the standpoint of the phase rule, the number of degrees of freedom which a system possesses is the number of variable factors, temperature, pressure and concentration of the components, which must be fixed in order to define completely the condition of the system. E.g. if the pressure and temperature of a given mass of gas are fixed, the volume of the gas must have a definite value. The system has, therefore, two degrees of freedom.

In statistical mechanics (e.g. the theory of specific heats of gases) a degree of freedom means an independent mode of absorbing energy by movement of atoms. Thus a monatomic gas has three translational degrees of freedom. Polyatomic molecules have in addition vibrational and rotational degrees of freedom.

dehumidification The removal of condensable water vapour from a vapour-gas mixture by condensation, absorption or adsorption. The term is most frequently applied to the removal of water vapour from air.

Dehumidification may be effected by cooling. Where small quantities of dry air are required dehumidification can be carried out using chemical absorbents, e.g. calcium chloride, phosphorus pentoxide, sulphuric acid, etc.

dehydrated castor oil, DCO Polyunsaturated carboxylic acid glyceride, typically a 9,11-octadecadienoic acid derivative, prepared by heating castor oil to 250–300°C in the presence of catalysts. Used to form sebacic and capryl derivatives (by cleavage at the C=C bond) and for paints and varnishes.

dehydro A prefix for an organic compound indicating the presence of less hydrogen. Often synonomous with anhydro-*.

dehydroacetic acid, 6-methylacetopyranone, $C_8H_8O_4$. Colourless needles; m.p. 109°C, b.p. 270°C. Prepared by boiling acetoacetic ester for several hours and then distilling off the unchanged ester. It is formed from keten by polymerization in the presence of $ZnBr_2$.

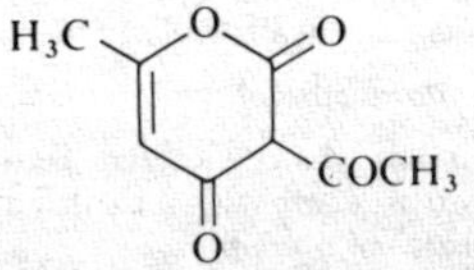

The parent acid is unstable and always reverts to the lactone form shown. It is reconverted to acetoacetic ester by boiling alcoholic potassium hydroxide. Reduced by hydrogen iodide to dimethylpyrone.

dehydrocyclization In petroleum cracking* or reforming* processes some of the alkanes may undergo dehydrocyclization yielding aromatics and hydrogen, e.g.:

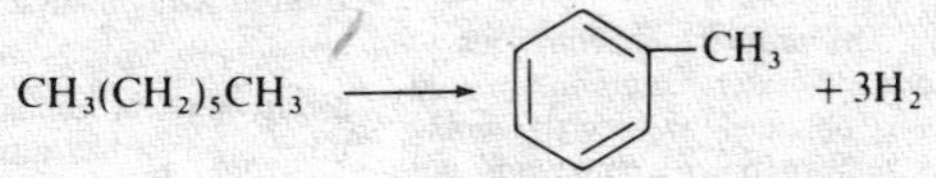

dehydrocyclization

dehydrogenases *See* enzymes.

dehydrogenation The process whereby the hydrogen content of a molecule is decreased and the degree of unsaturation is increased, e.g. cyclohexane ⟶ benzene. It is usually accomplished by catalysts such as platinum or palladium on charcoal or by heating the compound with selenium (which yields hydrogen selenide). The technique was frequently used in degradative studies to establish the basic skeleton of unknown natural products, e.g. Diels hydrocarbon from cholesterol.

deliquescence The property of having a water vapour pressure lower than that of the surrounding air so that water is absorbed (to give a solution).

deliquescent substance A material showing deliquescence.

delocalization A description of the normal situation for molecules or ions in which a bonding electron cannot be associated only with one atom but must be considered delocalized over the whole group. Applied particularly to aromatic molecules where in, e.g. benzene, the six π electrons are delocalized over the six carbon atoms. The extra stability of the molecule over a situation with three localized double bonds is called the delocalization energy.

delphinin The anthocyanin* pigment of the larkspur and many other plants. On hydrolysis it gives delphinidin, two molecules of glucose and *p*-hydroxybenzoic acid.

delphinine An alkaloid related to aconitine*.

Delrin Polymethanal.

delta bonding Lateral overlap between two orbitals (e.g. d orbitals) such that there are four regions of overlap.

delta value, δ A measure of shift in nuclear magnetic resonance*.

demasking *See* masking.

demeton-methyl *See* Metasystox; demeton-*S*-methyl.

demeton-*S*-methyl, *S*-(ethylthio)ethyldimethyl phosphorothiolate, $C_6H_{15}O_3PS_2$, $(CH_3O)_2P(O)SCH_2CH_2SCH_2CH_3$. A contact and systemic insecticide and acaricide.

demulsibility The property enabling oils to separate from oil–water emulsions. Good demulsibility is obtained by special refining processes to remove polar compounds not removed in normal refining.

demulsification Emulsions of oil and water formed in various operations may require special techniques to separate the two liquid phases. This process of demulsification can be effected by either physical or chemical means. The chemical methods depend on the opposing action of different emulsifiers on one another. Some emulsions can be broken by mechanical treatment while others can be broken electrically by a high voltage current. Some emulsions coagulate in the presence of electrolytes.

denaturants Substances added to dutiable articles to render them unfit for human consumption, e.g. denatured alcohol. *See* methylated spirits.

denaturation The irreversible change in solubility and other properties of proteins that occurs when a protein in solution is heated, shaken or treated with acid. Good examples are the precipitation of milk with vinegar (ethanoic acid) and the scrambling and beating of eggs. *See* proteins.

dendrite A tree-like crystal formation formed because crystal growth proceeds at different rates in certain directions from the initial crystallization point. Found particularly in alloys, and ice crystals. Many salts can be induced to crystallize as dendrites (*see* dendritic salt).

dendritic salt NaCl crystallized as dendrites.

It is less dense than normal rock salt, does not cake so easily, dissolves more quickly.

denitrification The return of fixed nitrogen, generally nitrates and ammonium salts, to the atmosphere generally as N_2 and N_2O.

denitrogenation The removal of nitrogen compounds, particularly from gas oil cracking feedstocks, to prevent the ammonia which would be formed neutralizing the acid catalysts used. Denitrogenation can be effected catalytically.

dense media separation Separation of solids of different density by means of a liquid of intermediate density which allows a 'float and sink' separation. Liquids of suitable density can be obtained by suspending finely-divided solids in water. Used for separation of mineral products such as coal, ores or aggregates.

densitometer An apparatus for measuring the intensities of e.g. lines on a photographic plate. Used in X-ray analysis and spectrometric analysis.

density-gradient column A graduated glass tube, filled with a mixture of solutions such that a density gradient is obtained throughout the length of the tube. Used for rapid determination of densities of small particles of solid.

D-2-deoxyribose, desoxyribose, $C_5H_{10}O_4$. The sugar isolated by hydrolysis of DNA. Colourless crystals, m.p. 91°C, soluble in water.

$$
\begin{array}{c}
\text{CHO} \\
| \\
\text{CH}_2 \\
| \\
\text{H---C---OH} \\
| \\
\text{H---C---OH} \\
| \\
\text{CH}_2\text{OH}
\end{array}
$$

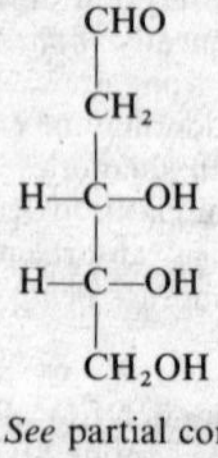

dephlegmator *See* partial condenser.

depilatories Materials used for hair removal. The most widely used is calcium thioglycollate in an alkaline medium at pH 12·3.

depot fat The reserve of fat in animals, deposited in adipose tissue and which usually undergoes metabolism slowly.

depression of freezing point When a solute is added to a pure solvent the freezing point of the solvent is lowered by an amount which is proportional to the number of solute particles in the solution. The depression caused by 1 mole of solute in 1 litre of solvent is called the molecular depression constant (K_f). K_f has a definite value for a particular solvent.

Solvent	*Molecular depression constant (°C/mole)*
water	1·86
ethanoic acid	3·90
benzene	4·90
cyclohexane	20·5

depsides Simple compounds with tannin-like properties formed by the condensation of the carboxylic group of one phenolcarboxylic acid with a hydroxyl group of a similar acid. Thus:

$$HO{\cdot}C_6H_4{\cdot}COOH + HO{\cdot}C_6H_4{\cdot}COOH \longrightarrow HO{\cdot}C_6H_4{\cdot}COO{\cdot}C_6H_4{\cdot}COOH$$

They are called didepsides, tridepsides, polydepsides, etc., depending on the number of phenol residues they contain. Obtained from lichens, present in tea.

dermatan sulphate *See* chondroitin.

derris Obtained from the root of the Far-Eastern shrub *Derris elliptica*, and originally used as a fish poison; contains rotenone. It is applied as a dust, and is used as an insecticide.

desalination *See* water.

desensitization A process used in photography to induce a loss of sensitivity in the silver halide emulsion over part of its spectral sensitivity range. Often used during development to allow subsequent processing to be carried out in a bright light but without affecting the latent image or inhibiting development.

desiccant A material used for drying. Some desiccants can react chemically with specimens being dried. Used in a desiccator. In order of effectiveness common desiccants are $CaCl_2 < CaO < NaOH < MgO < CaSO_4 < H_2SO_4 <$ silica gel $< Mg(ClO_4)_2 < P_2O_5$.

desmosterol, 24-dehydrocholesterol, $C_{27}H_{44}O$. A close biogenetic precursor of cholesterol. It is found at low levels in many animal tissues and was first isolated from chick embryo and from rat skin.

desmotropism Tautomerism.

destructive distillation The distillation of organic solids or liquids in which the substance being heated partly or wholly decomposes during distillation, leaving a solid or viscous liquid in the still. Volatile liquid products are collected after condensation. Typical destructive distillation processes involve the carbonization of coal, peat, wood or oil shale. *See also* carbonization.

desulphurization The removal of sulphur compounds from petroleum fractions using hydrogen in the presence of a catalyst. Also known as *hydrofining*, *hydrodesulphurization* and *hydrotreating*.

detergent oil Lubricating oil containing detergent or dispersant additives used in internal combustion engines in order to stabilize solid contaminants in a finely-divided state, and so prevent abrasion and accumulation of deposits. Additives used include alkaline earth metal salts of petroleum sulphonic acids, alkyl phenol sulphides, etc.

detergents Water-soluble, surface-active agents capable of wetting a variety of surfaces, removing greasy and oily deposits and retaining the dirt in suspension for ease of rinsing. While soap acts as a detergent, a large range of synthetic detergents, derived from petroleum, is now available. These detergents contain *hydrophilic* groups such as sulphate, sulphonate or polyether which confer water solubility. Long hydrocarbon chains – *hydrophobic* groups – enable them to dissolve oily materials.

Anionic detergents, probably the most important group, are typified by alkyl aryl sulphonates which ionize in water to produce a large anion. *Cationic* surface-active agents, which produce a large cation, are less used as detergents but find some application in the textile and dyeing industries and as adhesion and emulsifying agents. They are often quaternary ammonium compounds or derivatives of fatty acid amides. *Non-ionic* detergents generally consist of condensation products of alcohols or phenols with ethylene oxide. European production 1976 3·2 megatonnes.

detonating gas A mixture of H_2 and O_2 (molar ratio 2:1) produced by electrolysis of water and which explodes violently to reform H_2O on ignition.

detonation A type of explosion* typified by a much-increased reaction rate as well as a very high velocity percussion wave and a high pressure rise. Very high localized temperatures are produced.

This type of explosion is important in engines where detonation can vary from an incipient stage, often referred to as pinking, to a more severe condition known as knocking*.

detoxication, detoxification A term for the biochemical transformations which occur during the metabolism of foreign organic compounds by animals. The process does not always result in decreased toxicity.

deuterium, D. At.no. 1, at.wt. 2·01. An isotope of hydrogen which, because of the relatively large difference in masses, gives markedly different physical properties including rates of reactions to deutero compounds – bracketed figures below are for hydrogen analogues. D_2, b.p. −249·7 (−252·8°C); D_2O, b.p. 101·4 (100·0°C), m.p. 3·8 (0·0°C), ionic product at 25°C $1{\cdot}1 \times 10^{-15}$ ($1{\cdot}0 \times 10^{-14}$). Ordinary hydrogen compounds contain 1 part in 4500 D. D_2O (heavy water) is obtained by preferential evaporation of H_2O or preferential electrolysis of H_2O. Deuterium compounds are used as tracers in chemical and biological investigations; as solvents, etc., when protons are under investigation (e.g. n.m.r.), D_2O is used as a moderator in atomic reactors.

deuton, deuteron, ${}^{2}_{1}D^{+}$. The nucleus of the deuterium atom. Used in linear accelerators for bombarding other nuclei.

Devarda's alloy, 45% Al, 50% Cu, 5% Zn. Used to detect and analyse for nitrates.

developed colours *See* ingrain dyestuffs.

developers, photographic *See* aminophenols, developers, colour development. Solutions which cause reduction of exposed silver halide grains to silver.

development centres The areas of a silver halide crystal in which exposure to light has induced latent images and where, on development, reduction to metallic silver will start.

devitrification The crystallization of a glass (a supercooled liquid) which then becomes opaque.

'Dewar' benzene, bicyclo[2·2·0] hexadiene, C_6H_6. Colourless liquid with alkenic properties. Many substituted derivatives are known, the preferred method of preparation being the addition of an alkyne to a cyclobutadiene.

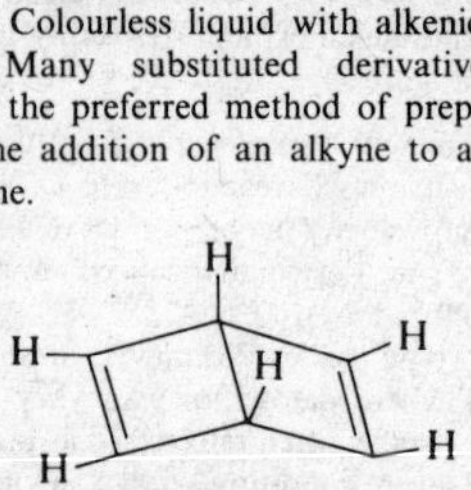

'Dewar' benzene is a valence isomer of benzene, to which it reverts on heating.

Heavily substituted valence bond isomers of benzene with prism-like structures ('prismanes') are now known.

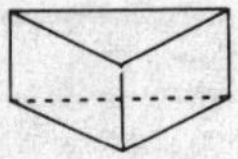

dewaxing Removal of wax from lubricating oil stock.

dexamphetamine The dextrorotatory isomer of amphetamine*.

dextran A mucilaginous polymer of glucose made by certain bacteria. It has mol.wt. between 30 000 and 250 000, and has both 1-4 and 1-6 linkages with branches about every five units, giving it considerable resistance to enzymic hydrolysis. Manufactured by the fermentation of sucrose. It can be used as a plasma substitute in blood transfusions, as an emulsifying and thickening agent, and as a stabilizer for ice creams. For cross-linked dextran *see* Sephadex.

The sodium salts of sulphuric acid esters of dextran are used as anticoagulants for the same purposes as heparin.

dextrins Intermediate products formed during the hydrolysis of starch to sugars. They are strongly dextrorotatory. Manufactured by the action of heat and hydrolytic agents on wet or dry starch and used as adhesives.

dextrose *See* glucose.

diacetin *See* acetins.

diacetone *See* diacetone alcohol.

diacetone alcohol, 2-hydroxy-2-methylpentan-4-one, $C_6H_{12}O_2$, $HOC(CH_3)_2CH_2COCH_3$. A colourless, odourless liquid; b.p. 166°C. Manufactured by treating propanone with lime, or sodium or barium hydroxides. It is readily decomposed by acids and strong alkalis. It is used as a lacquer solvent, particularly for brushing lacquers, as it evaporates slowly. It is also used as a dye-solvent in the printing trade.

diacetoneamine, $C_6H_{13}NO$, $Me_2C(NH_2)CH_2C(O)Me$. Colourless, strongly basic liquid; b.p. 25°C/0·2 mm. Obtained from mesityl oxide and ammonia or ammonia gas into boiling propanone.

diacetyl, biacetyl, $CH_3COCOCH_3$. A greenish-yellow liquid with a powerful odour; b.p. 88°C. It occurs in bay and other essential oils, and also in butter. It is formed in small amounts when tartaric acid, sugar and starch are charred by heat. Prepared from vinylacetylene and a solution of $HgSO_4$ in H_2SO_4 with decomposition of the product with HCl; oxidation of 2,3-butyleneglycol with oxygen at 270°C in the presence of a Cu catalyst; and the distillation of a solution of acetoin in 50% $FeCl_3$. It is volatile in steam. Reduced to 2,3-butyleneglycol by yeast; oxidized by hydrogen peroxide to ethanoic acid. Forms dimethylglyoxime with hydroxylamine. Hardening results from the addition of diacetyl to photographic gelatin.

diacetylenes, dialkynes Hydrocarbons containing two alkyne linkages in the molecule. They are mobile, colourless liquids with odours resembling that of ethyne. Their properties are similar to those of the acetylenes. Prepared by coupling copper or Grignard derivatives of monoacetylenes in the presence of $CuCl_2$. Diacetylene (buta-1,3-diyne) itself is conveniently obtained by treating 1,4-dichlorobut-2-yne with sodamide in liquid ammonia.

diagonal relationship A similarity (in solubility and thermal stability) between compounds of elements of the main groups related to one another diagonally (Li and Mg; Be and Al). Results from a similarity in size in combining forms.

dialin The trivial name given to each isomer (two) of dihydronaphthalene.

diallylmelamine, 2-diallylamine-4,6-diamino-*s*-triazine *See* cyanoguanidine.

dialysis A process used for the purification of colloidal sols. The comparatively large particles (1–100 pm) of the sol, in contrast to ions and molecules, cannot diffuse through a semipermeable membrane. By enclosing the impure sol within such a membrane (dialyser), which is supported in pure (preferably flowing) water, impurities of a crystalloid nature are gradually removed. The function of the kidneys in mammals is to dialyse the blood, removing impurities such as urea from the haemoglobin sol.

diamagnetism The property of all substances that causes repulsion of the substance from a magnetic field. Originates in the interaction of the electronic charge and the field and is an additive property of the atoms and groups present. Order of magnitude 10^{-3} that of paramagnetism so that the effect is swamped by the presence of unpaired electrons.

1,2-diaminoethane, ethylenediamine, $C_2H_8N_2$, $H_2NCH_2{\cdot}CH_2NH_2$. Colourless liquid which fumes in air and has a strong ammoniacal odour; m.p. 11°C, b.p. 116°C. Absorbs carbon dioxide and water from the air to give ethylenediamine carbamate. Manufactured by heating 1,2-dichloroethane with ammonia under pressure in the presence of CuCl. Forms a crystalline hydrochloride, and oil-soluble soaps with fatty acids; these are used as detergents and emulsifying agents. It is also used as a solvent for certain vat dyes and for casein, shellac and resin. Forms stable complexes with transition metals, in which it acts as a chelating agent. Used in textiles, paper, coatings, films and adhesives, in emulsifying agents (after condensation with carboxylic acids), rubber formulation.

1,6-diaminohexane, hexamethylene diamine, $H_2N{\cdot}(CH_2)_6{\cdot}NH_2$. Colourless solid when pure; m.p. 41°C, b.p. 204°C. Manufactured by the electrochemical combination of two molecules of acrylonitrile* to adiponitrile followed by catalytic reduction, or by a series of steps from cyclohexanone via adipic acid. Used in the production of Nylon [6, 6].

1,5-diaminopimelic acid, $C_7H_{14}N_2O_4$, $HO_2CC(H)(NH_2)(CH_2)_3C(H)(NH_2)CO_2H$. M.p. at least 305°C. The D, L and *meso* forms are all isolated from hydrolysates of bacterial proteins. It is an intermediate in the biosynthesis of lysine in many bacteria.

diamond A crystalline form of carbon, and one of the hardest and most infusible substances known; *d* 3·520. In a crystal of diamond every carbon atom is linked tetrahedrally to four others. The structure has C–C 1·54 Å, equal to that found in saturated hydrocarbons.

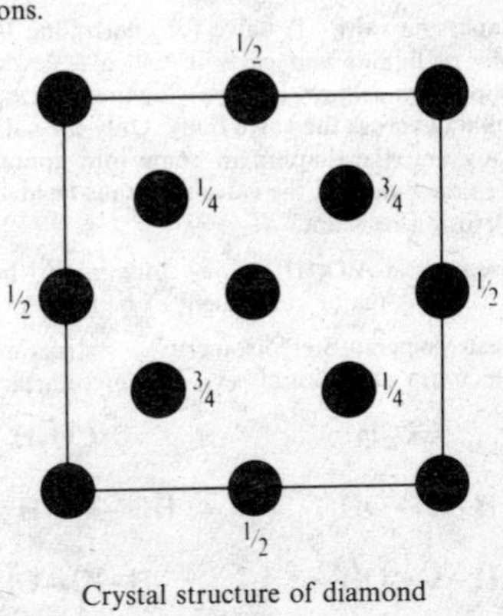

Crystal structure of diamond

This mineral, which in its transparent varieties is a most beautiful and costly gem, has been known from early times.

Important sources, e.g. in South Africa, are the basic igneous rock known as kimberlite and also alluvials. The valuable diamond may be colourless or faintly coloured, but must be transparent. The heavily coloured forms, known as carbonado or bort, are of no value as gems, but are used for rock drills, for lathe tools, and when powdered for cutting and polishing clear diamonds. In cutting, the natural crystalline form is obliterated and an artificial shape, which gives rise to a large amount of internal reflection producing the 'fire' of the stone, produced. The diamond is extremely hard, and stands highest in Mohs's scale of hardness. It possesses a high refractive index, and dispersive power. It is relatively much more transparent to X-rays than are its imitations. World production 1979 48 MKT, 62·5% southern Africa including Zaire.

Diamond is very resistant to chemical reagents; potassium dichromate and sulphuric acid attack it with the formation of CO_2. It burns in air or oxygen at 700°C to CO_2 leaving scarcely any ash; some forms of bort may leave as much as 4·5% ash.

Very high temperature and pressure on graphite in the presence of a metal catalyst gives synthetic diamonds big enough for many industrial uses.

diamorphine, heroin, $C_{21}H_{23}NO_5$. Diacetylmorphine. *See* morphine. Used in the form of its hydrochloride which is soluble in water and alcohol. M.p. 231–232°C.

diaphragm cell A cell for electrolysis of brine in which the anode and cathode compartments are separated by a diaphragm.

diaphragm valve A valve for controlling the flow of liquids and gases in which a flexible diaphragm is pressed down against a ridge running across the valve body. Only the valve body and the diaphragm come into contact with the fluid and the valve may thus be made corrosion-resistant.

diaspore, α-AlO(OH). *See* aluminium hydroxide. A major constituent of bauxite.

diastereomers Stereoisomeric structures which are not enantiomers. E.g. mesotartaric

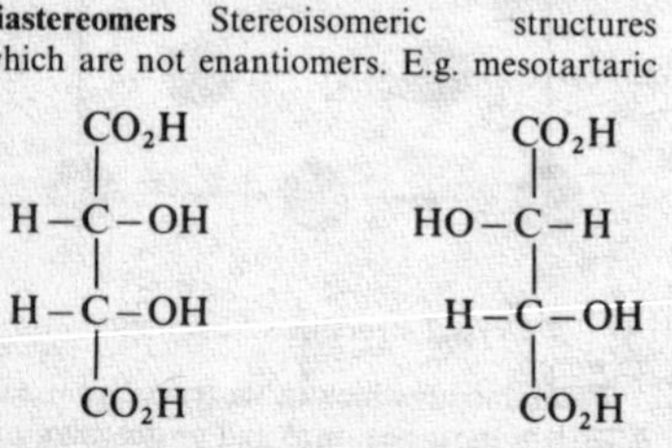

Mesotartaric acid   *d- or l*-Tartaric acid

acid and either one of the optically-active tartaric acids are said to be diastereomeric and differ only in the configuration at one carbon atom. Unlike enantiomers, diastereomers usually have substantially different chemical and physical properties.

diastereotopic *See* enantiotopic.

diatomic molecule A molecule with two atoms, e.g. N_2, HCl.

diatomite, kieselguhr A hydrated silica, the remains of diatoms. The particles are hollow and the material is used as a filler and an absorbent (nitroglycerine gives dynamite). Other uses are for polishing, filtration, decolourizing and for heat-insulation. Tripolite is a compact form. World use 1976 1·5 megatonnes.

Diazald A trade name for *N*-methyl-*N*-nitro-

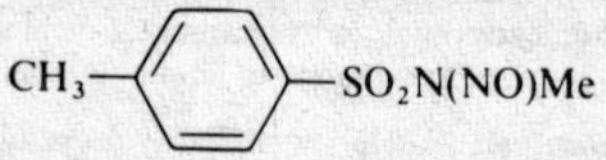

rosotoluene-*p*-sulphonamide. A white crystalline solid; m.p. 60°C. A convenient and easily stored source of diazomethane, which can be liberated by alkaline saponification.

diazepam, valium, 7-chloro-2,3-dihydro-1-methyl-5-phenyl-2H-1,4-benzodiazepin-2-one,

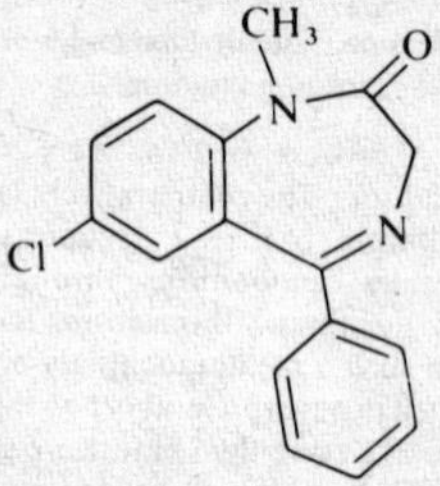

$C_{16}H_{13}ClN_2$. White plates; m.p. 125°C. Diazepam is one of several benzodiazepines which are very widely used as minor tranquillizers for allaying anxiety, as hypnotics or, in sufficiently high dosage given intravenously, as pre-anaesthetic sedatives.

diazinon, diethyl 2-isopropyl-4-methyl-6-pyrimidinyl phosphorothionate A compound which shows high activity against insects, particularly house flies.

diazoacetic ester *See* ethyl diazoethanoate.

diazoamino-compounds Prepared by condensation of a diazonium salt with a primary or secondary amine in the presence of sodium ethanoate, e.g.

$$C_6H_5N_2Cl + NH_2 \cdot C_6H_5 =$$
$$C_6H_5 \cdot N_2 \cdot NH \cdot C_6H_5$$
diazoaminobenzene

The diazoamino-compounds are usually yellow in colour, and do not dissolve in acid; they can usually be isolated and crystallized without decomposition. When treated with HNO_2 two molecules of diazonium salt are formed. Form an azo compound when warmed with an amine and its hydrochloride, e.g.

$$PhN=NNHPh \longrightarrow PhN=N-\!\!\!\left\langle\;\right\rangle\!\!\!-NH_2$$
$$\longrightarrow \left\langle\;\right\rangle\!\!-N=N-\!\!\!\left\langle\;\right\rangle\!\!\!-OH$$

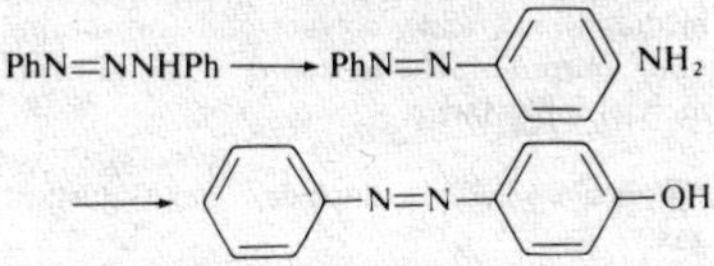

Some diazoamino-compounds are suspect mutagenic agents.

diazo compounds An indiscriminate name covering the diazoalkanes, diazonium salts and azo compounds.

diazo dyestuffs Azo dyestuffs with $-N_2-$ groups.

diazomethane, N_2CH_2. A yellow gas with a peculiar musty odour. It is very poisonous and highly explosive. Very soluble in ether. Prepared by treating a mixture of nitroso-methylurea and ether with cold concentrated potassium hydroxide solution, or better from Diazald* and alkali. The resulting ethereal solution is a convenient form in which to handle diazomethane. It is a very reactive substance, and forms methyl chloride with hydrochloric acid, acetonitrile with hydrogen cyanide and methyl esters with organic acids. It reacts with acyl chlorides to give diazoketones, and with esters of unsaturated organic acids to give pyrazoline carboxylic acids. With aldehydes a mixture of a methyl ketone and an alkene oxide is formed; the proportions depend upon the conditions of the reaction and upon the aldehyde used. Alcohols react slowly or not at all, but phenols form methyl ethers. On photolysis it produces carbene (methylene), $:CH_2$, which forms cyclopropanes with alkenes. It is a useful reagent in organic syntheses. *See* Arndt–Eistert synthesis.

diazonium compounds An important class of compounds containing the $RN=NX$ group with R aryl.

The diazonium salts are by far the most important diazo-compounds. These are salts derived from the base $R-N=N-OH$, e.g. benzene diazonium chloride, $PhN_2{}^+Cl^-$.

They are prepared by the action of HNO_2 on aromatic amines. The amine is dissolved in excess of mineral acid and sodium nitrite is added slowly until a slight excess of HNO_2 is present. The reaction is usually carried out in ice-cold solution. The solution then contains the diazonium salt of the mineral acid used, anhydrous diazonium salts of unpredictable stability may be precipitated with complex anions like $PF_6{}^-$, $SnCl_6{}^{2-}$ $BF_4{}^-$.

The diazonium salts usually decompose when warmed with water to give a phenol and nitrogen. When treated with CuCl, CuBr, KI, the diazo group is replaced by chlorine, bromine or iodine respectively (Sandmeyer reaction). A diazonium sulphate and hydroxylamine give an azoimide. The diazonium salt of anthranilic acid (2-aminobenzoic acid) decomposes to give benzyne*.

The most important reaction of the diazonium salts is the condensation with phenols or aromatic amines to form the intensely coloured azo compounds. The phenol or amine is called the secondary component, and the process of 'coupling' with a diazonium salt is the basis of manufacture of all the azo dyestuffs. The entering azo group goes into the *p*-position of the benzene ring if this is free, otherwise it takes up the *o*-position, e.g. diazotized aniline coupled with phenol gives benzeneazophenol. When only half a molecular proportion of nitrous acid is used in the diazotization of an aromatic amine a diazo-amino compound is formed.

Used in reprographic processes, particularly the ozalid blue-print paper processes.

dibasic acid An acid with two replaceable hydrogen atoms which may yield two series of salts, e.g. H_2SO_4 gives $NaHSO_4$ and Na_2SO_4.

1,2,5,6-dibenzanthracene, $C_{22}H_{14}$. Crystallizes in silvery leaflets, m.p. 262°C. A polycyclic aromatic carcinogen obtained from coal tar. The 7,8 derivative, m.p. 196°, is also known.

dibenzenechromium, $C_{12}H_{12}Cr$. Brown-

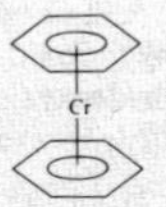

black crystals, m.p. 284–285°C, rapidly oxidized by air. Formed from $CrCl_3$, C_6H_6, $AlCl_3$ and reducing agents. The prototype π-arene complex with a sandwich structure with the two benzene rings symmetrically bonded to the metal.

dibenzoyl peroxide M.p. 107°C, PhC(O)OOC(O)Ph. Decomposes on moderate heating to give radicals and catalyses polymerization reactions, anti-Markownikoff additions (e.g. of HBr) across alkenes and radical halogenation of alkanes. Used commercially in the preparation of a wide range of styrene-based polymers. *See* peroxides, organic.

dibenzyl *See* 1,2-diphenylethane.

diborane, B_2H_6. *See* boron hydrides.

dibromochloropropane, 1,2-dibromo-3-chloropropane, $CH_2BrCHBrCH_2Cl$. A heavy amber liquid with a mildly pungent odour prepared by the addition of bromine to propenyl chloride. It is used as a soil fumigant and as a nematocide. Can cause sterility in humans.

1,2-dibromoethane, ethylene dibromide, $BrCH_2{\cdot}CH_2Br$. A colourless liquid with a sweet odour, m.p. 10°C, b.p. 132°C. Manufactured by passing ethene through bromine or bromine and water at about 20°C. Chemical properties similar to those of 1,2-dichloroethane; when heated with alkali hydroxides, vinyl bromide is formed. Used extensively in petrols to combine with the lead formed by the decomposition of lead tetraethyl, as a fumigant for stored products and as a nematocide.

2,6-di-*t*-butyl-4-methylphenol, 2,6-di-tert-butyl-*p*-cresol, ionol butylated hydroxytoluene, BHT. White solid, m.p. 69·5–70°C, used as an anti-oxidant especially to inhibit peroxide formation in ethers. A potent inactivator of lipid-containing viruses. Skin irritant.

dibutyl phthalate *See* phthalic esters.

dicarbollides *See* carboranes.

dicarboxylic acids Organic acids containing two carboxyl (–COOH) groups. They form both acidic and neutral salts and esters; some give anhydrides by loss of a molecule of water between the two carboxyl groups. Many occur naturally as the free acid or an ester. They are generally prepared by the oxidation of a glycol, hydroxy-acid or hydroxyaldehyde, or by hydrolysis of a dinitrile or cyano-acid.

dichlorobenzenes, $C_6H_4Cl_2$.

o-dichlorobenzene, 1,2-dichlorobenzene, b.p. 179°C.

m-dichlorobenzene, 1,3-dichlorobenzene, b.p. 172°C.

p-dichlorobenzene, 1,4-dichlorobenzene, m.p. 53°C, b.p. 174°C.

The *o*- and *p*-isomers are manufactured by the direct chlorination of benzene in the presence of iron as a catalyst, the resulting mixture being separated by fractional distillation. The *m*-isomer may be obtained by isomerization of the *o*- or *p*-compound in the presence of a catalyst.

p-dichlorobenzene is the most important of the three compounds, being used largely as a moth repellant and air deodorant. *o*-dichlorobenzene is used as a dye intermediate, insecticide, solvent and for various other purposes. The *m*-isomer is not of commercial importance.

$\beta\beta'$-dichlorodiethyl sulphide. *See* mustard gas.

dichlorodifluoromethane, CCl_2F_2. B.p. −30°C. Manufactured by the action of HF on CCl_4 using $SbCl_5$ as a catalyst. Known commercially as Freon-12 or Arcton-12. Widely used as a refrigerant and aerosol propellant. It is much less toxic than carbon tetrachloride.

1,2-dichloroethane, ethylene dichloride, $CH_2Cl{\cdot}CH_2Cl$. Colourless liquid with an odour like that of chloroform; b.p. 84°C. It is an excellent solvent for fats and waxes. Was first known as 'oil of Dutch chemists'. Manufactured by the vapour- or liquid-phase reaction of ethene and chlorine in the presence of a catalyst. It reacts with anhydrous ethanoates to give ethylene glycol diethanoate and with ammonia to give ethylenediamine, these reactions being employed for the manufacture of these chemicals. It burns only with difficulty and is not decomposed by boiling water.

Most of the ethylene dichloride produced is utilized for the manufacture of vinyl chloride, which may be obtained from it by pyrolysis or the action of caustic soda. Large quantities are also used in anti-knock additives for gasoline. As a solvent it has been displaced by trichloroethylene* and tetrachloroethylene*. U.S. production 1978 4·75 megatonnes.

dichloroethenes, dichloroethylenes, $C_2H_2Cl_2$. There are three of these compounds.

1,1-dichloroethene, 1,1-dichloroethylene, asymmetric dichloroethene, $CH_2{=}CCl_2$, colourless liquid, b.p. 37°C. Prepared by heating 1,1,1- or 1,1,2-trichloroethene with excess lime at 70–80°C. Polymerizes readily to an insoluble solid.

1,2-dichloroethene, acetylene dichloride,

cis *trans*

exists in two forms, *cis* and *trans*, both formed when acetylene tetrachloride (1,1,2,2-tetrachloroethane) is heated with water and zinc at 100°C. The product contains about 80% of the *cis*-form. When a mixture of equal volumes of ethyne and chlorine is passed over activated carbon at 40°C the *trans*-form is the chief product. *Cis*-form, b.p. 60°C; *trans*-form, b.p. 49°C. They are insoluble in water but are miscible with hydrocarbon solvents and are very similar in properties to trichloroethene; not affected by moisture or alkalis. They are used as substitutes for ether in fat extraction and as solvents, e.g. for rubber.

dichloroethylenes *See* dichloroethenes.

$\beta\beta$-dichloroethyl ether, di(2-chloroethyl) ether, $(ClCH_2CH_2)_2O$. A colourless liquid with a chloroform-like odour; b.p. 178°C. Manufactured by treating ethylene chlorohydrin ($CH_2Cl{\cdot}CH_2OH$) with chlorine and excess ethene at 80°C; or by heating ethylene chlorohydrin with strong sulphuric acid at 100°C. Reacts with amines to give morpholines and with sodium tetrasulphide to give rubber-like plastics. It reacts with fused potassium hydroxide to give divinyl ether. Used as a solvent in the Chlorex process for dewaxing mineral oils, and for removing grease from textiles. Used for the manufacture of surfactants and elastomers.

sym-dichloroisopropyl alcohol *See* glycerol dichlorohydrin.

dichloromethane, methylene chloride, CH_2Cl_2. A colourless liquid with a chloroform-like odour; b.p. 41°C. Prepared by heating chloroform with zinc, alcohol and hydrochloric acid; manufactured by the direct chlorination of methane. Decomposed by water at 200°C to give methanoic and hydrochloric acids. Largely used as a solvent for polar and non-polar substances, particularly for paint removal (30%), dissolving cellulose acetate and degreasing (10%). It is more stable than carbon tetrachloride or chloroform especially towards moisture or alkali. It is somewhat toxic. U.S. production 1981 280000 tonnes.

dichloromethyleneammonium salts, phosgenammonium, phosgene iminium, $[Cl_2C{=}NR_2]^+X^-$. Prepared from Cl_2 and $[Me_2NC(S)S^-]_2$ or the corresponding thiocarbamoyl chloride. Reactive electrophiles used in synthesis, e.g. RCH_2COCl plus $[Cl_2C{=}NMe_2]^+$, gives $(Me_2N)ClC{=}C(R)COCl$.

1,2-dichloropropane *See* propylene dichloride.

2,3-dichloropropyl alcohol *See* glycerol dichlorohydrins.

dichroism Pleochroism* in a uniaxial crystal.

dichromates Compounds containing the $[O_3CrOCrO_3]^{2-}$ anion. *See* chromates.

dicyandiamide *See* cyanoguanidine.

dicyclohexylcarbodiimide, $C_{13}H_{22}N_2$.

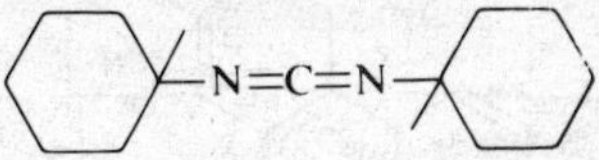

Crystalline solid; m.p. 35–36°C, b.p. 154–156°C, prepared by oxidizing *N,N′*-dicyclohexylthiourea with HgO in carbon disulphide solution, also obtained from cyclohexylamine and phosgene at elevated temperatures. Used as a mild dehydrating agent, especially in the synthesis of peptides from amino-acids. Potent skin irritant.

dicyclopentadiene Colourless solid; m.p.

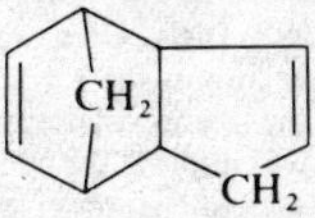

32°C, b.p. 170°C (decomp.), has a characteristic odour. It is the Diels–Alder product of cyclopentadiene* reacting with itself, the *exo*-form being formed most rapidly but the *endo*-form is thermodynamically favoured. At temperatures above 150°C a retro-Diels–Alder reaction occurs and cyclopentadiene monomer is regenerated; *see* diene reactions.

dicyclopentadienyl compounds *See* cyclopentadienylides.

dieldrin The common name for the insecticidal product containing not less than 85% of 1,2,3,4,10,10,(hexachloro-6,7-epoxy-1,4,4a,5,6,7,8,8a-octahydro-*exo*-1,4-*endo*-5,8-dimethanonaphthalene ($C_{12}H_8Cl_6O$). M.p. 175–176°C. The epoxy derivative of aldrin from which it is prepared by epoxidation with peracetic or perbenzoic acid. It has a high contact and stomach toxicity to most insects, and is very persistent. Stored in the bodies of birds and mammals.

dielectric constant The force F between two electric charges e, separated by a distance r in a vacuum, is given by $F = e^2/r^2$. In any other medium this relationship becomes $F = e^2/Dr^2$ where D is the dielectric constant of the medium. The dielectric constant is a measure of the polarity of the medium, some typical values being 1·00 (air); 1·013 (water as steam); 1·874 (hexane); 5·94 (chlorobenzene); 15·5 (liquid ammonia) and 80 (water at 20°C).

Diels–Alder reaction *See* diene reactions.

Diels' hydrocarbon, γ-methylcyclopenteno-phenanthrene, $C_{18}H_{16}$. The fundamental

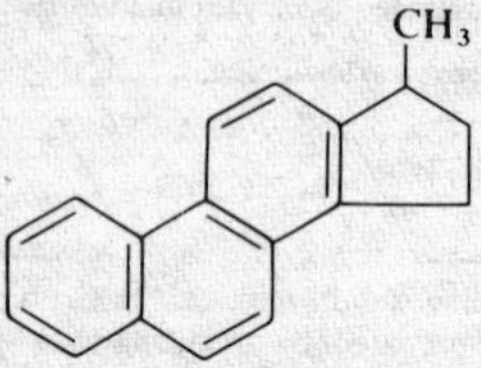

hydrocarbon on which the structure of the sterols and related compounds is based. It can be obtained by heating cholesterol with selenium.

dienes Organic compounds containing two carbon to carbon double bonds.

diene reactions, Diels–Alder reaction The Diels–Alder reaction is the 1,4-addition of an alkene or alkyne (dienophile) across a conjugated diene. An example is the addition of propenal to buta-1,3-diene to give Δ⁴-tetrahy-

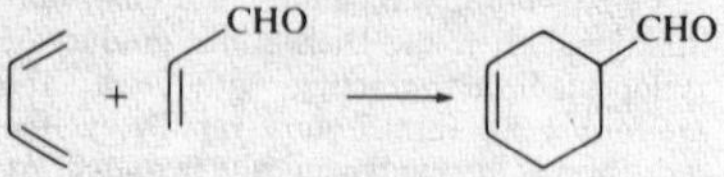

drobenzaldehyde. The dienophile should be substituted with electron-attracting groups such as –C(O)OR, –CN, $-CF_3$, $-NO_2$, to enhance its reactivity, and the conjugated diene should have a *cis*-configuration. Its utility in synthesis arises from the stereospecific course it follows, i.e.

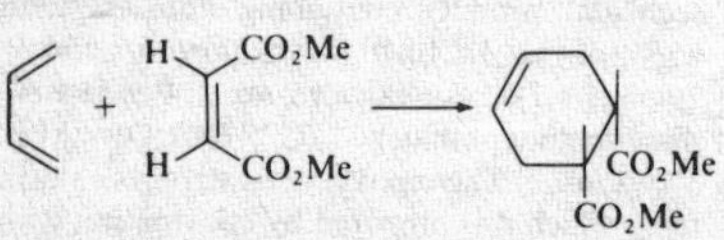

the addend from dimethylmaleate is also *cis*. Addition to cyclic dienes, e.g. cyclopentadiene, may give *endo* or *exo* products; often the *endo* adduct is produced first but slowly isomerizes to the more stable *exo* adduct; e.g. for maleic anhydride and cyclopentadiene.

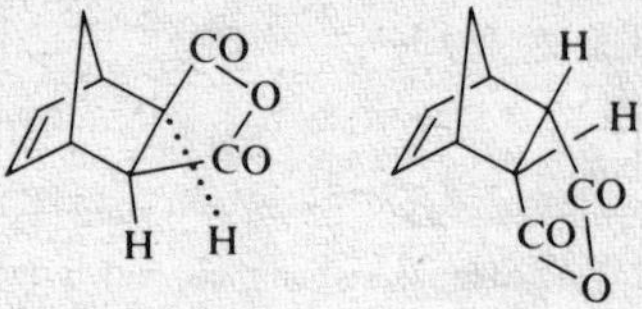

exo adduct *endo* adduct

dienoestrol, $C_{18}H_{18}O_2$. Colourless crys-

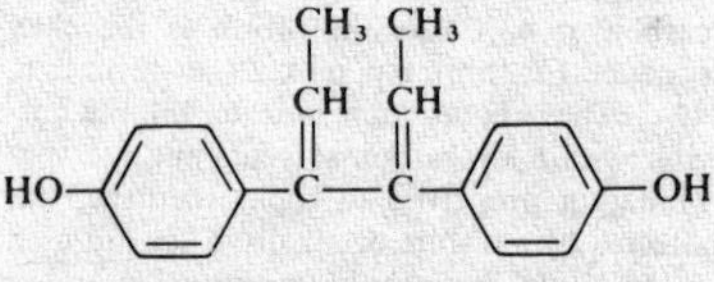

talline powder; m.p. 233°C, insoluble in water, soluble in alcohol and in organic solvents. Prepared by reducing *p*-hydroxypropiophenone and dehydrating the pinacol thus obtained.

It has oestrogenic properties (*see* oestradiol) and is used for the same purposes as stilboestrol*.

diesel fuel Fuel used for diesel (compression ignition) engines. The composition varies but is near that of gas oil*. Fuel for diesel engined road vehicles is commonly known as Derv.

Dieterici's equation A modification of van der Waals' equation, in which account is taken of the pressure gradient at the boundary of the gas. It is written

$$p(v-b) = RTe^{-a/RTV}.$$

where a and b are constants characteristic of each gas; at relatively low pressures it is identical with van der Waals' equation.

diethanolamine *See* ethanolamines.

diethylamine *See* ethylamines.

diethyldithiocarbamic acid, $Et_2NC(S)SH$. Typical dithiocarbamic acid. Gives a brown complex with copper salts, soluble in CCl_4, used in the determination of copper. The Zn^{2+} and $Et_2NH_2^+$ salts are used as accelerators in rubber vulcanization.

1,4-diethylene dioxide *See* dioxan.

diethyleneglycol, 2,2′-dihydroxydiethyl ether, $C_4H_{10}O_3$, $(HOCH_2CH_2)_2O$. A colourless and almost odourless liquid; b.p. 244°C. Readily absorbs moisture from the atmosphere. It is obtained as a by-product in the manufacture of ethylene glycol by the hydration of ethylene oxide. Manufactured by heating ethylene glycol with ethylene oxide. It is a solvent for cellulose nitrate, but not for the acetate. Used as a softening agent for textile fibres, as a solvent for certain dyes and as a moistening agent for glues, paper and tobacco. Its esters and ethers are used as solvents and plasticizers in lacquers.

diethyleneglycol dinitrate (DEGN), $C_4H_8N_2O_7$, $O(CH_2CH_2ONO_2)_2$. Formed by the nitration of diethyleneglycol. Viscous, colourless, odourless liquid; m.p. −11·5°C. Used to replace nitroglycerin as a propellant.

diethyleneglycol monoethylether, carbitol, $C_6H_{14}O_3$, $CH_3CH_2OCH_2CH_2OCH_2CH_2OH$. A colourless liquid with a pleasant odour; b.p. 202°C. Miscible with water and most organic solvents. Manufactured by heating ethylene oxide with ethylene glycol monoethyl ether under pressure. Used in the preparation of nitrocellulose lacquers and laminated glass.

diethylenetriamine, $NH_2CH_2CH_2NHCH_2CH_2NH_2$. B.p. 207°C. Important polyamine with uses similar to those of diaminoethane.

diethyl ether *See* ether.

diethyl oxalate, oxalic ester, $C_6H_{10}O_4$, $(C(O)OEt)_2$. Colourless liquid; b.p. 185°C. Prepared by distilling a mixture of oxalic acid, alcohol and carbon tetrachloride. Condenses with other esters in the presence of sodium. Used in organic syntheses.

diethyl oxide *See* ether.

diethyl phthalate *See* phthalic esters.

diethyl succinate *See* ethyl succinate.

diethyl sulphate, $(C_2H_5O)_2SO_2$. A colourless liquid with a faint ethereal odour; b.p. 208°C (decomp.). Manufactured by leading ethene under pressure into 100% sulphuric acid. It is decomposed by boiling water. Reacts with hydroxy-compounds and amines to give ethers, esters, amines and imines. Used as an ethylating agent in organic syntheses. Highly toxic by skin contact.

diethyl sulphide, ethyl sulphide, $(CH_3CH_2)_2S$. A colourless liquid with an ethereal odour when pure; usually it has a strong garlic-like odour; b.p. 92°C. Prepared by the action of KHS on ethyl chloride or potassium ethyl sulphate. When heated at 400–500°C it forms thiophene.

differential scanning calorimetry *See* thermal analysis.

differential thermal analysis *See* thermal analysis.

differential titration Analysis of a sample which contains two or more similarly reacting species which are differentiated, e.g. use of different bases or indicators.

diffraction pattern A beam of X-rays or electrons passed through a crystalline solid produces a diffraction pattern on a photographic plate. The pattern is due to the diffraction of the X-rays or electrons in a manner analogous to the diffraction of light by a diffraction grating. Reflected X-rays or electrons from crystalline materials also give diffraction patterns. The diffraction patterns are used in structure determination.

diffusion In any gaseous mixture or liquid solution which is kept at a uniform temperature, the composition eventually becomes uniform throughout the system, no matter what was the original distribution. This is explained on the basis of the molecular motion that is diffusion.

Graham showed that the rate of diffusion of different gases through a porous diaphragm was inversely proportional to the square roots of their densities; this is the basis of a method of separation of gases, and has been applied successfully to the separation of hydrogen and deuterium.

A diffusion mechanism is also used in dialysis* as a means of separating colloids from crystalloids. The rate of diffusion of molecules in gels is practically the same as in water, indicating the continuous nature of the aqueous phase. The diffusion of gases into a stream of vapour is of considerable importance in diffusion pumps.

Diffusion may also take place, although slowly, between solids.

diffusion pump A non-mechanical pump used for the production of high vacua. Gas molecules are forced through a fine jet by molecular bombardment with a stream of oil or mercury vapour from a boiler. Once through the jet the heavy vapour is condensed and returned to the boiler while the expelled gas is removed by a mechanical backing pump.

difluoromethane, CH_2F_2. B.p. −52°C. A gas used as an azeotropic mixture with other fluorocarbons as a refrigerant.

digestion The process of ageing (recrystallization) of a precipitate to obtain it in a form more suitable for filtration.

digilanids The primary glycosides of digitalis.

digitalin, $C_{36}H_{56}O_{14}$. M.p. 229°C. Composed of digitalose, glucose and digitaligenin. Its action is that of digitalis.

digitalis The dried leaves of *Digitalis purpurea*, the common foxglove. It is a valuable source of the cardiac glycosides. The glycosides* contain several sugars including the unusual digitalose* and digitoxose*, linked to steroidal aglucones. The glycosides cause a failing heart to beat more slowly, more regularly and more strongly. Digoxin* is now used in treating cardiac failure, rather than the extracts of digitalis.

digitalose, 3-methyl-D-fucose, $C_7H_{14}O_5$.

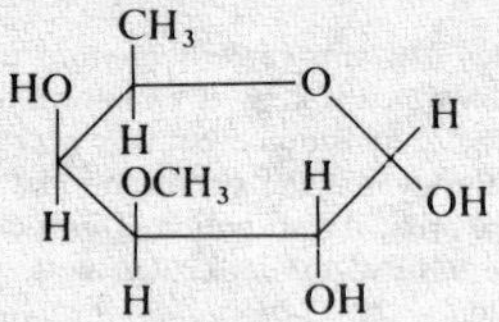

The sugar constituent of digitalin; m.p. 106°C.

digitonides Molecular complexes formed between digitonin and a variety of hydroxylic compounds.

digitoxigenin, $C_{23}H_{34}O_4$. A cardiac aglucone obtained by hydrolysing digitoxin*.

digitoxin, $C_{41}H_{64}O_{13}$. The chief constituent of the mixture of glycosides obtained from digitalis. It is composed of three digitoxose* units and digitoxigenin*. M.p. 263°C. It is used for the same purposes as digitalis.

digitoxose, $C_6H_{12}O_4$. The 2-desoxy sugar

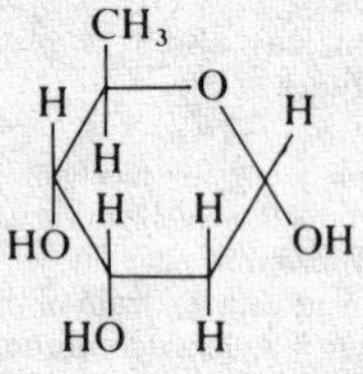

present in several of the cardiac glycosides of digitalis.

diglyme, $C_6H_{14}O_3$, $CH_3{\cdot}O{\cdot}CH_2{\cdot}CH_2{\cdot}O{\cdot}CH_2{\cdot}CH_2{\cdot}O{\cdot}CH_3$. Dimethyl ether of diethylene glycol. A colourless liquid; b.p. 160°C. Typical ether, useful as a high-temperature solvent.

digol The trivial name for diethyleneglycol*, $O(CH_2CH_2OH)_2$.

digoxin, $C_{41}H_{64}O_{14}$. A glycoside obtained from the leaves of *Digitalis lanata* which contains three digitoxose* units and digoxigenin. Colourless crystals, m.p. 265°C. The cardiac glycoside most frequently used in treating heart failure.

dihedral angle The angle relating two parts of a molecule. In H_2O_2 the dihedral angle in the solid is 94°.

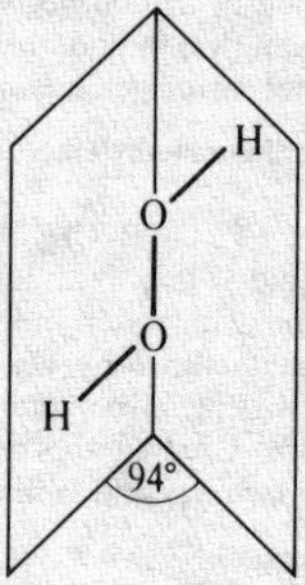

dihydroazirine, aziridine, ethyleneimine,

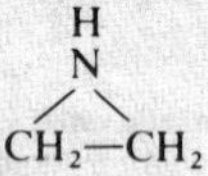

C_2H_5N. Colourless liquid with strong ammoniacal smell; b.p. 56°C. Miscible with water and strongly basic. Prepared commercially from 2-aminoethanol. Pure dry aziridine is comparatively stable but it polymerizes explosively in the presence of traces of water. Carbon dioxide is sufficiently acidic to promote polymerization.

Used industrially to cross-link hydroxylic polymers, polyethyleneimine. Possesses some carcinogenic properties. Polyethyleneimine is a hygroscopic liquid used in paper manufacture to confer wet strength and in textiles. *N*-alkylated derivatives also form useful polymers.

2,3-dihydropyran, C_5H_8O. B.p. 85°C. Pre-

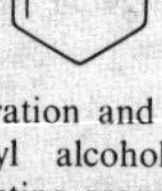

pared by dehydration and rearrangement of tetrahydrofurfuryl alcohol over alumina. Used as a protecting group* in organic synthesis it forms tetrahydropyranyl ethers by reaction with alcohols under very mild acid catalysis. These ethers are cleaved easily with dilute acid after the desired reaction has been accomplished. Has also been used to protect thiol, carboxyl and certain >N−H groups.

dihydroxyacetone, $HOCH_2C(O)CH_2OH$. Crystallizes as the monomer, m.p. about 80°C, but changes to the dimer on standing, m.p. about 115°C. Obtained by the action of certain *Acetobacter* on glycerol. It is a strong reducing agent. Used in preparations for application to the skin to simulate a sun-tan.

1,2-dihydroxybenzene, catechol, pyrocatechol, $C_6H_6O_2$. Colourless crystals; m.p. 105°C, b.p. 240°C. Strong reducing agent.

Obtained by fusing *o*-benzenedisulphonic acid with NaOH. It is used as a photographic developer, for preparation of dyes and pharmaceuticals, and as an antioxidant.

1,3-dihydroxybenzene *See* resorcinol.

1,4-dihydroxybenzene *See* hydroquinone.

dihydroxybutanes *See* butylene glycols.

2,2′-dihydroxydiethyl ether *See* diethylene-glycol.

1,2-dihydroxyethane, ethylene glycol, $HOCH_2{\cdot}CH_2OH$. Colourless, odourless, rather viscous hygroscopic liquid having a sweet taste, b.p. 197°C. Manufactured from ethylene chlorohydrin and $NaHCO_3$ solution, or by the hydration of ethylene oxide with dilute sulphuric acid or water under pressure at 195°C. Used in anti-freezes and coolants for engines (50 %) and in manufacture of polyester fibres (e.g. Terylene) and in the manufacture of various esters used as plasticizers. U.S. production 1979 1 900 000 tonnes.

dihydroxymalonic acid *See* mesoxalic acid.

3,4-dihydroxyphenylalanine, Dopa, $C_9H_{11}NO_4$. M.p. 282°C (decomp.). The naturally occurring substance is laevorotatory. It is an amino-acid isolated from various plant sources, but not found in the animal body. It is formed from tyrosine as the first stage in the oxidation of tyrosine to melanin. It is used in the treatment of Parkinson's disease.

1,2-dihydroxypropane, propylene glycol, $CH_3{\cdot}CHOH{\cdot}CH_2OH$. A colourless, almost odourless liquid. It has a sweet taste, but is more acrid than ethylene glycol; b.p. 187°C. Manufactured by heating propylene chlorohydrin with a solution of $NaHCO_3$ under pressure. It closely resembles dihydroxyethane in its properties, but is less toxic. Forms mono- and di-esters and ethers. Used as an anti-freeze and in the preparation of perfumes and flavouring extracts, as a solvent and in mould growth inhibition. U.S. production 1978 250 000 tonnes.

diimide, diimine, HN=NH. A reagent generated *in situ* by oxidation (Cu^{2+}/air, H_2O_2, HgO) of hydrazine, for the reduction of symmetrical multiple bonds (C=C, C≡C, N=N) by stereospecific *cis*-addition of hydrogen. Polar double bonds (C=O, C≡N, NO_2) are reduced less easily. In the absence of a substrate for reduction diimide disproportionates to nitrogen and hydrazine.

3,5-diiodotyrosine, iodogorgic acid, $C_9H_9I_2NO_3$. Pale, straw-coloured needles, m.p. 198°C. Found in certain marine organisms, such as corals and sponges, and in the thyroid gland.

diisobutyl ketone, isovalerone, 2,6-dimethyl-4-heptanone, $(CH_3)_2CHCH_2COCH_2CH(CH_3)_2$. Obtained by reduction of phorone, b.p. 169°C. Very good solvent for aldehyde resins.

2,4-diisocyanatotoluene, toluene-2,4-diisocyanate, 2,4-tolylenediisocyanate, T.D.I., Nacconate 100 A lachrymatory liquid; b.p. 251°C. Manufactured from phosgene and 2,4-diaminotoluene. Used for preparing polyurethane foams and other elastomers by reaction with polyhydroxy compounds. Produces skin irritation and causes allergic eczema and bronchial asthma.

diisopropyl ether *See* isopropyl ether.

diisopropylideneacetone *See* phorone.

diketen, $C_4H_4O_2$. A colourless highly

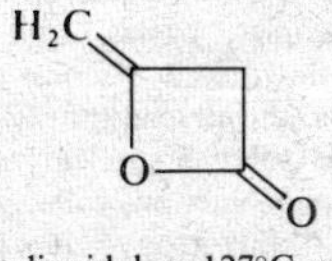

lachrymatory liquid; b.p. 127°C, m.p. −6·5°C. Manufactured from propanone via keten. It reacts with alcohols and amines to give acetoacetic esters and amides respectively.

diketones Organic compounds containing two keto (>C=O) groups. They are classified according to the number of carbon atoms separating the two keto groups. α- or 1,2-diketones are formed from their mono-oximes, the isonitrosoketones, by boiling with dilute sulphuric acid. The aliphatic α-diketones are yellow oils with pungent odours, while the aromatic diketones are crystalline solids. They react characteristically with *o*-phenylene-diamines to give quinoxalines; they form mono- and di-oximes with hydroxylamine and osazones with hydrazines. β- or 1,3-diketones R·CO·CH_2·CO·R are obtained by the reaction between an ester and a ketone in the presence of metallic sodium or sodamide. They show acidic properties and form metallic derivatives, many of which are insoluble in water but soluble in organic solvents. These diketones can exist in keto and enol forms. They react with phenylhydrazine to give pyrazoles, and with hydroxylamine to give iso-oxazoles. They form very stable derivatives (*see*, e.g., acetylacetonates) with metals. γ- or 1,4-diketones (R·CO·CH_2·CH_2·CO·R) are formed by treating sodium acetoacetic ester with α-bromoketones and hydrolysing the product. They readily form ring compounds.

2,5-diketopiperazine, glycine anhydride,

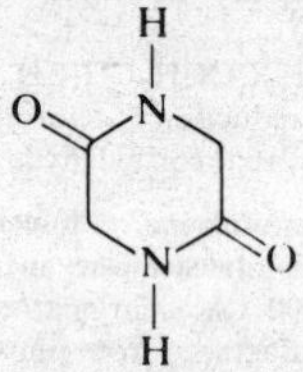

$C_4H_6N_2O_2$. Sublimes 260°C; sparingly soluble in water; hydrolysed by alkalis or mineral acids to glycylglycine. It and substituted diketopiperazines are formed by the condensation of amino-acids, and are obtained in small quantities on the hydrolysis of proteins.

dilatancy Certain pastes or suspensions (e.g. starch moistened with water) seem to become harder when they are subjected to shearing; in extreme examples, pastes, which are 'wet' enough to flow as a viscous liquid when allowed to stand, will temporarily crumble like a dry, brittle solid if suddenly stirred. This rheological phenomenon, called dilatancy, is the opposite of thixotropy*. It is explained by the friction of the particles against one another; when they are left undisturbed, the particles roll over one another and settle down into minimum volume, but when the medium is subjected to shear, they are forced into a less dense packing and the available fluid is insufficient to fill the interstices. Hence, the material seems dry and hard.

dilatometer An apparatus for measuring small changes of volume of a liquid, solution or solid immersed in a liquid. It usually consists of a cylindrical glass bulb with attached capillary tube, and change of volume of the contents of the bulb is noted by observing the movement of the meniscus in the capillary. Dilatometry is useful for determining transition temperatures, rates of reaction, polymerization, etc.

dilauryl peroxide, $(CH_3(CH_2)_{10}CO)_2O_2$. *See* peroxides, organic.

dilute solution A dilute solution is one which contains only a small amount of solute.

dilution, infinite With electrolyte solutions the equivalent conductivity (Λ) increases as the solution is diluted. If Λ is extrapolated to zero concentration, the limiting value, Λ_0, is called the equivalent conductivity at infinite dilution.

dimedon, dimethyldihydroresorcinol, 5,5-dimethyl-1,3-cyclohexanedione, $C_8H_{12}O_2$. Greenish-yellow needles or prisms, m.p. 148–149°C, sparingly soluble in cold water. Gives crystalline condensation products with aldehydes.

dimefox, bis(dimethylamino)phosphorofluoridate, bis(dimethylamino)fluorophosphine oxide, $C_{14}H_{12}FN_2OP$, $(Me_2N)_2P(O)F$. An extremely poisonous, persistent, systemic insecticide, which is usually supplied to the roots of plants.

dimer A compound formed by the addition polymerization of two molecules of a monomer.

dimethoate, *O,O*-dimethyl(*N*-methylcarbamoylmethyl)phosphorothiolothionate, Rogor, $C_5H_{12}NO_3PS_2$, $(MeO)_2P(S)SCH_2CONHMe$. A systemic insecticide.

1,2-dimethoxyethane, glyme, $CH_3OCH_2CH_2OCH_3$. A water-miscible colourless liquid, b.p. 83°C, used in selected reactions as an alternative to diethylether.

dimethyl acetylenedicarboxylate, $C_6H_6O_4$, $CH_3OOC{\cdot}C{\equiv}C{\cdot}COOCH_3$. A lachrymatory liquid; b.p. 195–198°C, which is extensively used as a dienophile in Diels–Alder reactions.

***cis*-dimethylacrylic acid** *See* tiglic acid.

dimethylamine *See* methylamines.

3-(dimethylaminomethyl)-indole, gramine Colourless solid; m.p. 132–134°C. An important intermediate for preparing indole derivatives, produced by treating indole* under Mannich reaction* conditions with methanal and dimethylamine.

3-dimethylaminophenol, $C_8H_{11}NO$. Colourless needles; m.p. 87°C, b.p. 265–268°C. It is prepared by NaOH fusion of dimethylaniline-3-sulphonic acid, or by heating a mixture of resorcinol, dimethylamine sulphate and dimethylamine solutions.

Other 3-(alkylamino)phenols are commonly used, and are prepared by analogous methods. They are used as intermediates in the preparation of dyestuffs.

4-dimethylaminopyridine, 4-pyridinamine Polymerization catalyst particularly for urethanes.

N,N-dimethylaniline, $C_8H_{11}N$, $PhNMe_2$. Colourless oil of characteristic smell; b.p. 193°C.

Prepared by heating aniline with methanol and a little sulphuric acid at 215°C.

Used as a solvent and as a second component in a few azo-dyes.

2,3-dimethylbutadiene, βγ-dimethylbutadiene, $C_6H_{10}, H_2C{=}CMe{\cdot}CMe{=}CH_2$. Colourless liquid; b.p. 69·5°C. Manufactured from propanone by conversion to pinacone and passing its vapour over heated $KHSO_4$. Slowly changes to a rubber-like substance; the change is hastened by metallic sodium or peroxides. Used in the manufacture of artificial rubber.

dimethylformamide, DMF, $HC(O)NMe_2$. Colourless liquid; b.p. 153°C, m.p. −61°C. It is an excellent solvent for a large variety of organic and inorganic materials and is widely used as a reaction medium. It can act as a catalyst in some substitution, elimination and addition reactions; synthetic applications include syntheses of aldehydes (with phosphorus oxychloride) and amines (Leuckart reactions). It is used as a solvent for lacquers, adhesives, dyes, etc.

dimethylglyoxime, diacetyl dioxime,

$$\begin{array}{l} CH_3{\cdot}C{:}NOH \\ \quad\;\; | \\ CH_3{\cdot}C{:}NOH \end{array}$$

$C_4H_8N_2O_2$. Colourless needles; m.p. 234°C, sublimes at 215°C. Prepared by the action of hydroxylamine on diacetyl, or by treating methyl ethyl ketone with ethyl nitrite and hydrochloric acid to which gives diacetyl monooxime which with sodium hydroxylamine monosulphonate gives dimethylglyoxime. It slowly polymerizes; condenses with *o*-phenylenediamine to give quinoxaline derivatives. Forms a dark-red crystalline nickel salt. Under appropriate conditions dimethylglyoxime can also be used for the detection and estimation of Bi, Cu, Co and Pd. Also used for rot-proofing of some fibres.

1,1-dimethylhydrazine, unsymdimethylhydrazine, Me_2NNH_2 M.p. −58°C, b.p. 64°C. A highly inflammable hygroscopic yellow liquid which fumes in air. Prepared from Me_2NH and NH_3 in the presence of catalysts. Used in rocket fuels. Corrosive to skin and mucous membranes.

dimethyl phthalate *See* phthalic esters.

dimethyl sulphate, $(CH_3O)_2SO_2$. A colourless and odourless liquid. Its vapour is extremely poisonous, the liquid is poisonous by absorption through the skin, b.p. 188°C. Prepared from methanol and chlorosulphonic acid. A good methylating agent, it reacts with amines and ammonia to give methylamines, with phenols and with organic acids to give methyl ethers and with organic acids to give methyl esters.

dimethyl sulphide, CH_3SCH_3. A colourless liquid; b.p. 37°C. Prepared by methylation of methane thiol. It has a pungent odour and is used for detecting pipe leakages by mixing with the gas stream. Forms complexes with transition metals.

dimethyl sulphite, $(CH_3O)_2SO$. A colourless liquid with a faint odour; b.p. 126°C. Prepared by boiling methanol with thionyl chloride. Decomposed by boiling water to methanol and SO_2. It reacts with organic acids, amines and alcohols to give methyl esters, methylamines and methyl ethers. Used as a methylating agent in organic syntheses.

dimethyl sulphoxide, DMSO, Me_2SO. A colourless, odourless solid; m.p. 18°C, b.p. 189°C, a powerful broad spectrum solvent for a wide variety of inorganic and organic reactants. Saturated aliphatic hydrocarbons are virtually insoluble in DMSO. Having low toxicity it can be used in biology and medicine, especially for low-temperature preservation. The chemical reactions are those of a typical lower aliphatic sulphoxide. With sodium hydride a salt of the anion $[CH_3SOCH_2]^-$ is produced (dimsyl sodium). The anion has several synthetic uses being a powerful nucleophile. Produced by the oxidation of dimethyl sulphide, obtained from Kraft pulping liquors.

dimethyl terephalate *See* phthalic esters.

dimorphism The existence of a substance in two crystal forms (*see* polymorphism).

dimsyl sodium *See* dimethyl sulphoxide.

1,3-dinitrobenzene, *m*-dinitrobenzene, $C_6H_4N_2O_4$. Forms colourless crystals; m.p. 90°C, b.p. 302°C. Very poisonous.

It is prepared by the direct nitration of

benzene or nitrobenzene, only small quantities of the *o*- and *p*-dinitro-compounds being produced. It is used as an intermediate in the preparation of *m*-nitroaniline and *m*-phenylenediamine.

4,6-dinitro-*o*-cresol, DNOC, $C_7H_6N_2O_5$. A yellow crystalline solid; m.p. 86°C, with insecticidal, ovicidal and herbicidal properties.

dinitrogen, N_2. The term used to denote the N_2 group, particularly in complexes, e.g. $(Ru(NH_3)_5N_2)Cl_2$.

dinitrogen oxide, N_2O. *See* nitrogen oxides.

dinitrogen tetroxide, N_2O_4. *See* nitrogen oxides.

2,4-dinitrophenylhydrazine, $C_6H_6N_4O_4$. Violet-red crystalline solid, m.p. 197°C (decomp.). Prepared from 2,4-dinitrochlorobenzene and hydrazine in boiling alcohol. It is an important reagent for aldehydes and ketones (Brady's reagent).

2,4-dinitrotoluene, $C_7H_6N_2O_4$. Colourless needle-shaped crystals; m.p. 71°C. Prepared by the nitration of the mixture of mononitrotoluenes obtained by mildly nitrating toluene. Used for the preparation of the corresponding mono- and di-amino-compounds, the latter compound being the precursor of toluene-2,4-di-isocyanate which is used in polyurethane formation.

dinoseb *See* DNBP.

diolefins Hydrocarbons containing two double bonds. They are divided into three classes, according to the position of the double bonds. Compounds of the type $-CH{=}C{=}CH-$ are called allenes after the simplest member of the series. They are said to have 'cumulated' double bonds. They react in most cases normally, that is, each double bond is unaffected by the proximity of the other; they are converted by heating into the isomeric acetylenes. Compounds of the type $-CH{=}CH-CH{=}CH-$ are said to have 'conjugated' double bonds and react somewhat differently from the other diolefins. For instance, bromine or hydrogen is often added so that a product of the type $-CHBr-CH{=}CH-CHBr-$ is formed. Also, these hydrocarbons participate in the Diels–Alder reaction (*see* diene reactions). They show a tendency to form rubber-like polymers. Hydrocarbons not falling into these two classes are said to have 'isolated' double bonds and each reacts normally. *See* olefins and butadienes.

diorite A widely distributed igneous rock composed of plagioclase feldspar and hornblende. Used as road metal.

diosgenin, $C_{27}H_{42}O_3$. M.p. 204–207°C. A

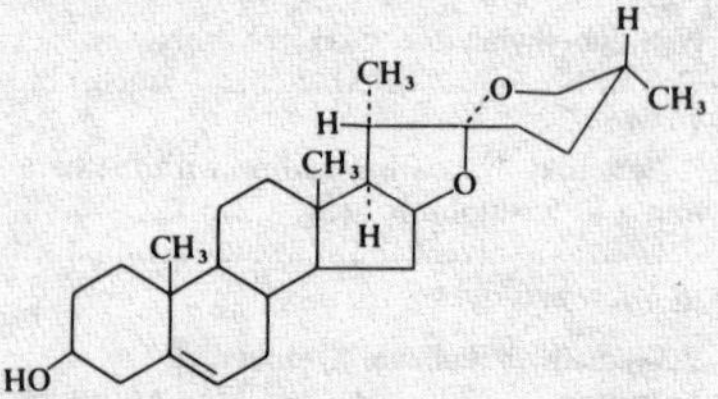

steroid sapogenin found abundantly in *Dioscorea* spp. (Mexican yams), and used in very large amounts as one of the principal starting materials for the commercial manufacture of steroid hormones.

dioxan, 1,4-diethylene dioxide A colourless

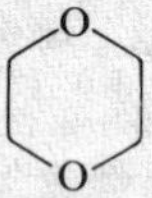

liquid with a faint and not unpleasant odour; b.p. 101°C. Manufactured by heating ethylene glycol with concentrated sulphuric acid or by passing ethylene oxide over solid $NaHSO_4$ at 120°C. Toxic. Forms complexes. Used as a solvent for cellulose acetate, resins, waxes and many organic substances.

dioxin The incorrect popular contraction of 2,3,7,8-tetrachlorodibenzo-*p*-dioxin*.

dioxygen, O_2. Formally the oxygen molecule but used for dioxygen complexes and dioxygenyl, $(O_2)^+$, salts.

dioxygenyl, $(O_2)^+$. *See* dioxygen.

dipentaerythritol, $C_{10}H_{22}O_7$, $[(CH_2OH)_3CCH_2]_2O$. Formed during the formation of pentaerythritol*; m.p. 222°C; generally present in pentaerythritol. Used in the preparation of drying oils, resin esters and alkyd resins; in the formation of plasticizers and fire retardant compositions.

dipentene *See* limonene.

diphenyl, $C_{12}H_{10}$. Volatile, colourless solid which forms large lustrous plates, m.p. 70·5°C, b.p. 254°C. Benzene vapour passed repeatedly through an iron tube heated to 720°C produces much diphenyl and polyphenyls. Bromobenzene when heated with copper powder (Ullman reaction) and the action of $CrCl_3$ or

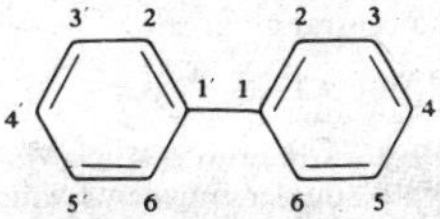

$CuCl_2$ on phenyl magnesium halides also provide reasonable quantities of diphenyl. Uses include its action as a fungistat during shipment of apples and oranges, and as a heat transfer agent (dowtherms) mixed with diphenyl ether and terphenyls.

diphenylamine, $C_{12}H_{11}N$, Ph_2NH. Colourless leaflets; m.p. 54°C, b.p. 302°C. Prepared by heating aniline and aniline hydrochloride at 200°C. It is only weakly basic, its salts with mineral acids being hydrolysed by water. It is slightly acidic, and gives an N-potassium salt. It is used as a second component in monoazo dyestuffs. Also used as a redox indicator. In sulphuric acid the colourless/violet colour change occurs between +0·76 and +0·86 volts. Diphenylamine derivatives are used as stabilizers for rocket propellants and polymers (particularly rubber).

1,2-diphenylethane, dibenzyl, bibenzyl, $C_{14}H_{14}$, $PhCH_2CH_2Ph$. Colourless crystals; m.p. 52°C, b.p. 284°C. It can be obtained by the action of sodium or copper upon benzyl chloride, by the action of $AlCl_3$ upon benzene and 1,2-dichloroethane or by heating benzoin or benzil. Used in the treatment of scabies.

diphenylguanidine, $C_{13}H_{13}N_3$, PhNHC(NH)NHPh. Colourless needles; m.p. 147°C. Prepared by the action of PbO and ammonia on thiocarbanilide. Extensively used as a rubber accelerator, particularly when mixed with other accelerators, such as zinc diethyldithiocarbamate.

diphenylpicrylhydrazyl A stable dark

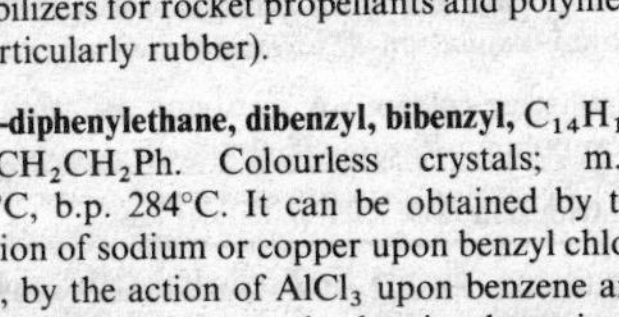

violet coloured free radical*; m.p. 137–138°C prepared by condensation of 1,1′-diphenylhydrazine with picryl chloride (2,4,6-trinitrochlorobenzene) and oxidation of the resulting orange hydrazine with PbO. Used as a dehydrogenating agent and an inhibitor of radical chain reactions.

diphos Commonly used abbreviation for a diphosphine.

diphosphines An important group of chelating agents including $Ph_2PCH_2CH_2PPh_2$, $Me_2PCH_2CH_2PMe_2$, etc. Prepared from R_2PNa and dihalides. Abbreviated diphos.

diphosphopyridine nucleotide, DPN *See* nicotinamide adenine dinucleotide.

dipole moment In a heteronuclear diatomic molecule, because of the difference in electronegativities of the two atoms, one atom acquires a small positive charge (q_+), the other a negative charge (q_-). The molecule is then said to have a *dipole moment* whose magnitude $\mu = qd$, where d is the distance of separation of the charges. With polyatomic molecules the net dipole moment is the vector sum of the dipole moments of the individual bonds within the molecule. Thus symmetrical molecules, e.g. CCl_4, may contain polar bonds but possess no net dipole moment. Measurements of dipole moments may be used to give information about the structure of complex molecules.

di-n-propyl ketone *See* butyrone.

dipterex Dimethyl(1-hydroxy-2,2,2-trichloroethyl)phosphonate, $C_4H_8Cl_3O_4P$, $(MeO)_2P(O)CH(OH)CCl_3$, an insecticide.

dipyridyl, $C_{10}H_8N_2$. Prepared by the action

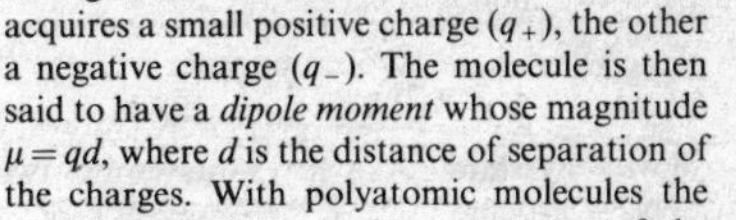

22′-dipyridyl 44′-dipyridyl

of sodium on pyridine with subsequent oxidation by air of the disodium salt formed. A mixture of isomers (22′-, 44′-, etc.) is formed. The quaternized 44′ derivative (Paraquat) and 22′ derivative (Diquat) are important as herbicides*. The 22′ isomer (bipy) is an important chelating agent; m.p. 70–73°C, b.p. 273°C.

Diquat, ethylene dipyridylium bromide,

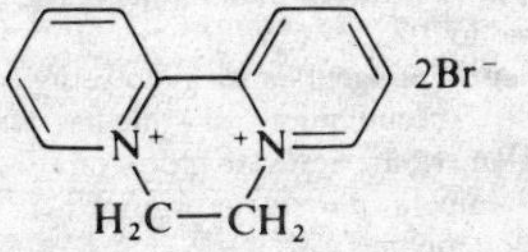

$C_{12}H_{12}Br_2N_2$. Yellow crystals; m.p. 335–340°C, containing water of crystallization; soluble in water. Made from 2,2′-dipyridyl and ethylene dibromide. A powerful herbicide which is rendered inactive on contact with the soil. *See also* paraquat.

diradical A molecule possessing two separate free electrons, e.g.

$$(C_6H_5)_2\dot{C}{\cdot}C_6H_4{\cdot}[CH_2]_4{\cdot}C_6H_4{\cdot}\dot{C}(C_6H_5)_2.$$

direct dyes Water soluble dyestuffs, generally azo dyes although some contain carboxyl groups. Dyeing is direct from a solution of the dyestuff.

disaccharides Sugars derived from monosaccharides by the elimination of a water molecule from two monosaccharides.

disacryl A hard acrolein polymer*. *See* propenal.

disazo dyestuffs Azo dyestuffs with two $-N_2-$ groups.

discharge tube A vessel, usually glass, which contains two metal electrodes and in which the passage of electricity occurs through a gas or vapour (usually at low pressure).

disinfectants Materials which destroy micro-organisms but not spores. Chlorine and hypochlorites (chlorates(I)) are examples.

dislocation A dislocation is normally a line defect in a crystal lattice although some other types are known.

dismutation reaction A particular type of metathesis reaction, e.g. $2CH_2{=}CHR$, giving $CH_2{=}CH_2$ and $CHR{=}CHR$. This reaction often occurs in the presence of catalysts.

disperse dyes A class of water-insoluble dyes generally used from an aqueous suspension the dyestuff having a high affinity for the fibre. Used particularly for nylons and synthetic fibres. The main types of dispersed dyestuffs are anthraquinone and aminomonoazo compounds.

dispersing agent A substance used in the production of emulsions or dispersions of immiscible liquids or liquids and solids. The dispersing agent may lower the interfacial surface tension (surface tension depressant) or increase the viscosity of the continuous phase (protective colloid).

Dispersing agents, such as polyethylene polyamide succinimides or methacrylate-type copolymers, are added to motor oils to disperse 'low-temperature sludge' formed in spark-ignition engines.

disproportionation A process in which a compound of one oxidation state changes to compounds of two or more oxidation states, e.g. $2Cu^+ \longrightarrow Cu + Cu^{2+}$ in aqueous solution. Alternatively a redistribution of groups around a central atom, e.g.

$$2PF_4Cl \longrightarrow PF_3Cl_2 + PF_5.$$

dissociation The process whereby a molecule is split into simpler fragments which may be smaller molecules, atoms, free radicals or ions. Particularly used in connection with photochemical, thermal and ionic (electrolytic) dissociation. The extent of dissociation is measured by the dissociation constant, K. For the process $AB \rightleftharpoons A + B$

$$K = \frac{[A][B]}{[AB]}$$

where the square brackets denote the concentrations (activities) of the species.

dissymmetric Chiral*.

distillation The separation of a liquid from a solid or another liquid by vaporization followed by condensation. The distillation can be carried out at atmospheric pressure or reduced pressure. Volatile material can be distilled from low temperature baths. *See* fractional distillation.

distillation column A rectifying or fractionation column. *See* rectification.

distribution law *See* partition law.

disulfoton, diethyl S-(2-(ethylthio)ethyl) phosphorothiolothionate, $C_8H_{19}O_2PS_3$, $(EtO)_2P(S)SCH_2CH_2SEt$. A systemic insecticide and acaricide.

disulphiram, tetraethylthiouram disulphide, $C_{10}H_{20}N_2S_4$, $Et_2NC(S)S{\cdot}SC(S)NEt_2$. Drug used in treating chronic alcoholism.

diterpene Unsaturated hydrocarbons of the empirical formula, $C_{20}H_{32}$, which may be considered as dimers of the terpenes. They may contain one or more closed carbon rings. The term is also applied to their simpler derivatives. They are mostly vegetable products. Abietic acid is a diterpene derivative.

dithiocarbamates, $M^+(R_2NCS_2)^-$. Prepared by the action of amines on CS_2 in the presence of, e.g., NaOH. Form complexes with metals (*S*-bonded, used in analysis, e.g. Cu^{2+}). Transition metal salts are used as fungicides and Zn salts are used in rubber vulcanization. Oxidized to thiuram disulphides*.

dithiolene ligands SCR·CRS groups chelated to metals. Known with many different R (e.g. CF_3, CN, $\frac{1}{2}C_6H_4$) the two sulphur atoms are at the correct positions for chelation. Dithiolene complexes frequently have trigonal-prismatic (rather than octahedral) geometry and the complexes readily undergo oxidation and reduction.

dithionous acid, hydrosulphurous acid, hyposulphurous acid, $H_2S_2O_4$. An acid known only in solution, unstable and a powerful reducing agent. Salts prepared by reducing sulphites. Dithionites contain the $(O_2S-SO_2)^{2-}$ ion, eclipsed structure.

dithiothreitol, 2,3-di-hydroxybutane-1,4-dithiol, $C_4H_{10}O_2S_2$, $HSCH_2CH(OH)CH(OH)CH_2SH$. A useful water soluble reagent for preserving thiols in the reduced state, and for reducing disulphides quantitatively to dithiols.

dithizone, diphenylthiocarbazone, $C_{13}H_{12}N_4S$. M.p. 165–169°C. A blue-black solid. Prepared from phenylhydrazine and CS_2. Used in chloroform solution as an extraction agent for most heavy metals, used particularly for the extraction and estimation of Pb.

dithranol, 1,8-dioxyanthranol, 1,8,9-trihydroxyanthracene, $C_{14}H_{10}O_3$. A yellow powder, m.p. 174–178°C. It is prepared by the reduction of 1,8-dioxyanthraquinone, and used medicinally for the treatment of psoriasis and other skin diseases.

diverse salt effect Effect of a salt without common ions on precipitation because of the effect of the ions on the activity of the precipitating ions.

divinyl acetylene, 1,5-hexadien-3-yne, $CH_2=CHC\equiv CCH=CH_2$. A colourless liquid which turns yellow on exposure to the air; it has a distinct garlic-like odour; b.p. 83·5°C. Manufactured by the controlled, low-temperature polymerization of acetylene in the presence of an aqueous solution of copper(I) and ammonium chlorides. It is very dangerous to handle, as it absorbs oxygen from the air to give an explosive peroxide. When heated in an inert atmosphere, it polymerizes to form first a drying oil and finally a hard, brittle insoluble resin. Reacts with chlorine to give a mixture of chlorinated products used as drying oils and plastics.

divinyl ether $C_4H_6O,(CH_2=CH)_2O$. Colourless liquid; b.p. 28°C. Prepared from $(ClCH_2CH_2)_2O$ with fused KOH in a NH_3 atmosphere or in ethylene glycol at over 200°C. Readily oxidized by air, slowly polymerizes to a jelly.

DME Dropping-mercury electrode*.

DMF, $Me_2NC(O)H$. Dimethyl formamide*.

DMSO, Me_2SO. Dimethyl sulphoxide.

DMT Dimethyl terephthalate. *See* phthalic esters.

DNA *See* nucleic acids.

DNBP, dinoseb, 2-*sec*-butyl-4,6-dinitrophenol, $C_{10}H_{12}N_2O$. Yellow solid, m.p. 42°C. A powerful insecticide and herbicide.

DNOC *See* 4,6-dinitro-*o*-cresol.

dodecahedral co-ordination Co-ordination by

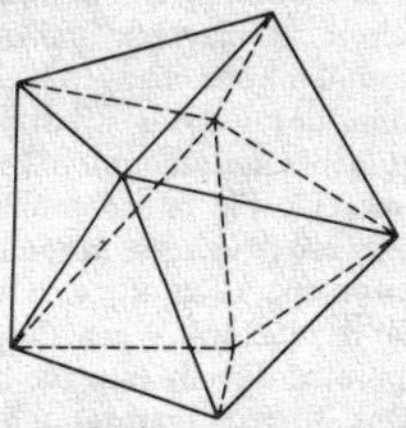

eight ligands at the corner of a dodecahedron. Found in the $[Mo(CN)_8]^{4-}$ ion. A dodecahedron has twelve triangular faces and consists of two pentagonal pyramids sharing four corners with two faces at the open edges.

dodecanedioic acid $HOOC(CH_2)_{10}COOH$ Plasticizer.

dodecylphenol, $C_{18}H_{30}O$. Made from phenol and propene tetramer. Reacts with ethylene oxides to give detergents and surfactants.

Doebner–von Miller reaction Condensation of an aromatic amine with an aldehyde or ketone in the presence of hydrochloric acid to form a quinoline derivative. A general method, thus aniline and ethanal give 2-methylquinoline (quinaldine) and *p*-phenetidine.

dolerite A widely distributed igneous rock composed essentially of plagioclase feldspar and augite. Used for road metal.

dolichols *See* polyprenols.

dolomite, $MgCO_3{\cdot}CaCO_3$. The term is loosely used for all mixtures of the two carbonates in the very common 'dolomite rock' or magnesian limestone, which is used for building purposes and for the basic linings of Bessemer converters and open-hearth steel furnaces. It is one of the most important raw materials for the production of magnesium and its salts.

Donnan membrane equilibrium This concerns the distribution of ions on each side of a membrane separating two portions of a solution of an electrolyte, e.g. NaCl, in water; on one side of the membrane a polyelectrolyte, e.g. a protein, is introduced, the molecules and ions of which cannot pass through the membrane. The distribution of the NaCl on the two sides of the membrane will be unequal and a membrane potential will be established. The theory of such systems, which are of importance in biology, was fully worked out by Donnan.

donor *See* co-ordinate bond.

dopa *See* 3,4-dihydroxyphenylalanine.

doping The name given to the practice whereby electrical, magnetic and other properties of a solid may be changed by incorporation of impurities in the host lattice. E.g. the incorporation of small amounts of Li^+ or Ga^{3+} into the nickel oxide lattice results in a respective decrease or increase in conductivity of the latter. Incorporation of e.g. B in Si forms useful semi-conductors. When the dopant is not incorporated directly into the lattice, but rather forms discrete aggregations within it, the process is referred to as domain doping.

d orbitals Atomic orbitals of azimuthal quantum number 2.

double bond Atoms can share two pairs of electrons forming a double bond (two covalent bonds). Ethene,

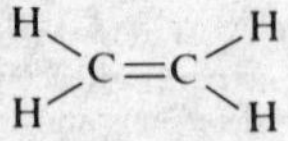

has a double bond. A double bond is stronger than a single bond although more reactive. Formally a double bond generally arises from overlap of p orbitals of two atoms which are already united by overlap of s orbitals. *See* pi bond, sigma bond.

double layer *See* electrical double layer.

double refraction Having different refractive indices in different crystallographic directions.

double salt A salt formed by crystallization from a solution of two or more components that has distinct properties in the solid but in solution behaves as a mixture of the components. E.g. iron(II) ammonium sulphate*, $FeSO_4,(NH_4)_2SO_4,6H_2O$. Double salts are either weak, easily dissociated complexes or the solid lattice has an extensive interaction, e.g. hydrogen bonding, not found in the components.

doublet Term used in n.m.r. spectroscopy to denote a signal split into an equal doublet by coupling to a further nucleus with spin $\frac{1}{2}$. Also used in group theory to denote particular energy states, e.g. with s = 2.

Dow process The process for the extraction of magnesium from sea-water by precipitation of $Mg(OH)_2$ by $Ca(OH)_2$ followed by solution of the hydroxide in hydrochloric acid.

dowtherm Mixtures of high-boiling organic substances used for purposes of heat exchange.

drier, dryer Plant or equipment for drying a material. Also metallic soaps* added to paint and varnishes to hasten drying.

Drikold A commercial name for solid CO_2.

drilling fluids Fluids pumped through a drill bit to cool the bit, pick up cuttings and seal the wall of the borehole. Fluids vary from dirty water to very viscous solutions. Additives include, e.g, barytes (to increase density), clays particularly bentonite (to increase viscosity), phosphates, tannins or lignite material (to reduce viscosity) and cotton seed hulls, mica or wood fibres (to seal the borehole).

dropping-mercury electrode, DME An electrode consisting of a column of mercury passing through a fine capillary and emerging in the solution as fine drops to give a continuously fresh surface. Used in polarography*.

dry cleaning The process of cleaning fabrics using solvents which dissolve dirt at low temperatures but do not swell fibres. The most widely used cleaning fluids are the highly inflammable hydrocarbons and chlorinated hydrocarbons, particularly C_2Cl_4.

dry ice Solid CO_2.

drying Removal of small quantities of water from a sample by physical or chemical means. Removal of the water from a solid may be speeded up by warming the sample. In freeze-drying* the water is removed at low temperature and low pressure. Drying may also refer to the removal of solvents other than water.

drying equipment Gases can be dried using equipment such as absorption* columns, while liquids may be dried by distillation, etc.

For solids various equipment and techniques are used.

drying oils Liquids which are oxidized by air to dry, hard resins. The oils consist mainly of unsaturated triglycerides (linseed, tung, safflower oils, dehydrated castor oil (DCO) and also of unsaturated hydrocarbon polymers, e.g. polybutadiene $(-CH_2CH{=}CHCH_2-)_n$. Used in coatings, paints, varnishes, enamels and lacquers. Oxidation is photochemical in nature and involves formation of C–O–C cross linkages.

dry point In a laboratory distillation the thermometer reading observed when the last drop of liquid evaporates from the lowest point in the flask.

DSC Differential scanning calorimetry. *See* thermal analysis.

DTA Differential thermal analysis. *See* thermal analysis.

dulcin, sucrol, *p*-ethoxyphenylurea, $C_9H_{12}N_2O_2$. Colourless crystals; m.p. 171–172°C. Used as a sweetening agent.

dulcitol, $C_6H_{14}O_6$. The alcohol from galac-

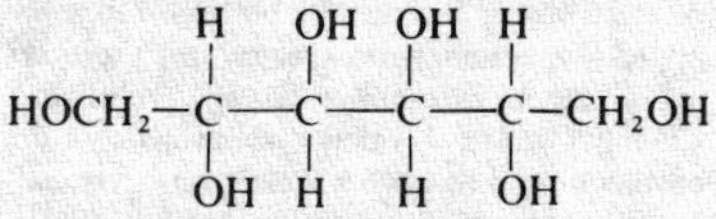

tose, m.p. 198°C, of wide occurrence in plants.

Dulong and Petit's law The product of the atomic weight and the specific heat of a metal is constant of value approximately 6·2. Although not true for all metals at ordinary temperatures, these metals and several non-metals approximate to the law at high temperatures.

Dumas' method A method for determining vapour density by determining the weight of a known volume of vapour. Also a method for determining the nitrogen content of a compound by oxidizing the compound with copper oxide and then reducing the combined nitrogen with copper to nitrogen gas, which can be measured.

duralumin An important aluminium alloy.

durene, 1,2,4,5-tetramethylbenzene, C_9H_{14}. Colourless solid; m.p. 79°C. Forms a quinone, duroquinone, a phenol, durenol and a carboxylic acid, durylic acid. Oxidized to pyromellitic dianhydride.

duriron An iron alloy containing Si (14%), Mn (2%), C (1%), S (0·1%). Particularly resistant to acids and used in chemical plant.

dutch metal A Cu–Zn alloy prepared in leaf form in imitation of gold-leaf.

Dy Dysprosium.

dyestuffs Intensely coloured compounds applied to a substrate, e.g. fibre, paper, cosmetic, hair, to give colour. React by absorption, solution, bonding. Pigments* retain their identity more closely on bonding to the substrate. Colours generally originate in electronic transitions; most dyestuffs are organic in nature but are frequently applied together with inorganic species. Classified in the colour index* by chemical nature; also classified by their method of application as acid, basic, direct, disperse, azo, sulphur, vat or fibre reactive.

dynamite A mixture of nitroglycerine with other substances to make the explosive safer to handle and to have specific explosive properties.

dysprosium, Dy. At.no. 66, at.wt. 162·50, m.p. 1409°C, b.p. 2335°C, *d* 8·5. A lanthanide*. The metal is hcp. A Dy metal foil is used to measure neutron fluxes. Dy compounds are used in lasers and phosphors.

dysprosium compounds Most compounds are in the +3 state Dy^{3+}(f^9 yellow green) $\longrightarrow$ Dy (−2·35 volts in acid) and are typical lanthanide compounds. Lower halides, DyX_2, are formed by DyX_3 plus Dy – it is not clear whether they are +2 compounds or $Dy^{3+}2X^{-}e$. Cs_3DyF_7, containing Dy(IV) is obtained using fluorine.

E

earth *See* alkaline earth metals.

ebonite, hard rubber, vulcanite A hard, black, inert material made by vulcanizing mixtures with a rubber:sulphur mixture near 2:1 and with less than 4% of the sulphur uncombined.

ebullioscopy Determination of molecular weight from rise in boiling point of solution.

ecdysone, $C_{27}H_{44}O_6$. An insect hormone,

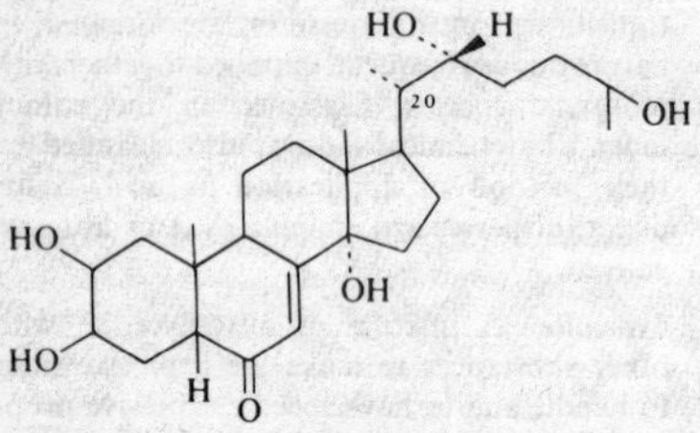

required together with 'juvenile hormone'* for larval moulting. Crystalline ecdysone from silk-worm pupae has the structure shown.

ecgonine, 3-hydroxy-2-tropanecarboxylic acid,

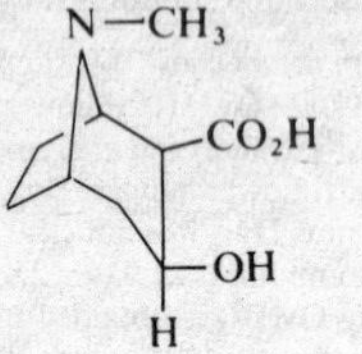

$C_9H_{15}NO_2$. Colourless crystalline material; m.p. 203°C. The major portion of the cocaine molecule, from which it may be obtained by hydrolysis with acid. Benzoylation and methylation reconvert it to cocaine. Forms a stable hydrochloride, m.p. 246°C. *See* cocaine.

eclipsed *See* conformation.

Edeleanu process An extraction process utilizing liquid sulphur dioxide for the removal of aromatic hydrocarbons and polar molecules from petroleum fractions.

EDTA Ethylenediaminetetra-acetic acid*.

EFA Essential fatty acids*.

effective atomic number rule Applied to transition metal carbonyls and to many organometallic derivatives. It is frequently found that the total number of electrons available to the central atom from its own electrons, by donation (lone pairs, pairs, etc. of electrons from π-bonding systems), by covalent bonding (each covalent bond to a substituent or another metal contributes 1 electron), and taking into account charge add up to the total number of electrons of the next noble gas. Does not apply to paramagnetic species and there are many other exceptions to the rule. E.g. $ClMn(CO)_5$.Mn 7 valence electrons, 1 from covalent bond to Cl, 5×2 electrons from carbonyls. Total 18 electrons, krypton structure.

efflorescence The loss of water by a salt hydrate arising because the pressure of water over the hydrate is greater than the partial pressure of water in the atmosphere.

effusion The passage of a gas through an orifice which has a diameter smaller than the mean free path of the gas molecules. The rate of effusion is proportional to the area of the orifice and to the mean velocity of the molecules (and hence to the reciprocal of the square root of the molecular weight of the gas). This principle is employed experimentally for investigating low vapour pressures and the molecular weight of gaseous species (e.g. in high-temperature equilibria); the apparatus used is called a Knudsen cell.

Eigen function In wave mechanics, the Schrödinger equation may be written using the Hamiltonian operator H as

$$H\psi = E\psi$$

Only certain energy values (E) will lead to solutions of this equation. The corresponding values of the wave functions are called Eigen functions or characteristic wave functions.

einstein A unit of radiant energy. If every molecule in a gram molecule of a substance absorbs one quantum of light, the total energy absorbed is $Lh\nu$, where L is the Avogadro number, h is Planck's constant, and ν the frequency of the light. $L = 6{\cdot}06 \times 10^{23}$; $h = 6{\cdot}548 \times 10^{-27}$ erg sec., and therefore for red light of frequency 4×10^{14}, one einstein $= 1{\cdot}588 \times 10^{11}$ ergs.

einsteinium, Es. At.no. 99, ^{253}Es (20 days) and ^{254}Es (270 days) may be formed by multiple neutron irradiation of Am, Pu and Cm; it is purified by ion exchange chromatography. The metal has not yet been prepared.

einsteinium compounds The element shows oxidation states of +2 and +3, the +3 state being that of a typical tripositive actinide, $EsCl_3$, EsOCl and Es_2O_3 are known. $EsBr_3$ is reduced to $EsBr_2$ by hydrogen.

Einstein's law of photochemical equivalence Before a chemical reaction can be induced by light, the light must be absorbed by some of the reactant molecules. Einstein proposed that each reacting molecule must be excited by absorption of one quantum of light. This principle is not easy to test, however, since secondary reactions occurring after the primary reaction may cause the destruction of many more of the molecules of the absorbing substance. In addition, excited molecules may lose energy by such processes as fluorescence or by undergoing deactivating collisions. In consequence the yield of reaction products may only be a fraction of that calculated from Einstein's law. The law is, however, considered to hold for the primary act of light absorption.

ejector A simple non-mechanical device used for the pumping of liquids and gases.

elaidic acid, *trans*-9-octadecenoic acid, $C_{18}H_{34}O_2$, $CH_3{\cdot}[CH_2]_7{\cdot}CH{:}CH{\cdot}[CH_2]_7{\cdot}COOH$. Crystallizes in plates; m.p. 46·5°C. It is the *trans*-isomer of oleic acid, and is prepared from it by treatment with isomerizing agents, of which 0·3% selenium at 150–220°C is the most effective. The product is an equilibrium with about 66% elaidic acid and 34% oleic acid.

elastase A proteinase from mammalian pancreas which hydrolyses elastin. It acts on peptide bonds between amino-acids which possess non-polar (alkyl) side chains.

elastin The protein of elastic tissue, ligaments and arterial walls. It is structurally related to collagen, but contains about 30% each of glycine and leucine and about 15% of proline.

elastomers Materials with rubbery properties, i.e. capable of rapid recovery from large deformation, generally synthetic materials. Elastomers include rubber, synthetic rubbers (poly-*cis*-1-4-isoprene and polybutadienes) and such materials as polychloroprene and butadiene copolymers.

electrical double layer An electrical double layer is generally formed at the interface between two phases. One phase acquires a net positive charge and the other a net negative charge. Thus in simple terms a negatively-charged dropping mercury electrode would be surrounded by an ionic sheath in which the proportion of positive ions decreases and that of negative ions increases with increasing distance from the interface.

electric cells *See* batteries.

electrochemical equivalent The weight of an element liberated from its ions or converted into them by unit quantity (1 coulomb) of electricity.

electrochemical series A series in which the elements are arranged in decreasing magnitude of their standard oxidation potentials. The use of the series allows some qualitative generalities to be drawn. Thus, for example, the higher the metal is in the series the greater the tendency for it to be oxidized. Similarly if a metal is added to a solution of another metal ion, which is below it in the electrochemical series, the metal ion will be displaced, e.g. zinc metal will displace copper from solutions of copper salts. The following series shows some of the more common elements in order (oxidation potentials in brackets):

Li	(3·04)	Na	(2·71)	H_2	(0·00)
Cs	(2·92)	Mg	(2·38)	Cu	(−0·34)
Rb	(2·92)	H^-	(2·23)	I^-	(−0·54)
K	(2·92)	Al	(1·71)	Ag	(−0·80)
Ba	(2·90)	Mn	(1·03)	Au	(−1·42)
Sr	(2·89)	Zn	(0·76)	F^-	(−2·87)
Ca	(2·76)	Cr	(0·74)		

electrochemistry The branch of chemistry concerned with electrolysis and other similar phenomena occurring when a current is passed through a solution of an electrolyte, or concerned with the behaviour of ions in solution and the properties shown by these solutions.

electrochromatography *See* electrophoresis.

electrocyclic reaction A term used to describe a highly stereospecific type of multicentre reaction involving either:

(1) The formation of a single σ-bond by

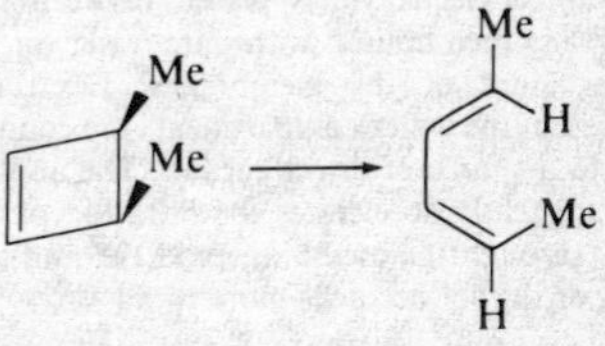

joining the ends of a linearly conjugated π-electron system, e.g. the ring opening of *cis*-1,2-dimethylbut-3-ene to the *cis*, *trans*-hexa-2,4-diene,

or (2) The opposite of this, e.g.

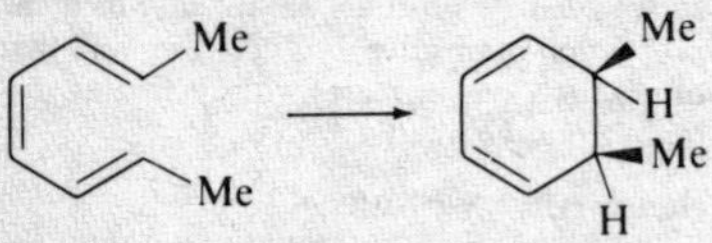

electrode An electro-conducting body which, when placed into an electrolyte, exhibits a certain electrical potential with respect to the bulk of the electrolyte. The material(s) used to introduce an electrical potential to a solid or solution.

electrode potential An electrode of any element in contact with a solution of its ions exhibits a potential called the electrode potential. The magnitude of the potential depends upon both the element and the concentration of the solution. Under conditions when no net current flows between the electrode and the solution the system is said to behave reversibly and the potential is termed the reversible electrode potential. For an element in contact with a solution of its ions at unit activity the electrode potential measured relative to the hydrogen electrode is termed the standard electrode potential (E°). The sign given to the electrode potential is arbitrary; values are usually quoted as 'oxidation' potentials, i.e. the electrode potential for the process $M \longrightarrow M^{n+}$. A positive electrode potential indicates that the oxidation process tends to occur spontaneously. A negative oxidation potential for $M \longrightarrow M^{n+}$ indicates that the process $M^{n+} \longrightarrow M$ tends to occur spontaneously.

If an appreciable current flows between the electrode and the solution, thus disturbing the reversible thermodynamic equilibrium conditions, the electrode is said to be polarized and the system is then operating under irreversible conditions.

electrodialysis In the dialysis of colloidal sols containing electrolytes, the removal of the latter is facilitated by means of an electric field. The colloidal solution containing the unwanted electrolyte is placed between two dialysing membranes, with pure water on the other sides of the membranes. Electrodes placed in the water compartments are connected to a source of direct current. The applied e.m.f. greatly accelerates the diffusion of the ions through the membrane and the purification of the sol becomes more rapid.

electrodispersion When an arc is struck between two metal electrodes under the surface of a liquid, particles of the metal are torn off and these remain colloidally dispersed in the liquid. Using this method colloidal dispersions of most metals were prepared by Bredig. A more refined electrodispersion process is obtained if the arc is maintained within a silica tube and the vaporized metal blown through a small hole in front of the arc into the dispersion medium using a stream of inert gas.

electrokinetic potential Frequently called the zeta (ζ) potential, it is the potential difference across the diffuse part of a double layer, i.e. between the rigid solution layer and the mobile part of the solution adjacent to the bulk solution. *See* electrical double layer.

electrokinetics Electrophoresis is one of several effects, collectively known as electrokinetic phenomena, which are met in systems with electrical double layers at the interface between a solid and a liquid (generally aqueous) or between one liquid and another. The other effects are electro-osmosis, streaming potential and sedimentation potential. They all arise from a partial separation of the fixed and mobile parts of the electrical double layer.

electrolysis The process of decomposing a substance, usually in solution or as a melt, by the passage of an electric current.

electrolysis, laws of *See* Faraday's laws of electrolysis.

electrolyte A substance which undergoes partial or complete dissociation into ions in solution, and thus acts as a conductor of electricity. Electrolytes with dissociation constants* greater than about 10^{-2} are called strong electrolytes.

electrolyte dissociation The process of the formation of ions in solution from an added solute. In the case of solid solutes with an ionic lattice (e.g. NaCl) there is complete dissociation into ions in dilute solution. For other types of solute (e.g. ethanoic acid) there is an equilibrium between the ions and the undissociated molecules.

electrolytic oxidation An oxidation process effected by electrolysis.

electrolytic reduction A reduction process effected by electrolysis.

electromagnetic radiation Oscillating electric and magnetic disturbances are propagated through space as waves of electromagnetic radiation. The different forms of electromagnetic radiation, which have different wavelengths and energies, include visible, infra-red and ultra-violet light as well as radiowaves and X-rays. All forms of electromagnetic radiation are propagated through a vacuum with the same velocity of $2{\cdot}98 \times 10^8$ m sec^{-1}. Electromagnetic radiation interacts with molecules according to the frequency (energy of the radiation) u.v. – visible – near i.r. (electronic transitions), i.r. (vibrations), microwave (rotations), radio frequencies (nuclear spins), X-rays (removal of electrons).

electrometer An instrument for detecting and measuring the magnitude of an electric charge.

electrometric titration Analytical method for following titrations by observing the e.m.f. of an inert electrode immersed in the solution. Of special use in titrating coloured solutions.

electron The electron is the ultimate indivisible negative charge, or particle of negative electricity, and forms an integral part of every atom. The electron has a mass of $\frac{1}{1837}$ that of a proton.

electron affinity The energy released when an electron is added to a neutral atom in the gaseous state, i.e.

$$A(g) + e \longrightarrow A^-(g) + \text{energy}$$

Electron affinities may be estimated using a Born–Haber cycle.

Electron affinities (kJ mol⁻¹)

			H				He
			73				−54
Li	Be	B	C	N	O	F	Ne
57	−66	15	123	−31	141	333	−99
Na	Mg	Al	Si	P	S	Cl	
21	−67	26	135	60	196	348	
						Br	
						340	
						I	
						297	

electron compounds Compounds formed by transition metals and one of the main group elements in which the structure of particular phases depends on the ratio of the total valency electrons available to the total number of atoms in the simplest empirical formula (making assumptions such that Fe, Co, Ni group elements have 0, Cu group 1 valence electrons). The ratios are 3:2 (β-phase bcc), 21:13 (γ-phase complex), 7:4 (ε-phase, hcp).

electron-deficient compounds Compounds in which a filled-shell configuration cannot be written for each atom if it is assumed that each bond between two atoms contains two electrons. The bonding in these compounds can be rationalized in terms of multicentre bonds in which two electrons may bind many atoms.

electron density In the investigation of crystal structure by X-ray diffraction the electron density distribution which is calculated is mapped out. Using such diagrams accurate bond lengths may be derived.

In theories of bonding the term is often used to indicate the probability of finding an electron at a particular point.

electron diffraction When a beam of electrons is allowed to interact with atoms or molecules, because of the wave properties of the electrons, they are diffracted. Electron diffraction is used to study molecular structure, particularly in the gas phase and the investigation of surfaces.

electronegativity A measure of the tendency of each atom in a stable molecule to attract electrons to itself. Thus in covalently bonded hydrogen chloride, HCl, the electrons are much closer to the chlorine than to the hydrogen; the chlorine being the more electronegative atom. Numerical values of electronegativity have been calculated by Pauling from bond dissociation energies, ionization energies and electron affinity values using the equations:

$$D_{A-B} = \tfrac{1}{2}(D_{A-A} + D_{B-B}) + 23(x_A - x_B)^2$$

and

$$125x_A = (\text{ionization energy} - \text{electron affinity})_A$$

where x is the electronegativity and D the bond dissociation energy. In general, electronegativity values increase from left to right across each row of the periodic table and decrease down each group. Some typical values are:

H				
2·1				
Li	C	N	O	F
1·0	2·5	3·0	3·5	4·0
Na	Si	P	S	Cl
0·9	1·8	2·1	2·5	3·0
				Br
				2·8
				I
				2·5

electron exchange Electron transfer* without chemical change.

electronic configuration (*See table pp. 153–4.*) (1) Atoms. The electronic configuration of an atom is the basis of the properties of the atom:

the chemical properties are determined by the electronic configuration of the outer shells. Each electron has four characteristic quantum numbers which are interrelated:

a. Principal quantum number, represented by n and having values 1, 2, 3 ...; this denotes the overall size and energy of the electron.

b. The azimuthal quantum number, represented by l and having values 0, 1, 2 ... to $n-1$. This denotes the angular momentum and thus shape of the most probable electron distribution corresponding to the shape of the orbital. Azimuthal quantum numbers with $l=0$, 1, 2 and 3 are also called s, p, d and f respectively.

c. A magnetic quantum number, represented by m, and having values $-l$, $-l+1$, ... 0 ... $l-1$, l. This denotes the orientation of the orbital. For any value of n there are one s orbital, three p orbitals, five d orbitals and seven f orbitals.

d. A spin quantum number $s = \pm\frac{1}{2}$.

An s orbital is spherically symmetrical and can contain a maximum of two electrons with opposed spins. A p orbital has a solid figure-of-eight shape; there are three equivalent p orbitals for each principal quantum number; they correspond to the three axes of rectangular coordinates.

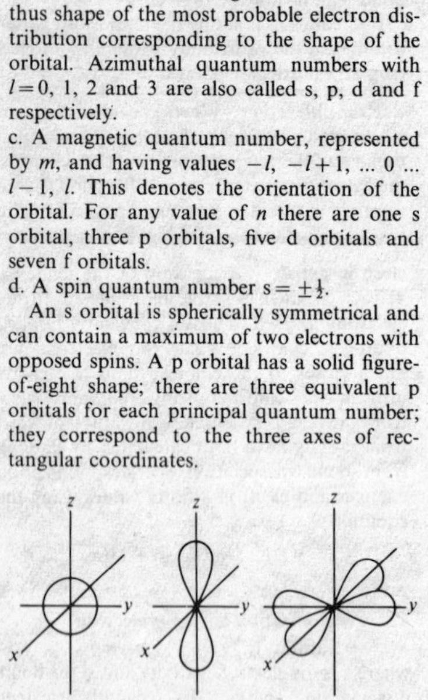

s orbital p_z orbital $d_{x^2-y^2}$ orbital

The d and f orbitals have more complex shapes; there are five equivalent d orbitals and seven equivalent f orbitals for each principal quantum number, each orbital containing a maximum of 2 electrons with opposed spins.

The above definitions must be qualified by stating that for principal quantum number 1 there are only s orbitals; for principal quantum number 2 there are only s and p orbitals; for principal quantum number 3 there are only s, p and d orbitals; for higher principal quantum numbers there are s, p, d and f orbitals.

To arrive at the electronic configuration of an atom the appropriate number of electrons are placed in the orbitals in order of energy, the orbitals of lower energy being filled first (Aufbau principle*), subject to the proviso that for a set of equivalent orbitals – say the three p orbitals in a set – the electrons are placed one in each orbital until all the orbitals are half filled (Hund's rule*). The order of energy of orbitals for most nuclei is 1s < 2s < 2p < 3s < 3p < 4s < 3d < 4p < 5s < 4d < 5p < 6s < 4f < 5d < 6p < 7s. . . . In any atom no two electrons may have all quantum numbers the same (Pauli exclusion principle*).

(2) Molecules. The electronic configurations of molecules can be built up by direct addition of atomic orbitals (LCAO method) or by considering molecular orbitals which occupy all of the space around the atoms of the molecule (molecular orbital method).

electronic transition In an atom or molecule the electrons have certain allowed energies only (orbitals). If an electron passes from one orbital to another an electronic transition occurs and there is emission or absorption of energy corresponding to the difference in energy of the two orbitals.

electron pair A pair of electrons in one orbital with opposed spins; a lone pair. Often the pair of electrons in a covalent or co-ordinate bond.

electron spin Properties of electrons can only be explained in terms of the electrons having spin, $s = \pm\frac{1}{2}$.

electron spin resonance, esr In the absence of any external fields, all spatial orientations of the electron spin are equally probable. On the application of a magnetic field, however, the spins are aligned either parallel or antiparallel to the direction of the magnetic field vector. The energy difference between these two states is $g\beta H$ where g is the Landé factor, β is the Bohr magneton and H is the strength of the applied magnetic field. Boltzmann's law governs the distribution of spins between the two states, and at normal temperatures there is a slight excess in the lower energy state.

In an electron spin resonance spectrometer, transitions between the two states are brought about by the application of the quantum of energy $h\nu$ which is equal to $g\beta H$. The resonance condition is defined when $h\nu = g\beta H$ and this is achieved experimentally by varying H keeping the frequency (ν) constant. Esr spectroscopy is used extensively in chemistry in the identification and elucidation of structures of radicals.

electron transfer reactions Reactions which involve transfer of electrons from one group to another, i.e. oxidation and reduction.

electron transport chain The collection of flavoproteins*, ubiquinones and cytochromes in mitochondria which is a coupled oxidation-reduction system. The passage of electrons

Electronic configuration of the elements

		At.no.	*1s*	*2s*	*2p*
hydrogen	H	1	1		
helium	He	2	2		
lithium	Li	3	2	1	
beryllium	Be	4	2	2	
boron	B	5	2	2	1
carbon	C	6	2	2	2
nitrogen	N	7	2	2	3
oxygen	O	8	2	2	4
fluorine	F	9	2	2	5
neon	Ne	10	2	2	6

			1s	*2(sp)*	*3s*	*3p*	*3d*	*4s*	*4p*
sodium	Na	11	2	8	1				
magnesium	Mg	12	2	8	2				
aluminium	Al	13	2	8	2	1			
silicon	Si	14	2	8	2	2			
phosphorus	P	15	2	8	2	3			
sulphur	S	16	2	8	2	4			
chlorine	Cl	17	2	8	2	5			
argon	Ar	18	2	8	2	6			
potassium	K	19	2	8	2	6		1	
calcium	Ca	20	2	8	2	6		2	
scandium	Sc	21	2	8	2	6	1	2	
titanium	Ti	22	2	8	2	6	2	2	
vanadium	V	23	2	8	2	6	3	2	
chromium	Cr	24	2	8	2	6	5	1	
manganese	Mn	25	2	8	2	6	5	2	
iron	Fe	26	2	8	2	6	6	2	
cobalt	Co	27	2	8	2	6	7	2	
nickel	Ni	28	2	8	2	6	8	2	
copper	Cu	29	2	8	2	6	10	1	
zinc	Zn	30	2	8	2	6	10	2	
gallium	Ga	31	2	8	2	6	10	2	1
germanium	Ge	32	2	8	2	6	10	2	2
arsenic	As	33	2	8	2	6	10	2	3
selenium	Se	34	2	8	2	6	10	2	4
bromine	Br	35	2	8	2	6	10	2	5
krypton	Kr	36	2	8	2	6	10	2	6

			1s	*2(sp)*	*3(spd)*	*4s*	*4p*	*4d*	*5s*	*5p*
rubidium	Rb	37	2	8	18	2	6		1	
strontium	Sr	38	2	8	18	2	6		2	
yttrium	Y	39	2	8	18	2	6	1	2	
zirconium	Zr	40	2	8	18	2	6	2	2	
niobium	Nb	41	2	8	18	2	6	4	1	
molybdenum	Mo	42	2	8	18	2	6	5	1	
technetium	Tc	43	2	8	18	2	6	6	1	
ruthenium	Ru	44	2	8	18	2	6	7	1	
rhodium	Rh	45	2	8	18	2	6	8	1	
palladium	Pd	46	2	8	18	2	6	10		
silver	Ag	47	2	8	18	2	6	10	1	
cadmium	Cd	48	2	8	18	2	6	10	2	
indium	In	49	2	8	18	2	6	10	2	1
tin	Sn	50	2	8	18	2	6	10	2	2
antimony	Sb	51	2	8	18	2	6	10	2	3
tellurium	Te	52	2	8	18	2	6	10	2	4

Electronic configuration of the elements

		At.no.	1s	2(sp)	3(spd)	4s	4p	4d	5s	5p
iodine	I	53	2	8	18	2	6	10	2	5
xenon	Xe	54	2	8	18	2	6	10	2	6

			(1,2,3)	4s	4p	4d	4f	5s	5p	5d	6s
caesium	Cs	55	28	2	6	10		2	6		1
barium	Ba	56	28	2	6	10		2	6		2
lanthanum	La	57	28	2	6	10		2	6	1	2
cerium	Ce	58	28	2	6	10	2	2	6		2
praseodymium	Pr	59	28	2	6	10	3	2	6		2
neodymium	Nd	60	28	2	6	10	4	2	6		2
promethium	Pm	61	28	2	6	10	5	2	6		2
samarium	Sm	62	28	2	6	10	6	2	6		2
europium	Eu	63	28	2	6	10	7	2	6		2
gadolinium	Gd	64	28	2	6	10	7	2	6	1	2
terbium	Tb	65	28	2	6	10	9	2	6		2
dysprosium	Dy	66	28	2	6	10	10	2	6		2
holmium	Ho	67	28	2	6	10	11	2	6		2
erbium	Er	68	28	2	6	10	12	2	6		2
thulium	Tm	69	28	2	6	10	13	2	6		2
ytterbium	Yb	70	28	2	6	10	14	2	6		2
lutetium	Lu	71	28	2	6	10	14	2	6	1	2

			(1,2,3,4)	5s	5p	5d	6s	6p	7s
hafnium	Hf	72	60	2	6	2	2		
tantalum	Ta	73	60	2	6	3	2		
tungsten	W	74	60	2	6	4	2		
rhenium	Re	75	60	2	6	5	2		
osmium	Os	76	60	2	6	6	2		
iridium	Ir	77	60	2	6	7	2		
platinum	Pt	78	60	2	6	9	1		
gold	Au	79	60	2	6	10	1		
mercury	Hg	80	60	2	6	10	2		
thallium	Tl	81	60	2	6	10	2	1	
lead	Pb	82	60	2	6	10	2	2	
bismuth	Bi	83	60	2	6	10	2	3	
polonium	Po	84	60	2	6	10	2	4	
astatine	At	85	60	2	6	10	2	5	
radon	Rn	86	60	2	6	10	2	6	
francium	Fr	87	60	2	6	10	2	6	1
radium	Ra	88	60	2	6	10	2	6	2

			(1,2,3,4)	5s	5p	5d	5f	6s	6p	6d	7s
actinium	Ac	89	60	2	6	10		2	6	1	2
thorium	Th	90	60	2	6	10		2	6	2	2
protactinium	Pa	91	60	2	6	10	2	2	6	1	2
uranium	U	92	60	2	6	10	3	2	6	1	2
neptunium	Np	93	60	2	6	10	5	2	6		2
plutonium	Pu	94	60	2	6	10	6	2	6		2
americium	Am	95	60	2	6	10	7	2	6		2
curium	Cm	96	60	2	6	10	7	2	6	1	2
berkelium	Bk	97	60	2	6	10	8	2	6	1	2
californium	Cf	98	60	2	6	10	10	2	6		2
einsteinium	Es	99	60	2	6	10	11	2	6		2
fermium	Fm	100	60	2	6	10	12	2	6		2
mendelevium	Md	101	60	2	6	10	13	2	6		2
nobelium	No	102	60	2	6	10	14	2	6		2
lawrencium	Lr	103	60	2	6	10	14	2	6	1	2
rutherfordium	Rf	104	60	2	6	10	14	2	6	2	2

along this chain from reduced NAD (usually generated by the citric acid cycle*) to eventually reduce oxygen is accompanied by the production of ATP. *See also* oxidative phosphorylation.

electro-osmosis If a direct current is passed through a tube of liquid which is fitted with two electrodes, one on either side of a diaphragm, the potential difference between the diaphragm and the liquid manifests itself by a movement of the liquid towards one or other of the electrodes. This phenomenon, which is akin to electrophoresis*, is termed electro-osmosis. It is modified by the presence of acids, bases and salts. The hydrogen-ion concentration at which there is no movement of liquid with respect to the membrane is termed the isoelectric point. *See* colloid.

electrophilic reagents These act by acquiring electrons, or a share in electrons, from a foreign molecule; examples are the bromonium ion, Br^+ and the nitronium ion, $NO_2{}^+$. The reagent can frequently only accommodate the extra electrons by undergoing fission, thus acidity is a special case of *electrophilicity*, the affinity for external electrons in general.

electrophilic substitution The exchange of an atom or group in a molecule for an entering electrophile according to

$$R-X+E^+ \longrightarrow R-E+X^+$$

The nitration, sulphonation and Friedel–Crafts acylation of aromatic compounds (e.g. benzene) are typical examples of electrophilic aromatic substitution.

electrophoresis The migration of charged particles, colloidal particles or ions through a solution under an electric field. Variation of pH can stop movement at the isoelectric point*. In electro-osmosis there is a constant flow of liquid relative to a stationary surface. Electrophoresis is used in analysis, particularly in biochemical applications (ionography, zone electrophoresis, electrochromatography) for both identification and separation.

electrophoretogram The separated species on, e.g., a column after electrophoresis.

electroplating The deposition of metals from solution in the form of a layer on other metals or, e.g., plastics, by passage of an electric current. A metallic article to be plated is made one electrode in a bath containing the other metal as aquo-ions or other complexes. Current density, pH, concentration, etc. all have a very marked effect on the adhesion and texture of the deposited metal. Among metals used for electroplating are Ag, Cr, Ni, Zn.

electrostatic precipitators Plants for the removal of fine suspended matter from a gas that depend for their action on the ionization of the gas between two highly charged electrodes. The ions so formed attach themselves to the dispersed particles, conferring a charge, with the result that the latter then migrate to the appropriate electrode.

electrovalent bond, polar bond Bonding by electrostatic attraction.

electrovalent compounds Compounds in which the major binding force is electrostatic attraction between positive and negative ions. The lattice does not contain discrete molecules and the ions are packed together to occupy space most efficiently (generally close packing of anions). Electrovalent compounds are distinguished from covalent compounds by the conductivity of melts and solutions in polar solvents, low volatility, solubility in polar solvents.

element A substance which cannot be further divided by chemical methods. The basic substances which build up chemical compounds. An element is defined by its atomic number (nuclear charge and electronic configuration). The table on pp. 153–4 shows the elements and their normal, ground-state, electronic configurations.

elementary particles The fundamental particles found in nature, e.g. proton, neutron, electron, positron.

elements, abundance of The relative abundance of elements depends on whether measurements are made on the crust, meteorites, core, etc. For the earth's crust (E) abundance of the lighter elements (expressed relative to $Si = 10^6$) are:

	E	A
Na	105 000	0·51
Mg	68 000	11·25
Al	300 000	1·07
Si	10^6	13·84
P	3 430	0·08
S	1 640	2·74
K	62 000	0·06
Ca	80 000	1·07
Ti	9 000	0·06
Mn	1 950	0·13
Fe	126 000	38·8
Ni	100	2·70
O	$2·96 \times 10^6$	27·17

(A) represents the average % composition of the earth by weight.

elements of symmetry The symmetry elements, centres, axes, planes of symmetry, present in a molecule, crystal lattice or crystal. Together with the arrangement of atoms the elements of symmetry spell out the space-group* of a crystal or the point-group of a molecule.

elevation of boiling point The boiling point of a solvent is raised by the presence of a solute. For small concentrations of the solute the rise in the boiling point is proportional to the number of solute particles present in the solution. The elevation caused by 1 mole of solute in 1 litre of solvent is termed the molecular elevation constant. (*See* colligative properties.)

elixirs Pharmaceutical solutions, frequently containing alcohol, used as sweetening or flavouring agents for drugs.

elutriation The process of separating a material into fractions of various sizes by allowing it to settle against an upward moving stream of fluid, usually air or water.

emanation An obsolete name for radon.

embonic acid *See* pamoic acid.

emerald The grass-green variety of beryl; contains some Cr^{3+}.

emerald green, Paris green, Schweinfürter green Basic copper ethanoate arsenates(III) prepared from verdigris (or other basic copper salt), sodium arsenate(III) and ethanoic acid. Used in insecticides for spraying fruit trees. Readily decomposed to soluble arsenic compounds so use is very restricted.

emery Impure α-Al_2O_3 (corundum) containing some iron oxide (magnetite). Used as an abrasive and polishing agent.

emetine, $C_{29}H_{40}N_2O_4$. Alkaloid extracted

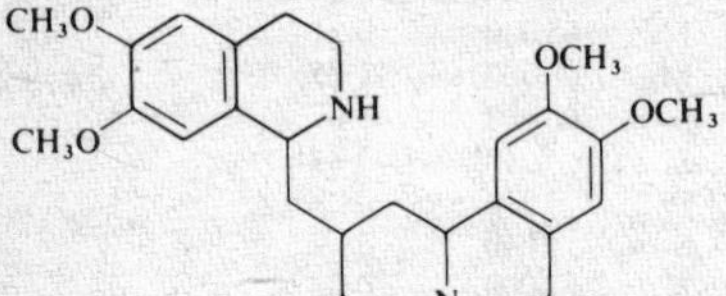

from ipecacuanha or obtained by the methylation of cephaeline, another ipecacuanha alkaloid. A white amorphous powder; m.p. 74°C. It is a powerful emetic used against amoebic dysentry.

emission spectroscopy An analytical technique in which atoms in excited states in flames or electric arcs are examined by their emission spectra. Emission spectroscopy is also used to investigate the energy levels of atoms in ground and excited states.

emulsification The preparation of a suspension, or emulsion, of one liquid in another.

emulsifier
(1) Another name for an emulsifying agent.
(2) Any machine for producing an emulsion.

emulsifying agent Dilute suspensions of oil in water behave as typical hydrophobic sols, and it is not possible to increase the concentration of oil unless a stabilizing material is added which decreases the interfacial tension. Such substances, known as emulsifying agents, are generally long-chain compounds containing a hydrophilic (carboxyl or sulphonate) group at one end of the molecule; these become orientated at the interface, the hydrophilic end projecting into the water. There are also solid emulsifying agents (e.g. carbon black) which possess widely different angles of contact for the two phases.

An emulsifying agent generally produces such an emulsion that the liquid in which it is most soluble forms the external phase. Thus the alkali metal soaps and hydrophilic colloids produce O/W emulsions, oil-soluble resins the W/O type (*see* emulsion).

The commonest emulsifying agents are the soaps, which have the disadvantage of coagulating in hard waters. For edible and pharmaceutical emulsions gums of various sorts are employed. Of recent years, large numbers of synthetic organic emulsifying agents have been developed. Many of these are sulphonates or quaternary ammonium compounds.

emulsion An emulsion is a disperse system in which both phases are liquids; generally one of the liquids is water or an aqueous solution, and the other an oil or other water-immiscible liquid. With a given pair of liquids (e.g. an oil and water) two distinct types of emulsion are possible according to which forms the dispersion medium. Emulsions in which an oil is dispersed in water are termed oil-in-water (O/W) emulsions; those in which the water is the disperse phase are water-in-oil (W/O) emulsions. It is possible to have any phase ratio in an emulsified system. Thus naturally-forming emulsions such as that of the condenser water of an engine may contain only a small fraction of oil, while semi-solid emulsion pastes may contain 95% of oil dispersed in water. In order to stabilize all but the most dilute emulsions, it is necessary to have present a third substance, the emulsifying agent*, to stabilize the system.

Emulsions have many industrial uses; they

are used widely as foods and pharmaceutical preparations, cosmetics, horticultural and insecticide sprays, as oil-bound water paints, as lubricants, for spraying of roads, etc.

emulsion stabilizers Compounds which are weakly adsorbed on the grain surface of a photographic emulsion where they displace the chemical sensitizer and which prevent the appearance of fog. *See* antifoggants.

enamels, vitreous enamel A glassy coating fused to metal at high temperatures, acting as a protective and decorative surface.

enamines Vinylamines; those having a hydrogen atom on nitrogen are usually unstable

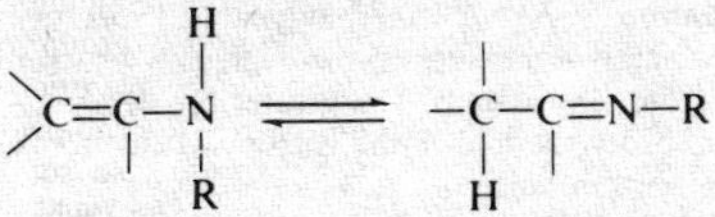

and rearrange to the corresponding imine, but those with no hydrogen on nitrogen are stable and important synthetic intermediates. These are usually easily prepared from ketones

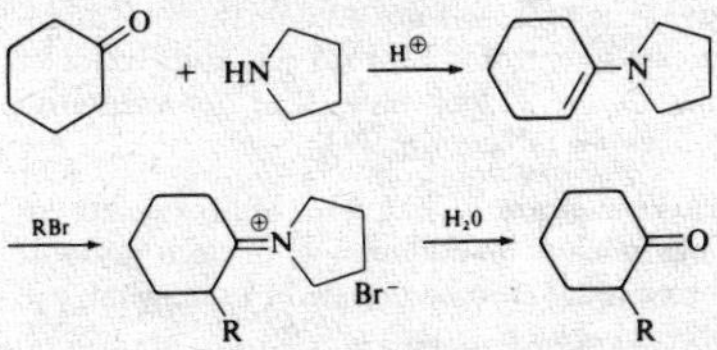

with an α-hydrogen and a secondary amine, e.g. pyrrolidine, with an acid catalyst. Since many undergo C-alkylation smoothly, and the substituted enamine can be cleaved easily back to a ketone, a facile substitution of the α-position of ketones is possible.

enantiomers Isomers differing in their configuration at a chiral atom.

enantiomorphic Possessing neither a plane of symmetry nor a centre of symmetry. Such a molecule cannot be brought into coincidence with its mirror image. The two enantiomorphs (the molecule and its mirror image) rotate the plane of polarized light in opposite directions. The interrelation between the two isomers is enantiomorphism.

enantiotopic The two identical ligands *a* attached to a prochiral* centre in [*Caabc*] are said to be enantiotopic if their separate replacement by a group different from *a*, *b* or *c* gives rise to a pair of enantiomers [*Cadbc*] and to be diastereotopic if diastereomers are obtained on replacement. E.g.

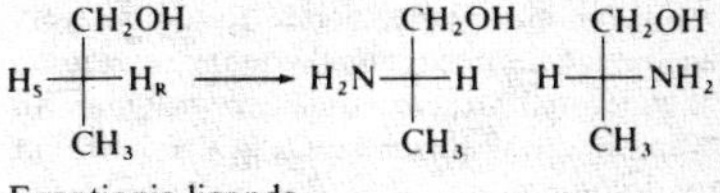

Enantiopic ligands at a prochiral centre

(enantiomers)

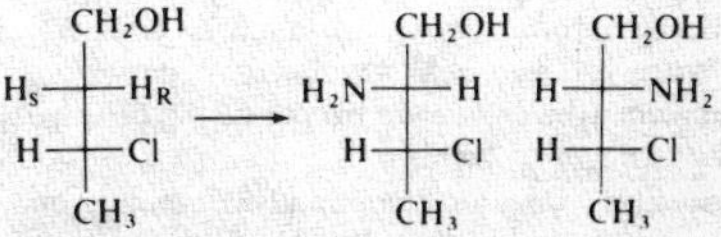

Diastereotopic ligands at a prochirol centre

diastereoisomers

enantiotropy The reversible transformation of one form or one allotrope into another at a definite transition temperature, e.g. red $HgI_2 \rightleftharpoons$ yellow HgI_2 at 126°C.

enclosure compounds *See* clathrate compounds.

endo- A prefix used to describe the orienta-

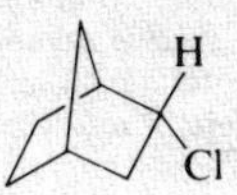

endo-norbornyl chloride

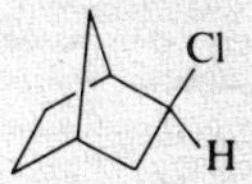

exo-norbornyl chloride

tion of atoms or groups with respect to the rest of the molecule, especially with polycyclic molecules, but not restricted to these classes. It indicates that the substituent is facing the inside (convex side) of the molecule. In certain instances such identification may be doubtful. The stereochemically related isomer is given the prefix *exo*-.

endopeptidases A class of peptide-cleaving enzymes which act at points in the interior of polypeptide chains rather than at the terminal groups. Important examples are trypsin, chymotrypsin and pepsin – all concerned in digestion. *See also* exopeptidases.

endothermic reaction A reaction in which heat is absorbed. According to the laws of thermodynamics a spontaneous endothermic reaction must be accompanied by an increase in entropy of the system and surroundings.

end point The stage in a titration when the indicator (in its widest sense) undergoes maximum change (colour, etc.) for a small amount of added titrant. The end point corresponds to a definite hydrogen or other ion concentration or redox potential. It does not necessarily correspond to the equivalence point.

end point, final boiling point In a laboratory distillation the maximum thermometer reading obtained in the test and usually coinciding with evaporation of all the liquid in the flask, but *see* decomposition point.

endrin The insecticidal product containing not less than 92% by weight of 1,2,3,4,10,10-hexachloro-6,7-epoxy-1,4,4a,5,6,7,8,8a-octahydro-1,4-*endo*,*endo*-5,8-dimethanonaphthalene (the *endo-endo* isomer of the principal constituent of dieldrin*).

enediols Organic compounds containing one carbon-carbon double bond and two hydroxyl groups.

energy levels Energy taken up by a molecule, say by the absorption of light, may cause an electronic transition; it may increase the energy of rotation of the molecule as a whole, or may increase the energy of vibration of the nuclei of the constituent atoms relative to one another. The changes all occur in accordance with the quantum hypothesis, i.e. the changes in energy are not continuous, but take place only in definite steps. Thus, every electronic orbital is associated with a specific energy value, or energy level. So the rotational and vibrational states have only certain restricted energy values, or energy levels. Diagrammatically, these are usually represented as a series of horizontal, parallel straight lines separated by distances proportional to the differences in energy of the various states.

enkephalins Two pentapeptides, *met*-enkephalin (H-tyr-gly-gly-phe-met-OH) and *leu*-enkephalin (H-tyr-gly-gly-phe-leu-OH) which occur naturally in the brain and have similar analgesic and other properties to the opiates*, morphine*, codeine*, pethidine, etc.

enols, enolic compounds The tautomeric forms of certain ketones, particularly those with an α-CH group and containing the >C=C−OH grouping. This grouping is also present in phenols, and enols resemble the phenols in several of their reactions. The hydrogen atom of the −OH group is acidic and is replaceable by sodium and other metals. The copper compounds of 1,3-diketones (acetylacetonates) and β-keto esters are of this type. The sodium compounds react with alkyl and acyl chlorides to give derivatives of the type $\underset{\mathrm{R}}{\overset{|}{\mathrm{C}}}\!-\!\overset{|}{\mathrm{C}}\!=\!\mathrm{O}$. The hydrogen of the −OH group can be replaced by an acyl group if the reaction is carried out in pyridine solution. *See* isomerism.

enthalpy (*H*) A thermodynamic state function defined as $H = U + PV$, where U is the internal energy and P and V are the pressure and volume respectively. For any process which occurs at constant pressure the heat absorbed or evolved is equal to the enthalpy change if the only work is pressure/volume work, i.e. $\Delta H = \Delta U + P\Delta V$ for a process at constant pressure. In chemical reactions the enthalpy change (ΔH) is related to the changes in free energy (ΔG) and entropy (ΔS) by the equation $\Delta G = \Delta H - T\Delta S$. Enthalpy is usually expressed in units of kilojoules per mole. For values *see* table on p. 393.

entrainment The carrying forward by a stream of gas or vapour of fine liquid droplets.

entropy (*S*) A thermodynamic quantity which is a measure of the degree of disorder within any system. The greater the degree of disorder the higher the entropy. Thus, for a given substance

$$S_{\text{gas}} > S_{\text{liquid}} > S_{\text{solid}}$$

Any change taking place which results in an increase in entropy has a positive entropy change (ΔS). Most spontaneous thermodynamic processes are accompanied by an increase in entropy. Entropy has units of joules per degree K per mole. For representative values *see* table on p. 393.

enyl complexes Complexes of organic groupings in which an odd number of electrons must be considered to be donated to the metal, e.g.

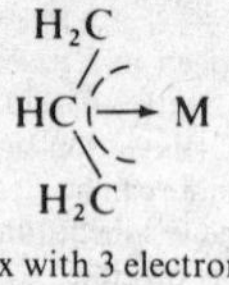

a π-allyl complex with 3 electrons donated (trihapto, 3 atoms bonded).

enzymes Proteins which catalyse reactions with a high degree of specificity and efficiency. Enzymes are present in all living organisms and are responsible for catalysing most of the reactions which take place in a cell, e.g. the hydrolysis of fats, sugars and proteins and their re-synthesis. They also catalyse the many forms of oxidation and reduction which provide energy for the cell.

Enzymes are obtained from plants, animals and micro-organisms by extraction with a suitable solvent, preferably after the cell structure has been destroyed by drying or grinding. They can be purified by precipitation and re-solution and by fractional absorption and elution. Many enzymes have been obtained crystalline.

Enzymes often need for their activity the presence of a non-protein portion, which may be closely combined with the protein, in which case it is called a prosthetic group, or more loosely associated, in which case it is a coenzyme. Certain metals may be combined with the enzyme such as copper in ascorbic oxidase and selenium in glutathione peroxidase. Often the presence of other metals in solution, such as magnesium, are necessary for the action of particular enzymes.

One outstanding feature of enzymes is that many of them are highly specific. Those that act on carbohydrates are particularly so, the slightest change in the stereochemical configuration of the molecule being sufficient to make a particular enzyme incompatible and unable to effect hydrolysis.

Enzymes act by lowering the overall activation energy of a reaction sequence by involving a series of intermediates, or a mechanism, different from the spontaneous uncatalysed reaction.

Most enzymes work best within a narrow pH range and are susceptible to a wide variety of compounds which inhibit or sometimes promote the activity. The majority of enzymes work most efficiently at around 40°C and at higher temperatures are rapidly destroyed.

Enzymes are classified in terms of the reactions which they catalyse and were formerly named by adding the suffix 'ase' to the substrate or to the process of the reaction. In order to clarify the confusing nomenclature a system has been developed by the International Union of Biochemistry and the International Union of Pure and Applied Chemistry (*see* 'Enzyme Nomenclature', Elsevier, 1973). The enzymes are classified into divisions based on the type of reaction catalysed and the particular substrate. The suffix 'ase' is retained and recommended trivial names and systematic names for classification are usually given when quoting a particular enzyme. Any one particular enzyme has a specific code number based upon the new classification.

Six main divisions of enzymes are recognized.

(1) *Oxidoreductases*. Enzymes catalysing redox reactions. The substrate which is oxidized is regarded as the hydrogen donor. This group includes the trivially named enzymes, dehydrogenases, oxidases, reductases, peroxidases, hydrogenases and hydroxylases.

(2) *Transferases*. Enzymes which transfer a group, possibly a methyl group or a glycosyl group, from one compound to another. The name of the group transferred is usually incorporated into the recommended name. Often the donor is a cofactor (coenzyme). Examples are the aminotransferases and the methyltransferases.

(3) *Hydrolases*. Enzymes catalysing the hydrolytic cleavage of C–O, C–N and C–C bonds. The systematic name always includes hydrolase but the recommended name is often formed by the addition of 'ase' to the substrate. Examples are esterases, glucosidases, peptidases, proteinases, phospholipases. Other bonds may be cleaved besides those cited, e.g. during the action of sulphatases and phosphatases.

(4) *Lyases*. These enzymes cleave C–C, C–O, C–N and other bonds by elimination leaving double bonds or conversely add groups to double bonds. This group includes decarboxylases, hydratases, dehydratases and some carboxylases.

(5) *Isomerases*. These catalyse the structural or geometric changes within a molecule. The division includes racemases, epimerases, *cis-trans*-isomerases, tautomerases and mutases.

(6) *Ligases* (*synthetases*). Enzymes catalysing the joining together of two molecules coupled with the hydrolysis of a pyrophosphate bond in ADP or a similar triphosphate. They include some carboxylases and many enzymes known as synthetases.

Often there is more than one enzyme catalysing the same reaction.

Enzymes are used in baking, cheese manufacture, wine-making, brewing and distillation, pharmaceuticals, leather tanning, paper manufacture, adhesives, sewage disposal, animal feeds and in detergents.

ephedrine, $C_{10}H_{15}NO$, PhCH(OH)CH(NHMe)Me. Colourless crystals containing about $\frac{1}{2}H_2O$; m.p. 40°C, b.p. 225°C. It is obtained from various species of *Ephedra*, or may be synthesized. Its pharmacological action resembles that of adrenaline. It is used to relieve catarrh, asthma and hay fever.

epicamphor, β-camphor
Colourless crystals; m.p. 182°C.

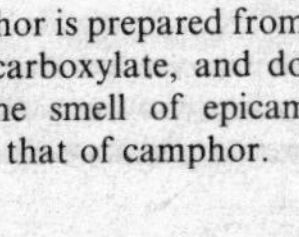

(–)Epicamphor is prepared from methyl-(+)-bornylene-3-carboxylate, and does not occur naturally. The smell of epicamphor differs slightly from that of camphor.

epichlorohydrin, 3-chloropropylene oxide, 2-chloro-1,3-epoxypropane,

$$CH_2Cl$$
$$HC$$
$$H_2C \diagdown O$$

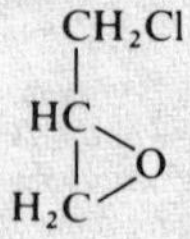

C_3H_5ClO. Colourless liquid with an odour resembling that of chloroform; b.p. 115–117°C. Prepared by treating glycerol dichlorohydrins with solid NaOH at 25–30°C, allyl chloride plus chlorine water, or chlorination of acrolein. Reduced by sodium amalgam to propenol oxidized by nitric acid to β-chlorolactic acid. Reacts with alcohols in the presence of potassium hydroxide to give diethers of glycerol. Used in the manufacture of glycerol, glycerol ethers, triethyl oxonium salts and epoxy resins.

epimerism A type of isomerism shown by substances which contain several asymmetric centres but differ in the configuration of one only of these.

In carbohydrate chemistry the term is restricted to sugars or their derivatives which differ only in the orientation of the groups attached to the carbon atom next to the potential aldehyde group of the sugar or to the corresponding group in the sugar derivative. Thus D-gluconic acid,

$$HOH_2C-\overset{\displaystyle H}{\underset{\displaystyle OH}{C}}-\overset{\displaystyle H}{\underset{\displaystyle OH}{C}}-\overset{\displaystyle OH}{\underset{\displaystyle H}{C}}-\overset{\displaystyle H}{\underset{\displaystyle OH}{C^*}}-COOH$$

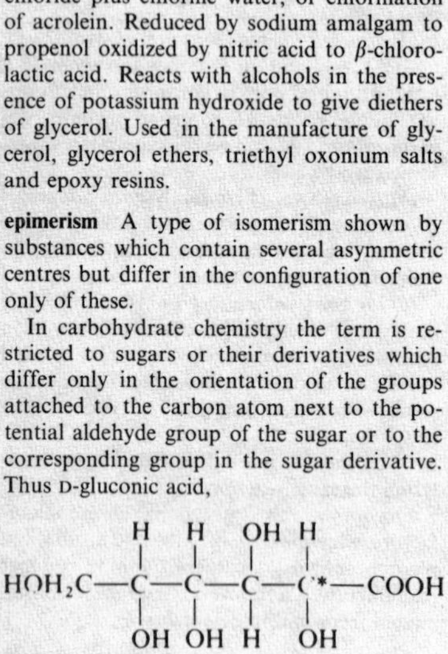

and D-mannonic acid.

$$HOH_2C-\overset{\displaystyle H}{\underset{\displaystyle OH}{C}}-\overset{\displaystyle H}{\underset{\displaystyle OH}{C}}-\overset{\displaystyle OH}{\underset{\displaystyle H}{C}}-\overset{\displaystyle H}{\underset{\displaystyle H}{C^*}}-COOH$$

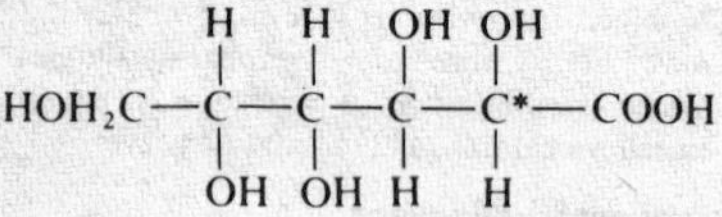

are epimeric at the starred atoms and each may be partially converted into the other by heating in pyridine or quinoline, such a process being called epimerization. The sugars themselves are epimerized to some extent when treated with dilute alkalis, but side-reactions also occur.

epimerization See epimerism.

epinephrine See adrenaline.

epoxidation The addition of an oxygen bridge across an alkene bond, achieved in special cases by oxygen and a catalyst, but more usually by a per-acid* such as perbenzoic acid or peracetic acid, or substituted analogues of these acids.

1,2-epoxide polymers Polymers of ethylene oxide, oxirane, CH_2–CH_2 and its derivatives
$$\diagdown O \diagup$$
by oxyalkylation with compounds in the presence of Lewis acids or bases or by catalytic polymerization over catalysts. The polymers are polyethylene glycols and range from liquids (lubricants, binders, solvents) to very high mol.wt. solids (for sheet, film, etc., thickeners, binders). Other epoxides, e.g. $CF_2 \cdot CF_2 \cdot O$ can also form useful polymers. See epoxy resins.

epoxy A prefix indicating a $\overset{\displaystyle O}{C-C}$ group in a molecule, e.g. $\overset{\displaystyle O}{CH_2-CH_2}$, epoxyethane.

1,2-epoxypropane See propylene oxide.

epoxy resins Polyethers formed by condensation of e.g. epichlorohydrin with polyols, such as bisphenol A, or by epoxidation of Diels–Alder adducts with peroxy compounds. The polyethers need curing agents to convert to resins. The resins are thermosetting, tough, adhesive, chemically resistant and have favourable electrical properties. They are used as adhesives (16%) protective coatings (48%) and in composites (22%). U.S. production 1981 135 000 tonnes.

Epsom salts, $MgSO_4,7H_2O$. See magnesium sulphate.

equatorial See conformation.

equilenin, $C_{18}H_{18}O_2$. M.p. 259°C (decomp.).

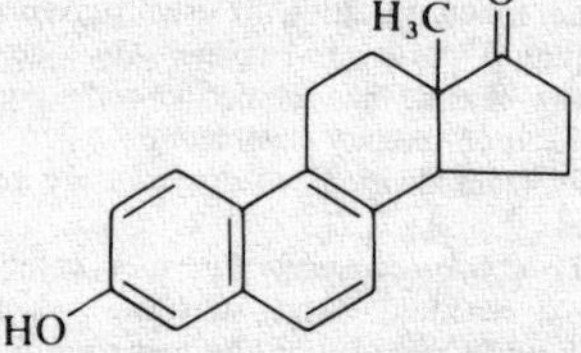

An oestrogen* found in the urine of pregnant mares. It is not so physiologically active as equilin.

equilibrium constant According to the law of mass action, for any reversible chemical reaction $aA + bB \rightleftharpoons cC + dD$, the equilibrium constant (K) is defined as

$$K=\frac{[C]^c[D]^d}{[A]^a[B]^b}$$

where [A] . . ., etc. are the active masses of A . . ., etc., more accurately termed the thermodynamic activities.

In thermodynamic terms the equilibrium constant is related to the standard chemical potential by the equation

$$\Delta\mu^\circ = -RT\ln K,$$

where R is the gas constant.

For gas reactions where the gases are assumed to follow ideal behaviour this equation becomes $\Delta G^\circ = -RT\ln K_p$, where K_p is defined in terms of the partial pressures of reactants and products. Thus for the general reaction above,

$$K_p = \left(\frac{p_C{}^c \cdot p_D{}^d}{p_A{}^a \cdot p_B{}^b}\right).$$

equilibrium diagram A simplified boiling point diagram which shows graphically, for a liquid mixture, the composition of the vapour or solid which is in equilibrium with the liquid. Widely used in metallography to show the equilibrium between liquid and solid alloys.

equilibrium, metastable If a system, capable thermodynamically of undergoing a spontaneous change, persists without change, it is said to be in a state of metastable equilibrium. The change requires initiation, e.g. a mixture of hydrogen and oxygen at 298 K is capable of undergoing spontaneous reaction but reaction only occurs when initiated by a spark or addition of a catalyst.

equilin, 3-hydroxy-1,3,5(10),7-oestratetraen-17-one, $C_{18}H_{20}O_2$. M.p. 238–240°C. A substance with the properties of a female sex hormone found in the urine of pregnant mares.

equivalence point The stage in a titration when the reactants and products are present in equivalent amounts according to the stoicheiometry of the reaction. This may not correspond exactly to the end point*.

equivalent, chemical
See chemical equivalent.

equivalent conductivity The specific conductance* multiplied by the volume (ml) which contains 1 g equivalent of the electrolyte.

equivalent proportions, law of A law stating that substances (elements and compounds) react together in the mass ratio of their equivalents. The law does not have wide application in its simple form because of variable oxidation states and non-stoicheiometries.

Er Erbium.

erbium, Er. At.no. 68, at.wt. 167·26, m.p. 1522°C, b.p. 2510°C, *d* 9·01. A lanthanide*. The metal is hcp. Used as an additive to improve the workability of vanadium; the oxide is used in colouring glass.

erbium compounds Erbium forms a single series of typical lanthanide compounds* in the +3 state E^{3+} (f^{11} rose) $\longrightarrow$ Er(−2·30 volts in acid).

Erdmann's salt $NH_4[Co(NH_3)_2(NO_2)_4]$.

ergine, lysergamide The amide of lysergic acid*.

ergometrine, $C_{19}H_{23}N_3O_2$. Water-soluble alkaloid of ergot*; m.p. 195–197°C (decomp.).

ergosterol, (245)-24-methyl-5,7,22-cholestatrien-3βol M.p. 163°C. First isolated from ergot of rye and occurring in many fungi and yeasts. Uv. radiation gives a number of products, amongst which is vitamin D_2*. *See* sterols.

ergot Sclerotium of the fungus *Claviceps purpurea*, which develops in the ovary and replaces the grain of rye. Ergot owes its pharmacological activity almost entirely to the alkaloids ergotoxine*, ergometrine* and ergotamine*.

ergotamine, $C_{33}H_{35}N_5O_5$. One of the ergot alkaloids*.

ergothioneine, thiolhistidinebetaine
Crystallizes with $2H_2O$; m.p. 290°C

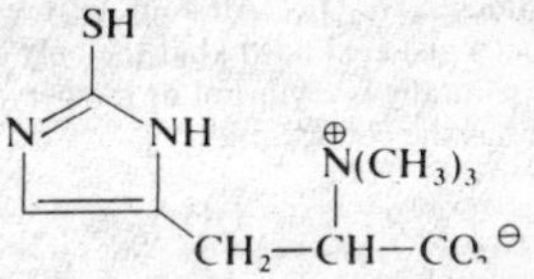

(decomp.). A base present in ergot and in mammalian blood corpuscles. It is a biological transmethylating agent.

ergotoxine A mixture of three of the ergot alkaloids*, ergocornine, $C_{31}H_{39}N_5O_5$, ergocrystine, $C_{35}H_{39}N_5O_5$ and ergokryptine, $C_{32}H_{39}N_5O_5$. All these have molecules built up of lysergic acid, D-proline, L-phenylalanine and other compounds.

eriochrome black T A useful complexometric reagent for Ca^{2+} and Mg^{2+} (used for determining water hardness). It has three acid sites, two of which are involved in colour changes. In Mg^{2+} titration at pH 10 with EDTA the colour changes from red to blue.

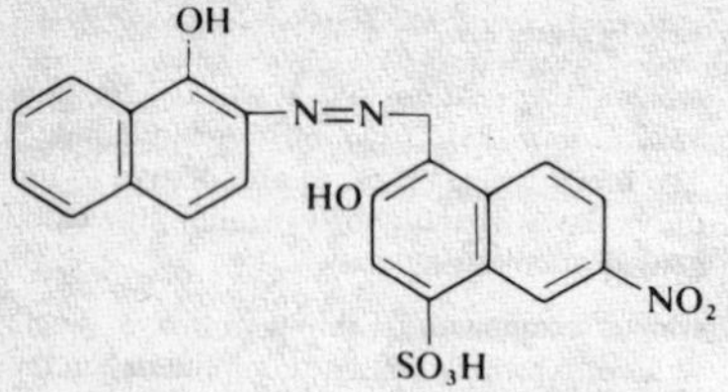

Erlenmeyer's rule This originally postulated that all unsaturated alcohols of the types

$$C-CHOH \quad \text{and} \quad {>}C-COH-C{<}$$

were unstable and that any reaction leading to them would give instead the aldehyde or keto forms

$$>CH-CHO \quad >CH-CO-C{<}$$

Later findings have modified the rule, especially when the carbon atoms concerned are connected with carbonyl- or carboxyl-groups, but it still holds good for monohydric unsaturated alcohols.

erubescite, Cu_3FeS_3. Cu ore.

erucic acid, cis-13-docosenoic acid, $C_{22}H_{42}O_2$, $CH_3 \cdot [CH_2]_7 \cdot CH{:}CH \cdot [CH_2]_{11} \cdot COOH$. M.p. 35°C. An unsaturated fatty acid belonging to the oleic acid series, occurring as glycerides in rape oil and other vegetable oils. It is the *cis*-isomer, the *trans*-isomer being brassidic acid.

erythritol, $C_4H_{10}O_4$. All four tetrose alcohols (tetritols) are known, but the only one to occur naturally is erythritol or mesoerythritol which has the configuration

$$HOH_2C-\underset{\underset{OH}{|}}{\overset{\overset{H}{|}}{C}}-\underset{\underset{H}{|}}{\overset{\overset{OH}{|}}{C}}-CH_2OH.$$

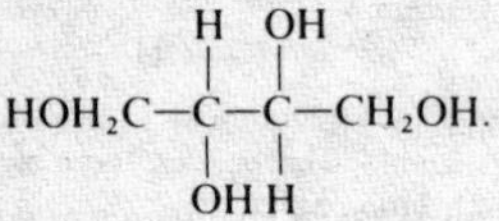

It is found in lichens and in some algae. It has m.p. 120°C, is very soluble in water and is about twice as sweet as sucrose. It is a reference compound upon which the *erythro* nomenclature is based.

The other tetritols were formerly called *d*- and *l*-erythritol, but have been renamed L- and D-threitol respectively. A racemic form can also be prepared.

erythro- A prefix used to distinguish one diastereomer from the other *threo* form which is based upon an analogy to erythrose and

threose. The erythro diastereomer is defined as the one in which at least two sets of similar substituents on adjacent asymmetric carbons are in the eclipsed configuration.

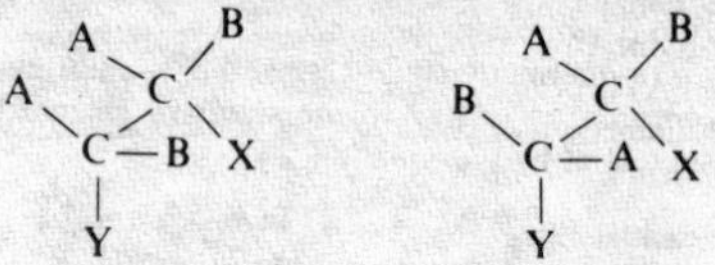

Erythro-form   *Threo*-form

A Newman projection of the above *erythro* form is

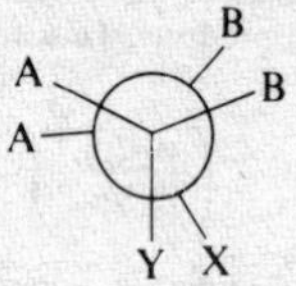

erythromycin, $C_{37}H_{67}NO_{13}$. An important macrolide antibiotic*, produced by *Streptomyces erythreus*.

Erythromycin is active against gram-positive and certain gram-negative bacteria, also against *Rickettsia* and spirochaetes. It is used for patients who are allergic to or do not respond to treatment with penicillins or tetracyclines.

erythrose, $C_4H_8O_4$. A tetrose sugar, both the

$$HOH_2C-\underset{\underset{H}{|}}{\overset{\overset{OH}{|}}{C}}-\underset{\underset{H}{|}}{\overset{\overset{OH}{|}}{C}}-CHO$$

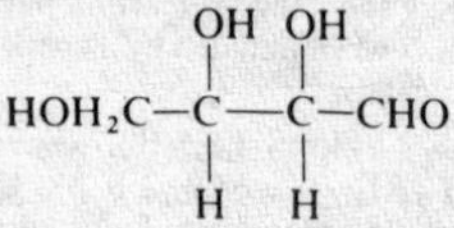

D-and L- forms of which have been prepared. It has been obtained as a liquid, very soluble in water and alcohol.

Es Einsteinium.

ESCA (electron spectroscopy for chemical analysis) *See* photo-electron spectroscopy.

esr Electron spin resonance*.

essential fatty acids, EFA A group of unsaturated acids which must be in the diet for healthy growth. A deficiency of EFA leads to lack of growth and skin lesions and necrosis. Essential fatty acids must possess an all-*cis* arrangement of double bonds in a 'methylene-interrupted' polyene system. The principal EFA are linoleic acid*, arachidonic acid, linolenic acid and γ-linolenic acid*. These acids are essential as the body cannot synthesize them fast enough to meet needs. EFA are abundant in the vegetable oils and to a lesser extent in fats of some animals or of marine origin. Adult man does not seem to suffer from EFA deficiency, but some infants fed on skimmed milk with a low fat content develop eczema.

essential oils More or less volatile oils extracted from plants. Many essential oils are refined, separated and purified and used as sources of compounds.

esterases *See* enzymes.

esterification The process of formation of esters by the combination of an acid with an alcohol.

esters Organic compounds formed by the union of an acid and an alcohol with elimination of water. They are volatile liquids or low-melting solids and are usually insoluble in water but soluble in alcohol or ether. Many esters have characteristic fruity odours and occur naturally in fruit. As the reaction between an acid and an alcohol is reversible, it cannot be carried to completion unless the water produced is removed. In some cases it is sufficient to have a large excess of alcohol present to secure a good yield of ester; in other cases it is usual to add a liquid such as benzene or carbon tetrachloride to the mixture and to distil off the water in the form of a low-boiling mixture. The reaction between alcohol and acid is very slow at ordinary temperatures, but the ester is formed much more rapidly at higher temperatures and in the presence of a small amount of an acid such as sulphuric, hydrochloric or benzenesulphonic acid. Esters are also produced by the action of an alcohol on an acid chloride or anhydride; amides also react with alcohols to give esters in the presence of boron trifluoride. Some esters may be obtained by the action of an aldehyde on aluminium ethoxide or isopropoxide. Methyl and ethyl esters are often conveniently obtained by treating the sodium salt of the acid with methyl or ethyl sulphates. Diazomethane reacts with organic acids to give methyl esters. Esters can also be prepared by heating an alcohol with the methyl or ethyl ester of the acid; in this case an exchange of alcohols occurs and methanol or ethanol is formed. Esters are completely decomposed to acid and alcohol by heating with dilute sodium hydroxide solution. They react with ammonia to give amides; with ketones in the presence of sodamide to give 1,3-diketones and with Grignard reagents to give tertiary alcohols. They are reduced by a mixture of sodium and ethanol to the alcohol corresponding to the acid of the ester. Used as solvents, flavouring essences and perfumes; they are also used in many chemical processes.

estradiol *See* oestradiol.

estriol *See* oestriol.

estrogens *See* oestrogens.

estrone *See* oestrone.

Et Ethyl, C_2H_5-.

Ètard's reaction A direct method for the production of aromatic aldehydes by the oxidation of the methylated homologues of benzene by chromyl chloride, CrO_2Cl_2. When benzene homologues with longer side chains are treated with CrO_2Cl_2 a ketone is the usual product, but with ethylbenzene a mixture of phenylacetaldehyde and acetophenone is obtained.

ethanal, acetaldehyde, $CH_3{\cdot}CHO$. Colourless liquid with a characteristic odour; b.p. 20·8°C. Miscible with water, alcohol and ether; insoluble in concentrated calcium chloride solutions. Manufactured by (1) vapour phase oxidation or dehydrogenation of ethanol over a catalyst; (2) vapour phase oxidation of butane (with methanal and methanol as by-products); (3) direct oxidation of ethene over a copper(II) chloride/palladium(II) chloride catalyst (Wacker process*). Prepared on the small scale by the oxidation of ethanol with potassium dichromate and sulphuric acid. Compounds are formed with many metallic salts; the crystalline sodium or potassium hydrogen sulphite compounds may be used to isolate and purify ethanal. It is oxidized to ethanoic acid and reduced to ethanol. Polymerization readily occurs to give paraldehyde or metaldehyde: when heated with alkalis, a brown resin is formed. Aldol is formed by the interaction of two molecules of ethanal. Ethanal is used almost exclusively as a chemical intermediate, three compounds, ethanoic acid, ethanoic anhydride and *n*-butanol accounting for the bulk of its consumption. Other industrial chemicals made from it include 2-ethylhexyl alcohol, pentaerythritol and chloral.

ethanamide, acetamide, $CH_3CO{\cdot}NH_2$. Crystallizes in long white needles which absorb water and finally liquefy. The crude substance has a strong odour of mice; m.p. 82°C, b.p. 222°C. Prepared by the dry distillation of ammonium ethanoate or by the action of ammonia on ethyl ethanoate. Weakly basic.

ethane, $CH_3 \cdot CH_3$. Colourless, odourless gas; forms an explosive mixture with air. B.p. −89°C. It occurs in natural gas. May be prepared by reduction of ethene or ethyne by hydrogen under pressure in the presence of a nickel catalyst, or by the electrolysis of a solution of potassium ethanoate. It has the general properties of the paraffins. Used in low-temperature refrigeration plant.

ethanenitrile *See* acetonitrile.

ethanoates, acetates Salts or esters of ethanoic acid* (acetic acid).

ethanoic acid, acetic acid, glacial acetic acid, CH_3COOH. Colourless liquid with a pungent, irritating odour; m.p. 16·6°C, b.p. 119°C.

Manufactured by the liquid-phase oxidation of ethanal at 60°C by oxygen or air under pressure in the presence of manganese(II) ethanoate, the latter preventing the formation of perethanoic acid. Another important route is the liquid-phase oxidation of butane by air at 50 atm. and 150–250°C in the presence of a metal ethanoate. Some ethanoic acid is produced by the catalytic oxidation of ethanol. Fermentation processes are used only for the production of vinegar.

Ethanoic acid will attack most metals and can form acidic, basic and normal salts. It can, however, be handled in e.g. stainless steel equipment. About half the ethanoic acid produced is used as ethanoic anhydride for the manufacture of cellulose ethanoate. Large quantities are also used for the manufacture of vinyl ethanoate and various solvents. Because a number of manufacturing processes yield large amounts of dilute acid its recovery is a problem of considerable importance. Simple rectification is expensive and hence various other processes such as azeotropic and extractive distillation and liquid–liquid extraction are used. U.S. production 1978 1 240 000 tonnes. Production of ethyl ester (85%) 100 000 tonnes.

ethanoic anhydride, acetic anhydride, $[CH_3C(O)]_2O$. Colourless liquid with a pungent odour; b.p. 139·5°C. Hydrolysed to ethanoic acid by boiling water. Manufactured by bubbling air through a mixture of ethanal and ethanoic acid in the presence of a catalyst; or by reacting keten, derived by the cracking of ethanoic acid or propanone with ethanoic acid. It reacts with compounds containing an −OH, −SH or −NH group to give acetyl (ethanoyl) derivatives which are useful for characterization. Largely used for the production of cellulose ethanoate, also used for the manufacture of vinyl ethanoate and aspirin. U.S. production 1978 680 000 tonnes.

ethanol, ethyl alcohol, alcohol, spirits of wine, CH_3CH_2OH. Colourless liquid with a pleasant odour, b.p. 78·3°C. Miscible with water with evolution of heat and contraction in volume; pure ethanol absorbs water vapour. Many gases are more soluble in it than in water. Some inorganic salts and many organic compounds are soluble in ethanol. It occurs only rarely in nature, except as a result of the fermentation of sugary plant juices by yeasts, and less often by certain bacteria and moulds. Alcohol was formerly manufactured almost exclusively by the fermentation of materials containing starch and sugars, but this method is now relatively unimportant. Most is at present made by the catalytic hydration of ethene, or by the hydrolysis of the mono- and di-ethyl sulphates formed by absorbing ethene in 90% sulphuric acid. The ethene is obtained from refinery gases or other petroleum fractions by cracking. Because ethanol forms an azeotrope with water it is not possible to obtain a product containing more than 95·6% alcohol by weight by straightforward fractionation of an aqueous solution. For the manufacture of 100% ethanol azeotropic distillation* is employed. Ethanol is oxidized to ethanal or ethanoic acid; with nitric acid a variety of products, including glycollic and oxalic acids are formed. Ethanolates (ethoxides) are formed by the action of sodium, calcium, aluminium and some other metals on ethanol. It reacts with acids to give esters. With sulphuric acid it yields ether, ethene or ethyl hydrogen sulphate. Bleaching powder converts it to chloroform, while chlorine gives chloral. Ethanol is used as a starting point for the manufacture of other chemicals, principally ethanal, in foodstuffs and as a solvent. U.S. production 1981 600 000 tonnes. Its pharmacological effects are basically those of a central depressant, low doses having an apparently stimulant effect due to the removal of normal inhibitory influences.

ethanolamines The three ethanolamines are low-melting, colourless solids which very readily absorb water and form viscous liquids; they have distinct ammoniacal odours and are strong bases.

Monoethanolamine, 2-aminoethyl alcohol, 2-hydroxyethylamine, $HOCH_2CH_2NH_2$. M.p. 10·5°C, b.p. 171°C.

Diethanolamine, di-(2-hydroxyethyl)amine, $(HOCH_2CH_2)_2NH$. M.p. 28°C, b.p. 217°C/150 mm.

Triethanolamine, tri-(2-hydroxyethyl)amine, $(HOCH_2CH_2)_3N$. M.p. 21°C, b.p. 277°C/150 mm.

All these compounds are manufactured by

heating ethylene oxide under pressure with concentrated aqueous ammonia. A mixture of the three is obtained, the proportion of each depending on the ammonia/ethylene oxide ratio used, and this is separated by fractional distillation.

The ethanolamines form soaps with fatty acids which are almost neutral in reaction and soluble in benzene. These are of great commercial importance, being used as detergents, emulsifying agents and in the preparation of cosmetics, toiletries, bactericidal and herbicidal products. Monoethanolamine is widely used for removing acid constituents such as carbon dioxide and hydrogen sulphide from natural gas streams. These acid gases are then removed by steam stripping. U.S. production 1978 160 000 tonnes.

ethanoyl chloride, acetyl chloride, CH_3COCl. Colourless liquid with a pungent odour; fumes in moist air, producing ethanoic and hydrochloric acids. B.p. 55°C. Reacts with water and other hydroxyl containing compounds. Prepared by the distillation of a mixture of ethanoic acid and PCl_3 or $POCl_3$. It is used to prepare ethanoyl (acetyl) derivatives of hydroxy- and amino-compounds.

ethene, ethylene, $CH_2=CH_2$. Colourless gas with a faint ethereal odour which occurs in natural gas, crude oil and coal gas. M.p. −169°C, b.p. −105°C. It is manufactured in larger quantities than any other organic chemical, most of it being derived from the vapour-phase cracking of ethane, various petroleum fractions and crude oil. It is also obtained from ethanol by vapour-phase dehydration using an activated alumina catalyst at 350°C. May be prepared by dropping alcohol on to syrupy phosphoric acid heated to 220°C. It is inflammable and forms an explosive mixture with air. Readily absorbed by concentrated sulphuric acid to give ethyl hydrogen sulphate; if the reaction is carried out under pressure, diethyl sulphate is largely formed. Absorbed by solutions of potassium permanganate to give ethylene glycol: by dilute chlorine water to give ethylene chlorohydrin. Reduced by hydrogen to ethane. Reacts with hydrogen bromide and hydrogen iodide at 100°C to give ethyl bromide and iodide; hydrogen chloride does not react under these conditions. With ammonia under pressure the ethylamines are formed. Reacts with water at 450°C, or at lower temperatures in the presence of catalysts, to give ethanol. Reacts with oxygen and a palladium catalyst to give ethanal (*see* Wacker process).

It is given off in minute amounts by ripe tomatoes and apples and has the property of accelerating the ripening of fruit. Also used in the manufacture of glycol (20%), ethylene oxide (15%), ethanol, styrene (14%), dichloroethane, ethyl chloride, vinyl esters, e.g. vinyl ethanoate and vinyl ethers, vinyl chloride (10%). When polymerized under high pressure, or at lower pressures in the presence of a catalyst, it gives the important plastic polyethylene* (40%). U.S. production 1982 14 megatonnes.

ethene polymers, polyethene, polyethylene, polythene Probably the most important polymeric substances. Polymerization is effected by high-pressure free radical processes over metal-oxide catalysts, low pressure polymerization using Ziegler catalysts. The polymers can be produced crystalline (high density) and non crystalline (low density). Used in fabrication (houseware, toys, etc.), pipe, wire coating, process machinery, textiles. Chlorination gives materials used for flooring, and electrical insulation. Chlorosulphonation (introduction of SO_2Cl groups) gives very durable materials. U.S. production of polyethene 1981 5·9 megatonnes (low density 65%).

ethenyl, vinyl The group $CH_2=CH-$.

ether, diethyl ether, diethyl oxide, $(CH_3CH_2)_2O$. Colourless liquid with a pleasant, characteristic odour; very volatile, its vapour forms an explosive mixture with air; b.p. 34·5°C. It is manufactured by passing ethanol vapour into a mixture of 92% ethanol and 78% sulphuric acid at 128°C. May also be produced as a by-product in the manufacture of alcohol from ethene. Commercial ether usually contains small amounts of water, ethanol, ethanal and the explosive peroxide. It is a comparatively inert compound and is an excellent solvent for a great many organic substances. Oxidized by nitric acid to ethanoic acid. Reacts with strong sulphuric acid to give ethyl hydrogen sulphate and with hydrogen iodide to give ethyl iodide. Chlorine reacts violently with ether at ordinary temperatures but forms various chloro-ethers at low temperatures. Crystalline co-ordination compounds are formed with some metallic salts. Can be employed for the production of general anaesthesia by inhalation and is generally a safe anaesthetic but it must be pure. Ether acts as a carminative to the stomach, and is administered in the form of spirit of ether as a restorative in collapse. It is also injected hypodermically for the same purpose. Medical and pharmaceutical uses account for a relatively small proportion of ether consumption, and it is mainly used as a chemical intermediate and as a solvent for oils, fats, waxes and alkaloids.

ethers Organic compounds of the type R–O–R′, where R and R′ are alkyl or aryl. Formed by heating the sodium derivative of a hydroxy-compound with an alkyl or aryl halide; or by treating the hydroxy-compound with an alkyl halide in presence of silver oxide. Methyl and ethyl ethers are conveniently prepared by the action of dimethyl and diethyl sulphates on a hydroxy-compound in presence of sodium hydroxide solution. Diazomethane will react with phenols in ethereal solution to give methyl ethers. The simpler aliphatic ethers are manufactured by the action of sulphuric acid on the appropriate alcohol or olefin. They are usually liquids with pleasant odours, but some aromatic and the higher aliphatic ethers are crystalline solids. They are insoluble in water but soluble in alcohol and diethyl ether. The commonest ether, diethyl ether, is usually called simply 'ether'.

ethinylation The reaction between ethyne and organic substrates to produce substances having an intact ethynic bond; e.g. the reaction of methanal and ethyne to give butyne-1-4-diol, $HOCH_2-C{\equiv}C-CH_2OH$.

ethisterone, ethinyltestosterone, $C_{21}H_{28}O_2$. Has the properties of progesterone and is given orally for treating functional uterine haemorrhage. Norethisterone, 19-norethisterone, m.p. 201–206°C is one of the progestational steroids used in the contraceptive pill.

ethoxy The CH_3CH_2O- group; $EtO-$.

ethoxyl The group CH_3CH_2O-; also written $EtO-$.

ethyl The group $CH_3{\cdot}CH_2-$; often written C_2H_5- and $Et-$.

ethyl acetate *See* ethyl ethanoate.

ethyl acetoacetate, acetoacetic ester, ethyl 3-oxobutanoate, $CH_3COCH_2COOC_2H_5$, $CH_3C(OH){:}CHCOOC_2H_5$. Colourless, mobile liquid with a pleasant odour; b.p. 181–182°C. Prepared by the action of sodium or sodium ethoxide on ethyl ethanoate, or of ethanol on diketen. It is the classical example of *keto-enol tautomerism*, and normal preparations of the ester contain 92·6% keto and 7·4% enol form. The proportion of the two forms is altered by change of temperature or by solution in solvents. Separation of the keto and enol forms has been effected by distillation from quartz or Pyrex apparatus: the pure components revert to the original equilibrium mixture on standing. Metallic derivatives are formed with sodium and other metals; these are of the type $CH_3C(ONa){:}CHCOOC_2H_5$; they react with alkyl halides to give alkyl acetoacetates $CH_3CO{\cdot}CHR{\cdot}COOC_2H_5$. These also form sodium derivatives, reacting with alkyl halides in a similar manner to give dialkyl esters. Ethyl acetoacetate and its alkyl derivatives react with strong alkalis to give ethanoic and alkylethanoic acids respectively. Dilute alkalis or acids react to give ketones. With many nitrogen compounds it reacts to form nitrogen-containing rings: thus urea gives methyl uracil; hydrazines give methyl pyrazolones; aniline gives methyl quinoline. Prolonged boiling causes decomposition with loss of alcohol and formation of dehydroacetic acid. It is a valuable means of synthesizing a wide variety of compounds and is used in the manufacture of phenazone.

ethyl acrylate, ethyl propenoate, $C_5H_8O_2$, $CH_2{:}CHCOOCH_2CH_3$. Colourless liquid; b.p. 101°C. Manufactured by treating ethylene chlorohydrin with sodium cyanide and heating the β-hydroxypropionitrile so formed with ethanol and sulphuric acid. Forms a colourless resin on standing. Used in the manufacture of synthetic resins, but the methyl ester is more commonly employed.

ethyl alcohol *See* ethanol.

ethylamines Organic compounds in which one or more of the hydrogen atoms of ammonia are replaced by ethyl groups. Colourless liquids with strongly ammoniacal odours; they are inflammable and burn with a yellow flame. Manufactured by passing ethene or ethanol vapour with ammonia under pressure over heated catalysts. The relative amounts of the three amines formed can be controlled by varying the proportions of ammonia to ethene or alcohol. They are all three strongly basic in character and show the typical properties of aliphatic amines.

Ethylamine, monoethylamine, $CH_3CH_2NH_2$. B.p. 19°C. Prepared by reduction of acetonitrile or by heating ethyl chloride with alcoholic ammonia under pressure. It is a strong base and will displace ammonia from ammonium salts. Forms a crystalline hydrochloride and also crystalline compounds with various metallic chlorides.

Diethylamine, $(CH_3CH_2)_2NH$. B.p. 55·5°C. Forms a crystalline ½ hydrate. Prepared by the action of a boiling solution of sodium hydroxide on nitrosodiethylaniline. Forms crystalline compounds with many metallic chlorides.

Triethylamine, $(CH_3CH_2)_3N$. Oily liquid;

b.p. 89°C. Prepared by heating ethylamine with an alcoholic solution of ethyl chloride under pressure. Readily oxidized by potassium permanganate.

ethylation Processes by which an ethyl group is added to a compound. In aliphatic chemistry this involves the substitution of the hydrogen atom of a hydroxyl, amino or imino group, and produces an ether or a secondary or tertiary amine respectively. In aromatic chemistry it may also mean the substitution of one of the hydrogen atoms of the ring by the ethyl group; this is carried out by the Friedel–Crafts reaction.

ethylbenzene, $PhCH_2CH_3$. B.p. 136°C. Recovered from mixed xylenes from petrochemical processes, prepared by addition of ethene to benzene, or, on a laboratory scale by the Friedel–Crafts reaction between C_2H_5Cl and benzene. Catalytically dehydrogenated, particularly in the presence of steam, to styrene. U.S. production 1978 3 800 000 tonnes.

ethyl *n*-butyl ketone, 3-heptanone, $C_2H_5CO(CH_2)_3CH_3$. Made by dehydrogenation of 3-heptanol; b.p. 148°C. Used as a solvent for lacquers and synthetic resins.

ethyl carbamate *See* urethane.

ethyl chloride, monochloroethane, CH_3CH_2Cl. Colourless liquid with an ethereal odour; burns with a green-edged flame; b.p. 12·5°C. Manufactured by reacting hydrogen chloride with ethene at 40°C in the presence of $AlCl_3$ using dichloroethane as solvent, or by the catalytic or photochemical chlorination of ethane. Reacts with ammonia at 100°C to give ethylamine hydrochloride; converted to ethanol by heating with KOH. Its principal use is in the manufacture of lead tetraethyl. Having a low b.p. it is useful for the production of local anaesthesia; it is applied in the form of a fine spray, and its rapid evaporation freezes the part to which it is applied. U.S. production 1978 245 000 tonnes.

ethyl chlorocarbonate *See* ethyl chloroformate.

ethyl chloroformate, chloroformic ester, chlorocarbonic ester, ethyl chlorocarbonate, $ClCOOCH_2CH_3$. Volatile liquid with an unpleasant odour; b.p. 94–95°C. Prepared by the addition of ethanol to cooled phosgene. It is very reactive and unites with many organic compounds containing $-OH$ groups to give carbethoxy-derivatives. Reacts with ethanol to give diethyl carbonate and with ammonia to form urethane. Used in organic syntheses.

ethyl citrate, triethyl citrate, $C_{12}H_{20}O_7$. Colourless oil with a bitter taste; b.p. 185°C/17 mm. Typical ester. Used as a fixative in perfumery.

ethyl cyanoacetate *See* ethyl cyanoethanoate.

ethyl cyanoethanoate, cyanoacetic ester, ethyl cyanoacetate, $C_5H_7NO_2$, $NC{\cdot}CH_2COOCH_2CH_3$. Colourless liquid; b.p. 207°C, 97°C/16 mm. Prepared by boiling cyanoacetic acid with ethanol and sulphuric acid. Reacts with ammonia to give cyanoacetamide. Like acetoacetic ester, it will react with sodium and the product will combine with halogen compounds such as alkyl bromides. It may be used to replace malonic ester or acetoacetic ester in many syntheses.

ethyl diazoethanoate, ethyl diazoacetate, diazoacetic ester, $C_4H_6N_2O_2$, $N_2CHCOOCH_2CH_3$. Yellow oil; b.p. 84°C/61 mm. Prepared by treating a solution of ethyl glycinate hydrochloride with sodium nitrite and sulphuric acid at 0°C. Decomposes explosively if distilled at atmospheric pressure, or if treated with concentrated hydrochloric or sulphuric acid. When treated with alkalis it forms nitrogen-containing ring compounds. Forms glycollic acid on boiling with water. Reacts with alcohols to give alkoxy-ethanoic esters, and with aldehydes to give β-keto-esters. Esters of unsaturated acids react with it, forming pyrazoline carboxylic esters; these lose nitrogen on heating to give derivatives of cyclopropane. Used in organic syntheses.

ethylene *See* ethene.

ethylene chlorohydrin *See* 2-chloroethyl alcohol.

ethylenediamine *See* 1,2-diaminoethane.

ethylenediaminetetra-acetic acid, EDTA, $(HO_2CCH_2)_2N{\cdot}CH_2{\cdot}CH_2{\cdot}N(CH_2CO_2H)_2$. An important compound, which owes its use to its sequestering properties. A multidentate chelating agent. Forms complexes with most elements.

ethylene dibromide *See* 1,2-dibromoethane.

ethylene dichloride *See* 1,2-dichloroethane.

ethylene dinitrate, ethylene glycol dinitrate, $C_2H_4N_2O_6$, $O_2NOCH_2{\cdot}CH_2ONO_2$. A colourless liquid with a sweetish taste and no appreciable odour; b.p. 105°C/19 mm; explodes at 215°C when rapidly heated. Manufactured by treating ethylene glycol with a mixture of fuming nitric and sulphuric acids at a low temperature or by passing ethene into a cold mixture of nitric and sulphuric acids. Used as an explosive of the dynamite type; it is mixed with nitroglycerin to reduce the freezing-point of dynamite.

ethylene glycol *See* 1,2-dihydroxyethane.

ethylene glycol dinitrate *See* ethylene dinitrate.

ethylene glycol monobutyl ether, butyl Cellosolve, $C_6H_{14}O_2$, $CH_3CH_2CH_2CH_2OCH_2CH_2OH$. Colourless liquid with a pleasant odour; b.p. 171°C. Manufactured by heating ethylene oxide with 1-butanol in the presence of nickel sulphate as a catalyst. Used as a solvent in brushing lacquers.

ethylene glycol monoethyl ether acetate, Cellosolve acetate, $C_6H_{12}O_3$, $CH_3CH_2OCH_2CH_2OOCCH_3$. Colourless liquid with a pleasant ethereal odour; b.p. 156°C. Manufactured by heating ethylene glycol monoethyl ether with ethanoic acid. Used as a solvent in nitrocellulose lacquers.

ethylene glycol monoethyl ether, Cellosolve, $CH_3CH_2OCH_2CH_2OH$. A colourless liquid with a pleasant odour; b.p. 135°C. Manufactured by heating ethylene oxide with ethanol and a catalyst, or by treating ethylene glycol with diethyl sulphate and sodium hydroxide. Used extensively as a solvent in nitrocellulose lacquers.

ethylene glycol monomethyl ether, methyl Cellosolve, $C_3H_8O_2$,$CH_3OCH_2CH_2OH$. Colourless liquid with a pleasant odour; b.p. 124°C. Manufactured by heating ethylene oxide with methanol, either under pressure or in the presence of a catalyst. Also obtained by treating ethylene glycol with dimethyl sulphate and sodium hydroxide. Used as a solvent for cellulose acetate lacquers and for certain dyes and resins. U.S. production 1976 1 630 000 tonnes.

ethyleneimine *See* dihydroazirine.

ethylene oxide, 1,2-epoxyethane, oxirane $CH_2{\cdot}CH_2O$ Colourless gas with a sweet odour which is somewhat lachrymatory; b.p. 10·5°C. Manufactured by heating ethylene chlorohydrin with $Ca(OH)_2$ or NaOH solution; or by the direct oxidation of ethene at 250–300°C using a silver catalyst. It forms an explosive mixture with air. Reacts with water in the presence of sulphuric acid to give ethylene glycol and polyethylene glycols; with alcohols and phenols to give ethers of glycol; and with hydrochloric acid to give ethylene chlorohydrin. It reacts with many primary and secondary amines to give ethanolamines; with organic acids to give monoesters of ethylene glycol, and with acid anhydrides to give diesters. It is reduced by hydrogen to ethanol, and is converted to ethanal by heating at 200–300°C in the presence of alumina. Its principal use is for polymerization to 1,2-epoxide polymers*. Also used as an intermediate in the manufacture of ethylene glycol, polyethylene glycols, glycol ethers, ethanolamines and similar compounds. It is also used as a fumigant. U.S. production 1980 2·3 megatonnes.

ethyl ethanoate, ethyl acetate, acetic ester, acetic ether, $CH_3CO_2C_2H_5$. Colourless liquid; b.p. 77°C, with a pleasant fruity odour. Manufactured from ethanol, ethanoic acid and concentrated sulphuric acid, or from ethyne via ethanal with an aluminium alkoxide as catalyst. It is an extremely useful solvent, especially for cellulose-type varnishes and adhesives; also used for cosmetic applications and as an artificial essence.

Used for the synthesis of ethyl acetoacetate*

ethyl fluid *See* anti-knock additives.

ethyl formate *See* ethyl methanoate.

2-ethylhexyl alcohol, octyl alcohol, $C_8H_{18}O$, $CH_3(CH_2)_3CHEtCH_2OH$. B.p. 181°C. Manufactured by heating 1-butanol with KOH and boric oxide at 270–300°C. Used as an anti-foaming agent, as a dispersing agent in the pigment and ceramic industries and as a means of introducing the 2-ethylhexyl group into other products. Its esters, e.g. with phthalic, stearic and phosphoric acids, are used as plasticizers. The acrylate on copolymerization with other monomers gives plastics which are 'internally' plasticized. Other esters are used as lubricants, bactericides, fungicides and insecticides.

ethyl hydrogen sulphate, ethylsulphuric acid, $C_2H_6O_4S$,$(EtO)(HO)SO_2$. Oily acidic liquid. Soluble in water and slowly hydrolysed by it to ethanol and sulphuric acid. Prepared by passing ethene into concentrated sulphuric acid or by heating ethanol and sulphuric acid. Gives ethene when heated alone, and diethyl sulphate when heated with ethanol at 140°C. Forms crystalline metallic salts which are soluble in water.

ethylidene The group $CH_3{\cdot}CH<$.

ethylidene diacetate, acetaldehyde diacetate, ethanal diethanoate, $CH_3CH(OOCCH_3)_2$. Colourless liquid with a characteristic heavy odour; m.p. 18·9°C, b.p. 169°C. Manufactured by passing ethyne into glacial ethanoic acid containing $HgSO_4$. The temperature is kept above 70°C; at lower temperatures vinyl ethanoate is also formed. It is slowly decomposed by water to give ethanal and ethanoic acid. Decomposed when heated above 150°C to give ethanoic anhydride and ethanal; some vinyl ethanoate is formed. It is an intermediate product in the manufacture of ethanoic anhydride from ethyne.

ethyl lactate, $C_8H_{10}O_3$, $CH_3CH(OH)C(O)OEt$. A colourless liquid with a pleasant odour, b.p. 154°C. Manufactured by distilling a mixture of (±)-lactic acid, ethanol and benzene in the presence of a little sulphuric or benzenesulphonic acid. It is a solvent for cellulose nitrate and acetate and also for various resins. Used as a lacquer solvent.

ethyl methanoate, ethyl formate, $HCOOCH_2CH_3$. Colourless liquid with the odour of peach-kernels; b.p. 54°C. Prepared by boiling ethanol and methanoic acid in the presence of a little sulphuric acid; the product is diluted with water and the insoluble ester separated and distilled. Used as a fumigant and larvicide for dried fruits, tobacco and foodstuffs. It is also used in the synthesis of aldehydes.

ethyl silicate, 'silicon ester', $Si(OC_2H_5)_4$. Thin mobile liquid; with a mild ester-like odour; m.p. −77°C, b.p. 168°C. Made by reacting alcohol with silicon tetrachloride. Slowly decomposed by water to give ethanol and silicic acid. It is used for waterproofing stonework, for making precision castings and for bonding.

ethyl succinate, diethyl succinate, $C_8H_{14}O_4$, $EtO_2CCH_2 \cdot CH_2CO_2Et$. Colourless liquid with a faint odour; b.p. 218°C. Manufactured by heating succinic acid with alcohol and sulphuric acid. Typical ester. Used as a fixative in perfumery.

ethylsulphuric acid *See* ethyl hydrogen sulphate.

ethyl vinyl ether, $CH_3CH_2OCH{=}CH_2$, C_4H_8O. B.p. 35°C. Prepared commercially by addition of ethyne to ethanol in the presence of catalysts. Readily polymerized, high mol.wt. polymers are used as adhesives, coatings, films; low mol.wt. polymers are used as plasticizers and resin-modifiers. Used in manufacture of glutanaldehyde and as an anaesthetic.

ethyne, acetylene, C_2H_2, $HC{\equiv}CH$. Colourless gas which when pure has a pleasant smell. M.p. −82°C, b.p. −84°C.

The compound has very wide explosive limits in air (2·3% and 80% ethyne by volume) and apart from this the liquid, and also the gas if the pressure exceeds 2 atm., are explosive even in the absence of air. For storage and transportation steel cylinders containing propanone and a porous material such as diatomaceous earth are used.

The original method for the manufacture of ethyne, the action of water on calcium carbide, is still of very great importance, but newer methods include the pyrolysis of the lower paraffins in the presence of steam, the partial oxidation of natural gas (methane) and the cracking of hydrocarbons in an electric arc.

Ethyne is the starting point for the manufacture of a wide range of chemicals, amongst which the most important are acrylonitrile, vinyl chloride, vinyl acetate, ethanal, ethanoic acid, tri- and perchloro-ethylene, neoprene and polyvinyl alcohol. Processes such as vinylation, ethinylation, carbonylation, oligomerization and Reppe processes offer the possibility of producing various organic chemicals cheaply. Used in oxy-acetylene welding.

Like all terminal alkynes, the ethynic hydrogen atoms are acidic in the sense that they are removed by suitable strong bases such as sodamide or a Grignard reagent to give $RC{\equiv}CNa$ or $RC{\equiv}CMgX$. (For ethyne R = H, although a second mole of base will also replace this hydrogen atom.) Many metals will give salts (*see* acetylides) which may be used for purification purposes, e.g. silver acetylide, $AgC{\equiv}CH$, a white shock-sensitive solid, insoluble in water, regenerates ethyne on treatment with mineral acids.

Eu Europium.

eucalyptol *See* 1,8-cineole.

eucarvone, $C_{10}H_{14}O$. Colourless oil; b.p. 88°C/10 mm. Obtained from carvone by treatment with HBr followed by alkali. Has an odour resembling menthone.

eugenol, 4-allyl-2-methoxyphenol, $C_{10}H_{12}O_2$. Colourless liquid; b.p. 254°C. The chief constituent of oil of cloves, it is also present in oil of bay and oil of cinnamon. It is used in dentistry as an antiseptic dressing and mixed with zinc oxide as a temporary filling. It is a strong antiseptic and local anaesthetic.

europium, Eu. At.no. 63, at.wt. 151·96, m.p. 822°C, b.p. 1597°C, *d* 5·26. A lanthanide*. The metal is bcc and is used as a neutron absorber in reactor technology. Eu compounds are used as red-phosphors in colour television.

europium compounds Europium forms compounds in the +3 and +2 oxidation states,

Eu^{3+} (f^6 pale pink) ⟶ Eu −2·41 volts in acid. Europium(III) salts are typical lanthanide derivatives. Europium(II) salts are pale yellow in colour and are strong reducing agents but stable in water. EuX_2 are prepared from $EuX_3 + Eu$ (X = Cl, Br, I) or $EuF_3 + Ca$; $EuCl_2$ forms a dihydrate. $EuSO_4$ is prepared by electrolytic reduction of Eu(III) in sulphuric acid. Eu(II) is probably the most stable +2 state of the lanthanides

$$Eu^{3+} \longrightarrow Eu^{2+}\ (-0{\cdot}43\text{ volts})$$

eutectic When a mechanical mixture of two substances shows a sharp melting point, the mixture is said to be a eutectic mixture. The temperature at which this occurs is called the eutectic temperature.

eutectic point The point on the phase diagram of a mixture of substances which represents the melting point and the composition of the eutectic mixture. The presence of eutectic points in mixtures of substances may be taken to indicate the formation of compounds between the components of the mixture in many, although by no means all, cases.

eutrophication Excessive growth of algae and higher plants on natural waters because of increased nutrients, particularly phosphates, etc., from sewage, detergents and fertilizers.

EVA plastics Copolymers of ethene and vinyl acetate (vinyl ethanoate).

evaporation All liquids and solids have a characteristic vapour pressure depending on the temperature. In a closed vessel, after a sufficient time has elapsed for equilibrium to be established, as many molecules leave the liquid surface to form vapour as return to it from the vapour phase to form liquid. In an open vessel, however, no such equilibrium is set up and if the molecules of vapour immediately over the liquid surface are removed, further liquid molecules vaporize to take their place. In this way the liquid bulk is continuously diminished and the process of evaporation occurs. In the chemical and process industries the term normally refers to the removal of water from a solution.

evaporators Equipment for the concentration of solutions by evaporation. Many types are available.

EXAFS Extended X-ray absorption fine structure spectroscopy. A spectroscopic technique which can determine interatomic distances very precisely.

exchange reaction A reaction which occurs without chemical change. Usually applied to reactions involving isotopic replacement, e.g.

$$^{16}O_2 + 2H_2{}^{18}O \rightleftharpoons {}^{18}O_2 + 2H_2{}^{16}O$$

or $H_2 + D_2 \rightleftharpoons 2HD$. Because of differences in zero point energies caused by the differing masses of the isotopes, the equilibrium constants in such reactions do not have a value of unity. Use of this fact is made in exchange reactions for the separation of isotopes.

exciplex A molecular charge transfer complex formed by the complexing of a donor (or acceptor) in an excited state with an acceptor (or donor) in its ground state. An exciplex is only stable in an energetically excited state.

excited state A state of an atom, molecule, etc., when the species has absorbed energy and become excited to a higher energy state as compared with the ground state. The excitation may be electronic, vibrational, rotational, etc.

exclusion principle In any atom no two electrons may have the same set of all four quantum numbers.

exo *See* endo.

exopeptidases A class of peptide-cleaving enzymes which act on the terminal amino-acid residues. They are further divided into carboxypeptidases acting on the carboxyl end, and aminopeptidases acting on the amino end of the peptide. *See also* endopeptidases.

exothermic reaction A reaction in which heat is evolved.

expanded plastics *See* cellular plastics.

explosion Rapid combustion in homogeneous fuel/air mixtures with the flame front* passing through the mixture from the source of ignition. Explosions may occur in constant pressure or constant volume situations. In the former there are only local pressure increases at the flame front where the reaction is completed but there are no significant pressure increases in the passageway or shaft. In constant volume explosions the flame front spreads very rapidly across the container or cylinder and high turbulence creates rapid flame propagation, giving an irregular flame front and high pressure increases.

explosive cladding A coating process in which detonation imports a high velocity to the coating material which then collides with the material to be coated under such conditions that bonding, e.g. metallic bonding, occurs. Used e.g. in the cladding of the interior surfaces of autoclaves.

explosives Substances or mixtures which, when submitted to shock, friction, sparks, etc., undergo rapid decomposition with the production of a considerable quantity of heat and large volumes of gas. The three main classes of explosives are:

*Propellants**. Explosives which burn at a steady speed and can be detonated only under extreme conditions.

Initiators. Explosives such as mercury fulminate and certain metallic azides which are extremely sensitive to mechanical shock, and are accordingly used in small quantities in detonators to initiate the explosion of larger masses of less sensitive material.

High explosives. Those explosives which normally burn without undue violence when ignited in an open space, but which can be detonated by a sufficiently large sudden mechanical or explosive shock.

expression The compression of a two-phase solid–liquid system in order to separate the liquid from the solid which is retained in the pressing equipment. Expression may be used instead of filtration where slurries are too viscous for pumping, or as an alternative to solvent extraction in oil production from vegetable seeds. Various forms of batch or continuous press or roller mills are used industrially.

extender A compounding ingredient, frequently only a filler, for rubber.

external indicators A virtually obsolete series of indicators used outside the reaction vessel; the indicator is used by transferring a drop of solution from the vessel to a drop of the indicator.

extinction coefficient (α) A quantity which is characteristic of a medium which absorbs light. The extinction coefficient (α) is defined as the reciprocal of the layer thickness, measured in centimetres, in which the intensity of radiation is reduced to $\frac{1}{10}$ of its original value. The extinction and absorption coefficients (K) are related; $\alpha = 0{\cdot}4343\,K$. *See* Beer's law.

extract In solvent extraction a portion of the feed is preferentially dissolved by the solvent and recovered by distilling off the solvent. This constitutes the extract.

extraction The removal of soluble material from a solid mixture by means of a solvent, or the removal of one or more components from a liquid mixture by use of a solvent with which the liquid is immiscible or nearly so. *See* leaching, liquid–liquid extraction.

extractive distillation If two substances have boiling points close together, or form an azeotrope, separation by normal fractionation is not feasible. Extractive distillation is a method whereby a third substance is added which decreases the volatility of one compound relative to the other, and makes separation by distillation possible.

extranuclear electrons *See* electronic configuration.

F

F Fluorine.

F centre An anionic site in a crystal occupied only by an electron, e.g. NaCl plus Na vapour gives blue $Na_{1+x}Cl$ containing F centres.

face-centred cubic The cubic close-packed lattice (ccp)*. *See* face-centred lattice.

face-centred lattice A lattice having an atom at the face-centre of each unit cell.

fac isomer *See* isomerism.

FAD Flavin-adenine dinucleotide.

fahl ore, Cu_3SbS_3. Tetrahedrite.

Fajans' method The titration of Cl^- with Ag^+ using fluorescein as an adsorption indicator. At the end point the precipitate becomes red.

Fajans' rules Ionic compounds are most readily formed by:

(1) ions with low charge (e.g. Na^+ more than Al^{3+});

(2) by large cations (Cs^+ more than Na^+);

(3) by small anions (Cl^- more than I^-).

Faraday effect A plane polarized light wave may be considered to consist of an electrical component and a magnetic component which are mutually perpendicular. When such a wave of plane polarized light is passed through a solid or liquid, which is placed in a magnetic field, the plane of polarization is rotated. The phenomenon is called the Faraday effect. Positive rotations are observed when the direction of rotation is the same as that of the magnetizing current.

Faraday's laws of electrolysis These state:

(1) the amount of decomposition during an electrolysis is proportional to the quantity of current passed;

(2) for the same quantity of electricity passed through different solutions the extent of decomposition is proportional to the chemical equivalent of the element or group liberated.

Faraday, unit The electrical charge of one mole of electrons. 1 Faraday = 96 487 coulombs.

farnesol, $C_{15}H_{26}O$, 3,7,11-trimethyldodeca-*trans*-2-*trans*-6,10-trienol. Sesquiterpene alcohol found in many essential oils. B.p.120°C/0·3 mm. Has a characteristic smell and is of value in perfumery. Oxidation gives the

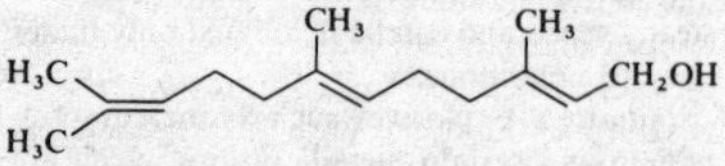

aldehyde farnesal. Farnesol is of importance in the biosynthesis of polyterpenoids including steroids.

Farnesol pyrophosphate is an immediate precursor of squalene*, the key intermediate in steroid and triterpenoid biogenesis, which arises from the coupling of two farnesol pyrophosphate molecules or of C_{15} units derived therefrom. The numerous types of sesquiterpenoid carbon skeletons represent various modes of cyclization of farnesol (sometimes with rearrangement) and it is probable that farnesol pyrophosphate is also the source of these compounds.

Farnesol derivatives also have 'juvenile hormone' activity.

fats

$(R'COO)CH_2{\cdot}CH(OOCR^2)CH_2(OOCR^3)$.

The fats are esters of fatty acids with glycerol with the general formula shown; R^1, R^2 and R^3 may be the same fatty acid residue, but in general the fats are mixed glycerides, each fatty acid being different. The fatty acids present in the greatest quantity in fats are oleic, palmitic and stearic acids. The term 'oil' is usually applied to those glycerides which are liquid at 20°C, and the term 'fat' to those that are solid at that temperature.

Fats are hydrolysed to glycerol and fatty acids by boiling with acids and alkalis, by superheated steam and by the action of lipases. If alkalis are used for hydrolysis, the fatty acids combine with the alkalis to form soaps. Alkaline hydrolysis is therefore sometimes called saponification.

When the fats are heated above 250°C they decompose with the production of acrolein, the intense smell of which is one of the best methods for detecting fats. The extraction of fats from tissues is most conveniently carried out by extraction with ether or some other solvent.

The fats are essential constituents of the food of animals, although conversion of carbohydrates to fats in the animal body does occur. They are partially absorbed from the gut as fats to the lymphatic system and partially hydrolysed by lipases and absorbed as fatty acids which are carried direct to the liver. They are split up in the body principally by oxidation at the *β*-carbon atom.

fatty acids The fatty acids are monobasic acids containing only the elements carbon, hydrogen and oxygen, and consisting of an alkyl radical attached to the carboxyl group. The saturated fatty acids have the general formula $C_nH_{2n}O_2$. Methanoic and ethanoic acid are the two lowest members of this series, which includes palmitic acid and stearic acid. There are various series of unsaturated fatty acids:

The oleic acid series, $C_nH_{2n-2}O_2$, with one double bond, of which acrylic acid is the lowest member.

The linoleic acid series, $C_nH_{2n-4}O_2$, with two double bonds.

The linolenic acid series, $C_nH_{2n-6}O_2$, with three double bonds.

There also exist natural fatty acids with four or more double bonds, fatty acids with hydroxy groups in the molecule, and certain cyclic fatty acids.

The lower members of the series are liquids soluble in water and volatile in steam. As the number of carbon atoms in the molecule increases, the m.p. and b.p. rise and the acids become less soluble in water and less volatile. The higher fatty acids are solids, insoluble in water and soluble in organic solvents.

The fatty acids occur in nature chiefly as glycerides (*see* fats), which constitute the most important part of the fats and oils, and as esters of other alcohols, the waxes. The naturally occurring fatty acids are mostly the normal straight-chain acids with an even number of carbon atoms.

Fe Iron.

Fehling's solution A solution of copper sulphate, sodium potassium tartrate and NaOH used for estimating and detecting reducing sugars.

feldspars A group of aluminosilicates* with framework structures with Al or Si in tetrahedral co-ordination. The most common constituents of igneous rocks. Subdivided into two groups depending on the detailed structural types. Used in the ceramic and enamelling industries. World use 1976 3·3 megatonnes.

felspars *See* feldspars.

fenchenes, $C_{10}H_{16}$. A group of generally bicyclic terpenes derived chemically from fenchone, in most cases with intramolecular rearrangement, of such a character that their carbon skeletons differ from those of fenchone.

fenchone, $C_{10}H_{16}O$. A dicyclic ketone of camphor-like smell, the (+)-form of which is an important constituent of fennel oil and of certain lavender oils; the (−)-form is found in thuja oil.

fenchyl alcohol, $C_{10}H_{18}O$. A secondary alco-

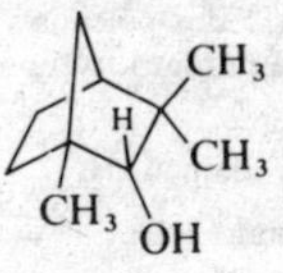

endo-α-

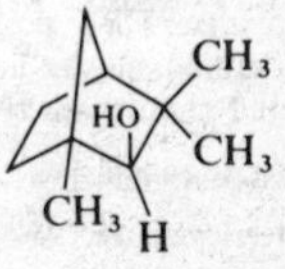

exo-β-

hol produced by the reduction of fenchone with sodium and alcohol. Both (+)- and (−)-fenchyl alcohols are each separable into two stereoisomers, *endo-α-* and *exo-β-*. Fenchyl alcohol can also be obtained by the hydration of (−)-pinene with a mixture of ethanoic and sulphuric acids.

Fenton's reagent An aqueous solution of $FeSO_4$ or other Fe^{2+} salt and hydrogen peroxide used for oxidizing polyhydric alcohols.

FEP plastics Copolymers of tetrafluoroethene and hexafluoropropene. Inert and corrosion resistant as Teflon* but can be processed by melt techniques.

fermentation Controlled microbial action to give useful products. Ethanol, lactic acid, ethanoic acid, gluconic acid, glutamic acid and many other amino-acids are produced by fermentation. Many pharmaceuticals, e.g. sterols, are produced by partial fermentation.

fermium, Fm. At.no. 100. ^{257}Fm (80 days) may be formed by multiple neutron irradiation of Am and Cm, it is purified by ion exchange chromatography. The metal has not yet been prepared and as the element is so radioactive its preparation is unlikely.

fermium compounds No solid fermium compounds are known. The element exhibits +2 and +3 oxidation states in solution. Fermium(II) has been obtained by reduction of Fm(III) with Mg metal and stabilized in an $SmCl_2$ lattice.

ferrates Formally oxyanions of iron.

Ferrates(VI), $[FeO_4]^{2-}$. Red-purple salts ($Fe(OH)_3$ in alkali plus Cl_2; anodic oxidation of Fe in alkali). Stable in basic solution but strongly oxidizing in acid solution. Na and K salts are soluble in water.

Ferrates(IV), e.g. Sr_2FeO_4. Mixed metal oxides.

Ferrates(III), e.g. M^IFeO_2, $M^{II}Fe_2O_4$ (spinels). Mixed metal oxides. Hexahydroxyferrates, $M_3^{II}(Fe(OH)_6)_2$ are formed from $M(OH)_2$ and iron(III) salts.

Ferrates(II), e.g. $Na_4[Fe(OH)_6]$ are formed in alkaline solutions.

ferredoxins Non-haem iron-containing proteins isolated from a number of bacteria and plants. They possess low redox-potentials. Ferredoxins are involved in nitrogen fixation and photosynthesis.

ferric compounds Iron(III) compounds.

ferricyanides *See* cyanoferrates (III).

ferrite Bcc iron.

ferrites Incorrect nomenclature. Mixed metal oxides of Fe(III). Used in electronics industry for any useful magnetic iron oxide.

ferritin A soluble protein which is important in the absorption and storage of iron in the body.

ferrocene, di-π-cyclopentadienyl iron, bis(h⁵-cyclopentadienyl) iron, $C_{10}H_{10}Fe$. M.p.

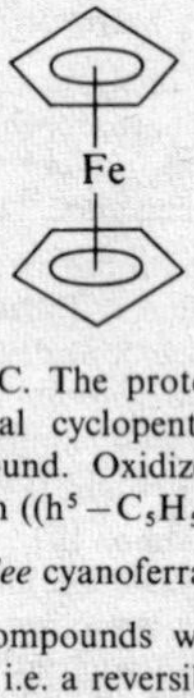

174°C, b.p. 249°C. The prototype of metallocenes, a typical cyclopentadienylide*. A sandwich compound. Oxidized to the blue ferricinium cation $((h^5-C_5H_5)_2Fe)^+$.

ferrocyanides, *See* cyanoferrates (II).

ferroelectrics Compounds which show dielectric hysteresis, i.e. a reversible spontaneous polarization due to non-cancellation of the elementary dipoles in a crystal. Examples include the perovskite $BaTiO_3$ and KH_2PO_4.

ferromagnetism Magnetic behaviour which does not follow the Curie–Weiss law and is field strength dependent. Has large positive magnetic susceptibility (χ) and the magnetic moment (M) does not fall to zero when the applied magnetic field is removed. All the magnetic moments within a crystallite can be considered parallel and the orientations in different crystallites become aligned in a magnetic field. Above the Néel point (temperature), magnetic behaviour is normal.

ferrous Iron(II) compounds.

fertilizers Although any substance which increases production when added to the soil could be called a fertilizer, usually contain N, P or K. Ammonium sulphate is the most important nitrogenous fertilizer; sodium nitrate, ammonium nitrate, ammonia and urea are also used. A mixture of ammonium nitrate and calcium carbonate is sold under the name of Nitro-chalk. The most important fertilizers containing phosphorus are the superphosphates*.

fibre reactive dyestuffs Reactive dyestuffs which form a direct chemical link to the OH groups of cellulose fibres and thus form very resistant colours.

fibres Materials which may be spun and are used in textiles, etc. Natural fibres include wool, cotton, etc. Inorganic fibres including asbestos, glass-wool and other materials used for insulation and packing are now becoming available. Synthetic fibres are based either on modified natural materials – cellulosics, e.g. rayon (270), cellulose acetate (140) – or on completely synthetic derivatives – polyesters (1700), nylons (1150), glass fibres (420), acrylics (330), polyolefines (300). All figures are U.S. 1978 production in megatonnes.

Fick's law of diffusion A law relating the rate of diffusion of a substance in a given direction to the gradient of its concentration.

Thus

$$N_A = -D_{AB}\frac{dC_A}{dz}$$

where N_A is the molar rate of diffusion/unit area; D_{AB} is the diffusivity of substance A in substance B; C_A is the molar concentration of A; z is the distance in the direction of diffusion.

fictile Molecules which rearrange very easily. *See* fluxional. Eg. $Fe_3(CO)_{12}$.

Fieser's solution An aqueous alkaline solution of sodium anthraquinone β-sulphonate (silver salt) reduced with sodium dithionite, $Na_2S_2O_4$, and used as a scrubbing solution for partially removing O_2 from, e.g., N_2.

filament wound plastics *See* reinforced plastics.

fillers Materials incorporated into polymeric compounds (plastics and rubbers) to improve the general properties, to introduce particular characteristics, or to reduce the cost even though certain physical properties may be somewhat impaired. In rigid materials, fibrous fillers (woodflour, woodpulp, cotton, glass-fibre, asbestos) are incorporated to improve the impact strength. Other fillers are used to increase hardness (slate powder), to improve electrical properties (mica) or to change the density ($BaSO_4$). In synthetic rubbers, fillers are classified as reinforcing fillers, inert fillers or diluents. In flexible thermoplastics, smaller proportions of inert fillers are used to reduce the cost and sometimes to aid processing.

Fillers are also used in the preparation of paper to improve the properties of the material and are added to bituminous materials such as plastics and road surfacing materials.

films Formally any material deposited or used in a thin section or in layer form. Film is generally deposited whereas sheet is rolled. Vapour deposition techniques include sputtering, evaporation. Films can be of metals, oxides, etc., but are most generally thermoplastic resins, cellophane, polyethene, polypropene, polystyrene, polyvinyl chloride, etc. Films of such polymeric organics are used as free film for wrapping, photographic images, etc.

filter Apparatus or plant for filtering operations.

filter-aids Materials such as kieselguhr which are added to a slurry to increase the rate of a filtration operation.

filter press A relatively simple and widely-used type of plant for filtration under pressure. The *chamber press* has a number of recessed plates which are covered with filter cloths, the plates then being clamped together. Feed enters through a central channel and a solid cake is built up in each of the chambers, filtrate being discharged through an outlet in the body of the press.

filtration The process of separating a solid from a liquid or gas by use of a membrane or medium (e.g. filter paper, glass sinter) which will allow the fluid to pass through but not the solid. Industrial filtration uses sand beds or perforated surfaces covered by cloth or wire gauze.

fine structure *See* multiplet.

fireclay A refractory clay consisting mainly of aluminium silicate, suitable for the manufacture of furnace bricks, crucibles, etc., and melting at over 1600°C.

fire extinguishers Materials which smother and extinguish, or at least prevent propagation of flame. Water is the most widely used extinguisher. Other important extinguishers are $NaHCO_3$ solution, CO_2-foam formed from a carbonate and acid, liquids such as $ClBrCH_2$, CH_3Br, CCl_4 (although these can give toxic by-products in use) and solids such as $NaHCO_3$, $KHCO_3$, $NH_4H_2PO_4$.

fire retardant An additive which causes resistance to burning when in contact with a high-energy ignition source. *See* flame resistant materials.

Fischer–Hepp rearrangement The nitrosamines of aromatic secondary amines when treated with hydrochloric acid give nuclear substituted nitrosoamines. Among the benzene derivatives, if the *para* position is free the –NO group displaces the hydrogen atom there; in naphthalene derivatives it enters the 1-position:

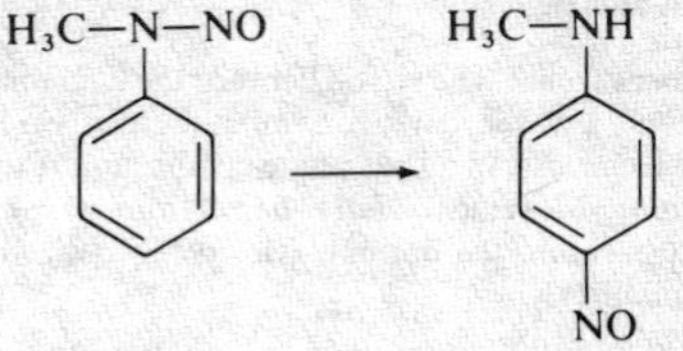

Fischer projection A method of representing three-dimensional structures in two-dimensional drawings in which the chiral atom(s) lies in the plane of the paper. The two enantiomeric forms of glyceraldehyde are represented as

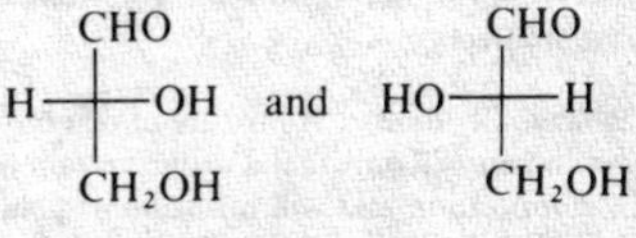

See sawhorse projections.

Fischer–Speier esterification method The catalytic use of a strong mineral acid to effect ester formation between an alcohol and an organic acid. The acids generally used are hydrochloric and sulphuric.

Fischer–Tropsch reaction The catalytic reaction of hydrogen and carbon monoxide (synthesis gas*) to produce high-molecular weight hydrocarbons.

The reaction produces a mixture of gaseous, liquid and solid paraffinic hydrocarbons.

fission *See* atomic energy.

fission products The products of the fission of the heavy nuclei. Fission is generally an asymmetric process, maximum yields occurring for products of mass ~ 90 and mass ~ 140, the distribution of products depending upon the energy of the bombarding particle. Fission products are generally intensely radioactive.

fixation The removal of unwanted silver halide from a photographic material by solubilization after development. The most widely used fixers are thiosulphates which form water-soluble complex salts with silver halides. Other inorganic fixers include thiocyanates, cyanides, ammonia and concentrated potassium iodide. Organic fixers include thiourea and allylthiourea (thiosinamine). Fixing baths may also include hardeners (chromium alum, potassium alum) to toughen the final gelatin layer.

flame emission spectroscopy Emission spectroscopy in which the emitting atom is present in a flame.

flame front The region between the luminous zone and the dark zone of unburned gases which exists in all gas phase combustion reactions. Movement of the flame front is used to measure the burning velocity* of gaseous mixtures.

flame resistant materials, flame retardants, flame-proofing materials Compounds added to fabric to reduce tendency to flame and to burn by glowing. The materials must be resistant to laundering and cleaning. Borates, phosphates, nitrogen compounds (e.g. diammonium phosphate), bromine, chlorine, Sb_2O_3, $Al_2O_3,3H_2O$ all have good flame resistant effects.

flammability limits A mixture of combustible gas and air will only ignite within certain limits of composition and will not support combustion if there is too little or too much combustible present.

flash-back In gas burners having a stationary flame it is essential that the flame speed of the combustible mixture should be balanced by the velocity of the mixture in the burner tube or flame ports. If the flame speed is too high there will be flash-back of the flame onto the gas injector tube. This can be caused by an increase of air/fuel ratio.

flash distillation A distillation technique involving very rapid removal of solvent. Used in desalination of sea-water.

flash photolysis A technique used to study very fast reactions which, usually, involve atoms or radicals in the gas phase. A powerful flash of light of very high energy (usually *c.* 10^5 joules) is passed through the reaction medium for a few microseconds thus causing dissociation. Intermediates in the reaction may be studied by following changes in absorption of a monochromatic light source with time. This technique has the advantage that relatively high concentrations of reaction intermediates are produced compared with those obtained with normal illumination methods.

flash point The lowest temperature at which the application of a small flame causes the vapour above a flammable liquid to ignite when the liquid is heated in a standard cup under prescribed conditions.

flavin-adenine dinucleotide (FAD) The active enzyme or prosthetic group of a large group of flavoprotein enzymes. These enzymes take part in a variety of metabolic oxidation-reduction reactions. They may act directly on the substrate or indirectly by reoxidizing reduced NAD or NADP.

The reduction of flavin in FAD is accompanied by loss of the characteristic yellow colour. The reduction-oxidation of flavoproteins can thus be followed spectrophotometrically.

flavin mononucleotide, FMN Riboflavin-5′-phosphate, the coenzyme component of some flavoprotein enzymes. It is the prosthetic group of Warburg's 'old yellow enzyme' which catalyses the oxidation of NADPH.

flavone, 2-phenylchromone, 2-phenyl-γ-benzopyrone, $C_{15}H_{10}O_2$. Colourless needles; m.p.

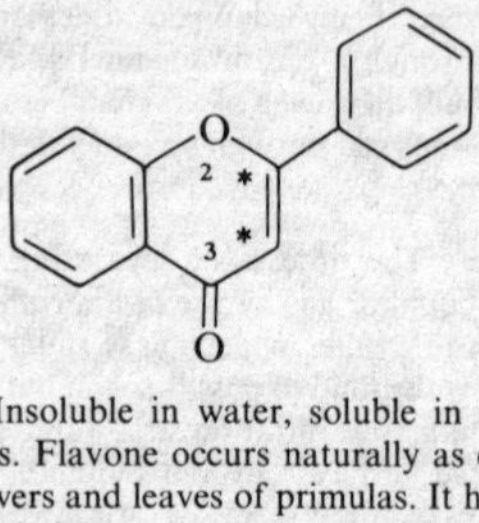

97°C. Insoluble in water, soluble in organic solvents. Flavone occurs naturally as dust on the flowers and leaves of primulas. It has been prepared from *o*-hydroxyacetophenone and benzaldehyde.

The reduction product formed by adding 2H across the starred double bond is called *flavanone*.

flavone glycosides Widely distributed in plants. The sugar residue is either glucose or rhamnose, occasionally a biose sugar. Further isomerization is possible according to which hydroxyl is concerned in the attachment of the sugar. They are, in general, colourless. They include apiin, acaciin, diosmin, lutusin, orobosin.

flavones A group of yellow pigments chemically related to flavone; they occur widely in the plant kingdom. The flavones are found uncombined, as glycosides and in association with tannins.

flavonol glycosides Contain derivatives of flavonol attached to sugar. They include fustin, galangin, kaempferitrin, robinin, datisein, quercitrin, quercetin, incarnatrin, quercimeritrin, serotin, etc. Very widely distributed in plants. Flavonol is 3-hydroxyflavone.

flavoproteins A group of proteins containing riboflavin as their prosthetic group which act as oxygen carriers in biological systems.

flint A compact massive quartz. Used in porcelain manufacture.

flocculation A term often used synonymously with coagulation and precipitation to describe the destruction of a colloidal sol. Sometimes coagulation is used to describe the aggregation part of the process and flocculation the resulting sedimentation of the sol in accordance with Stokes's law.

Florisil Trade name of a magnesium silicate used for chromatography*, especially of fluorine-containing compounds.

flotation A method of separation of ores in aqueous suspension by formation of a moderately stable foam by addition of a small quantity of a suitable reagent followed by aeration. The particles of ore collect in the liquid-air interfaces of the bubbles, the foam may then be removed from the top of the cell, allowed to break down and the ore concentrate recovered.

fluidization The suspension of a mass of solid particles in an upward-flowing stream of gas or liquid. The resulting mixture assumes many of the properties of a liquid, e.g. it can transmit hydrostatic forces and solid objects of smaller density will float in it.

fluidized bed A bed or mass of particles maintained in a state of fluidization*. Fluidized beds are of importance industrially with an ever-widening range of applications in various forms of gas-solid contacting.

Fluon *See* polytetrafluoroethene and fluorine-containing polymers.

fluorene, $C_{13}H_{10}$. Colourless shining flakes usually showing a violet fluorescence; m.p.

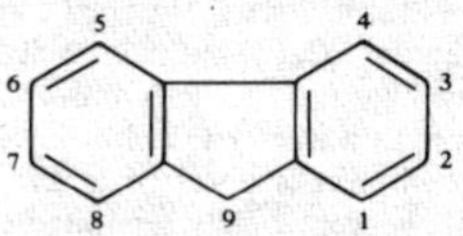

116°C, b.p. 294–295°C. Insoluble in water, easily soluble in organic solvents.

Prepared synthetically by reduction of diphenylene ketone.

Fluorene can be nitrated, sulphonated and chlorinated. Oxidation gives fluorenone (9 O).

fluorescein, $C_{20}H_{12}O_5$. Red crystals with green iridescence. M.p. 314–316°C (decomp.). It is made by heating phthalic anhydride with resorcinol. In alkaline solution it shows intense green fluorescence. It is a dyestuff and is used for colouring liquids in various instruments and by Water Authorities for determining the route of water supplies, leakages, etc.

fluorescence Many substances absorb electromagnetic radiation, e.g. light or X-rays, of a certain wavelength. Part of the energy thus absorbed is re-emitted as radiation of longer wavelength. This process is known as fluorescence. The absorption of the original radiation causes an electron to be promoted to a higher energy level in a molecule or atom of the absorbing substance. This excited electron then returns to its original level in a series of steps, each corresponding to an intermediate energy level. X-ray fluorescence is extensively used in the analysis of geological rock specimens. Fluorescence is used as a conventional analytical technique – *see* fluorimetry.

fluorescence indicator An indicator* which alters the intensity of, or colour of, its fluorescence according to e.g. the pH or redox potential of the solution. Used in the titration of deeply coloured or turbid solutions.

fluorescent brightening agents, optical brighteners Additives for detergents, paper or textiles which absorb white light and re-emit in the violet-blue. This re-emission counteracts the faint yellow colour of bleached materials.

fluorides The salts of HF. *See also* fluorine chemistry.

fluorimetry An analytical technique using the fluorescent properties of the substance to be estimated.

fluorinating agent A substance used to introduce F or replace e.g. H by F. Examples are F_2, KF, HF.

fluorine, F. At.no. 9, at.wt. 18·9984, m.p. −219·62, b.p. −188·14. The lightest and most reactive of the halogens. The chief commercial ore is fluorite*, CaF_2, but many silicates contain some fluorine. The element is obtained by electrolysis of a KF−HF melt using a carbon anode; hydrogen is liberated at the cathode and must be kept rigidly separated from the fluorine. Fluorine can be handled in metal (copper, nickel, monel) apparatus and stored as a gas in cylinders or as liquid fluorine. Elementary fluorine consists of F_2 molecules but they have a relatively low dissociation energy and react with most elements (metals tend to be coated by a coherent film of fluoride). Fluorine gas, HF and fluorides are toxic in large amounts but essential to life and beneficial in the prevention of dental caries. Fluorine compounds are used in the preparation and handling of metals (e.g. U, Al), in inert plastics, refrigerants, aerosol propellants, toothpaste, water treatment. Much of fluorine is used as HF. World production 1978 1·8 megatonnes.

fluorine chemistry Fluorine is the lightest halogen, electronic configuration $1s^22s^22p^5$. It forms a single series of compounds in the −1 oxidation state (E° −2·82V). High oxidation state compounds are covalent (e.g. WF_6, SF_6) low oxidation state compounds are ionic (e.g. NaF, CaF_2). In covalent fluorides the co-ordination number of F is 1 but higher co-ordination numbers are known in ionic fluorides. The F^- ion is a good complexing agent and can bridge (the bridging angle frequently approaches 180°). Fluorine forms compounds with all of the elements except He, Ne, Ar (compounds are formed here in excited states). It tends to bring out the highest **oxidation** states of the element and to reduce **metal–metal** bonding.

fluorine halides *See* halogen fluorides (under Cl, Br,I).

fluorine oxides *See* oxygen fluorides.

fluorine-containing polymers Inclusion of fluorine in polymers affects the properties of the polymers greatly, particularly in conferring stability to corrosion. Polytetrafluoroethene* $-(CF_2)-_n$, Teflon, PTFE is a highly crystalline solid prepared by free-radical polymerization of C_2F_4. It does not flow and fabrication uses powder metallurgy techniques. It is stable to above 250°C and is used where toughness, or good electrical insulation are required. Polyhexafluoropropene $-(CF(CF_3)CF_2)-_n$ is much more costly but is more readily fabricated. Polychlorotrifluoroethene $-(CF_2CFCl)-_n$, is almost as inert as PTFE but can be fabricated by moulding and extrusion, electrical properties are inferior to those of PTFE. Vinylidene fluoride polymers $-(CF_2CH_2)-_n$ and particularly co-polymers with $CF_3CF{=}CF_2$ (Viton) and $CClF{=}CF_2$ (Kel−F) are still resistant. Polyvinylfluoride $-(CH_2CHF)-_n$ is used in films and coatings particularly in the construction industries. Perfluoropolyethers have applications as inert fluids.

fluorite, fluorspar Mineral CaF_2. The white

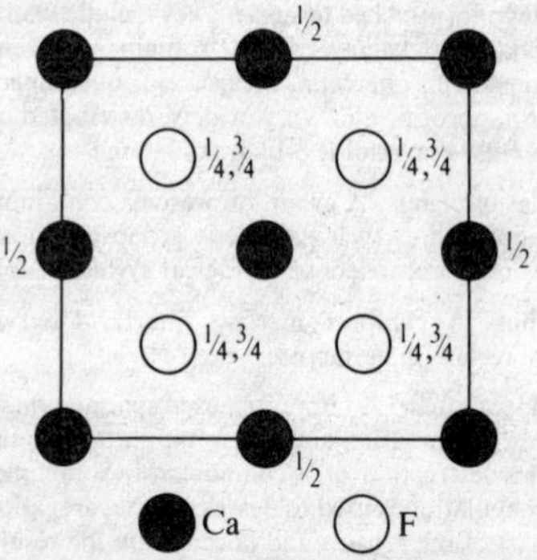

cubic crystals are often coloured yellow, blue, green or violet and have been used for making ornaments (blue-john). It is the chief commercial source of fluorine and fluorides (e.g. HF) and is also used in glasses, enamels and as a metallurgical flux. The crystal structure of fluorite has a cubic close packed array of F^- with each Ca^{2+} in tetrahedral holes. This structure is adopted by many oxides, fluorides and alloys, e.g. CeO_2 and $PbMg_2$.

fluoroacetates, fluoroethanoates Derivatives of monofluoroacetic acid, CH_2FCOOH. The acid itself, its methyl ester and all derivatives broken down in the body to the acid, are highly toxic substances, behaving as convulsant poisons with a delayed action. They are almost odourless, are difficult to detect chemically and are very stable.

Sodium fluoroacetate, which is not volatile and not irritating to the skin, is used as a rodenticide. It is made from CH_2ClCO_2Et and KF, which react to give ethyl fluoroacetate, which is then hydrolysed with NaOH in methyl alcohol.

fluoro acids Complex fluoro salts are formed by most elements but the free acids are, in general, not known (e.g. HF and BF_3 combine only in the presence of a base, e.g. NH_3). Salts are prepared by the action of F_2 on a mixture of alkali and other metal halide, by the action of HF, BrF_3, ClF_3, on appropriate mixtures of salts. The complex halide salts decompose on heating. The distinction between complex fluorides, which definitely contain complex anions, and mixed metal fluorides, e.g. $KMgF_3$, K_2UF_6, which do not contain discrete fluoro-anions is not clear-cut structurally.

fluoroapatite, $Ca_5(PO_4)_3F$. Naturally occurring phosphate rock. Of importance in teeth where it resists dental decay and in the production of fertilizers.

fluoroborates Salts of fluoroboric acid*.

fluoroboric acid, HBF_4. Formed, with $B(OH)_3$, from BF_3 plus water. Fluoroborates contain the tetrahedral $(BF_4)^-$ ion which is only very weakly basic.

fluorocarbon polymers *See* fluorine-containing polymers.

fluorocarbons A class of organic compounds in which some or all of the hydrogen atoms have been replaced by fluorine. They have very similar freezing and boiling points to the corresponding hydrocarbons but are heavier and have high viscosities and low surface tensions. They are often unreactive and have great thermal stability. Fluorocarbons can be made by fluorination of hydrocarbons with fluorine or inorganic fluorinating agents such as AgF_2 and CoF_3 or by the replacement of another halogen, e.g. chlorine, by the action of an alkali halide in a solvent of high dielectric constant. They may also be prepared by fluorination in an electrolytic cell using anhydrous hydrogen fluoride and the substance to be fluorinated as electrolyte. Derivatives of hexafluorobenzene, C_6F_6, show some of the typical properties of aromatic compounds. Fluorocarbon oils, greases and inert dielectrics are used under conditions in which ordinary materials would be attacked. Perfluoro and perfluorochloro polymers are used as plastics (*see* fluorine-containing polymers). Chlorofluorocarbons are used as refrigerants and aerosol agents and propellants. Many fluorine-containing molecules are valuable surfactants. Incorporation of fluorine has major effects on pharmaceutical activity.

fluorochromates Salts, $MCrO_3F$, containing the tetrahedral $(CrO_3F)^-$ ion (HF on chromates(VI)).

1-fluoro-2,4-dinitrobenzene, 2,4-dinitrofluorobenzene, DNF, Sanger's reagent, $C_6H_3FN_2O$. Yellow liquid; m.p. 26°C, b.p. 137°C/2 mm. Reagent introduced by Sanger for identification of the terminal amino group of peptides with which it forms an *N*-2,4-dinitrophenyl derivative under mild conditions.

fluoroethanoic acid, CH_2FCO_2H. *See* fluoroacetates.

fluoroform The trivial name for CF_3H, trifluoromethane.

fluorophosphonates *See* fluorophosphoric acids.

fluorophosphoric acids A series of anions $(PF_6)^-$, $(PO_2F_2)^-$, $(PO_3F)^{2-}$ are formed from PF_5 and HF and by hydrolysis of PF_5, etc. Hexafluorophosphates, M^IPF_6, have solubilities similar to perchlorates (BrF_3 on MF plus P_2O_5). Monofluorophosphates, (fluorophosphonates), $M^I{}_2PO_3F$ (H_3PO_4 in conc. HF) resemble sulphates. Esters (Ag_2PO_3F plus RX) are exceedingly toxic and inhibit cholinesterase action causing muscular constriction. Derivatives are used as insecticides and proposed as nerve gases. Salts are used in toothpaste.

fluorosilicates Salts containing $(SiF_6)^{2-}$ (octahedral) and $(SiF_5)^-$ (trigonal bipyramidal) species; the free acid is unknown but solutions formally containing H_2SiF_6 are known (SiO_2 plus aqueous HF) and are formed in superphosphate manufacture from fluorine-containing phosphates.

fluorosulphates, MSO_3F. *See* fluorosulphuric acid.

fluorosulphuric acid, HSO_3F. B.p. 163°C. Stable acid (distillation of KHF_2 and oleum). Fluorosulphates are formed from the acid or by the action of peroxydisulphuryl difluoride, $FS(O)_2OOS(O)_2F$ (F_2 on SO_3), on the elements. They are similar to perchlorates. HSO_3F is a major constituent of various strongly acidic mixtures, e.g. $HSO_3F{\cdot}SbF_5$, super acids.

fluorspar *See* fluorite.

Fluothane A trade name for halothane*.

flux

(1) An additive assisting in fusion – e.g. $CaCO_3$ in iron-smelting – although the limestone also helps with slag formation.

(2) A substance enabling two pieces of metal to be joined without oxidation.

fluxional Molecules which rearrange so easily on a particular time scale (e.g. n.m.r.) that the rearrangement process can be detected are termed fluxional or stereochemically non-rigid. Example PF_5 which although it is a trigonal bipyramid shows only one type of fluorine by ^{19}F n.m.r. spectroscopy. The rearrangements can often be shown and studied by reduction of temperature. Molecules which are fluxional over all available temperatures are termed fictile*.

fly ash, Na_2CO_3. Sodium carbonate.

Fm Fermium.

FMN *See* flavin mononucleotide.

foamed plastics *See* cellular plastics.

foams A foam is a coarse dispersion of a gas in a liquid, most of the phase medium being gas with the liquid in thin laminar sheets between the gas bubbles. Foaming does not occur with pure substances and it is developed by agitation of liquids and gases in the presence of stabilizing agents. These are generally surface-active substances (e.g. lauryl alcohol) which greatly lower the surface tension of the solution and which are themselves strongly adsorbed at the gas/solution interface. This adsorption process must take place very rapidly if break up of the foam is to be avoided. The persistence of different foams ranges from a few seconds (e.g. amyl alcohol solution) to hours and even months (soap and protein solutions).

Foams are used industrially and are important in rubber preparations (foamed-latex) and in fire fighting. The foam floats as a continuous layer across the burning surface, so preventing the evolution of inflammable vapours. Foams are also used in gas absorption and in the separation of proteins from biological fluids. *See* anti-foaming agents.

folic acid, pteroyl-L-glutamic acid, vitamin B_c Folic acid and its derivatives (mostly the tri- and heptaglutamyl peptides) are widespread in nature. It is a specific growth factor for certain micro-organisms, but in animals the intestinal bacteria provide the small quantities needed for growth. The coenzyme forms are actually the reduced products of folic acid. The main function of 5,6,7,8-tetrahydrofolate(THFA) is as a carrier of a C_1 (methanoate) unit in the biosynthesis of purines, serine and glycine. The nitrogen atoms at 5 and 10 positions are the reactive sites of the molecule. Green leaves are especially rich in this vitamin.

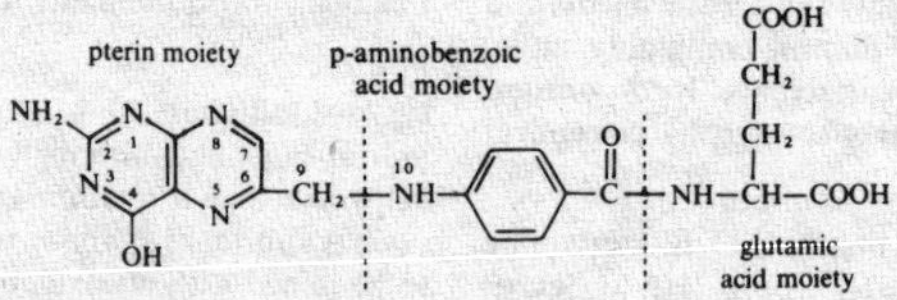

folic acid

food additives Chemicals added to foodstuffs to preserve, improve flavour and appearance. Amongst types of food additives are anti-caking agents (e.g. calcium phosphates), anti-foaming agents (silicones), bleaching agents (peroxide), colouring agents, emulsifying, gelling and thickening agents (e.g. seaweed colloids), enzymes (e.g. carbohydrases), firming agents (calcium salts), flavouring agents and enhancers (e.g. sodium glutamate), nutrients (iodine, fluoride, vitamins), pH adjusters (acids, bases), preservatives (anti-microbials, e.g. benzoic acid, SO_2, etc.), release agents (acetylated monoglycerides to prevent sticking to containers), starch modifiers (e.g. ethanoic anhydride).

force constant A measure of the restoring force to vibration in a bond. Given by

$$\nu = \frac{1}{c}\sqrt{\frac{f}{4\pi^2\mu}}$$

where ν is the vibrational frequency, (cm^{-1}), f is the force constant, c is the velocity of light and μ is the reduced mass of the oscillator A−B(g). The stronger the bond the greater the force constant.

formaldehyde, HCHO. *See* methanal.

formalin *See* methanal.

formamide, $HCONH_2$. *See* methanamide.

formates *See* methanoates.

formation constant Stability constant*.

formic acid, HCOOH. *See* methanoic acid.

formol titration A method for estimating carboxyl groups in amino-acids and free carboxyl groups in proteins, devised by Sörensen. Methanal reacts with the amino-groups of most amino-acids to give neutral methylene derivatives. Thus in the presence of methanal the carboxyl groups can be estimated by titration with alkali.

formoxy The group –O·CHO.

formyl The group C(O)H. Systematically the methanoyl group.

formylation The introduction of a formyl, C(O)H, group. Formylation of aromatic compounds may be compared with acylation and allows the introduction of an aldehyde group. Reactions due to Gattermann, Gattermann–Koch*, Reimer–Tiemann (chloroform heated with sodium phenoxide) and Vilsmeier (aromatic compound, phosphorus oxychloride and N-methylformanilide) are suitable methods of formylation. Ethyl methanoate or ethyl orthoformate react with carbanions to give formyl-derivatives.

fourier transform analysis, FT analysis The use of fourier transform techniques to obtain information from the interaction of electromagnetic radiation with scatterers, e.g. X-rays with solids to give structure, i.r. or other radiation in an interferometer to give an i.r. spectrum; n.m.r. techniques.

Fr Francium.

fractional crystallization The process of separating the components of a mixture containing two or more materials by making use of the change in solubility with temperature.

fractional distillation The process of separating a mixture into a series of fractions of different volatilities by means of distillation.

fractionating column *See* rectification.

francium, Fr. At.no. 87. Radioactive element which occurs as a member of the various natural decay series; m.p. approx. 27°C, b.p. approx. 680°C. ^{223}Fr (half-life 21 minutes) is the most stable isotope. Fr is an alkali metal, electronic configuration. $7s^1$ and its chemistry is known mainly from tracer studies. It forms a single series of Fr^+ compounds.

Franck–Condon principle According to this principle the time required for an electronic transition in a molecule is very much less than the period of vibration of the constituent nuclei of the molecule. Consequently, it may be assumed that during the electronic transition the nuclei do not change their positions or momenta. This principle is of great importance in discussing the energy changes and spectra of molecules.

Frasch process A process for obtaining sulphur by passing superheated water down a shaft to liquefy sulphur which is blown to the surface with compressed air.

free energy A thermodynamic state function represented by G (after Willard Gibbs). The free energy change (ΔG) in any system is related to the enthalpy and entropy changes by the equation:

$$\Delta G = \Delta H - T\Delta S.$$

ΔG is a measure of the maximum amount of *useful* work which may be obtained from the change under consideration. In any system the value of the free energy change determines the position of equilibrium in that at equilibrium the free energy is a minimum, i.e. $\Delta G = 0$ at equilibrium. For substances in their standard states, i.e. 1 mole at 1 atmosphere pressure and a specified temperature, the change in free energy is called the *standard free energy change*, ΔG°_T. This latter quantity is of considerable importance in that it determines the thermodynamic feasibility of a reaction. For a reaction to be thermodynamically feasible and thus for more products to be formed than reactants at equilibrium, the value of ΔG° must be negative, i.e. there must be a decrease in standard free energy from reactants to products. ΔG and ΔG° are usually expressed in kilojoules per mole for chemical reactions. For values *see* p. 393.

free radicals Molecules or ions with unpaired electrons and hence generally exceedingly reactive. The most stable free radicals are NO, NO_2 and molecular oxygen, O_2. Organic free radicals have a complete range of stability, those of relatively long life like triphenylmethyl which exist in equilibrium with their dimeric forms in solution; and those of relatively short life like methyl, which demand special techniques for their study. By the action of silver on triphenylmethyl bromide Gomberg obtained a product of empirical formula $(C_6H_5)_3C$ which behaved as a highly unsaturated molecule, and in solution proved to be an equilibrium mixture of $(C_6H_5)_3C\cdot$ and $(C_6H_5)_3C-C(C_6H_5)_3$. Numerous aryl-substituted ethanes have subsequently been prepared, and the degree of their dissociation is found to depend on (a) the nature of the solvent, (b) the concentration and temperature of the solution, and (c) the size and nature of the substituent groups.

The free radicals which have only a transient existence, like $\cdot CH_3$, $\cdot C_2H_5$ or $\cdot OH$, and are therefore usually met with only as intermediates in chemical reactions, can usually be prepared and studied directly only at low pressures of the order of 1 mm, when they may be transported from the place of preparation in a rapidly streaming inert gas without suffering too many collisions with other molecules. They may be frozen and stabilized in glasses and other matrices. Free radicals may be detected and studied by esr* spectroscopy.

freeze-drying Removal of water (or other solvent) from the frozen solid at e.g. −40° to 10°C at low pressure. Costly but avoids damage to sensitive materials; used for biological specimens and some foodstuffs.

freezing mixtures A mixture used for small-scale refrigeration consisting of salt, ice and water which freezes well below 0°C. Dry ice or liquid nitrogen and organic solvents are also used to prepare freezing mixture. *See* low temperature baths.

freezing point depression The depression caused by dissolving 1 mole of the solute in 1 litre of solvent is termed the molecular depression or cryoscopic constant of the solvent. *See* colligative properties.

Frenkel defect *See* defect structures.

Freons Trade name for a group of chlorofluorocarbons. They can be made by treating a suitable chlorinated hydrocarbon with hydrogen fluoride in the presence of an inorganic chloride such as $SbCl_5$ or with SbF_3/SbF_5. They are inert substances which have been used extensively as solvents, refrigerants and inert dispersing agents in aerosols. Individual members are distinguished by numbers; the more important Freons include the following: Freon 12, dichlorodifluoromethane CCl_2F_2 (b.p. −30°C); Freon 21, dichlorofluoromethane $CHCl_2F$ (b.p. +9°C); Freon 142, 1-chloro-1,1-difluoroethane CH_3CClF_2. Also called Geons, Genetrons, or Arctons. Use is being curtailed because of long-term effects on the upper atmosphere.

frequency, ν The number of wavelengths passing a fixed observer per second. Thus ν depends on the velocity of the wave motion and upon its wavelength. A frequency (wave number) used in spectroscopy is the number of complete waves per cm (cm^{-1}), the reciprocal of the wavelength. Frequency is proportional to energy.

Freundlich isotherm The empirical relationship between the amount of a substance adsorbed and the concentration of the solute

$$\frac{x}{m} = kc^{1/n}$$

where x is the amount absorbed, m the weight of adsorbent, c the concentration or pressure, and k and n constants. The values of the constants k and n vary with the system; n is usually between 0·2 and 0·7. The Freundlich equation is successful with adsorption systems of very variable characteristics, such as adsorption from solution and adsorption of gases. It expresses the rapid increase of adsorption with concentration at low concentrations, but is not so successful at higher concentrations.

Friedel–Crafts reaction In its broadest sense this reaction covers the acylation and alkylation of aromatic compounds by an electrophilic reagent. E.g. an acid chloride RCOCl and benzene will react together in the presence of a Lewis acid to give a ketone $RCOC_6H_5$. Alkyl halides, alcohols, and under certain conditions, alkenes, may be used in the place of the acyl halide and yield the alkyl-substituted aromatic system. The Lewis acid is usually anhydrous $AlCl_3$, but HF, BF_3, $FeCl_3$, $SnCl_4$ may also be used; these serve to generate the acylium ion $(RCO)^+$ or alkyl carbenium ion R^+ which is the electrophile. Alkyl group rearrangement may occur during reaction, e.g. *n*-propyl bromide may yield an isopropyl substituent. The reaction is useful for the synthesis of hydrocarbons and ketones, and is used industrially.

frontier orbital symmetry The theory that site and rate of reaction depend on geometries and relative energies of the highest occupied molecular orbital (HOMO) of one reactant and the lowest unoccupied molecular orbital (LUMO) of the other.

froth flotation *See* flotation.

D-fructose, $C_6H_{12}O_6$. Crystallizes in large needles; m.p. 102–104°C. The most common ketose sugar. Combined with glucose it occurs as sucrose and raffinose; mixed with glucose it is present in fruit juices, honey and other products; inulin and levan are built of fructose residues only. In natural products it is always in the furanose form, but it crystallizes in the pyranose form. It is very soluble in

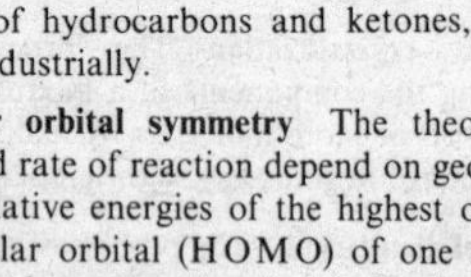

pyranose form

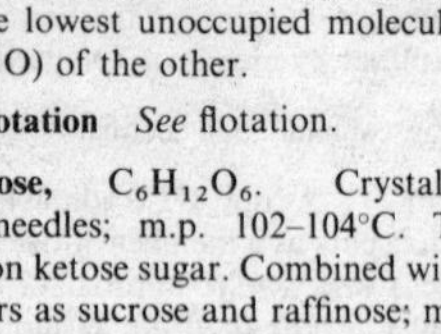

furanose form

water, is twice as sweet as glucose and in many respects behaves as does glucose.

L-fucose Crystallizes in microscopic needles;

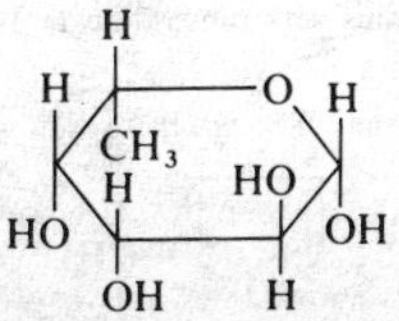

m.p. 145°C. The methyl pentose of gum tragacanth, of the blood group polysaccharides, and of seaweed, where its polymer fucoidin constitutes the cell walls.

fuel cell A cell, similar to a galvanic cell, which generates electricity directly by the electrochemical conversion of gaseous or liquid fuels fed to the cell as required. Fuel and oxidant are supplied to the respective electrodes which are often porous and activated by catalysts.

Fuels which have been used include hydrogen, hydrazine, methanol and ammonia, while oxidants are usually oxygen or air. Electrolytes comprise alkali solutions, molten carbonates, solid oxides, ion-exchange resins, etc.

Batteries of fuel cells have been used in spacecraft but, in general, the potentially high efficiency of the fuel cell has not been realized industrially.

fuel oils A general term applied to petroleum oil used for the production of power or heat. Fuel oils are normally divided into two categories: category 1 for engine fuels and category 2 for burner fuels. Motor and aviation gasolines are normally classified separately.

fuels Materials used for producing energy, particularly heat. They include fuel oils, coal, natural gas, synthetic gases, rocket fuels (propellants), uranium and plutonium (nuclear fuels).

fuller's earth Formerly any detergent clay for fulling cloth; later one used mainly for its adsorptive properties in decolourizing oils, fats and waxes, e.g. for lubricants, soaps and margarine.

fulminates *See* fulminic acid.

fulminic acid, carbonyl oxime (misnomer), **carbyloxime,** C=NOH. Unstable volatile compound with odour resembling HCN; polymerizes to metafulminuric acid. Hg(I) and Ag(I) salts are formed from the metal plus excess HNO_3 and added C_2H_5OH. These fulminates are explosive and shock sensitive and are used as detonators. Fulminic acid is an isomer of cyanic acid.

fulvenes Hydrocarbons of the general structure

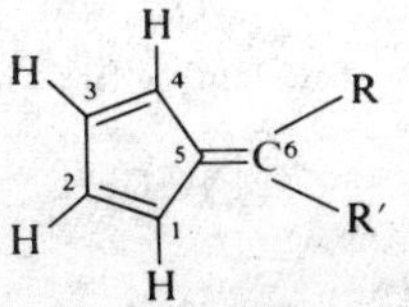

ture shown, where R and R′ are generally hydrogen, alkyl or aryl groups. They are yellow-orange to deep red-violet compounds. The compounds are prepared by base-catalysed condensation of cyclopentadiene and the appropriate keto-compound: they behave as conjugated dienes and peroxidize easily. The parent hydrocarbon fulvene (R=R′=H) is known and is isomeric with benzene.

fumarase The enzyme catalysing the reaction between fumaric acid and water to give (−)-malic acid in the citric acid cycle*.

fumaric acid, *trans*-butenedioic acid, $C_4H_4O_4$. Crystallizes in colourless needles or prisms;

HOOC·CH
‖
HC·COOH

m.p. 300–302°C in sealed tubes; sublimes above 200°C in open vessels. It can be manufactured by the isomerization of maleic acid by heating in the presence of catalysts, or made from carbohydrate sources by the action of *Rhizopus nigricans* or other moulds. When heated to 230°C it is converted to maleic anhydride; heated in sealed vessels with water at 150–170°C, it forms (±)-malic acid. Used as a food acid.

fumaryl The group

—OC·CH
‖
HC·CO—

fungicides Chemicals that inhibit fungal attack, e.g. in agriculture, on wood, plastics, etc. Important fungicides include sulphur, polysulphides and sulphur-containing chemicals (e.g. dithiocarbonates), heavy metal (Cu, Sn, Hg, Ni) compounds.

furaldehyde *See* furfural.

furan, furfuran, C_4H_4O. Colourless liquid; b.p. 32°C. Prepared by heating furoic acid at its boiling point. Forms resins in the presence of mineral acids, but is unaffected by alkalis. Reduced by hydrogen to tetrahydrofuran.

furanose Those forms of the sugars which

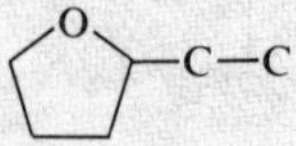

contain a ring of 4 carbons and oxygen and in contrast with the pyranose sugars* are characterized by their instability. They exist in α and β forms.

furazans, 1,2,5-oxadiazoles Formed by the

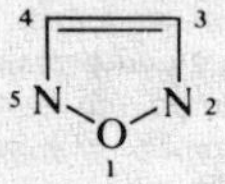

action of NaOH on the dioximes of α-diketones.

furfural, furfuraldehyde, 2-furaldehyde,

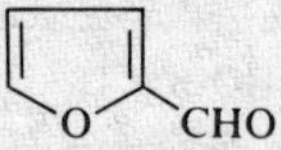

$C_5H_4O_2$. Colourless liquid with a peculiar odour; darkens on exposure to light and air; b.p. 162°C. It occurs in many essential oils and in fusel oil. Manufactured by heating corncobs, oat hulls or other pentose-containing material with steam under pressure at 180°C and fractionally distilling the liquor. Undergoes the Cannizzaro reaction with alkalis to give furoic acid and furfuryl alcohol. Reacts with ammonia to give furfuramide. Oxidized by sodium chlorate(V) in the presence of V_2O_5 to give fumaric acid. Forms resins with phenol, aniline and propanone. Used as a solvent for decolourizing rosin and in the solvent extraction of mineral oils.

furfuraldehyde *See* furfural.

furfuran *See* furan.

furfuryl- The group

furfuryl alcohol, $C_5H_6O_2$. Colourless liquid; b.p. 170–171°C. Prepared by the reduction of furfural or with furoic acid by the action of 30% NaOH solution on furfural in the cold. Forms resins with mineral acids. It is poisonous.

furfurylidene The group

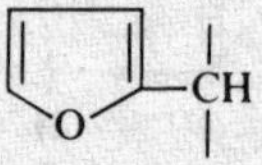

furoic acid, α-furoic acid, furancarboxylic acid, pyromucic acid, $C_5H_4O_3$. Colourless crys-

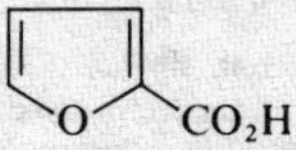

talline substance, m.p. 125–132°C (sublimes at 100°C), b.p. 230–232°C. Prepared, together with furfuryl alcohol, by the action of 30% NaOH solution on furfural in the cold. Forms soluble sodium and potassium salts. It is decomposed at its b.p. into furan and carbon dioxide.

furoyl The groups

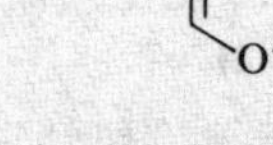

α- or 2-

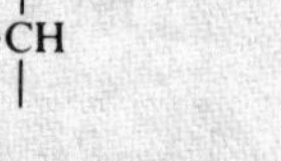

β- or 3-

The α- or 2-configuration is the more common.

furyl- The groups

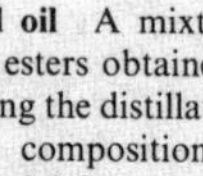

α- or 2-

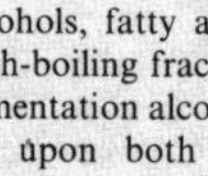

β- or 3-

The α- or 2-configuration is the more common.

fusel oil A mixture of alcohols, fatty acids and esters obtained as a high-boiling fraction during the distillation of fermentation alcohol. The composition depends upon both the material fermented and the yeast used. Used as a source of amyl and isobutyl alcohols.

fusidic acid, $C_{31}H_{48}O_6$. An antibiotic isolated from the fermentation froth of *Fusidium coccineum*.

G

Ga Gallium.

Gabriel's reaction The conversion of a halogen compound into the corresponding amino compound by treatment with potassium phthalimide and subsequent hydrolysis of the intermediate phthalimide compound. Thus chloroethanoic acid gives glycine. The method is of general application and has the great advantage of giving a pure primary amine free from mixture with secondary and tertiary products.

gadolinium, Gd. At.no. 64, at.wt. 157·5, m.p. 1311°C, b.p. 3233°C, *d* 7·89. A lanthanide*. The metal is hcp (to 1262°C) and bcc (to m.p.). The largest use of Gd is in ferrites and as the selenide for electronic equipment. Gd compounds are extensively used as catalysts, particularly olefine polymerization catalysts.

gadolinium compounds Gd forms a single series of typical lanthanide compounds* in the +3 state Gd^{3+} (f^7 colourless) $Gd^{3+} \longrightarrow Gd$ (−2·40 volts in acid). GdI_2 ($GdI_3 + Gd$). Gd_2Cl_3 and GdCl ($GdCl_3$ plus Gd) have metal–metal bonding.

galactans Polysaccharides which on hydrolysis give galactose. They occur in wood and in many algae. The most important galactan is agar*.

D-galactose, $C_6H_{12}O_6$. Crystallizes in the pyranose form; m.p. 118–120°C (monohydrate), 165·5°C (anhydrous). An isomer of glucose which is fairly widely distributed in plants. It is a constituent of raffinose and stachyose, of hemicelluloses, of pectin, of gums and mucilages, and of some glycosides. In animals it forms half the lactose molecule and is the sugar found in the brain. Chemically it is very similar to glucose. It has the structure

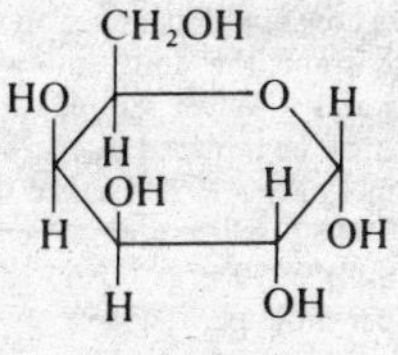

L-Galactose is a constituent of agar and other seaweed polysaccharides, and of linseed mucilage.

galena, PbS. A common lead ore, black or bluish-grey with a metallic lustre. It has the NaCl structure.

gallates Anionic Ga-containing species formed by dissolving Ga, Ga_2O_3, $Ga(OH)_3$ in excess base. Mixed oxides are also known.

gallic acid, 3,4,5-trihydroxybenzoic acid,

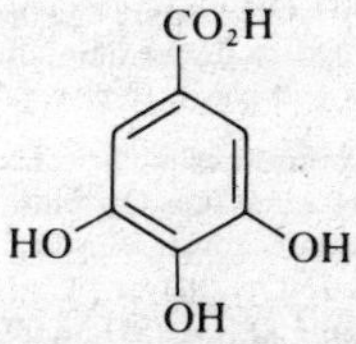

$C_7H_6O_5$. Colourless crystals with one molecule of water, m.p. 253°C, sparingly soluble in water and alcohol. It occurs free in woody tissue, in gall-nuts and in tea, and is a constituent of the tannins, from which it can be obtained by fermentation or by acid hydrolysis. It gives a blue-black colour with Fe^{3+} and is used in the manufacture of inks. On heating it gives pyrogallol.

Gallic acid has a limited use as a dyestuffs intermediate and as a source of medicinals.

gallium, Ga. At.no. 31, at.wt. 69·72, m.p. 29·78°C, b.p. 2403°C, *d* 5·907. Found in traces in Zn and Al minerals and in germanite (0·7% Ga). The element, metallic in appearance, has a complex structure; it is obtained by electrolysis of Ga salts. It dissolves in acids (and alkalis) and becomes coated with a thin coating of oxide on heating in air. Used in doping semi-conductors, light emitting diodes (GaAs; GaP), electrical devices, phosphors. World use 1976 14 000 kg.

gallium chemistry Gallium is a Group (III) element, electronic configuration $4s^24p^1$. Its main oxidation state is +3 ($E°$ for $Ga^{3+} \longrightarrow Ga$ − 0·56 volts in acid) and there is an extensive aqueous chemistry of the Ga^{3+} ion (generally octahedrally co-ordinated). Halides, e.g. $(GaCl_3)_2$, are dimeric and predominantly covalent. Organic derivatives, GaR_3, are monomeric and $[R_2Ga]^+$ complexed species are stable even in aqueous solution, e.g. $[Me_2Ga(H_2O)_2]^+$, $[Me_2GaOH]_4$. Complexes are readily formed. Gallium(I) compounds are relatively unstable and occur in $GaCl_2$, GaS, etc.; contain Ga_2 units.

gallium halides Gallium(III) fluoride, GaF_3 is ionic and forms GaF_3,H_2O. Complexes, e.g. $(NH_4)_3GaF_6$, are formed in aqueous HF solution (decomposes to GaF_3). Gallium(III) chloride, $[Cl_2Ga(\mu\text{-}Cl)_2GaCl_2]$ is covalent and gives GaOCl with water. The bromide and iodide are similar to the chloride. $GaCl_3$ plus Ga gives $Ga^I[Ga^{III}Cl_4]$ (also Br and I) which are strong reducing agents.

gallium oxide, Ga_2O_3. Formed by heating the nitrate, sulphate, hydroxide; has various forms similar to those of Al_2O_3. Forms a hydroxide (NH_4OH on Ga^{3+} solution). Both oxide and hydroxide are amphoteric, dissolving in acids and bases (to give gallates*).

gallium oxyacid salts These are all gallium(III) salts (Ga_2O_3 plus acid) which decompose on heating to Ga_2O_3. Known salts include $Ga(NO_3)_3,9H_2O$, $Ga(NO_3)_3,3H_2O$, $Ga(ClO_4)_3,6H_2O$, $Ga_2(SO_4)_3,18H_2O$.

gallium sulphides Ga_2S_3 (Ga plus S), GaS (Ga_2S_3 plus H_2 contains Ga_2 units) and Ga_2S (Ga plus H_2S) are known.

galvanizing The coating of steel with Zn for protection by dipping into molten Zn or by electrodeposition (cold galvanizing). Corrosion occurs preferentially at the Zn.

gamma rays (γ-rays) Electromagnetic radiation of very short wavelength, that is very hard X-rays. γ-Rays have great penetrating power and are emitted in most nuclear disintegrations.

gammexane *See* BHC.

gangliosides Sphingolipids composed of sphingosine, fatty acid, at least one sugar and sialic acid. Found in brain, nervous tissues and erythrocytes.

gangue The clay and other silicate material associated with ores.

garnets A group of silicates, $M_3^{II}\ M_2^{III}(SiO_4)_3$ (M^{II} = Ca, Mg, Fe; M^{III} = Al, Cr, Fe), containing discrete SiO_4^{4-} groups. Some garnets are used as gemstones with colours ranging from yellow-red to deep-red and also emerald green. Garnet sand is used as an abrasive. World use 1976 28 000 tonnes.

gas The gaseous state is the most diffuse state of matter in which molecules have almost unrestricted motion. A substance for which the volume increases continuously and without limit as the pressure is continuously reduced. A vapour is distinguished from a gas in that a gas is above the critical temperature of the substance.

gas absorption The solution of one component of a gaseous mixture in a liquid or the absorption on to a surface. Selective absorption is employed to separate components of mixtures of gases or vapours, e.g. the removal of NH_3 from NH_3–air mixtures by absorption in water.

gas analysis A mixture of gases may be analysed by e.g.:

(1) Absorbing each constituent preferentially by means of a reagent, e.g. CO_2 in KOH solution, CO in acid or alkaline CuCl.

(2) Exploding or burning the gas with oxygen or air and measuring both the change in volume and amount of waste gases formed by absorption.

(3) Titration as in volumetric analysis, e.g. NH_3 and H_2S.

(4) Absorbing the gas on a substance which can be weighed both before and after absorption.

(5) Changes in thermal conductivity, e.g. carbon dioxide in flue gas.

(6) Measurement of i.r. or u.v. spectra.

(7) Vapour phase chromatography.

(8) Measurement of magnetic susceptibility.

gas calorimeter Equipment for the determination of the calorific value* of a fuel gas by burning at a known rate or for small amounts of combustible gas combustion or explosion of a known volume of gas.

gas chromatography, gas–liquid chromatography (glc), vapour-phase chromatography (v.p.c.). *See* chromatography.

gas chromatography mass spectroscopy, GCMS. Gas chromatography with direct injection of the separated components into a mass spectrometer.

gas constant By combining Boyle's and Charles's laws, we arrive at a general equation of the form $pv = kT$, where p is the pressure, v the volume and T the temperature of the gas, while k is a constant depending on the amount of gas used, and the units employed for expressing p and v. When 1 mole of gas is considered, k may be replaced by R, and the equation becomes $pv = RT$. R is called the gas constant. $R = 0{\cdot}08204$ litre-atmos. per degree, or $8{\cdot}31432/(\text{mol})^4 \times 10^7$ ergs per degree, or 1·987 cal per mole per degree.

gas hydrates Clathrates formed by water with

atomic or small-molecular weight gases and some small organic molecules. The solid water forms open cages with the gases in holes. E.g. $Cl_2,7{\cdot}3H_2O$.

gasification Conversion of a solid or liquid feedstock to a lower molecular weight gaseous fuel of lower carbon to hydrogen ratio. *See* gasification of solid fuels, gasification of oil.

gasification of oil Petroleum feedstocks can be gasified to produce town gas*, synthesis gas* or substitute natural gas* (SNG). Depending on the feedstock the complete process may involve the cracking* of large molecules followed by further gasification involving a reforming reaction which reduces the carbon to hydrogen ratio of the feedstock. This may be effected by direct addition of hydrogen (hydrogasification) or by hydrogen generation from steam within the system (steam reforming). Alternatively the carbon content may be reduced by forming CO_2, coke or char which can be removed from the system.

gasification of solid fuels The conversion of the combustible matter in solid fuels to fuel gases. Various processes have been employed, ranging from the production of producer gas*, blue water gas*, carburetted water gas* to the more sophisticated Lurgi process*.

The main reactions involved are the exothermic producer gas reactions, which are basically

$$C + O_2 \longrightarrow CO_2$$

$$CO_2 + C \rightleftharpoons 2CO$$

and the endothermic water gas reaction

$$C + H_2O \rightleftharpoons CO + H_2$$

If air (or oxygen) and steam are both passed through a high-temperature bed of coal or coke these reactions can be balanced, thus controlling the bed temperature and the fusion of the ash. In the higher pressure Lurgi process the gas obtained is high in methane, formed in reactions such as

$$C + 2H_2 \rightleftharpoons CH_4$$

$$CO + 3H_2 \rightleftharpoons CH_4 + H_2O$$

gas laws The laws which describe the behaviour of gases. These are Boyle's law, Charles's law, Gay-Lussac's law of gaseous volumes and Avogadro's hypothesis.

gas oil A petroleum distillate fraction intermediate between kerosine and light lubricating oil. It is used as a domestic and industrial fuel oil and as a feedstock for cracking processes. Slightly more viscous grades are used as diesel fuels.

gasoline A complex blend of hydrocarbon fractions used as a fuel for motor and aviation purposes. The overall composition may vary, depending on the processes available, but the main component is generally a straight-run naphtha* from the primary distillation of crude oil. This may be blended with stabilized natural gasoline, cracked gasoline* from catalytic cracking of heavier fractions or from the by-products of chemical processes, components from conversion processes such as isomerization and catalytic reforming. Aviation gasoline in particular may also contain branched-chain alkanes from alkylation processes*.

The various additives used to obtain the required properties include anti-knock compounds, anti-oxidants, metal deactivators, anti-icing additives, etc.

gastrin A hormone of the gastro-intestinal tract: at physiological concentrations it stimulates gastric acid secretion, but it is inhibitory at high concentrations. Hog gastrin is

Tyr·Gly·Pro·Tyr·Met·(Glu)$_5$·Ala·Tyr·Gly·
Tyr·Met·Asp·PheNH_2

Gattermann–Koch reaction Formylation* of an aromatic hydrocarbon to yield the corresponding aldehyde by treatment with CO, HCl and $AlCl_3$; at atmospheric pressure CuCl is also required. The reaction resembles a Friedel–Crafts acylation since methanoyl chloride, HCOCl, is probably involved.

Gattermann's reaction A variation of the Sandmeyer reaction*; copper powder and hydrogen halide are allowed to react with the diazonium salt solution and halogen is introduced into the aromatic nucleus in place of an amino group.

Gattermann synthesis A method for the synthesis of aromatic hydroxyaldehydes. E.g. $AlCl_3$ is used to bring about the condensation of phenol with a mixture of gaseous hydrochloric acid and hydrocyanic acid; an aldimine hydrochloride is formed and on hydrolysis gives *p*-hydroxybenzaldehyde

gauche An isomer with non-eclipsed groups

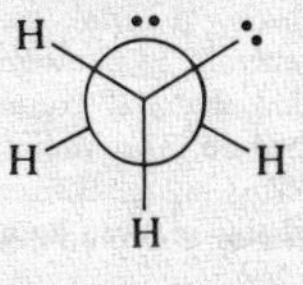

with low-pairs unaligned but not *trans*, e.g. hydrazine.

gaultherin, $C_{18}H_{26}O_{12}$. The parent glycoside of methyl salicylate, it is widely distributed in plants. It is a glycoside of the disaccharide, primeverose.

$$C_5H_9O_4{\cdot}O{\cdot}C_6H_{10}O_4{\cdot}O{\cdot}C_6H_4{\cdot}CO_2CH_3$$

Gay-Lussac's law When gases combine they do so in volumes which are in a simple ratio to each other, and to that of the product if it is also gaseous. Thus 1 vol of nitrogen combines with 3 vol of hydrogen to form 2 vol of ammonia. The law is only approximately true and only strictly holds for ideal gases.

Charles's law, which is a quite different law, is also sometimes referred to as Gay-Lussac's law.

GCMS Gas chromatography–mass spectroscopy*. *See* chromatography.

Gd Gadolinium.

Ge Germanium.

Geiger counter A device used to detect and measure amounts of radioactivity.

gel Hydrophilic colloids are capable under certain conditions, such as lowering of temperature, of partially coagulating to a mass of intertwining filaments which may enclose the whole of the dispersion medium to produce a pseudo-solid but easily deformable mass or jelly. Such gels may, in some instances, preserve a rigidity when they contain as little as 1% of disperse phase, and generally appear to be heterogeneous under the microscope. Gels are sometimes classified as hydrogels, alcogels, etc., according to whether the dispersion medium is water, alcohol, etc.

Inorganic gelatinous precipitates have much the same type of structure as coherent gels, and can often be made to form true gels on careful choice of conditions. Gels are sometimes divided into elastic and rigid gels. Gelatin is an example of the first type, silica gel of the second, but this classification is not strict, as even silica has a certain amount of elasticity. Gelatin gel may be converted back to the sol on heating and is thus a reversible gel; silica will not liquefy to a sol on any simple treatment, and is said to be irreversible. *See* colloids.

gelatin hardeners Materials used in hardening gelatin, generally in photography.

Photographic material containing gelatin can be hardened during manufacture; the process involves cross-linking between the gelatine polypeptide chains induced by hardener.

gelatins Boiling collagen* in dilute acids yields gelatins, which are roughly the constituent peptides of collagen. Gelatins swell in cold water, but are insoluble in it. They dissolve in hot water to give very viscous solutions. A solution of gelatins containing more than 1% gelatin, solidifies to a jelly on cooling. They are rich in the amino-acids glycine, proline, lysine, hydroxyproline and hydroxylysine. Gelatins are manufactured from bones and hides, and are used in cooking, in photography and in the manufacture of glue.

gel filtration An extremely useful technique in biochemistry for separating molecules according to size. Small molecules diffuse into the pores of a gel (e.g. Sephadex), while large molecules are excluded and so are quickly eluted through a column of the gel with a continuous flow. The size of the molecule (and therefore the approximate molecular weight) can be determined by using gels of the appropriate cross-linkage. Gel filtration is commonly used to determine the molecular weight of proteins or in studies of drug–protein binding. It is a powerful tool in protein purification.

gelignite A form of dynamite used as an explosive.

gem- A prefix indicating that particular substituents are attached to the same atom, e.g. $F_3SiSiCl_2SiF_3$, gem-dichlorohexafluorotrisilane.

gemstones Materials which because of their appearance and variety are used in jewellery and decoration. Synthetic stones simulate natural gemstones. α-Alumina (corundum) gives ruby (with Cr^{3+}) and sapphire (Fe plus Ti). Spinel, titania, $SrTiO_3$ are also used but are virtually all artificial. Diamonds can be synthesized but are mainly used for cutting.

genetic code The relationship between the sequence of amino-acids in a protein and the sequence of nucleotides in a nucleic acid which is the 'blue-print' for protein synthesis. Each amino-acid is coded by a triplet of nucleotides, e.g. phenylalanine is coded for by the pyrimidine bases uracil–uracil–uracil in messenger RNA and lysine by adenine–adenine–adenine. Most amino-acids are coded by more than one triplet (a codon) and in these cases the code is said to be 'degenerate', although it is in fact very useful in combating the effects of mutation. Some codons code for 'initiate' and 'terminate' protein synthesis. The genetic code is apparently universal and is the basis of the hereditary information of nucleic acids (*see also* nucleic acids and ribosomes).

Genetron *See* Freons.

genins Where a glycoside is composed of a carbohydrate group combined with a compound related in structure to the sterols, the non-carbohydrate portion of the molecule is called a 'genin'. Genins include digitalis, the toad venoms and the saponins, many of which have high cardiac activity. The latter are called cardiac aglucones.

gentiobiose, $C_{12}H_{22}O_{11}$. The sugar of amygdalin and a number of other glycosides, including gentianose, 6-[β-D-glucopyranosido]-D-glucose. M.p. 190–195°C.

geometrical isomerism *See* isomerism.

Geons *See* Freons.

geranial *See* citral.

geraniol, 3,7-dimethyl-2,6-octadien-1-ol A terpene alcohol; b.p. 229–230°C. Found in a very large number of essential oils. It can also be manufactured from turpentine. It is a colourless oil, smelling of roses, and is unstable in air. It forms citral-a on oxidation. By treatment with gaseous HCl geraniol is converted into limonene.

germanates Oxy compounds of germanium(IV) showing resemblances to simple silicates. E.g. $SrGeO_3$ (cyclic $[Ge_3O_9]^{6-}$) and Mg_2GeO_4 (orthogermanate). Hydrated GeO_2 dissolves in excess base to give $[Ge(OH)_6]^{2-}$ and other oxide-hydroxide anions and cations.

germanes, germanium hydrides Monogermane, GeH_4, b.p. −88°C; Ge_2H_6, b.p. −29°C, and Ge_3H_8 are formed from GeO_2 and $LiAlH_4$ or $NaBH_4$. Higher germanes, to Ge_9H_{20}, are formed by electric discharge on GeH_4. Germanes are oxidized in air. An extensive chemistry of GeH_3- and related groups is known. Germanes have potentialities as sources of very pure Ge.

germanium, Ge. At.no. 32, at.wt. 72·59, m.p. 937·4°C, b.p. 2830°C, *d* 5·35. Germanium occurs in sulphides, particularly zinc ores and is obtained from certain flue dusts. It is converted to volatile $GeCl_4$ which is hydrolysed with water and GeO_2 and reduced to the element with H_2 at 600°C. It has a structure close to that of diamond; it is oxidized by heating in air and combines with chlorine on heating. Its main use is as a semi-conductor but it is also used in phosphors. Forms alloys which expand on freezing and which are used in precision casting. Used in special glasses. World production 1976 800 tonnes.

germanium chemistry Germanium is a Group IV element, electronic configuration $4s^24p^2$. The normal oxidation state is +4 in covalent $GeCl_4$, GeH_4 and ionic GeO_2. Complexes, e.g. $[GeF_6]^{2-}$, are known mainly with octahedral co-ordination. The +2 state is quite strongly reducing. Organogermanium compounds of Ge(IV) are similar to the corresponding Si or Sn compounds.

germanium halides
Germanium fluorides. GeF_4, m.p. −15°C (heat on $BaGeF_6$, Ge plus F_2) is hydrolysed by water, forms $[GeF_6]^{2-}$ in aqueous HF. GeF_2 is polymeric (Ge plus GeF_4 at 100–300°C).

Germanium chlorides. $GeCl_4$, m.p. −49°C, b.p. 86°C (Ge plus Cl_2) is hydrolysed by water. $GeCl_2$, colourless reducing agent ($GeCl_4$ plus Ge).

Bromides and iodides are similar to the chlorides.

germanium hydrides *See* germanes.

germanium oxides *Germanium(IV) oxide,* GeO_2, white solid (Ge plus O_2), forms germanates*. The hydrated oxide is not a simple hydroxide. *Germanium(II) oxide,* GeO, a yellow indefinite material formed by hydrolysis of $GeCl_2$.

germanium sulphides GeS_2 and GeS are known.

getter A chemical for removing impurities. KF acts as a getter to remove HF. Li will remove O, S, etc. from Cu and Cu-alloys.

***g* factor** The proportionality factor in the relation between the magnetic moment μ and the number of unpaired electrons

$$\mu = g\sqrt{s(s+1)}$$

Sometimes called the gyromagnetic ratio. For a free electron $g = 2{\cdot}003$.

gibberellic acid, gibberellin A_3, $C_{19}H_{22}O_6$. M.p. 233–235°C. A plant-growth promoting metabolite of *Gibberella fujikuroi.* Gibberellic acid is the most important member of a family of compounds called gibberellins.

gibberellins *See* gibberellic acid.

Gibbs' equation of surface concentration This equation relates the surface tension (γ) of a solution and the amount (Γ) of the solute adsorbed at unit area of the surface. For a single non-ionic solute in dilute solution the equation approximates to

$$\Gamma = -\frac{c}{RT}\frac{d\gamma}{dc}$$

(where c is the concentration per unit volume and R is the gas constant).

Qualitatively the equation shows that solutes which lower the surface tension have a positive surface concentration, e.g. soaps in water or amyl alcohol in water. Conversely solutes which increase the surface tension have a negative surface concentration.

Gibbs free energy *See* free energy.

Gibbs–Helmholtz equation This equation relates the heats and free energy changes which occur during a chemical reaction. For a reaction carried out at constant pressure

$$\Delta G = \Delta H - T\Delta S$$

where ΔG is the Gibbs free energy change, ΔH is the enthalpy change and ΔS, the entropy change. For a reaction carried out at constant volume the corresponding equation is

$$\Delta A = \Delta E + T\left(\frac{\mathrm{d}A}{\mathrm{d}T}\right)_{\mathrm{v}} = \Delta E + T\Delta S$$

where ΔA is the change in Helmholtz free energy and ΔE is the change in internal energy.

Using this equation it is possible to calculate heats of reaction from the variation of ΔG with temperature.

gibbsite, γ-$Al(OH)_3$. *See* aluminium hydroxides.

Girard's reagents Quaternary ammonium salts of the type $Me_3NCH_2CONHNH_2{}^+X^-$ which form water-soluble compounds with aldehydes and ketones, and are therefore separable from other neutral compounds; the aldehyde or ketone may be subsequently regenerated after separation.

glass In general a supercooled liquid which forms a non-crystalline solid. Specifically a hard, brittle, amorphous material which is usually transparent or translucent, and resistant to chemical attack. The common lime–soda glass used for bottles and jars is a supercooled mixture of sodium and calcium silicates, prepared by fusing a mixture of sand, Na_2CO_3, lime or $CaCO_3$. Numerous special varieties of glass are now made, in which the silica is partly or (infrequently) wholly replaced by another acidic oxide, such as B_2O_3 or P_2O_5, and the Na is replaced by K (potash glass), Li, an alkaline earth metal, or Pb. 'Crown glass' contains K_2O or BaO as the basic constituent; 'flint glass', used largely for optical purposes, contains PbO. When glass is heated to the softening point for a prolonged period the constituents begin to crystallize and the glass 'devitrifies', or becomes opaque and more brittle. Coloured glasses are obtained by adding small quantities of certain metallic oxides (or occasionally other compounds) to the melt. Glass can be toughened by rapid cooling of the surface or by chemical treatment of the surface.

glass electrode *See* ion-selective electrodes. Used for determination of particular ions including pH.

glass fibre Glass filaments in the form of filaments, fabric, chopped fibre, added to thermoset resins to confer great strength. U.S. production 1977 360000 tonnes.

glauber's salt Sodium sulphate*.

glc, gas–liquid chromatography *See* chromatography.

globin A protein, belonging to the histone class. Combined with haem it forms the respiratory pigment haemoglobin.

globulins *See* proteins.

glove box, inert atmosphere box An enclosed space, generally with a glass or plastic viewing port and gloves, used for the manipulation of moisture or oxygen sensitive materials. The purging may be static, e.g. by a tray of P_2O_5; or by circulation through a series of adsorbants.

glucans Polymers of D-glucopyranose; β-glucans being polymers of β-D-glucopyranose. Included are cellulose, lichenin and other materials that are constituents of cell walls.

gluconic acid, dextronic acid, $C_6H_{12}O_7$

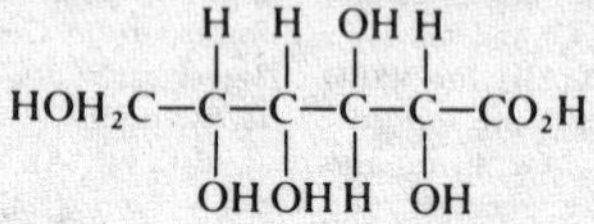

Colourless crystals; m.p. 125°C, soluble in water and alcohol. In aqueous solution forms equilibrium with its lactones. Gluconic acid is made by the oxidation of glucose by halogens, by electrolysis, by various moulds or by bacteria of the *Acetobacter* groups.

D-glucose, dextrose, $C_6H_{12}O_6$. The most common hexose sugar. It is present in many plants, and is the sugar of the blood. It is a constituent of starch, cellulose, glycogen, sucrose and many glycosides, from all of which it can be obtained by hydrolysis with acids or enzymes.

Like all hexoses it can exist in a number of forms.

(1) *Aldehyde*

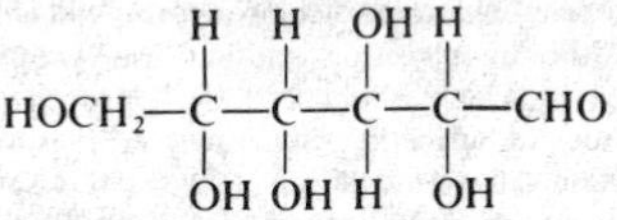

It exists in this form only in solution, though stable derivatives of the aldehyde structure are known. The optical antipode of D-glucose in which the positions of every H and OH are transposed is L-glucose.

(2) *α- and β-Glucopyranose*

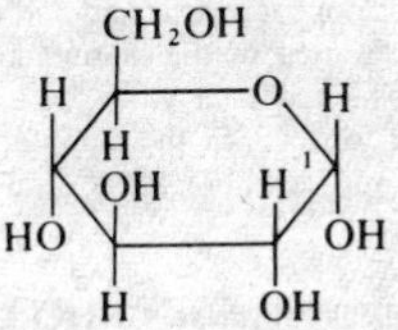

Carbon atom 1 in this formula is asymmetric and two stereoisomers therefore exist, depending on whether the OH group is written below (α) or above (β) the plane of carbon atoms. Both forms crystallize either as the monohydrate or anhydrous.

(3) *α- and β-glucofuranose*

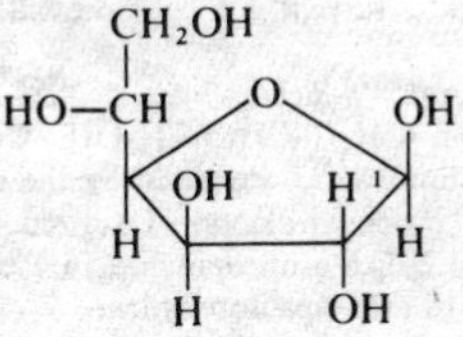

These are unstable and known only in solution. The ethyl glucosides can be obtained crystalline and other derivatives are known.

Ordinary glucose is α-glucopyranose monohydrate; m.p. 80–85°C and $[\alpha]_D$ +113·4°. In solution it gives a mixture with the β form with $[\alpha]_D^{20}$ +52·5°. It is manufactured from starch by hydrolysis with mineral acids, purification and crystallization, and is widely used in the confectionery and other food industries. It is about 70% as sweet as sucrose.

glucosidase *See* maltase.

glucosides *See* glycosides.

glucuronic acid, $C_6H_{10}O_7$. M.p. 165°C. An oxidation product of glucose in which the primary alcohol group is oxidized to carboxyl; it contains the pyranose ring, and exists in α and β forms. The animal organism combines toxic substances with glucuronic acid to excrete them in the urine. It is an important constituent of hemicelluloses and plant gums.

glucuronidase The enzyme responsible for splitting glucopyranouronides, and hence responsible for the formation and splitting of the conjugates that glucuronic acid forms with many substances in the mammalian body.

glue A colloidal mixture of proteins, built up from amino-acids prepared from animal or fish waste (skins, bones, tendons, etc. containing collagen) treated first with milk of lime, acidified and then heated with water at about 60°C. Glue is related to gelatine. An aqueous solution is used in flocculation, recovery of suspended particles. Hide glue (80%) forms a gel in glycerol which can be used in print rollers and in shock absorbing equipment. On addition of methanol the gels become stable to oil and grease.

glutamic acid, α-aminoglutaric acid, $C_5H_9NO_4$, $HO_2C{\cdot}CH_2{\cdot}CH_2{\cdot}CH(NH_2){\cdot}CO_2H$. M.p. 211–213°C. Glutamic acid is one of the acidic amino-acids and is present in large quantities among the products of protein hydrolysis. It can be produced from wheat gluten or beet sugar molasses by hydrolysis, but is usually manufactured by the fermentation of a carbohydrate in the presence of ammonium salts. Its monosodium salt has a meaty flavour and is used as a condiment.

glutamine, $C_5H_{20}N_2O_3$, $H_2N(O)C{\cdot}CH_2{\cdot}CH_2{\cdot}CH(NH_2){\cdot}CO_2H$. Needles; m.p. 184–185°C. It is the monoamide of glutamic acid, and is widely distributed in plants, especially in seedlings of the *Cruciferae* and *Caryophyllaceae*, and in the roots of the beet, the carrot and the radish.

glutaraldehyde, $HC(O)CH_2CH_2CH_2CHO$. Cross-linking agent for proteins and, commercially, for polyhydroxy resins. Also used for sterilization. Formed by hydrolysis of 3,4-dihydro-2-ethoxy-2H-pyran (from acrolein and ethyl vinyl ether).

glutaric acid, pentanedioic acid, $C_5H_8O_4$, $HOOCCH_2CH_2CH_2COOH$. M.p. 97–98°C, b.p. 302–304°C. Prepared by treating 1,3-dichloropropane with sodium cyanide and heating the product with NaOH. Forms an anhydride on heating at 230–280°C.

glutathione, glutamylcysteinylglycine, GSH, $C_{10}H_{17}N_3O_6S$. M.p. 190–192°C (decomp.).

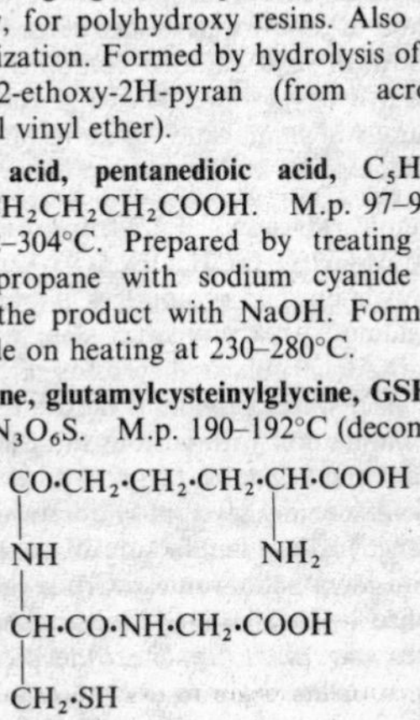

A tripeptide, which is very soluble in water and stable to heating. Glutathione is an important constituent of all cells in which it is usually the major non-protein thiol. Many enzymatic reactions specifically involve glutathione either as a substrate or as a product. Glutathione can act as an oxygen carrier. The enzyme glutathione reductase reduces the oxidized form GSSG back to the reduced form (GSH). Glutathione is also involved in some S-transferase reactions and in glyoxylase* action.

glutelins *See* proteins.

gluten A mixture of proteins obtained from wheat dough by washing out the starch. It contains mostly the two proteins gliadin and glutelin.

glyceraldehyde, glyceric aldehyde, 2,3-dihydroxypropanal, $C_3H_6O_3$, $OHC{\cdot}CH(OH){\cdot}CH_2OH$. Optically active. D-glyceraldehyde is a colourless syrup. May be prepared by mild oxidation of glycerol or by hydrolysis of glyceraldehyde acetal (prepared by oxidation of acrolein acetol). DL-glyceraldehyde forms colourless dimers, m.p. 138·5°C. Converted to methylglyoxal by warm dilute sulphuric acid. The enantiomers are the reference compounds for the configurations of all derived longer chain sugars.

glyceric acid, 2,3-dihydroxypropanoic acid, $C_3H_6O_4$, $HO_2C{\cdot}CH(OH){\cdot}CH_2OH$. An uncrystallizable syrup; it occurs in optically active forms. Prepared by oxidation of glycerin with nitric acid.

glyceric aldehyde *See* glyceraldehyde.

glycerides Esters of glycerol. Classified as mono-, di- and tri-glycerides according to the number of acid radicals combined with the three hydroxyl groups. The tri-glycerides occur naturally in animal and vegetable fats and oils. *See* fats.

glycerol, glycerin, 1,2,3-trihydroxypropane, $CH_2OH{\cdot}CH(OH){\cdot}CH_2OH$, $C_3H_8O_3$. Normally obtained as a colourless, odourless, viscous liquid with a very sweet taste. M.p. 20°C, b.p. 182°C/20 mm. It absorbs up to 50% of its weight of water vapour. It occurs (glycerides) in combination with various fatty acids in all animal and vegetable fats and oils.

Commercially glycerin is obtained as a by-product in the manufacture of soap, and by various synthetic routes. Crude glycerin is purified by distillation. The various synthetic routes start with propene. One proceeds via propenyl chloride, dichlorohydrin, epichlorohydrin to glycerin, another via propenyl alcohol which is oxidized by hydrogen peroxide to glycerin. Some glycerin is also obtained by fermentation of sugars. It is a good solvent for many organic and inorganic compounds. It reacts with hydrochloric acid to form chlorohydrins, and with nitric acid to give nitroglycerin. When heated with sulphuric acid or potassium hydrogen sulphate, acrolein (propenal) is formed. It is oxidized to a variety of products including glyceraldehyde, dihydroxyacetone, glyceric acid and oxalic acid. Glycerol is used in the manufacture of synthetic resins and ester gums, as a moistening agent for tobacco, in the manufacture of explosives and cellulose films, and has many other uses.

glycerol dichlorohydrins, $C_3H_6Cl_2O$. *Glycerol α-dichlorohydrin, sym-dichloroisopropyl alcohol, 1,3-dichloro-2-hydroxypropane,* $CH_2Cl{\cdot}CHOH{\cdot}CH_2Cl$. Colourless liquid with an ethereal odour; b.p. 174–175°C. Prepared by passing dry HCl into glycerin containing 2% ethanoic acid at 100–110°C. Converted to α-epichlorohydrin by KOH. Used as a solvent for cellulose nitrate and resins.

Glycerol β-dichlorohydrin, 2,3-dichloropropanol, $CH_2Cl{\cdot}CHCl{\cdot}CH_2OH$. Colourless liquid, b.p. 182°C. Prepared by the chlorination of propenyl alcohol. Oxidized by nitric acid to 1,2-dichloropropionic acid. Reacts with NaOH to give epichlorohydrin.

glycerol monochlorohydrins, $C_3H_7ClO_2$. *Glycerol α-monochlorohydrin, 3-chloropropylene glycol, 1,2-dihydroxy-3-chloropropane,* $CH_2Cl{\cdot}CHOH{\cdot}CH_2OH$. Colourless, rather viscous liquid; b.p. 139°C/18 mm. Prepared by passing dry HCl into glycerin containing 2% ethanoic acid at 105–110°C. Reacts with nitric acid to give a dinitrate which is used in the manufacture of low-freezing dynamites.

Glycerol β-monochlorohydrin, 2-chlorotrimethylene glycol, 1,3-dihydroxy-2-chloropropane, $CH_2OH{\cdot}CHCl{\cdot}CH_2OH$. Colourless liquid; b.p. 146°C/18 mm. It is obtained in small amounts in the preparation of the α-chlorohydrin.

glycerophosphoric acid, 3-phosphoglyceric acid Lecithins* are fatty acid esters of glycerophosphoric acid derivatives. Commercially glycerophosphoric acid is used to prepare the medicinal glycerophosphate salts, e.g. the calcium salt.

glyceryl The group

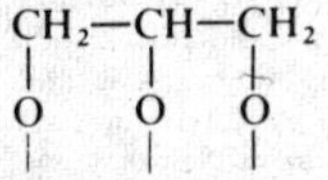

glyceryl trinitrate *See* nitroglycerine.

glycin *See* aminophenols.

glycine, aminoethanoic acid, glycocoll, $H_2N{\cdot}CH_2{\cdot}COOH$. Crystallizes in colourless prisms; m.p. 260°C (decomp.), turning brown at 228°C. It has a sweet taste. Glycine is the simplest of the amino-acids and is a hydrolysis product of proteins. It can be made synthetically by the action of ammonia on chloroethanoic acid and from proteins, such as gelatin, by hydrolysis with acids. Glycine can be synthesized in the animal body.

glycocholic acid, cholylglycine, $C_{26}H_{43}NO_6$. *See* bile salts.

glycogen, $(C_6H_{10}O_5)_X$. The reserve carbohydrate of the animal cell. The molecule is built up of a large number of short chains with 12, or sometimes 18, α-glucose units joined by 1–4 links, the chains being cross-linked (as in amylopectin) by α-1-6 glucoside links. Molecular weight about 4 million. It is a white amorphous powder with no reducing properties. It is broken down in the digestive system by glucosidases to glucose, but in the cells it is broken down to and built up from glucose-1-phosphate by means of phosphorylases.

glycol *See* 1,2-dihydroxyethane.

glycollic acid, hydroxyethanoic acid, hydroxyacetic acid, $CH_2OH{\cdot}COOH$. Colourless crystals, m.p. 80°C. Occurs in the juice of the sugar cane and beets. Prepared by boiling a concentrated aqueous solution of sodium monochloroethanoate. Also produced by electrolytic reduction of oxalic acid. Forms an anhydride when heated at 100°C. Used in textile and leather processing and cleaning (metals and dairy sanitation).

glycols Dihydric alcohols derived from aliphatic hydrocarbons by replacement of two hydrogen atoms by hydroxyl groups. They are colourless liquids. 1,2-Glycols are obtained by oxidation of olefins with potassium permanganate or lead tetra-acetate; by heating olefin chlorohydrins with weak alkalis or by heating alkane dihalides with sodium hydroxide.

glycolysis The metabolic breakdown of carbohydrates in living organisms: it may occur in the presence of oxygen ('aerobic glycolysis') or in its absence ('anaerobic glycolysis'). Anaerobic glycolysis of glucose ultimately yields two moles of ATP per mole of glucose, together with lactate, which can undergo further energy-yielding degradation in the presence of oxygen. Aerobic glycolysis yields pyruvate, which in turn may give rise to acetyl-coenzyme A for the citric acid cycle *. Pyruvate may also undergo other metabolic transformations, e.g. amination to give alanine.

glycoproteins *See* proteins.

glycosidases Enzymes which split off glucoses (glucosidases) or other sugars from glycosides.

glycosides Sugar derivatives in which the hydroxyl group attached to carbon 1 is substituted by an alcoholic, phenolic or other group. The term glucoside is now used for those glycosides which contain glucose as the sugar, while glycoside refers to all compounds whatever the constituent sugar. The non-sugar portion of the molecule is termed the aglycone. The simplest glucoside is methylglucoside, $C_6H_{11}O_5{\cdot}O{\cdot}CH_3$. As the 1 carbon atom is asymmetric there are two isomers distinguished as α and β. The natural glucosides nearly all belong to the β-series and are hydrolysed by β-glucosidases. Glycosides as a class are colourless, crystalline, bitter substances. Various functions have been ascribed to them. Probably most substances in plants are at some stage combined with sugar.

glycylglycine, diglycine, $C_4H_8N_2O_8$, $NH_2{\cdot}CH_2{\cdot}CO{\cdot}NH{\cdot}CH_2{\cdot}CO_2H$. Decomposes at 260–262°C. The simplest of the dipeptides, it is formed by the action of glycyl chloride on glycine or by hydrolysing diketopiperazine with hydrochloric acid.

glyme *See* 1,2-dimethoxyethane.

glyoxal, ethanedial, biformyl, $CH(O){\cdot}CHO$. Crystallizes in yellow prisms; its vapour is green and burns with a violet flame; m.p. 15°C, b.p. 51°C. Readily polymerizes on standing in the presence of moisture; the aqueous solution contains the dihydrate. It can be manufactured by oxidation of ethylene glycol with air using a copper oxide catalyst. Used to harden photographic gelatin.

glyoxaline *See* imidazole.

glyoxydiureide *See* allantoin.

glyoxylases The enzymes that catalyse the formation of lactic acid from methylglyoxal and are present in yeast, animal tissues and plants. Glutathione (GSH) is a coenzyme.

glyoxylate cycle An anaplerotic sequence* supplementing the tricarboxylic acid cycle in micro-organisms and in plants. Isocitric acid is cleaved to succinate and glyoxylate, the latter being used to form malic acid by combination with acetyl-coenzyme A.

glyoxylic acid, CH(O)·COOH A thick syrup, rather difficult to crystallize; m.p. 98°C. Widely distributed in plant and animal tissues. Prepared by electrolytic reduction of oxalic acid. Forms salts of the type $(HO)_2CHCOOM$. Also reacts as an aldehyde. Condenses with urea to give allantoin. Reduced to glycollic acid.

glyptals Alkyd resins*.

Gmelin, Handbuch der anorganische Chemie The definitive reference series in inorganic chemistry although the complete periodic table is not covered.

gold, Au. At.no. 79, at.wt. 196·9665, m.p. 1064·4°C, b.p. 2807°C, *d* 19·3. Gold occurs native, typically as small particles in quartz but also as deposits after erosion of the quartz and sometimes as nuggets. Extraction is with potassium cyanide in the presence of peroxide or air to give the $[Au(CN)_2]^-$ ion. Gold is also found in the slime after the electrolytic purification of copper. Gold is a bright yellow ccp metal. It is a good conductor of heat and electricity. Colloidal gold is formed by reducing gold chloride solution with, e.g., hydrazine. The colour of the colloid depends on the particle size. Reduction with a mixture of $SnCl_2$ and $SnCl_4$ gives Purple of Cassius, a colloidal gold oxide with adsorbed Au used in the manufacture of ruby glass. Gold is not attacked by oxygen or single acids (except H_2SeO_4). It dissolves in aqua regia and in halogen solutions and is attacked by fluorine. The main use of gold is in monetary systems but it is also used in jewellery, as an i.r. reflector, in electrical contacts and conductor systems, and its compounds in the treatment of rheumatoid arthritis. ^{198}Au is used as a radiation source (gold grains). World production 1976 48 tonnes.

gold chemistry Gold is an element of the copper group, electronic configuration $5d^{10}6s^1$. It shows oxidation states +5, +3, +2, +1. The +5 state is found only in AuF_5 and hexafluoroaurates(V). The common states are +3 (generally square planar) and +1 (generally linear). Both +3 and +1 states form extensive ranges of complexes particularly with soft bases. Organometallic compounds are readily formed. Gold(II) complexes are formed with S-ligands but many compounds which are apparently Au(II) derivatives contain Au(I) and Au(III). Gold forms stable metal–metal bonds with other metals and cluster compounds, e.g. $Au_{11}I_3L_7$ or $_8$ (L = phosphines) are formed on reduction.

gold cyanides Important in the extraction of the metal. *Gold(I) cyanide,* AuCN (HCN on AuOH) gives $KAu(CN)_2$. *Gold(III) cyanide,* $Au(CN)_3$, (acid on $KAu(CN)_4$) forms the $[Au(CN)_4]^-$ ion ($AuCl_3$ plus KCN).

gold grains ^{198}Au encapsulated in platinum used as a radiation source.

gold halides Gold(III) fluoride (BrF_3 on Au;

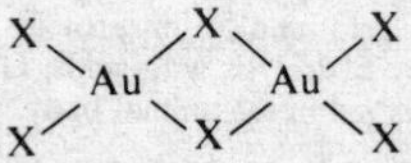

decompose AuF_3·BrF_3 by heating) is readily hydrolysed by water. Fluoroaurates(III), $MAuF_4$, are formed by the action of BrF_3 on a mixture of MF and Au. Fluoroaurates(V) containing $[AuF_6]^-$ species are formed by the action of strong fluorinating agents, e.g. KrF_2. Red brown AuF_5 is formed by heating KrF_2·AuF_5.

The remaining gold(III) halides AuX_3 (X = Cl, Br, I) (Au plus Cl_2 or Br_2, $AuCl_3$ plus KI) have dimeric structures with planar co-ordination about Au. All form haloaurates(III), $MAuX_4$, with excess halide and HX. Gold(I) halides, AuX (X = Cl, Br, I) are formed by gently heating the trihalides. They are insoluble in water but decompose in water to Au plus AuX_3. Gold(I) halides generally form linear complexes, particularly with phosphines and CO; complex halides $[AuX_2]^-$ (X = Cl, Br) are formed by reduction of $[AuX_4]^-$.

gold number A number used to define the efficiency of operation of protective colloids based upon the amount of protective colloid preventing the colour change of a gold sol from red to blue on coagulation by electrolytes.

gold, organometallic compounds Gold(I) compounds RAuL (L = sulphide, phosphine, isocyanide) and gold(III) alkyls, e.g. Me_3AuPPh_3 and $[Me_2AuBr_2]^-$ are known.

gold oxides Gold(I) oxide, Au_2O, is obtained from $[AuBr_2]^-$ and alkali. It decomposes above 200°C. An ill-defined gold(III) hydroxide is soluble in excess of alkali to give the $[Au(OH)_4]^-$ ion.

Goldschmidt process The preparation of sodium methanoate by pressurizing CO and

NaOH at approx. 200°C. Pyrolysis then gives sodium oxalate.

Goldschmidt reaction The use of Al powder for the reduction of metal oxides.

gold, standard Pure gold is too soft for use as ornaments or for coinage, and is alloyed with copper or silver, or both. The fineness is expressed either in parts per thousand, or in carats. Pure gold is 24 carat fine, and the five standard alloys of 22, 18, 15, 12 and 9, i.e. parts of gold in 24 of alloy, are legalized.

gold sulphides Grey Au_2S and black Au_2S_3 are formed by the action of H_2S on gold derivatives. Both decompose to the metal on heating.

gonadotropic hormones Hormones influencing the activity of the gonads, i.e. sex organs. The term is not applied to the steroid sex hormones but to a variety of peptide hormones, produced in the anterior lobe of the pituitary gland and elsewhere.

Gouy balance A balance for the determination of magnetic susceptibility. The sample is weighed in and out of a magnetic field and the susceptibility is calculated from the difference in weights.

graft copolymers *See* block copolymerization.

Graham's law of diffusion This law states that the rates at which two gases diffuse are inversely proportional to their densities, i.e.

$$\frac{\text{rate A}}{\text{rate B}} = \sqrt{\frac{\rho\text{B}}{\rho\text{A}}}$$

Graham's salt, $NaPO_3$. A polymeric metaphosphate. *See* phosphates.

gram atom The quantity of an element numerically equal to the atomic weight expressed in grams.

gram equivalent The equivalent weight of a substance expressed in grams.

gramicidin An antibiotic* produced by the soil bacterium, *B. brevis*, which is highly bactericidal *in vitro* and *in vivo* against bacteria.

gramine *See* 3-(dimethylaminomethyl)-indole.

gram molecular volume The volume occupied by the gram molecule of an element or compound in the gaseous state. According to Avogadro's hypothesis, under the same conditions of temperature and pressure, all gases have the same gram molecular volume. At stp the gram molecular volume is equal to 22·414 litres.

gram molecule The quantity of a compound, or element, equal numerically to the molecular weight expressed in grams. The weight in grams of one mole of a substance.

granite An igneous rock extensively used as a building stone and road metal.

granulation A general term for any process for producing granules. Granulation is frequently carried out by compacting a fine material and then crushing it, and thus employs techniques of both size enlargement and size reduction.

graphite, plumbago, black lead A form of

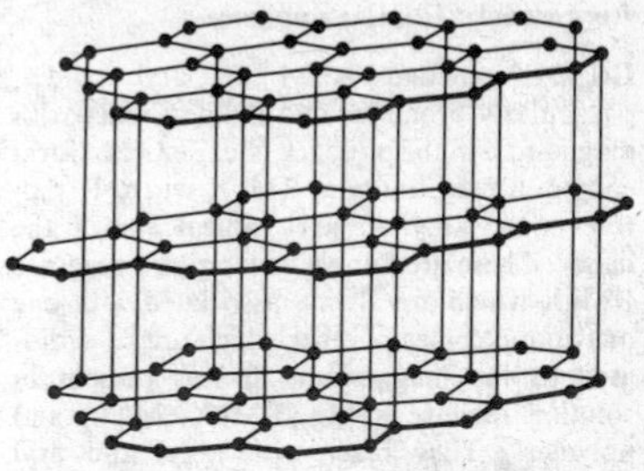

carbon occurring and formed in crystalline and microcrystalline forms. The structure has parallel hexagonal sheets of carbon atoms. The C–C distance within the sheets is 1·42 Å, the distance found in aromatic rings. The C–C distance between sheets is 3·40 Å and there is high conductivity along sheets, easy cleavage and ready incorporation of other substances between the layers (graphite compounds, lamellar compounds, intercalation compounds). Used as a lubricant, refractory, pencils, etc., electrical components, paints, carbonizing steels. Very pure graphite (coke from crude oil plus a pitch binder heated to 2800°C) is used as a moderator and reflector in nuclear reactors. World production 1976 460 000 tonnes.

graphite compounds Compounds formed by penetration of reactants between the layers of graphite. Graphite oxide, approx. C_2O (graphite plus HNO_3 or $KMnO_4$) and graphite fluoride (graphite plus F_2) are non-conducting because of break-up of the aromatic systems of the sheets (graphite fluoride is an efficient lubricant, also used with Li in batteries). In the conducting intercalation compounds various species are inserted between the sheets. The reactivity of the intercalated compound is profoundly modified, e.g. hydrolysis of fluorides is reduced.

gravimetric analysis Methods of analysis involving precipitation, etc. which end up with a final weighing of a stable derivative.

graviton An elementary particle: the particle which in wave form is a gravitational field.

greases *See* lubricating greases.

'green' acids The name given to sulphonic acids which are present in the acid sludge* phase following treatment of lubricating oil fractions with oleum or sulphuric acid. The sulphonic acids in the oil phase are termed 'mahogany acids'* used as emulsifiers, rust-proofing agents and lubricant additives. The water-soluble green acids are not used.

green vitriol Iron(II) sulphate.

Grignard reaction Alkyl and aryl halides, particularly bromides and iodides, react with magnesium in the presence of dry ether to form compounds of the type RMgX where R represents the alkyl or aryl radical and X the halide. These are known as Grignard reagents RMgX which are always associated with one or two molecules of ether which are co-ordinated to the magnesium. Species present in solution include R_2Mg, RMgX, MgX_2 and polymers. They react with alkyl and aryl halides to give hydrocarbons and with metal halides to give organometallics. Esters are formed when the reagents combine with orthoformic, methanoic or chloromethanoic esters; carboxylic acids by combination with solid carbon dioxide. Aldehydes give secondary alcohols, while ketones give tertiary alcohols. Amides and nitriles give ketones. Water and dilute acids react to give hydrocarbons. The immediate result of the reaction is often a complex magnesium compound which is decomposed by dilute acids to give the final product.

Grignard reagents *See* Grignard reaction.

grinding *See* crushing and grinding.

Grottius–Draper Law This law states that only light which is absorbed by a substance is effective in inducing a chemical change. It is not correct to consider that all the light absorbed brings about reaction, since some of it may be re-emitted as fluorescence or as heat. It is not necessary to absorb the light directly by the reacting substances; it is possible, as in photosensitization, to absorb the light by an inert substance which then transfers the stored energy to the reactants as thermal energy.

ground state The lowest energy electronic, vibrational, or rotational state of an atom, molecule or ion.

group A vertical group of elements in the normal form of the periodic table*. Within a group there are distinct chemical resemblances and gradations in properties.

group theory A mathematical method which considers the effect of a group of operators (e.g. symmetry elements, crystal field) on properties. Used in calculations on structure, spectra, magnetic susceptibility, etc.

growth hormone *See* somatropin.

guaiacol, 2-methoxyphenol, $C_7H_8O_2$. M.p. 32°C, b.p. 205°C. It is a constituent of guaiacum resin, and occurs in beechwood tar.

Guaiacol has a very characteristic odour and burning taste; its medicinal properties are identical with those of creosote. Used in the preparation of vanillin and paparvarin and for denaturing alcohol.

guaiazulene, *S*-guaiazulen, 1,4-dimethyl-7-isopropylazulene, $C_{15}H_8$. Blue plates, m.p. 31·5°C. Obtained by the dehydrogenation of guaiol with sulphur. If the dehydrogenation is carried out at a higher temperature using selenium an isomeric product, 2,4-dimethyl-7-isopropylazulene (*Se*-guaiazulene), is obtained.

guaiol, $C_{15}H_{26}O$. A crystalline alcohol; m.p. 93°C, obtained from the wool oil of *Bulnesia sarmienti*, Lorenz. On heating with sulphur it gives guaiazulene.

guanase The enzyme deaminating guanine to xanthine.

guanidine, iminourea, CH_5N_3, $(H_2N)_2C{=}NH$. Deliquescent crystals. Forms many salts, e.g. nitrate, m.p. 214°C. It can be prepared by oxidizing guanine or arginine, the usual method of preparation is by heating ammonium thiocyanate to give guanidine thiocyanate and obtaining the base from this. Guanidine derivatives are used in pharmaceutical manufacture, dyestuffs, explosives and resins.

guanine, 6-hydroxy-2-aminopurine, $C_5H_5N_5O$. An amorphous powder, soluble in acids and alkalis. It is present in all animal and vegetable tissues as a constituent of the nucleic acid portion of nucleoproteins.

guanosine *See* nucleosides.

gulose *See* hexose.

gum acacia, gum arabic Obtained as an exudate from acacia trees. It is the calcium salt of arabin, a complex polysaccharide built of glucuronic acid, arabinose, rhamnose and galactose units. It is used in pharmacy as an emulsifying and suspending agent and in the manufacture of plasters, in the food industry, and as an adhesive.

gums The true plant gums, e.g. gum acacia, gum tragacanth, are the dried exudates from various plants, obtained when the bark is cut or other injury is suffered. They are soluble in water to give very viscous colloidal solutions, sometimes called mucilages* and are insoluble in organic solvents. They are complex polysaccharides, each containing several different sugar molecules and uronic acid groups. Used as thickening agents in the food and pharmaceutical industries etc.

In the petroleum industry the term gum refers to the dark coloured polymer formed by the oxidation of certain unsaturated compounds of cracked or reformed gasolines.

guncotton *See* cellulose nitrate.

gunpowder The oldest and best-known explosive. As a propellant it has been replaced by modern smokeless powders, but it is still an important explosive used in primers, fuses and pyrotechnics. Ordinary black gunpowder consists of about 75% KNO_3, 15% charcoal and 10% sulphur. It is granulated during manufacture, grains of different sizes with different burning rates being produced, and these are polished by rotating in closed drums. Relatively insensitive to shock but is very easily ignited.

gutta-percha A naturally occurring polymeric material, isomeric with rubber but having the *trans* configuration. Now chiefly used as a rubber additive. Obtained from the latex of various tropical trees.

gypsum Mineral $CaSO_4,2H_2O$ also known as selenite. Used in the production of plaster of Paris, to control the rate of setting of Portland cement and as a filler. World production 1976 60 megatonnes.

gyromagnetic ratio For a nucleus the gyromagnetic ratio is equal to the ratio of the magnitude of the magnetic moment (μ) and the angular momentum (I), i.e.

$$\mu = gI \quad \text{where} \quad I = h\sqrt{I(I+1)},$$

I being the nuclear spin quantum number.

H

h *See* hapto.

H Hydrogen.

Ha Hahnium.

Haber process The process for the direct synthesis of ammonia* from N_2 and H_2 over a catalyst.

haem, $C_{34}H_{32}FeN_4O_4$. The combination of protoporphyrin IX and iron in the iron(II) state. Structurally it resembles haematin with the iron in the Fe(II) state and with no chlorine. Haem serves as the prosthetic group of haemoglobin*, cytochromes* and some peroxidases*. The iron may be oxidized to Fe(III) during reactions involving cytochromes or peroxidases, i.e. it becomes haematin.

haematin *See* haem and haemin.

haematite, α-Fe_2O_3. An important iron ore, normally blood-red in colour but forms black crystals (specular iron). Has the corundum structure with c.p. O^{2-} and octahedrally co-ordinated Fe.

haematoporphyrin *See* porphyrins.

haemin, $C_{34}H_{32}ClFeN_4O_4$. The Fe(III)

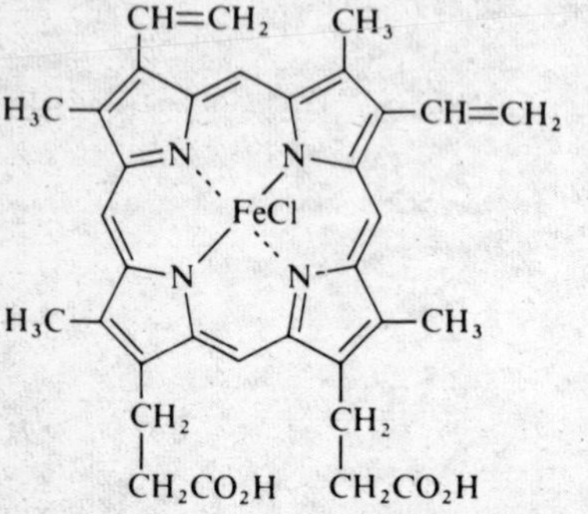

equivalent of haem*. It possesses one residual positive charge and so the isolated complex is neutralized by an anion such as chloride (shown above). In alkaline solution the chloride is replaced by hydroxide to give haematin.

haemocyanin The copper-containing respiratory pigment of molluscs and crustaceans.

haemoglobin The respiratory pigment of the blood of vertebrates. It is a conjugated protein consisting of the iron-porphyrin compound haem, combined with the basic protein globin. Combines loosely with oxygen to form oxyhaemoglobin, which readily loses oxygen again on exposure to a vacuum or an atmosphere devoid of oxygen. Haemoglobin combines with oxygen in the lungs, and carries it in the arteries to the tissues, where the oxygen pressure is low, and where it is reduced, returning to the lungs via the veins. Oxyhaemoglobin is scarlet in colour: reduced haemoglobin is of a purplish colour. Haemoglobin also combines very easily with carbon monoxide, forming carboxyhaemoglobin*, haemoglobin also plays a part in regulating the acidity of blood and in the carriage of carbon dioxide; muscle haemoglobin acts as a respiratory catalyst.

hafnium, Hf. At.no. 72, at.wt. 178·49, m.p. 2227°C, b.p. 4602°C, *d* 13·31. An element of Group IV, electronic configuration $6s^25d^2$. Occurs with Zr in baddeleyite* and zircon*. The metal is hcp. Has very similar properties to Zr and is separated by ion-exchange. In contrast to Zr it is a strong neutron absorber and is used as a control rod material for nuclear reactors. Used in alloys with, e.g., W and Ta. World production 1968 54 tonnes.

hafnium compounds Hafnium compounds are very similar to zirconium compounds* and the chemistry is mainly that of the +4 state.

hahnium, Ha. Element 105. *See* post-actinide elements.

halazone, $C_7H_5Cl_2$, p-$HO_2CC_6H_4SO_2NCl_2$. White powder; m.p. 213°C. Used for the sterilization of drinking water.

half-life, half-value period, $t_{\frac{1}{2}}$. The time taken for the concentration of a substance to fall to half its initial value. For radioactive elements the half-life is $0{\cdot}69 \times 1/\lambda$ where λ is the decay constant. The term is used for transient species (e.g. free radicals) and first order chemical reactions.

half reaction Most generally applied to redox reactions; the half reaction refers to the stoicheiometry of the reaction of one of the species involved, e.g.

$$MnO_4^- + 8H + 5e \longrightarrow Mn^{2+} + 4H_2O$$

half-value period *See* half-life.

half wave potential, $E_{\frac{1}{2}}$ The midpoint of the polarographic wave of an electrochemically reversible reaction. $E_{\frac{1}{2}}$ is characteristic of the species under investigation.

halibut-liver oil Obtained from the liver of the halibut, *Hippoglossus hippoglossus*. Rich in vitamin A and used in the treatment of vitamin A deficiency diseases.

halides Fluorides, chlorides, bromides and iodides; derivatives of the halogens(-1).

Hall effect If a current (I) is passed through a conducting crystal in a direction perpendicular to that of an applied magnetic field (H), the conductor develops a potential (V) between the faces which are mutually perpendicular to both the direction of the current and the magnetic field. This is known as the Hall effect; the magnitude of the potential difference is given by

$$V = \frac{IH}{Ned}$$

where N is the number of current carriers, e is the electronic charge and d, the distance of separation of the two faces across which the potential is developed.

haloamines Halogen derivatives of ammonia. Chloramine, $ClNH_2$, is formed from NaOCl and NH_3; m.p. −66°C. Difluoroamine, F_2NH; b.p. 24°C, ($H_2NC(O)NF_2$ (F_2 plus aqueous urea then H_2SO_4), and H_2NF, a low yield product from the electrolysis of fused NH_4HF_2 are NF_3 derivatives. All of the haloamines are explosive.

haloform reaction A reaction used for the chemical recognition of methylcarbonyl (acetyl) and methylcarbinol structures which give chloroform with NaOCl or the more easily recognized yellow iodoform with NaOI. *See* iodoform reaction.

haloforms The trihalogen derivatives of methane, CHF_3, $CHCl_3$, $CHBr_3$ and CHI_3.

haloform test A colour reaction for detecting any of the haloforms or dihalocarbenes. A haloform or a haloform precursor is added to a mixture of 10% sodium hydroxide and pyridine. Haloforms cause the mixture to turn pink to bright blue-red at room temperature.

halogenated rubbers *See* chlorinated rubbers. Other halogen analogues have little stability.

halogenation A process resulting in the incorporation of halogens by addition or substitution. Among organic compounds, alkenes add halogens, interhalogens and hydrogen halides across the C=C bond; some aromatic compounds react with halogens by substitution for hydrogen atoms in the presence of Lewis acid catalysts like $AlBr_3$, $SnCl_4$, BF_3, $FeBr_3$. *See also* fluorinating agent.

halogens The elements fluorine, chlorine, bromine, iodine and astatine of Group VII.

halonium ions Disubstituted halogen cations, RX^+R', where R,R′ may be alkyl or aryl groups and intermediates in electrophilic halogen additions. May be isolated as, e.g., SbF_6^- salts from media of low nucleophilicity. They are good alkylating (and arylating) agents.

halothane, Fluothane, $CHBrCl{\cdot}CF_3$. Colourless, mobile, heavy liquid, with an odour resembling chloroform and a sweet, burning taste. Non-inflammable; b.p. 49–51°C. Halothane is a general anaesthetic administered in oxygen, often along with dinitrogen oxide, N_2O. It possesses little acute toxicity, but habitual exposure to its vapour may, in rare cases, lead to liver damage.

Hammett equation A correlation between the structure and reactivity in the side chain derivatives of aromatic compounds. Its derivation follows from many comparisons between rate constants for various reactions and the equilibrium constants for other reactions, or other functions of molecules which can be measured (e.g. the i.r. carbonyl group stretching frequency). For example the dissociation constants K_a of a series of para substituted (O_2N-, $MeO-$, $Cl-$, etc.) benzoic acids correlate with the rate constant k for the alkaline hydrolysis of para substituted benzyl chlorides. If log K_a is plotted against log k, the data fall on a straight line. Similar results are obtained for meta substituted derivatives but not for orthosubstituted derivatives.

The equation

$$\log k = \rho \log K + C$$

expresses this relationship, and when the ring substituent is hydrogen, k_0 is the rate of hydrolysis of benzyl chloride and K_0 is the dissociation constant of benzoic acid. The equation

$$\log \frac{k}{k_0} = \rho\sigma \quad \text{where } \sigma = \log \frac{K}{K_0}$$

may be derived and is known as the Hammett equation. It is usual to maintain the term $\log K/K_0$ as referring to the dissociation of benzoic acids in water at 25°C, and to relate other measurable entities to it. The term σ is called the substituent constant since the strength of the benzoic acid depends upon the nature of the substituent. A more positive value of σ infers a more electron-attracting nature of the substituent. Negative σ values indicate electron-donating substituents.

The term ρ is a reaction constant and is mathematically evaluated for a particular reaction by plotting log k/k_0 against σ. The slope of the straight lines is ρ, and reflects the sensitivity of the reaction under study to effects of substituents. The value of ρ is obviously affected by temperature, solvent changes, etc.

The equation does not hold without exceptions even for meta and para substituents, especially when resonance interactions from substituents are possible.

Hammick and Illingworth's rules A set of empirical rules governing the course of aromatic substitution. If a mono-substituted benzene derivative has the formula C_6H_5XY, where X is attached directly to the nucleus and Y is an atom or group attached to X, then if:

(1) Y is in a higher group in the periodic table than X

(2) Y is in the same group as X but of lower atomic weight

XY is *meta* directing. If:

(3) Y is in a lower group than X

(4) X and Y are atoms of the same element

(5) XY is just a single atom

XY is *ortho* and *para* directing.

Hydrogen must be considered as a member of Group I: in such cases as $-CHO$ the hydrogen must be ignored and only the remaining element considered. *See also* Crum Brown's rule.

Hantzsch synthesis The formation of pyridine derivatives by the condensation of ethyl acetoacetate with ammonia and an aldehyde. Also applied to similar syntheses of pyrroles.

hapto The number of atoms in a ligand bonded to the acceptor. E.g. ferrocene, $(C_5H_5)_2Fe$ contains symmetrically bonded pentahapto C_5 rings, designated $(h^5-C_5H_5)$ or $(\eta^5-C_5H_5)$.

hard and soft acids and bases Bases may be arbitrarily divided into hard, non-deformable, e.g. F^-, and soft, polarizable, potentially π-bonded, e.g. PPh_3. Acids may be divided similarly into hard, non-polarizable, e.g. Na^+ and soft, polarizable, potentially π-bonded, e.g. Pt^{2+}. Hard and soft acid base theory (HSAB) postulates greatest stability for hard acid-hard base and soft acid-soft base pairs, e.g. Si^{4+} forms its most stable halide complexes with F^-; Pt^{4+} with I^-.

hardness The resistance of a material to pressure applied to a small area, i.e. resistance to crushing, abrasion, indentation or stretching.

Hardness is assessed on an empirical ten-point scale, diamond (10) is the hardest substance known, talc (1) is soft.

hardness of water The property conferred on water by the presence of alkaline earth salts which prevent formation of a lather with soaps. Temporary hardness ($Ca(HCO_3)_2$) is removed by boiling; permanent hardness, other Ca and Mg salts, is removed by ion exchange (permutite process) or by detergents.

Hartree–Fock orbital Relatively accurately calculated orbital shapes.

HCH Hexachlorocyclohexane. *See* BHC, gammexane.

hcp Hexagonal close-packed. *See* close-packed structures.

He Helium.

heat exchangers Units where heat is transferred from one fluid stream to a second fluid stream, resulting in one being heated and the other being cooled, without change of state in either case. Heat exchangers account for a large part of chemical plant investment.

heating oil A general term describing fuel oils used in domestic and industrial heating. Often applied to kerosines or gas oils.

heat of atomization The amount of heat required to dissociate one mole of an element into its gaseous atoms.

heat of combustion The amount of heat evolved when one mole of a substance is burned in oxygen at constant volume.

heat of crystallization The heat evolved when unit weight of solute crystallizes from an infinite quantity of a saturated solution.

heat of dissociation The amount of heat required to dissociate 1 mole of a compound into its constituent elements, or into certain specified smaller molecules or other species.

heat of formation, standard (ΔH_f^0) The heat evolved or absorbed when 1 mole of a substance is formed from its constituent elements at 1 atm. pressure and a stated temperature, commonly but not necessarily 298 K. By definition the standard heat of formation of an element in its standard state is taken arbitrarily to be zero.

See table on p. 393 for ΔH_f^0 values of some of the more common compounds.

heat of reaction The amount of heat (usually expressed in kilojoules) absorbed or evolved when specified amounts (usually 1 mole of the product) react under constant pressure conditions. For exothermic reactions the convention is adopted that the enthalpy change, the heat of reaction, is negative. Thus, e.g.

$$C(solid) + \tfrac{1}{2}O_2(gas) \longrightarrow CO(gas), \quad \Delta H = -110{\cdot}88\ kJ$$

110·88 kJ of heat are evolved when 1 mole of solid carbon reacts with ½ mole of oxygen gas at constant pressure.

It is essential to specify the physical states of the reactants and products, since there may be additional heat changes associated with changes in state.

heat transfer media Fluids used to convey heat from the place of generation to a location where it is required. The heat may be transferred as sensible heat, latent heat or both.

Steam is by far the most widely used medium, useful up to about 475 K. Up to about 700 K organic liquids such as the dowtherms* and mineral oil may be used. Mercury and molten salts, such as the eutectic* mixture of sodium nitrite, sodium nitrate and potassium nitrate may be used up to 875 K, while above this temperature air and flue gases must be used.

heat transmission oils Thermally stable oils, usually medium viscosity distillates, employed as circulated heat transfer media for industrial purposes.

heavy hydrogen Deuterium, ^2_1H.

heavy spar, $BaSO_4$. *See* barium sulphate.

heavy water, deuterium oxide, D_2O. *See* deuterium. Used to form deutero derivatives for use as tracers, the preparation of ^1H n.m.r. solvents and as a moderator in reactors.

Heisenburg uncertainty principle For small particles which possess both wave and particle properties, it is impossible to determine accurately both the position and momentum of the particle simultaneously. Mathematically the uncertainty in the position Δx and momentum Δp are related by the equation

$$\Delta p \times \Delta x = h/2\pi \quad (h = \text{Planck's constant})$$

helium, He. At.no. 2, at.wt. 4·00260, m.p. −272·2°C, b.p. −268·934°C. One of the noble gases (5×10^{-4}% in air but up to 7% in some natural gases). Found in some radioactive minerals as a product of radioactive decay (the α-particle is an He^{2+} ion). Separated from natural gas by liquefaction of the other gases. Used as an inert atmosphere for arc welding and for Ti, Zr, Si, Ge production, as a coolant in nuclear reactors, with 20% O_2 as an atmosphere for divers, for pressurizing liquid-fuel rockets, for filling balloons and in gas lasers. U.S. demand 1980 1000 M cu ft. He has electronic configuration $1s^2$ and has no normal chemistry although excited species containing bound He are formed in discharge tubes. Liquid He has no triple point and cannot be solidified at atmospheric pressure. Liquid He_{II} is a liquid which exhibits superconductivity and can flow over the edge of vessels. It changes to He_I at 2·2 K.

Helmholtz free energy The maximum amount of energy available to do work resulting from changes in a system at constant volume. *See* free energy and Gibbs–Helmholtz equation.

hemicelluloses The easily hydrolysed portion of cellulose; a series of sugars, hexoses and pentoses. Rather ill-defined and ill-differentiated from cellulose.

hemihedral forms Those forms in any crystal system which show the full number of faces required by the symmetry of the system are called holohedral forms. When only half the number of faces found in the holohedral form are present, the form is said to be hemihedral.

hemimellitene 1,2,3-trimethylbenzene, $C_6H_3(CH_3)_3$.

hemimorphite, $(OH)_2Zn_4Si_2O_7,H_2O$. An important zinc ore containing $(Si_2O_7)^{6-}$ units. Formerly called calamine or electric calamine (particularly U.S. usage).

hendecane The alternative accepted name for undecane $C_{11}H_{24}$, $CH_3(CH_2)_9CH_3$, a colourless liquid, b.p. 195°C.

Henderson–Hasselbach equation A simplified version of the relationships used in calculations on buffer solutions.

$$pH = pK_a + \log \frac{\text{fraction neutralized}}{\text{fraction unneutralized}}$$

Henry's law The mass of gas which is dissolved by a given volume of a liquid at constant temperature is directly proportional to the pressure of the gas. The law is only obeyed provided there is no chemical reaction between the gas and the liquid.

heparin A compound obtained from the liver that inhibits the clotting of blood. Its chief use is in treating thrombosis. It is an acetylated sulphuric ester of a complex carbohydrate containing glucosamine and glucuronic acid.

heptachlor, 1,4,5,6,7,8,8-heptachloro-3a,4,7,7a-tetrahydro-4,7-methanoindene, $C_{10}H_5Cl_7$. White solid; m.p. 95–96°C, soluble in organic solvents. An insecticide similar to chlordane*. Used to control cotton boll weevil.

heptalene The trivial name given to the hy-

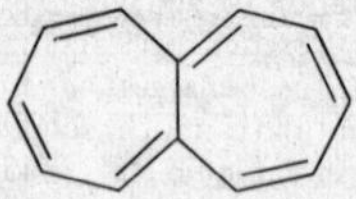

drocarbon $C_{12}H_{10}$. Obtained as an air-sensitive, reddish-brown liquid which is rapidly polymerized by oxygen or by warming to 50°C. Heptalene is an anti-aromatic* system.

heptamethyldisilazane, $Me_3SiNMeSiMe_3$. Used to prepare =NMe derivatives from halides.

***n*-heptane,** C_7H_{16}, $CH_3[CH_2]_5CH_3$. Colourless inflammable liquid; b.p. 98°C. Occurs, together with other isomeric hydrocarbons, in petroleum. Obtained by distillation of petroleum. Eight other paraffin hydrocarbons of the formula C_7H_{16} are possible. Has the general properties of the paraffins. Used with isooctane in defining the knock-rating of petrols.

heptanoic acid *See* oenanthic acid.

4-heptanone *See* butyrone.

heptose A carbohydrate with seven carbon atoms.

herbicides Weedkillers, often selective. *See* e.g. 2,4-D, 2,4-DB.

heroin *See* diamorphine.

Hess's law Sometimes called the law of constant heat summation, it states that the total heat change accompanying a chemical reaction is independent of the route taken in reactants becoming products. Hess's law is an application of the first law of thermodynamics to chemical reactions.

heteroauxin Indole-3-ethanoic acid.

heterocyclic Compounds which contain a closed ring system in which the atoms are of more than one kind, for example, pyridine, thiophen and furan.

heterogeneous catalysis A catalysed reaction where the catalyst is in a different phase from the reactants, e.g. a solid catalyst in a reaction between liquid or gaseous reactants. Such reactions are generally considered to involve adsorption and reaction of the reactants at the catalyst surface.

heterogeneous reaction A reaction which occurs between substances in different phases, e.g. between a gas and a liquid.

heteroleptic Having more than one substituent, e.g. SF_5Cl.

heterolytic reaction A reaction occurring with bond breaking so that the electrons forming the broken bond are shared unequally between the fragments formed, e.g.

$$A-B \longrightarrow A^+ + :B^-$$

Alternatively a reaction between a species with a pair of electrons and a species with a vacant orbital to form a covalent bond.

heteronuclear molecule *See* homonuclear molecule.

heteropoly acids Formally acids giving polymeric oxide- and hydroxide-bridged anions containing different metal and non-metal species. The free acids often do not exist. Examples are the heterotungstates.

HETP *See* theoretical plate.

hexaborane (10), B_6H_{10}; **hexaborane (12),** B_6H_{12}. *See* boron hydrides.

hexachlorobenzene, C_6Cl_6. Colourless crystals; m.p. 227°C, b.p. 326°C. Prepared by exhaustive chlorination of benzene with Cl_2 in the presence of $FeCl_3$ and commercially by the action of Cl_2 on hexachlorocyclohexane in C_2Cl_6. Used in preparative work (C_6Cl_5OH), for the preparation of C_6F_6 and its derivatives, as a wood preservative and in seed dressing.

hexachlorobut-1,3-diene, C_4Cl_6, $Cl_2C=CCl-CCl=CCl_2$. Colourless viscous liquid; b.p. 210–211°C. Principally used as a dispersant material for obtaining i.r. spectra of solids in mulls. Skin irritant.

hexachlorocyclopentadiene, C_5Cl_6. Liquid; b.p. 244°C, reacts with norbornadiene [(2.2.1]-bicycloheptadiene) to give the Diels–Alder adduct aldrin*.

hexachloroethane, C_2Cl_6. Colourless solid; m.p. 187°C, sublimes on heating. Manufactured by chlorination of *s*-tetrachloroethane in the presence of $AlCl_3$. With alkali at about 200°C it gives oxalic acid, and with antimony pentachloride above 450°C it gives carbon tetrachloride. Catalytic reduction gives *s*-tetrachloroethane.

hexadecane, cetane, $C_{16}H_{34}$,$CH_3{\cdot}[CH_2]_{14}{\cdot}CH_3$.M.p. 18°. A straight chain alkane.

hexafluorobenzene, C_6F_6. The simplest of a series of benzene derivatives having fluorine atoms in place of hydrogen. A colourless mobile liquid; m.p. 5·2°C, b.p. 80°C, thermally stable at over 500°C, which undergoes nucleophilic substitution to give pentafluorophenyl derivatives. Prepared C_6Cl_6 plus KF in a polar solvent. Readily substituted, e.g. KNH_2 gives $C_6F_5NH_2$.

hexafluorometallates Derivatives containing $[MF_6]^{n-}$ ions. Formed by many elements in the 3, +4, +5 oxidation states. Some derivatives, e.g. $[PF_6]^-$, $[SiF_6]^{2-}$, $[GeF_6]^{2-}$, are stable in aqueous solution and are available as solutions of strong acids. The majority of hexafluorometallates are immediately hydrolysed by water. Prepared by use of HF (aqueous or anhydrous), F_2, BrF_3.

hexafluoropropene, C_3F_6, $CF_3CF{=}CF_2$. B.p. −29°C. Monomer used to form polymers (only under rather drastic conditions) or copolymers with C_2F_4 and vinylidene fluoride, $CH_2{=}CF_2$. Hexafluoropropene may be prepared by thermal decomposition of $CF_3CF_2CF_2CO_2Na$ or is prepared commercially by low pressure pyrolysis of C_2F_4.

hexagonal close-packing *See* close-packed structures.

hexagonal system The crystal system* with a 6-fold axis as principal axis of symmetry. The unit cell is taken as in the trigonal system*. E.g. NiAs.

hexahelicene, phenanthro[3,4-*c*]phenanthrene,

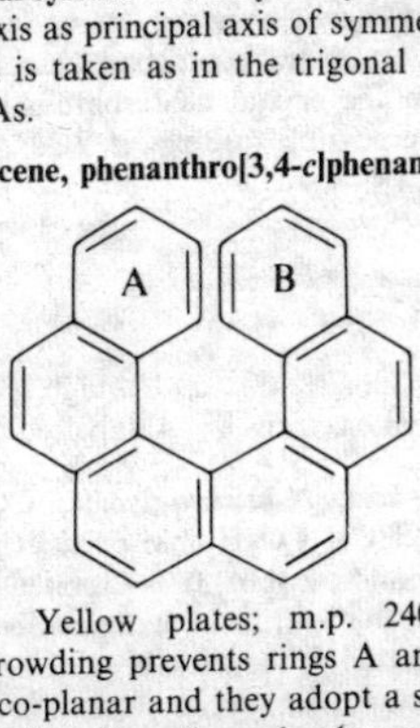

$C_{28}H_{16}$. Yellow plates; m.p. 240–242°C. Steric crowding prevents rings A and B becoming co-planar and they adopt a staggered configuration. Since the molecular geometry is similar to the first turn of a helix the molecule exists as an enantiomeric mixture (right-hand helix and left-hand helix). The enantiomers have been separated by resolution using a π-acid complex.

hexahydrocresols *See* methylcyclohexanol.

hexahydrophenol *See* cyclohexanol.

hexalin *See* cyclohexanol.

hexametaphosphates Derivatives of Graham's salt $(NaPO_3)_n$. *See* phosphates.

hexamethylbenzene, $C_{12}H_{18}$, $C_6(CH_3)_6$. Colourless crystalline solid; m.p. 164°C, b.p. 264°C. Prepared by the reaction of methyl chloride with benzene, toluene, or preferably pentamethylbenzene in the presence of $AlCl_3$. It is oxidized by potassium permanganate to mellitic acid.

hexamethyldisilazane, $(CH_3)_3SiNHSi(CH_3)_3$. B.p. 125·5°C. Prepared from trimethylchlorosilane and ammonia. Used for forming trimethylsilyl derivatives of carboxyl, hydroxyl, amino and thiol groups and especially useful for protecting these groups since the Me_3Si- group is cleaved by water. Used in peptide synthesis. A convenient method of forming volatile, non-polar derivatives of e.g. amino-acids which are then amenable to gas–liquid chromatography. The reagent is also used for siliconizing g.l.c. columns by direct injection into the gas stream. Used to prepare $=NH$ derivatives from halides.

hexamethylenediamine *See* 1,6-diaminoethane.

hexamethylenetetramine *See* hexamine.

hexamethyl phosphoramide *See* hexamethylphosphoric triamide.

hexamethylphosphoric triamide, hexamethylphosphoramide, HMPA, HMPT, $[(CH_3)_2N]_3PO$. M.p. 4°C, b.p. 232°C, dielectric constant 30 at 25°C. Can be prepared from dimethylamine and phosphorus oxychloride. Used as an aprotic solvent, similar to liquid ammonia in solvent power but easier to handle. Solvent for organolithium compounds, Grignard reagents and the metals lithium, sodium and potassium (the latter metals give blue solutions).

At temperatures of 220–240°C it functions as an efficient, neutral dehydrating agent, amides yielding nitriles and alcohols yielding alkenes.

hexamine, hexamethylenetetramine, methenamine, 1,3,5,7-tetra-aza-adamantane, urotropine, $C_6H_{12}N_4$. White crystals, sublime at

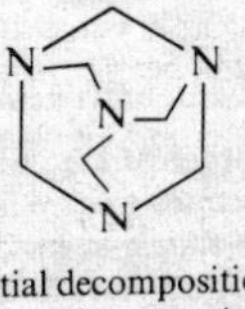

263°C with partial decomposition. Synthesized from methanal and ammonia. Hexamine is used as starter fuel for camping stoves, as an

accelerator in vulcanizing rubber and in the manufacture of resins. It is administered orally in the treatment of urinary infections.

hexan-2,5-dione, acetonylacetone, $CH_3COCH_2CH_2COCH_3$. Colourless liquid which becomes yellow on standing; b.p. 191°C. Obtained by boiling 2,5-dimethylfuran with dilute sulphuric acid. It readily condenses with a variety of substances to give derivatives of furan, thiophen and pyrrole, and is a solvent for cellulose acetate.

hexanes, C_6H_{14}. Five isomeric hexanes are known, b.p. 60–80°C. *n*-Hexane, $CH_3[CH_2]_4CH_3$, a colourless liquid, b.p. 69°C, which is insoluble in water, is the most important isomer, and is used principally as a solvent, particularly for oil-seed extraction.

hexanols, $C_6H_{11}OH$.
n-Hexanol, b.p. 156°C, fatty alcohol used in plasticizers. Prepared from ethanal by an aldol condensation or via $hexyl_3Al$.

4-Methyl-2-pentanol, b.p. 132°C, a solvent for nitrocellulose, urea-formaldehyde and alkyd resins. It is also used in ore flotation.

2-Ethyl-1-butanol, b.p. 147°C, is used in synthesis and as a solvent for printing inks and surface coatings.

hexobarbitone, $C_{12}H_{16}N_2O_3$. Colourless

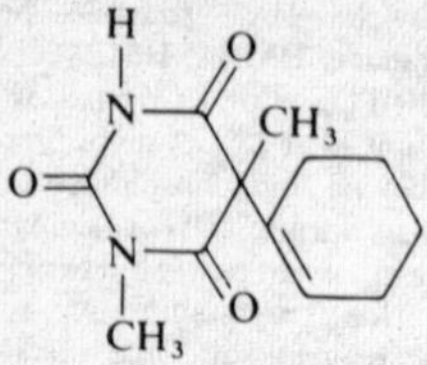

crystals m.p. 146°C. It is a rapidly absorbed hypnotic. The water-soluble sodium salt – hexabarbitone soluble – is injected intravenously to produce general anaesthesia of short duration.

hexoestrol, $C_{18}H_{22}O_2$ $(p\text{-}HOC_6H_4CHEt)_2$ Colourless crystals, m.p. 185–188°C. Prepared from stilboestrol dimethyl ether, or anethole. It resembles stilboestrol* in its oestrogenic activity.

hexone Methyl isobutyl ketone.

hexose A carbohydrate with six carbon atoms. The hexoses are by far the most important of the simple sugars, and almost all the polysaccharides are built up of hexose units. Three aldohexoses, D-glucose*, D-mannose* and D-galactose*, are common in plants, either in the free state or as components of polysaccharide molecules. Other less important aldohexoses are D-idose, D-gulose, D-talose, D-allose and D-altrose.

The four ketohexoses are fructose*, sorbose*, allulose and tagatose. *See* glucose for example of isomerization between open chain and cyclic structures in a typical hexose molecule.

hexyl The group $C_6H_{13}-$.

Heyrovsky–Ilkovic equation The relation between the half wave potential*, $E_{\frac{1}{2}}$, for a polarogram, the current, i, for the potential at the dropping mercury electrode, E_{DME}, the number of electrons, n, involved in the electrochemical reaction (id is the diffusion current).

$$E_{\text{DME}} = E_{\frac{1}{2}} + \frac{0{\cdot}059}{n} \log \frac{id - i}{i}$$

Hf Hafnium.

Hg Mercury.

high alumina cement *See* aluminous cement.

high spin state A term used in transition-metal chemistry to denote that the compound has the maximum number of unpaired electrons consistent with the electronic configuration and stereochemistry. Most commonly used for octahedral complexes where, depending upon the crystal field splitting two spin states are possible for, e.g., d^6 (Co^{3+}, Fe^{2+})

low spin – 0 unpaired electrons

high spin – 4 unpaired electrons

hippuric acid, N-benzoylglycine, $C_9H_9NO_3$, $PhC(O)NHCH_2CO_2H$. M.p. 187°C; soluble in water and alcohol. It is excreted in small quantities in the urine of mammals as a means of elimination of the toxic benzoic acid which is conjugated with glycine in the kidney.

histamine, 4-(2-aminoethyl)-imidazole, $C_5H_9N_3$. A base, formed by the bacterial degradation of histidine, and present in ergot and in many animal tissues, where it is liberated in response to injury and to antigen–antibody* reactions. If injected it causes a condition of shock with dilatation of many blood vessels, loss of plasma from the capillaries to the tissues and a rapid fall in blood pressure. It is normally prepared from protein degradation products.

histidine, α-amino-β-imidazolpropionic acid, $C_6H_9N_3O_2$. M.p. 277°C. The naturally occurring substance is laevorotatory. Histidine is one of the basic amino-acids occurring in the hydrolysis products of proteins, and particularly of the basic proteins, the protamines and histones. It is an essential constituent of the food of animals.

histones *See* proteins.

HMPA, HMPT *See* hexamethylphosphoric triamide.

Ho Holmium.

Hoesch synthesis A variation of the Gattermann synthesis of hydroxy-aldehydes, this reaction has been widely applied to the synthesis of anthocyanidins. It consists of the condensation of polyhydric phenols with nitriles by the action of hydrochloric acid (with or without $ZnCl_2$ as a catalyst). This gives an iminehydrochloride which on hydrolysis with water gives the hydroxy-ketone.

Hofmann conversion of amides Amides react with solutions of chlorine or bromine in excess sodium hydroxide (NaOX) to give primary amines containing one carbon atom less than the original amide. The halogen first replaces one of the hydrogen atoms of the amido group to give a chloro- or bromo-amide which reacts with alkali to give an isocyanate; this decomposes to give an amine and carbon dioxide. The reaction is used in the preparation of anthranilic acid. Overall

$$RCONH_2 \longrightarrow RNH_2$$

Hofmeister series *See* lyotropic series.

holmium, Ho. At.no. 67, at.wt. 164·93, m.p. 1470°C, b.p. 2720°C, *d* 8·80. A lanthanide*. The metal is hcp. There are at present no uses for this element.

holmium compounds Holmium forms a single series of typical lanthanide compounds* in the +3 state

Ho^{3+}(f^{10} brown-yellow) ⟶ Ho
(−2·32 volts in acid).

holohedral forms *See* hemihedral forms.

HOMO *See* frontier orbital symmetry.

homo- Used as a prefix in organic chemistry it indicates a difference of $-CH_2-$ in an

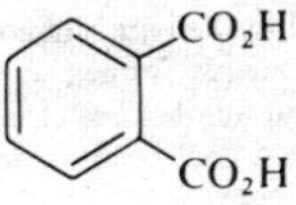

phthalic acid

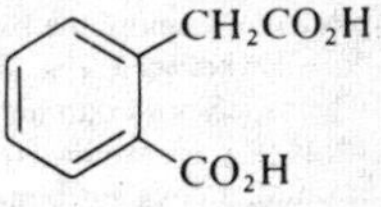

homophthalic acid

otherwise similar structure. It is applicable to aliphatic, alicyclic and aromatic compounds.

homocyclic Compounds which contain a closed ring system in which all the atoms are the same. As the atoms are nearly always carbon atoms, the term carbocyclic is often used. These compounds can be subdivided into aromatic or benzenoid compounds and alicyclic compounds, such as cyclohexane.

homogeneous catalysis The process which occurs when the catalyst is in the same phase as the reactants, e.g. the acid-catalysed hydrolysis of an ester. Recently the term has been used more loosely to describe processes where the catalyst and one of the reactants are in the same phase, e.g. the hydrogenation of liquid olefins catalysed by solutions of transition metal complexes.

homogeneous combustion *See* cool flames.

homogeneous reaction Reactions which occur between substances in the same phase, e.g. between gases or liquids. A reaction between two solids is not usually regarded as a homogeneous reaction.

homogenizer A type of colloid mill.

homogentisic acid, $C_8H_8O_4$. M.p. 152–

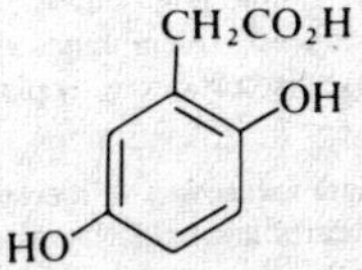

154°C. Excreted in the urine in the rare hereditary disease alkaptonuria. Homogentisic acid is easily oxidized in the air to dark-coloured polymeric products, so that urine from patients with alkaptonuria turns gradually black. It is formed from tyrosine and is an intermediate in tyrosine breakdown in the body. Alkaptonuria is due to the absence of the liver enzyme which cleaves the aromatic ring.

homoleptic Having only one type of substituent, e.g. MoF_6 or $Ti(CH_3)_4$.

homolytic reaction A reaction such that the

electrons forming the bond broken are shared equally between the fragments formed, or when species containing odd numbers of electrons (e.g. free radicals) react with each other forming a covalent bond. E.g.

$$Br_2 \longrightarrow 2Br\cdot$$

$$Br\cdot + CH_3CH_2\cdot \longrightarrow CH_3CH_2Br$$

homonuclear molecule A molecule consisting of identical atoms or groups of atoms, e.g. O_2, N_2 or Cl_2. A heteronuclear molecule contains different atoms, e.g. HCl, NO, CO.

homopolar bond A covalent bond*.

homopolar crystal A crystal with all bonds homopolar (covalent), e.g. diamond, zinc blende.

hormones Hormones are compounds secreted directly into the blood stream by certain organs of the body. They are then carried to other organs where they exert a definite physiological action. Hormones influence the rate at which reactions proceed. The tissues which specialize in the secretion of hormones are called endocrine glands. The thyroid gland secreting thyroxine is one example. Some hormones, however, are secreted by tissues which have other functions, e.g. the sex hormones secreted by the testis and ovary and insulin which is secreted by certain parts of the pancreas. Hormones also often affect the functioning of other hormone-producing tissues so that effects can be complex and integrated. The exact effects of hormones at a molecular level are still far from certain. Hormones do not fall into any one particular chemical class, they may be steroids, peptides or relatively simple compounds like adrenaline. Substances resembling the animal hormones are also produced by plants, e.g. the auxins.

hot carbonate processes Processes for the removal of carbon dioxide from gases by extraction with hot solutions of sodium or potassium carbonate. A significant amount of hydrogen sulphide may be removed at the same time.

hot working Deformation of metals and alloys carried out above the recrystallization temperature. There is no hardening and reductions in section can be made by processes such as hot rolling or extrusion.

HPAN *See* Krilium.

HPLC High pressure liquid chromatography.

Hudson's isorotation rule For a pair of sugars having the α- and β-constitutions, the molecular rotations may be represented as (A + B) and (−A + B) where A represents the contribution of the 1-carbon atom and B that of the rest of the molecule. The sum of these values, 2B, is characteristic of the sugar structure, and the difference, 2A, is characteristic of the end grouping. Though not strictly accurate this rule is useful on an empirical basis.

Hudson's lactone rule The sign of a molecular

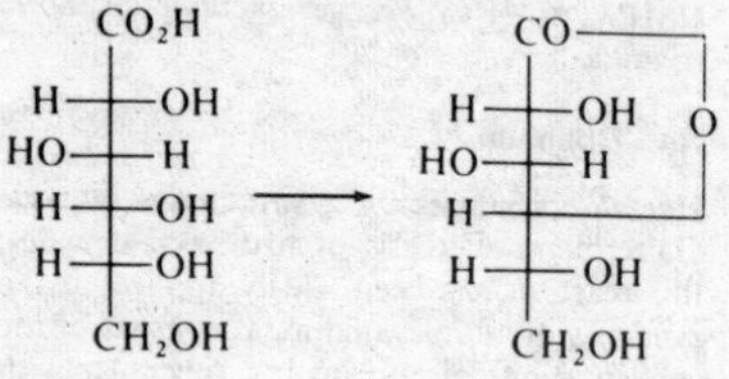

dextrorotatory

rotation in gamma lactones derived from aldose acids may be predicted by observing the configuration of the gamma (4) carbon atom. Using the planar projection formula, if the hydroxyl in position four is to the right, the lactone is dextrorotatory; if to the left it is laevorotatory. This rule gives reasonably good agreement between theory and observation.

humectant A substance, which because of its hygroscopic* nature, can be used to preserve the moisture content of materials, e.g. glycerol or sorbitol in confectionery or food, glycerol in tobacco.

Hume-Rothery's rule The statement that the phase of many alloys is determined by the ratios of total valency electrons to the number of atoms in the empirical formula. *See* electron compounds.

humic acids A group of aromatic acids of high molecular weight and unknown structure that are present in soil, and can also be obtained from peat and brown coal or after refining sugar beet. The colouring matter of most surface water. They may be derived from lignins.

humidification The evaporation of a liquid into a gas is known as humidification and the gas is said to become humidified.

humidity The humidity of moist air may be expressed several ways:

The *absolute humidity* is the mass of water vapour per unit mass of dry air.

The *percentage humidity* is the ratio of the amount of water vapour present per unit mass of dry air to the amount the air could hold if saturated at the same temperature, expressed as a percentage.

The *relative humidity* is the ratio of the partial pressure* of the water vapour in the air to the partial pressure of water vapour in the air when saturated at the same temperature. This ratio is usually expressed as a percentage.

The humidities for systems other than air and water are defined in an analogous manner.

humus The characteristic organic constituent of the soil. It is a dark-coloured amorphous material and is formed by the microbiological decomposition of plant materials, chiefly lignin and proteins. It occurs in the soil in company with clay, forming a colloidal clay-humus complex.

Hund's rules Rules which describe the electronic configuration of degenerate orbitals in the ground state. The electronic configuration will have the maximum number of unpaired electrons, i.e. p^3, ↑ ↑ ↑ rather than ↑↓ ↑ _. The ground state will have the maximum orbital angular momentum.

hyaluronic acid A collective name for mucopolysaccharides from the vitreous humor of the eye, the umbilical cord and synovial fluid. It is composed of units of D-glucuronic acid and *N*-acetyl-D-glucosamine.

hyaluronidase A glycolytic enzyme which hydrolyses hyaluronic acid and chondroitin. It acts by hydrolysing the link between glucuronic acid and the glucosamine moiety (or galactosamine in chondroitins).

hybridization The process whereby atomic orbitals of differing type but similar energies are combined to yield a set of equivalent hybrid orbitals. These hybrid orbitals do not in fact exist and the process of hybridization is simply a mathematical operation in the formation of the molecular orbitals by combining the atomic orbitals of the constituent atoms of a molecule. E.g. in methane, which contains four equivalent C–H bonds, it may be considered that the C-atom 2s and 2p orbitals are hybridized to give four equivalent sp^3 orbitals which are then each combined with an H-atom 1s orbital.

hydantoin, glycolylurea, $C_3H_4N_2O_2$. Colourless needles; m.p. 220°C. Soluble in alcohol, sparingly soluble in water. Prepared by the condensation of glycine with potassium cyanate and boiling the hydantoic acid so formed with hydrochloric acid. It is present in beet molasses. Many substituted hydantoins have been prepared.

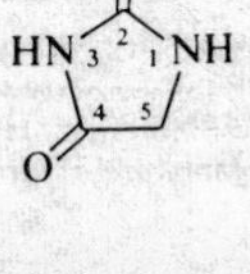

hydnocarpic acid, $C_{16}H_{28}O_2$. M.p. 59–60°C.

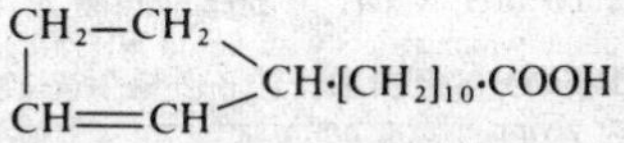

A fatty acid occurring as glycerides in chaulmoogra oil and other vegetable oils.

hydracrylic acid, β-lactic acid, 2-hydroxypropionic acid, $CH_2OH{\cdot}CH_2COOH$. An uncrystallizable syrup. Manufactured by treating ethylene cyanohydrin with NaOH. Converted to propenoic acid on heating. Used in the manufacture of propenoic esters.

hydrates Many compounds have crystallized water additional to that required for a simple stoicheiometry. Water can be bonded to cations by co-ordinated bonds from the oxygen or to anions by hydrogen bonding. In non-ionic derivatives dipole interactions are also important. The gross structure of many materials is determined by the bonding of the water of hydration.

hydration Ions (and many other species) in aqueous solutions are solvated by water and are said to be hydrated. The proton in aqueous solution is generally written as $[H_3O]^+{}_{aq}$ and three or four other water molecules are associated with the proton. Hydration occurs either by interaction of the lone-pairs of electrons in water with a cation or by hydrogen bonding with anions. Secondary hydration spheres also exist.

hydraulic cement *See* cement.

hydraulic conveying The conveying of particulate solids (e.g. coal, china clay, wood pulp) as a slurry through a pipeline.

hydraulic fluids Fluids used to transmit power and pressure. Most hydraulic fluids are based on low-viscosity mineral oils but for reasons of economy or safety aqueous-base fluids are also used.

hydrazides, $MNHNH_2$. The sodium and other alkali metal derivatives are formed from M, MNH_2, or MH and hydrazine. $NaNHNH_2$ explodes with oxygen or above 100°C. It cleaves a C=C bond and, e.g., PhCH=CHMe gives PhMe and MeCH=NNH₂ (a hydrazone). More generally derivatives of hydrazine.

hydrazine, N_2H_4. M.p. 1·4°C, b.p. 114°C. Has structure $H_2N{\cdot}NH_2$ in the gauche form. Manufactured from NH_3 or urea and NaOCl or Cl_2 in the presence of a ketone and gelatin (Raschig process). Forms an azeotrope with water, anhydrous N_2H_4 is obtained by distillation over NaOH or precipitation of the sulphate which reacts with liquid NH_3 to give $(NH_4)_2SO_4$ and N_2H_4. Hydrazine is a weak base giving hydrazinium salts, e.g. $(N_2H_5)Cl$ with strong acids. Aqueous solutions can be oxidizing giving $(NH_4)^+$ in acid with Ti^{3+}(E° +1·27 volt) in slow reactions but are more generally reducing giving N_2(E° acid +0·23 volt; E° alkaline +1·15 volt). Forms complexes but generally acts only as a monodentate ligand. Burns in oxygen, reacts with halogens. Used for removing O_2 from boiler-feed water, etc. and in the manufacture of hydrazides. Organic derivatives have many uses including use as high-energy fuels, blowing agents for foam plastics, antioxidants, herbicides.

hydrazinium salts *See* hydrazine.

hydrazobenzene, 1,2-diphenylhydrazine, $C_{12}H_{12}N_2$, PhNH·NHPh. Colourless plates; m.p. 131°C, insoluble in water, easily soluble in organic solvents. In moist air or in alcoholic solution it oxidizes spontaneously to azobenzene. It does not form salts with acids but is converted into benzidine by an intramolecular change. It gives two molecules of aniline when treated with a strong reducing agent.

It is prepared by reduction of nitrobenzene with iron and NaOH. It is also prepared by an electrolytic reduction of nitrobenzene. It is widely used for the preparation of benzidine.

hydrazoic acid, azoimide, N_3H, NNNH. M.p. −80°C, b.p. 37°C. Prepared from sodium azide and acid or $(N_2H_5)^+$ plus nitrous acid, HNO_2. Heavy-metal salts, azides, are used as detonators, alkali metal salts are stable and azides are used synthetically in organic chemistry.

hydrazones The derivatives formed by con-

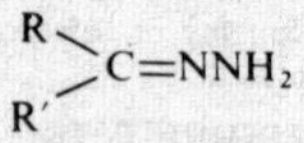

densation between an aldehyde or ketone and hydrazine. Substituted hydrazines, especially 2,4-dinitrophenylhydrazine, are used for preparing crystalline derivatives for chemical identification of keto-compounds.

hydrides There are several distinct types of hydrides.

(1) *Salt-like.* These are the hydrides of the most electropositive elements (e.g. Na) and contain H^- ions.

(2) *Covalent.* Formed by most of the non-metals and transition metals. This class includes such diverse compounds as methane, CH_4 and iron carbonyl hydride, $H_2Fe(CO)_4$. In many compounds the hydrogen atoms act as bridges. Where there are more than one hydride sites there is often hydrogen exchange between the sites. Hydrogens may be inside metal clusters.

(3) *Complexes.* These derivatives contain complex anions which may be considered as derived from co-ordination of an H^- ion to a metal or non-metal. Examples are the BH_4^- and ReH_9^{2-} ions.

(4) *Transition metal hydrides.* These are formed by hydrogen uptake by the metal. The phases are often non-stoicheiometric.

Hydrides frequently function as hydrogenation catalysts.

hydriodic acid, HI. An aqueous solution of hydrogen iodide*.

hydroboration The *cis*- addition of B−H bonds across the double bonds of olefins. Thus diborane, B_2H_6, reacts with ethene to give $B(C_2H_5)_3$. Breakdown of the alkyl borane with acid gives alkanes and with hydrogen peroxide gives alcohols (the orientation is the opposite to that which would result from the direct addition of water across the double bond).

hydrobromic acid, HBr. An aqueous solution of hydrogen bromide*.

hydrocarbon resins Thermoplastic polymers of mol.wt. less than 2000 obtained by cracking petroleum and from turpentine. Used in drying oils, with rubber and as plasticizers. Class includes coumarone-indene resins, petroleum resins, cyclopentadiene resins, terpene resins.

hydrocarbons This term includes all compounds of carbon and hydrogen only. They are subdivided into aliphatic and cyclic hydrocarbons according to the arrangement of the carbon atoms in the molecule. The aliphatic hydrocarbons are again subdivided into paraffins, olefins, diolefins, etc., according to the number of double bonds in the molecule. The cyclic hydrocarbons are subdivided into aromatic hydrocarbons and cycloparaffins.

hydrochloric acid, HCl. An aqueous solution of hydrogen chloride*. A saturated solution of HCl contains about 43% HCl and gives a constant-boiling mixture. Hydrochloric acid is used widely in the chemical industry, in the food industry (glucose, monosodium glutamate), in metallurgy and in the oil industry. Extremely corrosive and handled in glass or plastic equipment or in apparatus using special alloys (Ta, Ni–Mo). U.S. production 1978 2 500 000 tonnes.

hydrochlorides Salts formed by organic bases with HCl (for weak bases prepared in non-basic solvents, e.g. toluene). Also salts containing the $[ClHCl]^-$ anion.

hydrochlorinated rubber Material used in films produced by passing HCl gas into a solution of rubber in benzene. Contains *c.* 30% Cl.

hydrocortisone, cortisol, 17-hydroxycorticosterone, $C_{21}H_{30}O_5$. A cortisone* with an 11-OH group. White crystals; m.p. 217–220°C. Made synthetically from naturally occurring steroids, such as diosgenin or hecogenin, the 11-hydroxy group being introduced where necessary by microbiological means. It is also used as its 21-esters for local treatment of inflammatory and allergic conditions.

hydrocracking Processes which involve both the catalytic cracking and hydrogenation of petroleum fractions, yielding high quality gasolines.

hydrocyanic acid, HCN. *See* hydrogen cyanide.

hydrodealkylation A catalytic process generally used for the production of benzene from alkyl aromatics, toluene being the most common feedstock. Methane is also formed.

hydrodesulphurization *See* desulphurization.

hydrofining *See* desulphurization. Used, under milder conditions, as a sweetening process* or to remove alkenes for improvement of stability in light petroleum fractions.

hydrofluoric acid, HF. Aqueous solution of hydrogen fluoride. The system has a maximum boiling point at about 36% HF. Used widely in industry as a fluorinating agent for, e.g., cryolite (Al production), UF_4 and UF_6 production; etching glass, cleaning stainless steel, etc. Extremely corrosive and handled in monel, Teflon or platinum equipment.

hydroforming *See* catalytic reforming.

hydroformylation A particular type of carbonylation reaction. Under the influence of a cobalt or rhodium catalyst (e.g. $HCo(CO)_4$), an alkene will react with carbon monoxide and hydrogen to produce an aldehyde or a primary alcohol, depending upon conditions. *See* oxo reaction.

hydrogasification The process of converting crude oil or oil fractions to substitute natural gas* (SNG).

hydrogen, H. At.no. 1, at.wt. 1·00794, m.p. −259·14°C, b.p. −252·87°C. The lightest element, occurs free in some natural gases and is a widespread constituent of water, minerals and natural organic materials. The element is prepared by electrolysis of water, or industrially by steam reforming hydrocarbons (C_nH_{2n+2} plus H_2O over catalyst to give CO plus H_2; then further catalyst on CO plus H_2O to give H_2 plus CO_2 followed by purification by scrubbing CO_2) and also results during catalytic reforming of petroleum. Occurs as H_2 (*see* ortho-hydrogen); explodes over wide concentration ranges with oxygen on sparking and reacts with halogens. Combines with most elements to give hydrides*. Used industrially in metallurgy (37%), ammonia synthesis, hydrogenation of fats and oils, methanol synthesis, hydrodealkylation, etc. U.S. production 1980 96×10^9 cu.ft.

hydrogenation A specific method of reduction in which hydrogen is added to a substance by the direct use of gaseous hydrogen. The process is normally carried out with the aid of a catalyst and proceeds more rapidly at high pressures. Hydrogenation is of great industrial importance and is widely used in petroleum refining and in the petroleum chemical industry. It is also used in the refining of fatty oils and for many other chemical processes. Also used with coal to give hydrocarbons.

hydrogen bond Hydrogen attached to an electronegative element, e.g. O,F, gives a highly polarized bond. If this bond is directed towards another atom, E, with a lone-pair of electrons, a weak bond is formed, known as a hydrogen bond, which may be represented O – H – E. Such bonds cause the self-association of water and alcohols, and form the basis of ions such as $[HF_2]^-$. Hydrogen bonds are readily formed between hydroxyl and amino groups and oxygen, nitrogen and halogen atoms; they may be intermolecular or intramolecular. They are important in determining the arrangement of polypeptide and carbohydrate chains in proteins and cellulose, in adhesion and in hydrates.

hydrogen bromide, HBr. M.p. −88°C, b.p. −67°C. Colourless gas obtained from elements in presence of a catalyst (charcoal or Pt) or from Br_2, red phosphorus, and H_2O (HBr is the only volatile product). Gives hydrobromic acid in water (also prepared Br_2 plus H_2O with SO_2 or H_2S). Used in formation of bromides, as a brominating and reducing agent.

hydrogen chloride, HCl. M.p. −115°C, b.p. −85°C. Hydrogen chloride is formed as a by-product in chlorination of hydrocarbons (90%), from H_2 plus Cl_2 and NaCl plus H_2SO_4; much is used as its aqueous solution hydrochloric acid*. Hydrogen chloride is a colourless gas with an irritating smell. Forms hydrates ($1H_2O$ and $2H_2O$) and reacts with many metals (forms lower oxidation state chlorides, *see* chlorine). Used in the production of organic chlorides and as a condensing agent in organic chemistry. U.S. production 1979 2·9 megatonnes.

hydrogen cyanide, hydrocyanic acid, prussic acid, HCN. M.p. −13°C, b.p. 26°C. Colourless, completely miscible with water, burns in air, smell of bitter almonds. Manufactured from methane, air, NH_3 over a catalyst at 1000°C; also formed from metal cyanides and H_2SO_4, dehydration of formamide. Polymerizes on standing. Weak acid, gives salts, cyanides*; forms cyanides, nitriles RCN and isocyanides, isonitriles RNC. Used in organic and synthetic fibre synthesis and as a fumigant. HCN is a very powerful poison.

hydrogen electrode The standard electrode against which other electrode potentials are measured. The electrode consists of a piece of platinum foil, covered with a layer of active platinum black, which is immersed in a solution containing hydrogen ions whilst hydrogen gas is bubbled over the surface. For solutions in which the concentration of hydrogen ions is equal to unit activity and for a gas pressure of 1 atm. the electrode potential is said to be normal and is arbitrarily assigned a value of zero. Although other electrode potentials are quoted relative to the standard hydrogen electrode, the latter is seldom used in practice for this purpose. Usually electrode potentials are measured against a standard calomel electrode which is then referred to the 'hydrogen scale'.

hydrogen fluoride, HF. M.p. −83°C, b.p. 19·5°C. Colourless, strongly fuming, associated liquid prepared from KHF_2 or CaF_2 and H_2SO_4. Gives hydrofluoric acid* with water. Anhydrous HF is a good solvent for both inorganic salts and organic materials, it must be handled in metal (mild steel or monel) or Teflon or Kel-F apparatus. Used as a fluorinating agent for producing chlorofluorocarbons, uranium processing and as an alkylation catalyst in the petrochemical industry. Gives fluorides as salts. The hydrogen difluoride, $[HF_2]^-$, ion is also formed. Used for aluminium production (32%), fluorocarbons (30%), alkylation (5%), uranium processing (5%), steel manufacture (5%). U.S. production 1979 340 000 tonnes.

hydrogen halides, HF, HCl, HBr and HI. Hydrogen fluoride stands apart from the other hydrogen halides because of its strong hydrogen bonding and consequent association. HI is the strongest acid in water and the most readily dissociated. All form halide salts but the solubility pattern of the fluorides is frequently different from that of the other halides (e.g. AgF soluble, AgX insoluble; AlF_3 insoluble, AlX_3 soluble, tend to be dimeric).

hydrogen iodide, HI. M.p. −54°C, b.p. −36°C. Colourless gas fairly readily dissociated to H_2 and I_2. Very soluble in water to give an acid solution – forms 2, 3, $4H_2O$ hydrates. Prepared from H_2 and I_2 with catalyst or I_2 plus red phosphorus plus H_2O. Aqueous hydriodic acid solution prepared I_2 plus H_2O plus H_2S.

hydrogen-ion concentration Hydrogen-ion concentrations are measured as the number of gram ions of hydrogen ions present per litre of solution. Since these concentrations are usually small, the concentration is generally expressed as the pH of the solution; the pH being the logarithm of the reciprocal of the hydrogen-ion concentration, i.e. $pH = -\log[H^+]$. The pH can be measured using a glass electrode or, less accurately, by coloured indicators.

hydrogen ions Chemically free hydrogen ions are rarely formed and dissociation to H^+ occurs only when the resultant ion can be complexed. Thus HCl gas is undissociated but gives $[H(H_2O)_n]^+Cl^-$ and $[NH_4]^+Cl^-$ with H_2O or NH_3.

hydrogenolysis The cleaving of a chemical bond by hydrogen, usually in the presence of a hydrogenation catalyst.

hydrogen overvoltage In electrolytic processes where hydrogen is evolved at the cathode, the potential of the latter is usually more negative than that formally required for hydrogen liberation. The cathode, when thus behaving as an irreversible electrode, is said to exhibit an overpotential or overvoltage. The potential difference between the irreversible cathode and the reversible hydrogen electrode in the same solution is the hydrogen overvoltage at the particular cathode.

hydrogen peroxide, H_2O_2. M.p. −0·4°C, b.p. 150°C, *d* 1·45. Colourless syrupy liquid prepared by oxidation of an anthraquinol with oxygen and subsequent reduction of the quinone with hydrogen; also prepared electrolytically via persulphates. The pure liquid is stable but aqueous solutions are unstable and alkalis and dust catalyse decomposition to oxygen and water. Often used as an aqueous solution. Concentration expressed in volume of O_2 that can be liberated: 6% = 20 vol. Can act as a powerful oxidizing agent (E° acid solution +1·77 volts) when converted to H_2O and as a reducing agent (E° +0·68 volts) when converted to O_2 (e.g. by $KMnO_4$). The parent compound of peroxides. Forms adducts, perhydrates, with some salts. Used in chemicals (33%), pollution control (19%), textiles (17%), paper (13%); free H_2O_2 is used as a propellant. U.S. use 1979 110 000 tonnes.

hydrogen persulphides, sulphanes, H_2S_x (x = 2–6). Formed from polysulphides and acid followed by cracking. Decompose to H_2S and S.

hydrogen sulphide, sulphuretted hydrogen, H_2S. M.p. −85°C, b.p. −61°C. Colourless, very poisonous gas produced by decaying animal or vegetable matter and present in some mineral waters. Prepared FeS plus dilute acid or *in situ* by hydrolysis of thiols. Forms sulphides with most metals, particularly in solution where it acts as a weak acid.

hydrolith, CaH_2. Calcium hydride*.

hydrolysis A term used to signify reactions due to the presence of the hydrogen or hydroxyl ions of water. E.g. when salts are dissolved in water the resulting solution is often either slightly acidic or alkaline due to hydrolysis. Thus, sodium ethanoate yields an alkaline aqueous solution due to hydrolysis; some of the ethanoate ions combine with the hydrogen ions of the water to form the undissociated ethanoic acid, thereby leaving an excess of hydroxyl ions in solution. The reaction may be represented as

$$Na^+ + CH_3COO^- + H_2O \rightleftharpoons Na^+ + OH^- + CH_3COOH$$

and the equilibrium constant, K, for this reaction is given by

$$K_1 = \frac{[OH^-][CH_3COOH]}{[CH_3COO^-][H_2O]}$$

However, since the water is present in large excess its concentration may be regarded as being constant and hence

$$K_2 = \frac{[OH^-][CH_3COOH]}{[CH_3COO^-]}$$

K_2 is called the hydrolysis constant for sodium ethanoate. Hydrolysis occurs when salts involving weak acids or bases are dissolved in water. It is often also found with metal ions in solution. The ion $[M(H_2O)_n]^{x+}$ dissociates to the hydroxy species $[M(H_2O)_{n-1}(OH)]^{(x-1)+}$.

In organic chemistry the term is used to describe the conversion of an ester to an acid and an alcohol (saponification), the addition of the elements of water to a molecule, e.g. the conversion of a nitrile to an amide

$$R{\cdot}CN + H_2O \longrightarrow RCONH_2$$

and the displacement of a group or radical by hydroxyl, e.g.

$$RCH_2Cl + H_2O \longrightarrow RCH_2OH + HCl.$$

All hydrolysis reactions are catalysed by acid, or base, or both.

hydrometer A simple device for measuring the density or specific gravity of a liquid, consisting of a cylindrical float whose lower end is conical and weighted and whose upper end terminates in a narrow tube. When placed in a liquid the instrument floats upright with the tube partially immersed, the immersion depending on the liquid density; the latter may be read off from a scale on the tube.

Because of their widespread use for simple process control, hydrometers are frequently calibrated, not in specific gravity, but in some units related to it, which bear (or bore at one time) some relationship to the concentration being so measured.

hydroperoxides A class of compounds containing the hydroperoxyl group, HO·O−, the simplest of which is hydrogen peroxide. Organic hydroperoxides are produced by aerial oxidation of many organic compounds, and decompose, frequently with extreme ease, to give alcohols, carbonyl compounds and other products.

hydrophilic colloid Colloidal sols of proteins, carbohydrates, soaps and other complex organic materials are marked by considerable stability and are not coagulated by small concentrations of electrolyte such as are sufficient to destroy hydrophobic or suspensoid sols such as those of metals or their sulphides. This stability is due to the existence of a sheath of dispersion medium (water) which protects the hydrophilic (water-loving) particles. The essential difference between hydrophilic and hydrophobic colloids is that the former are not coagulated at the isoelectric point.

Hydrophilic sols are characterized by a high viscosity, and often set to gels on cooling; it is possible to prepare them containing a high proportion of the dispersed colloid. They are either soluble monomolecular substances (e.g. proteins, polymers) or monomolecular substances which aggregate in solution to form colloidal micelles (e.g. soaps, dyes).

hydrophobic bonding The association of non-polar molecules or groups in aqueous media, resulting from the tendency of water molecules to exclude non-polar species.

hydrophobic colloids Sols of gold, arsenic, etc., are stable owing to the mutual repulsion arising from the like charges on all the colloidal particles. Such sols are not hydrated. On the addition of small concentrations of electrolytes, the charges on the particles become neutralized and coagulation takes place. Such sols are termed hydrophobic (water-hating), and have a viscosity equal to, or only slightly greater than, water. Hydrophobic sols are coagulated at the isoelectric point. It is normally only possible to prepare hydrophobic sols in low concentration.

hydroquinone, $C_6H_6O_2$, *p*-$C_6H_4(OH)_2$. Colourless prisms; m.p. 170°C, b.p. 285°C/730 mm. With $FeCl_3$ it is oxidized to quinone. It is best prepared from quinone by reduction with sulphur dioxide. It is used as a photographic developer and antioxidant.

hydrosilylation The addition of an Si–H bond across an olefinic bond to form an organosilane (the reaction is frequently catalysed by Pt salts) which may be used in silicone production.

hydrosulphurous acid, $H_2S_2O_4$. *See* dithionous acid.

hydroxides Compounds containing inorganic OH^- groups. Alkali hydroxides are strongly basic.

hydroxonium ions, $[H(H_2O)_n]^+$. *See* hydration of ions.

hydroxy- A prefix used in organic chemistry to denote that a substance contains a hydroxyl (–OH) group.

hydroxyacetone *See* hydroxypropanone.

hydroxyapatite, $Ca_5(OH)(PO_4)_3$. *See* apatite.

hydroxyethylamines *See* ethanolamines.

hydroxyhalides Basic halides of the metals. *See* basic salts.

hydroxylamine, H_2NOH. Colourless solid which explodes on heating. M.p. 33°C. Weak base prepared by reduction of NO_3^- or NO_2^- electrolytically, with SO_2, or with H_2 and catalyst. The solid decomposes slowly on keeping but salts, e.g. (hydroxylamine hydrochloride, $NH_2OH{\cdot}HCl$) are fairly stable. Forms oximes. Acts as a reducing agent or as oxidizing agent. Used in formation of caprolactam for Nylon and as a free radical scavenger.

hydroxylation The introduction of a hydroxyl group into an organic compound. It may be achieved chemically, e.g. by the action of lead tetraethanoate, potassium permanganate or permethanoic acid on an olefin, or, in particular with steroids, biochemically by the use of appropriate enzymes.

hydroxyl ion, OH^-. The anion formed by ionic dissociation of water and present in hydroxides. Combines with H^+ to give water. Present in solutions of bases.

2-hydroxy-3-naphthoic acid, $C_{11}H_8O_3$. M.p. 216°C. Prepared by heating sodium 1-naphtholate at 250°C under a pressure of 4 atm. of carbon dioxide. It is used to prepare the anilide, Naphtol AS, and also is used in dyestuffs production.

hydroxyproline, 5-hydroxypyrrolidine-2-carboxylic acid, $C_5H_9NO_3$. M.p. 270°C. The laevorotatory form is a constituent of connective-tissue proteins, e.g. collagen and elastin, but not of other proteins.

hydroxypropanone, acetol, hydroxyacetone, acetone alcohol, pyroracemic alcohol, pyruvic alcohol, $C_3H_6O_2$, $CH_3CO{\cdot}CH_2OH$. Colourless liquid with a pleasant odour; b.p. 145–146°C. Prepared by the action of chloropropanone on potassium ethanoate, also formed by the action of certain *Acetobacter* on propene glycol. Tends to polymerize.

8-hydroxyquinoline, oxine, C_9H_7ON. Light brown needles, m.p. 75–76°C. Forms insoluble complexes with metals. The solubilities of the derivatives vary with pH, etc. and hence oxine is widely used in analysis. Used for estimating Mg, Al, Zn and many other metals. Many oxinates are extracted and the metal is estimated spectrophotometrically. Derivatives, e.g. 2-methyl tend to be specific, for, e.g., Mg^{2+}. Copper derivatives are used as fungicides.

hygroscopic Substances which absorb water from moist air and form either a moist solid or solution. E.g. $MgCl_2$, P_2O_5. Hygroscopic substances can often be used as desiccants.

hyoscine,(−)scopolamine, $C_{17}H_{21}NO_4$.

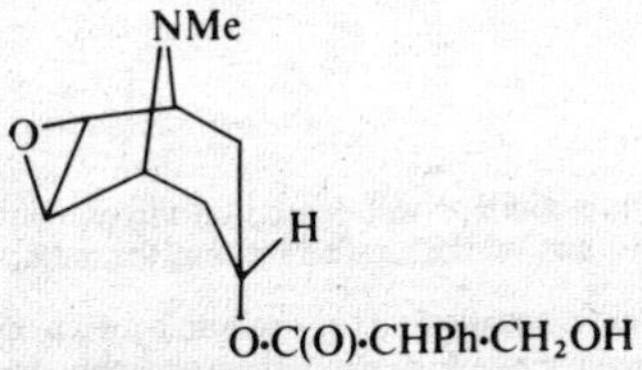

Obtained as a syrup from plants of the *Solanaceae* family. Intensely poisonous, its action resembles that of atropine. Sedative in small doses.

hyoscyamine, $C_{17}H_{23}NO_3$. Colourless needles, m.p. 108·5°C. *l*-Form of atropine.

hyperchromic An increase in absorption intensity. The opposite is hypochromic.

hyperconjugation Alkyl groups interact electronically with unsaturated systems, e.g. the benzene nucleus, to which they are attached, in a similar way to unsaturated groups. The magnitude of this effect, called hyperconjugation, depends on the number of hydrogen atoms attached to the carbon atom immediately adjacent to the unsaturated system. Specifically it indicates C−H conjugation. Postulated to account for the stabilization of aliphatic carbonium ions, and ease of certain elimination reactions.

hyperons Elementary particles of mass 2185 times that of the electron. Hyperons are found in cosmic radiation.

hypo, $Na_2S_2O_3$. Sodium thiosulphate*. Used in photography.

hypobromates, M[BrO]. Salts of HOBr. *See* bromates.

hypobromous acid, HOBr. *See* bromates.

hypochlorites, M[ClO]. Salts of HOCl. *See* chlorates.

hypochlorous acid, HOCl. *See* chlorates. Used as powerful oxidizing and bleaching agents. *See* bleaching powder.

hypochromic *See* hyperchromic.

hypofluorites Covalent compounds containing −OF groups (e.g. F_2O, CF_3OF, SF_5OF). Prepared by the action of fluorine on an oxygen-containing compound. Ionic hypofluorites, containing OF^-, are not known as stable compounds. HOF is a very unstable substance formed from reaction of F_2 with ice.

hypoiodates, M[OI]. *See* iodates.

hypophosphoric acid, $H_4P_2O_6$. Prepared from $NaClO_2$ and red phosphorus. Salts contain $[O_3P-PO_3]^{4-}$ ions. Stable in aqueous alkali but converted to phosphate by fused NaOH. A tetrabasic acid, forms tetra-alkyl esters.

hypophosphorous acid, H_3PO_2, $H_2P(O)OH$. Crystalline solid; m.p. 26°C. The sodium salt, NaH_2PO_2, is formed from yellow P and NaOH. The acid and its salts are powerful reducing agents giving hydrides, or low oxidation state compounds. The acid is monobasic and liberates H_2 and PH_3 on heating with NaOH.

hyposulphurous acid, $H_2S_2O_4$. *See* dithionous acid.

hypoxanthine, 6-oxypurine, $C_5H_4N_4O$. M.p. 150°C (decomp.). Sparingly soluble in cold water, more so in hot. A breakdown product of adenine in animals.

hypsochromic Shifts absorption to shorter wavelength, higher frequency, higher energy. *See* bathochromic.

hysteresis Processes which occur mainly in solids in which the response obtained on increasing the variable differs from that on reducing the variable. Occurs e.g. during the magnetization of iron, during the absorption of moisture by textiles.

I

I Iodine.

ice The solid form of water. The transition point between liquid and solid water at one atmosphere pressure is defined as 0° on the Centrigrade scale, also equal to 0° Celsius. This freezing temperature is lowered by increasing the pressure. By crystallization of water under greatly increased pressure it is possible to obtain forms of solid water which are physically distinct from ordinary ice (Ice I). The molecules in ice are held together by hydrogen bonds.

icosahedral Having the symmetry of an icosahedron, a solid figure with twelve apices; each face is triangular. Icosahedral units are found in many boron derivatives, e.g. $B_{12}H_{12}^{2-}$.

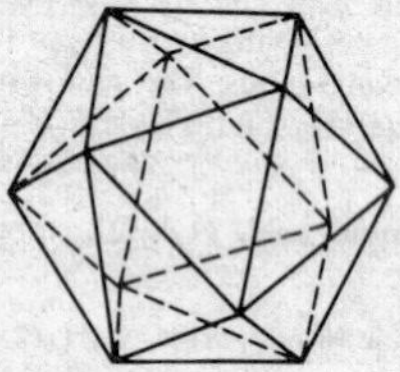

ideal gas A gas which obeys exactly the gas laws, i.e.

$$PV = nRT$$

In practice, no gas shows ideal behaviour, although helium, hydrogen and nitrogen approximate to ideality at high temperatures and low pressures.

ideal solution A solution, liquid or solid, in which the thermodynamic activity of each component is proportional to its mole fraction, i.e. a solution obeying Raoult's law. Such a solution would be formed from its components with zero heat of mixing, zero volume change and the ideal entropy of mixing. Only mixtures containing closely similar substances, e.g. isotopic elements, form ideal solutions, although many mixtures show behaviour which approximates to ideality (e.g. benzene and toluene).

idose A hexose*.

ignition limits *See* flammability limits.

ignition temperature The minimum temperature at which a fuel can be ignited. The temperature obtained will depend on the technique used and on the air–fuel ratio of the mixture.

Ilkovic equation The relation between diffusion current, i_d, and the concentration c in polarography which in its simplest form is $i_d = k'c$.

ilmenite, $FeTiO_3$. A black Ti ore occurring in massive form and in sands. The structure has cp oxygen with Fe and Ti in octahedral co-ordination.

imidazole, $C_3H_4N_2$. Also called glyoxaline or

iminazole. The ring is present in polybenzimidazoles*.

imides In organic chemistry imides are nitrogen-containing ring compounds containing the group

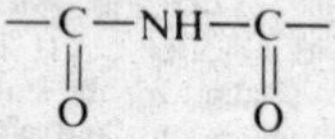

They are formed by heating dibasic acids or their anhydrides with ammonia. The hydrogen atom of the NH group is acidic and can be replaced by a metal. Mild hydrolysis breaks the ring to give the half amide of the acid. *See* succinimide and phthalimide.

In organic chemistry the term refers to compounds containing the NH_2^- ion or the >NH group. These are prepared by the action of heat on amides or by metathetical reactions in liquid ammonia. The heavy metal imides are explosive.

imines Compounds containing an >NH group. May be cyclic (e.g. dihydroazirine) or linear.

iminium salts Salts containing cations in which the charge delocalization includes a nitrogen atom.

imino The name given to the group >NH where the nitrogen atom is part of a ring system or is united by both bonds to a single atom. Such compounds as R·CO·NHR are usually regarded as substituted amides and not as imino compounds.

impact resistance A measure (generally

approximate) of the capacity of a substance to withstand shock or impact under particular test conditions.

improper rotation, S_n. A symmetry element consisting of rotation about an axis through $2\pi/n$ followed by reflection through a plane perpendicular to the axis of rotation.

IMS Industrial methylated spirits.

In Indium.

inconel *See* nickel alloys.

indan, 2,3-dihydroindene B.p. 117°C. The first reduction product of indene.

indanthrine dyestuffs An important series of dyestuffs based on amino derivatives of anthraquinone.

indene, C_9H_8. Colourless liquid, b.p. 188°C,

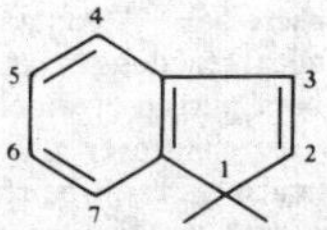

m.p. −2°C, obtained from petroleum fractions. It is an acidic hydrocarbon and yields a sodium salt. The reactions of the 5-membered ring resemble those of cyclopentadiene in forming fulvenes, and pentahapto metal derivatives.

indene resins *See* coumarone and indene resins.

indican, indoxyl-3-glucoside, $C_{14}H_{17}NO$. *See* indoxyl. Also the potassium salt of indoxyl-3-sulphuric acid, the form in which toxic indoxyl is excreted as one of the ethereal sulphates.

indicator A substance which allows the progress of a chemical change to be followed, but generally a substance which marks a precise stage in a chemical reaction, usually by a change of colour, or fluorescence of a solution, or by the alteration in the nature of a precipitate. E.g. methyl orange, an indicator used in the neutralization of acids with bases, is red at pH less than 3·1 and gradually changes through orange to a full yellow colour as the pH increases to 4·4.

indigo *See* indigotin.

indigoid dyes Vat dyes and pigments containing indigo-like groupings and giving similar colours. Contain indole or thionaphthene systems.

indigotin, $C_{10}H_{10}N_2O_2$. A very important and long known dyestuff (woad). Originally

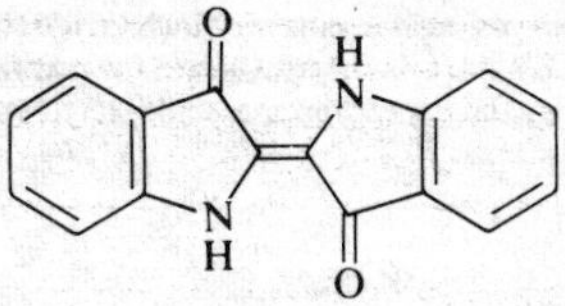

obtained from indican (the glucoside of indoxyl) but now prepared from anthranilic acid and chloroethanoic acid to give phenylglycine-*o*-carboxylic acid which with KOH and $NaNH_2$ gives indoxylic acid, decarboxylated to give indoxyl and oxidized to indigotin by air. Other syntheses involve initial reaction of anthranilic acid and chloroethanoic acid or methanal bisulphite compound and KCN. The dark blue indigotin is reduced by sodium dithionite to the soluble leuco-compound, indigotin-white which is used in dyeing and then reoxidized to indigotin. Sulphonated indigo dyestuffs are also known.

indium, In. At.no. 49, at.wt. 114·82, m.p. 156·61°C, b.p. 2080°C, *d* 7·31. Occurs in some specimens of zinc blende; precipitated from solution electrolytically or with Zn. The metal is tetragonal with a distorted cp structure. The metal is oxidized on strong heating in air and dissolves in acids. Used in low melting alloys. InP is used in electric motors and other In compounds are used in electronic components and semi-conductors. Indium compounds are toxic. World production 1976 45 tonnes.

indium chemistry Indium is a Group III element, electronic configuration $5s^2 5p^1$. It shows the normal group oxidation state of +3 ($E°In^{3+} \longrightarrow In$ in acid solution −0·34 volts) and a relatively stable +1 state ($E°In^{+} \longrightarrow In$ in acid solution −0·25 volts) which disproportionates in aqueous solution. The chemistry of the +3 state is similar to that of Al(III).

indium halides Indium forms InX_3 (X = F, Cl, Br, I), (X_2 on In). InF_3, $InCl_3$ and $InBr_3$ are fairly ionic. InI_3 is $I_2In(\mu\text{-}I)_2InI_2$. $InF_3,3H_2O$ and $InCl_3,4H_2O$ are known. Complex species $[InX_6]^{3-}$ are formed with F and Cl. InX (X = Cl, Br, I) (InX_3 plus In) and intermediate halides $InCl_2$ ($In^+[In^{III}Cl_4]$), In_2Cl_3 ($In_3^{I}[In^{III}Cl_6]$), In_4Cl_5 and In_4Cl_7 are also known.

indium oxides In_2O_3 (In plus O_2) is also formed by dehydration of $In(OH)_3$ (In^{3+} plus OH^-). In_2O (heat on In_2O_3) is yellow when hot, red when cold and is readily oxidized to In_2O_3.

indium oxyacid salts $In(NO_3)_3,5H_2O$, $In(ClO_4)_3,11H_2O$, $In_2(SO_4)_3$ are formed from $In(OH)_3$ and the appropriate acid. The sulphate forms alums.

indole, C_8H_7N. M.p. 52°C, b.p. 253–254°C.

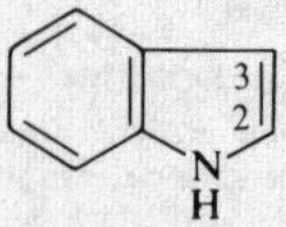

Occurs in coal tar, in various plants and in faeces, being formed by the action of the intestinal bacteria on tryptophan. It can be prepared by the action of acid on the phenylhydrazone of pyruvic acid to give indole-2-carboxylate which can be decarboxylated to indole.

indole-3-acetic acid, $C_{10}H_9NO_2$. M.p. 164–165°C. Has auxin* properties. It is the active constituent of horticultural 'rooting hormone' preparations.

indolizine (pyrrocoline) ring system The

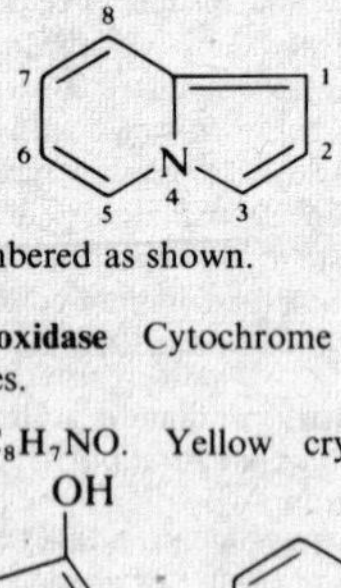

system numbered as shown.

indophenol oxidase Cytochrome oxidase. *See* cytochromes.

indoxyl, C_8H_7NO. Yellow crystals, m.p.

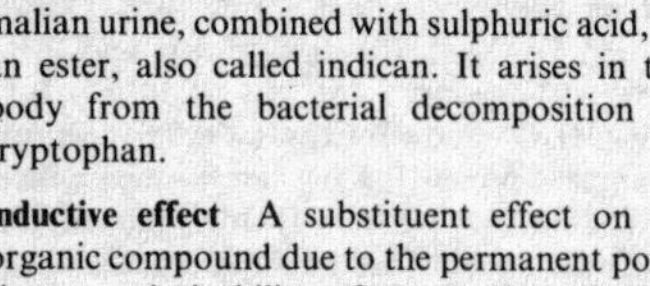

85°C, soluble in water and alcohol. It occurs in woad as the glucoside indican, and in mammalian urine, combined with sulphuric acid, as an ester, also called indican. It arises in the body from the bacterial decomposition of tryptophan.

inductive effect A substituent effect on an organic compound due to the permanent polarity or polarizability of groups is called an inductive effect. It may be visualized as the inducement of charge either towards or away from a group or atom, with a resulting dipole. Such a hypothesis can account for the increased acidity of trichloroethanoic acid over ethanoic acid.

industrial methylated spirits (IMS) Ethanol rendered non-potable by mixing 95% alcohol with wood naphtha.

inelastic neutron scattering Bombardment of a crystal by a beam of monochromatic neutrons results in a scattering of the latter due to collision with the crystal nuclei. Since the nuclei are not of infinite mass and they indulge in limited motion because of their thermal energy, such collisions cannot be regarded as truly elastic. The impinging neutron may either transfer some of its energy to the crystal vibrations or gain energy from these vibrations. Such collisions involving energy transfer are said to be inelastic and they result in scattered neutrons which are of different wavelength from the incident beam. The inelastic scattering of neutrons gives valuable information regarding crystal structure and vibrations within the crystal.

inert atmosphere box A container for carrying out chemical reactions in which the atmosphere has been purged of water, oxygen and nitrogen by recycling over active absorbants for the particular gases to be removed. Used for handling very sensitive materials; modifications can be used for handling reactive metals, e.g. Ti at high temperatures.

inert complex A complex which participates in reactions, particularly ligand exchange reactions, either very slowly or sufficiently slowly that the rates of reaction can be readily followed. $[Co(NH_3)_6]^{3+}$ is an inert complex. Compare labile complex.

inert gases A semi-obsolete name for the noble gases*.

inert gas formalism The octet rule*; the effective atomic number rule*.

inert pair effect An apparent correlation between the existence of oxidation states n and (n − 2) and the s^2 configuration for the latter for the heavy p block elements (n corresponds to the group-oxidation state). E.g. Pb shows oxidation states Pb(IV) and Pb(II).

infra-red radiation, i.r. Radiation of frequency less than about 10^{13} per sec ($10\,000\ cm^{-1}$), i.e. of wavelengths greater than about 8000 Å. It is invisible to the human eye, but possesses considerable penetrating power, and photographic plates may be sensitized towards it, so that it finds application in long-distance photography. Heat-seeking devices and devices for night viewing rely on detection of i.r. radiation. Many forms of detector are known. Interatomic vibrations in molecules cause absorption in the i.r. region.

infra-red spectroscopy Vibrations within molecules show characteristic absorption bands in the infra-red region of the spectrum. Infra-red spectroscopy provides valuable information about molecular structure, gives information on groups present and on molecular symmetry. The technique has been highly developed and is widely used as a routine tool in analysis and research.

infusions Dilute solutions containing the water-soluble components of vegetable drugs prepared by aqueous extraction of the drugs.

ingrain dyestuffs Insoluble dyestuffs formed directly on the fibre from soluble components.

inhibitor A general term for any compound which will inhibit, i.e. slow down or stop, a reaction, generally by preventing propagation of chain reactions or by passivation of surfaces, etc. The term is often used to describe any additive which will prevent a particular tendency during working operations, e.g. anti-oxidants, anti-corrosion agents, etc.

initial boiling point In a laboratory distillation the thermometer reading observed when the first drop of condensate falls from the lower end of the condenser tube.

initiators *See* explosives.

inks Colouring matter dissolved or dispersed in a solvent or carrier; on drying the colouring matter is bonded to the substrate.

inner sphere mechanism A mode of redox reaction in which the complexes form an intermediate with a shared ligand.

inorganic chemistry The chemistry of all the elements other than carbon. Simple carbon compounds, e.g. carbides, carbonates, carbonyls and cyanides are often regarded as inorganic: the organic derivations of the metals and metalloids are considered as organometallic*.

inosinic acid, hypoxanthine ribonucleotide, $C_{10}H_{13}N_4O_8P$. One of the first purine

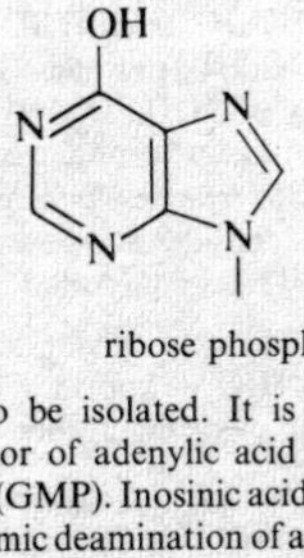

ribose phosphate

nucleotides to be isolated. It is the biosynthetic precursor of adenylic acid (AMP) and guanylic acid (GMP). Inosinic acid may be prepared by enzymic deamination of adenylic acid.

inositol, $C_6H_{12}O_6$. Hexahydroxycyclohex-

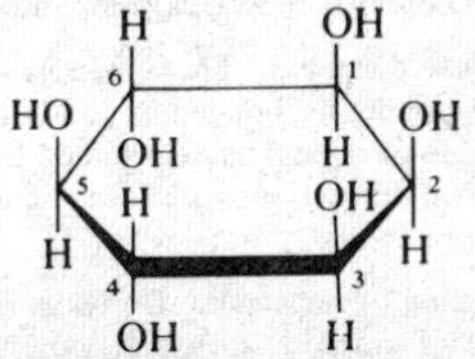

anes. Nine stereoisomers are possible, four occur naturally, but only one, *meso*-or *i*-inositol (formula shown), is widely distributed. It is present in yeast, in plants as phytic acid ester and as a constituent of both plant and animal phosphatides. M.p. 225–226°C, with $2H_2O$ m.p. 218°C. Has a sweet taste. *i*-Inositol is an essential growth factor for certain micro-organisms and its absence stops the growth of hair in rats and mice, so it is classified in the vitamin B group. It can most easily be obtained from corn steep liquor by precipitation as calcium phytate.

insecticides Materials used to control insects by poisoning (chrysanthemic acids, contact poisons, systemic poisons) by use of attractants or repellants (fumigation-acrylonitrile, CS_2, etc.).

insertion reaction A common type of reaction in which an incoming molecule is inserted between previously bonded atoms, e.g. $[(OC)_5MnCH_3]$ plus CO gives $[(OC)_5MnC(O)CH_3]$. The inserted molecule contained a multiple bond or a lone pair.

insoluble azo dyes *See* ingrain dyestuffs.

instability constant The reverse of a stability constant*; the measure of dissociation.

insulator A substance with a very low electrical conductivity. In insulators there is a wide separation between completely filled and completely empty electronic energy levels. Most pure solid ionic substances are insulators although impurities or imperfections may introduce semi-conductivity*.

insulin The hormone, secreted by the Islets of Langerhans in the pancreas, which maintains the balance of glucose metabolism. The absence of insulin leads to the disease *diabetes mellitus*.

Insulin is built up of two polypeptide chains, 'A' of 21 amino-acids and 'B' of 30 amino-acids, linked by two disulphide bridges.

integral tripack material Colour films and papers in which the three separate emulsions (for violet-blue, green and red) are coated on the same support in layers and after exposure and colour development* give dyes of complementary colour (yellow, magenta and cyan).

interatomic distances The distances between the nuclei of atoms, bonded or non-bonded, in a molecule or crystal as determined by X-ray or electron diffraction, spectroscopic or other physical methods.

intercalation compounds Derivatives, particularly of graphite, in which molecules are accommodated in holes or between layers in the lattice. The resulting compounds are easily handled and stored (e.g. graphite-SbF_5, used as a fluorinating agent; graphite-$FeCl_3$ stable to water.

interfacial angles, constancy of The angle between any particular pair of faces of a crystal is constant and characteristic of the substance of the crystal. It does not depend on the size of the crystal.

interferon A small molecular weight protein (about 30 000) produced by cells in response to the presence of foreign double stranded RNA (e.g. viruses). Interferon then acts on uninfected cells stimulating them to produce a protein which prevents replication of the RNA in these cells.

interhalogen compounds Compounds formed by the halogens with one another. The known compounds are: ClF, BrF, ClBr, ICl, IBr, ClF_3, BrF_3, IF_3, ICl_3, ClF_5, BrF_5, IF_5 and IF_7. All these compounds are formed by direct combination of the two halogens. They are readily volatile and exceedingly reactive, especially if the compound in question contains fluorine. No interhalogen compounds are yet known which contain more than two different halogen atoms. Oxide derivatives are also known, e.g. IOF_5, ClO_3F. The interhalogens form related anions and cations, e.g. BrF_3 gives $[BrF_2]^+$ (with SbF_5) and $[BrF_4]^-$ (with KF). The anions, polyhalide anions, can contain more than two halogens, e.g. $[IBrCl]^-$.

internal compensation A type of optical activity in which a molecule contains two asymmetric centres which produce equal and opposite rotation of the plane of polarization, with the result that the substance is optically inactive.

internal energy (*E*) For any system in a given state it is possible to ascribe to it a definite amount of energy. This energy is termed the internal energy of the system and includes all forms of energy other than those arising from the position of the system in space. The internal energy (E) is related to the enthalpy (H) by the equation

$$H = E + PV$$

where P is the pressure and V the volume.

interstitial compounds Compounds in which the structure may be considered to be derived by inserting small species (H, B, C, N, B) into the holes of a cp metallic lattice (particularly transition elements).

invar *See* nickel alloys.

invariant system A system which, when at equilibrium, possesses no degrees of freedom. *See* phase rule.

inverse spinel *See* spinel.

inversion *See* stereospecific reactions; Walden inversion, invertase. The symmetry operation, *i*; reflection through a point.

inversion temperature Most gases when rapidly expanded experience a cooling effect (the Joule–Thomson effect). Below 193 K hydrogen also shows this effect, but above this temperature it behaves abnormally and is warmed by expansion. This temperature of 193 K is called the inversion temperature. Helium shows similar behaviour to hydrogen with an inversion temperature of 33 K.

invertase, sucrase The enzyme which hydrolyses sucrose to a mixture of glucose and fructose. Invertase has been used in making syrups for the confectionery trade.

invert sugar The laevorotatory mixture of glucose and fructose obtained by hydrolysing sucrose.

iodates Salts containing iodine oxyanions. *Iodates*(I), M[OI], hypoiodates. Formed I_2 plus HgO; the acid is very unstable; iodates(I) are similar to chlorates(I). Iodates(III) are not known. *Iodates*(V) MIO_3. The free acid, HIO_3, is the only known free halic acid (I_2 plus conc. HNO_3) dehydrates to I_2O_5. Strong oxidizing agents (IO_3^- plus I^- gives I_2). The $[IO_3]^-$ ion is pyramidal. *Iodates*(VII), periodates, IO_4^- from HIO_4, $[IO_6]^{5-}$ from H_5IO_6. Formed by electrolysis of HIO_3 which gives $[IO_6]^{5-}$ derivatives, e.g. NaH_4IO_6,H_2O; with base gives MIO_4. Again strong oxidizing agents (potential $[IO]^- \longrightarrow I^- + 0{\cdot}49$ volts; $[IO_3]^- \longrightarrow I^- + 1{\cdot}085$ volts; $H_5IO_6 \longrightarrow [IO_3]^- + 1{\cdot}7$ volts).

iodic acids The acids giving rise to iodate* anions. Only HIO_3 is known in the free form. All are strong oxidizing agents.

iodimetry The use of iodine in titrations, generally titrated with thiosulphate and converted to I^- using starch (blue to colourless) as indicator. Used indirectly in estimating oxidizing agents which liberate I_2 from acid KI solution.

iodine, I. At.no. 53, at.wt. 126·9045, m.p. 113·5°C, b.p. 184·35°C, *d* 4·94. Occurs as I^- in brines, as iodates(V) in caliche and in sea-water and seaweeds. Liberated by oxidation (of I^-) or reduction (of $(IO_3)^-$). The element is a black solid giving a violet vapour (I_2 molecules). Simple solutions (CCl_4) are violet, solutions in which there is appreciable charge transfer (C_2H_5OH, KI/H_2O) are brown. The element reacts with other halogens (to give interhalogen compounds), and bases (to give iodates). Used in disinfectants (22%), medically (e.g. for hyperthyroidism), synthesis, photography, food supplements, electric light bulbs (quartz-halogen lamps). Iodine is an essential element in the diet. U.S. use 1982 3000 tonnes.

iodine chemistry Iodine is a halogen element, electronic configuration $5s^2 5p^5$. Its most stable oxidation states are diamagnetic −1 (e.g. I^-) ($\frac{1}{2}I_2 \longrightarrow I^- -0{\cdot}5V$) +1 (ICl, also complexed I^+ cations (e.g. $[Ipy_2]^+NO_3$ $AgNO_3$ plus I_2 plus pyridine) +3 (ICl_3), +5 (IF_5, HIO_3), +7 (IF_7, KIO_4). Iodine thus shows considerably more cationic character than Cl or Br and forms some oxysalts (e.g. $(IO)_2SO_4$ (I_2O_5 plus I_2 plus H_2SO_4), $I(O_2CCH_3)_3$ (I_2 plus $(CH_3CO)_2O$ plus fuming nitric acid).

iodine chlorides Iodine monochloride, ICl, b.p. 97°C, and iodine trichloride, ICl_3, yellow-orange solid dissociated to ICl and Cl_2 above room temperature. *See* interhalogen compounds.

iodine fluorides Iodine trifluoride, IF_3, yellow solid. Iodine pentafluoride, IF_5, m.p. 9°C, b.p. 100°C. Colourless liquid used as mild fluorinating agent and solvent. Square pyramid. Iodine heptafluoride, IF_7. B.p. 4°C. *See* interhalogen compounds.

iodine oxides Iodine pentoxide, I_2O_5 is the most important oxide (heat on HIO_3) and regenerates HIO_3 with water. It is white, polymeric and a strong oxidizing agent (e.g. CO gives CO_2, I_2 being liberated quantitatively). The other oxides I_2O_4, I_4O_9, I_2O_7 are also polymeric.

iodobenzene, PhI, C_6H_5I. Colourless liquid, b.p. 188°C. The iodine atom is very reactive, and is easily removed by metals. With chlorine an oxidized compound phenyliodine dichloride ($C_6H_5ICl_2$) is formed, the iodine being tripositive as in iodine trichloride.

Iodobenzene is prepared by direct iodination of benzene with iodine and nitric acid under reflux or from diazotized aniline and potassium iodide.

iodoform, triiodomethane, CHI_3. Yellow crystalline solid having a powerful and characteristic odour, m.p. 119°C. Volatile in steam. Manufactured by the electrolysis of a solution of an iodide in dilute alcohol or propanone. Reacts with KOH to give methylene iodide. Decomposed slowly by light and air to CO_2, CO, I_2 and H_2O.

iodoform reaction A test used to detect the presence of a $CO{\cdot}CH_3$ group (or groups which may be converted to this under the reaction conditions). Iodine in alkali converts the acetyl group into iodoform.

iodogorgic acid *See* 3,5-diiodotyrosine.

iodomethane *See* methyl iodide.

iodometric methods Analytical methods involving iodimetry*.

ion An atom or group of atoms which has gained or lost one or more electrons, and thus carries a negative or a positive charge. Ions are present in solutions of electrolytes in solvents of high dielectric constant. They are generally complexed by the solvent. They are also present in melts and in gaseous ionic materials. Ions are present in gases which have been subjected to suitable electric disturbances such as a high-tension discharge, or a beam of electrons.

The word is also used to denote a unit in a solid crystal of an electrovalent compound such as NaCl in which each Na^+ is electrically attracted by the surrounding six Cl^- and each Cl^- is electrically attracted by the surrounding six Na^+. The structure of such crystals is termed ionic to indicate that the crystal is not an aggregate of independent molecules.

ion exchange When an ionogenic surface* is in contact with water or other ionizing solvent an electrical double layer* is set up. The counter-ions are partially free and can readily be exchanged for others of the same sign supplied by soluble electrolytes. Thus the sodium in certain silicates can be replaced by potassium by exhaustively washing the material with a solution of KCl (*see* base exchange). It is exploited particularly with ion-exchange resins. These are synthetic insoluble cross-linked polymers carrying acidic or basic side-groups which have high exchange capacity and have many applications, including water-treatment, extraction, separation, analysis and catalysis. Using a 'mixed bed' of anion and cation exchange resins the electrolytes can be removed from a water solution, and very pure water obtained.

ion exchange chromatography *See* chromatography.

ion exchange resin *See* ion exchange.

ion exclusion The medium inside the pores of an ion-exchange resin is subject to the Donnan membrane equilibrium. It therefore contains a lower concentration than does the external solution of any electrolytes that may be present in the latter. This is 'ion exclusion'. On the other hand, non-electrolytes are not subject to the Donnan effect. Hence, a partial separation of electrolytes from non-electrolytes can be obtained; by using a chromatographic technique complete separation can be effected in favourable cases.

ionic atmosphere The Debye–Hückel theory of interionic attraction shows that in solutions ions of one charge type are surrounded by an atmosphere of ions predominantly of the opposite charge type. The effect felt by the central ion is the same in all directions. When an electric field is applied in the solution, however, the central ion moves in one direction whereas the ionic atmosphere moves in the opposite sense giving rise to an asymmetric distribution of ions about the central ion. This effect is important in studies of electrical conductivities of solutions.

ionic product The product of the concentrations of the ions of a substance in solution. This factor is a constant for weakly ionized substances, because the concentration of undissociated material may be taken to be constant, if it is present in a large excess. *See* solubility product.

ionic radii The effective radius of ions in crystal lattices. *See* table under radii, p. 339

ionic strength For an electrolyte solution the ionic strength is given by

$$I = \tfrac{1}{2}\Sigma_i m_i z_i^2$$

where z is the charge on the ionic species and m is the molal concentration, the summation being continued over all the different ionic species in the solution.

ionization chamber A device for measuring the absolute intensity of a beam of X-rays or ionizing particles.

ionization energies, ionization potentials The energy required to remove an electron from a free atom or ion in the gaseous state. Normally measured in electron volts or kilojoules per mole.

Ionization energies (eV) for selected atoms and ions:

H	13·598	Sc	6·54
He	24·587	Ti	6·82
He^+	54·42	V	6·74
Li	5·39	Cr	6·77
Li^+	75·60	Mn	7·43
Be	9·32	Fe	7·87
B	8·30	Fe^+	16·18
C	11·26	Fe^{2+}	30·65
C^+	24·38	Co	7·86
C^{2+}	47·88	Co^+	17·06
N	14·53	Co^{2+}	33·50
N^+	29·60	Ni	7·64
N^{2+}	47·44	Ni^+	18·15
O	13·62	Ni^{2+}	35·20
O^+	35·12	Cu	7·72
O^{2+}	54·90	Zn	9·39
F	17·42	Ga	6·00
Ne	21·56	Ge	7·88
Ne^+	41·08	As	9·81
Ne^{2+}	63·45	Se	9·75
Na	5·14	Br	10·56
Na^+	47·30	Kr	14·00
Mg	7·65	Rb	4·18
Mg^+	15·04	Sr	5·69
Al	5·99	Y	6·25
Al^+	18·83	Zr	6·95
Si	8·15	Nb	6·77
P	10·49	Mo	7·06
S	10·36	Tc	7·28
S^+	23·41	Ru	7·36
Cl	12·97	Rh	7·46
Ar	15·75	Pd	8·33
Ar^+	27·63	Ag	7·57
K	4·34	Cd	8·99
K^+	31·71	In	5·79
Ca	6·11	Sn	7·34
Ca^+	11·87	Sb	8·64
Te	9·01	Os	8·70
I	10·45	Pt	9·00
Xe	12·13	Au	9·22
Xe^+	23·39	Hg	10·43
Cs	3·89	Tl	6·11
Cs^+	25·10	Pb	7·42
Ba	5·21	Ra	5·28
Ba^+	10·00	Th	7·50
La	5·61	U	6·19
Ta	7·88	Np	6·16
W	7·98	Pu	5·71
Re	7·87		

($1\,eV = 0{\cdot}16 \times 10^{-18}\,J \equiv 96{\cdot}37\,kJ\,mol^{-1}$)

ionization, heat of The amount of heat required to split one mole of an electrolyte into its constituent ions. For water, the heat of ionization is 57·3 kJ mol^{-1}.

ionization of water Pure water has a very low conductivity and is only very slightly ionized. The concentrations of hydrogen and hydroxyl ions, which are equal, are 10^{-7} g. ions l^{-1}. The ionic product for water is $[H^+][OH^-]=10^{-14}$ at 25°C, but increases rapidly with temperature. The pH of pure water is clearly 7. Any solution with pH >7 contains excess hydroxyl ions and is alkaline; those with pH <7 are acidic, containing excess hydrogen ions.

ionization potential *See* ionization energies.

ionogenic surface The stability of a colloid particle is due to the charge it carries. This charge is ionic in nature and if the stabilizing ions on the surface are due to the ionization of the particle wall material, the surface is said to be ionogenic.

ionol *See* 2,6-di-*t*-butyl-4-methylphenol.

ionone, $C_{13}H_2O$. Pseudoionone (*see* citral) treated with sodium hydrogen sulphate, dilute nitric acid, strong sulphuric acid or phosphoric acid, gives a mixture of α- and β-ionone. These products are of commercial importance, being powerful odorants, smelling strongly of violets.

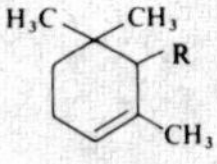

α-ionone

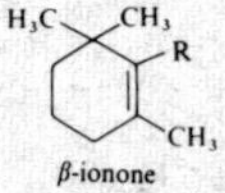

β-ionone

R = –CH=CHOCH$_3$

α-Ionone can be separated from β-ionone by distilling the mixed sodium hydrogen sulphite compounds in steam when the β-ionone distils, leaving the α-compound, which yields α-ionone by treatment with alkali. α-Ionone has b.p. 123–124°C/11 mm. It has a sweeter smell, more closely resembling that of violets, than the β-compound.

ion-pair When the ions in solution are close together, the approximations made in the Debye–Hückel theory no longer hold and the energy of the mutual attraction of ions is great enough to give rise to the formation of ion-pairs. This effect is more pronounced in mixed and non-aqueous solvent systems and there is evidence for triple ion-pairing in some instances.

ion-pair partition *See* phase-transfer chemistry.

ion-selective electrodes An electrode which responds specifically to one ion present in solution and develops a potential according to the concentration of that ion. Used for metals and for anions. The electrode contains a standard solution and a membrane or other conductor placed in contact with the solution to be examined. Glass electrodes are widely used ion-selective electrodes.

ions, hydration *See* hydration.

Ir Iridium.

i.r. Infra-red.

iridates *See* iridium oxides.

iridium, Ir. At.no. 77, at.wt. 192·22, m.p. 2410°C, b.p. 4130°C, *d* 22·65. One of the platinum metals with similar properties to Rh. For separation *see* ruthenium. It occurs in osmiridium and as a native Pt alloy. The metal is ccp. It is used principally as an alloying material for Pt, as an alloy with Os for pen tips and in catalysts.

iridium chemistry Iridium is the heaviest member of the cobalt group, electronic configuration $5d^76s^2$. It shows oxidation states from +6 (IrF_6) to −1 $[Ir(CO)_4]^-$. The +6 and +5 states are strongly oxidizing and the +4 and +3 states the most stable; the +4 state for Ir is more stable than for Rh. Ir(III) and (IV) ammines and other N-bonded complexes are very stable. The +2 state is uncommon but the +1 state (generally square planar) often with phosphine or carbonyl ligands is well known (e.g. Vaska's compound*) but undergoes oxidative addition to give Ir(III). Ir(0) exists as carbonyls, e.g. $Ir_2(CO)_8$ and phosphine complexes.

iridium halides The known fluorides are yellow IrF_6 (Ir plus F_2) m.p. 44°C, b.p. 53°C; yellow $(IrF_5)_4$ (Ir plus F_2); IrF_4 (Ir plus IrF_6); and black IrF_3 (heat on IrF_5). Fluoroiridates containing IrF_6^-, IrF_6^{2-}, and IrF_5^{2-} species are known. Green-blue $IrCl_3$ (Ir plus Cl_2 at 600°C) is the only known anhydrous chloride; $IrBr_3$, IrI_3 and hydrates are known. Chloroiridates(IV) $[IrCl_6]^{2-}$ (Ir compounds plus Cl_2 plus HNO_3) are reduced to $[IrCl_6]^{3-}$.

iridium oxides Black IrO_2 is formed by heating Ir in air. A hydrated form and hydrated Ir_2O_3 are precipitated from Ir(IV) or Ir(III) solutions with alkali; the latter is readily oxidized to IrO_2. Mixed oxides, e.g. $CaIrO_3$, are formed by solid state reactions; iridates are not stable in solution.

iron, Fe. At.no. 26, at.wt. 55·847, m.p. 1535°C, b.p. 2750°C, *d* 7·87. Iron was known in Egypt in 3000 B.C. and was used in Europe about 1000 B.C. Fe is found in most clays, sandstones and granites, meteoritic iron contains Co and Ni. The chief ores are haematite (Fe_2O_3), brown haematite (hydrated Fe_2O_3), magnetite (Fe_3O_4), siderite or spathic iron ore ($FeCO_3$), iron pyrites (FeS_2) and chalcopyrite ($CuFeS_2$). The metal is bcc below 960°C, ccp to 1401°C, bcc to 1535°C. Iron is used extensively as a metal for fabrication, either as iron or as steel.

Iron is usually prepared from oxide or carbonate ores, from which S, As, etc. have been removed by roasting in air, by reduction with carbon. The ore is mixed with coke and $CaCO_3$ and heated in a blast furnace, the maximum temperature of which is about 1300°C. The major acidic impurities are removed as slag (calcium silicate, aluminate, etc.) and the molten crude metal run off into pigs; pig-iron contains 2–4% of carbon with a little P, S and Si. If the Si content is high, the C is present almost entirely as graphite; on remelting and casting, such an alloy gives grey cast iron. If the Si content is low, the carbon is present as cementite, Fe_3C, and gives white cast iron on casting. The sulphur content governs that of Mn, since excessive sulphur forms MnS, which is appreciably soluble in the slag. The cast irons are too brittle for many purposes. Wrought iron is made by melting cast iron with scrap-iron in a reverberatory furnace (puddling furnace) lined with Fe_2O_3, which oxidizes C, Si, etc., most of which are then removed by rolling. Cast iron melts at about 1200°C. Wrought-iron, which is tough, fibrous and malleable, melts at above 1500°C, though it softens at about 1000°C. It is not hardened by quenching.

Steels contain *c.* 1·5% carbon and other alloying elements, from pig-iron the C content must be reduced, from wrought-iron the C content must be increased. Above 900°C steel is a solid solution of C in γ-Fe (austenite). On slow cooling, austenite gives cementite Fe_3C and an Fe–C solution; below 690°C the γ-Fe gives a eutectic of ferrite (*c.* 0·06% C in α-Fe) and cementite. The eutectic is known as pearlite and is soft. Quenching austenite from 900°C to 150° gives martensite (supersaturated C in α-Fe) which is very hard but may be tempered (reduced hardness but still tough) by reheating martensite to 200–300°C.

In case hardening C and N are introduced into the iron by heating in a N_2-NH_3 atmosphere or by immersing the steel in molten $NaCN/Na_2CO_3$ at about 900°C.

Iron is converted into steel by the Bessemer, Siemens–Martin or electrical processes. Iron can be rendered passive by contact with concentrated HNO_3. World production iron ore 1976 520 megatonnes.

iron acetates *See* iron ethanoates.

iron alums, $MFe^{III}(SO_4)_2,12H_2O$. *See* alums. Iron alum, ferric alum is the ammonium salt.

iron ammonium sulphate, Mohr's salt, $(NH_4)_2Fe(SO_4)_2,6H_2O$. Pale green crystalline solid used as a standard reducing agent (e.g. for $KMnO_4$). More stable than $FeSO_4$.

iron bromides *Iron(II) bromide,* $FeBr_2$, occurs as hydrates (Fe plus HBr). *Iron(III) bromide,* $FeBr_3$, formed anhydrous from Fe plus Br_2 as dark red crystals. $FeBr_3,6H_2O$ is dark green.

iron buff Hydrated iron(III) oxide. Important pigment used for black dyeing of silk and for khaki dyeing of cotton, in paints and enamels.

iron carbides Of importance in metallurgy – *see* iron.

iron carbonate, $FeCO_3$. Only the iron(II) derivative known (white powder from Fe^{2+} plus Na_2CO_3 in absence of air) very readily oxidized. Occurs naturally as siderite, spathic iron ore, chalybite.

iron carbonyls Iron forms $Fe(CO)_5$ (trigonal bipyramidal), $Fe_2(CO)_9$ and $Fe_3(CO)_{12}$. $Fe(CO)_5$ is formed from Fe plus CO at 200°C and 300 atm., m.p. −20°C, b.p. 102°C. Light gives orange $Fe_2(CO)_9$ and heat gives dark green $Fe_3(CO)_{12}$. These are parent compounds of wide ranges of carbonyl derivatives in which CO is replaced by other ligands, e.g. PPh_3, butadiene, etc. They react with acetylenes to give products of telomerization of the C_2 groups and insertion of CO and Fe. $Fe(CO)_5$ gives $(Fe(CO)_4)^{2-}$ with base which reacts with acids to give $H_2Fe(CO)_4$. Halide, nitrosyl, organo derivatives are also known.

iron chemistry Iron is a first row transition element, electronic configuration $3d^64s^2$ which shows a wide variety of oxidation states and co-ordination numbers. The most stable oxidation state is +2 (ferrous) (in e.g. $FeCl_2$, $K_4Fe(CN)_6$) although +3 (ferric) (e.g. $FeCl_3$, $K_3Fe(CN)_6$) is also important. Many complexes are known for these states but these are labile and frequently difficult to prepare (except low spin compounds, e.g. cyanoferrates).

$E°\ Fe^{3+} \longrightarrow Fe^{2+}$ +0·77 volts, $E°\ Fe^{2+} \longrightarrow Fe$ −0·09 volts (acid solution). Other oxidation states include −2 $(Fe(CO)_4)^{2-}$,$OFe(CO)_5$; +1 $(Fe(H_2O)_5$-

$NO)^{2+}$; +4 $FeH_4(PR_3)_3$; +6 $(FeO_4)^{2-}$. The latter state is strongly oxidizing. The co-ordination can increase over 6 in the +2 and +3 states (e.g. Fe^{III}-EDTA complex).

iron chlorides Iron(III) chloride, $FeCl_3$. Dark red crystals (Fe plus Cl_2) forms hydrates but hydrolysed to brown derivatives in water. Forms complex chlorides, e.g. K_2FeCl_5,H_2O and $(NH_4)_2FeCl_5,H_2O$ (ammonio-chloride of iron). Iron(II) chloride, $FeCl_2$ white solid (Fe plus HCl gas) forms many complexes including $FeCl_2{\cdot}6NH_3$, $FeCl_2,4H_2O$ (also 6, 2 and $1H_2O$), $KFeCl_3,2H_2O$.

iron, complex cyanides *See* cyanoferrates. Derivatives of cyanoferrates, e.g. pentacyanonitrosylferrates, nitroprussides, containing $[Fe(CN)_5NO]^{2-}$ are also known.

iron ethanoates, iron acetates Iron(II) ethanoate, $Fe(O_2CCH_3)_2$, pale green crystals. Iron(III) forms the $[Fe_3O(CH_3CO_2)_6(H_2O)_3]^+$ cation and some other hydrolysed species.

iron fluorides Iron(III) fluoride, FeF_3. Only slightly soluble in water. White salt (HF on Fe_2O_3). Forms hydrates and complex hexafluoroferrates containing $[FeF_6]^{3-}$ ions. Iron(II) fluoride, FeF_2. Prepared Fe plus HF gas forms $4H_2O$ hydrate and, in the absence of oxygen, perovskites, e.g. $KFeF_3$.

iron hydroxides *See* iron oxides.

iron iodide, iron(II) iodide, FeI_2, 9, 6, 5, 4 and $1H_2O$. Prepared from Fe and I_2. Soluble in water, oxidized to I_2 and Fe(III) derivatives in air.

iron nitrates Iron(III) nitrate, $Fe(NO_3)_3,9H_2O$ also $6H_2O$. The pure salt is pale amethyst but impure salts are first colourless and then brown because of hydrolysis. Iron(II) nitrate, $Fe(NO_3)_2,6H_2O$, unstable material (Fe plus HNO_3) also forms $9H_2O$ hydrate.

iron, organic derivatives Iron carbonyls readily form π-bonded olefine complexes by displacement of CO and acetylenes form both σ- and π-bonded derivatives. Simple σ-bonded iron organometallics are relatively unstable but h^5-cyclopentadienyl derivatives, e.g. ferrocene*, and h^4-cyclobutadiene derivatives are stable.

iron oxides and hydroxides These substances form a complex interrelated system. They are of importance as dyes and pigments (e.g. iron buff, sienna, umber, rouge, ochre). Oxides all show non-stoicheiometry. Fe_2O_3 dark brown has two forms, one, haematite, with the corundum structure, the other (γ-form) with a structure related to the spinel structure and closely related to Fe_3O_4, magnetite (inverse spinel, $Fe^{3+}(Fe^{2+},Fe^{3+})O_4$) and $FeO(Fe_{1-x}O)$ has a cp oxide structure (NaCl structure). Fe_2O_3 is only soluble in acids with difficulty and becomes almost insoluble if strongly ignited. FeO is difficult to prepare (heat on FeC_2O_4) and decomposes to Fe_3O_4 and Fe below 570°C. It dissolves readily in acids.

Hydroxide $Fe(OH)_3$ (Fe^{3+} plus OH^-) has definite existence and there are many ill-defined hydrates used as pigments. FeOOH has two forms goethite and lepidocrocite. Colloidal $Fe(OH)_3$ is easily obtained as a deep red sol. Many Fe(III) hydroxy complexes are known. $Fe(OH)_2$ may be formed from Fe^{2+} and OH^- in the absence of O_2 but it is very readily oxidized.

Iron oxides react with other metal oxides to give ferrates*.

iron(III) perchlorate, $Fe(ClO_4)_3,6H_2O$.

iron phosphates Iron(III) phosphate, $FePO_4,2H_2O$, is precipitated from solution but is soluble in dilute acids; complex phosphates are known (the complexing of Fe by $PO_4{}^{3-}$ is used to affect redox potentials during Fe^{2+}/Fe^{3+} titrations). Medicinal iron phosphate is a dark blue powder containing iron(II and III) phosphates and iron(III) hydroxides.

iron sulphates
Iron(III) sulphate, $Fe_2(SO_4)_3$, 12, 10, 9, 7, 6, 3 and 0 H_2O. Formed from Fe(III) salts plus $(NH_4)_2SO_4$. Decomposes to Fe_2O_3 and SO_3 on heating. Sulphato complexes and alums are also known.

Iron(II) sulphate, $FeSO_4,7H_2O$, ferrous sulphate, green vitriol, copperas. Forms 5 and 1 hydrates and a white anhydrous salt. Obtained from the waste liquor resulting from pickling steel in H_2SO_4. Slowly oxidized in air; forms more stable double sulphates, e.g. $M_2Fe(SO_4)_2,6H_2O$, e.g. iron ammonium sulphate.

iron sulphides FeS_2 occurs naturally as pyrites and marcasite, both contain $S_2{}^{2-}$ anions. Grey FeS (Fe^{2+} plus H_2S or $(NH_4)_2S$, Fe plus S) has the nickel arsenide structure. Pyrrhotite, Fe_7S_8, is an Fe-deficient version of FeS. It seems unlikely that Fe_2S_3 exists as such. Some complex sulphides appear to be formed.

iron thiocyanates Iron(III) thiocyanate complexes are deep red species formed from Fe^{3+} and NCS^- and used for the detection of either constituent. $Fe(NCS)_2,3H_2O$ is known but is unimportant.

Irving–Williams order The stabilities of series of complexes with different metals fall into the sequence
$Mn^{2+} < Fe^{2+} < Co^{2+} < Ni^{2+} < Cu^{2+} > Zn^{2+}$.
This is known as the Irving–Williams order and is general for many ligands.

isatin, $C_8H_5NO_2$. Yellowish-red prisms,

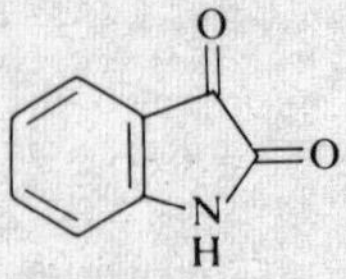

m.p. 200–201°C. Prepared by heating α-isatin-anilide with dilute mineral acid. It is used for the preparation of vat dyes by condensation with indoxyl (to give indirubin) or thio-indoxyl.

islands of nuclear stability *See* post-actinide elements.

isoamyl- The group $Me_2CH{\cdot}CH_2CH_2-$.

isoamyl alcohol *See* amyl alcohols.

isobars Atomic species of the same mass number (i.e. protons plus neutrons) but different chemical properties (i.e. protons or electrons).

isoborneol, $C_{10}H_{18}O$. M.p. 214°C. *See* borneol.

isobornyl chloride *See* bornyl chloride.

isobutane, 2-methylpropane, C_4H_{10}, $Me_2CH{\cdot}CH_3$. Colourless gas; m.p. −145°C, b.p. −10°C. Occurs in natural gas, and is produced in large amounts in the cracking of petroleum. It is used in refrigeration plant. For chemical properties *see* paraffins.

isobutyl-, 2-methylpropyl The group $Me_2CH_2CH_2-$.

isobutyryl The 2-methylpropanoyl-group, Me_2CHCO-.

isocrotonic acid *See* crotonic acids.

isocyanates Compounds containing the group $-N{=}C{=}O$. Alkyl isocyanates are prepared from dialkyl sulphates and potassium cyanate. They readily polymerize to isocyanuric esters and are formed as intermediates in the Hofmann decomposition of amides and the Curtius decomposition of azides. Aryl isocyanates are best prepared by the reaction of phosgene and an aromatic amine. PhNCO reacts with alcohols to give urethanes, with ammonia to give phenylurea and with aniline to give diphenylurea. Converted on heating to carbodiimides.

The polycondensation of di-isocyanates with polyhydric alcohols gives a wide range of polyurethanes which are used as artificial rubbers and light-weight foams, and have other important properties. Isocyanates are also used as modifiers in alkyd resins*.

isocyanides *See* isonitriles.

isodurene 1,2,3,5-Tetramethylbenzene, $C_6H_2(CH_3)_4$.

isodispersion Sols in which the dispersed particles, of colloidal dimensions, are all of the same size. Haemoglobin forms an isodispersion of particles all of mol.wt. 68 000.

isodrin The isomer of aldrin* obtained by the Diels–Alder* reaction between cyclopentadiene and 1,2,5,6,7,7-hexachloronorbornadiene. Used as an insecticide.

isoelectric point As dispersed substances in general are either positively or negatively charged according to whether positive (including hydrogen) ions or negative (including hydroxyl) ions are predominantly adsorbed, they also possess an isoelectric point at which the net charge on the particles is zero – positive and negative ions being adsorbed to an equal extent. At the isoelectric point the electrophoretic velocity is zero and the substance tends to flocculate. With hydrophilic substances which are protected from coagulation by possession of a sheath of water, the stability is minimum at the isoelectric point (*see* colloid, coagulation, zwitterion).

isoelectronic Molecules or ions which have the same number of electrons, e.g. CO and N_2.

isoleptic Having all ligands the same, e.g. $Fe(CO)_5$, $[Cu(NH_3)_6]Cl_3$.

isoleucine, 2-amino-3-methylvaleric acid, $C_6H_{13}NO_2$, $CH_3{\cdot}CH_2{\cdot}CHMe{\cdot}CHNH_2{\cdot}$-COOH. Colourless crystals, m.p. 284°C (decomp.). The naturally occurring substance is dextrorotatory. An amino-acid, occurring with leucine as a product of protein hydrolysis.

isomerases Enzymes catalysing isomerization of compounds, e.g. 3-phosphoglyceraldehyde to phosphodihydroxyacetone by phosphotriose isomerase. *See* enzymes. Mitases are isomerases which catalyse the movement of a radical from one part of the molecule to another, e.g. glucose-1-phosphate to glucose-6-phosphate.

isomerism Compounds possessing the same

composition and the same molecular weight, but differing in at least one of their physical or chemical properties, are said to be isomeric, and each is an isomer, or isomeride, of the others. The isomerism may be of several types.

(1) *Structural isomerism*, due to differences in the order in which the atoms are joined together.

a. Chain or nuclear isomerism, due to differences in the arrangement of e.g. carbon atoms in the molecule as in

$$CH_3CH_2CH_2CH_3 \quad \text{and} \quad CH_3CH\begin{array}{c}CH_3\\CH_3\end{array}$$

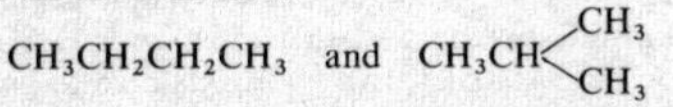

b. Position isomerism, due to differences in the position of some group or atom with regard to the carbon chain or ring, as in $CH_3CH_2CH_2Cl$ and $CH_3CHClCH_3$.

c. Functional group isomerism or metamerism, due to differences in the type of compound, as in CH_3COCH_3 and CH_3CH_2CHO. Tautomerism, or dynamic isomerism, is a special case of structural isomerism in which two isomers are directly interconvertible. The reversibility of the change is due to the mobility of a group or atom, which can move from one position to another in the molecule with rearrangement of a double bond. In certain cases it may be due only to a shifting of the double bonds.

The more common types of structure showing tautomerism are the *keto-enol type*,

$$>\overset{\overset{\displaystyle H}{|}}{C}-C=O \rightleftharpoons >C=C-OH$$

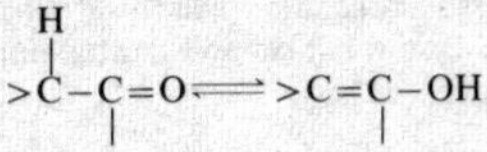

as in acetoacetic ester; the *three carbon* type,

$$>\overset{\overset{\displaystyle H}{|}}{C}-C=C< \rightleftharpoons >C=C-\overset{\overset{\displaystyle H}{|}}{C}<$$

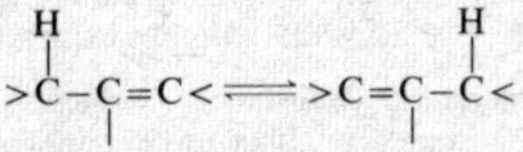

as in the substituted glutaconic acids; and the *amido-amidol* type,

$$-\overset{\overset{\displaystyle H}{|}}{N}-C=O \rightleftharpoons -N=C-OH$$

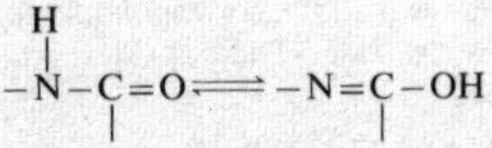

as in urea. These isomers are known as tautomers, or tautomerides, and exist in equilibrium in the liquid state or in solution. The speed of conversion of one form to the other varies widely in different cases, and may be so slow that it is possible to isolate both forms in a relatively pure state. There is no interconversion in the solid state. The term *desmotrope* is applied to those tautomers which can be separated from each other.

(2) *Space isomerism or stereo-isomerism*, due to differences only in the spatial orientation of the atoms in the molecule.

a. Optical isomerism due to asymmetry. Such isomers differ in the direction in which they rotate the plane of polarized light.

b. Geometrical, or *cis-trans* isomerism, due to different arrangements of dissimilar atoms or groups attached to two atoms joined by a double bond or forming part of certain ring structures. Thus the isomeric fumaric and maleic acids are

$$\begin{array}{cc}H-C-COOH & H-C-COOH\\ \| & \text{and} \quad \| \\ HOOC-C-H & H-C-COOH\end{array}$$

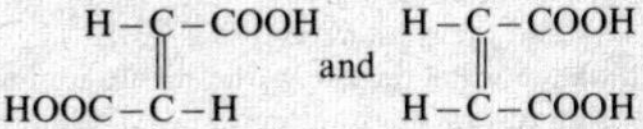

respectively. The presence of the double bond restricts the free rotation of the C atoms joined by it, and permits the existence of the two forms. Similarly, in certain ring structures which need not contain double bonds, the free rotation of the atoms is restricted and other groups or atoms attached to them may lie above or below the plane of the ring and give rise to geometrical isomers. This isomerism is not confined to the $>C=C<$ bond, but exists also in compounds having $>C=N-$ or $-N=N-$ bonds. In the case of the $>C=C<$ double bond or a carbon ring, the isomers are distinguished by the prefixes *cis-* and *trans-* according as two given groups or atoms lie on the same or on opposite sides of the plane of the double bond or ring. In the other cases, the prefixes *syn-* and *anti-* are used; thus the *syn-* aldoximes have the structure

$$\begin{array}{c}R-C-H\\ \| \\ N-OH\end{array}$$

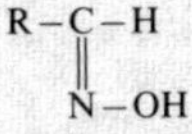

For ketoximes, the group which is *syn* or *anti* to the hydroxyl group must be specified. Difficulty in assigning the terms *cis* and *trans* arises when there is no pair of identical, or similar, groups substituting an alkene. Then the terms (E) [German *entgegen* = across] and (Z) [German *zusammen* = together] replace the terms *trans/cis*. In many cases, but not all, the term *cis* corresponds to (Z) and *trans* to (E). Under this new system the two ligands attached to each olefinic carbon atom are put in order of precedence by application of sequence rules* and the symbols (E) and (Z) are then used to describe those isomers in which the ligands of higher precedence are on the opposite side (E), etc. Maleic acid is then referred to as the (Z)-isomer and fumaric acid as the (E)-isomer. A similar nomenclature can be applied to oximes.

Stereoisomers are also found in inorganic complexes, e.g.

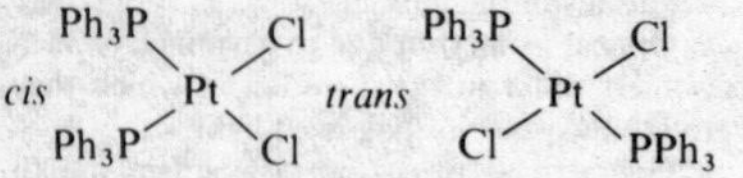

which have square planar co-ordination about the central platinum atom.

Other types of isomerism are:

Ionization isomerism, e.g. $[Pt(NH_3)_4Cl_2]Br_2$ and $[Pt(NH_3)_4Br_2]Cl_2$.

Mer and *fac* isomerism in e.g. CoL_3X_3.

Linkage isomerism, e.g. the nitrite ion can co-ordinate through either the oxygen or nitrogen atoms. Since co-ordination numbers in inorganic chemistry vary widely from the limited range adopted by carbon the possibilities for isomerism in inorganic derivatives are greater than in organic chemistry.

See conformation.

isomerism, nuclear The existence of atomic nuclei with the same atomic and mass number but having different nuclear energies and radioactive decay properties. E.g. ^{124}Sb has three isomers, ^{124}Sb, ^{124m}Sb, $^{124m_2}Sb$.

isomerization The conversion of a compound to an isomer of that compound.

isomorphism Having the same crystal structure (isostructural). Isomorphous substances will often induce crystallization of each other and may even crystallize on each other (e.g. violet chromium alum will crystallize on colourless potassium alum).

isonitriles, isocyanides, carbylamines Organic compounds, R–NC. They are colourless liquids, having a toxic and highly disagreeable odour. Prepared by treating alkyl halides with silver cyanide, but more conveniently by dehydration of formamides (RNH·CHO) with phosphorus oxychloride in the presence of base; the reaction between a primary amine, chloroform and alcoholic KOH generates a carbylamine which is a useful test, but of limited preparative value. They are reactive substances; acids cause decomposition to a primary amine and methanoic acid. Heating to 100–200°C causes isomerization to cyanides. They are oxidized by sulphur to *iso*thiocyanates and by HgO to *iso*cyanates. Reaction with carbonyl compounds and amines gives derivatives of α-amino-acids. Their stable complexes with transition metals are analogous to the metal carbonyls, e.g. $Ni(CNPh)_4$ and $Ni(CO)_4$.

isonitrosoketones Tautomeric monoximes of

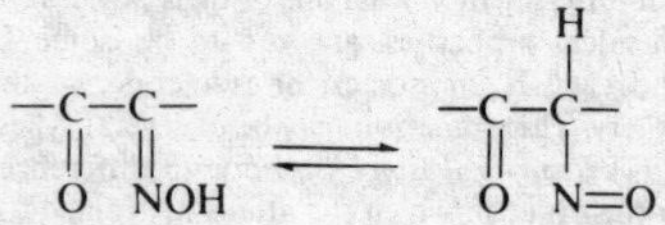

α-diketones: formed by treating ketones with amyl nitrite and hydrochloric acid. They are colourless crystalline solids. Dissolve in alkalis to give salts having an intense yellow colour. Form α-diketones when treated with H_2SO_4 or HNO_2. React with hydroxylamine to give dioximes of the diketones.

isoparaffins Aliphatic hydrocarbons having a branched carbon chain structure. They usually have higher anti-knock value than the corresponding straight chain normal paraffin. *See* octanes.

isophthalic acid, 1,3-benzene dicarboxylic acid Colourless crystals, m.p. 345–347°C with sublimation, obtained by oxidation of *m*-xylene.

isopiestic Solutions of salts, etc. which have the same partial pressure of solvent are termed isopiestic solutions. The isopiestic method of studying the osmotic properties of solutions is to allow solutions to come to isopiestic equilibrium, by diffusion of vapour from one to another, and with a solution of a standard substance in an enclosed vessel at a constant temperature; the concentrations of the isopiestic solutions are then determined. The method is slow but capable of attaining a high accuracy.

isopolyacids Formally acids which give polymeric oxide- and hydroxide-bridged anions containing one metal only. Formed by e.g. Mo (VI). The free acids often do not exist.

isopolymorphism Substances which have more than one crystalline form isomorphous with the crystalline forms of another substance are said to show isopolymorphism. Thus, the octahedral and rhombic forms of As_2O_3 are isomorphous with the corresponding forms of antimony trioxide. These oxides are isodimorphous.

isoprene, 2-methylbutadiene, $C_5H_8, H_2C = CMe{\cdot}CH = CH_2$. Colourless liquid with a penetrating odour; b.p. 34°C. It is obtained, together with other products, by the distillation of raw rubber, or by passing the vapour of turpentine oil through a metallic tube heated to redness. Manufactured from butanone and methanal in the presence of KOH to give 2-methyl-3-ketobutanol, and reducing this to 2-methylbutylene glycol. The glycol is converted to isoprene when its vapour is passed over a heated acid phosphate. It is also manufactured from propanone and ethyne in an ether solvent in the presence of powdered KOH. The resultant acetylenic alcohol is catalytically reduced to 1,1-dimethylallyl alcohol which is dehydrated by passing the vapour over alumina. It is converted to synthetic rubber by the action of catalysts (*see* isoprene polymers).

isoprene polymers Stereoscopic polymerization of isoprene to *cis*-polyisoprene, the synthetic analogue of natural rubber occurs with Ziegler–Natta catalysts or lithium alkyls. Used as an elastomer and replacement for natural rubber (as polyisoprene). *Trans*-polyisoprene is used in golf-ball covers. U.S. production of *cis*-polyisoprene 1978 59 000 tonnes.

isoprene rule This proposes that the carbon skeletons of terpenes are built up of isoprene units, so that C_4 of one unit is attached to C_1

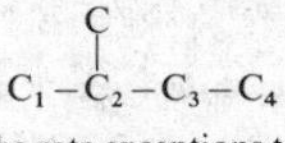

of the next. The rate exceptions to this rule can be accommodated by assuming that a methyl group migration occurs at some stage in the biogenesis. Isoprene itself is not the actual precursor, but the closely related compound mevalonic acid* is incorporated in high yield into carotenoids, squalene, cholesterol and other terpenoid compounds.

isopropanol *See* 2-propanol.

isopropanolamines, 2-propanolamines Alkanolamines with at least one $NCH_2CHOHCH_3$ grouping. Important materials include monoisopropanolamine $NH_2CH_2CHOHCH_3$, b.p. 159°C; di-isopropanolamine $NH(CH_2CHOHCH_3)_2$, b.p. 248°C; triisopropanolamine $N(CH_2CHOHCH_3)_3$, b.p. 300°C. Manufactured from ammonia and propylene oxide. Used as weed-killers, as stabilizers for plastics, in detergents, alkanolamine soaps for sweetening natural gas and in synthesis.

isopropenyl acetate, $C_5H_8O_2$, $CH_2{=}C(CH_3){\cdot}O{\cdot}CO{\cdot}CH_3$. Colourless liquid; b.p. 96–97°C. Prepared from propanone and keten in the presence of sulphuric acid. The reaction is reversible, hence isopropenyl acetate finds application as an acetylating agent, being more efficient than keten itself. Gives ethanoates (acetates) with alcohols and amines, enol acetates with aldehydes and ketones, and acid anhydrides with carboxylic acids. The substance is heated with isopropenyl acetate and an acid catalyst, so that propanone is continuously removed by distillation.

Isopropenyl acetate gives spinnable fibres on copolymerization with vinyl chloride.

isopropyl The group CH_3CHCH_3.

isopropyl alcohol *See* 2-propanol.

isopropylbenzene *See* cumene.

isopropyl chloride, 2-chloropropane, C_3H_7Cl, $CH_3{\cdot}CHCl{\cdot}CH_3$. Colourless liquid with a pleasant odour; b.p. 36·5°C. Manufactured by treating isopropyl alcohol with hydrochloric acid in the presence of zinc chloride, or by passing the vapour of the alcohol and hydrochloric acid gas over a heated metallic chloride, such as magnesium chloride. It is used as a fat solvent.

isopropyl ethanoate, isopropyl acetate, $C_5H_{10}O_2$, CH_3COOPr^i. Colourless liquid with a fragrant odour; b.p. 88°C. Manufactured by leading propene into hot ethanoic acid containing sulphuric acid, or by heating isopropyl alcohol with ethanoic and sulphuric acids. Used as a solvent for cellulose nitrate and various gums.

isopropyl ether, C_6H_4O, $Me_2CHOCHMe_2$. Colourless liquid with sweet camphor-like odour, b.p. 68°C. Made by heating isopropanol with sulphuric acid. Used extensively as an industrial solvent particularly for dewaxing lubricating oils and for the extraction of fats and oils. U.S. production 4500 tonnes p.a.

isopropylidene The group

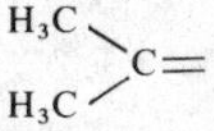

isopropylmethylbenzenes *See* cymenes.

44′-isopropylidenediphenol *See* bisphenol A.

isopropylnaphthalenese Important solvents.

isoquinoline, C_9H_7N. Crystallizes in tablets,

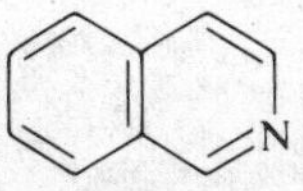

m.p. 24°C, b.p. 242°C. Odour like that of quinoline.

isosbestic point The spectra of transition metal complex ions often show a point or points at which extinction coefficients are equal. These points are referred to as isosbestic points. They are used in spectrophotometric studies of the formation or of kinetics of reactions in solution to investigate processes such as isomerization, which may occur during such reactions.

isostructural Having the same lattice type and crystal structure. E.g. K_2PtCl_6 and Rb_2PtCl_6.

isotactic polymers Stereospecific polymers of, e.g., $(-CH_2CHR-)_n$ with all the substituted carbons having the same stereo-configuration and, with a hypothetical straight chain, all the substituents on the same side of the chain. Prepared by polymerization using specific catalysts, often Ziegler catalysts. *Compare* syntactic polymers.

isothermal change A change in a system which occurs whilst the system is maintained at constant temperature by removing the heat produced, or by adding heat to compensate that absorbed during the change.

isothiocyanates Derivatives of the type RNCS. *See* thiocyanic acid.

isotones Atomic nuclei having the same number of neutrons but different mass numbers, e.g. ^{132}Xe and ^{133}Cs.

isotonic solutions Solutions which have the same osmotic pressure.

isotope Elements (or atoms) which have the same atomic number and, therefore, similar chemical properties, but different atomic weights are called isotopes, e.g. $^{16}_{8}O$ and $^{18}_{8}O$. In addition to stable isotopes, very many radioactive isotopes of elements are known. Although isotopes have similar chemical properties, their slight difference in mass causes slight differences in physical properties. Use of this is made in isotopic separation processes using techniques such as fractional distillation, exchange reactions, diffusion, electrolysis and electromagnetic methods.

isotope effect Reactions involving the breaking of a C−H bond are slower when that hydrogen is replaced by deuterium or tritium and are distinguished by a 'primary kinetic isotope effect'. This difference in rate of breaking of a C−H bond compared to a C−D bond is due to the greater vibrational energy at room temperature of a C−H bond by comparison with that of the C−D bond. Consequently less energy is required to break the C−H bond.

isotropic A substance which possesses identical properties in all directions in its crystal. Cubic substances and some non-crystalline glasses are isotropic.

itaconic acid, methylenesuccinic acid, $C_5H_6O_4$.

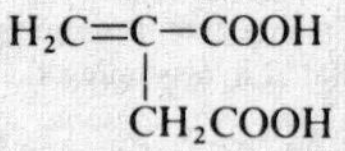

White crystals; m.p. 162–164°C. It can be prepared by the fermentation of sugar with the mould *Aspergillus terreus* or by heating citraconic anhydride with water at 150°C. Electrolysis of the potassium salt in solution gives allene. Itaconic acid is used as a comonomer in plastics; its esters are polymerized to lubricating oils and plasticizers.

J

J A measure of the coupling constant in nuclear magnetic resonance.

Jahn–Teller effect The Jahn–Teller theorem states that, when any degenerate electronic state contains a number of electrons such that the degenerate orbitals are not completely filled, the geometry of the species will change so as to produce non-degenerate orbitals. Particularly applied to transition metal compounds where the d^9 state is $d_\gamma^6 d_\varepsilon^3$. Cu(II) generally has the distorted octahedral co-ordination typical of the Jahn–Teller effect. Where a species is not distorted the vibrational spectra frequently show anomalous intensities if subject to the Jahn–Teller theorem.

jj coupling *See* Russell–Saunders coupling.

Jones reductor A tube containing zinc amalgam. Used for reduction of solutions (e.g. $Fe^{3+} \longrightarrow Fe^{2+}$) prior to titration or estimation.

Joule's law The internal energy of a gas depends only on its temperature (being independent of its pressure and volume). Like the other gas laws, it is only approximately true. At high pressures it is invalidated by the existence of intermolecular forces.

Joule–Thomson effect, Joule–Kelvin effect Most gases (except H and He) undergo cooling when they are expanded adiabatically through a throttle. This is called the Joule–Thomson effect. The change in temperature is proportional to the drop in pressure; the Joule–Thomson coefficient is the change of temperature per unit change of pressure. The effect, which is used for liquefying air, making solid carbon dioxide, etc. is related to departures from Boyle's law and Joule's law. H and He, which show a warming effect at ordinary temperatures, also show cooling if the experiment is conducted below their inversion temperatures.

juvenile hormones A widely distributed group of compounds which regulate larval development or reproduction of insects and are therefore considered to be insect hormones. Hormone activity is shown by some simple terpenoids such as the two *trans*-(Δ^6)-farnesols, their methyl ethers, farnesol and methyl farnesoate dihydrochloride; ecdysone* is also concerned in the larval moulting process. Knowledge of such compounds has important implications for insect control.

K

K Potassium.

kainite, $MgSO_4$, $KCl,3H_2O$. Occurs in the Stassfurt salt deposits as beds. Used as a source of K salts and as fertilizer.

kairomone A compound or mixture acting between different species which adaptively favours the receiving species, e.g. a chemical by which a predator recognizes its prey or by which an insect is attracted to its food plant. *See also* allomone, pheromone.

kaolin *See* china clay.

kaolinite, $Al_2(OH)_4Si_2O_5$. The most important constituent of china clay.

Kapustinskii equation For an ionic crystal composed of cations and anions, of respective charge z_+ and z_-, which behave as hard spheres, the lattice energy (U) may be obtained from the expression

$$U = \frac{Lz_+z_-e^2M}{R}\left(1-\frac{1}{n}\right);$$

where L is Avogadro's number; e, the electronic charge; R, the equilibrium internuclear distance; M, the Madelung constant and n an integer. The Kapustinskii equation is a simplification of the above expression in which it is assumed that the value of M/v, where v is the number of ions in one 'molecule', is constant for different structures and that n has an average value of 9. Thus, if r_+ and r_- are the thermodynamic radii of cation and anion respectively, and $R=(r_+ + r_-)$, the Kapustinskii equation states

$$U = \frac{0{\cdot}777N_0vz_+z_-e^2}{(r_+ + r_-)}$$

Karathane A trade name for 2,4-dinitro-6-(1-methylheptyl)phenyl crotonate, $C_{18}H_{24}N_2O_6$, a compound which has both acaricidal and fungicidal activity. It is a red-brown oil of high boiling point, insoluble in water but soluble in most organic solvents. Karathane is used for the control of powdery mildew, and is non-toxic to mammals.

Karl Fischer reagent A mixture of I_2 and SO_2 dissolved in pyridine – MeOH used as a titrant for water with which HI is liberated and the pH determined with a meter.

Karplus equation The relationship between the n.m.r. coupling constant J for vic-H atoms and the dihedral angle between the C–H bonds.

kayser A suggested unit of frequency 1 cm^{-1}. The kilo-kayser is 1000 cm^{-1}.

K capture *See* radioactivity.

Kel-F Trade name for a polymeric chlorotrifluoroethene, often copolymerized. May be a liquid or solid. Inert to chemical attack and a thermoplastic (Teflon cannot be moulded).

Kelvin scale *See* absolute temperature.

kephalin *See* cephalins.

kepone, $C_{10}Cl_{10}O$. A high-melting ketonic solid prepared by condensation of two molecules of hexachlorocyclopentadiene in the presence of SO_3, oleum or HSO_3Cl. An insecticide and fungicide but poisonous to mammals.

keratins An insoluble group of proteins of the scleroprotein class, found in the skin, hair, nails, horns, hoofs and feathers of animals. Keratins are insoluble in the usual protein solvents and are unattacked by pepsin or trypsin.

kernite, rasorite A hydrated sodium borate, $Na_2B_4O_7,4H_2O$, which occurs in California and is a major source of boron compounds. *See* borates, borax.

kerosine A petroleum fuel of boiling range approximately 410–570 K.

ketals Ketone acetals of the general formula

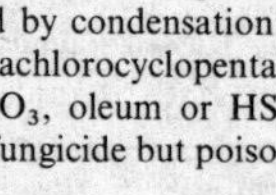

They are colourless liquids with characteristic odours, and are prepared by the condensation of ketones with alkyl orthoformates in the presence of alcohols, or by the reaction of acetylenes with alcohols in presence of HgO and BF_3. In some cases trichloroethanoic acid is used as the catalyst. They lose alcohol when heated and form vinyl ethers. Exchange of alcohol groups occurs when the ketals of the lower alcohols are boiled with alcohols of greater molecular weight. *See* acetals.

keten, $CH_2{:}CO$. A colourless gas, manufactured by passing propanone rapidly through metallic tubes heated at 550–800°C, or by heating ethanoic acid at 700–1000°C. It is very unstable and dimerizes spontaneously to diketen. Reacts with water to form ethanoic acid, with alcohols to give ethanoates, with ethanoic acid to give ethanoic anhydride and with amines to give acetyl derivatives. Used as an acetylating agent, particularly in the manufacture of cellulose acetate.

ketens Organic compounds containing the group $>C=C=O$. They are formed by the action of zinc powder on the acid bromide or chloride of an α-bromo- or α-chloro-fatty acid, or by heating the anhydride of a dibasic acid. Keten itself and other ketens containing the group $-CH=C=O$ are colourless substances, polymerized by quinoline and do not form addition compounds with substances containing a $>C=O$ or $>C=N-$ group. The remaining ketens are highly coloured and form addition compounds with quinoline and with substances containing $>C=O$ and $>C=N-$ groups. Both types react readily with water to give acids, with alcohols to give esters and with halogens and halogen acids.

keto- A prefix used to denote that the substance in question contains a carbon atom attached to an oxygen atom by a double bond and to two other carbon atoms by single bonds.

ketols Organic compounds containing both a keto- and an alcohol group. They are formed by oxidation of glycols or by condensation between two molecules of a ketone. They exhibit the typical properties of both alcohols and ketones.

ketomalonic acid *See* mesoxalic acid.

ketone bodies, acetone bodies The ketone bodies, acetone (propanone), $CH_3{\cdot}CO{\cdot}CH_3$, acetoacetic acid, $CH_3{\cdot}CO{\cdot}CH_2{\cdot}COOH$ and β-hydroxybutyric acid, $CH_3{\cdot}CHOH{\cdot}CH_2{\cdot}COOH$), are excreted in the urine in severe cases of diabetes.

ketones Organic compounds containing the $C{\cdot}CO{\cdot}C$ group. They include aliphatic, aromatic, cyclic and mixed ketones. The cyclic ketones have the carbon of the carbonyl group as part of a ring. They are formed by the dry distillation of the calcium or barium salts of acids, or by passing the acids over ThO_2 at 400°C. Ketones may be obtained by oxidation of secondary alcohols. Aromatic and mixed ketones are usually prepared by Friedel–Crafts reaction using acyl halides. Aliphatic ketones are usually liquids and aromatic ketones solids, they have ethereal or aromatic odours. They are reduced to secondary alcohols, but are resistant to oxidation and give mixtures of acids and other products when strongly oxidized. They form oximes with hydroxylamine, hydrazones with hydrazine and substituted hydrazines, and semicarbazones with semicarbazide. Isonitrosoketones are formed by the action of sodium nitrite.

ketose A ketose is a sugar containing a potential keto-(CO) group. The presence of the group may be obscured by its inclusion in a ring system. Ketoses are called ketopentoses, ketohexoses, etc., according to the number of carbon atoms they contain.

ketoximes Organic substances containing the group $>C=NOH$. They are formed by treating ketones with hydroxylamine. They are similar in properties to the aldoximes, but undergo the Beckmann rearrangement when treated with sulphuric acid. They are reduced to primary amines.

ketyls, e.g. $K^+(OCPh_2)^-$. Salts of radical anions of ketones, intensely coloured; dimerize readily.

Kevlar An aromatic amide fibre.

kieselguhr *See* diatomite.

kinase Any enzyme catalysing the transfer of phosphate from ATP to the indicated substrate, e.g. hexokinase (transfer to hexose).

kinetics, chemical The branch of chemistry which is concerned with the explanation of observed characteristics of chemical reactions (e.g. the variation of reaction velocities with pressure, temperature or concentration).

kinetic theory of gases The theory in which the properties of gases are derived by applying the laws of probability and of particle dynamics to a system in which the gas is assumed to consist of elastic particles possessing temperature-dependent, random motion.

Kipp's apparatus Equipment for the production of a gas by interaction of a liquid and a solid. It consists of three receptacles, the top is reservoir for the liquid and is connected to the bottom. The middle contains the solid and a tap for the gas. When gas is released the liquid rises and reacts with the solid, when the tap is closed the back pressure returns the liquid to the lower reservoir and reaction ceases. Once widely used for the production of H_2S (HCl and FeS) and CO_2 (HCl and $CaCO_3$).

Kirchhoff's equation If Q is the heat evolved when a process (physical or chemical) is carried out at temperature T, the heat which would be evolved if the same process were carried out at a different temperature can be calculated with the aid of Kirchhoff's equation, namely

$$dQ/dT = C_1 - C_2$$

where C_1 and C_2 are the total heat capacities of the system before and after the process respectively (e.g. reactants and products in a chemical reaction). Kirchhoff's equation is a direct consequence of the first law of thermodynamics.

Kjeldahl method An analytical method for the determination of nitrogen particularly in organic materials. The N is converted to NH_4^+ with conc. H_2SO_4 and catalysts. After neutralization the NH_3 is distilled off and estimated by titration after absorption.

knocking Both spark-ignition and compression-ignition engines may show a reduced power output accompanied by sharp knocking sounds. This is normally a function of both fuel composition and compression ratio. *See* anti-knock additives, knock rating.

knock rating The tendency for gasolines and diesel fuels to produce knocking* in service is designated by octane number* and cetane number.

Knudsen cell *See* effusion.

Kohlrausch equation This equation, which describes the behaviour of strong electrolytes on dilution, states that

$$\Lambda_\infty - \Lambda_v = kC^{\frac{1}{2}}$$

where Λ_∞ is the equivalent conductivity at infinite dilution, Λ_v that at volume v, k is a constant and C is the concentration of electrolyte. The equation is valid only at high dilutions.

Kolbe reaction The preparation of saturated or unsaturated hydrocarbons by the electrolysis of solutions of the alkali salts of aliphatic carboxylic acids. Thus, ethanoic acid gives ethane,

$$2CH_3CO_2^- \longrightarrow CH_3 - CH_3 + 2CO_2$$

and succinic acid gives ethene,

$$\begin{array}{ccc} H_2C-CO_2^- & & CH_2 \\ | & \longrightarrow & || \quad + \; 2CO_2 \\ H_2C-CO_2^- & & CH_2 \end{array}$$

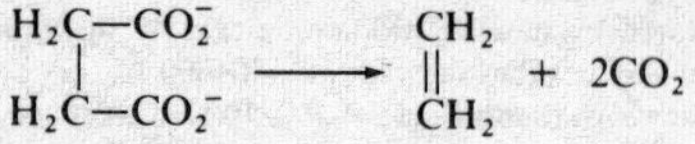

Kr Krypton.

Krebs's cycle *See* citric acid cycle.

Krilium The trade name of a soil conditioner. The solid form of Krilium has as active ingredient a copolymer of about equal molar proportions of vinyl ethanoate and the partial methyl ester of maleic acid. It may be formulated with lime, bentonite, etc. In aqueous form, Krilium contains a copolymer of about equal molar proportions of isobutene and ammonium maleamate. Other polymers are also used.

krypton, Kr, At.no. 36, at.wt. 83·80, m.p. −156·6°C, b.p. −152·3°C. One of the noble gases (% abundance $1·1 \times 10^{-3}$ % in air) obtained pure by fractional distillation of liquid air. Used (with Ar) for fluorescent lights, and in certain photographic flash lights. Electronic configuration $4s^2 4p^6$. Kr forms a limited range of chemical compounds and some clathrates. Linear KrF_2 (Kr plus F_2 in presence of an electrical discharge) forms adducts with Lewis acid fluorides, e.g. $KrF_2 \cdot 2SbF_5$.

kurchatovium Element 104. *See* post-actinide elements.

kurrol salt, $NaPO_3$. A long-chain metaphosphate formed by annealing Graham's salt. *See* phosphates.

kynurenine, 3-anthraniloyl-L-alanine, $C_{10}H_{12}N_2O_3$. An intermediate in the metabolic breakdown of tryptophan. This proceeds mainly through cleavage of the indole ring to give formylkynurenine, which is hydrolysed to kynurenine: this undergoes oxidation to 3-hydroxykynurenine, and side-chain cleavage at the carbonyl group gives 3-hydroxyanthranilic acid. This in part undergoes further ring fission with the formation of acyclic metabolites and of pyridine-carboxylic acids such as quinolinic and nicotinic acid. 3-Hydroxykynurenine is the biogenetic precursor of the ommochromes, eye pigments of insects and crabs.

L

La Lanthanum.

labile complex A complex which participates in very fast reactions, particularly ligand exchange reactions, generally within the time of mixing. E.g. $[Fe(H_2O)_6]^{3+}$, $[Co(H_2O)_n]^{2+}$. *Compare* inert complex.

lability A measure of the ease of replacement of attached groups in complexes and molecules. Cr^{3+}, Co^{3+} complexes are inert, many other complexes are relatively labile.

lactalbumin A protein, belonging to the albumin class, present in milk. It has a mol.wt. of about 17 500, and is very similar in its properties to the albumin of the blood. It contains the essential amino-acid cystine, which is only present in traces in casein, the chief protein of milk.

lactams Amino-acids when heated lose water to form lactams. The water is eliminated between the carboxyl group and the amino-group and a cyclic compound is formed. Also obtained by reduction of the imides of dicarboxylic acids or by isomerization of the oximes of cyclic ketones. Rings containing five, six and eight atoms can be obtained. The lactams are decomposed by heating with alkalis and the amino-acid is formed. They are colourless solids and are extremely poisonous. *See* caprolactam.

β-lactic acid Hydracrylic acid.

lactic acids, 2-hydroxypropanoic acids, $C_3H_6O_3$, $CH_3{\cdot}CHOH{\cdot}COOH$. The acids are colourless syrupy liquids which readily absorb moisture, and are formed by the fermentation of sugars by the *lactobacilli* and some moulds. When slowly distilled they lose water and form lactide. L-Lactic acid, sarcolactic acid, occurs in muscle, being formed by the breakdown of carbohydrate; m.p. 25–26°C. D-Lactic acid; m.p. 18°C, b.p. 122°C/14 mm. Lactic acid is manufactured by the fermentation of pure sugars or of various sugar-containing materials such as starch hydrolysates, and by the hydrolysis of lactonitrile formed by reacting ethanal with HCN. The major use is in the food and beverage industries, where it is used as an acidulant and for the manufacture of a bread additive. It is also used as a chemical intermediate, in textile finishing and in leather tanning.

lactide, $C_6H_8O_4$. Colourless crystals. Prepared by the slow distillation of concentrated solutions of lactic acids. L-Lactide, m.p.

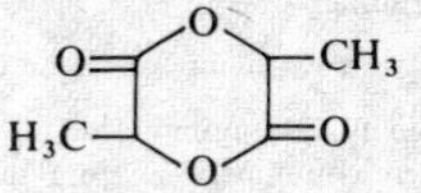

95°C, b.p. 150°C/25 mm. Prepared from L-lactic acid. It is partially converted to lactic acid by water. D-Lactide is similar. DL-Lactide crystallizes in colourless needles, m.p. 124·5°C, b.p. 142°C/8 mm. Obtained from DL-lactic acid.

lactides When α-hydroxy-fatty acids are heated they lose water and form lactides. These contain the group

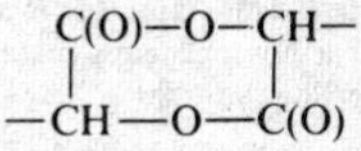

and are formed by the interaction between two molecules of the hydroxy-acid by elimination of water. They are decomposed by heating with water to regenerate the original acid.

lactones Anhydrides formed by intramolecular elimination of water between the hydroxyl and carboxyl groups of hydroxy-acids. Since they are ring compounds containing carbon atoms, the ease with which they are formed depends upon the number of atoms in the ring to be formed. The commonest lactones are those of γ- and δ-hydroxyacids, since these these have rings containing five and six atoms respectively. They are usually formed spontaneously in concentrated solutions of the acids, or by heating such solutions with small amounts of sulphuric acid. They are usually crystalline solids which are partially decomposed by water with regeneration of the hydroxy-acid.

lactose, $C_{12}H_{22}O_{11}$. Milk sugar, first discovered in 1615, occurs in the milk of all animals. Human milk contains 6%, cow's milk, 4%. It is manufactured by the evaporation of whey.

It is glucose-4-β-galactoside. Like glucose it

gives rise to two series of isomeric derivatives and is optically active. Lactose is hydrolysed by the enzyme lactase.

laevorotatory *See* optical activity.

laevulinic acid, levulinic acid, laevulic acid, 4-oxopentanoic acid, $C_5H_8O_3$, $CH_3COCH_2CH_2COOH$. Colourless crystals; m.p. 33–35°C, b.p. 245–246°C. Prepared by heating cane sugar or starch with concentrated hydrochloric acid. Reduced to γ-valerolactone. Also reacts as a ketone. Used in cotton printing.

laevulose An old name for fructose.

lake asphalts Naturally-occurring mixtures of bitumen with mineral and organic matter.

lakes Insoluble pigments obtained by precipitating natural and artificial colouring matters (dyestuffs, dye-wood extracts, cochineal, etc.) on to suitable bases in the presence of Al^{3+}, salts of Mg^{2+}, Zn^{2+}, Sn^{4+} or other metals, and compounds like tannic acid commonly used as mordants in dyeing processes.

lambda point, λ point the temperature of transition between He_I and He_{II}.

Lambert's law This states that layers of equal thickness of a homogeneous material absorb equal proportions of light

$$I = I_0 e^{-Kd},$$

where I is the intensity of the transmitted light, I_0 that of the incident light, d the thickness of the layer and K a constant, characteristic of the substance, known as the absorption coefficient; K also depends on the wavelength of the light employed. When solutions are considered, it is clearly desirable to modify this expression to include the concentration of the absorbing molecules. This modification is embodied in Beer's law*.

lamellar compounds An alternative name for intercalation compounds.

laminarin An extract from brown seaweeds. It is a white powder with the formula $(C_6H_{10}O_5)_n$, consisting of about 20 β-D-glucopyranose units linked through carbon atoms 1 and 3; branch linkages may also be present.

laminates *See* reinforced plastics.

lamp black A soft black carbon pigment obtained by incomplete combustion of natural gas or petroleum (formerly seen in paraffin lamp chimneys). A soft grade of carbon black* used extensively in inks and paints.

Lande *g* factor *See* *g* factor.

langmuir An arbitrary unit used as a measure of adsorption. 10^{-6} torr adsorbant exposed to a surface for 1 second.

Langmuir adsorption isotherm A theoretical equation, derived from the kinetic theory of gases, which relates the amount of gas adsorbed at a plane solid surface to the pressure of gas in equilibrium with the surface. In the derivation it is assumed that the adsorption is restricted to a monolayer at the surface, which is considered to be energetically uniform. It is also assumed that there is no interaction between the adsorbed species. The equation shows that at a gas pressure, p, the fraction, θ, of the surface covered by the adsorbate is given by:

$$\theta = \frac{bp}{1+bp},$$

where b is a constant called the adsorption coefficient, it is the equilibrium constant for the adsorption process.

Although still used the Langmuir equation is only of limited value since in practice surfaces are energetic inhomogeneous and interactions between adsorbed species often occur.

lanoceric acid, $C_{30}H_{60}O_4$. M.p. 104–105°C. A saturated dihydroxy-fatty acid occurring combined in wool fat.

lanolin Crude preparation of cholesterol and its esters obtained from wool fat. It is a pale yellow tenacious substance melting at about 37°C. Lanolin itself is anhydrous lanolin mixed with about 30% of water. In medicine it is used either alone or mixed with soft paraffin, lard or other fat as a base for many ointments, in which lipid-soluble drugs are administered to the skin. It readily emulsifies with water, making it suitable also as a base for creams.

lanosterol, isocholesterol, $C_{30}H_{50}O$. M.p.

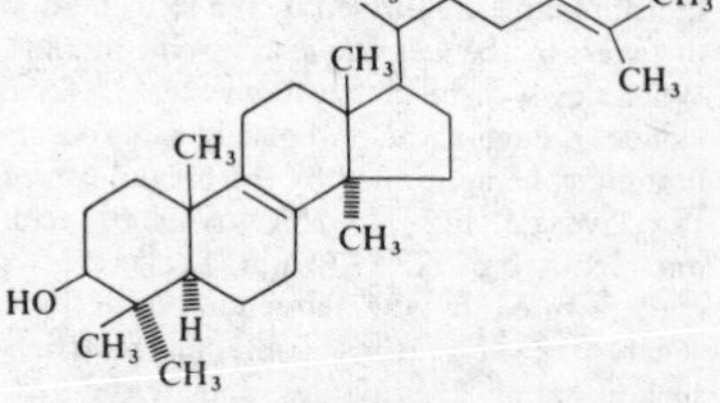

138–139°C. A triterpenoid or trimethylsterol, first found in the non-saponifiable material of wool wax. Lanosterol (4,4,14α-trimethyl-5α-cholesta-8,24-dien-3β-ol) is the precursor in animals and fungi of other sterols such as cholesterol*. It is formed by oxidative cyclization of squalene* followed by methyl group rearrangement.

lanthanides The elements from lanthanum to lutetium, in which the 4f orbitals are being filled. The elements lanthanum, cerium, praseodymium, neodymium, promethium, samarium, europium, gadolinium, terbium, dysprosium, holmium, erbium, thulium, ytterbium and lutetium. Except for Pm the elements are not uncommon – the normal source is monazite which contains 90% La, Ce, Pr, Nd in its lanthanide content. The elements are separated by chromatography or ion-exchange. The metals are prepared by $MF_3 + Ca$ at 1450–1700°C but for Sm, Eu and Yb the reaction $La + M_2O_3$ with volatilization of M is used. The metals are shiny and are attacked by water and acids and burn in oxygen or air to M_2O_3 or MO_2 and react in hydrogen to MH_2 and MH_3.

lanthanide compounds The normal oxidation state is M(III) and hydrides, oxides, hydroxides and halides MX_3 (MF_3 are insoluble in water) are formed by direct reaction of the elements or from solution. Salts of oxy-acids are generally hydrates with high co-ordination numbers for M. O-bonded ligands, particularly chelating ligands, form stable complexes (*see* shift reagents) and some N-bonded and halide complexes are known. Cyclopentadienyls, MCp_3 are stable; alkyls and aryls are relatively unstable. Particular lanthanides Ce, Pr and Tb form M(IV) compounds particularly MF_4 and MO_2, some other lanthanides also form +4 complex fluorides. Particular lanthanides Sm, Eu and Yb form ionic MX_2.

lanthanide contraction There is a regular decrease in ionic and atomic radii across the lanthanides. This means that radii of elements after the lanthanides are very similar to radii of corresponding elements before the series. Thus properties depending on ionic size, e.g. structure, solubility, of elements before and after the lanthanides are very similar, e.g. Mo and W.

lanthanum, La. At.no. 57, at.wt. 138·9055, m.p. 920°C, b.p. 3454°C *d* 6·14. The first lanthanide* (4f) element. The metal is used in steels and in Mg and Al alloys. La_2O_3 is used in glasses, as a glass polish, as a high temperature refractory and as a host matrix for fluorescent phosphors.

lanthanum compounds Lanthanum forms a single series of typical lanthanide compounds* in the +3 state La^{3+} (f° colourless)$\longrightarrow$La (−2·52 volts in acid). $La(NO_3)_3,6H_2O$ is used as an analytical reagent for ethanoate ions. LaI_2 is probably La^{3+}, $2I^-$, e^-.

lapis lazuli Mineral ultramarine*. Used as a semi-precious stone and known from antiquity.

laser A device which produces beams of monochromatic light of very great intensity in which the waves are coherent. The name is derived from Light Amplification by Stimulated Emission of Radiation.

Light emitted from an ordinary source, such as a filament lamp, is not of a single wavelength (non-monochromatic) and neither is it coherently phased, i.e. the wave fronts are out of step with one another. When an atom in its ground state (E_0) absorbs energy (photons) it is excited to a higher energy level (E_1). The energy absorbed may now be spontaneously released either immediately or after some time to yield the ground state of the atom once more. In a laser, however, the excited atom is struck by a photon of exactly the same energy as the one which would be emitted spontaneously. The excited atom is stimulated to emit a photon and return to the ground condition with the result that two photons of precisely the same wavelength are produced. The process can now be repeated throughout the system.

An example of a solid laser is the ruby crystal (Al_2O_3 containing about 0·05% chromium). Glasses containing neodymium are used for high-output lasers; gas lasers using helium–neon mixtures or caesium vapour have also been made. The most efficient lasers are those using semi-conductors, with for example gallium arsenide phosphide as the active crystal.

Due to the very high intensity of the laser beams and their coherent nature they may be used in a variety of ways where controlled energy is required. Lasers are used commercially for excitation with a specific energy, e.g. in Raman spectroscopy or isotope separation.

Lassaigne's test A general test for the presence of nitrogen, halogens or sulphur in an organic compound, a little of which is heated with a pellet of sodium in a hard-glass test tube; the hot tube is plunged into distilled water and the product ground. The presence of nitrogen is shown by the formation of Prussian blue on heating with a solution of iron(II) sulphate containing a trace of Fe(III) and hydrochloric acid. With halogen-containing substances sodium halide is formed, detectable with silver nitrate; while the presence of sulphur can be shown with sodium nitroprusside or lead ethanoate.

latent heat The heat absorbed or evolved during a change in the physical state of a substance. The heat effect in such changes is reversible. The heat absorbed when a solid is converted to liquid at the m.p. is called the latent heat of fusion, that absorbed when a liquid is converted to a gas at the b.p. and 1 atmosphere pressure is termed the latent heat of vaporization. Latent heats are usually measured in joules per mole. Latent heats are entropies of phase changes.

latent image The change occurring in silver halide grains which makes them developable after exposure to light.

latex A stable aqueous dispersion of a polymer. Originally applied to natural rubber latex from the bark of certain trees. Many synthetic rubbers and polymers (e.g. PVC, polyacrylates) are produced as latexes. Latexes may be used for direct manufacture of rubber and plastic goods, by dipping, moulding, spreading, electrodepositing and impregnating (e.g. textiles, adhesives, paints).

lattice The regular three-dimensional array of atoms in a crystal. A lattice is built up from unit cells*. Lattice constants are the unit cell dimensions and angles. The dimensions are generally in Å (Ångstrom) units.

lattice energy The energy required to break down one mole of an ionic lattice into its constituent ions in the gaseous state and infinitely separated from each other, i.e. the enthalpy change associated with the process

$$M^{+}X_{n}^{-}(\text{solid}) \longrightarrow M^{+}(g) + nX^{-}(g)$$

Lattice energies may be derived from the Born–Haber cycle* or calculated using the Kapustinskii equation*.

Laue pattern The symmetrical array of spots obtained on a photographic plate exposed to a non-homogeneous beam of X-rays after its passage through a crystal. The patterns constitute the earliest, although one of the most difficult, methods of investigating crystal structure by means of X-rays.

lauric acid, *n*-dodecylic acid, dodecanoic acid, $C_{12}H_{24}O_2$, $CH_3\cdot[CH_2]_{10}\cdot COOH$. Needles, m.p. 44°C, b.p. 225°C/100 mm. A fatty acid occurring as glycerides in milk, spermaceti, laurel oil, coconut oil, palm oil and other vegetable oils. The metal salts are widely used.

laurite Natural (Ru, Os)S_2; an important source of Ru and Os.

lauryl alcohol, dodecyl alcohol, 1-hydroxydodecane, $C_{12}H_{26}O$, $CH_3\cdot[CH_2]_{10}\cdot CH_2OH$. M.p. 24°C, b.p. 150°C/20 mm. Manufactured by reduction of ethyl laurate or glyceryl trilaurate by hydrogen under pressure in the presence of catalysts or more practically by reaction of C_2H_4 with $AlEt_3$ followed by hydrolysis. Converted to the sulphate for use as detergents, these generally contain higher homologues.

laves phases Alloys, e.g. $MgZn_2$, where structure is determined by size rather than by electronic configurations.

law of constancy of interfacial angles *See* interfacial angles, constancy of.

law of mass action *See* mass action, law of.

law of rationality of intercepts *See* rationality of intercepts, law of.

lawrencium, Lr. At.no. 102. Lawrencium isotopes have been formed by bombardment of Cf targets with ^{10}B and ^{11}B nuclei in a cyclotron or linear accelerator. It is the last element of the actinide (5f) series. It is only known in the +3 oxidation state, no solid compounds are known.

layer lattice Crystal structures in which well-defined layers, either simple or composite, of atoms may be distinguished. The forces between the atoms within the layers are stronger than those holding the layers together, leading to good cleavage parallel to the layers. Examples are graphite, boric acid, cadmium iodide and chloride, and molybdenum sulphide.

LCAO method A method of calculation of molecular orbitals based upon the concept that the molecular orbital can be expressed as a linear combination of the atomic orbitals.

leaching The extraction of a soluble material from an insoluble solid by dissolution in a suitable solvent. The soluble material may be either liquid or solid.

lead, Pb. At.no. 82, at.wt. variable depending upon source, approx. 207·2, m.p. 327·502°C, b.p. 1740°C, *d* 11·34. Lead is the ultimate product of the various radioactive-decay series. Occurs naturally as galena, PbS, and cerussite, $PbCO_3$ which are roasted in air to PbO and reduced to Pb either by the reaction between PbO and PbS (to give Pb plus SO_2) or with carbon or electrolytically. The silvery-white metal is ccp. Lead is used extensively in alloys, in storage batteries (25%), for covering electrical cable, in water and noise proofing, for

$PbEt_4$ (anti-knock agents), radiation and sound shields, in paints, high quality glass. World production 1980 4·0 megatonnes. Lead compounds are toxic and hazardous to health.

lead accumulator An accumulator consists of a number of plates of Pb, each alternate plate is connected to one wire and the remaining plates to another wire. The plates are covered with a mixture of Pb oxides and Pb sulphate, and are placed in dilute sulphuric acid. When a current is passed through the accumulator PbO_2 is formed on one set of plates and Pb on the other, and some H_2SO_4 is produced. When the accumulator is used to generate electricity H_2SO_4 acts on the Pb, the PbO and PbO_2 forming $PbSO_4$ and water to give a potential.

lead acetates *See* lead ethanoates.

lead–acid batteries *See* batteries, lead accumulator.

lead azide, $Pb(N_3)_2$. An initiator ($Pb(O_2CCH_3)_2$ and NaN_3). *See* explosives.

lead bromide, $PbBr_2$. M.p. 373°C, b.p. 916°C. Precipitated from cold water and soluble in hot. No lead(IV) bromide is known.

lead carbonate, $PbCO_3$. Occurs naturally as cerussite ($Pb(O_2CCH_3)_2$ plus $(NH_4)_2CO_3$). Forms basic carbonates, e.g. $Pb_3(OH)_2(CO_3)_2$, white lead, prepared from lead, air, CO_2, steam, ethanoic acid or electrolytically; used in paints.

lead chemistry An electropositive element of Group IV, electronic configuration $6s^26p^2$. Lead shows true cationic character in the +2 state in, e.g, PbF_2, but the +4 state is largely covalent and is quite strongly oxidizing. The +4 state forms series of complexes and some complexes, e.g. chlorides, and oxo-hydroxide complexes are formed by Pb(II). The lone-pair of electrons in Pb(II) has little stereochemical effect. Lead(IV) organic derivatives are stable and compounds with Pb–Pb bonds are formed. PbH_4 (acid plus Mg/Pb) is unstable.

lead chlorides
Lead(II) chloride, $PbCl_2$. Insoluble cold water, sparingly soluble hot. Matlockite is PbFCl. $PbCl_2$ forms many complex chlorides and also basic chlorides. Cassel yellow (approx. $PbCl_2{\cdot}7PbO$) is prepared by fusion of the constituents and is used as a pigment.

Lead(IV) chloride, $PbCl_4$, is a yellow liquid, m.p. −15°C, decomposes explosively at 100°C (H_2SO_4 on $(NH_4)_2PbCl_6$). Yellow hexachloroplumbates(IV), e.g. $(NH_4)_2PbCl_6$, are precipitated from the yellow solution resulting from PbO_2 in cold HCl.

lead chromates $PbCrO_4$ is precipitated from solutions of Pb^{2+} and $CrO_4{}^{2-}$ or $Cr_2O_7{}^{2-}$. In basic solution the precipitate is more orange (chrome yellow, chrome orange). Primrose chrome contains some $PbSO_4$. Cologne yellow is prepared by heating $PbSO_4$ with a $K_2Cr_2O_7$ solution.

lead ethanoates
Lead(II) ethanoate, $Pb(O_2CCH_3)_2{,}3H_2O$, sugar of lead. Soluble in water (ethanoic acid and ethanoic anhydride plus Pb_3O_4).

Lead(IV) ethanoate, $Pb(O_2CCH_3)_4$, (Pb(II)ethanoate plus Cl_2) is a powerful oxidizing agent which will convert vicinal glycols to aldehydes or ketones and 1,2-dicarboxylic acids into alkenes. Primary amides give ketones and amines give nitriles.

lead fluorides
Lead(II) fluoride, PbF_2, rutile structure, precipitated from aqueous solution.

Lead(IV) fluoride, PbF_4, strong fluorinating agent (PbF_2 plus F_2). Forms complexes containing the $[PbF_6]^{2-}$ ion.

lead hydroxides *See* lead oxides.

lead iodide, PbI_2. Golden-yellow, precipitated from solution. Forms complex iodides. No lead(IV) iodide is known.

lead nitrate, $Pb(NO_3)_2$. Colourless crystals (PbO plus HNO_3). Forms basic nitrates. Used in calico printing, as a mordant in dyeing and in the manufacture of chrome yellow pigment.

lead, organic derivatives Derivatives of lead(IV) made generally by alkylation of Pb(II) compounds e.g. R_3PbX and other derivatives are known. Tetramethyl lead, Me_4Pb, and tetraethyl lead, Et_4Pb, TEL, are extremely important anti-knock additives for petroleum (gasoline) prepared Na/Pb plus RCl; electrolysis of $Na(AlEt_4)$ with a lead anode and Hg cathode; electrolysis of RMgX with lead anodes. Organolead compounds are toxic.

lead oxides
Lead(II) oxide, PbO, exists in two forms as orange-red litharge and yellow massicot. Made by oxidation of Pb followed by rapid cooling (to avoid formation of Pb_3O_4). Used in accumulators and also in ceramics, pigments and insecticides. A normal hydroxide is not known but hydrolysis of lead(II) oxyacid salts gives polymeric cationic species, e.g. $[Pb_6O(OH)_6]^{4+}$ and plumbates are formed with excess base.

Trilead tetroxide, Pb_3O_4, red lead and other intermediate phases, e.g. Pb_7O_{11} and Pb_2O_3 are formed by heating PbO_n or PbO_2. Uses are as for PbO but extensively used in corrosion resistant paints.

Lead(IV) oxide, PbO_2. Chocolate brown (electrolytic oxidation of Pb(II) salts). Used as an oxidizing agent.

Plumbates(IV), e.g. K_2PbO_3, are formed by solid state reactions and plumbates containing $[Pb(OH)_6]^{2-}$ ions are formed from aqueous solution.

lead silicates   Silicates with low lead concentrations are used in glazing pottery and glass manufacture.

lead sulphate, $PbSO_4$.   Occurs naturally as anglesite, precipitated from Pb(II) solutions. Basic salts are known and are used as white pigments.

lead sulphide, PbS.   Black compound occurring naturally as galena, precipitated from aqueous solution. Used as rectifier.

lead tetraacetate   *See* lead ethanoates.

lead tetraethyl, TEL, Et_4Pb. *See* lead, organic derivatives.

lead tetramethyl, Me_4Pb. *See* lead, organic derivatives.

lean gas   A fuel gas defined as having a calorific value* of 100–450 Btu/ft³ (3·7–16·8 mJ/m³). Obtained from liquid and solid hydrocarbons by steam-refining in the presence of a catalyst at about 1075 K to yield mixtures of CO and H_2O or by partial oxidation at 1350–1650 K to give a mixture of CO, H_2 and CO_2 plus N_2 if air is the oxidant. *See* Lurgi process, producer gas, blue water gas.

leather   Animal and other skins treated by tanning*. Has a characteristic porous composition. Polymers which fairly closely simulate late leather are now available.

Leblanc process   Virtually obsolete process for Na_2CO_3. NaCl with H_2SO_4 gives Na_2SO_4 (HCl gas by-product); $[SO_4]^{2+}$ plus coke gives Na_2S which with $CaCO_3$ gives Na_2CO_3 plus CaS.

Le Chatelier's theorem or principle   When a constraint is applied to a system in equilibrium, the equilibrium will tend to move in such a way as to neutralize the effect of the constraint. Thus, e.g., in a reaction such as

$$N_2O_4 \rightleftharpoons 2NO_2$$

increasing the total pressure will cause the equilibrium to move towards the formation of more N_2O_4 thereby decreasing the total number of molecules in the system.

lecithin   A generic name for substances of the

$$
\begin{array}{l}
CH_2{-}OCOR^1 \\
| \\
CH{-}OCOR^2 \\
\quad\quad\quad O \\
| \quad\quad\quad \| \\
CH_2O{\cdot}P{-}OCH_2CH_2\overset{+}{N}(CH_3)_3 \\
\quad\quad\quad \| \\
\quad\quad\quad O
\end{array}
$$

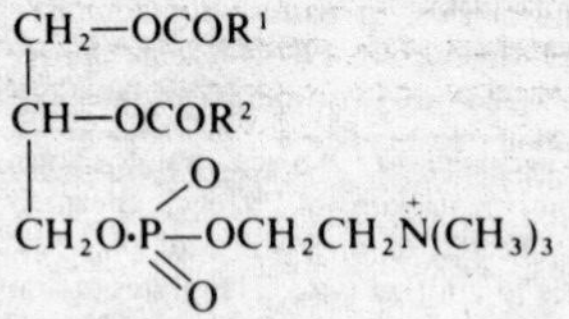

type shown, where R^1 and R^2 are fatty acid residues. Usually one acid is saturated and one unsaturated. Lecithins are found in every animal and vegetable cell and are an essential constituent of cells especially membranes. They swell up with water to give slimy emulsions or colloidal solutions. They melt at about 60°C.

Commercial lecithin is a mixture of phosphatides and glycerides obtained in the manufacture of soya bean oil. It gives a thick yellow emulsion with water, and is widely used in the food and other industries.

See also phospholipids.

Leclanché cell   *See* batteries.

LEED   *See* low energy electron diffraction.

lel   *See* chelate effect.

lepargylic acid   *See* azeleic acid.

leucine, α-aminoisocaproic acid, $C_6H_{13}NO_2$, $Me_2CHCH_2CH(NH_2)COOH$.   Colourless plates, m.p. 293–295°C (decomp.). Optically active. Leucine is one of the most common of the amino-acids obtained from proteins. It can be obtained by fractional crystallization of hydrolysed proteins.

Leuckart reaction   The conversion of ketones and aromatic aldehydes to primary amines by reaction with ammonium methanoate at a high temperature.

leuco compounds   Colourless products from dyestuffs. E.g. reduction of triphenylmethane dyes containing $[Ar_3C]^+$ ions gives colourless Ar_3CH. Leuco bases, carbinols, e.g. Ar_3COH, and other derivatives are known.

Lewis acid   *See* acid.

Lewis base   *See* base.

lewisite, 2-chlorovinyldichloroarsine, 1-chloro-2-dichloroarsino-ethene, $C_2H_2AsCl_3$, $ClCH{:}CH{\cdot}AsCl_2$.   A pale yellow liquid, m.p. $-13°C$, b.p. 190°C. It has a strong smell, resembling that of geraniums. War gas, systemic poison. It is hydrolysed by water and also destroyed by alkalis and by oxidizing agents. It can be manufactured by bubbling ethyne through a mixture of anhydrous arsenic trichloride and aluminium chloride.

Li Lithium.

libration In a crystalline lattice the rotational motions of molecules within the lattice may be restricted due to bonding with neighbouring atoms. The resulting rotational oscillations of the whole molecule are called librational modes.

licanic acid, $C_{18}H_{28}O_3$. White crystals.

$$CH_3{\cdot}[CH_2]_3{\cdot}[CH{:}CH]_3{\cdot}[CH_2]_4{\cdot}CO{\cdot}[CH_2]_2{\cdot}COOH$$

The naturally occurring isomer, α-licanic acid, has m.p. 74–75°C. It is unstable, and is readily isomerized to the β-form, m.p. 99·5°C. It occurs in oiticica and other oils which were previously used in the protective-coating industry.

Liebermann's reaction A colour-test for $-NO$ or $-OH$ groups. For the detection of an $-NO$ group a small quantity of the substance is dissolved in concentrated sulphuric acid and a crystal of phenol added. A blue-green colour develops on warming, this changes to red on pouring into water, and back to blue with excess of alkali. When the test is used for detecting a phenol, the substance and a crystal of sodium nitrite are dissolved in sulphuric acid and warmed. On dilution and addition of alkali many phenols give distinctive colours.

lift-off In gas burners having a stationary flame it is essential that the flame speed* of the combustible mixture should be balanced by the velocity of the mixture in the burner tube or flame ports. If the flame speed is too low or the mixture velocity too high the flame can lift-off the burner mouth or flame ports.

ligand A complexing group in co-ordination chemistry. Generally the entity from which electrons are donated. NH_3 is a ligand in $[Co(NH_3)_6]Cl_3$.

ligand field theory An extension of crystal-field theory.

light–heavy selectivity In solvent extraction the phenomenon in which the extractive power of the solvent is dependent on the molecular weight of the component extracted.

light scattering The scattering of light by suspended particles. Can be used to estimate molecular weights of macromolecules and to detect end-points when a precipitate is formed in the reaction.

light stabilizers *See* ultra-violet absorbers.

lignans A group of natural products obtained by ethereal or alcoholic extraction from the wood or exuded resin of the *Coniferae* and other plants, and characterized by the presence in the molecule of the 2,3-dibenzylbutane skeleton.

lignin A highly polymeric substance occurring with cellulose in lignified plant tissues. It is largely responsible for the strength of wood, which contains 25–30%, and it is extractable from wood pulp by the action of SO_2 and lime-water: it occurs (up to 6%) in sulphite waste liquors from paper mills. Lignin is a phenyl propene polymer of variable mol.wt. It is of commercial value as a source of vanillin, phenols and other aromatic chemicals. Used as a filler for plastics and as a dispersant, emulsifier, etc.

lignite, brown coal Immature coals intermediate in composition between peat and bituminous coals. They occur in thick seams, often quite near the earth's surface.

lignoceric acid, $C_{24}H_{48}O_2$. M.p. 84°C. Fatty acid present free and combined in many oils, fats and waxes but principally tall oil*.

ligroin Essentially petroleum ether; strictly that fraction of refined naphtha with b.p. 80–130°C. Contains aliphatic hydrocarbons and is used as a solvent.

lime *See* calcium hydroxide and calcium oxide.

limestone Rocks of sedimentary origin containing the remains of marine organisms or chemically precipitated or transported calcium carbonate. The pure mineral consists of $CaCO_3$ in the form of calcite, but is rarely found. Commercial limestone contains iron oxide, alumina, magnesia, silica and sulphur, with a CaO content of 22–56% and a MgO content of up to 21%. It is used as such as a fertilizer and for many other purposes, or is calcined to calcium oxide.

limewater *See* calcium hydroxide. Used as a mild antacid.

limiting density (of a gas) If a mass w g of a gas occupies v litres at 0° and pressure p atmospheres, the quotient w/pv is the density of the gas per unit pressure. Since gases do not obey Boyle's law accurately, w/pv varies with the pressure. The value of this quotient when p approaches zero is called the limiting density of the gas. It represents the density which the gas would possess as an ideal gas, i.e. a gas which obeys the gas laws perfectly.

limonene, citrene, carvene, $C_{10}H_{10}$. B.p. 176°C. An optically active terpene. Its racemic

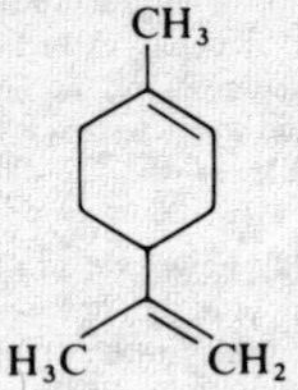

form is known as depentene. Very widely distributed.

linalool, $C_{10}H_{18}O$. B.p. 198–199°C. Present

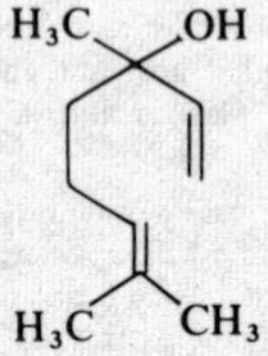

in many essential oils. It is readily converted into geraniol by acid reagents.

linamarin, acetonecyanhydrin, β-D-glucopyranoside, $C_6H_{11}O_5{\cdot}O{\cdot}C(CH_3)_2{\cdot}CN$. M.p. 143–145°C. Present in young flax plants and in *Phaseolus lunatus* and is also the glycoside of rubber seeds. Such glycosides may be regarded as primary materials for synthesis.

lindane The gamma isomer of BHC*, HCH*.

Linde process A high-pressure process for the production of liquid oxygen and nitrogen by compression to about 200 bar (20 MN/m²) followed by refrigeration and fractionation in a double column.

line spectra Absorption of energy can only occur between allowed energy levels. Absorption of energy by an atom can only lead to specific changes to excited states (e.g. with transfer of electrons from one energy level to others). The reverse of this process is generally accompanied by emission of energy of a single definite frequency corresponding to the energy involved in the process. The spectrum of an atom thus consists of a series of discrete lines, each line at a frequency corresponding to the appropriate electronic transition. The line spectrum is characteristic of an element in its atomic state.

linkage Bond*.

linoleic acid, *cis,cis*-9,12-octadecadienoic acid,

$C_{18}H_{32}O_2$. A fatty acid which is easily oxidized in air. It occurs widely, in the form of glycerides, in vegetable oils and in mammalian lipids. Cholesteryl linoleate is an important constituent of blood. The acid also occurs in lecithins. Together with arachidonic acid it is the most important essential fatty acid* of human diet.

linolenic acid, *cis,cis,cis*-9,12,15-octadecatrienoic acid, $C_{18}H_{30}O_2$. A fatty acid which

occurs in glycerides of linseed oil, amongst other oils. It also occurs widely in lecithin and serum triglycerides. γ-Linolenic acid, *cis,cis,-cis*-6,9,12-octadecatrienoic acid has been isolated from the seed oil of *Oenothera biennis*. Both acids can act as essential fatty acids*.

linseed oil Linolenic acid – containing oil of vegetable origin used as a drying oil, in alkyd resins and in margarine.

lipases Enzymes which act upon the fatty acid esters of glycerol (triglycerides, diglycerides and monoglycerides). They are therefore a class of hydrolases. *See also* phospholipases.

lipids These are natural substances of a fat-like nature. The exact definition, however, varies. Strictly they are fatty acids or derivatives, which are soluble in organic solvents and insoluble in water, e.g. the simple fats and waxes, and the phospholipids and cerebrosides. Many people would also consider such compounds as sterols and squalene to be lipids.

lipoproteins *See* proteins.

liquid crystals On heating a number of substances, e.g. *p*-azoxyanisole and ammonium oleate, a cloudy liquid is first formed which changes at higher temperature to a clear liquid, each transition occurring at a fixed temperature. The cloudy liquid has definite ordered structure and it is called a liquid crystal, or more generally, is said to be in a mesamorphic state; its viscosity, etc., are those of liquids. Substances which form liquid crystals are composed of molecules which possess a high degree of asymmetry – long thin molecules or flat planar ones – which tends to allow little rotation in the liquid state at low temperature. Liquid crystals exhibiting two-dimensional order are said to be in the cholesteric or smetic phases; those showing only one-dimensional order are nematic. Application of a voltage causes disorder; the basis of the extensive use of liquid crystals in display units. Cholesteric liquids scatter coloured, polarized light and can be made the basis of thermometer units. Spectroscopic studies in the presence of liquid crystals give much additional information over more simple techniques.

liquid–liquid extraction A method of extracting a desired component from a liquid mixture by bringing the solution into contact with a second liquid, the *solvent*, in which the component is also soluble, and which is immiscible with the first liquid or nearly so. Some of the component enters the solvent, forming an *extract*, while the solution that is left, the *raffinate*, is depleted in the component by this amount.

liquid oxygen explosives, LOX explosives Explosives based on liquid oxygen and a fuel, generally carbon black.

liquid paraffin A highly refined white oil used medicinally as a powerful laxative.

liquids, structure of The structure of a liquid is intermediate between the perfect disorder of molecules in a gas and the highly perfect three-dimensional order in a solid. The immediate environment of an atom or molecule in a liquid is not very different from that in a solid and the density of packing is only slightly lower, but the order is very local and no long-range order occurs.

liquidus curve The freezing point of a molten mixture of substances varies with the composition of the mixture. If the freezing points are plotted as a function of the composition, the line joining the points is called a liquidus curve. Such mixtures usually freeze over a range of temperature. If the temperature at which the last traces of liquid just solidify (assuming that sufficient time has been allowed for equilibrium to be established) are plotted against composition the resulting line is called a solidus curve.

liquified natural gas (LNG) Natural gas liquified at atmospheric pressure by cooling to about −160°C, normally for purposes of transportation.

liquified petroleum gas (LPG) Mixtures of C_3 and C_4 hydrocarbons obtained from petroleum refining operations. Commercially two mixtures are available; *commercial butane* containing about 85% butanes with some propane and pentanes and *commercial propane* containing about 92% propane along with some ethane and butanes.

These fuels are used for domestic and industrial purposes; being stored and transported as liquids in thick-walled spheres or cylinders. Liquefaction is obtained by compression or refrigeration.

Gross calorific values are high, being about 100 000 kJ/m^3.

lithia, Li_2O. *See* lithium oxide.

litharge, PbO. *See* lead oxide.

lithium, Li. At.no. 3, at.wt. 6·941, m.p. 180·54°C, b.p. 1347°C, *d* 0·531. The lightest alkali metal. The most important Li minerals are lepidolite, spodumene, petalite and amblygonite (all aluminosilicates). The metal is obtained by electrolysis of the fused chloride. It is a silver-white metal, bcc, which tarnishes rapidly in air and reacts with water, halogens, nitrogen, hydrogen. Lithium is used in low melting alloys, as a heat transfer medium. Lithium compounds are used extensively in organic synthesis (organolithiums), in ceramics and glasses (Li_2CO_3), in fluxes (LiF), in lubricating greases (LiOH), in air conditioning plant (LiCl and LiBr), in fungicides, foodstuffs, storage batteries and in Al production.

lithium aluminium hydride, lithium tetrahydroaluminate, $LiAlH_4$. A specific reducing agent widely used in organic chemistry and as an agent for the formation of aluminium derivatives and hydrides (e.g. of B, Si). *See* aluminium hydride.

lithium aluminium tri-*t*-butoxyhydride, $LiAl(OBu)_3H$. A selective reducing agent for converting RCOCl to aldehydes and reducing steroid ketones. *See* aluminium hydride.

lithium carbonate, Li_2CO_3. Sparingly soluble in water (only alkali carbonate to be so); forms soluble $Li(HCO_3)_2$ with excess CO_2, loses CO_2 on strong heating. Used in the manufacture of other lithium compounds.

lithium chemistry Lithium is an alkali metal, electronic configuration $1s^22s^1$ forming a single series of Li(I) compounds ($E°$ $Li^+ \longrightarrow Li$ −3·04 volts in acid). Li is exceptional amongst the alkali metals in forming an insoluble carbonate and fluoride (similar to Be^{2+}). Li compounds are predominantly ionic but covalent, and polymeric derivatives, particularly lithium alkyls, are known.

lithium chloride, LiCl. Very hygroscopic (Li_2CO_3 plus HCl) forms hydrates. Used in welding aluminium.

lithium fluoride, LiF. Relatively insoluble in water, precipitated from solution.

lithium hydride, LiH. Formed from Li plus

H_2 at 500°C. Stable crystalline compound, gives LiOH and H_2 with water (may ignite in moist atmosphere). Used as a source of H_2, as a reducing agent, and for the preparation of hydrides.

lithium hydroxide, $LiOH,H_2O$. Strong base (Li salt or ore plus $Ca(OH)_2$). Used in greases, storage batteries and absorbing CO_2.

lithium organic derivatives Of great use in industrial and laboratory synthesis. Prepared RCl in benzene or petroleum ether plus Li and also from R′Li on RH; R′Li plus RBr; Li plus R_2Hg. Solutions, of e.g. Bu^nLi are commercially available. Often polymeric, e.g. $(LiMe)_4$. With chelating amines, e.g. $Me_2NCH_2CH_2NMe_2$ form monomeric derivatives which are very active. In addition to synthesis, lithium organic derivatives are used in alkene-polymerization.

lithium oxide, lithia, Li_2O. White solid (Li plus O_2), gives LiOH with water.

lithium tetrahydroaluminate, $LiAlH_4$. *See* aluminium hydride.

lithography The mechanical printing process involving the use of a photographically or otherwise prepared printing plate, on which the image areas are oleophilic (non-aqueous, ink receptive) and the non-image areas are hydrophilic (water receptive, ink repellant). The plate is generally made of grained Al or Zn, on which is coated the light-sensitive material.

lithopone, Charlton white, Orr's white A mixture of $BaSO_4$ (70%) and ZnS (30%) prepared from BaS and $ZnSO_4$ solution and used after calcination as a pigment.

litmus A natural colouring matter obtained from lichens after oxidation in the presence of NH_3. Used as indicator (red acid, blue alkaline) largely in the form of litmus paper.

LNG Liquefied natural gas*.

lone pair A pair of electrons in a molecule which is not shared by two of the constituent atoms, i.e. does not take part in the direct bonding. Lone pairs can generally form coordinate bonds. They influence the stereochemistry of the molecule. Thus CH_4, NH_3 (one lone pair) and OH_2 (two lone pairs) all have structures based on a tetrahedral distribution of bonding and lone pairs about the central atom.

Loschmidt's number The number of molecules of an ideal gas in unit volume at stp; equal to $2{\cdot}687 \times 10^{19}$ per cm^3.

low energy electron diffraction (LEED) A technique used to investigate the two-dimensional surface structure of solids. The surface under investigation is bombarded with electrons of low energy, typically between 6 and 600 V, and the electrons diffracted by the surface atoms are collected on a fluorescent screen. Using this method it is possible to examine both surface structure and rearrangements which may occur following chemisorption and surface reaction.

low-spin state *See* high-spin state. The minimum number of unpaired electrons.

low temperature baths These are used for freezing and distilling gases. Liquid N_2 (−196°C), liquid O_2 (−183°C – very hazardous with any organic material), solid CO_2 (−78·5°C) are used together with various slush baths (liquid plus liquid N_2 or CO_2) −160°C isopentane; −140°C, 30–60 petroleum ether; −130°C pentane; −126°C methylcyclohexane; −116°C diethyl ether; −111°C CCl_3F; −95°C toluene; −60°C $CHCl{=}CCl_2$; −45°C C_6H_5Cl; −23°C CCl_4.

low temperature carbonization (LTC) *See* carbonization.

LOX explosives *See* liquid oxygen explosives.

LPG *See* liquefied petroleum gas.

LS coupling Russell–Saunders coupling*.

LSD Lysergic acid diethylamide. An hallucinogen.

LTC Low temperature carbonization*.

Lu Lutetium.

lubricating greases Solid or semi-fluid lubricants made by the incorporation of a thickening agent in a liquid lubricant. This structure gives resistance to shear forces and to temperature.

The base lubricant is usually a petroleum oil while the thickener usually consists of a soap or soap mixture. In addition they may contain small amounts of free alkali, free fatty acid, glycerine, anti-oxidant, extreme-pressure agent, graphite or molybdenum disulphide.

Greases may contain soaps of lithium, calcium, sodium, aluminium, etc., or they may be non-soap greases.

Non-soap greases using finely divided solids as thickeners are useful as lubricants at elevated temperatures. Materials used include organo-clays such as dimethyldioctyl-decylammonium bentonite (Bentone greases) or selected dyestuffs which produce brightly coloured greases.

lubricating oil Any liquid used for the purposes of lubrication. They consist mainly of mineral oils derived from crude petroleum although some vegetable oils are used in 'compounded' oils for special purposes. More recently synthetic fluids have been used, particularly for aviation turbines.

The main synthetic fluids used as special lubricants are esters, polyglycols, silicones, halogenated hydrocarbons and polyphenyl ethers.

luciferins Substances responsible for bioluminescence. Enzymatic oxidation is responsible for the characteristic luminescence of the firefly. The oxidation product is formed in an excited state which liberates excess energy as light rather than heat.

luminol, 5-aminophthalhydrazide Yellow crystals, m.p. 329–332°C. Treatment of this material with a number of oxidizing agents in the presence of alkali causes it to emit a blue light (chemiluminescence).

luminous paints Compounded from solid materials (such as calcium sulphide) which, by use of suitable methods of preparation and by addition of traces of heavy metals, can be made to exhibit phosphorescence.

LUMO *See* frontier orbital symmetry.

Lurgi coal gasification process A process involving the gasification of coal under pressure with mixtures of steam and oxygen. The reactions produce CO and H_2.

lutein, luteol, xanthophyll, $C_{40}H_{56}O_2$. It has the normal carotenoid structure (*see* carotene), with the rings:

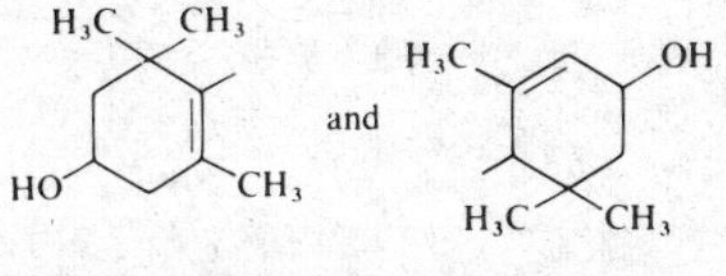

The most common of the xanthophyll pigments, it is present in all green leaves, in blossoms and in various animal sources. It crystallizes in violet prisms with one molecule of methanol; m.p. 193°C, soluble in organic solvents giving yellow solutions. It is related to α-carotene in the same way as zeaxanthin is to β-carotene.

lutetium, Lu. At.no. 71, at.wt. 174·967, m.p. 1656°C, b.p. 3315°C, *d* 9·84. The last of the lanthanide*(4f) elements. The metal is hcp. After neutron irradiation Lu nuclides emit pure β radiation which can be used in cracking, polymerization and other catalytic processes.

lutetium compounds Lutetium forms a single series of typical lanthanide compounds* in the +3 state

Lu^{3+} (f^{14} colourless) ⟶ Lu (−2·25 volts in acid).

lutidine, C_7H_9N. The lutidines are dimethylpyridines. The best known is 2,6-lutidine, which is isolated from the basic fraction of coal tar. It is an oily liquid, b.p. 144°C.

lycopene, $C_{40}H_{56}$. The red carotenoid pigment of tomatoes, rose hips and many other berries. Crystallizes from petroleum ether, m.p. 175°C.

Lyman series *See* Balmer series.

lyogels *See* xerogels.

lyophilic Colloidal substances in which the dispersed particles display a marked stability due to solvation are said to be lyophilic. *See* colloids.

lyophobic Colloidal substances in which the dispersed particles are unsolvated and hence easily coagulated. *See* colloids.

lyotropic series Although hydrophilic sols are not affected by small concentrations of electrolytes, they may be 'salted out' by the addition of certain salts which possess strong dehydrating properties. Citrates, tartrates and sulphates are very efficient in this connection; iodides and thiocyanates tend to disperse rather than coagulate. Thus silk or cellulose will peptize to colloidal sols in strong thiocyanate solutions but are reprecipitated by the addition of sulphates to the sols. The arrangement of the different anions in order of their salting-out efficiency is termed the lyotropic or Hofmeister series. This also has a connection with the coagulating power of salts on hydrophobic sols.

lysergic acid, $C_{16}H_{16}N_2O_2$. M.p. 238°C (de-

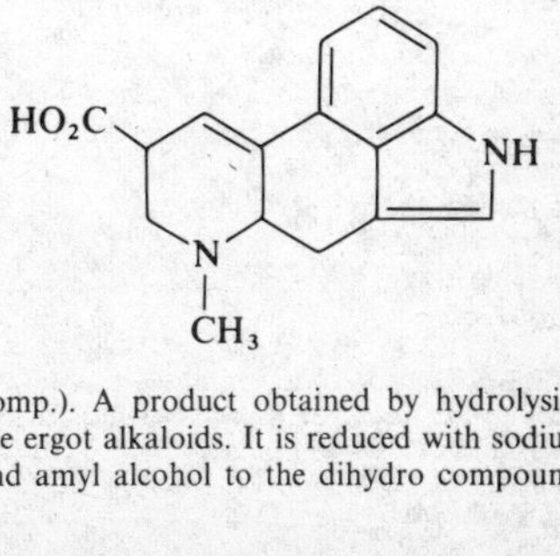

comp.). A product obtained by hydrolysing the ergot alkaloids. It is reduced with sodium and amyl alcohol to the dihydro compound.

The diethylamide is a strong hallucinogen.

lysine, α,ε-diaminocaproic acid, $C_6H_{14}N_2O_2$, $H_2N{\cdot}CH_2{\cdot}[CH_2]_3{\cdot}CH(NH_2){\cdot}COOH$. Colourless needles, m.p. 224°C (decomp.), very soluble in water, insoluble in alcohol. L-(+)-Lysine is one of the basic amino-acids occurring in particularly large quantities in the protamine and histone classes of proteins. It is an essential amino-acid, which cannot be synthesized by the body and must be present in the food for proper growth. It can be manufactured by various fermentation processes or by synthesis.

lysol A soapy solution containing 50% isomeric cresols, used as a disinfectant.

lysosomes Particles, occurring in cells, which contain hydrolytic enzymes. These enzymes would destroy many cell constituents if not kept separate. Damage to the lysosomal membrane can lead to release of those enzymes which attack the rest of the cell, as occurs in dying or damaged tissue.

lysozyme An enzyme which occurs especially in nasal mucosa, egg white, plant latex and various animal tissues. It has the ability to dissolve certain bacteria.

M

Macleod's equation

$$\gamma = K(D-d)^4$$

where γ is the surface tension, D the density of the liquid and d the density of the vapour.

macrolides Compounds with antibiotic activity made by *Streptomyces* spp. and characterized by having a macrocyclic ring. Many macrolides contain one or more unusual sugars in the molecule, including an amino sugar, usually desosamine. Others contain polyene chains. The macrolides, which include erythromycin*, are most active against gram-positive bacteria.

macromolecule Molecules of high molecular weight (usually greater than 10 000). Examples of this type of molecule are proteins and some polymers. The term macromolecular is also used to describe structures such as that of diamond, which may be regarded as being one large molecule.

Maddrell salt, $NaPO_3$. A long chain phosphate*obtained by heating $Na_2H_2P_4O_7$ to 230–300°C.

Madelung constant For an ionic crystal composed of cations and anions of respective change z_+ and z_-, the lattice energy U_0 may be derived as the balance between the coulombic attractive and repulsive forces. This approach yields the Born–Landé equation,

$$U_0 = \frac{LMz_+z_-e^2}{r_0}\left(1-\frac{1}{n}\right),$$

where L is Avogadro's number; r_0, the equilibrium internuclear distance; e, the electronic charge; n, an integer and M the Madelung constant.

Values of M depend upon the structure varying between 1·763 (CsCl) and 2·519 (fluorite structure).

magnesia Magnesium oxide.

magnesite, $MgCO_3$. A mineral form of magnesium carbonate used as a source of magnesium compounds and in the manufacture of refractories.

magnesium, Mg. At.no. 12, at.wt. 24·305, m.p. 648·8°C, b.p. 1090°C, *d* 1·74. Alkaline earth metal. The principal useful ores are dolomite $(Ca,Mg)CO_3$, carnallite $(KMgCl_3,6H_2O)$, kainite $(MgSO_4{\cdot}3Na_2SO_4)$, magnesite $(MgCO_3)$, schönite $(K_2Mg(SO_4)_2,6H_2O)$. Obtained from one of the above ores or from sea-water (*see* Dow process) or brines (precipitated as $Mg(OH)_2$). The metal is obtained by electrolysis of an $MgCl_2$-halide melt or by reduction of fused dolomite with ferrosilicon. The metal is silver-white, very light, hcp. Burns in air after ignition, reacts slowly with moist air to become covered with a film of oxide. Used in alloys (43%) and castings (20%), for deoxidizing and desulphurizing metals, in batteries and sacrificial corrosion. Used in glass, ceramics, fillers ($MgCO_3$, $MgCl_2$) in flocculating agents, catalysts, medicinal uses, sugar refining ($Mg(OH)_2$), refractories, cement, paper manufacture, (MgO and mixed magnesium oxides), in tanning, sizing, etc. ($MgSO_4$). World production Mg 1976 240 000 tonnes.

magnesium alloys Used on account of their light weight (aircraft industry) and for casting, forging, extrusion, etc.

magnesium carbonate, $MgCO_3$. Occurs naturally as magnesite and in dolomite. Addition of CO_3^{2-} to Mg^{2+} solutions gives basic carbonates which dissolve in excess CO_2 to give $Mg(HCO_3)_2$. Heating the solution to 50°C gives crystals of $MgCO_3$, $3H_2O$; the 5 and 1 hydrates are also known. Basic carbonates of approximate composition $3MgCO_3$, $Mg(OH)_2,4H_2O$ are used medicinally as antacids and mild laxatives.

magnesium chemistry Magnesium is an alkaline earth element, electronic configuration $3s^2$. It shows a single +2 oxidation state (E° $Mg^{2+} \longrightarrow Mg$ − 2·38 volts in acid solution). Compounds are largely ionic although there is major covalent character in Grignard reagents and in some other derivatives. Mg compounds generally have 6-co-ordinate Mg; $[Mg(H_2O)_6]^{2+}$ is not hydrolysed. Many complexes, particularly with O-and N-ligands are known. Mg complexes are of great importance in photosynthesis in plants (chlorophyll contains a MgN_4 group but there is additional interaction with other O atoms).

magnesium chloride, $MgCl_2$. Obtained from carnallite, KCl, $MgCl_2,6H_2O$ by fusion when the KCl is precipitated; cooling gives carnallite and $MgCl_2,6H_2O$. Dehydrated by heating in HCl gas. $MgCl_2$ has m.p. 708°C, b.p. 1412°C. The hygroscopic hexahydrate is used in moistening cotton threads in spinning, oxide chlorides (heat on hydrates) are used in cements. Fused $MgCl_2$ is electrolysed commercially to give Mg and Cl_2. $MgCl_2$ forms complexes including anionic chlorocomplexes, e.g. $(Et_4N)_2[MgCl_4]$.

magnesium halides Magnesium fluoride, MgF_2, is precipitated from aqueous solution. It has the rutile structure. Perovskites, e.g. $KMgF_3$, are known. The other halides are similar to the chloride* although complexes are not known. They have layer lattices.

magnesium hydroxide, $Mg(OH)_2$. Formed by hydration of MgO or precipitation from an Mg^{2+} solution with OH^-. Only slightly soluble in water to give an alkaline solution. Occurs naturally as brucite, and $Mg(OH)_2$ layers can occur in silicates – the chondrodites.

magnesium nitrate, $Mg(NO_3)_2$. Stable salt forms 9, 6 and 2 hydrates. Hydrates dehydrate to basic salts but $Mg(NO_3)_2$ can be obtained by heating hydrates in nitric acid vapour.

magnesium nitride, Mg_3N_2. Formed from elements above 300°C (used to remove N_2) hydrolysed to NH_3 and $Mg(OH)_2$.

magnesium, organic derivatives, RMgX, R_2Mg. *See* Grignard reagents.

magnesium oxide, magnesia, MgO. Occurs naturally as periclase. Prepared (m.p. 2640°C) by combustion of Mg in O_2 or ignition of $Mg(OH)_2$, $MgCO_3$ or $Mg(NO_3)_2$. Has NaCl structure. Hydrated (rate depends on whether previously ignited) to $Mg(OH)_2$. Used as a refractory, for heat insulation and as an antacid.

magnesium perchlorate, $Mg(ClO_4)_2,6H_2O$. Formed from solution. The anhydrous material (anhydrone) is used for drying gases.

magnesium phosphates, $Mg_3(PO_4)_2,8H_2O$. Precipitated from aqueous solutions. $MgNH_4PO_4$ is precipitated from solutions containing ammonium salts; used in estimation of Mg^{2+} and $PO_4{}^{3-}$.

magnesium silicates Many silicates contain Mg^{2+} Used medicinally to remove stomach acidity and (after hydrolysis to active SiO_2) to remove toxins.

magnesium sulphate, $MgSO_4$. Obtained (with $7H_2O$ – Epsom salts) from $MgCO_3$ and H_2SO_4; occurs naturally as kieserite or reichardite, $MgSO_4,H_2O$. Hydrates give $MgSO_4$ at 200°C. Epsom salts are used as a purgative, as a dressing for cotton goods and in dyeing.

magneson, *p*-nitrobenzeneazoresorcinol,

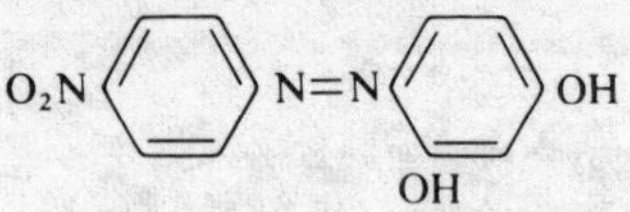

$C_{12}H_9N_3O_4$. A brownish-red powder, soluble in sodium hydroxide. Used for the detection and estimation of magnesium, with which it forms a blue lake in alkaline solutions.

magnetic moment For paramagnetic substances the magnetic moment (μ) is related to the magnetic susceptibility (χ) by the equation

$$\mu = 2{\cdot}84\sqrt{\chi_m(T-\theta)}$$

where χ_m is the molar susceptibility, T the absolute temperature and θ is a constant for the particular substance (called the Curie temperature).

magnetic optical rotatory dispersion *See* magnetic polarization of light.

magnetic polarization of light When a beam of plane polarized light is passed through a transparent medium which is placed in a magnetic field, the plane of polarization is rotated. This phenomenon is known as magnetic rotation, magnetic optical rotatory dispersion, the Faraday effect. The light beam must be travelling in the same direction as the magnetic lines of force. The angle of rotation (ω) is related to the length of the medium traversed (l) and to the magnetic field strength (H) by the equation: $w = øHl$, where ø is a constant known as Verdet's constant. Magnetic rotation is not connected with normal optical activity. The effect is shown by transparent substances whether or not they possess a centre of asymmetry.

magnetic quantum number *See* electronic configuration.

magnetic resonance *See* nuclear magnetic resonance, electron spin resonance.

magnetic separation Separation techniques involving magnetic properties. Materials, e.g. minerals, can be separated into ferromagnetic, magnetic, diamagnetic and non-magnetic fractions. Separation is generally carried out on crushed material of small particle size.

magnetic susceptibility Atoms or molecules may have a permanent magnetic moment or an induced moment due to the influence of a magnetic field. In general, the extent of magnetization of any substance is a function of the field in which that substance is placed. The magnetic susceptibility per unit volume, κ, is defined as the intensity of magnetization induced (I) divided by the field strength (H)

$$\kappa = I/H.$$

The molar magnetic susceptibility χ_m is given by the expression

$$\chi_m = \frac{M}{\rho} \cdot \kappa$$

where M is the molecular weight and ρ the density of the substance. Magnetic susceptibilities are usually determined using a Gouy balance but nuclear magnetic resonance methods may also be used. For a paramagnetic substance $1/\chi$ is proportional to the absolute temperature (Curie–Weiss law).

magnetic tape Used for recording information (sound, video, computer output, instrumental output). Consists of magnetic particles (generally γ-Fe_2O_3 but other materials e.g. CrO_2 also used). Information is imprinted via electrical signals converted to magnetic fields, which in turn affect the magnetization of the tape particles.

magnetite, Fe_3O_4. The mineral form of black Fe_3O_4 with the spinel structure. Used as an iron ore, a flux, a pigment for glazes and occasionally as a refractory material for lining furnaces for producing iron.

Magnus's green salt $[Pt(NH_3)_4][PtCl_4]$.

mahogany acids The sulphonic acids remaining in the oil phase following the refining of lubricating oil by oleum or sulphuric acid. *See* green acids.

malachite, $Cu_2(OH)_2CO_3$. A native green hydrated carbonate of copper used as a decorative stone as it takes a high polish. Produced artificially (green verditer) from Na_2CO_3 and a solution of $CuSO_4$ and used as an artist's colour.

Malathion, O,O-dimethyl-S-1,2-di(ethoxycarbonyl)ethyl phosphorothiolothionate, $C_{10}H_{19}O_6PS_2$. An organophosphorus

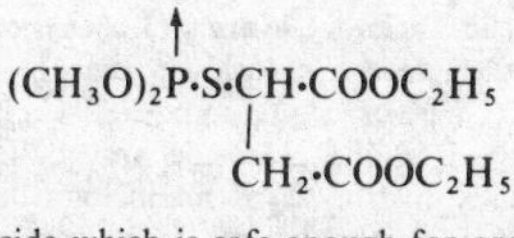

insecticide which is safe enough for ordinary garden use. B.p. 156–157°C/7 mm. It is manufactured by the condensation of O,O-dimethylphosphorothiolothionic acid and diethyl maleate in the presence of hydroquinone (as anti-polymerization agent). Malathion is effective against thrips, aphids and houseflies, and gives partial control against scale insects and red spider mite.

maleamic acid, $C_4H_5NO_3$. The half amide of

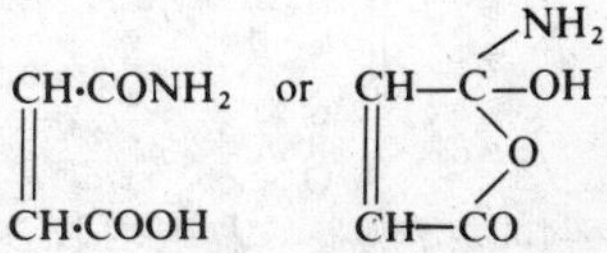

maleic acid; m.p. 172–173°C (decomp.). Its ammonium salt, which is made by treating maleic anhydride with ammonia, is used to make Krilium*.

maleic acid, *cis*-butenedioic acid, $C_4H_4O_4$.

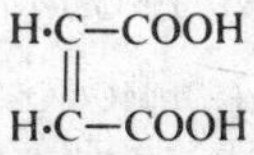

Colourless prisms; m.p. 130°C. Manufactured by treating maleic anhydride with water. It is converted to the anhydride by heating at 140°C. By prolonged heating at 150°C or by heating with water under pressure at 200°C, it is converted to the isomeric (trans) fumaric acid. Reduced by hydrogen to succinic acid. Oxidized by alkaline solutions of potassium permanganate to mesotartaric acid. When heated with solutions of sodium hydroxide at 100°C, sodium(±)-malate is formed. Used in the preparation of (±)-malic acid and in some polymer formulations.

maleic anhydride, $C_4H_2O_3$. Colourless; m.p.

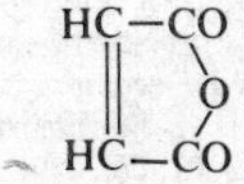

53°C, b.p. 200°C. Manufactured by the oxidation of benzene by air at 400–450°C over a vanadium pentoxide catalyst; furfural, crotonaldehyde and butenes may also be used as starting materials. Reacts with hot water to give maleic acid. Reduced by hydrogen in presence of catalysts to succinic anhydride. Reacts as a Diels–Alder dienophile with conjugated diolefins to give derivatives of Δ^4-tetrahydrophthalic anhydride. Forms resinous substances with terpenes. Its principal use is in the production of polyester resins and copolymers. It is also employed in the manufacture of alkyd resins for varnishes, drying oils, agricultural chemicals and fumaric acid. U.S. production 1978 150 000 tonnes.

maleic hydrazide, 3,6-dioxo-1,2,3,6-tetrahydropyridazine, $C_4H_4N_2O_2$. Colourless

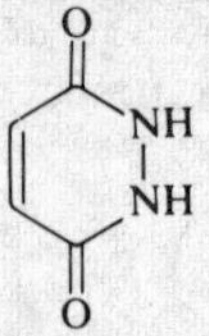

crystals; m.p. 296–298°C. Prepared by condensation of maleic anhydride with aqueous hydrazine sulphate. A plant growth regulator, it is toxic to grasses, and is used to retard the growth of some plants, e.g. hedges.

malic acid, hydroxysuccinic acid, $C_4H_6O_5$.

$$\begin{array}{c} H_2CCOOH \\ | \\ HOCHCOOH \end{array}$$

(−)-Malic acid crystallizes in colourless needles; m.p. 100°C. It occurs in many acid fruits, such as grapes, apples and gooseberries. It can be prepared by microbiological processes using various moulds or from (+)-bromosuccinic acid by the action of NaOH.

malleability The property which enables a body – particularly a metal – to be extended in all directions by hammering or rolling. The degree of malleability is gauged by the thinness of leaf or foil which it is possible to produce.

malonic acid, propanedioic acid, $C_3H_4O_4$, $CH_2(COOH)_2$. Colourless crystals; m.p. 136°C. Occurs in the mixed calcium salts obtained during the processing of sugar beet. Prepared by heating a solution of sodium cyanoethanoate with NaOH until no more ammonia is evolved. Decomposes above 140°C to give ethanoic acid. Forms crystalline acid and neutral salts. Reacts with aldehydes in presence of primary and secondary bases to give unsaturated substituted acids; some of these very readily lose CO_2 to give $\alpha\beta$-unsaturated fatty acids. Its esters are used in organic syntheses.

malonic ester, diethyl malonate, $C_7H_{12}O_4$, $CH_2(COOEt)_2$. A colourless liquid with a faint aromatic odour, b.p. 199°C. Insoluble in water; miscible with organic solvents. Manufactured by treating sodium monochloroethanoate with sodium cyanide in alkaline solution at 60°C. The resulting sodium cyanoethanoate is heated with alcohol and sulphuric acid to give malonic ester. It is converted to malonic acid and alcohol by boiling with water or dilute alkalis. Like acetoacetic ester, it reacts with sodium alkoxides or metallic sodium – a sodium atom displacing one of the hydrogen atoms of the $CH_2<$ group. This sodium derivative reacts with halogen compounds to give substituted malonic esters of the type $RCH(COOEt)_2$. With sodium followed by R′Cl disubstituted malonic esters $RR'C(COOEt)_2$. are formed. The substituted malonic esters are hydrolysed by alkali to the acids which are readily decarboxylated to substituted ethanoic acids. Malonic ester and the esters of the substituted malonic acids react with urea to give barbituric and substituted barbituric acids, which are important drugs.

malononitrile, $CH_2(CN)_2$. Prepared from NaCN and sodium chloroethanoate, b.p. 223°C, polymerizes above 120°C. Condenses with *o*-chlorobenzaldehyde to give *o*-chlorobenzal malononitrile, $ClC_6H_4C{=}C(CN)_2$, CS gas – a riot-control gas. Also used in the synthesis of vitamin B, and as a polar-additive for lubricating oils.

malonyl coenzyme A A highly reactive natural thioester; which is the chain-elongating agent in the biosynthesis of fatty acids.

malt Cereal grains which have been allowed to sprout and then are heated to destroy life. The sprouting activates the enzyme systems. Barley, wheat, rye, oats, sorghum, corn and rice are converted to malt. Malt is used in brewing, in food additives and as animal feed.

maltase The enzyme found in malt that is specific in splitting maltose into two molecules of glucose. Maltases are also found in the digestive systems of animals and in yeast but should be more correctly termed α-glucosidases.

maltose, $C_{12}H_{22}O_{11}$. A disaccharide which is present free in small quantities in barley grains and some other plants but is more commonly produced by the action of amylase on starch or glycogen. It is 4-[α-D-glucopyranosido]-D-glucopyranose; the α-form crystallizes as the monohydrate in colourless needles, m.p. 102–103°C. Used in brewing, soft drinks and foods.

mandelic acid, 1-hydroxy(1-phenyl)ethanoic acid, $C_8H_8O_3$. Colourless prisms; (±)- m.p. 118°C, (+)-or (−)- m.p. 133°C. Occurs combined in the glucoside amygdalin. Prepared by hydrolysis of mandelonitrile (benzaldehyde cyanohydrin). It is administered in large doses in the treatment of urinary infections.

manganates Double oxides containing manganese.

Manganates(VII), $[MnO_4]^-$, permanganates. Dark purple tetrahedral anion (electrolyte oxidation of $[MnO_4]^{2-}$. Powerful oxidizing agent
E° $(MnO_4)^-$ to MnO_2 in basic solution +1·68 volts;
E° $[MnO_4]^-$ to Mn^{2+} in acid solution +1·49 volts; used in volumetric analysis, *see* permanganate titrations. The free acid $HMnO_4$ and $HMnO_4,2H_2O$ are obtained at −75°C ($Ba(MnO_4)_2$ plus H_2SO_4), both are violent oxidizing agents; these are the acids of the oxide Mn_2O_7.

Manganates(VI), $(MnO_4)^{2-}$. Deep green tetrahedral ion (MnO_2 plus fused KOH with KNO_3, air, etc.). Only stable in basic solution.

Manganates(V), hypomanganates, $[MnO_4]^{3-}$. Deep blue ion (MnO_2 in conc. KOH; $[MnO_4]^-$ plus excess $[SO_3]^{2-}$). The salts are rapidly hydrolysed.

Manganates(IV), manganites. Mixed-metal oxides containing Mn(IV). Prepared by solid state reactions.

Manganates(III), again mixed-metal oxides; present in the spinel Mn_3O_4, $Mn^{II}Mn^{III}{}_2O_4$.

manganese, Mn. At.no. 25, at.wt. 54·9380, m.p. 1244°C, b.p. 1962°C, *d* 7·20. A transition element of Group VII. The principal ore is pyrolusite (MnO_2) but higher oxides (Mn_2O_3, Mn_3O_4) and the carbonate are also known. Occurs in nodules on the sea-bed. After roasting, manganese ores are reduced with Al or C; pure Mn is formed by electrolytic reduction. Mn metal is a soft grey metal with various low temperature forms which do not correspond to those adopted by other metals. Mn is a reactive metal combining with oxygen on heating in air, reacts slowly with water, combines with halogens, N_2, P, Si, S and C on heating. The major use of Mn is in steel (95%) as a deoxidizing and desulphurizing agent. Also used in alloys. Manganese compounds are used in dyes, paints, batteries (MnO_2), chemical processes, fertilizers, herbicides, fungicides. World production of Mn metal 1976 10 megatonnes.

manganese acetates *See* manganese ethanoates.

manganese alums Derivatives of manganese (III) sulphate prepared by electrolytic oxidation of Mn(II). *See* alums.

manganese borates Rather indefinite materials precipitated from Mn^{2+} solutions by borates and used as driers for linseed oil (catalysts for oxidation of the oil).

manganese bromide, $MnBr_2$. Forms hydrates (anhydrous by heating hydrates in stream of HBr gas). Very soluble in water.

manganese carbonate, $MnCO_3$. Occurs naturally as the mineral rhodochrosite, precipitated from Mn^{2+} solutions by use of $NaHCO_3$ solution saturated with CO_2. Normally precipitated as basic carbonates. Rapidly oxidized in air.

manganese carbonyl, dimanganese decacarbonyl, $Mn_2(CO)_{10}$. Formed by action of CO under pressure on a Mn(II) salt in the presence of a strong reducing agent. Golden-yellow crystals, structure $(OC)_5Mn-Mn(CO)_5$. Parent compound of series of manganese carbonyl derivatives, e.g. $Na[Mn^{-1}(CO)_5]$ ($Mn_2(CO)_{10}$ plus Na); $[Mn(CO)_5Cl]$ ($Mn_2(CO)_{10}$ plus Cl_2); $[(h^5-C_5H_5)Mn(CO)_3]$ (NaC_5H_5 plus $Mn_2(CO)_{10}$).

manganese chemistry Manganese is an electropositive transition element of Group VII, electronic configuration $3d^54s^2$. It shows a great range of oxidation states. Mn(II) shows a range of cationic chemistry

E° $Mn^{2+} \longrightarrow Mn$ −1·03 volts in acid

but in basic solution hydroxide or oxide species are formed and the tendency to oxide species increases with the higher oxidation states. Mn(II) and Mn(III) form extensive ranges of complexes. Mn(II) compounds are predominantly very pale pink in colour, most are spin-free (except cyanides). Manganese(III) compounds (brown) show Jahn–Teller distortion (E° $Mn^{3+} \longrightarrow Mn^{2+}$ in aqueous solution +1·51 volts). Manganese(IV) is confined to MnO_2, MnF_4 and some complexes. Mn(IV) is strongly oxidizing as are manganese(V), (VI) and (VIII) – *see* manganates. These compounds generally have tetrahedral co-ordination. Lower oxidation state compounds are formed with soft-ligands. Cyanide forms complexes with Mn(III), Mn(II). Mn(I) ($K_5Mn(CN)_6$ formed by reduction of $K_4Mn(CN)_6$) and Mn(0) ($K_6Mn(CN)_6$ formed by reduction of $K_5Mn(CN)_6$ with K in liquid NH_3) and CO from $Mn^0{}_2(CO)_{10}$ and $[Mn^{-1}(CO)_5]^-$. There is some metal–metal bonding but very little in comparison with Re. Manganese compounds are of importance in photosynthesis.

manganese chlorides The only stable chloride is manganese(II) chloride, $MnCl_2$. Forms hydrates. Forms range of complex chlorides, e.g. $KMnCl_3$ (perovskite), M_2MnCl_4 (tetrahedral – green or octahedral – pink), and K_4MnCl_6. Manganese(III) chloride, $MnCl_3$, is black and decomposes at −40°C (Mn(III) ethanoate plus HCl at $-100°C$). Red–brown complexes, e.g. K_2MnCl_5 (MnO_2 in HCl plus KCl) are known. Chloromanganates(IV), e.g. Rb_2MnCl_6, are deep red compounds formed MnO_2 in cold HCl plus MCl. Very unstable manganese(VII) and (VI) oxide chlorides, MnO_3Cl, $MnOCl_3$ (green) and MnO_2Cl_2 (brown) are formed from Mn_2O_7 and chlorosulphuric acid.

manganese cyanides Simple cyanides are not known but complex cyanides of manganese (III) (orange $[Mn(CN)_6]^{3-}$); manganese(II) (violet or yellow, e.g. $[Mn(CN)_6]^{4-}$); manganese(I) ($[Mn(CN)_6]^{5-}$); and manganese(0) ($[Mn(CN)_6]^{6-}$) are formed by the appropriate oxidation or reduction in the presence of cyanide. These are all spin-paired compounds.

manganese ethanoates Manganese(II) ethanoate, $Mn(O_2CCH_3)_2,4H_2O$, pink crystals. Manganese(III) ethanoate, $[Mn_3O(O_2CCH_3)_6]^+[O_2CCH_3]^-$. Red–brown (Mn(II) ethanoate plus $KMnO_4$ in CH_3CO_2H). Useful starting point for manganese(III) compounds, oxidizing agent and catalyst for decarboxylation.

manganese fluorides
Manganese(II) fluoride, MnF_2. Forms hydrates (MnF_2 obtained by heating NH_4MnF_3; tetrahydrate from aqueous solution). Solutions are hydrolysed. Insoluble perovskites, e.g. $KMnF_3$, formed from solution.

Manganese(III) fluoride, MnF_3, formed from F_2 on, e.g., $MnCl_2$. Purple–red, forms complexes, e.g. $K_2MnF_5.H_2O$, K_3MnF_6, hydrolysed by water.

Blue *manganese(IV) fluoride,* MnF_4 (Mn plus F_2), is immediately hydrolysed by water; complex hexafluoromanganates containing $[MnF_6]^{2-}$ ions are yellow (electrolytic oxidation of lower fluorides or use of BrF_3).

Green *permanganyl fluoride,* MnO_3F, is an explosive liquid ($KMnO_4$ plus HSO_3F).

manganese hydroxides Relatively indefinite compounds formed by precipitation. $Mn(OH)_2$ occurs as pyrochroite and has a layer lattice. Readily oxidized in air. MnO•OH occurs as groutite and manganite with two forms as AlO•OH and FeO•OH; an important source of manganese(III) compounds. Hydrated manganese(IV) oxides* are known but are not simple hydroxides.

manganese iodide, MnI_2. Forms hydrates.

manganese nitrate, $Mn(NO_3)_2$. Formed anhydrous from Mn plus N_2O_4. Forms hydrates. Decomposes on heating to MnO_2.

manganese, organic derivatives Simple manganese alkyls are not very stable but carbonyl derivatives, e.g. $CH_3Mn(CO)_5$ (CH_3I plus $NaMn(CO)_5$), are well known. Cyclopentadienyl derivatives, e.g. $[(h^5-C_5H_5)Mn(CO)_3]$, and $[(h^5-C_5H_5)_2Mn]$, and π-arene complexes, e.g. $[(h^6-C_6H_6)Mn(CO)_3]^+$ are stable.

manganese oxides The lower oxides are similar to those of iron, non-stoicheiometric phases are formed. The higher oxides form manganates*. MnO is stable up to $MnO_{1\cdot13}$ and has the sodium chloride structure. Grey or green formed by reduction of higher oxides on heating $Mn(OH)_2$ or $MnCO_3$ in the absence of air. Mn_3O_4 is the stable oxide in air above 940°C. It is black with a distorted spinel structure. It occurs naturally as haussmannite. Mn_2O_3 has two forms (one related to the spinel structure of Mn_3O_4). Occurs naturally as braunite formed by heating MnO_2 in air at 550–900°C. There are related hydroxy compounds, MnO•OH (*see* manganese hydroxides). Mn_5O_8 is $Mn_2^{II}Mn_3^{IV}O_8$. MnO_2 occurs as pyrolusite and as polianite and ramsdellite. The structural chemistry is extremely complex with MnO_6 octahedra sharing corners and some Mn^{4+} ions can be replaced by Mn^{2+} or related low oxidation state ions. Formed fairly pure by heating $Mn(NO_3)_2$. Powerful oxidizing agent (organic materials and Cl_2 Weldon process) and catalyst (e.g. decomposition of $KClO_3$). Used in dry cells, glass (to give amethyst tint). Mn_2O_7, oily molecular oxide formed from $KMnO_4$ and concentrated H_2SO_4. Very strong oxidizing agent, can explode.

manganese phosphates Manganese(II) phosphates are formed from aqueous solution – often by precipitation – e.g. $Mn_3(PO_4)_2,7H_2O$, NH_4MnPO_4,H_2O (latter used for estimation of Mn, weighed as $Mn_2P_2O_7$ after ignition).

manganese sulphates
Manganese(II) sulphate, $MnSO_4$ forms hydrates. Manganese(III) sulphate, $Mn_2(SO_4)_3$, formed from MnO_2 and H_2SO_4 or by electrolytic oxidation of $MnSO_4$ solution. Green mass, forms violet aqueous solution, hydrolysed by excess water. Forms alums and certainly contains sulphato complexes. Manganese(IV) sulphate, $Mn(SO_4)_2$, black crystals ($MnSO_4$ in H_2SO_4 plus $KMnO_4$) readily hydrolysed.

manganic salts Generally manganese(III) salts.

manganites Generally applied to manganates(IV)*.

manganous salts Manganese(II) derivatives.

mannans Polysaccharides made up of a chain

of mannose units. The best known is mannan A from the endosperm of the ivory nut which is used for making buttons, etc. Hydrolysed to mannose.

Mannich reaction The replacement of active hydrogen atoms in organic compounds (e.g. a hydrogen of a ketone or a hydrogen on the aromatic ring of a phenol) by aminomethyl or substituted aminomethyl groups. This process is also called aminomethylation, and may be accomplished by reacting 37% methanal with the desired amine, e.g. $(CH_3)_2NH$, and the compound RH with active hydrogen atom(s).

$$(CH_3)_2NH + CH_2O + RH \longrightarrow R{\cdot}CH_2{\cdot}N(CH_3)_2 + H_2O$$

D-mannitol, mannite, $C_6H_{14}O_6$. Mannitol,

$$CH_2(OH)\cdot\overset{H}{\underset{OH}{C}}\cdot\overset{H}{\underset{OH}{C}}\cdot\overset{OH}{\underset{H}{C}}\cdot\overset{OH}{\underset{H}{C}}\cdot CH_2(OH)$$

the alcohol corresponding to mannose, is widely distributed in plants and fungi. White crystalline powder with a sweet taste; m.p. 168°C. It can be made from glucose by electrolytic reduction or isolated from natural sources, e.g. seaweed or manna ash.

D-mannose, $C_6H_{12}O_6$. Obtained by hy-

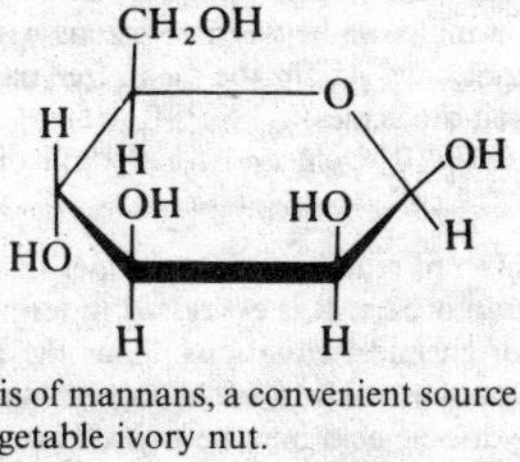

drolysis of mannans, a convenient source being the vegetable ivory nut.

manometers Instruments used to measure pressure differences between two points (one may be the atmosphere) in a liquid or a gas. In its simplest form consists of an upright transparent **U**-tube partially filled with liquid (often Hg); the difference in levels gives the pressure difference directly. Mechanical manometers (using the deflection of a diaphragm) and other devices are more costly but more precise methods of pressure measurement.

manufactured gas A general term for fuel gas obtained by a conversion process from solid or liquid hydrocarbons. *See* town gas.

marble, $CaCO_3$. A dense form of calcium carbonate* often used decoratively.

margarine Manufactured food product used in place of butter. Produced from vegetable oils, contains some polyunsaturated fats. There is about 20% aqueous phase (milk or water) together with food additives, vitamins and colour.

Markownikoff's rule The rule states that in the addition of hydrogen halides to an ethylenic double bond, the halogen attaches itself to the carbon atom united to the smaller number of hydrogen atoms. The rule may generally be relied on to predict the major product of such an addition and may be easily understood by considering the relative stabilities of the alternative carbenium ions produced by protonation of the alkene; in some cases some of the alternative compound is formed. The rule usually breaks down for hydrogen bromide addition reactions if traces of peroxides are present (anti-Markownikoff addition).

marsh gas Methane*.

Marsh's test for arsenic The arsenic-containing specimen is converted to volatile AsH_3 which is decomposed to a brown stain on heating. Estimation is by comparison of stains. Sb reacts similarly but the Sb stain is not soluble in NaOCl.

martensitic transformation Transformations involving lattice shear and no diffusion.

maser A device which involves the laser principle but employs microwave radiation.

masking An analytical technique involving addition of a masking agent which stops interfering species participating in a reaction. Demasking is the reverse process, e.g. breakdown of a cyanide complex with acid.

mass action, law of The rate at which a substance reacts is proportional to its active mass (concentration), and hence the velocity of a chemical reaction is proportional to the products of the concentrations of the reactants. Thus, in the reaction

$$A + B \rightleftharpoons C + D$$

the velocity of the forward reaction is

$$V_f = k_1[A][B]$$

where k_1 is a constant, and the square brackets denote concentrations. Similarly, for the backward reaction,

$$V_b = k_2[C][D]$$

At equilibrium, $V_f = V_b$

or $$k_1[A][B] = k_2[C][D],$$

hence $$\frac{k_1}{k_2} = \frac{[C][D]}{[A][B]} = K$$

The constants k_1 and k_2 are called velocity

constants, and K is called the equilibrium constant.

mass defect *See* packing fraction.

massicot, PbO. *See* lead oxides.

mass number The total number of particles (protons and neutrons) in the nucleus.

mass spectrograph In the mass-spectrometer the relative abundance of isotopes or other ionized species is determined by measuring the positive ion currents arriving at a fixed electrometer with various controlled magnetic fields and accelerating potentials. Resolution may be to 1 part in 50 000 and accuracy to better than 1 part in 10^6.

mass spectrometer *See* mass spectrograph.

mass spectrum The results obtained from a mass spectrometer are called mass spectra. They consist of a series of lines of varying intensity at different (m/e) ratios. For elements, the individual lines correspond to different isotopes. For compounds individual lines may be due to the presence of isotopic species, fragmentation of the molecule or ionization of the parent species in the mass spectrograph.

masurium An obsolete name for technetium*.

matches Practical method of producing flame. The usual method of flame production is a friction-initiated reaction between potassium chlorate and red phosphorus. The active materials are in a glue matrix and a glass powder is often used in the striker. The match may be wood or cardboard.

matlockite, PbFCl. *See* lead chlorides.

matrix isolation The formation and study of unstable species by reaction between two or more substrates generally in the atomic or small molecular form which are produced in a Knudsen cell* or, e.g., in a discharge and in which the reactive species are condensed out on a cold surface in the presence of a great excess of inert material (generally argon). Species such as metal clusters, actinide carbonyls, (M plus CO), PdC_6H_5 and uncomplexed $MgCl_2$ have been studied. Characterization is generally by spectroscopic methods.

matte The indefinite mixture of artificial sulphides produced by a smelting operation.

maximum co-ordination number *See* co-ordination numbers.

maximum covalency *See* co-ordination numbers.

maximum multiplicity, law of *See* Hund's rules.

MCD Magnetic circular dichroism. *See* optical rotatory dispersion.

MCPA, 2-methyl-4-chlorophenoxyacetic acid, Methoxone, $C_9H_9ClO_3$. Made by chlorination of *o*-cresol followed by reaction with chloroethanoic acid. White crystals, m.p. 118–119°C. As usually obtained, crude MCPA contains both 4- (60%) and 6- (40%) chloro-isomers, and is a light brown solid. Selective weedkiller.

MCPB, 4-(4-chloro-2-methylphenoxy)-butyric acid, $C_{11}H_{13}ClO_3$. A compound in itself harmless to plants, but when absorbed and translocated in the cells, $C_{11}H_{13}ClO_3$ is converted to a powerful herbicide, and results in the death of the plant. Acts as a selective weedkiller. Other butyric acid derivatives used commercially are 2,4-D_B and 2,4,5-T_B, the butyric acid analogues of 2,4-D and 2,4,5-T*.

Md Mendelevium.

MDI *See* methylene diisocyanate.

Me Methyl, CH_3-.

mean free path The average distance traversed by a molecule between collisions with another molecule. At stp the mean free path for hydrogen molecules is $1{\cdot}8 \times 10^{-6}$ cm, for nitrogen $9{\cdot}5 \times 10^{-6}$ cm and for carbon dioxide $6{\cdot}3 \times 10^{-6}$ cm.

mechanism of reaction The manner in which a reaction occurs. It is expressed in terms of a series of chemical equations. Thus the mechanism of the reaction of hydrogen peroxide with hydriodic acid may be written as

$$HI + H_2O_2 \longrightarrow H_2O + HOI$$
$$HOI + HI \longrightarrow H_2O + I_2$$

medicinal paraffin oil *See* liquid paraffin.

medicinal white oil *See* liquid paraffin.

MEK Methyl ethyl ketone. *See* butanone.

melamine, cyanuramide,2,4,6-triamino-1,3,5-triazine, $C_3H_6N_6$. Colourless; m.p. 354°C. It is manufactured by heating dicyandiamide, $H_2N{\cdot}C(NH){\cdot}NH{\cdot}CN$, either alone or in the presence of ammonia or other alkalis, in various organic solvents. Melamine is an important material in the plastics industry. Condensed with methanal and other substances it gives thermosetting resins that are remarkably stable to heat and light. U.S. production 1980 80 000 tonnes.

melamine resins *See* melamine.

melanin The brown pigment of hair, skin and eyes. It is a polymer arising by oxidation of tyrosine by the action of tyrosinase.

melibiose, 6-[β-D-galactopyranosido]-D-glucopyranose, $C_{12}H_{22}O_{11}$. Obtained with fructose by the hydrolysis of raffinose.

melissic acid, triacontanoic acid, $C_{30}H_{60}O_2$, $CH_3 \cdot [CH_2]_{28} \cdot COOH$. A fatty acid occurring in bees' wax, m.p. 94°C.

melissyl alcohol, myricyl alcohol, $C_{30}H_{62}O$, $CH_3 \cdot [CH_2]_{28} \cdot CH_2OH$. Colourless crystals; m.p. 87°C, soluble in organic solvents. It is present as melissyl palmitate in bees' wax.

mellitic acid, benzenehexacarboxylic acid, $C_{12}H_6O_{12}$. Colourless needles; m.p. 286–288°C. When heated it decomposes into pyromellitic anhydride, water and CO_2.

Occurs as the aluminium salt (honeystone) in some lignite beds. Prepared by oxidation of charcoal with concentrated nitric acid.

It condenses with resorcinol and aminophenols to give phthalein and rhodamine dyestuffs respectively. Esters are used in the formation of polyimides*.

melting point The temperature at which a solid changes into a liquid at a specified pressure. Pure solids usually have sharp well-defined melting points. Mixtures usually melt over a range of temperatures and have, therefore, ill-defined melting points.

membrane cell An electrolytic cell for the production of NaOH and Cl_2 from brine in which the anode and cathode are separated by a membrane. *See* Castner–Kellner cell, mercury cell, diaphragm cell.

membrane equilibrium *See* Donnan membrane equilibrium.

membrane hydrolysis When a colloidal electrolyte consisting, e.g. of giant anions and normal cations, is separated by a membrane from pure water, the cations are able to diffuse out but the colloidal anions are retained. For each positive ion which penetrates the membrane, a molecule of water will dissociate to provide a compensating OH^- ion, the corresponding H^+ ion tending to diffuse through the membrane to replace the original cation of the colloidal electrolyte. The liquid inside the dialyser thus becomes acid, that outside the dialyser alkaline. This process is termed membrane hydrolysis. *See also* colloids, colloidal electrolyte, membrane equilibrium.

mendelevium, Md. At.no. 101. ^{256}Md (1·5 hours) and ^{258}Md (2 months) have been formed by bombardment of Es with ^{4}He particles in a cyclotron or linear accelerator. The Md^{2+} ion is fairly stable in aqueous solution

$$Md^{3+} \xrightarrow{+0{\cdot}2v} Md^{2+}$$

No solid compounds are known.

***p*-menthadiene, 1-Δ$^{2:8(9)}$-menthadiene,**

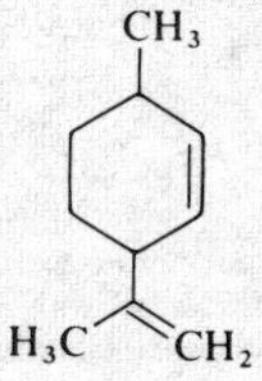

$C_{10}H_{16}$.
The most important menthadiene, an optically active monocyclic terpene found in chenopodium oil. Used in the manufacture of *p*-cymene.

menthol, $C_{10}H_{20}O$. An optically active

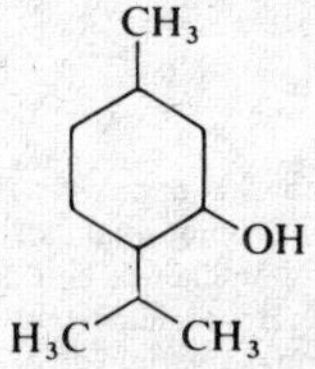

monocyclic terpene alcohol. A constituent of many oils. (−)-Menthol, m.p. 43°C, b.p. 216°C. (−)-Menthol and its stereoisomerides are obtainable by chemical reactions from a large number of terpene derivatives, for example, by the reduction of menthone, isomenthone or piperitone. (−)-Menthol can be converted into *p*-cymene and can be oxidized to menthone. It is used externally as an analgesic in rheumatism and by inhalation in the alleviation of nasal congestion and sinusitis.

menthone, $C_{10}H_{18}O$. An optically active monocyclic ketone. Obtained by oxidizing menthol. (−)-Menthone is present in some oils. When (−)-menthone is dissolved in strong sulphuric acid, and the mixture cooled, it is partially converted into (+)-*iso*menthone.

mercaptals, $RR'C(SR'')_2$. Acetals in which the oxygen atoms have been replaced by sulphur atoms. They are oily liquids with unpleasant odours. Formed by treating mercaptans with aldehydes or ketones in the presence of hydrochloric acid or zinc chloride.

They are stable compounds and are not decomposed by dilute acids or alkalis. They are frequently employed in synthetic organic chemistry for protecting* the carbonyl group.

mercaptans, thiols Organic compounds containing an $-SH$ group directly united to a carbon atom. The sulphur analogues of alcohols. They are liquids with strong and unpleasant odours. Found in crude petroleum. Prepared by the action of alkyl or aryl halides on KHS or by the reduction of the chlorides of sulphonic acids. Give complexes and metal derivatives, e.g. $Pb(SR)_2$, mercaptides. Oxidized by air or mild oxidizing agents to disulphides (RS•SR); nitric acid oxidizes them to sulphonic acids. React with aldehydes and ketones to form mercaptals and mercaptols. Long chain derivatives oxidized to give surfactants. Used in polymerization systems to control the length of the polymer chain. Also used as odorants, rubber vulcanizers, flotation agents and in the manufacture of various agricultural chemicals.

mercaptobenzthiazole, $C_7H_5NS_2$.

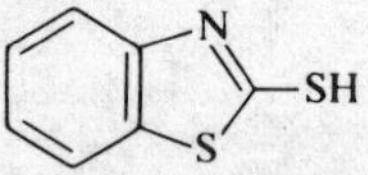

Crystalline powder, m.p. 174–179°C. Prepared by treatment of thiocarbanilide with sulphur, or by heating aniline, carbon disulphide and nitrobenzene. It is an important rubber accelerator, and on oxidation gives dibenzthiazyl disulphide, also a rubber accelerator.

mercapturic acids Derivatives of N-acetylcysteine which are excreted in the urine of animals following the ingestion of certain foreign organic compounds, such as alkyl halides and aromatic hydrocarbons.

mercuration The formation of Hg(II) derivatives of aromatic compounds by direct substitution, e.g. $Hg(O_2CCH_3)_2$ reacts with C_6H_6 to give $C_6H_5HgO_2CCH_3$. The resulting organomercurials are useful synthetic intermediates, e.g. for preparing halogeno derivatives.

mercuric Mercury(II) compounds.

mercurous Mercury(I) compounds.

mercury, Hg. At.no. 80, at.wt. 200·59, m.p. −38·87°C, b.p. 356·58°C, *d* 13·6. The only important mercury ore is cinnabar, HgS; in Hg extraction HgS is roasted in air to give Hg vapour. Mercury is a silver-white liquid metal with a rhombohedral structure derived from hcp. The metal is unattacked by dilute acids but dissolves in hot oxidizing acids. Hg is used in electrical equipment – switches, Hg-vapour lamps and also in pumps and thermometers and in $NaOH-Cl_2$ production. Uses of Hg compounds include pesticides, dental fillings, batteries, catalysis. Hg and many Hg compounds are toxic and tend to accumulate in higher animals. World production 1976 8400 tonnes.

mercury amalgams Intermetallic compounds of mercury. Some have definite compositions, e.g. $NaHg_2$, and are effectively alloys. Hg rich amalgams are liquid. Fe does not form amalgams and is used for containers for mercury. Used in dental stoppings, cadmium amalgam is the cathode in the Weston standard cell.

mercury chemistry Mercury is the heaviest Group 2B element, electronic configuration $5d^{10}6s^2$. The normal oxidation state is +2 (E° $Hg^{2+} \longrightarrow Hg$ +0·796 volts in acid solution). The Hg^{2+} ion is large and many compounds have high covalent character; 2-co-ordination, linear, is common but higher co-ordination is known. Mercury(I) (E° $\frac{1}{2}Hg_2{}^{2+} \longrightarrow Hg$ +0·798 volts in acid solution).

Exists as the $(Hg-Hg)^{2+}$ ion. Other polymercury cations, e.g. $Hg_3{}^{2+}$ (Hg plus AsF_5), $Hg_4{}^{2+}$, etc., are also known. All positive oxidation state compounds of Hg are readily reduced to the metal.

mercury chlorides

Mercury(II) chloride, $HgCl_2$, corrosive sublimate, m.p. 280°C, b.p. 302°C. Essentially covalent material (Hg plus Cl_2; Hg plus aqua regia). Forms complex halide ions, e.g. $(HgCl_4)^{2-}$ $(HgCl_3)^-$ in excess HCl and forms complexes. Very poisonous.

Mercury(I) chloride, Hg_2Cl_2, calomel, white solid precipitated from $Hg_2{}^{2+}$ solutions by Cl^- or from Hg plus $HgCl_2$. Used as an insecticide and as a purgative (must be free from $HgCl_2$) and in the calomel electrode.

mercury fulminate, $Hg(ONC)_2$. Grey crystals prepared by dissolving Hg in concentrated HNO_3 and pouring solution into C_2H_5OH. Used as a detonator. Safe if stored under water but relatively unstable in hot climates.

mercury iodides

Mercury(II) iodide, HgI_2. Scarlet (to 126°C) or yellow substance ($HgCl_2$ solution plus KI or Hg plus I_2). Forms complex iodides with excess iodide (Nessler's reagent).

Mercury(I) iodide, Hg_2I_2, pale green, is precipitated from a solution of a $Hg_2{}^{2+}$ salt by I^-.

mercury nitrates

Mercury(II) nitrate, $Hg(NO_3)_2$. Forms

hydrates with 8, 2, 1 and $\frac{1}{2}H_2O$ (Hg in concentrated HNO_3). Hydrolysed by water – forms basic nitrates.

Mercury(I) nitrate, $Hg_2(NO_3)_2$. Forms dihydrate from aqueous solution (Hg plus cold dilute HNO_3). Useful source of mercury(I) compounds.

mercury, organic derivatives Organomercury(II) derivatives, e.g. R_2Hg and RHgX are well established (e.g. RMgX plus mercury halide; C_6H_6 plus $Hg(O_2CCH_3)_2$ gives $C_6H_5HgO_2CCH_3$) but mercury(I) derivatives are unknown. Olefine and acetylene complexes are formed and are intermediates in reactions involving Hg as a catalyst. Organomercury compounds are used extensively to prepare other organometallic derivatives; they have a limited use as bactericides.

mercury oxides *Mercury(II) oxide,* HgO. Yellow–orange powder (Hg^{2+} solution plus excess NaOH). Decomposes to Hg and O_2 on heating. Many derivatives containing an oxygen co-ordinated to three or four mercury atoms are known. Mercury(I) oxide appears not to exist.

mercury sulphates

Mercury(II) sulphate, $HgSO_4,H_2O$. Colourless solid (Hg plus excess hot H_2SO_4). Readily hydrolysed to Hg_2OSO_4 (turpeth mineral).

Mercury(I) sulphate, Hg_2SO_4 formed by precipitation or excess Hg plus H_2SO_4. Hydrolysed to basic salts.

mercury sulphide, HgS. Occurs naturally as red cinnabar but a metastable black form (zinc blende structure) may be formed by precipitation or by grinding Hg plus S. Dissolves in alkali sulphide solutions to give thiomercurates. Formerly used as a red pigment – *see* vermilion red.

mer isomer *See* isomerism.

merocyanine dyes Polymethine dyestuffs comprising a nitrogen-containing heterocyclic system linked through a conjugated chain to an acidic ketomethylene heterocycle. Used as dyestuffs and as spectral sensitizers in photography.

mesaconic acid, methylfumaric acid, $C_5H_6O_4$.

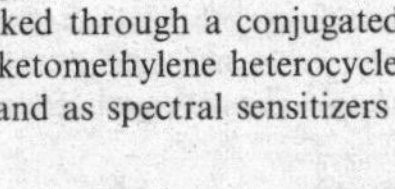

Colourless crystals; m.p. 204°C. Prepared by photochemical isomerization of citraconic acid in the presence of bromine.

mescaline, β-(3,4,5-trimethoxyphenyl)-ethylamine, $C_{11}H_{17}NO_3$. An alkaloid obtained from *Lophophora williamsii.* M.p. 35°C, b.p. 180°C/12 mm. It is a central nervous system depressant and causes visual hallucinations. Mescaline is the hallucinatory principle of peyote.

mesitite *See* breunnerite.

mesitylene, 1,3,5–trimethylbenzene, C_9H_{12}. Colourless liquid; b.p. 165°C. It occurs in crude petroleum. Prepared by adding concentrated sulphuric acid to cooled propanone, and then heating the mixture after it has stood for 24 hours; mesitylene can then be extracted from an alkaline solution.

mesityl oxide, isopropylidene acetone, 4-methyl-3-penten-2-one, $C_6H_{10}O$, $Me_2C = CHCOCH_3$. Colourless liquid; b.p. 129°C, with a strong peppermint-like odour. Prepared by distilling diacetone alcohol in the presence of a trace of iodine. Converted to phorone by heating in propanone with dehydrating agents such as sulphuric acid. It is a solvent for cellulose acetate and ethylcellulose and other polymers.

meso-ionic A compound is defined as meso-ionic if it is a five- or six-membered heterocycle which cannot be represented satisfactorily by any one covalent or polar structure, and possesses a sextet of electrons in association with all the atoms comprising the ring. *See* sydnones and nitron.

mesomerism A phenomenon encountered in molecules which may be formulated (in the conventional manner) in two or more ways that have the same spatial arrangement of the atoms and nearly the same potential energy (as, for example, in the Kekulé forms of benzene). In such cases the actual distribution of the electrons which go to make up the bonds is a weighted mean of the distributions corresponding to the various possible formulae. There is, however, no tautomerism or oscillation between the various forms. The electron distribution corresponds to an intermediate state. This phenomenon is termed mesomerism or resonance. It leads to an increase in stability.

mesomorphic state *See* liquid crystals.

meson, mesotron Mesons are subatomic particles with either a positive, negative or zero charge. Mass ca 200 times that of an electron and a mean lifetime of ca 10^{-7} seconds. Mesons are found in cosmic radiation and result from high energy nuclear collisions.

mesothorium Names used for Ra and Ac isotopes produced in natural decay series.

mesoxalic acid, ketomalonic acid, $C_3H_2O_5$. Occurs with $1H_2O$ probably as dihydroxymalonic acid, $(HO)_2C(CO_2H)_2$, m.p. 121°C. Prepared by the action of nitrous acid on malonic acid. Acts as a ketone towards hydroxylamine and phenylhydrazine; forms esters with alcohols which may be of the keto or dihydroxy forms.

mesyl The methane sulphonyl group, hence mesylate.

meta In meta-cresol (*m*-cresol, 3-hydroxy-

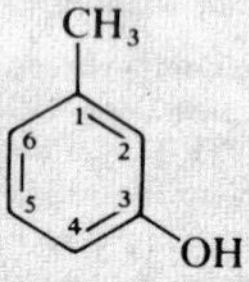

toluene), the CH_3 and OH groups are said to be in the meta position to each other. The same prefix is found in metaphosphoric acid and metabisulphites, but in these cases it is always written in full.

For the substance 'meta', *see* metaldehyde.

metabolism The chemical changes that occur in a living animal or plant. The metabolism of a substance refers to the changes that substance undergoes in an animal or plant. Metabolism is often sub-divided into anabolism, i.e. synthetic reactions, such as the manufacture of proteins and fats, and catabolism, i.e. destructive reactions, such as the breakdown of sugar to carbon dioxide and water, which release large amounts of energy.

metaboric acid, HBO_2 The intermediate product during the loss of water from boric acid.

metal A general term characterizing certain elements, e.g. gold, silver, copper, mercury, sodium, which have a characteristic lustrous appearance, which are good conductors of heat and electricity, and generally enter chemical reactions as positive ions, or cations. There are certain elements, such, for example, as tellurium, which have the physical properties of a metal and the chemical properties of a nonmetal. The distinction between the two groups is not sharp. The typical metallic structures are the body-centred cubic, bcc, the face-centred cubic (cubic close-packed), ccp, and the hexagonal close-packed structures, hcp. The following table of examples shows the widespread occurrence of these structures.

Structure	*Metal*
body-centred cubic (bcc)	Li, Na, K, Rb, Cs, Ba, β-Zr, V, Nb, Ta, α-Cr, Mo, α-W, α-Fe.
face-centred cubic (cubic close-packing, ccp)	Cu, Ag, Au, Ca (below 450°C), Sr, Al, β-La, β-Tl, Th, Pb, γ-Fe, β-Co, β-Ni, Rh, Pd, Ir, Pt.
hexagonal close packed (hcp)	Be, Mg, Ca (above 450°C), Y, α-La, α-Tl, Ti, α-Zr, Hf, β-Cr, Re, α-Co, α-Ni, Ru, Os.

A few metals have more complicated structures (noted under the elements concerned) whilst the structures of the metalloids (semi-metals) are intermediate between those of true metals and homopolar compounds. E.g. germanium and grey tin have the diamond structure; arsenic, antimony and bismuth have layer structures, and selenium and tellurium have hexagonal structures containing chains in which each atom has only two neighbours (compare the twelve equidistant neighbours in close-packed structures). Equally, certain alloys and compounds have all the properties of a metallic element. Bonding in metals is generally considered as being by loss of one or more electrons which form a macroscopic molecular orbital surrounding the close-packed metal ions.

metalation The removal of a relatively acidic proton from a compound and replacement by a metal atom, such as lithium, sodium or potassium (*see also* mercuration). The reaction is facilitated by the addition of chelating bases like N,N,N′,N′-tetramethylethylenediamine.

metal atom deposition *See* matrix isolation.

metal carbonyls Co-ordination derivatives between CO and a metal in which the carbon is generally bonded to the metal. Most transition metal and actinide elements form carbonyls, $M(CO)_x$ ($x = 1, 2, 3$, etc.) using matrix isolation techniques involving metal atoms and CO. These derivatives are only stable in matrices at low temperature. Most of the transition metals form stable carbonyls (MCl_n, $Macac_m$, etc., plus reducing agent plus CO under pressure). The carbonyl groups can also be bonded.

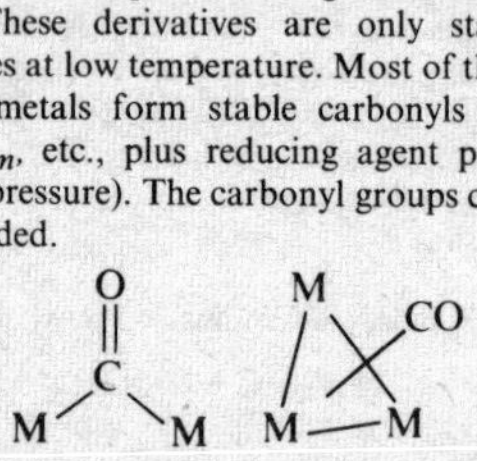

Simple stable carbonyls, except $V(CO)_6$, have an electronic configuration corresponding to the next noble gas. Carbonyl groups can be substituted by other unchanged ligands (e.g.

PPh_3, pyridine) organic groups (olefines, aromatic derivatives) and also form other derivatives, e.g. halides, hydrides, sulphides.

metal cluster compounds Compounds containing clusters of metal atoms linked together by covalent (or co-ordinate) bands.

metaldehyde, $(C_2H_4O)_n$ ($n=4$ or 6). A solid crystalline substance, sublimes without melting at 112–115°C; stable when pure; it is readily formed when ethanal is left in the presence of a catalyst at low temperatures, but has unpredictable stability and will revert to the monomer. It is used for slug control and as a fuel.

metallic conduction Whereas electrolytic conduction arises from the movement of ions and results in decomposition of the conductor, conduction in metals arises from the movement of electrons and does not cause decomposition. The conductivity of a metal may be explained by assuming that the 'sea' of electrons behaves as gas molecules observing the kinetic laws. Each atom in the metal contributes only one or two electrons to the energy state known as the 'conduction band'; these are the electrons which are free to move.

metallic soaps Water insoluble, alkaline earth, heavy metal, Li salts of long chain carboxylic acids used as driers*, in greases and heavy duty lubricating oils, fungicides (Cu and Zn salts), rubber, water proofing, cosmetics and pharmaceuticals (Zn and Mg stearates). Napalm is an Al soap $Al(OH)R_2$ used for gelling gasoline and as an incendiary.

metallizable dyes, mordant dyes Dyes that have a metal atom added *in situ* (chelated) after application to the fibre.

metallocenes The *bis*-(h^5-cyclopentadienylmetal) compounds $(C_5H_5)_2M$, formed by reaction between the cyclopentadienyl anion and suitable derivatives of the transition metals, e.g. M=Fe, ferrocene: M=Co, cobaltocene: M=Ni, nickelocene. Known for e.g., Ti, green: Cr, red: Fe, orange: Co, purple: Os, yellow. Only those metallocenes of the iron group (Fe, Ru, Os) are diamagnetic, the others exhibiting varying degrees of paramagnetism, e.g. Cp_2Co, one unpaired electron: Cp_2Mn, five unpaired electrons. Only the iron group metallocenes exhibit aromatic behaviour (*see* ferrocene) and are not particularly sensitive to oxidation. Many metallocenes may be oxidized to cations, e.g. $[(C_5H_5)_2Fe^+]$, ferricinium: $[(C_5H_5)_2Co^+]$, cobalticinium. Metallocene derivatives are readily formed.

metallochromic indicator A compound which forms a complex with a markedly different colour from that of the free indicator, e.g. NCS^- used with Fe^{3+}, eriochrome black T used with Mg^{2+}.

metalloids Elements which have properties between the obvious metals and obvious non-metals. The classification is arbitrary and often based on the structure and properties of the free element. E.g. As, Sb, Bi.

metallurgy The science of the extraction, working and use of metals and their alloys.

metal–metal bonds Covalent or co-ordinate bonds between metals in compounds, e.g. $(OC)_5Mn-Mn(CO)_5$. Often inferred from a low magnetic susceptibility, Raman spectroscopy or electron counting (effective atomic number rule*) but only established by X-ray structure determinations. Bond orders can be up to 4.

metal passivators In catalytic operations deposition of trace amounts of certain metals may passivate or poison the catalyst. Typical metal passivators are V and Ni present in petroleum fractions.

metal surface treatments All used metal objects are surface treated at some time. Oxides are removed chemically. Dirt and grease are removed with organic solvents particularly chlorocarbons or alkaline detergents. Surfaces are then pickled in acids (generally phosphoric).

metamerism A change in observed colour depending upon the nature of the illuminating light.

metaphosphates Phosphates* containing cyclic or chain linked PO_4^{3-} groups.

metastable state *See* equilibrium, metastable.

Metasystox A trade name for demeton-methyl, $C_6H_{15}O_3PS_2$, which is a mixture of O,O-dimethyl-2-ethylthioethyl phosphorothionate, and O,O-dimethyl-S-(2-ethylthioethyl) phosphorothiolate. Used as a pesticide on crops.

metathetical reaction An exchange reaction, e.g.

$$CF_3I + NaMn(CO)_5 \longrightarrow CF_3Mn(CO)_5 + NaI$$

or

$$2AgNO_3(aq) + BaCl_2(aq) \longrightarrow 2AgCl + Ba(NO_3)_2.$$

The driving force is generally the insolubility of one of the products in the reaction medium or the volatility of one of the products.

methacrylic acid, α-methylacrylic acid, 2-methylpropenoic acid, $C_4H_6O_2$,

$CH_2=CMeCOOH$. Colourless prisms; m.p. 15–16°C, b.p. 160·5°C. Manufactured by treating propanone cyanohydrin with dilute sulphuric acid. Polymerizes when distilled or when heated with hydrochloric acid under pressure, *see* acrylic acid polymers. Used in the preparation of synthetic acrylate resins*; the methyl and ethyl esters form important glass-like polymers.

methadone, amidone, 6-dimethylamino-4,4-diphenyl-3-heptanone, $C_{21}H_{27}NO$, $Me_2NCH(CH_3)CH_2CPh_2C(O)Et$. Prepared from 2-chloro-1-dimethylaminopropane by treating with diphenylmethyl cyanide in the presence of sodamide, and converting to a mixture of ketones from which the methadone is separated. It is converted to its hydrochloride, which is a colourless crystalline powder, soluble in water and in alcohol, m.p. 235°C. It is topologically similar to morphine. It is a powerful analgesic, which also depresses the cough centre with high oral effectiveness and long duration of action.

methanal, formaldehyde, HCHO. Colourless gas with a characteristic and pungent odour. Soluble in water to 52%: the commercial solution – formalin – contains about 37% formaldehyde. It is manufactured by passing methanol vapour and air over a heated metal or metal oxide catalyst, and to a lesser extent by the oxidation of paraffinic hydrocarbons. It reacts with water to give stable hydrates, called polymethylene glycols, of the type $(CH_2O)_n,H_2O$. In dilute solutions most of the formaldehyde is present as methylene glycol, $CH_2(OH)_2$ or CH_2O,H_2O. In more concentrated solutions, trimethylene glycol $(CH_2O)_3,H_2O$ is also present. When formalin is allowed to stand, flocculent masses of paraformaldehyde are precipitated. Its principal use is in the manufacture of urea- (30%), phenol- (20%), melamine-, acetal (10%) and polyformaldehyde resins. Large quantities are also used in the manufacture of ethylene glycol, pentaerythritol and hexamethylenetetramine.

It is a powerful germicide, whether in solution or vapour. Rooms may be disinfected by spraying the solution or vaporizing paraformaldehyde. Dilute solutions are used for sterilizing surgical instruments, as applications in certain skin diseases and for storing pathological specimens. U.S. production 1983 2·5 megatonnes (37% soln.).

methanamide, formamide, $HCONH_2$. A colourless, rather viscous, odourless liquid; it absorbs water vapour; m.p. 2·5°C, b.p. 210°C (decomp.). Manufactured by direct reaction of carbon monoxide and ammonia under pressure, or by distillation of ammonium methanoate. It is converted to methanoic acid or a methanoate by concentrated acids or alkalis. Forms compounds with metals and with acids. It and its derivatives are good solvents for many organic and inorganic compounds.

methanation The process of converting CO or CO_2 to methane by a catalytic reaction with hydrogen under pressure. Methanation is used industrially in several ways including (1) for removal of CO from the gas stream in ammonia synthesis, (2) the purification of hydrogen by removing CO and (3) conversion of CO and hydrogen to methane to increase the calorific value in town gas* or SNG* production.

methane, marsh gas, CH_4. A colourless, odourless gas; m.p. −184°C, b.p. −164°C. Liquid at −11°C under pressure. It is the chief constituent of natural gas associated with oil, formed by the decay of vegetable matter in marshy places; it occurs in coal mines, where explosive mixtures of it with air are known as firedamp, and in coal gas. It is manufactured by passing a mixture of one volume of CO and three volumes of hydrogen at atmospheric pressure over a nickel catalyst heated to 230–250°C. The gas mixture is obtained from blue water gas*, also obtained from the anaerobic digestion of sewage. Methane is chemically fairly inert but reacts explosively with chlorine at ordinary temperatures; at low temperatures methyl chloride is formed. Used in the manufacture of methyl and methylene chlorides, hydrogen, ammonia, carbon monoxide, the production of carbon black. Used as a fuel.

methanesulphonic acid, CH_3SO_2OH. M.p. 20°C, b.p. 167°C/10 mm, obtained by oxidation of dimethyl disulphide or by the catalytic reaction between methane and sulphur trioxide. Readily soluble in water, but almost insoluble in hydrocarbon solvents.

It is used as a catalyst in esterification, dehydration, polymerization and alkylation reactions. Converted by e.g., thionyl chloride, to methanesulphonyl chloride (mesyl chloride) which is useful for characterizing alcohols, amines, etc. as methanesulphonyl (mesyl) derivatives.

methanoates, formates Esters and salts of methanoic acid.

methanoic acid, formic acid, H·COOH. A colourless liquid which fumes slightly and has a penetrating odour; m.p. 8·4°C, b.p. 100·5°C. Occurs in sweat and urine, and in stinging nettles. Prepared by heating oxalic acid with glycerol, when formic acid distils off.

Manufactured by decomposing sodium methanoate, from the reaction of carbon monoxide and sodium hydroxide, with sulphuric acid; also made by the hydrolysis of methyl methanoate obtained by reacting carbon monoxide with methanol. It is used in textile dyeing and finishing, in leather tanning and as an intermediate for other chemicals. It is a good solvent for many organic and inorganic compounds and is a strong acid.

methanol, methyl alcohol, wood spirit, wood naphtha, CH_3OH. Colourless liquid with a spirituous odour; it is poisonous, small doses causing blindness; b.p. 64·5°C. Occurs as esters in various plant oils, such as wintergreen. It is manufactured by reacting together hydrogen and CO or CO_2, the reaction being carried out over a catalyst at 50–350 bar and 250–400°C. Some methyl alcohol is also produced by the partial oxidation of hydrocarbons. It is readily oxidized by air in presence of nickel or platinum to methanal. Reacts with sulphuric acid to give methyl hydrogen sulphate, dimethyl sulphate and dimethyl ether. Forms methyl chloride with hydrogen chloride in the presence of zinc chloride. When heated with sodalime it is converted to sodium methanoate and hydrogen. Reacts with sodium to give sodium methoxide. It is a good solvent for many inorganic salts and organic compounds. Used in the manufacture of methanal (45%), methanoic acid (10%), methyl chloride (5%), MTBE (5%) and many other organic compounds, also as a solvent and as a denaturant for ethyl alcohol. U.S. production 1983 1·4 B gals.

methenamine *See* hexamine.

methionic acid, methylenedisulphonic acid, $CH_4O_6S_2$, $H_2C(SO_3H)_2$. A colourless, crystalline solid which readily absorbs water vapour; decomposes on distillation. The potassium salt is prepared by heating methylene chloride with an aqueous solution of potassium sulphite under pressure at 150–160°C. The free acid is obtained by decomposing the sparingly soluble barium salt with sulphuric acid. The aryl esters are very stable, but the alkyl esters decompose on heating to give ethers. Resembles malonic acid in some of its reactions.

methionine, 2-amino-4-(methylthio)butanoic acid, $C_5H_{11}NO_2S$, $CH_3SCH_2CH_2CH(NH_2)CO_2H$. M.p. 283°C (decomp.). Soluble in water and alcohol. The naturally occurring substance is laevorotatory. Methionine is one of the natural sulphur-containing amino-acids, and is present in small quantities in the hydrolysis products of proteins. It is an essential constituent of the food of mammals and is particularly important in that it and choline are the only compounds in the diet known to take part in methylating reactions.

methoxy- A prefix used to denote that the substance contains a methoxyl, CH_3O-, group.

methyl The group CH_3-, Me–.

6-methylacetopyranone *See* dehydroacetic acid.

methyl acrylate, methyl propenoate, $C_4H_6O_2$, $CH_2{:}CHCOOCH_3$. Colourless liquid; b.p. 80°C, insoluble in water, soluble in organic solvents. It is manufactured (1) by treating ethylene chlorohydrin with sodium cyanide and heating the β-hydroxypropionitrile so formed with methanol and sulphuric acid, (2) by the pyrolysis of the ethanoate derived from methyl lactate, or (3) by the reaction of ethyne, CO and methanol in the presence of nickel carbonyl. It readily polymerizes to give colourless rubber-like poly(methyl acrylate). *See* acrylate resins.

methylal, methylformal, dimethoxymethane,

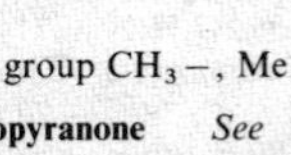

$C_3H_8O_2$. Colourless liquid with a pleasant odour; b.p. 42°C. It occurs in commercial formalin, and is made by heating polyoxymethylene with methanol in the presence of $FeCl_3$ by treating methylene chloride with sodium methoxide, or by treating a mixture of methyl alcohol and methanal with calcium chloride and a little hydrochloric acid. It is a good solvent, and is also used to replace methanol in many reactions.

methyl alcohol Methanol*.

methylamines Compounds of ammonia in which 1, 2 or 3 of the hydrogen atoms have been replaced by methyl groups. They are manufactured by the vapour phase reaction of methanol and ammonia at 350–450°C under pressure over an alumina catalyst. The product consists of a mixture of all three compounds; the proportion of the two higher amines may be increased, if so desired, by recycling the monomethylamine or reducing the ammonia/methanol ratio; the proportion of monomethylamine may be increased by diluting the reaction mixture with water or trimethylamine.

Monomethylamine, methylamine, CH_3NH_2. Colourless, inflammable gas with an ammoniacal odour, very soluble in water, m.p.

−92·5°C, b.p. −60°C. Occurs naturally in some plants, in herring brine and in crude bone oil. It may be prepared in the laboratory by heating formalin with ammonium chloride, when it is obtained as the crystalline hydrochloride. Monomethylamine is largely employed in the manufacture of herbicides, fungicides and surface-active agents.

Dimethylamine, C_2H_7N, $(CH_3)_2NH$. Colourless, inflammable liquid with an ammoniacal odour, mp. −96°C, b.p. 7°C. Occurs naturally in herring brine. Prepared in the laboratory by treating nitrosodimethylaniline with a hot solution of sodium hydroxide. Dimethylamine is largely used in the manufacture of other chemicals. These include the solvents dimethylacetamide and dimethylformamide, the rocket propellant unsymmetrical dimethylhydrazine, surface-active agents, herbicides, fungicides and rubber accelerators.

Trimethylamine, C_3H_9N, $(CH_3)_3N$. Colourless liquid with a strong fishy odour, miscible with water, m.p. −124°C, b.p. 3·5°C. It occurs naturally in plants, herring brine, bone oil and urine. It reacts with hydrogen peroxide to give trimethylamine oxide and with ethylene oxide to give choline; its commercial importance stems chiefly from this latter reaction.

The solutions of all the methylamines in water are alkaline, but the alkalinity decreases with the number of methyl groups.

methyl *n*-amyl ketone, $CH_3CO(CH_2)_4CH_3$. B.p. 152°C. Occurs in oil of cloves, manufactured by dehydrogenation of 2-heptanol. Used as a solvent for lacquers and resins.

methylated spirits *See* industrial methylated spirits.

methylation The name given to a process by which a methyl group is added to a compound. In aliphatic chemistry this involves substitution of the hydrogen atom of, e.g., a hydroxyl, amino or imino group, and produces an ether or a secondary or tertiary amine respectively. In aromatic chemistry it may also mean the substitution of one of the hydrogen atoms of the ring by the methyl group; this is carried out by the Friedel–Crafts reaction. Amines and amino-compounds may be methylated by heating with methanal in methanoic acid solution, or by heating with methyl iodide or dimethyl sulphate. For methylation of hydroxy compounds *see* ethers.

methyl bromide *See* bromomethane.

methyl butynol, 2-methyl-3-butyn-2-ol, $HC{\equiv}CC(Me)OHMe$. Colourless liquid, b.p. 104°C, obtained from ethyne and propanone. It is used as a corrosion inhibitor and as a chemical intermediate.

methyl cellosolve *See* ethylene glycol monomethyl ether.

methyl chloride *See* chloromethane.

methylcyclohexane, C_7H_{14}, $C_6H_{11}CH_3$. Colourless liquid, b.p. 101°C, obtained by hydrogenating toluene and used as a solvent.

methylcyclohexanol, methylhexalin, hexahydrocresol, $C_7H_{14}O$. The commercial product is a mixture of the 1,2;1,3;1,4 isomers: It is a colourless, rather viscous liquid, with a penetrating odour, boiling range 165–180°C. Manufactured by reduction of cresols with hydrogen in the presence of nickel catalysts. Oxidized to methylcyclohexanones. Used as a solvent for fats, oils, gums and waxes, and in nitrocellulose lacquers. Soaps containing it are important detergents. The esters are used as plasticizers in lacquers. The individual methylcyclohexanols are prepared in a similar manner from the pure cresol, and exist in *cis* and *trans* modifications. They are volatile in steam.

methylcyclohexanone, $C_7H_{12}O$. The commercial product is a mixture of the three isomers corresponding to methylcyclohexanols. It is a colourless liquid with a penetrating but not unpleasant odour, boiling range of 164–172°C. Soluble to 3% in water; miscible with alcohol and hydrocarbon solvents. Manufactured by passing the vapour of commercial methylcyclohexanol over a heated copper catalyst. Used as solvent for resins and gums in the manufacture of lacquers.

methyldemeton Metasystox*.

methylene The divalent group $=CH_2$. Prepared in the free state by the thermal or photochemical dissociation of keten or diazomethane, and is extremely reactive. *See* carbene.

methylene blue (Basic Blue 8) An important

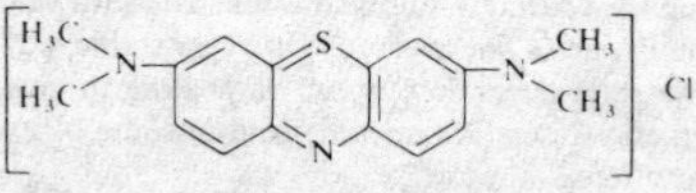

dyestuff prepared by boiling indamine with dilute acid or a solution of zinc chloride. It is used in dyeing fast fibres and in calico-printing and is of considerable value as a staining material in bacteriological work and microscopy and as a mild antiseptic.

methylene chloride *See* dichloromethane.

methylene diisocyanate, MDI, $CH_2(NCO)_2$. Intermediate used in manufacture of urethane foams.

methylenedisulphonic acid *See* methionic acid.

methyl ethyl ketone, MEK *See* butanone.

methyl fluorosulphate, methyl fluorosulphonate, $FS(O)_2OCH_3$. Colourless liquid, b.p. 94°C. Functions as a powerful methylating agent, even for amides and nitriles which are not attacked by conventional alkylating agents like dialkyl sulphates.

methyl glyoxal, pyruvic aldehyde, $C_3H_4O_2$, CH_3COCHO. Yellow liquid, b.p. 72°C, with pungent odour. The liquid readily polymerizes. Obtained by oxidation of propanone with SeO_2 and from the fermentation of sugar.

methylhexalin *See* methylcyclohexanol.

methyl iodide, iodomethane, CH_3I. Colourless liquid with characteristic odour; b.p. 43°C. Prepared by heating methanol and iodine with red phosphorus. Reacts with the hydrogen atoms of hydroxyl and amino-groups to give methoxy- and methylamino-compounds and hence used in organic chemistry for methylating these compounds. Also used for preparing methyl lithium and methyl Grignard reagents.

methyl isobutyl ketone, MIBK, 4-methyl-2-pentanone, hexone, $CH_3COCH_2CHMe_2$. Ketone used in solvent extraction, e.g. for Ta and mineral oils, and as a solvent for many polymers, e.g. acrylic esters, alkyds, polyvinylacetate. Prepared by selective hydrogenation of mesityl oxide.

methyl methacrylate, methyl α-methylacrylate, methyl 2-methylpropenoate, $C_5H_8O_2$, $CH_2=C(CH_3)CO_2Me$. Colourless liquid; b.p. 100°C. Manufactured by heating acetone cyanohydrin with methanol and sulphuric acid. It is usually supplied containing dissolved polymerization inhibitor, on removal of which it is readily polymerized to a glass-like polymer. *See* acrylate resins.

methylnaphthalenes Found extensively in crude oil and reformed petroleum; the various mono-and dimethyl-derivatives are all known and readily interconverted. Oxidized to naphthalene carboxylic acids which are used as dyestuff intermediates.

N-methyl-N′-nitro-N-nitrosoguanidine, $C_2H_5N_5O_3$. M.p. 118°C. Previously used as a precursor for the production of diazomethane* by treatment with 40% aqueous potas-

$$CH_3-N-\overset{\overset{\displaystyle NH}{\|}}{C}-N-NO_2$$
$$\quad\ \ \ | \qquad\qquad |$$
$$\quad\ \ NO \qquad\quad H$$

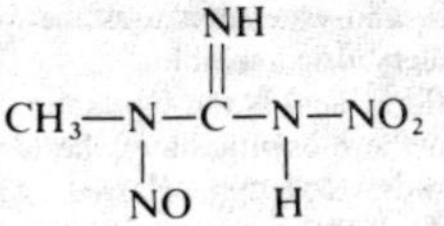

sium hydroxide, the compound is now known to be a potent mutagen for a wide spectrum of organisms and is accepted as carcinogenic.

methyl orange, 4-dimethylamino-4′-azobenzene sodium sulphonate, $C_{14}H_{14}N_3NaO_3S$. Orange crystals prepared by the action of dimethylaniline on diazotized sodium sulphanilate. It is used as an indicator, pH range 3·1 (red) to 4·4 (yellow).

N-methyl-2-pyrrolidone Biodegradable solvent particularly for resins.

methyl red, o-carboxybenzeneazodimethylaniline, $C_{15}H_{15}N_3O_2$. M.p. 181–182°C. An indicator, prepared by the action of dimethylaniline on diazotized anthranilic acid, pH range 4·4 (red) to 6·0 (yellow).

methyl salicylate, methyl (2-hydroxybenzoate), $C_8H_8O_3$. Colourless liquid, m.p. −8·6°C, b.p. 223°C. Exists in an almost pure state in the essential oils of wintergreen and sweet birch, but is mostly prepared by esterifying salicylic acid. It has a pleasant characteristic odour and is readily absorbed through the skin. It is used in perfumery and as a flavouring agent in food, drinks, dentifrices and cosmetics. It has the general medicinal properties of the salicylates, and is applied either alone or with other analgesics for easing the pain of lumbago, rheumatism and sciatica.

methyl succinic acid *See* pyrotartaric acid.

methyl-tert-butyl ether *See* MTBE.

methyl violet (Basic Violet 1) A violet dye obtained by the oxidation of dimethylaniline with $CuCl_2$. It consists of a mixture of the hydrochlorides of tetra-, penta- and hexamethyl-p-rosanilines. It is used in dyeing jute, for colouring methylated spirits, as a bacteriological stain and as an indicator.

metol *See* aminophenols.

mevalonic acid, $C_6H_{12}O_4$. D-Mevalonic acid,

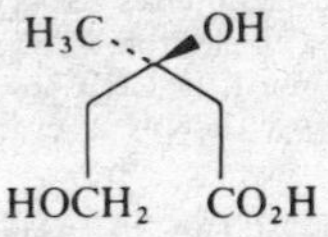

the 3(R) form of 3,5-dihydroxy-3-methylpentanoic acid, was found to replace ethanoate as an essential growth factor for *Lactobacillus acidophilus*, and was isolated as the δ-lactone from distillers' dried solubles.

D-Mevalonic acid is the fundamental intermediate in the biosynthesis of the terpenoids and steroids, together classed as polyisoprenoids. The biogenetic isoprene unit is isopentenyl pyrophosphate which arises by enzymic decarboxylation-dehydration of mevalonic acid pyrophosphate. D-Mevalonic acid is almost quantitatively incorporated into cholesterol synthesized by rat liver homogenates.

Mg Magnesium.

MIBK Methyl isobutyl ketone.

micas Aluminosilicates with layers of linked $(Si, Al)O_4$ tetrahedra similar to those of clays or talc. Readily split into thin insulating sheets; used as fillers and to impart lustre to wallpapers and paints. The most important micas are phlogopite, $KMg_3(OH)_2Si_3AlO_{10}$, and muscovite, $KAl_2(OH)_2Si_3AlO_{10}$. If there are hydrated ions between the layers (instead of K) 'hydrated micas' montmorillonite, vermiculite, normally classified as clays are found. Fluorophlogopite, $K_2Mg_6Al_2Si_6O_{20}F_4$ with other micas is widely used in the electrical industry. World use 1976 230 000 tonnes.

micelle A term used to describe the colloidal particles in aggregation colloids, e.g. soaps and dyes. A micelle is a submicroscopic aggregate of molecules. Occasionally the term micelle is used for polyelectrolyte particles.

Michaelis constant An experimentally determined parameter inversely indicative of the affinity of an enzyme for its substrate. For a constant enzyme concentration, the Michaelis constant is that substrate concentration at which the rate of reaction is half its maximum rate. In general, the Michaelis constant is equivalent to the dissociation constant of the enzyme-substrate complex.

Michael reaction A name originally applied in a restricted sense to the base catalysed addition of active methylene compounds (e.g. malonic ester) to activated unsaturated systems (e.g. $\alpha:\beta$ unsaturated ketones). It is now used in a broader sense to define the conjugate additions of nucleophiles across double bonds conjugated with electron-withdrawing groups such as carbonyl (in $CH_3CH{=}CHCOCH_3$), nitrile ($CH_2{=}CHCN$) (*see* cyanoethylation), nitro, etc.

Michler's hydrol, bis(p-dimethylaminophenyl)-carbinol, $C_{17}H_{22}N_2O$. Colourless crystals; m.p. 96°C.

Prepared by the oxidation of bis(p-dimethylaminophenyl)-methane by PbO_2 and glacial ethanoic acid. It can also be obtained by reduction of Michler's ketone with sodium amalgam in alcohol. It is used for the preparation of triphenylmethane dyestuffs by condensation with bases, phenols and other aromatic compounds.

Michler's ketone, bis(p-dimethylaminophenyl)-ketone, $C_{17}H_{20}N_2O$. Glistening leaflets; m.p. 172°C. Prepared by the action of phosgene on dimethylaniline at 100°C. Reduction with sodium amalgam in alcohol gives Michler's hydrol. Treatment with PCl_3 gives the dichloride. The dichloride is used commercially for the preparation of triphenylmethane dyes by condensation with bases, phenols and many other aromatic compounds.

microbalance A balance capable of detecting and measuring changes in weight of the order of 10^{-6} g or less. For ordinary quantitative analysis on the small scale, the microbalances in use generally weigh up to about 10 g, and are sensitive to a change of 10^{-6} g. For special purposes balances have been constructed which are sensitive to as little as 10^{-11} g, but the maximum load is correspondingly smaller than that of the ordinary analytical microbalance. Electro-microbalances are systems where the displacement is nullified electrically. Such balances are easier to operate and much more robust than conventional microbalances.

microcosmic salt, $NaNH_4HPO_4{,}4H_2O$. *See* sodium ammonium hydrogen phosphate.

microcrystalline wax *See* petroleum wax.

micron, μ One micron $= 10^{-4}$ cm.

microstate In any system of particles the total energy of the system may be distributed among the particles in any one of a large number of equally probable arrangements. Each of these possible arrangements is termed a microstate.

microwaves Electromagnetic radiation of frequency 10^{10} to 10^{12} Hz. Absorption produces rotation in molecules and microwave spectra* are used to determine precise molecular dimensions. Physical effects – heating used for food – and polymerization can be induced by microwaves.

microwave spectroscopy Microwaves are electromagnetic waves with wavelengths in the range 10^{-2}–1 m. Spectra are obtained by passing a microwave beam, produced by an oscillator using a Klystron electronic valve, through the gaseous sample; the transmitted beam being detected by a crystal receiver. The resolution of such a microwave spectrometer is vastly superior to the best infra-red instruments and wavelengths are measured to seven figures. The spectra correspond to transitions between levels very close in energy, and information is derived about pure rotational spectra. Microwave spectroscopy is being used for the identification of molecules in the interstellar medium.

milk of lime A suspension of $Ca(OH)_2$ in water.

millimicron, mμ One millimicron = 10^{-7} cm.

mineral colours Inorganic pigments (prussian blue, chrome yellow, iron buff, etc.), once used extensively when precipitated within fibres, now used for colouring paints, papers, plastics.

mineral dressing, ore dressing Concentration of the valuable constituents of a mineral by the application of purely physical processes. Techniques in use include dense media separation, froth flotation, electrostatic and magnetic separation, and jigging.

mineral solvents *See* white spirits.

minium, Pb_3O_4. *See* lead oxides.

miscibility A term used to denote the extent of mixing. Gases mix in all proportions and are said to be completely miscible. However liquids may be completely miscible, e.g. alcohol and water, only partially miscible, i.e. will only dissolve in each other to a limited extent, e.g. aniline and water; almost completely immiscible, e.g. water and mercury. The miscibility of liquids depends upon their chemical and physical similarities, in particular on their internal pressures. Liquids which have approximately equal internal pressures are miscible in all proportions.

mispickel, arsenopyrite, FeAsS. A whitish common mineral. A principal source of As.

mixed crystals Crystals deposited together from a solution containing two substances (which are normally isomorphous). Also termed solid solutions.

mixed indicator The use of a mixture of two or more indicators* in order to sharpen the colour change or to restrict the pH range (etc.) over which the colour change occurs.

mixed metal oxides Compounds which are formally derived from oxides but contain two or more metal species in an arbitrary ratio. Generally formed by heating mixtures of the appropriate oxy-salts. The structures are determined largely by close packing of the oxide ions and the metal atoms occupy octahedral or tetrahedral holes between the oxygen atoms. Very few mixed metal oxides contain discrete oxyanions.

Mn Manganese.

Mo Molybdenum.

MO Molecular orbital* (theory).

mobility, ionic According to Kohlrausch the equivalent conductivity at infinite dilution (Λ_∞) of a salt can be expressed as the sum of the equivalent ionic conductivities of the cation (λ_+) and anion (λ_-). The absolute velocities of migration of the ions, the ionic mobilities (u), which are the mean velocities of the ions in cm per sec for a potential gradient of 1 volt per cm are given by

$$u_+ = \frac{\lambda_+}{F} \quad \text{and} \quad u_- = \frac{\lambda_-}{F}$$

where F is the Faraday.

Mohr method Titration of Cl^- with Ag^+ in the presence of added CrO_4^{2-}. A red precipitate forms at the end point.

Mohr's salt, $(NH_4)_2SO_4{\cdot}FeSO_4{,}6H_2O$. Iron ammonium sulphate.

molality The concentration of a solution expressed in terms of the number of moles of solute dissolved in 1 litre of solvent. Not to be confused with molarity.

molar conductivity The specific conductance of a solution of an electrolyte multiplied by the volume in ml containing 1 mole of the electrolyte.

molar heat The amount of heat required to raise the temperature of 1 mole of a substance by 1°C, at either constant pressure or constant volume.

molarity The concentration of a solution expressed as the number of moles of the solute in 1 litre of solution.

mole The quantity of a substance which contains one gram formula weight of the substance. One mole of any substance contains $6{\cdot}023 \times 10^{23}$ (Avogadro's number) molecules or atoms. Formerly called one gram molecule.

molecular beam A beam of molecules in which all the molecules have velocities, and therefore energies, which fall into a very narrow range. Such beams may be used to measure molecular velocity distributions, molecular collision cross-sections and in kinetic studies.

molecular diameters Although very few molecules are in fact spherical, for the purpose of chemical kinetics they are assumed to be so. Some typical values are: H_2, $2{\cdot}38 \times 10^{-8}$ cm; O_2, $3{\cdot}19 \times 10^{-8}$ cm; NH_3, $3{\cdot}9 \times 10^{-8}$ cm; and C_6H_6, $6{\cdot}6 \times 10^{-8}$ cm.

molecular distillation A distillation process carried out at very low pressures, 1·3 N/m^2 or less. Under these conditions the mean free path of the molecules is of the same order as the distance between the surface of the liquid being distilled (*the distilland*) and the surface of the condensate (*distillate*), and they can therefore travel between the two with relatively few collisions. Because of this, distillation can proceed at a fast rate and the risk of thermal decomposition is minimized.

Molecular distillation is used in the separation and purification of vitamins and other natural products, and for the distillation of high-boiling synthetic organic compounds.

molecular orbitals The orbitals which belong to a group of atoms forming a molecule. The orbitals may be bonding, anti-bonding or non-bonding according to whether the presence of an electron tends to hold the molecule together, to cause disruption of the molecule or to have no bonding effect on the molecule. The total number of molecular orbitals must be equal to the total number of atomic orbitals used in the bonding. In general only outer electrons need be considered for molecular orbitals.

molecular refractivity *See* specific refractivity.

molecular sieve Zeolites* containing channels which can be used for the specific absorption of, e.g., water or gases. Sieves containing different pore sizes will absorb different molecules.

molecular spectrum *See* band spectrum and continuous spectrum.

molecular weight The weight of one molecule referred to the standard of $^{12}C = 12{\cdot}000$ amu. It is a ratio. The molecular weight is equal to the sum of the atomic weights of the constituent nuclei. The molecular weight expressed in grams is known as the gram molecular weight.

molecule The smallest particle of matter which can exist in a free state (*see* atom). In the case of ionic substances such as sodium chloride, the molecule is considered as NaCl, which exists as an ion-pair in the gas phase, although the solid consists of an ordered arrangement of Na^+ and Cl^- ions.

molecule, mass of The actual mass of a molecule can be obtained by dividing the gram molecular weight by Avogadro's number. Thus the mass of a nitrogen molecule is

$$\frac{28}{6{\cdot}02 \times 10^{23}} = 4{\cdot}65 \times 10^{-23}\,g$$

mole fraction The mole fraction (X_A) of a compound A in a mixture containing in addition only compounds B, C and D is the ratio of the number of molecules of A to the total number of molecules of A + B + C + D. In practice, the concentrations of A, B, C and D are expressed as fractions of the number of moles of each present. The sum of the mole fractions of the components of the system, i.e. $X_A + X_B + X_C + X_D$ is equal to unity.

Molisch's test A general test for carbohydrates. The carbohydrate is dissolved in water, alcoholic 1-naphthol added, and concentrated sulphuric acid poured down the side of the tube. A deep violet ring is formed at the junction of the liquids. A modification, the 'rapid furfural test', is used to distinguish between glucose and fructose. A mixture of the sugar, 1-naphthol, and concentrated hydrochloric acid is boiled. With fructose and saccharides containing fructose a violet colour is produced immediately the solution boils. With glucose the appearance of the colour is slower.

molybdates Molybdates are very similar to tungstates*. Isopolymolybdates $[Mo_7O_{24}]^{6-}$ and $[Mo_8O_{26}]^{4-}$ are well established. Heteropolymolybdates are similar to the corresponding tungstates. Used as corrosion inhibitors for metals, particularly in anti-freezes.

molybdenite, MoS_2. The principal ore of Mo. Has a layer lattice and is used as a lubricant.

molybdenum, Mo. At.no. 42, at.wt. 95·94, m.p. 2617°C, b.p. 4612°C. An element of Group VI electronic configuration $4d^5 5s^1$. Principal ore molybdenite, MoS_2, roasted to MoO_3, converted to ammonium molybdate, purified, heated to MoO_3 and reduced to Mo with hydrogen. The metal is bcc. The massive metal is lustrous and attacked only by $HNO_3 - HF$ or fused $KNO_3 - NaOH$ or Na_2O_2. Mo metal combines with oxygen at red heat and slowly assumes a blue patina in air. Used extensively in steels, particularly cutting steels for high-temperature use, and as a filament material. MoS_2 is an important solid lubricant and Mo compounds are used in pigments. World production Mo 1976 90 000 tonnes. Mo is a necessary trace element in animal diets and the active enzymes of nitrogen-fixing bacteria contain Mo and Fe. World use 1976 900 000 tonnes.

molybdenum blue *See* molybdenum oxides.

molybdenum carbonyl derivatives The parent carbonyl, $Mo(CO)_6$, is a colourless sublimable solid formed CO + reducing agent + $MoCl_5$. Carbonyl groups can be substituted by many other ligands to give, e.g. $Mo(CO)_4(PPh_3)_2$.

molybdenum chemistry Molybdenum is a typical element of transition Group VI and shows oxidation states from +6 to −2. There is little aqueous chemistry except that of complex oxyanions and some complex halides. MoF_6 and $MoCl_6$ are molecular and lower oxidation state halides polymeric with some Mo−Mo bonding. Carbonyl and phosphine derivatives are typical low oxidation state compounds. Complexes are formed with most donor atoms but particularly with *O*− and *S*− ligands in higher oxidation states. Cyanide and thiocyanate (N-bonded) complexes are well established. N_2 complexes are known and analogues are presumably formed during nitrogen fixation.

molybdenum halides The known fluorides are MoF_6 ($Mo + F_2$), b.p. 35°C; tetrameric MoF_5 ($Mo + MoF_6$), Mo_2F_9 ($Mo(CO)_6 + F_2$), MoF_4 (heat on Mo_2F_9) and MoF_3 ($Mo + MoF_6$). Complex fluorides containing Mo(VI), (V) and (III) are known containing ions such as MoF_6^-. The known chlorides are grey $MoCl_6$ ($MoO_3 + SOCl_2$), dimeric $MoCl_5$ ($Mo + Cl_2$), $MoCl_4$ ($MoCl_5 + C_2Cl_4$), $MoCl_3$ and $MoCl_2$ (reduction of $MoCl_5$). Both $MoCl_3$ (Mo_2 pairs) and $MoCl_2$ ($[Mo_6Cl_8]Cl_4$) contain metal–metal bonds. Complex chlorides containing e.g. $MoCl_6^-$ are known. Oxide halides are formed by hydrolysis of halides or by halogenation of oxides. The halides and oxide halides all form complexes with a range of ligands. Bromides and iodides, MoX_4, MoX_3 and MoX_2 are similar to chlorides.

molybdenum organometallic derivatives A few relatively stable molybdenum alkyls are known (e.g. $MoCl_3Me{\cdot}Et_2O$[$MoCl_4 + ZnMe_2$], Li_4-$[Mo_2Me_8]{\cdot}4Et_2O$ [$Mo(O_2CCH_3)_2 + LiMe$]). Cyclopentadienyl derivatives, e.g. $[(h^5-C_5H_5)$-$Mo(CO)_3]_2$ and many π-acetylene complexes are known and stable.

molybdenum oxides White MoO_3 (heat on the hydrated oxide formed by acidification of a molybdate), violet Mo_2O_5 ($MoO_3 + Mo$), brown–violet MoO_2 ($MoO_3 + H_2$). MoO_2 contains Mo−Mo bonds and non-stoicheiometric phases are also known. Mild reduction of hydrated MoO_3 gives a blue hydrous oxide, molybdenum blue. MoO_3 forms mixed oxides on fusing with other oxides – these contain linked MoO_6 octahedra rather than discrete anions. MoO_3 is formally the acid oxide of molybdates(VI)*.

molybdenum sulphides MoS_3, MoS_2, Mo_2S_3. MoS_2 has a layer structure which leads to its use as a solid lubricant.

molybic acids *See* molybdates.

monazite A mixed phosphate (Ce,La,Nd,Pr)PO_4 containing some thorium silicate (ThO_2 1–18%). The principle lanthanide ore.

Mond process A process for the extraction of nickel from nickel ores. *See* nickel.

monel A nickel alloy.

monoacetin *See* acetins.

monobasic acid An acid which has only one replaceable hydrogen and forms only one series of salts, e.g. HCl and H_3PO_2 ($HOP(O)_2H_2$).

monochloroethane, C_2H_5Cl. *See* ethyl chloride.

monochloroethylene, $CH_2=CHCl$. *See* vinyl chloride.

monochloromethane, CH_3Cl. *See* chloromethane.

monochromatic radiation Radiation of a single wavelength. *See* monochromator.

monochromator A device, prism, grating, single crystal for selecting a single wavelength from a wide wavelength range of radiation or particles. Used in spectrophotometers and in X-ray diffraction equipment.

monoclinic system The crystal system* with one 2-fold axis and/or one plane of symmetry. A unit cell has three unequal axes one perpendicular to the other two which are obliquely inclined. E.g. $Na_2CO_3,10H_2O$.

monodentate ligand A ligand with only one co-ordinately active lone pair of electrons which interact through a single atom, e.g. NH_3.

monolayers *See* unimolecular films.

monomer A simple molecule which can be polymerized or telomerized.

monosaccharides Sugars with the formula $C_nH_{2n}O_n$ (n = 5 or 6).

monosodium glutamate *See* sodium glutamate.

monotropy The thermodynamic stability of only one form of a substance. All other forms are metastable and tend to change to the stable form. Phosphorus has only the violet form stable; the common yellow form is metastable.

monovinylacetylene, buta-1-ene-3-yne, $CH_2{:}CH{\cdot}C{:}CH$. Colourless gas with a sweet odour; b.p. 5°C. Manufactured by the controlled low-temperature telomerization of ethyne in the presence of an aqueous solution of CuCl and NH_4Cl. Reduced by hydrogen to butadiene and, finally, butane. Reacts with water in the presence of $HgSO_4$ to give methyl vinyl ketone. Forms salts. Forms 2-chlorobutadiene (chloroprene) with hydrochloric acid and certain metallic chlorides.

montanic acid, octacosanoic acid, $C_{27}H_{55}COOH$. Glistening scales; m.p. 89°C. A straight-chain, saturated fatty acid, that occurs in many natural waxes.

montmorillonite A clay mineral, an aluminosilicate with variable composition. Has a high cation exchange capacity. Two varieties are known, one swelling and giving gels in water the other having marked absorptive properties. Bentonite* consists mainly of montmorillonite.

mordant A substance used to fix or develop the colour in dyeing.

mordant dyes *See* metallizable dyes.

morphine, $C_{17}H_{19}NO_3$ Colourless prisms with one molecule of water; m.p. (anhydrous) 254°C, with decomposition.

Morphine is the principal alkaloid of opium. It acts both as a base and as a phenol and reacts to form methylmorphine (codeine*) and diacetylmorphine (diamorphine* or heroin).

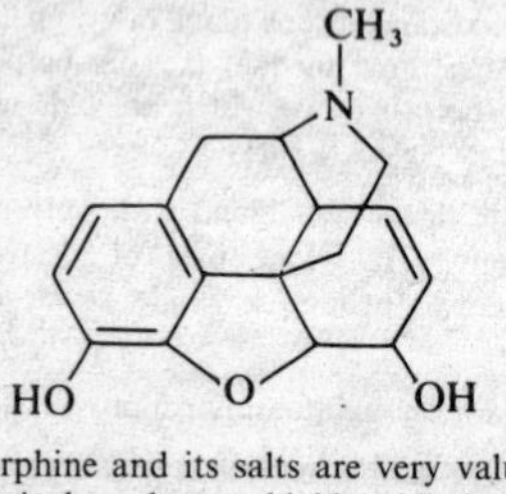

Morphine and its salts are very valuable analgesic drugs but are highly addictive. In addition to suppression of pain, morphine causes constipation, decreases pupillary size and depresses respiration. Only the (+)-stereoisomer is biologically active. They appear to produce their effects on the brain by activating neuronal mechanisms normally activated by the naturally occurring pentapeptides the enkephalins.

morpholine, C_4H_9NO. A colourless liquid

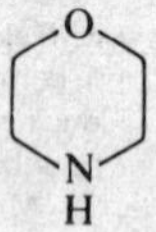

with a strong ammoniacal odour; b.p. 129°C. Manufactured by heating $\beta\beta'$-dichlorodiethyl ether with ammonia, or by heating diethanolamine with 70% sulphuric acid. It is a moderately strong base, forming soaps with fatty acids. Absorbs water and carbon dioxide from the air to give morpholine carbamate, a colourless, crystalline solid. Dilute solutions in water distil with little change in composition. It has a high solvent power for a very wide range of substances. Used as a solvent and for decreasing corrosion in boilers. Its soaps are used as emulsifying agents.

mosaic gold, SnS_2. *See* tin sulphides.

Mössbauer effect The resonance fluorescence by γ-radiation of an atomic nucleus, returning from an excited state to the ground state. The resonance energy is characteristic of the chemical environment of the nucleus and Mössbauer spectroscopy may be used to yield information about this chemical environment. Used particularly in the study of Fe, Sn and Sb compounds.

motor spirit Fuel with a boiling range of approximately 310–450 K used for internal combustion engines. *See* gasoline.

moulding The making of shaped objects in metals, plastics or other materials. Moulding can involve use of liquid material or may involve fashioning under pressure.

MSG Sodium glutamate.

MTBE Methyltert-butyl ether. Used as a gasoline additive.

mu-, μ- Designation of a bridging species. E.g. gaseous aluminium chloride is $[Cl_2Al(\mu\text{-}Cl)_2AlCl_2]$.

mucic acid, 2,3,4,5-tetrahydroxyhexanedioic acid, $C_6H_{10}O_8$. Colourless crystals, m.p.

$$HO_2C\cdot\overset{H}{\underset{OH}{C}}\cdot\overset{OH}{\underset{H}{C}}\cdot\overset{OH}{\underset{H}{C}}\cdot\overset{H}{\underset{OH}{C}}\cdot CO_2H$$

206°C (rapid heating). Manufactured by the oxidation of lactose or the galactans from wood with nitric acid. When heated with water it forms a soluble lactone. Converted to furoic acid by heat, and to allomucic acid, an optical isomer, by heating with pyridine. Distillation of the ammonium salt gives pyrrole. Used in the manufacture of pyrrole and other heterocycles.

mucilages Polysaccharides, generally containing galacturonic acid, xylose and arabinose residues which swell in water.

mucochloric acid, 2,3-dichloro-4-oxo-2-butenoic acid, $C_4H_2Cl_2O_3$. Colourless plates; m.p. 127°C. More correctly represented lactone, it is prepared from furfuraldehyde and chlorine in aqueous solution.

Used to harden gelatine.

mucopeptides Structural components of the cell walls of gram-positive and many gram-negative bacteria. They are insoluble polymers of *N*-acetylglucosamine and *N*-acetylmuramic acid, probably involving $\beta(1 \longrightarrow 4)$-glycosidic linkages; small peptide residues, usually composed of D- or L-alanine, D-glutamic acid, and either lysine or α,ε-diaminopimelic acid, are attached to the polymer. The antibacterial action of penicillin is ascribed to its inhibition of cell-wall synthesis, possibly by prevention of the interlinking of mucopeptide chains through the peptide residues, or by interference with biosynthesis of one of the essential constituents such as muramic acid.

mucoproteins *See* proteins.

muffle furnaces A furnace designed so that the charge does not come into contact with the hot combustion gases. This is effected by putting the charge in a retort or 'box' heating the outside walls.

Mulliken symbols The designators, arising from group theory, of the electronic states of an ion in a crystal field. A and B are singly degenerate, E doubly degenerate, T triply degenerate states. Thus a D state of a free ion shows E and T_2 states in an octahedral field.

mullite, $Al_6Si_2O_{13}$. The stablest aluminium silicate, formed by heating other aluminosilicates; used in refractories and glasses.

multicentre bond A bond formed by a bonding molecular orbital between three or more atoms containing a single pair of electrons. Such bonds occur in the boron hydrides.

multidentate ligand, polydentate ligand A ligand which has two or more sites at which it can co-ordinate. Such a ligand can form a bridge or can form a chelate compound. An example is $H_2NCH_2CH_2NH_2$.

multiple bonding Bonding between atoms in which there is overlap of orbitals in more than one position in space. The electrons may come equally from each atom or be derived from only one of the atoms (co-ordinate bond, back bonding). Generally the primary bond is a σ-bond with extra interaction by π- or δ- bonding.

multiple-effect evaporator In a simple evaporator, containing only one evaporator unit or *effect*, approximately 1 kg of steam is required to evaporate 1 kg of water. A multiple-effect evaporator contains several evaporator units in series, and the steam generated in one effect is fed to the calandria of the next, where it acts as a heating medium and produces further evaporation. Thus 1 kg of steam is able to evaporate, very approximately, *N* kg of water, where *N* is the number of effects.

multiple proportions, law of Originally proposed by Dalton in 1804, the law states that when two elements A and B combine to form more than one compound, the weights of B which combine with a fixed weight of A are in the proportion of small whole numbers. E.g. in dinitrogen oxide, nitrogen oxide and dinitrogen tetroxide the amounts of oxygen which combine with one part of nitrogen are in the proportion 1:2:4. This law is only of very limited value as very many examples of non-stoicheiometric compounds are known.

multiplet In a spectrum the individual lines often consist of two or more closely grouped fine lines, which make up the fine structure of the original line. Each such group, which constitutes a line in the ordinary spectrum, is called a multiplet. In an electronic spectrum this fine structure arises from the slightly differing amounts of energy an electron possesses due to its spin. The multiplicity is the maximum possible number of such energy values which an atom in a specified state can possess by virtue of the electron spin. A multiplet which arises from an electronic transition where there are two spin states is called a *doublet*; where there are three spin states a *triplet* is found.

In addition to this electron spin fine structure there are often still finer lines present. These are known as the *hyperfine* structure, which arises from the different weights of the isotopes of an element or from the spin of the nucleus.

In nuclear spectroscopy fine structure arises from coupling between nuclear spins.

muon A μ-meson. *See* meson.

muonium An electron plus a muon, formally analogous to hydrogen.

muriatic acid An old name for hydrochloric acid, still occasionally used in industry. Potassium chloride is sometimes called 'muriate of potash'.

muscone, 3-methylcyclopentadecanone,

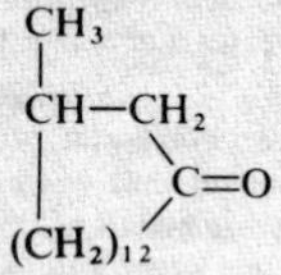

$C_{16}H_{30}O$. A yellow liquid; b.p. 330°C. The perfume base obtained from the scent glands of the Tibetan musk deer; also available by synthesis.

musk, artificial, $C_{11}H_{13}N_3O_6$.
(1) *2,4,6-Trinitro-3-tert-butyltoluene*, m.p. 97°C, yellow crystals. Used in perfumery and known as *musk baur*.

(2) *Musk ketone* is 4-*tert*-butyl-2,6-dimethyl-3,5-dinitroacetophenone, yellow crystals, m.p. 135–136°C.

(3) *Musk xylene* is 1-*tert*-butyl-3,5-dimethyl-2,4,6-trinitrobenzene, m.p. 105°C.

Both (2) and (3) are used in artificial perfumes.

mustard gas, 2,2′-dichlorodiethyl sulphide, $C_4H_8Cl_2S$, $(CH_2ClCH_2)_2S$. Colourless oily liquid with a faint garlic-like odour; m.p. 13–14°C, b.p. 215–217°C. Manufactured by treating S_2Cl_2 with ethene at 30–35°C. Decomposed violently by bleaching powder. It is a powerful vesicant and poison, and causes conjunctivitis and temporary blindness. It is an important agent in chemical warfare.

mustard oil *See* allyl isothiocyanate.

mutagens Chemical substances or physical agents which raise the frequency of mutation greatly above the normal level. Many chemical mutagens are carcinogens.

mutarotation The spontaneous change of optical activity of an optically active substance when it is dissolved in water, or some other solvent. Mutarotation is frequently catalysed by acids or bases, or by both.

mutases *See* isomerases and enzymes.

Mylar A polyester film.

myoglobin The oxygen-carrying protein of mammalian muscle. Myoglobin is used by muscle as a rapidly available source of oxygen.

myosin *See* actomyosin.

myrcene, $C_{10}H_{16}$,
$Me_2C{=}CHCH_2CH_2C({=}CH_2)CH{=}CH_2$.
An acyclic monterpene, b.p. 166–168°C, found in many essential oils, e.g. in verbena oil and oil of hops.

myricyl alcohol *See* melissyl alcohol.

myristic acid, *n*-tetradecanoic acid, $C_{14}H_{28}O_2$, $CH_3{\cdot}[CH_2]_{12}{\cdot}COOH$. Crystallizes in leaflets, m.p. 58°C, b.p. 250°C/100 mm. A fatty acid occurring as glycerides in milk and in large quantities in certain vegetable oils.

N

N Nitrogen.

Na Sodium.

nacrite *See* kaolinite.

NAD *See* nicotinamide adenine dinucleotide.

NADP *See* nicotinamide adenine dinucleotide phosphate.

napalm The aluminium soap of naphthenic* and palmitic acids used to gel gasoline. The thickened gasoline has been used for military flame throwers and incendiary bombs. *See* metallic soaps.

naphtha A term used to describe special boiling point* spirits (SBP) but also used as a description of light distillate feedstock for gas or petrochemical manufacture. The boiling range is generally about 40–150°C.

naphthacene ring system The system numbered as shown.

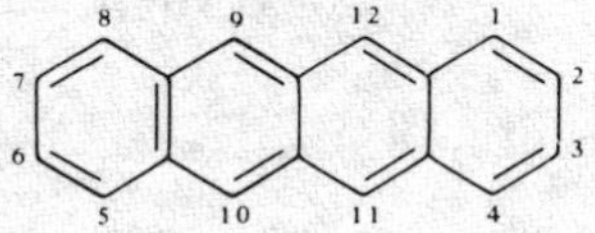

naphthalene, $C_{10}H_8$. Very volatile in steam, readily sublimes at low temperatures; m.p. 80°C, b.p. 218°C. White crystalline solid with a penetrating tarry smell; it burns with a very smoky flame. Obtained commercially from petroleum fractions by demethylation of methylnaphthalenes with H_2 at 750°C and 10–70 atm.

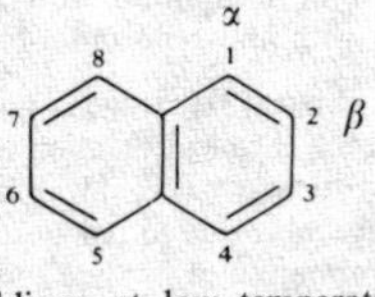

It is a typically aromatic compound and gives addition and substitution reactions more readily than benzene. Can be reduced to a series of compounds containing 2–10 additional hydrogen atoms (e.g. tetralin, decalin), which are liquids of value as solvents. Exhaustive chlorination gives rise to wax-like compounds. It gives rise to two series of monosubstitution products depending upon whether the substituent is in the 1- or 2-position. Readily nitrates and sulphonates to give valuable dyestuffs intermediates, the substituent entering chiefly into the 1-position except in the case of sulphonation at a high temperature (150°C) which gives a mixture containing 80% of naphthalene-2-sulphonic acid.

Most naphthalene produced is utilized in the manufacture of phthalic anhydride, for plasticizers, alkyd resins and polyesters. It is also used in the manufacture of 2-naphthol and insecticides. Naphthalene derivatives are of importance, particularly as dyestuff intermediates.

naphthalene-sodium, $[C_{10}H_8]^- Na^+$. Sodium metal reacts with naphthalene in certain ethers like THF* or diglyme* to give a deep green solution. The metal dissolves by transferring an electron to the naphthalene forming a radical anion. Used as a reagent for metalation reactions* for which it is better than sodium alone.

naphthalene sulphonic acids Naphthalene monosulphonic acids are obtained by direct sulphonation of naphthalene with concentrated sulphuric acid. A mixture of the two possible isomers is always obtained in proportions varying with the temperature of reaction. Fusion with NaOH gives the appropriate naphthol.

Naphthalene-1-sulphonic acid crystallizes with $2H_2O$, m.p. 90°C. Used for the preparation of 1,8- and 1,5-nitronaphthalene sulphonic acids.

Naphthalene-2-sulphonic acid crystallizes with $3H_2O$, m.p. 83°C. It is also used for the preparation of nitro-derivatives.

Naphthalene disulphonic acids are prepared by more prolonged sulphonation than in the preparation of the monosulphonic acids. Four isomeric acids are obtained.

Naphthalene trisulphonic acids can be obtained by more drastic sulphonation of naphthalene or its mono- and disulphonic acids. Only the 1,3,5-, 1,3,6- and 1,3,7-acids are obtained. The most important of the trisulphonic acids is the 1,3,6-acid which is used for the preparation of H-acid, a dyestuff intermediate.

naphthalic acid 1,8-Naphthalene dicarboxylic acid.

naphthenes Cyclo-alkanes; saturated hydrocarbons, containing at least one closed ring of carbon atoms, e.g. cyclohexane.

naphthenic acids Carboxylic acids obtained from crude petroleum. The water-soluble sodium salts of the acids are then decomposed with mineral acids. Many belong to a series of general formula $C_nH_{2n-1}COOH$ and are derivatives of cyclopentane, but more complicated alicyclic ring systems have also been found. The sodium salts are important detergents and emulsifying agents, while many other salts have importance – *see* metallic soaps.

naphthoic acids Naphthalene carboxylic acids. Prepared by oxidizing methylnaphthalenes.

1-naphthol, α-naphthol, $C_{10}H_8O$.
Colourless; m.p. 94°C, b.p. 278–280°C.

Usually prepared on the large scale by caustic soda fusion of sodium naphthalene-1-sulphonate, but can also be obtained by high-temperature alkaline digestion of 1-chloronaphthalene. Oxidized by acid permanganate or nitric acid to phthalic acid. Gives nitroso-compounds when treated with nitrous acid.

Can be coupled in the 4-position with diazotized bases to give a series of azo dyestuffs; more usually it is first sulphonated to give a valuable series of intermediates for solubilized azo dyestuffs. The nitro-1-naphthols are themselves dyestuffs, e.g. 2,4-dinitro-1-naphthol ('naphtol yellow').

2-naphthol, β-naphthol, $C_{10}H_8O$. White crystals, slightly pink when impure, m.p. 122°C, b.p. 285–286°C.

Prepared commercially by NaOH fusion of sodium naphthalene-2-sulphonate.

With nitrous acid it gives 1-nitroso-2-naphthol. It can also be chlorinated and sulphonated. Oxidized ultimately to phthalic acid on prolonged oxidation.

It couples readily with diazotized bases to give extensive series of azo dyes, and may also couple on the fibre to give ingrained dyes. These azo dyes have a hydroxyl group ortho to the azo group which is of special value for after-treatment with co-ordinating metals, e.g. after-chroming with dichromate. It gives rise to important series of dyestuff intermediates obtained by sulphonation and nitration.

Although 2-naphthol is employed chiefly for the manufacture of dyes, considerable quantities are used in the manufacture of anti-oxidants.

1-naphthylamine, α-naphthylamine, $C_{10}H_9N$. Colourless crystals; m.p. 50°C, b.p. 301°C. Basic and forms sparingly soluble salts with mineral acids. Prepared by the reduction of 1-nitronaphthalene with iron and a trace of hydrochloric acid or by the action of ammonia upon 1-naphthol at a high temperature and pressure.

1-Naphthylamine readily diazotizes and couples to aromatic hydroxylic or basic compounds. It was thus used as a first component in a number of important monoazo dyes, but its use has been severely curtailed because of its potent carcinogenicity. It sulphonates to give naphthionic acid (1-naphthylamine-4-sulphonic acid).

2-naphthylamine, β-naphthylamine, $C_{10}H_9N$. Crystallizes in lustrous leaflets; m.p. 112°C, b.p. 294°C. It is prepared by Bucherer's method; 2-naphthol is heated with a strong solution of ammonium sulphite and ammonia at 150°C.

It was used as an end component in a few azo-dyes, but this use has been discontinued because of its carcinogenic character.

1-naphthyl-N-methylcarbamate, $C_{12}H_{11}NO_2$.

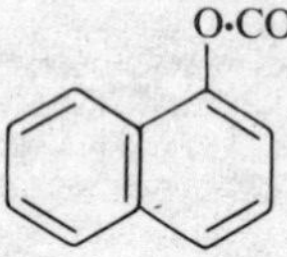

A contact insecticide with the trade name 'Sevin'. White solid, m.p. 142°C. It is prepared by reaction of 1-naphthol with methyl isocyanate or with phosgene and a base.

nascent hydrogen Hydrogen produced by electrolysis or chemical reaction and used *in situ* for reduction. The powerful reducing action is probably due to reaction at surfaces or to the formation of hydride species.

natrolite, $Na_2(Al_2Si_3O_{10}),2H_2O$. *See* zeolites.

natural gas Gas obtained from underground accumulations which may or may not be directly associated with crude oil. Such gas would normally contain at least 95% methane and other low-boiling hydrocarbons, the remainder being nitrogen and carbon dioxide. If oil is not present the gas is normally *dry*, i.e. methane is the main hydrocarbon constituent. When natural gas is associated with crude oil it frequently contains appreciable amounts of higher hydrocarbons, pentanes, hexanes, etc., and is referred to as *wet* gas. Used as a fuel and raw material.

Nb Niobium.

Ne Neon.

Néel point The temperature at which magnetic susceptibility becomes normal. *See* antiferromagnetism and ferromagnetism.

negative adsorption When adsorption at a surface occurs from a solution the concentration of some components may be higher at the surface than in the bulk, whilst with other components the concentration will necessarily be less at the surface than in the bulk. These latter components are said to undergo negative adsorption. According to the Gibbs' adsorption equation any solute which increases the interfacial tension of the system will be negatively adsorbed, e.g. the negative adsorption of sodium chloride at the air/water interface, which has been confirmed experimentally.

nematic liquid crystals *See* liquid crystals.

neodymium, Nd. At.no. 60, at.wt. 144·24, m.p. 1010°C, b.p. 3127°C, *d* 7·00. A typical lanthanide*. Nd is used in glasses and in capacitors.

neodymium compounds Stable Nd compounds are typical lanthanide compounds in the +3 state

Nd^{3+} (f^3 red) $\longrightarrow$ Nd (−2·44 volts in acid)

A +4 fluoride, Cs_3NdF_7, has been prepared by the action of F_2 on $CsCl + NdCl_3$. $NdCl_2$($NdCl_3$ + Nd) and NdI_2 are known but it is not certain that these compounds contain Nd^{2+} ions.

neon, Ne. At.no. 10, at.wt. 20·179, m.p. −248·67°C, b.p. −246·05°C. A relatively abundant noble gas (1·82 × 10^{-3}% of air). Obtained pure by fractionation of liquid air. Widely used in fluorescent tables (red neon lights), lighting and Geiger–Muller tubes, electrical equipment and in gas lasers. The electronic configuration is $2s^2 2p^6$ and Ne forms no normal chemical compounds.

neopentyl The group Me_3CCH_2–.

neophyl The group $PhMe_2CCH_2$–.

neoprene, polychloroprene Synthetic rubbers obtained by emulsion polymerization of chloroprene, $CH_2{=}CCl{\cdot}CH{=}CH_2$. The mol.wt. is controlled by addition of sulphur or sulphur-containing additives during the polymerization. Neoprene has high tensile strength, resilience and resistance to abrasion. It is resistant to solvents and is used in automotive goods, in wire and cable sheathing, road and building construction, and as a fibre binder for e.g. paper. U.S. production 1978 161 000 tonnes.

neoretinene *See* rhodopsin.

nephelauxetic effect The changes in the electronic (d $\longrightarrow$ d, etc.) spectra of complexes due to varying degrees of covalent character in the metal–ligand bonds. Measured as the ratio, β, of the Racah parameter B in the complex as compared with B in the free ion. The greater β the greater the ionic character in the bond.

nephelometry A method of quantitative analysis which involves the spectrophotometric estimation of the scattering of light by a colloidal suspension of a precipitate.

neptunium, Np. At.no. 93, at.wt. 237·0482, m.p. 640°C, b.p. 3902°C, *d* 20·45. Traces of Np occur naturally, being formed by neutron capture by natural uranium, but Np is normally prepared artificially by neutron irradiation of ^{238}U or ^{235}U [^{237}Np (2·2 × 10^6 years)]. Np is separated by selective oxidation and solvent extraction. The metal is formed by reduction of NpF_3 with lithium; there are six crystal forms. Neptunium is converted into ^{238}Pu for use as a power source.

neptunium compounds Neptunium is a typical early actinide element and forms compounds similar to those of uranium but also showing a strongly oxidizing +7 oxidation state.

$$NpO_2^{2} \text{ (pink)} \xrightarrow{+1\cdot14} NpO_2 \text{ (green)} \xrightarrow{+0\cdot74} Np^4 \text{ (yellow-green)} \xrightarrow{+1\cdot15} Np^3 \text{ (blue)} \xrightarrow{-1\cdot83} Np \text{ (volts in 1M acid).}$$

The metal dissolves in acid to give Np^{3+} oxidized by air to Np^{4+} and to NpO_2^+ by mild oxidants. Electrolytic oxidation gives green NpO_5^{3-} stabilized by large cations. NpF_6 (NpF_4 plus fluorine, m.p. 55°C orange) and NpF_5 (reduction of NpF_6 with iodine) are the only +6 and +5 halides. NpF_4 is formed by the action of HF and oxygen on NpO_2 ($NpCl_4$ and $NpBr_4$ are also known); all NpX_3 are formed from aqueous solution. The stable binary oxide is NpO_2. Complex fluorides of Np(V) and Np(IV) and complex chlorides of Np(IV) are formed from aqueous solution. The NpO_2^{2+} and NpX_4 derivatives form many complexes particularly with oxygen-donors.

neral *See* citral.

Nernst equation This equation relates the e.m.f. of a cell to the concentrations or, more accurately, the activities of the reactants and products of the cell reaction. For a reaction

$$aA + bB \rightleftharpoons cC + dD$$

the e.m.f., *E* of the cell is given by

$$E = E^\circ - RT\ln\left(\frac{a_C^c a_D^d}{a_A^a a_B^b}\right)$$

where E° is the standard e.m.f of the cell and the a_n are the activities of the reactants and products under a given set of conditions.

nerol, 3,7-dimethyl-2,6-octadien-1-ol, $C_{10}H_{18}O$. B.p. 225–226°C. A terpenic alcohol and a constituent of neroli, petit-grain and bergamot, and of many other essential oils. Nerol has a blander smell than its isomer, geraniol, and is more valuable as a constituent of perfumes.

nerolidol, $C_{15}H_{26}O$. B.p. 276°C. A sesquiterpene alcohol, the (+)-form of which is found in neroli oil and in Peru balsam. The corresponding (±)-compound has been obtained synthetically. On oxidation with chromic acid it yields farnesal.

nerve gases *See* fluorophosphoric acids.

nervone, $C_{48}H_{91}NO_8$. A cerebroside, crystallizing in needles from alcohol, m.p. 180°C. It hydrolyses to D-galactose, sphingosine and nervonic acid.

nervonic acid, *cis*-14-tetracosanic acid, $C_{24}H_{46}O_2$, $CH_3{\cdot}[CH_2]_7{\cdot}CH{:}CH{\cdot}[CH_2]_{13}{\cdot}CO_2H$. Crystalline powder, m.p. 43°C, obtained by hydrolysis of nervone*.

Nessler's reagent An alkaline solution of HgI_2 in KI used for detecting and estimating ammonia (brown colour or precipitate formed).

Nessler tubes Cylinders of thin glass, generally graduated, used for comparing turbidities and colours of solutions.

neuraminic acid, $C_9H_{17}NO_8$. An important

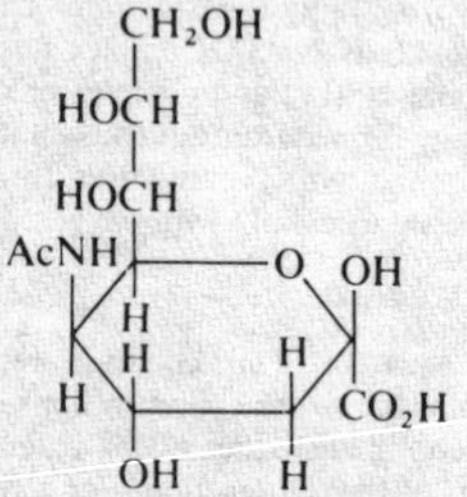

acetylneuraminic acid

amino sugar which is widely distributed in animal tissues and secretions especially in the form of the *N*-acyl derivatives (*see* sialic acids).

neurine, trimethylvinylammonium hydroxide, trimethylethenylammonium hydroxide, $C_5H_{13}NO$, $[Me_3NCH{=}CH_2]^+OH^-$. A liquid forming a crystalline trihydrate. It is present free and combined in brain and other animal and vegetable products and is formed as a product of putrefaction of lecithin. It can be prepared synthetically from choline and decomposes easily to trimethylamine.

neutralization, heat of The amount of heat evolved when 1 g equivalent of an acid is neutralized by 1 g equivalent of a base. For strong acids and strong bases in dilute solution the only reaction which occurs is $H^+ + OH^- \longrightarrow H_2O$ and the heat of neutralization has a constant value of 57·35 kJ.

neutrino An elementary particle without charge and with spin ½, postulated to account for the conservation of angular momentum in nuclear transformations. Neutrinos appear to have variable finite mass and to oscillate from one type to another.

neutron One of the fundamental particles in an atom. It has almost the same mass as a proton but is electrically neutral. A free neutron, obtained from an atomic pile, has a mean lifetime of 750 s; it decays to yield a proton and an electron. A neutron has magnetic moment 1·91 B.M. and interacts with atoms with unpaired electrons. Neutrons have been used for the determination of molecular structure by the diffraction which occurs when a substance is subjected to a neutron flux.

Neville-Winther acid The trivial name given to 4-hydroxy-1-naphthalene sulphonic acid.

Newlands's law *See* octaves, law of.

Newman projections *See* sawhorse projections.

Ni Nickel.

niacin *See* nicotinic acid.

nichrome A nickel alloy.

nickel, Ni. At.no. 28, at.wt. 58·69, m.p. 1453°C, b.p. 2732°C, *d* 8·90. Nickel is a transition element of Group VIII occurring naturally as mixed metal sulphide ores (e.g. millerite, pentlandite), as mixed arsenides, antimonides, sulphides, as garnierite and in the iron ore pyrrhotite, Fe_nS_{n+1}. The ore is roasted to NiO which is reduced to Ni with carbon. The details of the final purification depend upon the type of the ore but may involve electrolytic processes or conversion to $Ni(CO)_4$ at 50°C with CO followed by decomposition to pure Ni at

180°C (Mond process). The silver-white metal is ccp. The metal is not tarnished by air but is attacked by acids (except conc. HNO_3). It is resistant to F_2. It is used extensively in alloys, particularly in steels and cast iron and as a coinage metal. Used in glass (green) in catalysts (particularly for hydrogenation). Western world production 1981 662 000 tonnes.

nickel accumulator The nickel accumulator uses the reaction

$$Fe(s) + NiO{\cdot}OH + H_2O \longrightarrow Fe(OH)_2(s) + Ni(OH)_2$$

The reaction is carried out in an alkaline medium, usually potassium hydroxide with some lithium hydroxide. On charging, the $Fe(OH)_2$ is reduced to Fe and the $Ni(OH)_2$ oxidized to nickel(III) oxide. Its discharge-voltage is only about 1·2 V, and as compared with the lead accumulator it has a rather low current efficiency and is rather more expensive. It has some compensating advantages in lower weight and greater mechanical strength, for the electrode materials are compressed into perforated pockets of nickel-plated steel, and the accumulator can therefore withstand mechanical shocks and high rates of charge or discharge.

nickel alloys Have corrosion and scaling resistance but are expensive and so are confined to specialized applications. Examples are monel, hastelloy, michrome, inconel, nimonics, permalloy, invar, nilo.

nickel ammines Anhydrous Ni salts and Ni salts in the presence of aqueous ammonia form ammines, e.g. $[Ni(NH_3)_6]^{2+}$ (deep blue). Many of the ammines lose ammonia on exposure to air.

nickel ammonium sulphate, $Ni(NH_4)_2(SO_4)_2{,}6H_2O$. Blue-green crystalline material formed from a solution of the components. Used in electroplating.

nickel arsenide, NiAs. An important structural type, the As is hcp with the Ni in octahedral holes and the As in trigonal prismatic co-ordination. This structure is adopted by, e.g., CoTe, CrSb, FeS, NbS.

nickelates Nominally any anionic nickel-containing species; practically oxo species. Nickelates(IV) are ill defined (NiO plus alkali plus KNO_3 or O_2). Nickelates(III), $MNiO_2$, are prepared similarly.

nickel bromide, $NiBr_2$. Forms hydrates.

nickel carbonates Unstable ill-defined basic materials precipitated from solution. $NiCO_3{,}6H_2O$ ($NaHCO_3$ solution plus CO_2) and $NiCO_3$ ($CaCO_3$ plus $NiCl_2$) are both relatively unstable.

nickel carbonyl, $Ni(CO)_4$. M.p. −25°C, b.p. 43°C. Toxic material prepared Ni plus CO at less than 100°C. Decomposes to pure Ni at higher temperatures (Mond process for purifying Ni). Forms extensive ranges of substituted derivatives including some cluster-carbonyl derivatives.

nickel chemistry Nickel is an electropositive element of Group VIII, electronic configuration $3d^8 4s^2$. The most stable oxidation state is Ni(II) (E° $Ni^{2+} \longrightarrow Ni$ −0·24 volts in acid solution) (green or yellow octahedral; red planar; blue tetrahedral) which readily forms complexes. Ni(III) and (IV) are known for fluorides and oxides and some other Ni(III) complexes are known. Ni(I), e.g. $Ni(PPh_3)_3Cl$ is fairly stable. Ni(0), carbonyl, $Ni(PPh_3)_4$ and many organic derivatives and Ni(−1) is present in carbonylates.

nickel chloride, $NiCl_2$. Forms hydrates with octahedral nickel. Complex chlorides, e.g. blue $[NiCl_4]^{2-}$ (formed in ethanol), are known.

nickel cyanides Nickel(II) cyanide, $Ni(CN)_2$, is precipitated from a Ni^{2+} salt and CN^- in aqueous solution. Forms $Ni(CN)_2{,}4H_2O$ and many complexes including ammonia complexes with strong clathrating powers for aromatic hydrocarbons. Gives the $[Ni(CN)_4]^{2-}$ ion with excess CN^-; these can be reduced to red $[Ni_2(CN)_6]^{4-}$ (contains Ni – Ni bond) and $[Ni(CN)_4]^{4-}$ ions.

nickel dimethylglyoxime, $C_8H_{14}N_4NiO_4$. The normal form in which nickel is weighed in analysis. There is metal–metal bonding in the solid. The red complex is precipitated from alkaline solution.

nickel fluorides Nickel(II) fluoride, NiF_2, yellow-green, forms green $NiF_2{,}4H_2O$, and perovskites, e.g. $KNiF_3$. K_3NiF_6 and K_2NiF_6 are known.

nickel hydroxide Apple green $Ni(OH)_2$ is formed by precipitation from aqueous solutions of Ni(II) salts. It is not amphoteric. Black NiO·OH ($Ni(NO_3)_2$ solution plus Br_2 in aqueous KOH) and related compounds are useful oxidizing agents (*see* nickel accumulator).

nickel iodide, NiI_2. Forms hexahydrate. Similar to bromide.

nickel nitrate, $Ni(NO_3)_2{,}6H_2O$. Green

crystals from aqueous solution. $Ni(NO_3)_2$ is formed from Ni plus N_2O_4.

nickel, organic derivatives The most important nickel organic derivatives are the bis-π-allyls, e.g. $(h^3\text{-}C_3H_5)_2Ni$ (C_3H_5MgBr plus $NiBr_2$) which is a very active catalyst for cyclotelomerization of diolefines. Ni(0) complexes, e.g. $[Ni\{P(OEt)_3\}_4]$, catalyse olefine-diene coupling to give 1,4-hexadienes. Nickelocene, $(h^5\text{-}C_5H_5)_2Ni$, is an electron-rich compound and often reacts as if it had a free olefinic bond. Simple aryls, e.g. $(R_3P)_2NiX(Ar)$, are also known.

nickel oxides Nickel(II) oxide, NiO, has the NaCl structure (heat on oxyacid salt or $Ni(OH)_2$). Dissolves in acids. Higher nickel oxides are not known although indefinite oxide-hydroxides (e.g. NiO·OH – *see* hydroxides) are formed.

nickel perchlorate, $Ni(ClO_4)_2xH_2O$. Soluble nickel salt.

nickel sulphate, $NiSO_4xH_2O$. Fairly soluble nickel salt. Forms double sulphates, e.g. nickel ammonium sulphate.

nickel vitriol, $NiSO_4,7H_2O$. Hydrated nickel sulphate.

nicol prism A calcite crystal, cut in such a way that only one of the refracted rays emerges, the other being returned by total internal reflection in the direction of the light source. The two refracted rays produced when a beam of light enters a calcite crystal are plane-polarized in directions which are mutually at right angles, and the light emerging from a nicol prism is plane-polarized.

nicotinamide *See* nicotinic acid.

nicotinamide adenine dinucleotide (NAD) An extremely important and widespread coenzyme, formerly known as diphosphopyridine nucleotide (DPN). The coenzyme functions as a hydrogen transfer agent as indicated. The hydrogen transfer is stereospecific both to the substrate and to the coenzyme. The reduced form of the coenzyme may be designated as 'reduced NAD' or as 'NADH'. NAD is biosynthetically derived from nicotinamide or nicotinic acid*. It is involved in a large number of oxidation-reduction reactions catalysed by dehydrogenases*, which are usually highly specific for NAD or NADP (*see also* nicotinamide adenine dinucleotide phosphate).

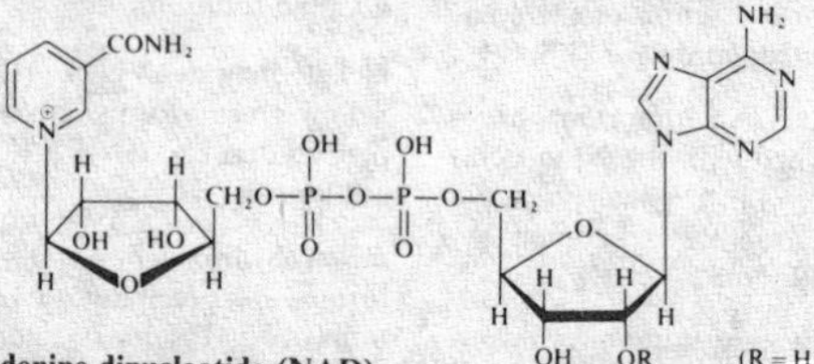

nicotinamide adenine dinucleotide (NAD)

nicotinamide adenine dinucleotide phosphate (NADP) An important coenzyme of the pyridine nucleotide class. The structure is that of NAD* except that the adenosine has an additional phosphate group at the 2′-position ($R = OPO_3H_2$). NADP was formerly known as triphosphopyridine nucleotide (TPN). Like NAD, NADP is concerned in hydride transfer reactions, but the two coenzymes function at different stages in metabolism: NAD – NADH is especially concerned in oxidative phosphorylation, while NADP – NADPH is particularly involved in biosynthesis.

nicotine, 3-(1-methyl-2-pyrrolidyl)pyridine, $C_{10}H_{14}N_2$. When pure nicotine is a colourless liquid, b.p. 247°C, but darkens on exposure to air and light. Crude nicotine contains small amounts of other alkaloids, but (–)-nicotine is the principal component.

It is used as an insecticide and usually manufactured from tobacco.

nicotinic acid, niacin, 3-pyridine carboxylic acid, $C_6H_5NO_2$. M.p. 232°C. It can be prepared by oxidizing nicotine with a variety of agents, or, more cheaply, from pyridine or quinoline. Nicotinic acid is an essential component of mammalian diet. It is the pellagra-preventing factor of vitamin B. The amide, nicotinamide is incorporated into nicotinamide adenine dinucleotide*.

nido Applied to open structures, e.g. $B_4C_2H_8$ (nest-like). *See* carboranes.

ninhydrin, 1,2,3-triketohydrindene hydrate, $C_9H_4O_3,H_2O$. Light brown crystals, losing water at 125–130°C, m.p. 242°C (decomp.). Ninhydrin can be prepared by oxidizing diketohydrindene with SeO_2. It gives a blue colour on heating with proteins, peptides and

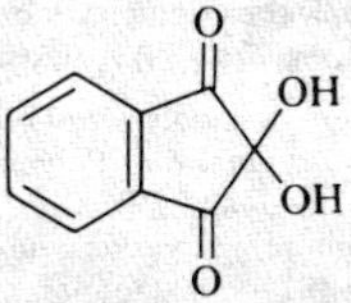

amino-acids and is much used for their detection, particularly as a spray reagent in paper chromatography.

niobates *See* niobium oxides.

niobite, (Fe,Mn)(Nb,Ta)$_2$O$_6$. *See* columbite.

niobium, Nb. At.no. 41, at.wt. 92·9064, m.p. 2468°C, b.p. 4742°C, *d* 8·57. An element of Group V, electronic configuration $5s^1 4d^4$. Once known as columbium, Cb. Apart from pyrochlorite, $CaNaNb_2O_6F$, all Nb minerals contain Ta although Nb is much more abundant. The important mineral is columbite–tantalite (Fe,Mn)(Nb,Ta)$_2$O$_6$. Ores are treated by NaOH fusion followed by acid washing to give mixed pentoxides. Nb is separated by liquid extraction from HF solution with isobutylmethylketone. The metal is prepared from Nb_2O_5 by reduction with carbon, it is bcc, remains lustrous in air but reacts with oxygen or steam at high temperatures and dissolves in HNO_3-HF. Nb is used in special steels and Nb/Ti alloys are superconductors. World use 1976 12 000 tonnes.

niobium chemistry Niobium chemistry is dominated by the +5 oxidation state although lower halides* which, with the exception of the fluorides, contain metal–metal bonds, and lower oxides are known. Hydrides (e.g. $NbH_{0\cdot6-0\cdot8}$, NbH_{2-x}), borides (NbB, Nb_3B_2), carbides (Nb_4C_3, NbC), and nitrides (Nb_2N, Nb_4N_5, Nb_5N_6, NbN) are readily formed. Alkoxides (e.g. $Nb(OEt)_5$ from $NbCl_5 + EtOH + NH_3$) and dialkylamides (e.g. $Nb(NMe_2)_5$ from $NbCl_5$ and $LiNMe_2$) form well-established classes of compounds. A relatively limited range of complexes, apart from complex halides, is known. Sulphides Nb_3S_4, NbS_2 are known.

niobium halides Known for all NbX_5 and NbX_4.NbX_5(Nb + X_2); NbF_5, m.p. 72°C, b.p. 235°C; $NbCl_5$, m.p. 205°C, b.p. 254°C. Hydrolysis gives oxide halides, e.g. $NbOF_3$ and $NbOCl_3$. Complexes are formed with many ligands and complex halides, particularly complex fluorides and oxide fluorides, e.g. NbF_6^-, $NbOF_5^{2-}$, are present in acid solution. NbX_4 (NbX_5 + Nb) are dark coloured solids with (except for NbF_4) some metal–metal bonding. NbI_4 is easily formed by heating NbI_5. Adducts are formed with various ligands. Lower halides, e.g. $[Nb_6Cl_{12}]Cl_{4/2}$, are formed by further reduction. They contain metal-cluster species and there is extensive chemistry based on $(Nb_6X_{12})^{n+}$ (n = 2, 3 or 4) and $Nb_4X_{11}^-$ species. NbF_3 has an ionic structure.

niobium organometallic chemistry Nb-alkyls and aryls, e.g. $MeNbCl_4$ and $NbMe_5(Me_2PCH_2CH_2PMe_2)$ are known although relatively unstable. Cyclopentadienyl derivatives $(h^5-C_5H_5)_2NbX_3$, $(h^5-C_5H_5)Nb(CO)_4$ are relatively stable as is $[Nb(CO)_6]^-$ ($NbCl_5$ + Na in diglyme + CO (pressure)). Tantalum compounds are very similar.

niobium oxides Nb_2O_5 is the most important oxide and is prepared by dehydration of the hydrated oxide formed by hydrolysis of halides. It is a dense, white, inert material. Reduction of Nb_2O_5 gives NbO_2 and NbO. The pentoxide forms niobates, e.g. $KNbO_3$ as mixed-metal oxides by fusion with other oxides or carbonates. Reduction gives bronzes*. Isopolyniobates, e.g. $[H_xNb_6O_{19}]^{(8-x)-}$ (x = 0,1,2) are formed by fusing Nb_2O_5 with alkali and dissolving the melt in water.

nitramide, H_2NNO_2. *See* nitrogen, oxyacids.

nitrates Salts of nitric acid*, HNO_3. All metallic nitrates are soluble in water, some, e.g. $UO_2(NO_3)_2$, are readily extracted into organic solvents. Formed, generally as hydrates, from nitric acid and metal oxides, hydroxides and carbonates. Anhydrous nitrates, e.g. $Cu(NO_3)_2$, formed by dissolving Cu in liquid N_2O_4 or reacting with N_2O_5; many adducts are formed with excess N_2O_4, they are soluble in organic solvents and are often volatile. Many nitrates are ionic but heavy metal nitrates and anhydrous nitrates have covalently bonded nitrate groups. On heating most nitrates decompose to the metal oxide and NO_2 and O_2 although $NaNO_3$ and KNO_3 give nitrites and oxygen and NH_4NO_3 gives N_2O and H_2O. Nitrates may be tested for by the 'brown ring' test; iron(II) sulphate is added to the cold solution to be tested, and concentrated sulphuric acid added down the side of the test tube; in the presence of a nitrate a brown or black ring is formed at the junction of the two liquids. This test is unreliable in the presence of bromide or iodide (which also give coloured rings); a sounder test and estimation of nitrate is by reduction to ammonia with Devarda's or Arnd's alloy and NaOH or by precipitating nitron nitrate. Nitric acid and the nitrates (particularly those of Na, K, Ca and

NH_3) are of great commercial importance; practically all modern explosives contain high proportions of ammonium nitrate or organic nitro-compounds. Nitrates are also employed as fertilizers.

nitrating acid Generally a mixture of nitric and sulphuric acids used for commercial nitration, e.g. in the production of nitroglycerin*.

nitration *See* nitrating acid, nitro-compounds.

nitrene group, RN=. The imino group. Nitrene derivatives, e.g. $ReCl_3(PR'_3)_2NR$ and $[WF_5NMe]^-$ are formed by condensation reactions.

nitric acid, HNO_3. M.p. −42°C, b.p. 83°C (decomp.). Extremely important acid produced industrially by oxidation of NH_3 over Pt to NO and H_2O followed by reaction of the NO with oxygen to NO_2 and further oxidation and absorption in water to give aqueous HNO_3. Forms constant boiling mixture (*c.* 68% HNO_3) and 1:1 and 1:3 hydrates. Concentrated with conc. H_2SO_4. Gives nitrates*. Gaseous nitric acid is $HONO_2$ with planar nitrogen. Used in the production of fertilizers (NH_4NO_3), explosives, dyestuffs, etc. Behaves as strong acid but also as a strong oxidizing agent (reduced to various nitrogen oxides or even to NH_3) especially to non-metals and organic materials. May be handled in stainless steel or Al equipment as oxide films are formed. U.S. production 1978 7·3 megatonnes.

nitric oxide, NO. *See* nitrogen oxides.

nitrides Compounds of nitrogen with other elements prepared by the action of N_2 or NH_3 on the element or by ammonolysis of halides. Electropositive elements form ionic nitrides formally containing N^{3-} ions (e.g. Li_3N), hydrolysed to NH_3. Transition elements form compounds that structurally can be considered to have N atoms in holes in metal lattices. These compounds are very hard, inert and high melting. The less electropositive elements form covalent nitrides – formally derivatives of NH_3 but often containing rings (e.g. BN, P_3N_5). Transition metal nitrides and BN, Si_3N_4, AlN are of importance as hard materials and for ceramics.

nitriding A method of surface hardening steel by treating it in an atmosphere of NH_3 which gives a surface layer of nitride. The steel must contain specific alloying elements, e.g. Al, Cr, Mo.

nitrido complexes Complexes containing the N^{3-} ligand, e.g. $[OsO_3N]^-$, the osmiamate ion.

nitrile rubber A copolymer of butadiene (55–80%) with acrylonitrile. Polymerized in an emulsion. An oil- and solvent-resistant rubber used in sealing rings, gaskets, hoses, etc. U.S. production 1978 73000 tonnes.

nitriles Organic cyanides containing the $-C{\equiv}N$ group. They are colourless liquids or solids, with peculiar but not unpleasant odours. They are formed by heating amides with P_2O_5 or by treating halogen compounds with NaCN. They are decomposed by acids or alkalis to give carboxylic acids containing the same number of carbon atoms as the nitrile. Reduced to primary amines. Aliphatic nitriles are valuable solvents for many metal complexes and salts, are used as chemical intermediates, fuel additives, lubricants, plasticizers, insect repellants, weed control, CO_2 removal from natural gas, electroplating additives.

nitrilotriacetic acid, $N(CH_2COOH)_3$. An important complexing (sequestering) agent. Forms particularly stable complexes with Zn^{2+}.

nitrites Salts containing the $[NO_2]^-$ ion. Salts of nitrous acid*. Organic O-bonded derivatives RONO. Alkali nitrites are used in bacon as preservatives.

nitroamines Organic compounds containing both nitro- and amino-groups, usually prepared by partial reduction of poly-nitro-compounds. They have some importance as dyestuffs intermediates.

***m*-nitroaniline, 1-amino-3-nitrobenzene,** $C_6H_6N_2O_2$. Yellow needles; m.p. 114°C, b.p. 285°C. It is prepared by reduction with sodium sulphide of one of the nitro-groups in *m*-dinitrobenzene. It is used as a first component in the preparation of azo-dyes.

***p*-nitroaniline, 1-amino-4-nitrobenzene.** Yellow crystals; m.p. 147°C.

Prepared by heating *p*-nitrochlorobenzene with concentrated aqueous ammonia in an autoclave at 170°C. It is also prepared by alkaline hydrolysis of *p*-nitroacetanilide or by nitrating and hydrolysing benzylideneaniline.

It is an important dyestuffs intermediate, being used as a first component in azo-dyes and also as a source of *p*-phenylenediamine.

2-nitroanisole, *o*-nitroanisole, $C_7H_7NO_3$. Colourless oil, m.p. 9°C, b.p. 273°C. Prepared by heating 2-nitrochlorobenzene with NaOMe or from 2-nitrophenol and MeCl. Important as a dyestuffs intermediate after reduction to *o*-anisidine with Fe and HCl.

nitroanthraquinones The direct nitration of anthraquinone results chiefly in substitution at the α-positions, 1-nitro- and 1,5- and 1,8-dinitroanthraquinones being obtained.

The compounds have limited value as dyestuffs intermediates.

nitrobenzene, $C_6H_5NO_2$, $PhNO_2$. Colourless, highly refractive liquid with characteristic smell; m.p. 6°C, b.p. 211°C.

It is manufactured by reacting benzene with a mixture of nitric and sulphuric acids. Most of the nitrobenzene produced is used to manufacture aniline, which is obtained by reduction. A considerable proportion is used as a raw material in the dyestuffs industry, either as nitrobenzene as such, or as aniline.

Further nitration gives *m*-dinitrobenzene; sulphonation gives *m*-nitrobenzene sulphonic acid. Reduction gives first azoxybenzene, then azobenzene and aniline depending upon the conditions. Used in the dyestuffs industry as such or as aniline.

nitrocellulose *See* cellulose nitrate.

nitrochlorobenzenes, $C_6H_4ClNO_2$.
1-Nitrochlorobenzene, needles, m.p. 32°C, b.p. 245°C.

4-Nitrochlorobenzene, prisms or leaflets, m.p. 83°C, b.p. 238°C.

A mixture of the two mononitrochlorobenzenes is prepared by nitration of chlorobenzene. Further nitration of the mixture or of either of the mononitro-compounds gives 2,4-dinitrochlorobenzene, m.p. 51°C, b.p. 315°C.

The nitrochlorobenzenes are valuable dyestuffs intermediates. The presence of the nitro-groups makes the chlorine atom very reactive and easily replaceable. Treatment with ammonia or dilute alkalis substitutes an amino- or hydroxy-group for the chlorine atom and gives a series of nitroanilines and nitrophenols.

nitro-compounds (aromatic) These form a large group of compounds which contain the nitro-group; NO_2.

Nitro-compounds are prepared by the direct action of nitric acid. The reaction is greatly facilitated if a mixture of nitric and sulphuric acid is used.

In the nitration of benzene, *m*-dinitro- and sym-trinitrobenzenes are obtained under more vigorous conditions. With naphthalene, 1-nitronaphthalene is the first product and further nitration gives a mixture of 1,5- and 1,8-dinitronaphthalenes; 2-nitronaphthalene is never obtained.

The nitro-hydrocarbons are neutral substances but when a nitro-group is introduced into a phenol or amine the acidic properties are greatly increased or the basicity decreased. The presence of a nitro-group also tends to make halogen atoms in the same molecule much more reactive.

On reduction, the nitro-compounds give rise to a series of products, e.g.

$PhNO_2 \longrightarrow$ nitrosobenzene, $PhNO \longrightarrow$

N-phenylhydroxylamine, PhNHOH and further reduction can give azoxybenzene, azobenzene, hydrazobenzene and aniline. The most important outlet commercially for the nitro-compounds is the complete reduction to the amines for conversion to dyestuffs. This is usually done in one stage with iron and a small amount of hydrochloric acid.

Some nitro-compounds are themselves coloured and can be used as dyestuffs, e.g. picric acid. In this case the nitro-group can be considered to be the chromophore. For aliphatic nitro-compounds *see* nitroparaffins.

nitro-dyes The coloured salts of various nitrophenols or sulphonic acid derivatives. Acid dyestuffs.

nitrogen, N. At.no. 7, at.wt. 14·0067, m.p. −209·86°C, b.p. −195·8°C. Colourless gas existing as N_2 (air contains 75·5% by weight, 78·1% by volume) and also as nitrates, ammonia and in proteins in animal and vegetable tissues (about 16% N by weight). Obtained industrially by fractionation of liquid air, obtained on a laboratory scale by heating NH_4NO_2 (a mixture of $NaNO_2$ and NH_4Cl) or barium or sodium azides. N_2 is fairly inert at room temperature (*see* nitrogen fixation) but combines with most electropositive elements on heating and with O_2 (under irradiation), H_2 (in presence of catalyst forms ammonia* in Haber process*). CaC_2 (to give calcium cyanamide), MOH plus carbon (to give cyanides). Active nitrogen is excited N_2 produced by an electric discharge. Used in ammonia synthesis and as an inert atmosphere. Liquid nitrogen is used as a refrigerant. U.S. production 1982 505 × 10^9ft³. World use of nitrogenous fertilizers (as N) 1976/77 45·8 megatonnes.

nitrogenase A complex of two metal-containing proteins present in nitrogen-fixing bacteria and blue–green algae.

nitrogen chemistry Nitrogen is an electronegative element of Group V, electronic configuration $1s^22s^22p^3$. Its only simple ionic chemistry is that of the nitride ion N^{3-}. The normal oxidation state is +3; pyramidal (rapid inversion) in 3-co-ordinate molecules, e.g. NH_3, 4-co-ordinate tetrahedral in e.g. $[NH_4]^+$. The +5 oxidation state does not show 5-co-ordination but is 4-co-ordinated in e.g. $[NF_4]^+$ and 3-co-ordinate in the planar nitrate, $[NO_3]^-$ ion.

nitrogen chlorides Nitrogen trichloride, NCl_3, explosive oily liquid formed from Cl_2 and NH_4Cl solution. $NHCl_2$ and NH_2Cl (mono and dichloroamine) are also formed. ClNO, nitrosyl* chloride, NF_2Cl and $NFCl_2$ are also known.

nitrogen-donors Tripositive nitrogen compounds, particularly ammonia and its derivatives, form extensive series of complexes. Nitrosyl complexes are N-bonded. $(NO_2)^-$ can be O- or N-bonded but the latter is more stable. Nitrates are O-bonded.

nitrogen fixation The use of atmospheric nitrogen in the manufacture of commercially important nitrogen compounds, nitric acid, ammonia and ammonium salts, and mainly used as fertilizers. Conversion to ammonia is effected by the Haber process, and nitric acid is obtained by oxidation of ammonia. Bacteria present in the roots of certain plants fix atmospheric nitrogen; so do *Azotobacter*, free-living soil micro-organisms, though the latter are inefficient and play only a small part in regulating the soil nitrogen balance (*see* nitrogenase). Some metal complexes interact with N_2 to give N_2 complexes which may be reduced to NH_3. The return of the fixed nitrogen to the atmosphere is denitrification*.

nitrogen fluorides *Nitrogen trifluoride*, NF_3, m.p. −206·7°C, b.p. −129°C, prepared by the electrolysis of fused ammonium hydrogen fluoride. It is a very inert substance, but the other nitrogen fluorides tend to explode, either spontaneously or in the presence of organic materials. Other nitrogen fluorides are *tetrafluorohydrazine*, F_2NNF_2, prepared by the action of Cu on NF_3 and *cis-* and *trans-difluorodiazine*, FN=NF. There is an extensive chemistry of the substituted nitrogen fluorides, e.g. nitrosodifluoroamine, F_2NNO.

nitrogen hydrides The simple hydride is ammonia, NH_3, but hydrazine, N_2H_4 is also known and derivatives of these, e.g. $[NH_4]^+N_3^-$, ammonium azide.

nitrogen iodides *Nitrogen tri-iodide*, NI_3, black powder formed from NH_3 gas over $KIBr_2$. Alcoholic solutions of I_2 and NH_3 give a black solid $NH_3{\cdot}NI_3$ (NI_4 tetrahedra) which is stable when moist but shock-sensitive when dry.

nitrogen mustards The name given to certain poly-(β-chloroethyl)-amines, $RN(CH_2CH_2Cl)_2$, in which R is an alkyl, alkylamine or alkyl chloride group. These have vesicant and other properties similar to those of mustard gas, including an action on cells resembling that of X-rays. They have been used in place of radiation therapy in the treatment of leukaemia, Hodgkin's disease and similar conditions.

nitrogen oxides
Dinitrogen oxide, nitrous oxide, N_2O. Colourless gas, m.p. −91°C, b.p. −88·5°C (heat on NH_4NO_3). Decomposes to N_2 and O_2 above 500°C; can be detonated. Linear molecule NNO. Used as a mild anaesthetic.

Nitrogen oxide, nitric oxide, NO. Colourless gas, paramagnetic with only slight tendency to dimerization, m.p. −164°C, b.p. −152°C. Prepared from Cu, and 8M HNO_3 and industrially by oxidation of ammonia over a catalyst. Decomposes at high pressures to N_2O and NO_2, reacts with O_2 (to NO_2) and halogens (to XNO). Removal of an electron gives the NO^+ (nitrosonium* ion); combines with transition metal derivatives to give nitrosyls*.

Dinitrogen trioxide, N_2O_3. Only stable in solid state (m.p. −102°C). Pale blue solid giving deep blue liquid; the gas contains some $ONNO_2$ molecules. Prepared from NO and O_2 or NO and N_2O_4 with freezing; decomposes to NO and NO_2.

Nitrogen dioxide, NO_2, and *dinitrogen tetroxide*, N_2O_4 are in equilibrium (NO_2 brown, paramagnetic N_2O_4 colourless, diamagnetic) with N_2O_4 in the solid (m.p. −11°C, b.p. 21°C). NO_2 is a bent molecule ONO, N_2O_4 is largely planar O_2NNO_2. Formed from NO plus O_2 or Cu plus HNO_3. The liquid (particularly in the presence of donor molecules such as $CH_3CO_2C_2H_5$) acts as a good solvent and anhydrous nitrates, e.g. $Cu((NO_3)_2$ (Cu plus N_2O_4) may be prepared in this solvent. Used as an oxidizing agent (nitrous fumes) and as an intermediate in the formation of nitric acid.

Dinitrogen pentoxide, N_2O_5. White solid (HNO_3 plus P_2O_5) readily decomposes to NO_2 and O_2, sublimes 32·5°C. In solid state $(NO_2)^+(NO_3)^-$; gaseous molecules O_2NONO_2.

Nitrogen trioxide, NO_3. Unstable white solid (N_2O_5 plus O_3).

nitrogen, oxyacids
Hyponitrous acid, $H_2N_2O_2$. A solution of the sodium salt is formed from $NaNO_2$ and Na

amalgam; $Ag_2N_2O_2$ is precipitated from solution, $H_2N_2O_2$ formed from $Ag_2N_2O_2$ and HCl in ether. The hyponitrite ion is *trans* $(O-N-N-O)^{2-}$.

Nitroxylic acid, H_2NO_2. Yellow Na_2NO_2 formed by electrolysis $NaNO_2$ in NH_3. Structure unknown; free acid unknown.

Nitramide, H_2NNO_2. A weak acid.

Nitrous acid, HNO_2. *See* separate entry.

Hyponitric acid, $H_2N_2O_3$, trioxo dinitrate(II). Sodium salt from H_2NOH in MeOH with $C_2H_5NO_2$; many complexes are known; the free acid is unstable.

Peroxonitrous acid, HOONO. Isomer of nitric acid (HNO_2 plus H_2O_2).

Nitric acid, HNO_3. *See* separate entry.

nitrogen sulphides *See* sulphur nitrides.

nitrogen trichloride *See* nitrogen chlorides.

nitrogen trifluoride *See* nitrogen fluorides.

nitroglycerin(e), glyceryl trinitrate, $C_3H_5N_3O_9$, $CH_2(ONO_2)\cdot CH(ONO_2)\cdot CH_2(ONO_2)$. Oily liquid prepared by treating glycerin with a mixture of nitric and sulphuric acids. When pure it is colourless, odourless, insoluble in water; m.p. 8°C. It is a very powerful and dangerous explosive and is never used alone due to its sensitivity. Used in propellants and dynamites. Used in medicine to give relief from chest pain in angina.

1-nitroguanidine, $H_2N\cdot C(:NH)\cdot NH\cdot NO_2$. White crystalline powder, m.p. 232°C; soluble in hot water. Usually manufactured from calcium carbide via calcium cyanamide, dicyandiamide and guanidine nitrate which is converted to nitroguanidine by the action of conc. H_2SO_4. It is used in some modern propellants to make them cooler and flashless.

nitron, $C_{20}H_{16}N_4$,PhN·N·NPh·CH·NPh. Used in ethanoic acid solution as a precipitating agent for nitrates. ClO_4^-, PF_6^-, ReO_4^-, WO_4^{2-} also form precipitates.

1-nitronaphthalene, $C_{10}H_7NO_2$. Yellow needles; m.p. 61°C, b.p. 304°C. Oxidized by chromic acid or potassium permanganate to 3-nitrophthalic acid; acid reducing agents give 1-naphthylamine. Readily nitrated, sulphonated and chlorinated.

Prepared by the direct nitration of naphthalene with a mixture of nitric and sulphuric acids. Its chief use was for the preparation of 1-naphthylamine and its derivatives.

nitronium salts Salts containing the NO_2^+ ion, e.g. $[NO_2]^+[WF_6]^-$. Prepared by the action of strong acids on concentrated nitric acid, by reactions involving nitrogen dioxide in bromine trifluoride or by NO_2 plus oxidizing agents. The nitronium ion is the active agent in the nitration of aromatic systems in mixtures of nitric acid and sulphuric acid or with nitronium fluoroborate.

nitroparaffins, $C_nN_{2n+1}NO_2$. Colourless, pleasant smelling but rather toxic liquids, sparingly soluble in water. They may be prepared by the action of silver nitrite on alkyl halides. The lower members are manufactured by the vapour phase reaction of propane with nitric acid at 400°C; a mixture of products results which may be separated by distillation. Nitroparaffins containing an α-hydrogen atom dissolve in aqueous alkali giving the salt of the *aci*-nitro compound, e.g.

$$CH_3NO_2 \longrightarrow [CH_2=NO_2]^- Na^+$$

which on treatment with concentrated mineral acids is hydrolysed to the corresponding aldehyde or ketone. Reduction of nitroparaffins with tin and hydrochloric acid gives the amine.

The lower nitroparaffins are used as propellants, as solvents and as chemical intermediates, e.g. nitromethane is an excellent solvent for polar materials especially metal salts.

***p*-nitrophenetole, 4-ethoxynitrobenzene,** $C_8H_9NO_3$. Yellow crystals; m.p. 58°C, b.p. 283°C. Prepared by the ethylation of *p*-nitrophenol with ethyl chloride. It is used for preparing *p*-phenetidine.

2-nitrophenol, *o*-nitrophenol, $C_6H_5NO_3$. Bright yellow needles; m.p. 45°C, b.p. 214°C. Prepared together with 4-nitrophenol by careful nitration of phenol. Sodium sulphide reduces it to 2-aminophenol which is used in dyestuffs and photographic processes.

4-nitrophenol, *p*-nitrophenol, $C_6H_5NO_3$. Colourless needles; m.p. 114°C. Prepared as 2-nitrophenol. Reduction with iron and hydrochloric acid gives 4-aminophenol.

nitroprussides Containing the $[Fe(CN)_5NO]^{2-}$ ion. *See* cyanoferrates.

nitrosamines Compounds having the general formula $\begin{matrix}R\\R'\end{matrix}\!\!>N\cdot NO$. Most are yellow oils or solids and arise from the interaction of nitrous acid with secondary or tertiary amines. The great majority are carcinogenic to experimental animals.

4-nitrosodimethylaniline, $C_8H_{10}N_2$. The free base forms dark green leaflets; m.p. 85°C. On standing it goes browner in colour. It is basic.

It reduces to 4-aminodimethylaniline and gives dimethylamine with hot sodium hydroxide solution. It is prepared by the action of nitrous acid upon dimethylaniline at 0°C.

It is used for the preparation of azines and oxazines by condensation with *m*-diamines and phenols.

nitroso dyes Dyestuffs containing nitroso-, NO, groups, generally quinone oximes prepared from nitrous acid and phenols or hydroxylamine on quinones. Acid dyes.

nitrosomethylurea, $NH_2CON(NO)CH_3$. A colourless crystalline solid; m.p. 123–124°C. Prepared by adding sodium nitrite solution to a solution of methylurea nitrate. It is unstable, carcinogenic and slowly decomposes at ordinary temperatures. Reacts with cold potassium hydroxide solution to give diazomethane, but this use has now been superseded by Diazald*.

nitrosonium hydrogen sulphate, chamber crystals, $NOHSO_4$. White solid; m.p. 73°C (decomp.). Prepared SO_2 and fuming nitric acid. Used in diazotization.

nitrosonium salts Salts containing the NO^+ ion which result from the reaction of nitrosyl chloride with metallic chlorides, e.g. $[NO]^+[AlCl_4]^-$, or from reactions in BrF_3.

nitrosonium tetrafluoroborate, $NOBF_4$. White solid; NOF plus BF_3. It is very hygroscopic; reaction with water gives oxides of nitrogen, but reactions with alcohols give nitrito esters. Aromatic primary amine hydrochlorides give diazonium tetrafluoroborates which may be pyrolysed to give aromatic fluoro derivatives. Secondary aliphatic amines react with it to give nitroso-amines.

4-nitrosophenol, $C_6H_5NO_2$. Greyish-brown leaflets, decompose at 124°C. It gives a red sodium salt. Prepared by the action of nitrous acid upon phenol or by the action of hydroxylamine hydrochloride upon *p*-benzoquinone.

It can act either as a nitrosophenol or as a quinone monoxime. Coloured derivatives, e.g. the sodium salt, are derived from the quinone form. Direct methylation gives a coloured compound which has been shown to be quinonemethoxime $O{:}C_6H_4{:}NOCH_3$; *p*-nitrosoanisole $CH_3O{\cdot}C_6H_4{\cdot}NO$ has been obtained by another route and is colourless.

p-Nitrosophenol condenses with bases to form indophenols, which are used for the production of sulphur colours.

nitrososulphuric acid, nitrosulphuric acid, nitrosulphonic acid Alternative names for nitrosonium hydrogen sulphate, $NOHSO_4$.

nitrosyl halides A series of covalent compounds, XNO, formally derivatives of $[NO]^+$ salts, prepared by direct interaction of X_2 and NO. Form nitrosonium salts with Lewis acid halides, act as strong oxidizing agents. Have bent molecules. FNO, m.p. −133°C, b.p. −60°C, colourless; ClNO, m.p. −62°C, b.p. −6°C, orange–yellow; BrNO, m.p. −56°C, b.p. 0°C. Other nitrosyl derivatives are known. Nitrosyl chloride is used in the production of caprolactam from cyclohexane.

nitrosyls Compounds containing NO bonded to metals through the nitrogen. In general similar to carbonyls but the M–N bonds are more stable to substitution than M–C bonds. For electron-counting assumed to have 3-electrons donated to the metal (i.e. NO^+ complexes). Formed direct by direct action of NO or indirectly by the action of nitrosyl halides. The iron group forms very stable nitrosyls, e.g. $Fe(CO)_2(NO)_2$, $[Fe(CN)_5NO]^{2-}$ (nitroprusside ion), $[Ru(NO)Cl_5]^{2-}$. Thionitrosyls, containing MNS groups are also known.

nitrosyl sulphuric acid, $NOHSO_4$. *See* nitrosonium hydrogen sulphate.

2-nitrotoluene, $C_7H_7NO_2$. Colourless liquid; b.p. 220°C. Volatile in steam, insoluble in water, miscible with most organic solvents.

It is prepared by the direct nitration of toluene as a 50–60% component of the mixture of isomers. Used for the preparation of *o*-toluidine.

3-nitrotoluene, $C_7H_7NO_2$. M.p. 16°C, b.p. 230°C. Prepared (3–4%) by separation from the mixture of isomers obtained on mononitrating toluene.

4-nitrotoluene, $C_7H_7NO_2$. Colourless crystals; m.p. 51–52°C, b.p. 238°C.

Prepared by mononitrating toluene, as a 40% constituent of the mixture of isomers. It is reduced to *p*-toluidine with iron and hydrochloric acid.

Further nitration of the 2- and 4- isomers yields 2,4-dinitrotoluene, yellow crystals, m.p. 71°C. This material is reduced to 2,4-diaminotoluene and treated with phosgene to give 2,4-diisocyanatotoluene, a precursor of polyurethanes.

nitrourea, $NH_2CONHNO_2$. Colourless crystalline solid; m.p. 158°C. Prepared by adding urea nitrate to concentrated sulphuric acid at 0°C and pouring the mixture on to ice. It is a strong acid, and forms salts. Not readily oxidized but can be detonated. Used in the preparation of semicarbazide.

nitrous acid, HNO_2. Unstable weak acid prepared by acidifying solutions of nitrites, e.g. KNO_2. The free acid is known in the gas phase (NO, NO_2 plus H_2O). Nitrites are formed by reduction of nitrates with C or Pb. Can act as oxidizing agent (to e.g. I^-, Fe^{2+}) more normally a reducing agent (converted to HNO_3). Forms both O- and N-nitrites with organic groups and in co-ordination compounds; the N-bonded nitro-derivatives are generally more stable. Organic thionitrites ONSR are also known.

nitrous fumes An NO-N_2O_4 mixture used in oxidation.

nitrous oxide, N_2O. *See* nitrogen oxides.

nitroxylic acid, H_2NO_2. *See* nitrogen, oxy-acids.

nitryl halides, XNO_2. Known for F and Cl; the parent compounds of nitronium salts, formally halogen nitro derivatives. FNO_2, m.p. $-166°C$, b.p. $-72°C$, colourless, prepared NO_2Cl plus AgF or N_2O_4 plus CoF_3.$ClNO_2$, m.p. $-145°C$, b.p. $-15°C$, colourless, prepared chlorosulphuric acid plus HNO_3. Both hydrolysed by water.

n.m.r. A widely used abbreviation for nuclear magnetic resonance*.

n.m.r. shift reagents Various lanthanide chelates, principally of β-diketones are useful for simplifying n.m.r. spectra because of their ability to cause spectacular changes in chemical shifts of nuclei adjacent to an electronegative substituent in a molecule. Europium complexes, added as solids or solutions, cause downfield shifts and praesodymium complexes result in upfield changes in resonances.

No Nobelium.

nobelium, No. At.no. 102, ^{255}No(3 min), is produced by the action of ^{12}C and ^{13}C nuclei on a curium target. The metal has not been prepared.

nobelium compounds Nobelium shows only +2 and +3 oxidation states. The +2 state (f^{14}) being most stable. No solid compounds have been isolated.

$$No^{3+} \xrightarrow{+1{\cdot}45V} No^{2+} \xrightarrow{+1{\cdot}64V} No$$

noble gases The elements helium, neon, argon, krypton, xenon and radon. All occur as minor constituents of the atmosphere and He is a component of some natural hydrocarbon gases. Separated by fractional distillation. Used (particularly He, Ar) to provide an inert atmosphere, e.g. for welding, and in electric light bulbs, valves and discharge tubes (particularly Ne). Liquid He is used in cryoscopy. The amounts of He and Ar formed in minerals by radioactive decay can be used to determine the age of the specimen. Xe and to a lesser extent Kr and Rn have a chemistry; the other noble gases do not form chemical compounds.

node A region of very low electron density in an atom or molecule as predicted from wave equations.

non-aqueous solution A solution in which the solvent is not water but may be organic (e.g. Et_2O) or inorganic (e.g. NH_3), protonic (e.g. EtOH), or non-protonic (e.g. BrF_3). Non-aqueous solvents may be self-ionized, e.g. $2NH_3 \rightleftharpoons NH_4^+ + NH_2^-$.

non-aqueous solvents *See* non-aqueous solution.

non-aqueous titrations Analytical procedures carried out in organic and other non-aqueous media because of insolubility of the reactants in water, or because some acids and bases cannot be titrated in water. Thus H_2SO_4 and $HClO_4$ can be titrated as a mixture in methylisobutyl ketone whilst amines can be titrated in 3-methylsulpholane.

non-ionic detergents *See* detergents.

non-polar molecule A molecule which has a zero dipole moment. Symmetrical covalent compounds, e.g. CCl_4, are non-polar.

non-soap greases *See* lubricating greases.

non-stoicheiometric compounds Compounds in which the ratios of the numbers of atoms present do not follow simple integers. Many transition metal oxides, silicates, etc. are non-stoicheiometric and the phenomenon is common.

nonylphenol, $C_{15}H_{20}O$. Made phenol plus nonylenes (propene trimer). Used to form phenolic resins (with aldehydes) and reaction with ethylene oxide to give surfactants.

nopinene *See* β-pinene.

nor- The most common use of the prefix nor- in the name of an organic compound is to indicate the loss of a CH_2 group, thus nor-nicotine has the formula $C_9H_{12}N_2$, while nicotine is $C_{10}H_{14}N_2$. The same convention is used with steroids. In terpene nomenclature nor- indicates the loss of all the methyl groups from the parent compound. With amino-acids however the use of nor- turns the trivial name of a branched chain compound to the isomeric straight chain compound, thus norvaline is 2-amino-*n*-valeric acid.

noradrenaline, norepinephrine, $C_8H_{10}NO_3$.

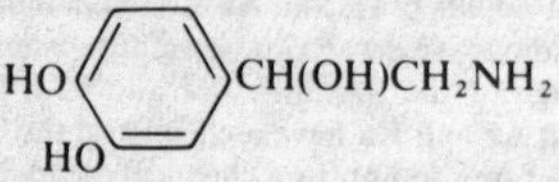

M.p. 103°C. Noradrenaline is released in the adrenal medulla with adrenaline, and also at the sympathetic nerve endings. Its release from a nerve fibre is followed by binding to a receptor molecule on the next nerve or muscle fibre, probably causing a change in the electrical charge of the receptor-cell membrane. Biosynthetically it normally serves as a precursor for adrenaline.

Acts to constrict small arteries, thereby increasing blood pressure and to contract smooth muscle. Used in cases of peripheral vasomotor collapse.

norbornadiene, bicyclo[2·2·1]hepta-2,5-diene,

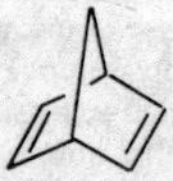

C_7H_8. A slightly coloured liquid, b.p. 90°C, obtained by a Diels–Alder reaction between cyclopentadiene and ethyne at approximately 150°C. At temperatures in excess of 450°C it rearranges to cycloheptatriene (tropilidene).

Nordhausen sulphuric acid *See* sulphuric acid, fuming.

norethisterone, 17-ethinyl-19-nortestosterone. *See* ethisterone.

normality *See* normal solution.

normal liquid A liquid in which the molecules have no tendency to associate. The simpler liquid hydrocarbons approximate to this type, while water, ethanoic acid and alcohol are examples of abnormal liquids.

normal solution A solution which contains 1 g equivalent of the substance in 1 litre of solution. A normal solution must have the equivalent referred to a specific reaction and this is often not done. A normal solution may be indicated by the letter N, thus, N-hydrochloric acid signifies a normal solution of hydrogen chloride (for titration with base).

normal temperature and pressure *See* standard temperature and pressure, ntp.

Np Neptunium.

NQR Nuclear quadrupole* resonance.

NTA Nitrilotriacetic acid*, $N(CH_2COOH)_3$.

ntp Normal temperature (298 K) and pressure (760 mm).

nuclear magnetic resonance A technique for studying nuclear magnetic moments*. The substance under test is placed in a very strong magnetic field which must be both uniform and steady. This produces a 'splitting' of the nuclear energy levels. The sample is then subjected to an additional, weak, oscillating magnetic field produced by passing r.f. current, at e.g. 30 MHz frequency, through a surrounding coil, and the frequency of this field is slowly 'scanned' over an appropriate range. At certain precise frequencies the nuclear magnets resonate with the field in undergoing transitions between the magnetic energy levels. The resonance is detected in a search coil and is amplified and recorded. The technique has assumed great importance for analysis of molecular structure because the resonance frequencies of a particular element are influenced by the electronic environment in which the atom is situated. Thus three different kinds of hydrogen atom are detected in ethanol, corresponding to those in the CH_3, CH_2 and OH groups. In addition to direct information from the resonance frequencies the nuclei couple together to give fine structure in the n.m.r. spectrum. The extent and nature of the coupling provides further information on molecular structure.

Information from an n.m.r. spectrum is classified into the chemical shift, δ (the relative shift from a standard [Me_4Si for ^1H, CCl_3F for ^{19}F] which is rendered independent of the field), and the coupling constants, J, which are determined directly from the spectra.

Complex n.m.r. spectra can be simplified by the use of n.m.r. shift reagents*.

nuclear paramagnetism The nuclei of many species have spin and the total angular momentum is given by

$$\frac{h}{2\pi}\sqrt{I(I+1)}$$

where I is the nuclear spin number ($I = 0, \frac{1}{2}, 1, \frac{3}{2} \ldots$). $I = 0$ corresponds to a nucleus without spin. Since nuclei are also charged they behave as magnetic dipoles and nuclear magnetic resonance techniques can be used to study nuclei with $I > 0$.

Isotope	*Spin no.* (I)	% *abundance*
^{1}H	$\frac{1}{2}$	99·98
^{2}H	1	0·0156
^{12}C	0	98·9
^{13}C	$\frac{1}{2}$	1·1
^{14}N	1	99·62
^{15}N	$\frac{1}{2}$	0·38
^{19}F	$\frac{1}{2}$	100
^{31}P	$\frac{1}{2}$	100

nuclear quadrupole moment The electric quadrupole moment of the nucleus. Affected by chemical environment and can be used to investigate structure (NQR spectroscopy).

nuclear spin From the hyperfine structure of certain spectra (*see* multiplet) it has been concluded that the nucleus has a property with the characteristics of spin. The angular momentum of nuclear particles is expressed in units of ($h/2\pi$). ($h=$ Planck's constant). *See* nuclear magnetic resonance.

nucleases Those enzymes which hydrolyse nucleic acids to oligonucleotides or mononucleotides. Nucleases are present in plasma, intestinal juice and the liver.

nucleic acids These are essential components of all living cells, as they carry the necessary hereditary information enabling highly specific proteins to be constructed. There are two types of nucleic acid; deoxyribose nucleic acid, DNA, is found in the nuclei of cells; ribose nucleic acid, RNA, is found mainly in the cytoplasm. They are polynucleotides; DNA is built of the nucleotides of the purines adenine and guanine and the pyrimidines cytosine and thymine. The DNA molecule consists of two helical chains coiled round the same axis, each chain consisting of alternate phosphate and sugar groups connected through carbons 3 and 5 of the pentose molecule, and with the bases attached to the pentose. The chains are held together by hydrogen bonds between the bases. Spatial considerations demand that of two linked bases one must be a purine and the other a pyrimidine. It is also assumed that only two specific pairs can bind, adenine with thymine and guanine with cytosine. Thus given a sequence of bases on one chain, that on the other is determined. If the chains are uncoiled and separated any new chain built on one of the originals has a standard sequence.

In RNA the sugar is ribose and uracil replaces thymine. Different varieties of RNA exist with widely different molecular weights. Messenger RNA conveys information in the form of the sequence of bases from DNA to RNA on the ribosomes, where proteins are built up from amino-acids on the RNA template. Each sequence of three bases, called a codon, codes for a specific amino-acid. Thus the order of amino-acids on a protein is ultimately based on the order of nucleotides on the DNA molecule. *See* genetic code.

Transfer or soluble RNAs are specific carrier molecules for amino-acids during protein synthesis on ribosomes with ribosomal RNA as the template. There is at least one t-RNA molecule for each amino-acid.

nucleide *See* nuclide.

nucleon A collective name for particles of mass number 1, i.e. protons and neutrons.

nucleophilic reagents Groups which act by donating or sharing their electrons. E.g. the hydroxyl anion, halide ions, OR^- and SR^- groups, and amines.

nucleophilic substitution Reactions proceeding in the broadest sense according to the equation

$$R-X+:N \longrightarrow R-N+X^- \qquad (1)$$

where R may be an alkyl, aryl, metal or metalloid group, and X and N may be a wide variety of both inorganic and organic anions: in addition N may be an uncharged compound with an unshared electron pair, e.g. amines, water. The mechanisms of these reactions have been investigated very extensively, and proceed in two ways: *S_n2 reactions* (substitution nucleophilic bimolecular) proceed formally according to step (1) since the rate-determining step of the reaction involves two particles, i.e. the kinetics are second order: in general S_n2 reactions involve a Walden inversion and there is a 5-co-ordinate intermediate in which both X and N are bonded to the same carbon of R. *S_n1 reactions* (substitution nucleophilic unimolecular) are dependent upon a preliminary cleavage of the R–X bond, step (2), in determining the rate

$$R-X \longrightarrow R^+ + X^- \qquad (2)$$

of the reaction: this is slow compared to step (3)

$$R^+ + N \longrightarrow R-N^+ \qquad (3)$$

These reactions follow first-order kinetics and proceed with racemisation if the reaction site is an optically active centre. For alkyl halides nucleophilic substitution proceeds easily: primary halides favour S_n2 mechanisms and tertiary halides favour S_n1 mechanisms. Aryl halides undergo nucleophilic substitution with difficulty and sometimes involve aryne* intermediates.

nucleoproteins *See* proteins.

nucleosides Glycosides of heterocyclic bases, in particular of purines and pyrimidines. They are crystalline substances, sparingly soluble in water. The nucleosides forming part of the molecule of ribose nucleic acid are the 9-β-D-ribofuranosides of adenine and guanine (adenosine and guanosine), and the 3-β-D-ribofuranosides of cytosine and uracil (cytidine and

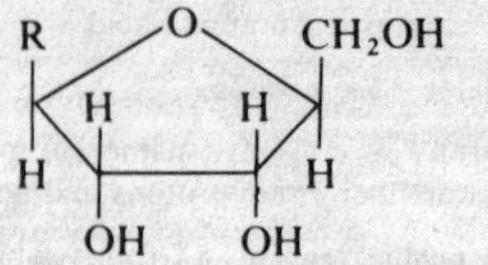

adenosin R =

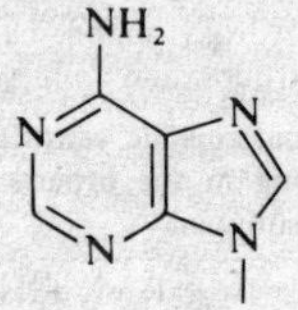

deoxycytidine R =

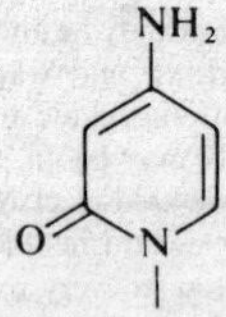

uridine). In deoxynucleic acid the sugar is deoxyribose and the uracil is replaced by thymine.

nucleotides Originally the term was applied to the phosphate esters of ribose and deoxyribose with purine or pyrimidine bases attached, which were identified as the constituents of nucleic acids*. Now the term nucleotide is used for phosphates of glycosides of all heterocyclic bases to encompass such compounds as adenosine triphosphate*, nicotinamide adenine dinucleotide* and other coenzymes.

nucleus, atomic The nucleus of an atom is the small but massive internal core of the atom in which is concentrated the weight and positive charge. The weight and any radioactive properties of the atom are associated with the nucleus, whereas chemical properties and u.v. and visible spectra are associated with the planetary electrons. The nucleus is built up from sub-atomic particles.

nuclide A species of atom characterized by the constitution of its nucleus. Commonly used in the context of radioactive decay. More than 1600 nuclides are known.

Nujol A trade name for a heavy medicinal liquid paraffin*. Extensively used as a mulling agent in spectroscopy.

Nylon A class of synthetic fibres and plastics, polyamides. Manufactured by condensation polymerization of α, ω-aminomonocarboxylic acids or of aliphatic diamines with aliphatic dicarboxylic acids. Also formed specifically, e.g. from caprolactam. The different Nylons are identified by reference to the carbon numbers of the diacid and diamine (e.g. Nylon 66 is from hexamethylene diamine and adipic acid). Thermoplastic materials with high m.p., insolubility, toughness, impact resistance, low friction. Used in monofilaments, textiles, cables, insulation and in packing materials. U.S. production 1983 1·1 megatonnes.

O

O Oxygen.

ob *See* chelate.

occlusion A term used to denote either the retention of a gas or of solids by a metal, or the absorption of an electrolyte by a precipitate. Generally used to indicate the penetration of the metal lattice by atoms or molecules of the occluded substance, sometimes with the formation of an interstitial compound (e.g. the occlusion of hydrogen by palladium).

ochre Earthy iron-containing pigments yellow to brownish red in colour.

ocimene, 3,7-dimethyl-1,3,6-octatriene, $C_{10}H_{16}$, $CH_2{:}CH{\cdot}C(CH_3){:}CH{\cdot}CH_2{\cdot}CH{:}C(CH_3){\cdot}CH_3$. An acyclic monoterpene. It is found in the oil obtained from the leaves of *Ocimum basilicum*, grown in Java, and probably in various other essential oils; b.p. 81°C/30 mm. On heating ocimene is isomerized to allo-ocimene, 2,6-dimethyl-2,4,6-octatriene, $(CH_3)_2C{:}CH{\cdot}CH{:}CH{\cdot}C(CH_3){:}CH{\cdot}CH_3$.

octahedral co-ordination Co-ordination of six

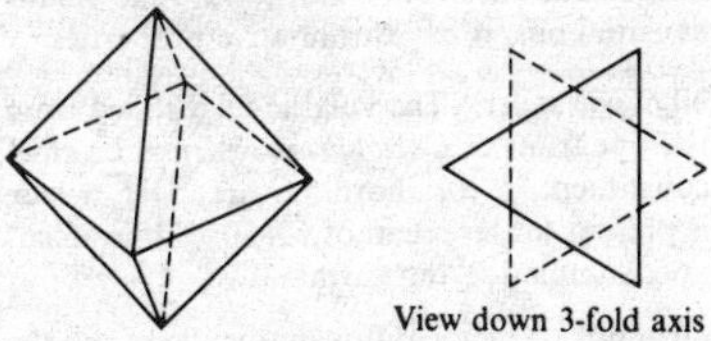

View down 3-fold axis

ligands which are disposed at the corners of an octahedron. SF_6 has octahedral co-ordination about sulphur. An octahedron has eight triangular faces.

octahedron *See* octahedral co-ordination.

octane number Motor or aviation gasolines are designated for knock rating* by the arbitrary scale of octane number. Iso-octane (2,2,4-trimethylpentane) is given a value of 100 and *n*-heptane a value of 0. The octane number of the gasoline is the percentage of iso-octane in a blend with *n*-heptane which has the same anti-knock performance as the gasoline under standard test conditions.

octanes, C_8H_{18}. Hydrocarbons of the paraffin series. There are eighteen possible paraffin hydrocarbons of this formula. They occur in petroleum and have boiling-points between 99°C and 125°C. The most important isomer is 2,2,4-trimethylpentane, $(CH_3)_3C{\cdot}CH_2{\cdot}CH(CH_3)_2$, usually called iso-octane. This is produced in large quantities by various methods from the butane-butylene fraction of the gas from the cracking of petroleum. It is a colourless liquid; b.p. 99°C. It has marked anti-knock properties and is used as a standard for determining the 'knock-rating' of petrols. Another isomer, 2-methylheptane, is sometimes, and more correctly, also called iso-octane. It has b.p. 117°C.

octanoic acid, caprylic acid, $C_8H_{16}O_2$, $CH_3{\cdot}[CH_2]_6{\cdot}COOH$. M.p. 16°C, b.p. 239°C. A fatty acid occurring free in sweat, as esters in fusel oil, from which it is most easily prepared, and as glycerides in the milk of cows and goats and in coconut and palm oils.

2-octanol, secondary octyl alcohol, capryl alcohol, $C_8H_{18}O$, $CH_3(CH_2)_5CHMeOH$. Colourless liquid with a strong odour; b.p. 179°C. Prepared by distilling castor oil with concentrated NaOH solution. Used as a foam-reducing agent. Esters of dibasic acids are used as plasticizers.

octant rule This relates the sign and the amplitude of Cotton effect* curves with the geometry and substitution of cyclohexanone derivatives and has been applied especially to steroids. It may also be extended to other cyclic ketones.

octaves, law of Newlands (1863–1864) found that when the elements were arranged in order of increasing atomic weights, the first element was similar to the eighth, the second to the ninth, the third to the tenth and so on. He termed this relationship the 'law of octaves'. *See* periodic law. The law is a natural consequence of the early part of the periodic table.

octet A group of 8 valence electrons (neglecting d electrons). Noble gases (except He) have 8 valence electrons and such arrangements have a certain stability in both ionic compounds, e.g. Na^+F^-, and covalent compounds, e.g. CCl_4. There are many compounds which do not have octets of electrons, e.g. most transition element compounds (except compounds for which the effective atomic number rule* holds), NO (paramagnetic), BF_3 (B has 6 electrons) and SF_6 (S has 12 electrons).

octyl alcohols Compounds with the formula $C_8H_{17}OH$. *See* 2-octanol.

***p-t*-octylphenol, diisobutylphenol,** $Me_3CCH_2CMe_2C_6H_4OH$. M.p. 81–83°C, b.p. 286–288°C. Made by alkylation of phenol. Forms oil-soluble resins with methanal (salts used as oil additives) and surfactants (with ethylene oxide).

oenanthic acid, *n*-heptanoic acid, $C_7H_{14}O_2$. An oily liquid, m.p. −9°C, b.p. 115–116°C/11 mm. It can be prepared by the oxidation of heptanal, and has growth inhibiting properties against micro-organisms. Present in some natural oils and waxes.

oestradiol, estradiol, 1,3,5(10)-oestratriene-3,17β-diol, $C_{18}H_{24}O_2$. M.p. 174°C. Found in the urine of pregnant mares and the ovaries of pigs. Oestradiol appears to act at a cellular level, by binding to the nucleus of cells in its target organ, where it seems to stimulate the synthesis of RNA.

oestriol, estriol, 1,3,5(10)-oestratriene-3,16α,17β-triol, $C_{18}H_{24}O_3$. M.p 283°C. An oestrogen present in the urine of pregnant women. It possesses female sex hormone activity but is not as active as oestrone.

oestrogens, estrogens The follicular female sex hormones. During pregnancy large quantities are excreted. They are found also in the urine of male members of the equine species. Oestradiol is the most potent of the naturally occurring oestrogens, and is probably the true hormone, of which oestrone, oestriol, equilin* and equilinin are metabolic products. Stilboestrol and hexoestrol* are synthetic products with oestrogenic activity that are much used medicinally, as they can be administered by mouth and are much cheaper than the naturally occurring oestrogens.

oestrone, estrone, 3-hydroxy-1,3,5(10)-oestratrien-17-one, $C_{18}H_{22}O_2$. M.p. 258°C. A female sex hormone.

oil Naturally occurring hydrocarbons, mainly aliphatic, used as a fuel after refining and as a feedstock for the petrochemical industry. Calorific value *c.* $4{\cdot}4 \times 10^7$ Jkg^{-1}. World production 2×10^9 tonnes per year. *See* crude oil.

Also used to describe an oily substance obtained from natural products, e.g. oil of anise.

oil of anise Commonly called oil of aniseed, it is obtained by steam distillation from star anise fruits, *Illicium verum*. It consists chiefly of anethole* and is used as a carminative and as an ingredient of cough lozenges and cough mixtures and as flavouring agent.

oil of camphor Two fractions of the natural oil of camphor are used medicinally.

(1) Light oil of camphor is an almost colourless fraction containing a small amount of camphor, about 30% of cineole and the remainder terpenes.

(2) Dark oil of camphor consists very largely of safrole.

Both are used externally in the preparation of liniments.

oil of caraway Obtained by distillation from the ripe fruits of *Carum carvi*. It contains about 60% of carvone, and is used to relieve flatulence, especially in infants.

oil of cinnamon The volatile oil distilled from the bark of *Cinnamonum zeylanicum*. The chief constituent of this oil is cinnamic aldehyde, $C_6H_5{\cdot}CH{:}CH{\cdot}CHO$ (50%). In the U.S. the rectified oil of cassia is known as oil of cinnamon. Oil of cinnamon is carminative and antiseptic.

oil of cloves The volatile oil distilled from cloves consisting largely of eugenol. It has antiseptic properties, and is an effective local anaesthetic used in dentistry. Like many essential oils, it has carminative properties.

oil of coriander The volatile oil distilled from the ripe fruits of *Coriandrum sativum*. Its chief constituent is coriandrol, $C_{10}H_{17}OH$, which is present to the extent of 65–80%. It is aromatic, stimulative and carminative.

oil of dill Volatile oil steam-distilled from the ripe fruits of *Anethum graveolens*. It contains about 50% of carvone*, to which it owes its medicinal properties. It is a carminative, and is especially useful for relieving flatulence in infants.

oil of peppermint The oil distilled from the fresh flowering tops of *Mentha piperita* containing about 50% of menthol together with menthyl esters and the ketone, menthone*. It is an aromatic carminative and relieves flatulence and colic.

oil of wintergreen *See* methyl salicylate.

olefine complexes Co-ordination compounds in which an olefine is bonded to a metal by overlap of the π-bonding orbitals of the olefin with a suitable orbital of the metal and by back-bonding from a suitable orbital of the metal into a π-anti-bonding orbital of the olefin. The olefin (or acetylene) is thus bonded sideways on to the metal. Olefin complexes are formed by many transition and non-transition metals and are intermediates during many reactions of olefins in the presence of metals. E.g.

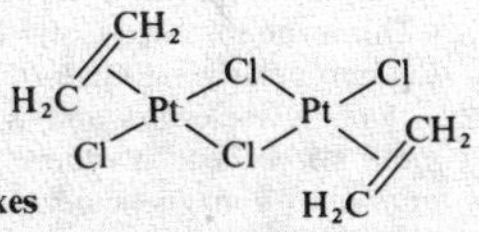

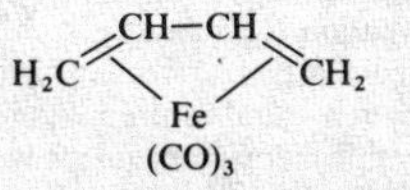

olefine complexes

olefin polymers *See* polyalkenes.

olefins *See* alkenes.

oleic acid, *cis*-9-octadecenoic acid, $C_{18}H_{34}O_2$, $CH_3[CH_2]_7{\cdot}CH{:}CH{\cdot}[CH_2]_7{\cdot}COOH$. Colourless liquid. B.p. 286° C/100 mm., insoluble in water. Oleic acid occurs naturally in larger quantities than any other fatty acid, being present as glycerides in most fats and oils. It forms one third of the total fatty acids of cow's milk. A crude grade from inedible tallow is used in the production of lubricants, detergents, resins and other products.

oleum *See* sulphuric acid, fuming.

oligomer A polymer composed from only a small number of monomeric units, hence oligomerization.

olivine, $9Mg_2SiO_4$, Fe_2SiO_4. A member of the silicates, $M_2{}^{II}SiO_4$ (M = Mg, Fe, Mn) containing discrete $SiO_4{}^{2-}$ anions. The Mg-rich olivines are used in the manufacture of refractory bricks.

'onium compounds Groups of compounds containing cations of the type R_XA^+ which are analogous to ammonium compounds. Ions include phosphonium R_4P^+, oxonium R_3O^+, sulphonium R_3S^+, iodonium R_2I^+.

open shell compound A species for which the total number of valency electrons does not correspond to a noble gas configuration or the octet* rule. E.g. BF_3 (6 outer electrons) SF_6 (12 outer electrons). In the transition series applied to compounds in which there is a partially filled d shell.

opium Dried latex which exudes from the unripe capsules of *Papaver somniferum* after the outer skin has been cut with a knife. It contains a large number of alkaloids, chief among which are morphine*, codeine* and papaverine.

Oppenauer oxidation The oxidation of secondary alcohols to ketones using aluminium *tert*-butoxide.

opsin *See* rhodopsin.

optical activity The property possessed by certain substances of rotating the plane of polarization of polarized light. The property is exhibited by certain solids and liquids (e.g. tartaric acid and some tertiary alcohols), and may also be observed in solutions of such substances, and in the vapour phase. It is associated with asymmetry either of a molecule, an ion or a crystal lattice. Molecules may be asymmetric due to having four different groups joined to one carbon atom, or from restricted rotation about a central bond.

Optically active substances are termed dextrorotatory or laevorotatory according to whether the plane of polarization of the light is rotated to the right or to the left with respect to the direction of incidence of the light. The prefixes *d*-and *l*-were once used to indicate which optical isomer was which, but now the correct prefixes are (+)- for dextrorotatory, and (−)- for laevorotatory compounds, with (±)- for racemic compounds. These signs are arbitrary as the rotation can change with wavelength of the light (Cotton effect*, optical rotatory dispersion). With carbohydrates and amino-acids the prefixes D- and L- are used to indicate configuration, not direction of rotation. The convention has been made that the configuration of D-glyceric aldehyde shall be represented by formula (1) and L-glyceric aldehyde by formula (2).

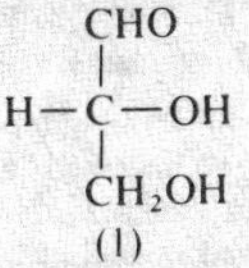

(1)

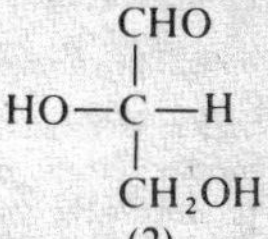

(2)

When the formulae of other sugars are written in this way those which have the configuration

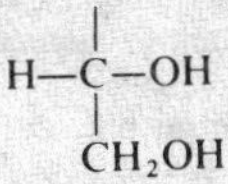

are given the prefix D-, while L- is used for those with the configuration

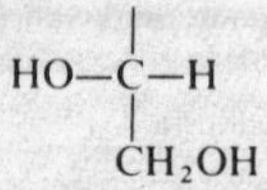

The same nomenclature has been adopted for amino-acids, the configurational family to which the α-carbon atom belongs being denoted by the prefixes D- and L-.

The R, S convention is a scheme which has largely superseded the D, L system to denote configuration about a chiral centre in a molecule. The convention allows unequivocal designation of the absolute configuration in a description of the positions in space of ligands attached to a chiral centre, in relation to an agreed standard of chirality like a right-hand helix.

Groups attached to the chiral centre are given an order of priority according to the sequence rules*. For an enantiomeric carbon compound the group of lowest priority is viewed through an imaginary triangle - - - - - formed by joining the other three groups which interpose themselves between the chiral centre and the observer. If the order of decreasing priority of the three groups on the triangle is clockwise the configuration is specified as (R) [*rectus*]; and (S) [*sinister*] if the decreasing priority is counter-clockwise.

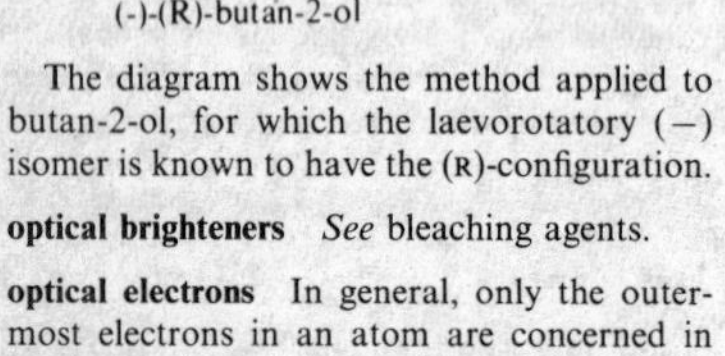

(-)-(R)-butan-2-ol

The diagram shows the method applied to butan-2-ol, for which the laevorotatory (−) isomer is known to have the (R)-configuration.

optical brighteners *See* bleaching agents.

optical electrons In general, only the outermost electrons in an atom are concerned in producing the emission or absorption of visible light. Such electrons are termed optical electrons.

optical exaltation An abnormal increase in the refractivity of an organic compound, over and above the additive amount to be expected from the bonds present, due to the presence of a system of conjugated double bonds.

optical purity The ratio of a substance's specific rotation to the specific rotation of the pure enantiomer.

optical rotatory dispersion, ORD A description of the changes in optical rotation of an optically active molecule, organic or inorganic, with the wavelength of the light. The variation of specific rotation with wavelength increases as an absorption band is approached, and the rotation may pass through a maximum or minimum when the absorption band is crossed. This is the so-called Cotton effect*. The effect is particularly sensitive to structural features of many molecules and is used in structural and stereochemical studies. ORD may be induced by a magnetic field, *see* magnetic polarization of light.

optoacoustic spectrometry A spectrophotometric technique whereby modulated electromagnetic radiation falling upon the sample causes heating when absorption occurs. The heating causes a pressure change in the cell containing the sample which is detected by a microphone. Potentially can be used to examine most regions of the electromagnetic spectrum.

orbital Loosely used to describe the geometrical figure which describes the most probable location of an electron. More accurately an allowed energy level for electrons. *See* electronic configuration.

orbital angular momentum, L. The total angular momentum of a set of electrons in an atom or ion. For filled and empty shells, L is zero.

orbital moment The component of the magnetic moment of an atom or ion resulting from the motion of the electron around the nucleus.

orbital splitting The raising of the degeneracy of a series of orbitals of equivalent energy, e.g. by placing in a crystal field*.

ORD Optical rotatory dispersion.

order of reaction The power to which the concentration terms are raised in the mathematical equation which describes the variation of the rate of a reaction with concentration of reactants at a given temperature. For example, if in the reaction of two substances A and B the rate of reaction is proportional to the concentrations of each reactant, the reaction is said to be first order in A and in B and second order overall. The order may be other than unity and in some cases may be non-integral. It is an experimentally determined quantity which does not necessarily have any relationship to the actual number of molecules involved in the reaction since many reactions proceed in a series of steps. The order refers only to the slowest step.

ore dressing *See* mineral dressing.

organic Until the early part of the nineteenth century chemical substances of animal or plant origin were designated as 'organic' and held to differ fundamentally from 'inorganic' substances of mineral origin in that a 'vital force' was required for their production. The 'vital force' theory was discredited in 1828 by Wöhler but the term 'organic' has remained. Organic chemistry is now the study of the compounds of carbon, whether they be isolated from natural sources or synthesized in the laboratory. Most, but not all, organic compounds contain H as well as C, while other common elements are O, N, the halogens, S and P. These and other elements are usually bound to carbon by covalent bonds as distinct from the ionic links typical of inorganic compounds. A few very simple carbon compounds, e.g. metallic carbonates, are considered as inorganic, and the large class of organometallic compounds is of interest to both inorganic and organic chemists.

organoboranes Compounds containing at least one direct carbon to boron bond. The commonest classes are the trialkylboranes R_3B and the dialkylboranes R_2BH. Developed as extremely useful organic synthetic reagents, they are decomposed by aqueous acid to alkanes and by alkaline peroxide to alcohols: the latter reaction provides a convenient method, via hydroboration, for the effective *cis*-hydration of alkenes in an anti-Markownikoff direction. Acid hydrolysis of the hydroboration product of internal acetylenes provides a convenient synthesis of *cis*-alkenes.

Methods of producing B–C bonds include hydroboration*, nucleophilic displacement at a boron atom in BX_3 (X= halogens or $B(OR)_3$) by e.g. a Grignard reagent, and a *pseudo*-Friedel–Crafts reaction with an aromatic hydrocarbon, BX_3, and $AlCl_3$.

organoborates Compounds formally containing BR_4^- anions although some derivatives, particularly lithium compounds, e.g. $LiBMe_4$, are polymeric. Most easily prepared from metal alkyls and trialkylboranes. Postulated as intermediates in reactions between organoboranes and nucleophiles. The BR_4^- ions react with electrophiles in synthetically useful reactions. There may be intermolecular transfer to E^+, e.g. RCOX reacts with BR'_3 to give ketones RCOR′, or intramolecular transfer particularly to the α-carbon, e.g.

$$NaR_3B-C{\equiv}CR' \xrightarrow{HCl} R_2BCR{=}CHR'$$

The organoboranes formed by such reactions may be decomposed to olefines.

organochlorine insecticides Highly active insecticides which, until recently, have been very widely used. Find their way into the bodies of animals and insects and are stored, principally in the fat. The application of organochlorine insecticides is now widely restricted.

organoelement compounds Compounds of an element containing organic residues bonded directly to the element. *See* organometallic compounds. Formed by all elements except for the noble gases.

organolithic Composites of organics and silicates.

organometallic compounds Compounds in which carbon atoms or organic groups are linked to metal or metalloid atoms. Metal isonitrile and carbonyl derivatives are considered as organometallic although carbides and cyanides are not generally so classified. Organometallic derivatives range from substantially ionic, e.g. $NaCH_3$, to covalent, e.g. $H_3CMn(CO)_5$. Widely used in synthesis (e.g. RMgX, Grignard reagents) and catalysis (e.g. Ziegler catalysts and aluminium alkyls). The only naturally occurring organometallic compound so far discovered is vitamin B_{12} coenzyme.

organometalloids Compounds containing a direct bond between one or more carbon atoms and one or more metalloid atoms.

organophosphorus compounds Compounds containing C–P bonds but often extended to esters and thio-esters. The most important application of organophosphorus compounds is in the field of pesticides.

Among the best known organophosphorus pesticides are malathion, parathion, schradan and dimefox.

Organophosphorus compounds are widely used as ligands in metal-complex chemistry, including some catalytic processes; triphenylphosphine is the commonest such ligand. Several classes of organophosphorus compounds are employed in organic synthesis, e.g. triphenylphosphine in the Wittig synthesis of alkenes and HMPA (hexamethylphosphoramide) as a solvent. Phosphate esters* are used for flame-proofing cotton and synthetic materials and as gasoline additives.

organosol A collodial dispersion in which an organic dispersion medium is used. Specific names such as alcosol, benzosol may be used to indicate that the dispersion medium is alcohol or benzene. Used for e.g. nitrocellulose, collodions, rubber solutions in toluene or naphtha.

organotin derivatives *See* tin, organic derivatives. Used as fungicides.

Orgel diagrams Simple graphs showing the relation between the energies of various electronic states and the crystal field splitting.

ornithine, αδ-diaminovaleric acid, $C_5H_{12}N_2O_2$, $NH_2{\cdot}CH_2{\cdot}CH_2{\cdot}CH_2{\cdot}CH(NH_2){\cdot}CO_2H$. M.p. 140°C. An amino-acid occasionally formed in the hydrolysis products of proteins and occurring in the urine of some birds as dibenzoylornithine. Ornithine is a precursor of arginine in plants, animals and bacteria.

L-ornithuric acid, *N,N*-dibenzoyl-L-ornithine *See* ornithine.

Orsat gas-analysis apparatus *See* gas analysis.

ortho In the case of disubstituted derivatives of benzene the prefix ortho has a precise structural significance. Thus ortho-cresol,

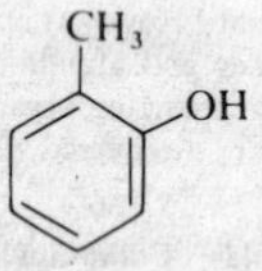

commonly written *o*-cresol, the CH_3 and OH groups are said to be in the ortho position to each other. Using numbers to designate substituent position *o*-cresol is 2-hydroxy-toluene.

The prefix is also used in orthophosphates, orthocarbonates, orthoformates and orthosilicates, which are derivatives of $PO(OH)_3$, the hypothetical $C(OH)_4$, $HC(OH)_3$ and $Si(OH)_4$ respectively; in these the word 'ortho' is always written in full.

orthoformic ester *See* triethyl orthoformate.

ortho-hydrogen The hydrogen molecule, H_2, exists in two forms because of differences in the relative spin of the nuclei of the two atoms. When both spin in the same direction we have *para-hydrogen*; when the spins are symmetrical, or in opposite directions, we have *ortho-hydrogen*. Ordinary hydrogen is a mixture of the two forms; at normal temperatures the composition being 25% para and 75% ortho. Almost pure para-hydrogen has been prepared by cooling ordinary hydrogen to a very low temperature in contact with charcoal, which acts as a catalyst for the conversion. The two forms have identical chemical properties, but slightly different physical properties.

orthorhombic system The crystal system* with a 2-fold axis at the intersection of two planes of symmetry or perpendicular to two other 2-fold axes. A unit cell has three unequal axes at right angles.

Os Osmium.

osazones Organic compounds containing the grouping

—C=N—NHAr

—C=N—NHAr

They are formed by treating α-diketones, α-hydroxyaldehydes, hydroxyketones, aminoaldehydes or aminoketones with arylhydrazines. Sugars can be identified by their osazones which have characteristic melting-points, formation times or crystal appearance.

osmates *See* osmium oxides.

osmiamates *See* osmium oxides.

osmic acid The name given to osmium tetroxide and its solution in water. This solution contains free OsO_4. In alkaline solution $[OsO_4(OH)_2]^{2-}$ ions are present.

osmiridium Native or synthetic alloys of osmium (15–40%) and iridium (50–80%). It is used for special purposes in which a very hard and incorrodible material is required, such as the tips of fountain pen nibs and sparking points.

osmium, Os. At.no. 76, at.wt. 190·2, m.p. 3045°C, b.p. 5027°C, *d* 22·6. Osmium is a platinum group metal. It occurs as a sulphide or as osmiridium (extraction *see* Ru). The metal is hcp and has very similar properties to Ru. Oxygen at high temperatures gives OsO_4. Os is used for hardening purposes in alloys with other platinum metals.

osmium chemistry Osmium is an element of the iron group, electronic configuration $5d^66s^2$. The chemistry is similar to that of Ru but there is greater stability of the higher oxidation states (halides and oxides); oxidation states from +8 to 0 are known. $Os(CO)_5$ and some polymeric carbonyls are formed and the organometallic chemistry is similar to that of Fe. Os(III) and Os(II) (predominantly octahedral) form extensive series of complexes with N-, P- and As-ligands. Osmium nitrosyls appear to be very stable as are nitrido derivatives (*see* oxide).

osmium halides Yellow OsF_7 (Os plus F_2 at 60°C and 400 atm) dissociates to OsF_6 very readily. Pale green OsF_6 (Os plus F_2) m.p. 32°C, blue tetrameric OsF_5 and yellow OsF_4 (both OsF_6 plus $W(CO)_6$) are the known fluorides. Oxide fluorides, e.g. $OsOF_5$ (Os plus F_2 plus O_2) and complex fluorides, e.g. $(OsF_6)^-$, $(OsF_6)^{2-}$ are known. The highest chloride is red $OsCl_4$ (Os plus Cl_2 at 600°C) it decomposes on heating to $OsCl_3$. Chloroosmates(IV) and (III), $(OsCl_6)^{2-}$ and $(OsCl_6)^{3-}$ are useful starting materials in Os chemistry. $OsBr_4$, $OsBr_3$ and OsI_3 and some oxide halides and oxide halide complexes are known.

osmium oxides Colourless OsO_4 (Os plus O_2 at high temperatures), m.p. 25°C, b.p. 100°C, is a toxic material with a penetrating odour. It is used as an oxidizing agent; alkenes give dihydroxy species. Dark coloured OsO_2 (Os plus OsO_4) has the rutile structure.

In strongly alkaline solution OsO_4 gives red osmates(VIII), $[OsO_4(OH)_2]^{2-}$, which can be reduced to pink or blue osmates(VI), $[OsO_2(OH)_4]^{2-}$. Osmates(VII) are formed by heating mixed oxides in oxygen. Osmates(VI) undergo substitution by anions, e.g. CN^- to give $[OsO_2(CN)_4]^{2-}$.

Osmiamates, e.g. $KOsO_3N$ ($K_2OsO_4(OH)_2$ plus ammonia) are stable and related species, e.g. $[Os^{VI}NCl_4]^-$, can be formed by substitution.

osmosis When a solvent and a solution are separated by a semi-permeable membrane, such as a piece of parchment or a plant or animal membrane, the solvent tends to flow spontaneously through the membrane into the solution. This process is known as osmosis. A similar flow is observed when two solutions of different chemical potential are separated by a semi-permeable membrane.

osmotic pressure The excess hydrostatic pressure which must be applied in order to counterbalance the process of osmosis. Thus a solution under this excess pressure can come to osmotic equilibrium with the same solvent on the other side of a membrane permeable only to the solvent molecules. For dilute solutions of non-electrolytes the osmotic pressure is equal to the pressure they would exert if they existed as gases under the same conditions of molecular concentration and temperature. For dilute solutions the osmotic pressure (π) is given by

$$\pi = \frac{cRT}{M}$$

where c is the concentration of solute, of molecular weight M, in grammes per unit volume. R is the gas constant and T, the absolute temperature.

Ostwald ripening A process of crystal growth in which a mixture of coarse and fine crystals of a substance are left in contact with a solvent. This results in a growth of the large crystals and the ultimate disappearance of the fine crystals.

outer sphere complexes Weak complexes in which there is minimal disturbance of the inner co-ordinated ligands. Postulated in many electron transfer reactions.

ovalbumin, egg albumin The chief protein constituent of white of egg. It can be prepared by precipitating the globulins from egg white by half saturation with ammonium sulphate and then precipitating the ovalbumin by adding ethanoic acid.

over-voltage, over-potential The excess potential, over and above that of the reversible electrode, that must be applied to an electrode for electrolysis to proceed at a measurable rate. E.g. at a platinized platinum electrode the evolution of hydrogen will occur practically at the reversible electrode potential. However, at a mercury electrode no hydrogen is evolved until an over-voltage of approximately 1 volt is reached. Over-voltage is of great practical importance in processes such as electro-plating, electro-analysis, electro-reduction, polarography, etc. *See* hydrogen over-voltage.

oxalates Salts or esters of oxalic acid.

oxalic acid, ethanedioic acid, $C_2H_2O_4$.

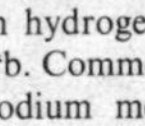

Crystallizes from water in large colourless prisms containing $2H_2O$. It is poisonous, causing paralysis of the nervous system; m.p. 101°C (hydrate), 189°C (anhydrous), sublimes 157°C. It occurs as the free acid in beet leaves, and as potassium hydrogen oxalate in wood sorrel and rhubarb. Commercially, oxalic acid is made from sodium methanoate. This is obtained from anhydrous NaOH with CO at 150–200°C and 7–10 atm. At lower pressure sodium oxalate formed; from the sodium salt the acid is readily liberated by sulphuric acid. Oxalic acid is also obtained as a by-product in the manufacture of citric acid and by the oxidation of carbohydrates with nitric acid in presence of V_2O_5.

When heated with sulphuric acid the acid is decomposed to CO, CO_2 and H_2O. Gives methanoic acid when heated with glycerin. Forms both acid and neutral salts. The normal potassium and ammonium salts are soluble in water; the corresponding acid salts are less soluble and the salts of other metals are only sparingly soluble. The calcium salt is very insoluble. Oxalic acid is used for metal cleaning and as a chemical intermediate. Because of its bleaching action it is widely employed for textile finishing and cleaning, and is also used for whitening leather.

A common domestic use is the removal of rust or blood stains from white materials, e.g. cotton.

oxamide, $C_2H_4N_2O_2$, $(CONH_2)_2$. Colourless crystalline solid from ethyl oxalate and concentrated ammonia. Forms cyanogen when heated with P_2O_5.

oxanthrol Oxanthrol (1) is tautomeric with

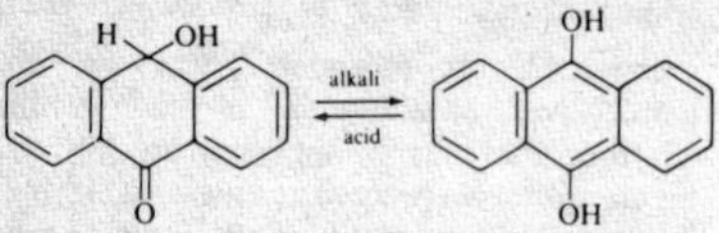

(1) (2)

anthrahydroquinone (2); both forms may be isolated according to experimental conditions from the zinc dust and alkali reduction of anthraquinone. Oxanthrol is a greenish-yellow compound, decomposing at 205°C, which dissolves in alkali giving a blood-red solution; on shaking in air the solution is rapidly decolourized and yellow anthraquinone separates, leaving an alkaline solution of hydrogen peroxide.

oxazole ring The ring numbered as shown.

oxetane polymers Polymers derived from oxetane $CH_2CH_2CH_2O$ many of the oxetane* derivatives being derived from pentaerythritol so that the repeating unit is $\text{-(}CH_2CR_2CH_2O\text{)-}_n$. Polymerization is with Lewis acids, particularly BF_3.

Poly (*3,3-bis* (*chloromethyl*) *oxetane*), chlorinated polyether ($R=CH_2Cl$) is widely used for injection moulding and coating.

oxetanes Derivatives of oxetane C_3H_6O. Formed by an intramolecular Williamson reaction on, e.g. pentaerythritol derivatives, pyrolysis of cyclic carbonate esters or photochemical addition of aldehydes or ketones to

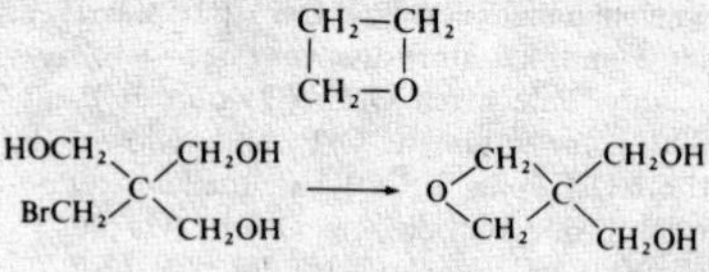

olefines. The commercially important 3,3-bis (chloromethyl)oxetane, b.p. 90°C, is prepared by a Williamson reaction. The oxetanes readily polymerize with Lewis acids to give useful oxetane polymers*.

oxidases Dehydrogenating enzymes that transfer hydrogen directly to molecular oxygen.

oxidation Any process by which the proportion of the electronegative constituent in a compound is increased, e.g. $Cu_2O \longrightarrow CuO$. In oxidation electrons are removed from the species oxidized, e.g.

$$Cu^+ \longrightarrow Cu^{2+} + e$$

and the oxidation state* is increased. In organic chemistry particularly the most obvious effect is to increase the proportion of oxygen, e.g. $C_2H_5OH \longrightarrow CH_3CO_2H$. For specific selective oxidizing agents *see* e.g. Oppenauer oxidation, selenium dioxide, chromium trioxide.

β-oxidation The degradation of fatty acids in biological systems to produce acetyl-CoA which is fed into the citric acid cycle to produce ATP, etc., eventually. It occurs by repeating a sequence of events which removes a 2 carbon fragment each time.

oxidation-reduction indicator *See* redox indicator.

oxidation-reduction potential *See* redox.

oxidation state The difference (+ve or −ve) between the number of electrons associated with an atom in a compound as compared with an atom of the element. In ions corresponds to the ionic charge, e.g. NaCl[Na (+1), Cl (−1)]; in covalent compounds to the formal change assuming an ionic structure taking account of electronegativity to determine charge, e.g. CCl_4 is $C^{4+}4Cl^-$ and thus C (+4). Co-ordinate bonds do not affect the oxidation state, e.g. $[Co(NH_3)_6]^{3+}$ contains Co (+3) and $Fe(CO)_5$ contains Fe (0). Elements are in the zero oxidation state. The oxidation state, Roman numerals, is used in systematic nomenclature, e.g. MnO_4^- is manganate(VII).

oxidative addition A type of reaction in which there is direct addition to an element or multiple bond with increase in oxidation state. E.g. addition of Cl_2 to PCl_3 gives PCl_5.

oxidative phosphorylation The process occurring along the respiratory or electron transport chain* by which ATP is generated through a series of coupled reactions from reduced coenzymes, such as NADH, produced in the citric acid cycle*.

oxide A compound of oxygen with another element. Oxides are divided into acidic oxides, which react with bases to form salts (e.g. SO_2, P_2O_5); basic oxides which react with acids to form salts (e.g. CuO, CaO); amphoteric oxides, which exhibit both basic and acidic properties (e.g. Al_2O_3); neutral oxides (e.g. CO, NO) which do not react with either acids or gases; peroxides, metallic derivatives of hydrogen peroxide; and superoxides containing O_2^- species. Free acids and gases cannot always be dehydrated to oxides. Many oxides (e.g. FeO) are non-stoicheiometric and contain metal atoms in two or more oxidation states. Most oxides have structures with close-packed oxide ions. Non-stoicheiometry may increase on heating (e.g. ZnO) and may cause marked colour changes. Fusion of two or more oxides often gives mixed metal oxides which frequently do not contain discrete oxyanions.

oxidized rubber, Rubbone Produced by oxidation of rubber with air in the presence of a catalyst. Can be vulcanized and chlorinated. Used for impregnation, electrical insulation and as an adhesive.

oxidizing agent A material which brings about oxidation and in the process is itself reduced. E.g. MnO_4^- in acid oxidizes $Fe^{2+} \longrightarrow Fe^{3+}$ and is itself reduced to Mn^{2+}.

oximes Organic compounds containing the group $=N\cdot OH$ united directly to a carbon atom. *See* aldoximes and ketoximes.

oxine 8-hydroxyquinoline.

oxirane The correct name for the oxygen-

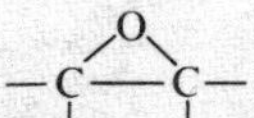

containing 3-membered heterocycle, more commonly called an epoxide.

oxitol *See* Cellosolve.

oxonium A positive ion containing a central oxygen atom. The hydroxonium ion, H_3O^+, exists in aqueous solutions of acid, and solid salts containing this ion may be prepared. The trimethyloxonium, Me_3O^+, and triphenyloxonium, Ph_3O^+, ions and their derivatives are also known. Trialkyloxonium salts are strong alkylating agents for organic and inorganic nucleophiles.

oxopyrazolines *See* pyrazolones.

oxo reaction A reaction which involves the addition of water gas or synthesis gas ($H_2 + CO$) to an alkene under pressure in the presence of a catalyst. The latter may be mixtures of oxides of cobalt, thorium and copper on kieselguhr, a cobalt carbonyl or rhodium phosphine complexes. The products are aldehydes which may be further hydrogenated using the same catalysts to primary alcohols. A modification using carbon monoxide and water instead of water gas yields carboxylic acids. A terminal alkene $R\cdot CH{=}CH_2$ usually yields a mixture of terminal aldehyde $R\cdot CH_2\cdot CH_2\cdot CHO$ and branched aldehyde $R\cdot\underset{\displaystyle CHO}{\underset{|}{C}}H\cdot CH_3$ depending upon catalyst and conditions. These may be converted to the mixed alcohols, especially where used for plasticizers or detergent manufacture. *See* carbonylation, hydroformylation.

oxyanions Anions which may be considered as being formed by co-ordination of oxide ions, O^{2-}, to metal or non-metal cations, e.g. the sulphate ion SO_4^{2-} is formally built up from S^{6+} and $4O^{2-}$ ions. Oxyanions are the normal species in water as complex fluoroanions are the normal species in anhydrous hydrogen fluoride.

oxyazo dyes Azo dyestuffs containing a hydroxyl group.

oxycyanogen, $(OCN)_2$. A pseudohalogen (KOCN plus Cl_2).

oxydemeton methyl, $C_6H_{15}O_4PS_2$. A systemic and contact insecticide.

oxygen, O. At.no. 8, at.wt. 15·9994, m.p. −218·4°C, b.p. −182·962°C. Colourless, odourless gas condensing to a pale-blue liquid and gas. Occurs in air (21% by volume) and by far the most abundant element (lithosphere 47% by weight, oceans 89%). Manufactured by distillation of liquid air and can be prepared in laboratory by heating $KClO_3$ with MnO_2. Reacts with most elements on heating to form oxides. Exists as paramagnetic O_2 and as ozone, O_3. Used in steel making (65%), in synthesis (e.g. HNO_3, H_2SO_4, ethylene oxide), in welding, in mining, explosives (*see* liquid oxygen explosives), in rocket fuels, sewage digestion. U.S. production 1980 415 billion cu. ft.

oxygen carrier A molecule which can react directly with molecular oxygen (to form a peroxide or superoxide derivative, generally a complex) and can subsequently use the oxygen in an oxidation reaction – generally of an organic substrate, e.g. haemoglobin.

oxygen cathode A platinum or gold electrode used in amperometry for measuring oxygen content by measuring the rate of cathodic reduction of O_2. The cathode is used with a reference electrode. A Clark electrode is a typical oxygen electrode.

oxygen chemistry Oxygen is an electronegative Group VI element, electronic configuration $1s^2 2s^2 2p^4$. Its predominant chemistry is in the -2 oxidation state in oxides* (covalent, or ionic), alkoxides (ROM compounds) and ethers. The M–O bond order may approach 2 in compounds such as F_3PO. The oxygen commonly is 2 (bent) co-ordinate but co-ordination numbers of 3 (R_3O^+ and TiO_2), 4 (Cu_4OCl_6), 6 (MgO) are known. O—O bonding is fairly unusual but is known in e.g. O_2, O_3 and peroxy compounds. The dioxygenyl $(O_2)^+$ cation (O_2 plus PtF_6 gives O_2PtF_6; O_2F_2 plus BF_3 gives O_2BF_4) shows that other oxidation states can exist. $[O_2]^-$ and $[O_2]^{2-}$ formally contain other oxidation states.

oxygen fluorides *Oxygen difluoride*, OF_2. Prepared by passing F_2 through dilute cooled NaOH solution. M.p. $-224°C$, b.p. $-145°C$. It is only slowly hydrolysed by water at ambient temperature, and is a strong oxidizing agent when sparked.

Other oxygen fluorides, O_2F_2, O_4F_2, O_5F_2 and O_6F_2, are prepared by passing an electric spark through a mixture of O_2, F_2 and sometimes O_3.

oxythallation Thallium(III) salts oxidize a wide variety of organic substrates under mild conditions in alcohol, dilute acid or diglyme solution. Unstable organothallium derivatives are produced as intermediates.

oxytocin A cyclic peptide hormone secreted by the posterior lobe of the pituitary gland. It stimulates contraction of the smooth muscle of the uterus. During lactation it promotes the ejection of milk.

ozalid process *See* blue print paper.

ozokerite A naturally occurring mineral wax which on refining yields a hard white wax known as ceresin*.

ozone, O_3. Allotrope of oxygen, m.p. $-193°C$, b.p. $-112°C$, blue in colour. Present in upper atmosphere and protects earth's surface from excess u.v. radiation. Prepared by action of electric discharge on O_2 and can be separated from O_2 by fractionation. Has bent molecule. Used for oxidation, sterilization and purification.

ozonides Products of the action of ozone on various classes of unsaturated organic compounds; have the structure shown. Prepared by passing ozonized air or oxygen into the

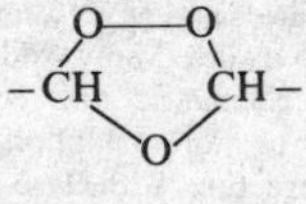

unsaturated compound dissolved in a suitable solvent. Most ozonides are thick oils with an unpleasant choking smell and are explosive. Readily decomposed, some by water, others by ethanoic acid or reducing agents. From an analysis of the decomposition products the position of the double bond in the original unsaturated compound can usually be determined. Inorganic ozonides, such as potassium ozonide, KO_3, also exist. These contain the O_3^- ion.

ozonizer Apparatus for the preparation of ozone by passing oxygen through an electrical discharge.

ozonolysis A procedure for the addition of ozone to an unsaturated organic compound, giving an ozonide which can be unstable and is not normally isolated but is decomposed *in situ*. E.g. an alkene gives 2 ketones

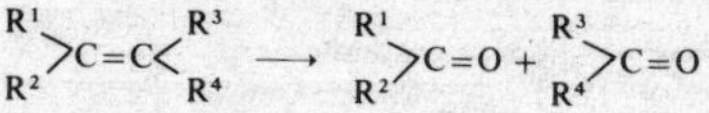

after decomposition of the ozonide with zinc and dilute aqueous acid.

P

P Phosphorus.

Pa Protactinium.

packed column, packed tower Used for gas-liquid contacting, for gas absorption*, for small-scale industrial distillation operations, and for liquid–liquid extraction*.

packing fraction The mass of an atom is not the simple sum of the masses of the elementary particles making up the atom, since some of the mass is converted to energy ($E=mc^2$) which is used to hold the nucleus together. The difference between the actual atomic mass and the mass number (the total mass of the protons and neutrons which constitute the nucleus) is termed the mass defect. The packing fraction is defined as the mass defect divided by the mass number. For the lightest and heaviest elements, the packing fraction is positive; intermediate elements have a negative packing fraction.

paint Covering material for wood, metal, etc., used to improve appearance and to protect the substrate against e.g. rust. The basic components of paints are pigment (decoration but also protects against sunlight), the non-volatile vehicle (drying oils, resins, etc.) which holds the pigment in place and the volatile vehicle (water, hydrocarbon solvents) which allows ease of application and eventually evaporates. European production 1976 4·7 megatonnes.

palladium, Pd. At.no. 46, at.wt. 106·42, m.p. 1552°C, b.p. 3140°C, *d* 12·02. The least noble of the platinum metals resembling Pt in its mechanical properties. Pd compounds are readily reduced to the metal which has the ccp structure. The metal is dissolved by conc. HNO_3 or hot H_2SO_4; it is oxidized in air at dull red heat. Used in alloys (white gold, Au plus Pd), as a catalyst (particularly for hydrogenation) and in electrical components. World demand 1976 2 565 000 troy oz.

palladium ammines An extensive range of palladium(II) complexes, generally 4-co-ordinate, containing $Pd-NH_3$ (or amine) bonds.

palladium black A finely divided precipitate obtained by reducing solutions of Pd salts. Used as a hydrogenation catalyst.

palladium chemistry Palladium is an element of the nickel group, electronic configuration $4d^{10}$. The chemistry is similar to that of Pt although the +6 and +5 oxidation states are missing. There are extensive ranges of complexes. Organometallic derivatives and hydrides are much less stable than for Pt. Pd forms an aquo-ion $[Pd(H_2O)_4]^{2+}$ and salts containing this ion are known.

palladium halides Palladium fluorides known are brick-red PdF_4 (PdF_3 plus F_2 under pressure), black PdF_3 ($Pd^{II}Pd^{IV}F_6$; $PdCl_2$ plus BrF_3) and violet PdF_2 (PdF_3 plus SeF_4). Fluoropalladates (IV), containing PdF_6^{2-} ions, are formed from BrF_3 solution, all palladium fluorides are hydrolysed by water. The only known chloride is $PdCl_2$ (Pd plus Cl_2), $PdCl_6^{2-}$ and $PdCl_4^{2-}$ species are formed in solutions containing hydrochloric acid. $PdBr_2$ and PdI_2 are also known. Extensive series of complexes are formed with PdX_2.

palladium hydrides Palladium absorbs hydrogen up to $PtH_{0 \cdot 6+x}$. Hydrido complexes, e.g. $[PtHBr(PEt_3)_2]$, are formed by reduction of the dihalides. Hydrido derivatives are presumably intermediates in hydrogenation reactions using Pd catalysts.

palladium organic derivatives Palladium alkyls and aryls are much less stable than those of Pt. Pd-olefine complexes and -allyl derivatives, e.g. $(C_3H_5PdCl)_2$ are well established. They are of major importance in catalysis, e.g. the Wacker process*.

palladium oxide The only known oxide is PdO formed by direct combination of the elements at about 800°C or by ignition of salts. It decomposes to the elements at 875°C. A hydrated form is precipitated from solution.

palmitic acid, hexadecanoic acid, $C_{16}H_{32}O_2$, $CH_3 \cdot [CH_2]_{14}COOH$. Crystallizes in needles, m.p. 63°C, b.p. 351°C. Palmitic acid is one of the most widespread fatty acids and occurs as glycerides in most animal and vegetable fats and oils and as esters of alcohols other than glycerol in various waxes. The best sources are Japan wax and palm oil. It forms 40% of the total fatty acids of cows' milk. A solid mixture of palmitic and stearic acids, 'stearine', is used for making candles, and the sodium and potassium salts of palmitic and stearic acids are soaps. Palmitic acid can be obtained by hydrolysing natural fats with superheated steam when it is the first fatty acid to distil over.

palygorskite A form of fuller's earth*.

pamoic acid, embonic acid, 2,2′-dihydroxy-1,1′-dinaphthyl-methane-3,3″-dicarboxylic acid, $C_{23}H_{16}O_6$. Used to form salts with organic bases; used in medicine where insolubility in water and lack of absorption from the alimentary canal are required, e.g. viprynium embonate.

panchromatic sensitization Extension of the sensitivity of photographic silver halide emulsions to the whole of the visible spectrum by addition of sensitizing dyes.

pantothenic acid, $C_9H_{17}NO_5$. An oil. It is

$HOCH_2{\cdot}CMe_2{\cdot}C{\cdot}CHOH{\cdot}CO{\cdot}NH{\cdot}CH_2{\cdot}CH_2$
|
COOH

required by higher animals as well as by some micro-organisms. This vitamin is present in many natural products and is a constituent of coenzyme A*.

paper An interwoven mat of hydrated cellulose fibres. Most types of fibrous cellulose material can be used in its preparation. The properties of the paper depend upon the degree of hydration and sub-division of the fibres. In order to make paper less porous and partially water-repellant it is sized*. Some inorganic materials, e.g. $BaSO_4$ may be added to the paper.

paper chromatography *See* chromatography.

para In para-cresol, *p*-cresol, the CH_3 and

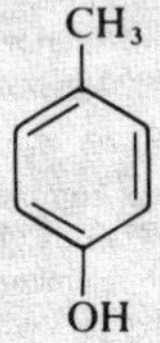

OH groups are in the para or 1,4 position to each other. This use of the prefix is confined to disubstituted benzene derivatives; in such cases as para-hydrogen and paraldehyde the prefix has no uniform structural significance and is always written in full.

parabens Alkylhydroxybenzoates used as food preservatives.

paracetamol, 4-acetamidophenol, $C_8H_9NO_2$. White crystalline powder; m.p. 169–172°C. It is very widely used orally as an analgesic and antipyretic.

parachor The parachor *p*, where γ is the surface tension, *M* the molecular weight and *D* and *d* the densities of the liquid and vapour

$$p = \gamma^{\frac{1}{4}} \frac{M}{D-d}$$

respectively, is an empirical measure of molecular volume and was widely used to determine the constitution of organic and inorganic compounds.

paracyanogen, $(CN)_x$. The white or brown

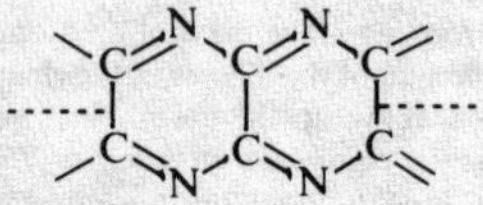

polymer formed when cyanogen is heated to 400°C.

paraffin *See* kerosine.

paraffin oil *See* kerosine.

paraffins *See* alkanes.

paraffin wax *See* petroleum wax.

paraformaldehyde, paraform, trioxymethylene A mixture of polymethylene glycols of the type $(CH_2O)_n{,}xH_2O$ where *n* is 6–50. It is a white, amorphous powder having the odour of methanal, m.p. 120–130°C. The commercial product contains 95% methanal and is obtained in white flocculent masses when solutions of methanal are evaporated or allowed to stand. When heated it is converted to methanal. Used as a convenient solid substitute for methanal.

para-hydrogen *See* ortho-hydrogen.

paraldehyde, $(C_2H_4O)_3$. A colourless, mobile

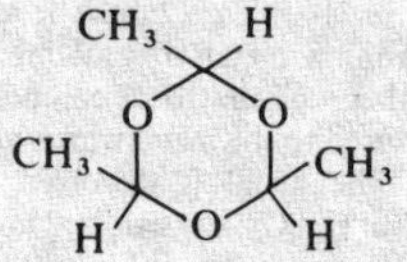

liquid, m.p. 12·5°C, b.p. 124°C, slightly soluble in water. Formed when ethanal is allowed to stand in the presence of a catalyst at moderate temperatures.

paramagnetism If when a substance is placed in a magnetic field it causes a greater concentration of the lines of magnetic force within itself than in the surrounding magnetic field, it is said to exhibit paramagnetism. Paramagnetism is associated with the presence of unpaired electrons in an ion or molecule. Magnetic susceptibility measurements of paramagnetic substances have contributed significantly to the theory of valency and knowledge of structure.

Paraquat, 1,1'-dimethyl-4,4'-bipyridylium dimethylsulphate or dichloride Both salts are

$$H_3C-N^+\!\!=\!\!\!\bigcirc\!\!-\!\!\bigcirc\!\!=\!\!N^+\!\!-CH_3 \quad 2Cl^- \text{ or}$$
$$2(CH_3OSO_3)^-$$

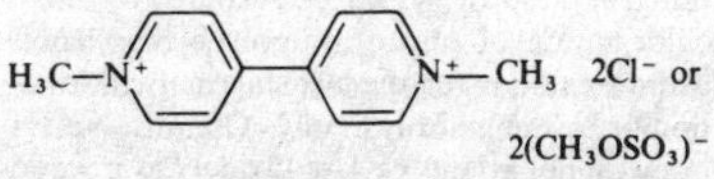

white crystalline solids, m.p. 175–180°C. Prepared by quaternization of 4,4'-bipyridyl with dimethyl sulphate or methyl chloride. Paraquat is a contact herbicide. It is poisonous by ingestion, causing irreversible lung damage.

parathion, diethyl-*p*-nitrophenylphosphorothionate, $C_{10}H_{14}NO_5PS$. An insecticide.

Paris green *See* emerald green.

Parkerizing A trade name for a process for phosphate treatment of steel. *See* phosphate coatings.

partial condenser A type of condenser. Vapour is partially condensed by heat exchange with another process stream, and the remainder is condensed in a *final condenser* using cooling water.

partial pressure In a mixture of gases or vapours each constituent can be considered to contribute to the total pressure that pressure which the same amount of the constituent would exert if it were present alone in a vessel of the same volume as that occupied by the mixture. This is called the partial pressure of the constituent.

parting The process of separating Au from Ag. As the chemical properties of the elements are very different this is an easy matter and methods used include dissolving the Ag in conc. H_2SO_4 or HNO_3 or the preferential reaction of Ag with Cl_2 to give AgCl.

parting agents *See* abherents.

partition column chromatography *See* chromatography.

partition law, distribution law In a heterogeneous system of two or more phases in equilibrium, the ratio of the activities (or less accurately the concentrations) of the same molecular species in the phases is a constant at constant temperature. This constant is termed the partition coefficient.

PAS 4-aminosalicylic acid.

Paschen series *See* Balmer series.

passivators *See* metal passivators.

passivity Many metals (e.g. Fe, Co, Ni, Cr) form a surface layer of oxide in contact with strong oxidizing agents (e.g. conc. HNO_3, chromic acid, H_2O_2). Further solution does not occur and many properties are altered. Other metals, e.g. Al and Mg normally have a tough surface film of oxide which prevents further attack so that although electropositive they do not dissolve in water. Passivity can also be conferred by fluorine which forms a layer of fluoride on, e.g., Cu or Ni.

Pasteur effect Yeast and other cells can break down sugar in the presence of oxygen (eventually to CO_2 and H_2O) or in its absence (to CO_2 and ethanol). The decomposition of sugar is often greater in the absence of oxygen than in its presence, i.e. the Pasteur effect. With oxygen, less toxic products (alcohol) are produced and the breakdown is more efficient in terms of energy production.

pasteurization The process of removal of some non-spore-producing organisms from milk, butter, beer by application of low heat (63°C for 30 minutes, 72°C for 20 seconds) or radiation.

patina The decorative and corrosion-resistant surface given to bronze, iron, etc. as an oxide film formed on warming the metal.

Patterson function The Fourier transform of observed intensities after diffraction (e.g. of X-rays). From the Patterson map it is possible to determine the positions of scattering centres (atoms, electrons).

Pauli exclusion principle In any atom no two electrons can have all four quantum numbers the same. *See* exclusion principle.

Pb Lead.

PCB Polychlorinated biphenyls.

Pd Palladium.

pearl ash *See* potassium carbonate.

peat A naturally-occurring, brown, fibrous mass of partially decayed plant material. The calorific values of peat on a dry basis range from about 16–20 MJ/kg. Used for firing power stations or can be converted to peat charcoal by low temperature carbonization*. World demand 1976 200 megatonnes.

pectic substances *See* pectin.

pectin A purified carbohydrate product obtained from the dilute acid extract of the inner portion of the rind of citrus fruits or from apple pomace. It is a mixture, the most important constituent of which is methyl pectate, which is the methyl ester of pectic acid, a high molecular weight polymer of D-galacturonic acid, which has the formula:

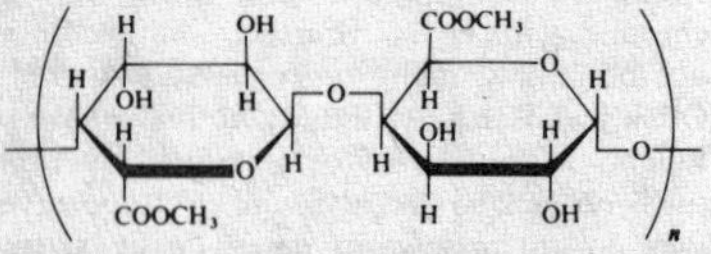

Pectin also contains araban and galactan. It is present in fruits, root vegetables and other plant products, and confers on jams their typical gelling property. Pectin is manufactured as a white powder, soluble in water, and used to assist the setting of jams and jellies, and for numerous other purposes. Low methoxyl pectins, with under 7% methoxyl, give firmer gels than pectins proper.

pelargonic acid, nonanoic acid, $C_9H_{18}O_2$, $CH_3\cdot[CH_2]_7\cdot COOH$. An oily liquid, with a rancid odour, m.p. 12·5°C, b.p. 253–254°C. It can be prepared by the oxidation of oleic acid or undecane. It has been found, with other fatty acids with odd numbers of carbon atoms, in small amounts in human hair.

pelargonin The anthocyan glycoside of the scarlet geranium, dahlia, gladiolus and many other plants. It is the 3,5-diglucoside of pelargonidin. Other anthocyans which contain pelargonidin are (1) callistiphin from red carnations and the purple red aster which is a 3-β-glucoside, (2) monardaein or salvianin from scarlet *Salvia splendens*, etc., which is a 3,5-diglucoside carrying in addition *p*-hydroxycinnamic acid and malonic acid.

Pelargonin-3-biosides are of widespread occurrence, e.g. in the nasturtium, scarlet runner bean, gloxinia. *See* anthocyanins.

penicillin The penicillins are a group of bactericidal antibiotics* containing the basic ring structure of penicillanic acid:

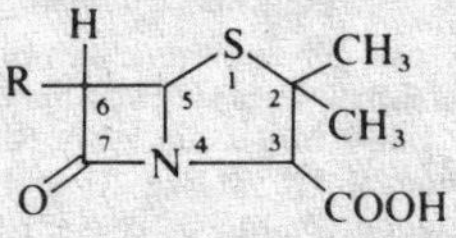

The first penicillin was benzylpenicillin (penicillin G, $R=C_6H_5\cdot CH_2\cdot CO\cdot NH-$), which is made by growing suitable strains of *Penicillium chrysogenum* on a carbohydrate medium.

Many semi-synthetic penicillins are made from 6-aminopenicillanic acid (6-APA, $R=NH_2$).

penicillinase An enzyme from some bacteria, which hydrolyses the β-lactam ring of penicillin to yield inactive penicilloic acid.

pentaborane (9), B_5H_9; **pentaborane (11),** B_5H_{11}. *See* boron hydrides.

pentachlorophenol, C_6Cl_5OH. Solid having a pungent odour when warm. Prepared by direct chlorination of phenol or polychlorophenols using a catalyst for the last stages of chlorination or by the hydrolysis of C_6Cl_6 with NaOH in methanol at 130°C. Used widely as a wood preservative and as a fungicide.

pentaerythritol, $C_5H_{12}O_4$, $C(CH_2OH)_4$. Colourless crystalline compound, m.p. 260°C. It is made by reacting ethanal and methanal in the presence of alkali. It is used in alkyd resin production, in the manufacture of its tetranitrate, and in the production of oxetanes. U.S. production 1978 56 000 tonnes. *See* dipentaerythritol and tripentaerythritol.

pentaerythritol tetranitrate, (PETN or penthrite), $C(CH_2\cdot O\cdot NO_2)_4$. Needles, m.p. 140–141°C. Made by nitrating pentaerythritol and is a very stable and extremely powerful and violent explosive. Mixed with TNT it forms pentolite. Used medicinally in tablet form to relieve the pain of angina pectoris.

pentanes, C_5H_{12}. There are three possible isomers of this formula; they occur in the low boiling fractions of petroleum from which they are prepared. They are inflammable liquids. $CH_3\cdot CH_2\cdot CH_2\cdot CH_2\cdot CH_3$, *n*-pentane, b.p. 38°C, is used as a standard illuminant in photometry.

pentanoic acids *See* valeric acids.

pentosans Hemicelluloses which give pentoses on hydrolysis. The most widely distributed are xylan and araban.

pentose A carbohydrate with five C atoms. Of the aldopentoses, both stereoisomers of arabinose and the D-forms of xylose and ribose occur naturally; lyxose does not occur naturally. There are four (two pairs of stereoisomers) possible ketopentoses.

pentose phosphate pathway A sequence of reactions of sugar phosphates that provides a source of hexoses, pentoses, heptoses and tetroses for a living cell. It is concerned in the initial stages of photosynthesis. A major function appears to be the generation of NADPH which is apparently preferred to NADH in biosynthetic reactions.

pepsin, pepsinogen The principal proteolytic

enzyme in the gastric juice of vertebrates. The enzyme splits links between the α-carboxyl group of a dicarboxylic amino-acid and the α-amino group of an aromatic acid.

peptidases Proteolytic enzymes. Some, called endopeptidases,* attack the internal links of a peptide chain, e.g. trypsin and chymotrypsin, and others, called exopeptidases,* attack at both ends of the chains.

peptides Substances composed of two or more amino-acids, and designated as di-, tri-, oligo-, poly-peptides, etc. according to the number of amino-acids linked by the peptide bond –CO·NH–. The simplest peptide is glycylglycine, $H_2N{\cdot}CH_2{\cdot}CO{\cdot}NH{\cdot}CH_2{\cdot}CO_2H$.

Naturally occurring peptides possessing important physiological activity include glutathione, gastrin and angiotensin; several pituitary hormones (e.g. oxytocin, vasopressin); and a number of antibiotics (e.g. bacitracin, polymyxin). Peptides also arise, and are frequently studied, as partial hydrolysis products of proteins. Methods for peptide synthesis are very highly developed. *See also* proteins.

peptization This term is sometimes used in a general way to imply the converse of coagulation, i.e. dispersion, especially when the process results in the formation of a colloidal sol. Peptization is, however, generally restricted to a chemical means of dispersion in which the colloidal particles are stabilized by the adsorption of charged ions. An example of peptization is the formation of silver iodide sols by shaking the neutral precipitate with a small excess of either KI or $AgNO_3$; with the former reagent the resultant sol is negative; with the latter, positive.

peptones Derived proteins of low mol.wt. formed by degradation of proteins by, e.g. enzymes.

per-acids Oxy-acids in which $[O-O]^{2-}$ or $[O-OH]^-$ groups have replaced oxide in oxy-anions (compounds containing the superoxide ion $[O_2]^-$ are formally not per-acids but the distinction is often difficult to make). The O–O group may be bonded end on

[M–O–O] or symmetrically 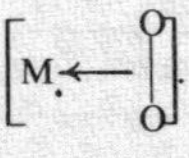.

Prepared by use of H_2O_2 or by electrolytic oxidation. Persulphuric acids, perborates, are of importance. (Permanganates, perchlorates and periodates are not salts of per-acids.) Organic per-acids are prepared similarly. The more stable per-acids such as perbenzoic acid, monoperphthalic acid convert olefines to epoxides and by hydrolysis to *trans*-glycols. Perethanoic and permethanoic acids give glycol monoesters with olefines. *See* peroxides, organic.

perborates, perborax Sodium perborates (borax plus Na_2O_2 or H_2O_2 plus NaOH). Quite stable and used as a bleaching agent and antiseptic.

perbromates, $M[BrO_4]$. *See* bromates (VII).

perbromic acid, $HBrO_4$. *See* bromates.

percarbonates Carbonates with varying amounts of H_2O_2 of crystallization are stable as solids and $2Na_2CO_3,3H_2O_2$ is used in laundering and for cleaning dentures.

perchlorates, $MClO_4$. *See* chlorates(VII).

perchloric acid, $HClO_4$. *See* chlorates(VII). Perchloric acid ($KClO_4$ plus H_2SO_4) is a colourless or yellowish strongly fuming liquid which can be distilled at low pressure and gives crystalline $HClO_4,H_2O$. The free acid can explode and frequently does so in the presence of organic materials. $HClO_4$ is a strong oxidizing acid and is sometimes used in analysis. Forms ionized salts.

perchloroethylene Trivial name for tetrachloroethene, $Cl_2C{=}CCl_2$. U.S. production 1978 330 000 tonnes. Used as cleaning fluid.

perchloryl fluoride, $FClO_3$. *See* chlorine oxide fluorides.

perewskites *See* perovskite.

perfluoroalkyl derivatives Organic derivatives in which all the H atoms, except those in the functional groups, have been replaced by fluorine. Thus C_6F_6 is perfluorobenzene; C_6F_5COOH is perfluorobenzoic acid.

Perfluoroalkyl derivatives have important technical uses, e.g. sulphonic acids as surfactants; introduction of perfluoroalkyl groups confers useful properties on many drugs.

perfume chemistry The broad contents of perfumes are essential oils (of plant origin), flower oils, natural extracts from resins, gums, etc., and fixatives (e.g. ambergris, civet) which render the perfume more lasting. Materials used in perfumery are both natural and synthetic.

peri A prefix sometimes used to denote the positional relationship between groups occupying the 1,8 positions on the naphthalene ring. As a special case the two groups may be part of a third ring. 1-Naphthylamine-8-sulphonic acid is sometimes referred to as peri acid.

periclase MgO either natural or formed by heating $Mg(OH)_2$; used in furnace linings.

period A series of elements in the periodic table, a period being elements from an alkali metal to a noble gas. The periods are (1) the very short period H, He, (2) two short periods Li–Ne and Na–Ar, (3) two long periods containing transition elements K–Kr and Rb–Xe, (4) a very long period containing the rare-earth elements Cs–Rn, (5) an unfinished period Fr–.

periodates *See* iodates(VII).

periodic acids *See* iodates(VII).

periodic law Mendeléeff (1869), independently of Newlands (octaves law), put forward the law that 'the properties of the elements are periodic functions of the atomic weights', i.e. by arranging the elements in order of increasing atomic weights, elements having similar properties occur at fixed intervals. This principle, with the modification that atomic numbers are used instead of atomic weights, forms the basis of the periodic table.

periodic table A table of the elements arranged in order of atomic numbers, the arrangement being such as to emphasize the chemical relationships between the elements and to emphasize the relationship between elements with similar configurations. *See* periodic law. *See table below.*

peritectic A type of solid–liquid phase equilibrium system, commonly found with binary alloys, in which the two components form limited ranges of solid solution in each other, although they are completely miscible in the liquid state. A peritectic system differs from a eutectic system in that the former shows no minimum in the freezing-point composition curve.

Perkin reaction A condensation between aromatic aldehydes and the sodium salts of fatty acids or their aromatic derivatives. The reaction between benzaldehyde and sodium ethanoate in the presence of ethanoic anhydride leads to sodium cinnamate

$$C_6H_5CHO + H_3C{\cdot}COONa \longrightarrow C_6H_5CH{:}CHCOONa$$

In general, condensation takes place at the α-carbon atom, leading to simple cinnamic acids or to their α-substituted derivatives. When possible, the anhydride corresponding to the sodium salt should be the condensing agent.

perlite Originally the name of a glassy volcanic rock, but now also applied to a light-weight product produced on a considerable scale by rapidly heating to their softening point crushed and graded perlites and similar glassy highly siliceous volcanic materials containing 2–6% of water. The particles 'pop' with the evolution of the water as steam, yielding a product with *d* approx. 1·2.

Used as a light-weight aggregate in heat and sound insulating plasters, as a loose fill and in

periodic table

s^1	s^2	d^1	d^2	d^3	d^4	d^5	d^6	d^7	d^8	d^9	d^{10}	p^1	p^2	p^3	p^4	p^5	p^6
H																	He
Li	Be											B	C	N	O	F	Ne
Na	Mg											Al	Si	P	S	Cl	Ar
K	Ca	Sc	Ti	V	Cr	Mn	Fe	Co	Ni	Cu	Zn	Ga	Ge	As	Se	Br	Kr
Rb	Sr	Y	Zr	Nb	Mo	Tc	Ru	Rh	Pd	Ag	Cd	In	Sn	Sb	Te	I	Xe
Cs	Ba	La	Hf	Ta	W	Re	Os	Ir	Pt	Au	Hg	Tl	Pb	Bi	Po	At	Rn
Fr	Ra	Ac															
			La	Ce	Pr	Nd	Pm	Sm	Eu	Gd	Tb	Dy	Ho	Er	Tm	Yb	Lu
			Ac	Th	Pa	U	Np	Pu	Am	Cm	Bk	Cf	Es	Fm	Md	No	Lw

light-weight concrete. World demand 1976 440 000 tonnes.

permanganates Salts containing the dark purple $[MnO_4]^-$ ion. $KMnO_4$ is the most important salt (MnO_2 (pyrolusite) plus KOH plus air at 300°C followed by treatment of $[MnO_4]^{2-}$ electrolytically or with CO_2 to give $[MnO_4]^-$ plus MnO_2). Used in metal treatment and as a bactericide. $NaMnO_4, 3H_2O$ is more soluble; a solution is used as a disinfectant (Condy's fluid). $AgMnO_4$ is fairly insoluble but is very reactive. *See* manganates(VII).

permanganate titrations On account of its intense colour and easy purification, potassium permanganate is an important volumetric oxidizing agent which requires no indicator even in 0·01 M-solution. When used, as in most estimations, in presence of sulphuric acid, it is reduced to Mn^{2+} (5 electrons). Solutions are standardized after having been boiled (or left to stand for some days) and then filtered. The best standards are sodium oxalate and pure (electrolytic) Fe; crystallized oxalic acid and $Fe(NH_4)_2(SO_4)_2,12H_2O$ are also frequently used but are less reliable. The solution may be used for the direct titration of Fe^{2+}, H_2O_2, $[Fe(CN)_6]^{4-}$ oxalate, suitable back titrations enable it to be used for the estimation of a great variety of oxidizing and reducing agents. In the titration of Fe^{2+}, Cl^- interferes but titration can be carried out in presence of an excess of manganese(II) sulphate. In some estimations permanganate is used in almost neutral solution, when hydrated manganese dioxide (or manganate(IV)) is formed. An important instance of this procedure is the determination of Mn(II) (the Volhard method); the titration is carried out in the presence of a Zn^{2+} salt and suspended ZnO.

permanganic acid, $HMnO_4$. *See* manganates(VII).

permonosulphuric acid, Caro's acid, $HOS(O)_2OOH$. Powerful oxidizing agent (H_2O_2 on conc. H_2SO_4 or HSO_3Cl; hydrolysis of $H_2S_2O_6$ with H_2SO_4). The pure acid can be obtained crystalline but the salts are unstable.

permutite process The process of softening hard water by replacing Ca^{2+} and Mg^{2+} by Na^+ by passage through a zeolite*. The zeolite is regenerated with NaCl solution.

perovskite Mineral $CaTiO_3$.

perovskite structure A structure adopted by many ABX_3 compounds in which A and B are positive ions and X is e.g. F^- or O^{2-}. The structure has A at the corners of a cube, B at the body centre and X (A and X are ccp) at the face centres. A has 12X neighbours, B has 6X (octahedral) neighbours. Examples are $KNiF_3$, $BaTiO_3$, $KNbO_3$. Many perovskites show distortions from the idealized cubic structure.

peroxidases The enzymes which destroy peroxides by transferring peroxide oxygen to oxidizable substances. They are especially abundant in plants, e.g. horseradish, and oxidize phenols to give coloured products. Like catalases, they often contain haematin. In animals glutathione peroxidase is very important. It reduces peroxides by transferring oxygen to reduced glutathione.

peroxides Compounds containing linked pairs of oxygen atoms – derivatives of H_2O_2. They yield H_2O with dilute acids. BaO_2 and Na_2O_2 are typical peroxides. PbO_2, NO_2, MnO_2 are sometimes erroneously termed peroxides.

peroxides, organic Organic peroxides are used as sources of free radicals for free radical polymerization, as curing agents for resins and elastomers and as cross-linking agents for polyolefines. The most commonly used are dibenzoyl peroxide (m.p. 106°C) and dilauroyl peroxide (m.p. 55°C) prepared from the acid chloride or anhydride and Na_2O_2 or H_2O_2. Other peroxides are similar. Peroxyethanoic acid (b.p. 36°C/30 mm) is used for epoxidation and bleaching. It is prepared from ethanoic acid and H_2O_2. Other peroxyacids are similar.

peroxychromium compounds H_2O_2 reduces $[CrO_4]^{2-}$ to Cr^{3+} but gives blue $CrO(O_2)_2$ which can be extracted into ether (test for Cr or after distillation of CrO_2Cl_2 for Cl^-). Other derivatives are also known and are relatively stable.

peroxyethanoic acid, $CH_3C(O)OOH$. *See* peroxides, organic.

perrhenates Rhenates(VII). *See* rhenium oxides.

Perspex Trade name for cast polymethylmethacrylate sheet – an acrylic resin.

persulphates Salts of persulphuric acid*.

persulphuric acid, perdisulphuric acid, $HOS(O)_2OOS(O)_2OH$. Dibasic acid formed as salts by electrolysis of sulphates at low temperatures and high current density. The acid and persulphates are strong oxidizing agents ($E°[S_2O_8]^{2-}$ to $SO_4{}^{2-}$ +2·01 volts in acid) but the reactions are often slow. *Compare* permonosulphuric acid.

perxenates Xenates(VIII).

perylene A polycyclic aromatic hydrocarbon.

pesticides *See* insecticides.

petalite, $(Li, Na)(AlSi_4)O_{10}$. Aluminosilicate containing 2–4·5% Li_2O used as a source of Li.

petrochemicals, petroleum chemicals Petroleum chemicals may be obtained from a variety of petroleum fractions, normally from processing of intermediates produced in primary conversion processes. The main basic raw materials are natural gas, refinery gases, liquid petroleum fractions and waxes, while the primary conversion processes include cracking, partial oxidation and pyrolysis.

petrol Motor fuel, gasoline or aviation fuel (U.K. usage). Originally a trade name.

petrolatum An unctuous salve-like residuum from non-asphaltic petroleum. It is available commercially under a wide variety of names and in various degrees of purity, i.e. petroleum jelly, soft paraffin, mineral jelly, paraffinum molle, etc. Two varieties, white and yellow, are used in pharmacy, and are used as a basis for medicaments. *See* petroleum wax.

petroleum A term used to describe oil products in general but also used for crude oil*. World total production 1976 2863 megatonnes.

petroleum coke The solid, carbonaceous material obtained as a by-product in thermal cracking* processes for the production of gasoline. Coke is also produced in *coking* processes, i.e. the severe cracking of residual pitches and tars into various petroleum products.

petroleum ether, light petroleum Lower aliphatic hydrocarbons of high volatility and a narrow distillation range. It is made in a number of grades; the two most common having distillation ranges 40–60°C and 60–80°C which contain C_5 and C_6 hydrocarbons respectively.

petroleum jelly *See* petrolatum.

petroleum wax Solid, waxy hydrocarbons obtained from crude oils, especially paraffinic crudes. They are often classified in two main types – paraffin wax and microcrystalline wax.

Paraffin waxes have a macrocrystalline structure and consist, largely, of *n*-alkanes of formulae $C_{20}H_{42}$ and upwards with some iso- and cycloalkanes.

Microcrystalline waxes, produced from heavy lubricating oil residues, have a microcrystalline structure and consist largely of iso- and cycloalkanes with some aromatics.

Petroleum wax is used in the manufacture of candles, polishes, ointments and for waterproofing purposes. Waxes are also used as a cracking feedstock for the production of 1-alkenes for conversion to detergents.

pewter (75% Sn, 25% Pb). An important tin alloy.

Pfeiffer effect The change in rotation of a solution of an optically active substance on the addition of a racemic mixture of an asymmetric compound.

Ph The phenyl group, C_6H_5-.

pH *See* hydrogen-ion concentration.

phaeophorbide *See* chlorophyll.

phaeophytin *See* chlorophyll.

phalloidins *See* amanitins.

phase Any part of a system, which is uniform in chemical composition and physical properties and separated from other homogeneous parts of the system by boundary surfaces. E.g. in a closed system containing ice, water and water vapour, there are three phases. Since all gases are completely miscible with each other, they can only exhibit one phase. With liquids and solids several phases are possible.

phase diagram A diagram showing the conditions of equilibrium between various phases. In one-component systems a pressure–temperature diagram is used; in binary systems a three-dimensional diagram relating pressure, composition and temperature, or two-dimensional diagrams relating composition and temperature at constant pressure or composition and pressure at constant temperature are used.

phase rule First derived by Willard Gibbs (1876), the phase rule relates the number of phases, P, the number of degrees of freedom, F, and the number of components, C, in a chemical system according to the equation

$$P+F=C+2$$

phase-transfer chemistry Reactions between heterogeneous systems (solid–liquid or liquid–liquid) brought about by addition of a catalytic amount of an agent which transfers a reactant across the interface into the system where reaction can occur. Thus R_4N^+ or R_4P^+ cations are used to transfer otherwise insoluble anions into non-polar media from aqueous media. Crown ethers* can be used to catalyse direct solid–liquid phase transfer of salts into non-polar solvents, e.g. KOH into toluene. Also called ion-pair partition.

α-phellandrene, 5-isopropyl-2-methyl-1,3-cyclohexadiene, $C_{10}H_{16}$. A colourless odoriferous oil, b.p. 175–176°C, which tends to decompose when distilled at atmospheric pressure. A monocyclic terpene. (+)-α-Phellandrene occurs in bitter fennel oil, gingergrass oil, and Ceylon and Seychelles cinnamon oil, and (−)-α-phellandrene in various eucalyptus oils.

β-phellandrene, 3-isopropyl-6-methylenecyclohexene, $C_{10}H_{16}$. A colourless, odoriferous oil; b.p. 171–172°C. An optically active monocyclic terpene. (+)-β-Phellandrene occurs in water-fennel oil, lemon oil and the oil from *Bupleurum fruticosum*, and (−)-β-phellandrene in Japanese peppermint oil and in the oil from *Pinus contorta*.

phenacetin, *N*-acetyl-*p*-phenetidine, $C_{10}H_{13}NO_2$. White crystals, m.p. 137–138°C. Prepared from phenol, via *p*-nitrophenol, *p*-nitrophenetole and *p*-phenetidine. It is used medicinally as an antipyretic analgesic similar to aspirin*. It has chronic toxicity towards the kidney.

phenacyl The trivial name for the group

$$C_6H_5-\overset{\overset{\large O}{\|}}{C}-CH_2-.$$

phenanthrene ring system The system numbered as shown.

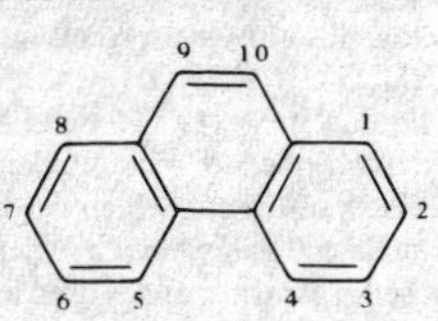

***o*-phenanthroline, 4,5-phenanthroline,**

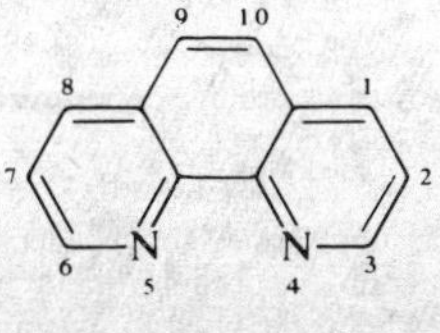

$C_{12}H_8N_2$. Made by heating *o*-phenylenediamine with glycerol, nitrobenzene and conc. sulphuric acid. It exists as a monohydrate, m.p. 94°C, or anhydrous, m.p. 117°C. It is used as a complexing reagent in the estimation of Fe^{2+} and in the preparation of numerous chelates of metals generally in low oxidation states.

***p*-phenetidine, 4-ethoxyaniline,** $C_8H_{11}NO$. Colourless oil, m.p. 2–4°C, b.p. 254°C. Its hydrochloride has m.p. 234°C, and is easily soluble in water.

It is usually prepared by the reduction of *p*-nitrophenetole with iron and hydrochloric acid.

It is used as a dyestuffs intermediate. Its acetyl derivative is phenacetin.

phenetole, ethoxybenzene, $C_8H_{10}O$, PhOEt. A colourless, aromatic liquid, b.p. 172°C. It is prepared by heating KOPh with ethyl iodide, or by adding ethanol to a mixture of phenol and P_2O_5 at 200°C.

phenidone *See* 1-phenyl-pyrazolid-3-one.

phenobarbitone, phenylethylbarbituric acid,

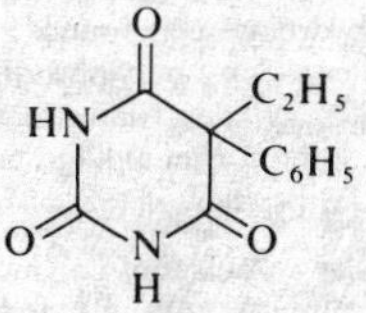

$C_{12}H_{12}N_2O_3$. White crystals, m.p. 174°C. Prepared by condensing the ethyl ester of phenylethylmalonic acid with urea. It is a more active hypnotic than barbitone. It and its sodium salt – soluble phenobarbitone – are used as sedatives and in treating epilepsy.

phenol, C_6H_6O, PhOH. Colourless crystals; m.p. 43°C, b.p. 183°C. Soluble in water at room temperature; larger amounts of phenol in water give a two-phase system of phenol solution floating on a lower layer of wet phenol. Miscible with water in all proportions at 84°C.

Obtained synthetically by one of the following processes: fusion of sodium benzenesulphonate with NaOH to give sodium phenate; hydrolysis of chlorobenzene by dilute NaOH at 400°C and 300 atm. to give sodium phenate (Dow process); catalytic vapour-phase reaction of steam and chlorobenzene at 500°C (Raschig process); direct oxidation of cumene (isopropylbenzene) to the hydroperoxide, followed by acid cleavage to propanone and phenol; catalytic liquid-phase oxidation of toluene to benzoic acid and then phenol. Where the phenate is formed, phenol is liberated by acidification.

Phenol is acidic and forms metallic salts. It is readily halogenated, sulphonated and nitrated. It is of great commercial importance and is used in quantity for the manufacture of phenolic and epoxy resins (40%), caprolactam (15%), bisphenol A (20%) and alkyl phenols (6%). Thus it is a starting material for making both precursors of Nylon 66. It is also used in the manufacture of various dyes, explosives, pharmaceuticals and perfumes. U.S. production 1981 1·2 megatonnes.

phenol aldehydes Aromatic aldehydes (R′O)ArCHO. Examples are anisaldehyde, ethylvanillin. Used in perfumery and flavouring, electroplating, metal complexing.

phenol ethers, ArOR (R aliphatic or aromatic). Examples are diphenyl ether, anisole. Occur in essential oils. Used as heat transfer media (Ph_2O) and antioxidants (with bulky substituents). Form stable resins.

phenol-formaldehyde resins Curable resins formed principally by phenol but also other aromatic hydroxyl compounds (resorcinol, cresols, xylenols, *p-t*-butylphenol, *p*-phenylphenol) with methanal under acid or basic conditions. Used for mouldings, coatings and bondings. U.S. production 1981 585 000 tonnes.

phenolic resins A class of polymers resulting from interaction of phenols and aldehydes (generally methanal). *See* phenol-formaldehyde resins.

phenolphthalein, $C_{20}H_{14}O_4$. Colourless

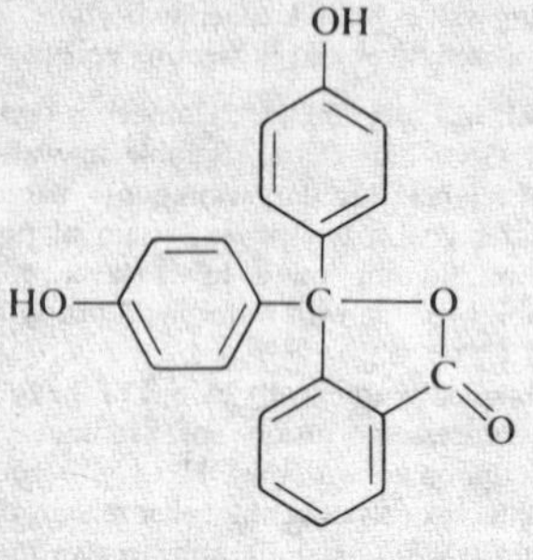

colourless form

crystals; m.p. 254°C. Prepared by heating phthalic anhydride and phenol in the presence of sulphuric acid. Extensively used as an indicator with pH range 8·3 (colourless) to 10·4 (red). The disodium salt (carboxylate and one phenyl group) is red but the trisodium salt is colourless. It is used medicinally as a purgative.

phenol red, phenolsulphonphthalein, $C_{19}H_{14}O_5S$. Sulphonic acid analogue of phenolphthalein Prepared by heating phenol with *o*-sulphobenzoic anhydride and used as an indicator with pH range 6·7 (yellow) to 8·3 (red). It is used as a test for renal function.

phenols Compounds containing at least one hydroxyl group attached directly to a carbon atom of an aromatic ring. They have a slight acidic nature, and are soluble in caustic alkalis forming metallic salts, the phenates. Their solubility in water increases with the number of hydroxyl groups. Polyhydric phenols have a strong reducing action. They are widely distributed in natural products, e.g. tannins, anthocyanins, tyrosine.

phenothiazine, $C_{12}H_9NS$. Colourless or light

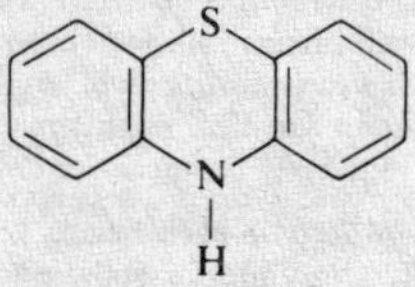

yellow crystals which darken to olive-green on exposure to light; m.p. 182°C. It is made by heating diphenylamine with sulphur, and is used as an anthelmintic in veterinary practice, but is poisonous to man.

Substituted phenothiazines form an important group of tranquillizers, e.g. chlorpromazine. Some, e.g. promethazine, have powerful antihistamic* activity.

phenoxyacetic acid herbicides *See* 2,4-D; MCPA.

phenoxyethanol, ethylene glycol monophenyl ether, $C_6H_5O{\cdot}CH_2{\cdot}CH_2OH$. A colourless, viscous liquid, b.p. 117°C/7 mm. Made by heating sodium phenate with ethylene chlorohydrin. It is an antiseptic which is particularly effective against *Pseudomonas pyocyaneus*, and certain other organisms, and is used in the local treatment of infected wounds.

phenoxy resins, polyhydroxyethers Linear

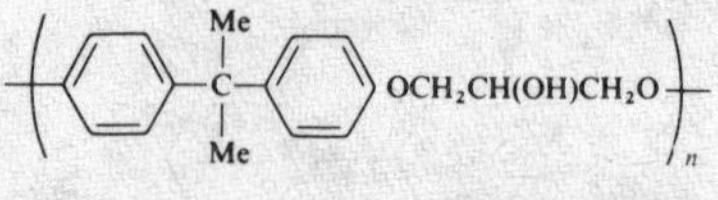

thermoplastics formed by condensation of epichlorhydrin, bisphenol A and NaOH in dimethylsulphoxide. Used for containers.

phenyl The group C_6H_5- or Ph. The radical appears to occur as an intermediate in many organic reactions in solution, e.g. in the decomposition of the benzene diazonium cation in the presence of Cu(I).

phenylacetic acid *See* phenylethanoic acid.

phenylalanine, α-amino-β-phenylpropionic acid, $C_9H_{11}NO_2$, $PhCH_2CHNH_2{\cdot}COOH$. Crystallizes in colourless leaves; m.p. 283°C. The naturally occurring substance is laevorotatory. It is one of the essential amino-acids that must be present in the food of animals.

***m*-phenylenediamine, 1,3-diaminobenzene,** $C_6H_8N_2$. Colourless crystals; m.p. 63°C, b.p. 287°C. Turns brown in air. Prepared by a one-stage reduction of *m*-dinitrobenzene with iron and hydrochloric acid.

Basic, forms a stable water-soluble dihydrochloride. Diazotization gives brown azo-dyes (Bismarck brown) owing to the coupling of the partially diazotized base with the excess of diamine. Is also used as an end component of many azo-dyes, readily coupling with one or two molecules of diazo compound.

***o*-phenylenediamine, 1,2-diaminobenzene,** $C_8H_8N_2$. Brown-yellow crystals; m.p. 103–104°C, b.p. 256–258°C. Its solutions reduce Ag^+ ions and it is used as a photographic developer. It is also used as a dye-precursor, for the synthesis of phenazine derivatives and for characterizing (*inter alia*) α-diketones.

***p*-phenylenediamine, 1,4-diaminobenzene,** $C_6H_8N_2$. White crystals; m.p. 147°C, b.p. 267°C, darken rapidly in air. Prepared by reducing *p*-nitroaniline or aminoazobenzene. Oxidizing agents convert it to quinone derivatives, hence it cannot be diazotized with nitric acid.

It is used for hair dyeing, as a rubber accelerator and as a photographic developer. Derivatives are extensively used as developers in colour photography.

phenylethanoic acid, $PhCH_2CO_2H$. M.p. 76°C, b.p. 266°C. Occurs naturally as esters, manufactured by hydrolysis of benzylcyanide. Used as an aqueous solution of its potassium salt as a penicillin precursor also in amphetamine production.

2-phenylethyl alcohol, $PhCH_2CH_2OH$. B.p. 220°C. Has a faint rose odour, occurs free or combined in many essential oils (e.g. orange juice). Made by reduction of ethyl phenylacetate with Na in alcohol or better by Friedel–Crafts reaction between benzene and ethylene oxide. Used extensively in perfumery (the ethanoate is also used).

phenylethyl ethanoate *See* phenylethyl alcohol.

***N*-phenylglycine,** $C_8H_9NO_2$. White crystals, m.p. 127°C. It can be prepared: (1) By the condensation of aniline with chloroethanoic acid. (2) By heating aniline, NaOH, methanal and KCN. (3) By heating aniline, lime and trichloroethene.

It is an important intermediate in the indigo industry. When phenylglycine is fused with sodamide a good yield of indoxyl is obtained. The indoxyl is easily oxidized to indigo.

***N*-phenylglycine-*o*-carboxylic acid, (2-carboxyphenyl)aminoethanoic acid,** $C_9H_9NO_4$. Sandy powder; m.p. 207°C (decomp.).

Prepared by the action of chloroethanoic acid upon anthranilic acid. It can also be prepared by the action of KCN and methanal on anthranilic acid, followed by hydrolysis with NaOH of the nitrile of anthranilic acid first produced.

It is used for the preparation of indigo into which it is readily converted on alkaline fusion.

phenylhydrazine, $C_6H_8N_2$, $PhNHNH_2$. A colourless, refractive oil; b.p. 240–241°C, with slight decomposition. It rapidly becomes brown when exposed to air. It forms a stable hydrochloride which can be crystallized unchanged from hot water. It is very poisonous by skin absorption and inhalation of the vapour.

It is prepared commercially by treating benzene diazonium chloride with sodium sulphite and then reducing the mixture with zinc dust and ethanoic acid. It can also be prepared by reduction of benzene diazonium chloride with $SnCl_2$.

Phenylhydrazine is a strong reducing agent; it reduces Fehling's solution in the cold and reduces aromatic nitro compounds to amines. With ketonic groups, hydrazones are formed. Compounds containing the grouping $-CHOHCO-$, e.g. glucose, fructose, etc., react with two molecules of phenylhydrazine to give osazones.

Phenylhydrazine condenses with acetoacetic ester to give a pyrazolone derivative which on methylation gives phenazone. The sulphonic acid similarly gives rise to the tartrazine dyestuffs. It is used to make indole derivatives by the Fischer process.

phenylhydrazine-4-sulphonic acid, $C_6H_8N_2O_3S$. Colourless needles, with $\frac{1}{2}H_2O$. Prepared by reducing diazotized sulphanilic acid with an excess of sodium sulphite. It is a typical hydrazine in its reactions with ketones, and with acetoacetic ester. The latter reaction gives rise to the tartrazine dyestuffs, and is much used commercially.

phenylhydrazones Compounds containing

the PhNH−N=C= group. Formed by treating aldehydes or ketones with phenylhydrazine in ethanoic acid solution. They are crystalline solids having definite melting-points and are usually sparingly soluble. Decomposed by heating with hydrochloric acid to regenerate the aldehyde or ketone. React with sodium ethoxide to give hydrocarbons, and with zinc chloride to give derivatives of indole. Formerly used for the identification of aldehydes and ketones, 2,4-dinitrophenylhydrazones give better derivatives.

phenyl isocyanate, isocyanatobenzene, C_6H_5NCO, PhNCO. A pungent lachrymatory almost colourless liquid; m.p. −33°C, b.p. 162°C. Used as a dehydrating agent and for characterization of alcohols. Prepared from aniline and phosgene in the presence of hydrogen chloride.

phenylmagnesium halides *See* Grignard reagents.

phenylmercuric ethanoate, $C_8H_8HgO_2$, $PhHgO{\cdot}OCCH_3$. White crystalline substance; m.p. 152–153°C. Prepared from benzene and mercury (II) ethanoate in glacial ethanoic acid. A powerful fungicide and selective herbicide, it has high mammalian toxicity.

1-phenylpyrazolid-3-one, phenidone, $C_9H_{10}N_2O$. Colourless crystalline solid, m.p. 121°C. Made by reacting phenylhydrazine with ethyl acrylate to obtain the hydrazide which cyclizes to the product. Its major commercial importance is as a photographic developing agent, being used particularly in conjunction with hydroquinone.

phenytoin, 5,5-diphenylhydantoin, $C_{15}H_{12}N_2O_2$. Used as its sodium salt, which is a white hygroscopic powder. Unstable, readily absorbing carbon dioxide and liberating phenytoin. Made by treating α-bromodiphenylacetylurea with alcoholic ammonia. It has a mild hypnotic and strong anticonvulsant action, and is used in the treatment of grand-mal and focal epilepsy.

pheromones Substances which are secreted externally by one individual and which elicit a specific response from other individuals of the same species. E.g. insect sex attractants such as bombykol (hexadeca-10-*trans*,12-*cis*-dien-1-ol) from the silk moth; trail markers of ants; and 'Queen substance' of the honey bee (9-oxodec-2-*trans*-enoic acid).

philosopher's stone *See* transmutation.

phloroglucinol, 1,3,5-trihydroxybenzene, $C_6H_6O_3$. Colourless crystals; $2H_2O$ from water. Water lost at 100°C. Anhydrous compound, m.p. 200–219°C. The aqueous solution gives a blue–violet colour with Fe^{3+}, reduces Fehling's solution and absorbs oxygen, but less readily than pyrogallol.

It occurs in many natural glycosides. It can be prepared by fusing resorcinol with NaOH and is manufactured from trinitrotoluene via trinitrobenzoic acid and triaminobenzene.

It reacts chemically as if it were either 1,3,5-trihydroxybenzene or the triketone of hexahydrobenzene. The trialkyl derivative of the hydroxy-form or the trioxime of the keto-form can be obtained from phloroglucinol by reacting in a suitable manner. In certain cases derivatives of the two forms are interchangeable. Used in printing, photography, adhesives, pharmaceuticals.

pH meter An instrument used for measuring the pH of solutions, suspensions, etc., generally with the aid of a glass electrode. It consists essentially of an electrometer-valve potentiometer with an adjustment for backing off the asymmetry potential of the electrode. Both 'null' type, which is the more accurate, and 'direct-reading' instruments are available.

phorate, diethyl S-(ethylthiomethyl)phosphorothiolothioate, $C_7H_{17}O_2PS_3$, $(EtO)_2P(S)SCH_2SEt$. A systemic insecticide and acaricide.

phorone, di-isopropylideneacetone, 2,6-dimethyl-2,5-heptadiene-4-one, $C_9H_{14}O$, $Me_2C{=}CHCOCH{=}CMe_2$. Yellow liquid having a camphor-like odour; m.p. 28°C, b.p. 198·5°C. It is formed when propanone is saturated with HCl and allowed to stand. Resembles camphor in many of its properties and is a solvent for cellulose nitrate. Used to prepare diisobutyl ketone (reduction).

Phosdrin, dimethylmethyl-2-crotonylphosphate, $C_7H_{13}O_6P$, $(MeO)_2P(O)OCMe{=}CHCO_2Me$. A trade name for a highly active systemic insecticide; both *cis* and *trans* isomers are active.

phosgene, Cl_2CO. *See* carbonyl derivatives.

phosgenimmonium salts, phosgene iminium salts *See* dichloromethyleneammonium salts.

phosphagen *See* creatine phosphate.

phosphatases A group of enzymes of outstanding importance. They hydrolyse the many different combinations of phosphoric acid with organic substances, for example the sugar phosphates. The enzymes also act synthetically, and they function independently of the living cell.

phosphate coatings Commonly applied to

steels and other alloys in order to give improved corrosion resistance. Such coatings provide an excellent keying surface for painting and also improve the retention of lubricants on the surface.

phosphate esters, $(RO)_3PO$. Derivatives formed by phosphoric acids – particularly H_3PO_4 – and alkyl or aryl alcohols. Tricresylphosphate, $(MeC_6H_4O)_3PO$ (a mixture of all isomers) formed Cl_3PO on technical cresol, is important as a plasticizer, flame-retardant agent for fabrics and as a hydraulic fluid. $(MeO)_2(PhO)PO$, $(BuO)_3PO$, $(MeO)_2$(xylylO)PO, $(MeO)_3PO$) are used as gasoline additives. Phosphate esters (e.g. $(BuO)_3PO$) are extensively used in solvent extraction processes. Tris(2,3-dibromopropyl)-phosphate has been used as a fabric flame retardant but is carcinogenic. The genetic substances DNA and RNA are phosphate esters and many phosphates are of great biological importance.

phosphates Formally any salt of a phosphorus oxyacid *. Generally applied to salts of H_3PO_4.

phosphatides *See* phospholipids.

phosphazenes Linear or cyclic compounds containing $(P{=}N)_x$ groups. *See* phosphonitrilic derivatives.

phosphides Compounds of the elements with phosphorus. The non-metals form covalent materials (e.g. S). The most electropositive elements form ionic derivatives, e.g. Na_3P rapidly hydrolysed to PH_3 by water. The transition elements form derivatives containing P_n groups (e.g. CdP_2) or hard materials with complex structures.

phosphinates Alkyl derivatives of $HOP(O)H_2$.

phosphine, PH_3. *See* phosphorus hydrides.

phosphinites Derivatives of $R_2P(H)O$.

phosphites, $(RO)_3P$. Derivatives of phosphorous acid, prepared PCl_3 plus ROH or ArOH. Used as intermediates in organic synthesis; with R′X, $(RO)_3P$ undergoes the Michaelis–Arbuzov reaction to $(RO)_2P(O)R'$. Form complexes with transition metals. $(RO)_3P$ have been used as anti-oxidants in lubricating oils. Metal derivatives of phosphorous acid are also termed phosphites – *see* phosphorus oxyacids.

phosphocreatine *See* creatine phosphate.

phospholipases Enzymes which effect the cleavage of phosphatides either at the carboxylate or at the phosphate linkages.

phospholipids, $R'C(O)OCH_2{\cdot}CH(OC(O)R'')$-$CH_2OP(O)_2OR'''$. A group of substances of a fat-like nature which are essential components of living cells. They contain phosphorus and usually nitrogen and on hydrolysis give fatty acids, phosphoric acids and bases such as choline (from R‴). The phospholipids consist of a number of different species usually with glycerol as the 'back-bone'. They include lecithins, cephalins and sphingomyelins. Phospholipids are amphoteric and mainly exist as components of cell membranes. Lecithins, cephalins and related compounds are also more specifically known as phosphatides as they are derivatives of phosphatidic acid.

phosphomolybdates Salts of the hetero-phosphomolybdic acids (*see* molybdates) but particularly of $H_3PMo_{12}O_{40}$. The yellow ammonium salt is used for the estimation of Mo or P.

phosphonates Derivatives such as $RP(O)(OR)_2$.

phosphonites Derivatives such as $RP(OR)_2$; a dialkyl alkylphosphonite.

phosphonitrilic derivatives, phosphazenes,

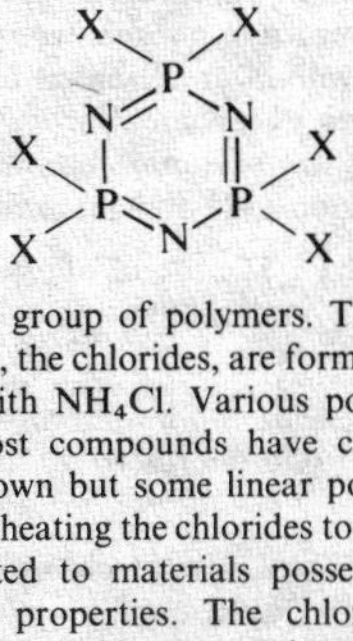

$(PNX_2)_n$ A group of polymers. The simplest compounds, the chlorides, are formed by heating PCl_5 with NH_4Cl. Various polymers are known; most compounds have cyclic structures as shown but some linear polymers are known. On heating the chlorides to 300°C they are converted to materials possessing some rubber-like properties. The chlorine atoms may be substituted by many other groups. The rings have low hydrolytic stability.

phosphonium salts White crystalline salts analogous to the ammonium halides, e.g. $[PH_4]I$. Formed by the action of dry hydrogen halides on phosphine, which functions as a weak base. Phosphonium salts are readily dissociated into PH_3 and the HX, phosphonium chloride not being stable at normal pressures and temperatures. Organophosphonium salts, e.g. $[(C_6H_5)_3PH]^+[BF_4]^-$, are much more stable than the unsubstituted derivatives; the stability increases with increasing substitution of hydrogen by alkyl or aryl groups. Triphenylalkyl phosphonium salts having the general formula $[(C_6H_5)_3PCHR^1R^2]^+X^-$ are used to make phosphoranes (ylids) $(C_6H_5)_3PC\langle{}^{R^1}_{R^2}$ employed in the Wittig reaction.

phosphoproteins *See* proteins.

phosphoranes Derivatives generally covalent, of 5-covalent phosphorus, i.e. R_5P, Ph_5P. The Wittig intermediates $RR'C{=}PR''_3$ can also be included under this heading.

phosphorescence Some substances absorb light and as a result are induced to emit radiation themselves. When this emission ceases when the irradiation is stopped, the process is called fluorescence; when the emission continues after the source of irradiation is removed, the process is termed phosphorescence. Phosphorescence may be regarded as fluorescence in which the atoms or molecules of the absorber are capable of existing in metastable states.

phosphoric acids Oxyacids containing phosphorus(V) – *see* phosphorus oxyacids. Prepared phosphates plus H_2SO_4 or P oxidized to P_2O_5 hydrolysed to H_3PO_4. The salts of phosphoric acids, phosphates, occur naturally (particularly calcium phosphate, apatite) and organic derivatives are essential constituents of living tissues. Phosphates, particularly superphosphates*, are of importance as fertilizers. U.S. production 1982 10·2 megatonnes.

phosphorimetry The use of phosphorescence in analysis. A non-routine method used for trace concentrations.

phosphorous acids Phosphorus oxyacids* containing phosphorus(III) particularly H_3PO_3.

phosphors Substances which exhibit phosphorescence. Doped oxide lattices, e.g. $Ca_5(F,Cl)(PO_4)_3$:Sb, Mn are used in fluorescent lamps. YVO_4:Eu is a red phosphor used in colour television tubes, on stamps to assist with automatic sorting. Zn_2SiO_4:Mn i.r. stimulated phosphors are used for night vision. Organic phosphors are used as optical brighteners*.

phosphorus, P. At.no. 15, at.wt. 30·97376, m.p. (white) 44·1°C, b.p. (white) 280°C, *d* 1·82. Phosphorus is a non-metal of Group V. It occurs naturally as calcium phosphate and as apatite* and fluoroapatite*. Phosphorus is obtained from $Ca_3(PO_4)_2$ by fusion with sand and coke in an electric arc furnace; P distils out and is condensed under water. There are many allotropes but three main forms. The most reactive form is white phosphorus (stored under water) which contains P_4 molecules. It is purified by zone melting. Heating for a long period at a high temperature gives red phosphorus. It is a polymeric material. Heating white phosphorus under high pressure gives black phosphorus with a graphite-like structure. The vapour contains P_4 and P_2 species. White phosphorus is spontaneously inflammable, black phosphorus is almost non-inflammable. P dissolves in alkalis but only in oxidizing acids.

Phosphorus compounds are very important as fertilizers (world use 1976/77 27·3 megatonnes as P_2O_5) but are widely used in matches, pesticides, special glasses and china ware, alloys (steels, phosphor bronze), and metal treating (10%), detergents (40%), electrical components (e.g. GaP), foods and drinks (15%). Phosphates are an essential constituent of living organisms. U.S. production of phosphorus 1982 372 000 tonnes.

phosphorus chemistry Phosphorus is an element of Group V, electronic configuration $3s^23p^3$. The stable oxidation states are +5, (5-co-ordinate covalent), e.g. PF_5, 6-co-ordinate complexes (e.g. $[PF_6^-]$), 4-co-ordinate tetrahedral, e.g. $[PCl_4]^+$, $[PO_4]^{3-}$ and +3 (mainly 3-co-ordinate pyrimidal, e.g. PH_3). Catenation occurs readily and is frequently based on P_4 tetrahedra although cyclic phosphines (e.g. $[CF_3P]_{4\ \text{and}\ 5}$) are known. Cationic species, e.g. $P_4{}^{2+}$ are formed in oxidizing species such as oleum.

phosphorus halides Halides of the types PX_5, PX_3, POX_3, PSX_3 and P_2X_4 are known. Pentahalides are known for fluorine, b.p. −75°C (P plus F_2; PCl_5 plus NaF or AsF_3 (trigonal bipyrimidal)) Cl, decomp. 167°C, (P plus excess Cl_2 $[PCl_4]^+[PCl_6]^-$), Br, $[PBr_4]^+Br^-$. POX_3 result from partial hydrolysis. The trihalides (all X) are pyrimidal and are prepared by PCl_3 plus NaF or P plus a deficiency of X_2. (PF_3, b.p. −101°C, PCl_3, b.p. 76°C); P_2X_4(F,Cl,I) by coupling reactions or use of electrical discharges. PX_3 are very weak Lewis acids but good Lewis bases (e.g. $[Ni(PF_3)_4]$) forming very stable complexes. PX_5 are good acceptors. The phosphorus halides are used in the synthesis of phosphorus compounds, particularly phosphate and phosphite esters. PF_5 is hydrolysed to fluorophosphoric acids.

phosphorus hydrides Phosphine, PH_3, m.p. −133·5°C, b.p. −88°C. Formed water on Ca_3P_2 or yellow phosphorus plus NaOH solution. Decomposes at 450°C. Weak base (*see* phosphonium salts). As generally prepared spontaneously inflammable because of the presence of P_2H_4, b.p. 52°C. Alkyl and aryl phosphines can contain P–H bonds but are more stable than PH_3 (particularly aryl derivatives). Higher phosphines probably exist.

phosphorus, organophosphorus acids The practice of naming organophosphorus compounds has been largely unsystematic. Since 1952 a nomenclature has been in use which considers organophosphorus compounds to be derivatives of phosphorus hydrides, oxyacids and oxides. Some of the parent compounds are unstable and cannot be isolated.

Acid		*Ester, etc.*
$P(OH)_3$		
phosphor*ous* acid	⟶	phosph*ite* e.g. $(CH_3O)_3P$ trimethyl phosphite
$HP(OH)_2$		
phosphon*ous* acid	⟶	phosphon*ite* e.g. $C_6H_5P(OC_2H_5)_2$ diethylphenyl phosphonite
$H_3CP(OH)_2$		
methylphosphonous acid	⟶	methylphosphonite
$H_2P(OH)$		
phosphin*ous* acid	⟶	phosphin*ite* e.g. $(CH_3)_2P-OEt$ ethyl dimethylphosphinite
$HP(O)(OH)_2$		
phosphon*ic* acid	⟶	phosphon*ate*
$H_3CP(O)(OH)_2$		
methylphosphonic acid	⟶	methylphosphonate e.g. $CH_3P(O)(OC_6H_5)_2$ diphenyl methylphosphon*ate*
$H_2P(O)OH$		
phosphin*ic* acid, hypophosphorous acid	⟶	phosphin*ate* e.g. $(C_6H_5)_2P(O)OCH_3$ methyl diphenylphosphinate
$OP(OH)_3$		
phosphor*ic* acid	⟶	phosphate e.g. $(C_2H_5O)_3P{=}O$ triethylphosph*ate*

phosphorus oxides Phosphorus forms a series of oxides the most important of which are P_4O_{10} and P_4O_6.

Phosphorus(V) oxide, P_4O_{10}, phosphorus pentoxide. There are several forms (P burned in excess O_2). P_4O_{10} has a great affinity for water and is used as a drying and dehydrating agent (e.g. HNO_3 gives N_2O_5) it is itself converted to phosphoric acids (*see* phosphorus oxyacids). P_4O_{10} reacts with metal oxides to give phosphates. Its structure has a P_4 tetrahedron with O along each edge and on each P. P_4O_9, P_4O_8 and P_4O_7 are intermediate oxides (heat on P_4O_6) with similar structures to P_4O_{10}.

Phosphorus(III) oxide, P_4O_6, phosphorus trioxide, m.p. 24°C, b.p. 174°C. A waxy material (burn P in deficiency of O_2). It burns in excess O_2 to P_2O_5, reacts with, e.g. Cl_2 to $POCl_3$ and dissolves in water to give phosphorus(III) oxyacids. The structure is similar to that of P_4O_{10} but without the terminal oxygens.

phosphorus oxyacids Because phosphorus shows oxidation states +5 and +3 and can in addition share oxygens (particularly between tetrahedral atoms) there is a range of phosphorus oxyacids, *compare* phosphorus, organophosphorus acids. POH groups have ionizable hydrogen, P–H groups are non-ionizable.

Hypophosphorous acid, H_3PO_2, $H_2P(O)OH$. A monobasic acid. $Ba(H_2PO_2)_2$ is formed when white phosphorus is dissolved in $Ba(OH)_2$ solution. H_3PO_2 and its salts are strong reducing agents.

Phosphorous acid *, H_3PO_3, $HP(O)(OH)_2$. A dibasic acid (PCl_3 plus cold water), strong reducing agent.

Hypophosphoric acid, $H_4P_2O_6$, $HP(O)(OH)(\mu\text{-}O)P(O)(OH)_2$ although can be tetrabasic (Na salt from NaOCl on red phosphorus).

Phosphoric acids are derivatives of P_4O_{10}. Condensed acids and anions are known. Orthophosphoric acid, H_3PO_4, $OP(OH)_3$, gives tetrahedral anions. Pyrophosphoric acid, $H_4P_2O_7$, $(HO)_2P(O)(\mu\text{-}O)P(O)(OH)_2$, a tetrabasic acid. Metaphosphates (HPO_3) are obtained by dehydration of H_3PO_4 at 300°C. Many species are known; a cyclic $(P_3O_9)^{3-}$ ion and various chain phosphates, e.g. Maddrell's salt, Kurrol's salt, Graham's salt, all $(NaPO_3)_x$, are well characterized. So-called sodium hexametaphosphate is a long chain compound used in water-softening – chelates Ca^{2+} ions. Short chain $[P_3O_{10}]^{5-}$, $[P_4O_{13}]^{6-}$ are also known.

Polyphosphates are used as detergents (calgon) but disposal of the residual phosphate causes major problems. Phosphates are of importance as flame-proofing agents. *See also* phosphorus.

phosphorus sulphides Yellow solids P_4S_3, P_4S_4, P_4S_5, P_4S_7 and P_4S_{10} formed from red phosphorus and sulphur. The structures are based on a P_4 tetrahedron with some bridging S atoms and, in P_4S_7 and P_4S_{10}, terminal S atoms. P_4S_3 is used in matches; P_4S_{10} in the preparation of organo-sulphur compounds (particularly pesticides and oil additives).

phosphorylases Enzymes that catalyse the addition of phosphate across glycosyl or related bonds.

phosphoryl halides, $P(O)X_3$. *See* phosphorus halides.

phosphotungstates Heteropolytungstates. *See* tungstates.

photoacoustic spectroscopy A spectroscopic technique in which a sample is irradiated with chopped monochromatic radiation and the absorption of the radiation causes a temperature fluctuation which is transmitted to the gas in contact with the sample and which is in turn transmitted to a microphone. Used to measure the optical absorption spectra of solids and liquids, including opaque materials.

photochemistry The investigation of chemical reactions brought about by the action of light. Only light which is absorbed by the system can produce chemical effects; any light which passes through the material or is scattered is necessarily ineffective. Consequently, it is essential to determine the absorption spectrum of the substances concerned and to identify the spectral regions responsible for the reaction. Absorption of light is a quantum process. The primary step in a photochemical reaction is therefore the absorption of a quantum of light energy by a particular atom, molecule or ion which is thereby raised to an excited state containing an excess quantum of energy equal to $h\nu$ (where h is Planck's constant and ν is the frequency of the light absorbed). Only electronic excitation can bring about chemical reaction and therefore radiation in the far i.r. region is ineffective. Photochemical reactions are generally noticed with visible or u.v. radiation, the latter being the more powerful. Radiation in the far u.v. is powerful enough to break any chemical bond: e.g. oxygen gas is dissociated into oxygen atoms, the photochemical reaction being written

$$O_2(g) + h\nu \longrightarrow 2O$$

However, not all photochemical reactions involve actual dissociation of molecules into atoms; many proceed by way of electronically excited molecules. An excited molecule may return to its ground state within about 10^{-4} sec while emitting its absorbed energy as fluorescence or passing it on to neighbouring molecules as thermal energy; in these cases no photochemical reaction is produced. In some cases, however, the excited molecule is capable of undergoing reaction, the light supplying the activation energy needed to induce reaction.

After the primary step in a photochemical reaction, the secondary processes may be quite complicated, e.g. when atoms and free radicals are formed. Consequently the quantum yield, i.e. the number of molecules which are caused to react for a single quantum of light absorbed, is only exceptionally equal to exactly unity. E.g. the quantum yield of the decomposition of methyl iodide by u.v. light is only about 10^{-2} because some of the free radicals formed re-combine. The quantum yield of the reaction of $H_2 + Cl_2$ is 10^4 to 10^6 (and the mixture may explode) because this is a chain reaction.

The absorbed light may act as catalyst for a spontaneous reaction, but in other cases it may supply energy to make possible a reaction which, without light, would be thermodynamically impossible. In some cases, such a reaction reverses itself by thermal reaction (e.g. if left in the dark) and, hence, during irradiation a *photostationary state* is reached.

The light producing a photochemical reaction is most commonly absorbed by one of the reactants, but many examples are known where energy absorbed by another species is passed to the reactants; this is the phenomenon of *photosensitization*.

photoconduction The phenomenon observed when certain substances, e.g. selenium, which are poor conductors of electricity in the dark, become good conductors when exposed to light. Photoconduction arises from the absorption of light energy causing electrons to be promoted from the valence band of the solid into the conduction band. It is the basis of some commercial photocopying processes.

photodissociation The process which occurs when a molecule on absorbing a quantum of light energy undergoes dissociation to smaller molecules, radicals or atoms. E.g. propanone undergoes photodissociation to methyl radicals and carbon monoxide.

photo-electric cells Exposure of certain elements, particularly the alkali metals, to light of suitable wavelength results in the emission of electrons. This phenomenon is known as the photo-electric effect. Since the intensity of emission is proportional to the intensity of the incident light, the effect can be made the basis of an instrument for measuring light intensity.

photo-electron spectroscopy Bombardment of an atom with photons of a specific energy results in the emission of an electron of energy E, where $E = (E_i - E_b)$, E_i and E_b being the respective energies of the photon and of binding of the ejected electron. The technique may be used for measuring binding energies of valence electrons or for inner core electrons. Inner core spectroscopy, more commonly termed ESCA (electron spectroscopy for chemical analysis) or X-ray photo-electron spectroscopy (XPS), generally uses Al($k\alpha$) 1486 eV as the photon source. Ultra-violet photo-electron spectroscopy using He^{I}, 21·2 eV and He^{II} 40·8 eV radiation is used for examining valence shell electrons.

photographic developers Chemical reducing agents which specifically reduce silver halide grains in the photographic plate which have been exposed to light.

photographic gelatin The medium for silver halide emulsion used in photography. It is a protective colloid for the emulsion grains but also enhances sensitivity.

photoluminescence A general term used to describe the emission of light as a result of an initial absorption of light. The phenomena of fluorescence and phosphorescence are examples of photoluminescence.

photolysis The decomposition or reaction of a substance or substances on exposure to light. In general wavelengths between 2000 Å and 8000 Å are suitable.

photon A quantum of light energy. The energy of a photon is given by $E = h\nu$, where ν is the frequency of the radiation and h is Planck's constant.

photosensitization A process in which a photochemical reaction is induced to occur by the presence of a substance (the photosensitizer) which absorbs the light but itself is substantially unchanged at the end of the reaction, the absorbed light energy being passed on to the main reactants. E.g. when hydrogen is exposed to light of wavelength 2536 Å, no absorption of the light occurs, and the hydrogen remains completely unaffected. If mercury vapour is added to the hydrogen, the mercury atoms are excited. When such an excited mercury atom collides with a hydrogen molecule, it can transfer some of its energy to the hydrogen, and cause it to dissociate into atoms. The hydrogen has apparently been made sensitive to the light which it does not absorb.

photostationary state In a chemical reaction system the term stationary state is used to indicate a state of equilibrium under specified conditions. In a photochemical reaction, a stationary state is attained when the rate of removal of reactants by light is equal to the rate of recombination or reaction of the products to yield the reactants. This condition is referred to as a photostationary state.

photosynthesis The process by which green plants build up their carbon compounds from atmospheric carbon dioxide using light as the energy source.

phototropy The reversible change of colour of certain substances when exposed to light of a certain wavelength.

phrenosin, $C_{48}H_{93}O_9$. M.p. 212°C. A cerebroside* present in brain and nervous tissue. On hydrolysis it gives D-galactose, sphingosine and cerebronic acid*.

phthalamide, $C_8H_8N_2O_2$, 1,2-$C_6H_4(CONH_2)_2$. Colourless crystals; on heating to 200–210°C melts (decomp.) into phthalimide and ammonia.

Prepared by stirring phthalimide with cold concentrated ammonia solution. Hydrolysed to phthalic acid with dilute acids. Dehydration with ethanoic anhydride gives first *o*-cyanobenzamide and then phthalonitrile.

phthalic acid, $C_8H_6O_4$, 1,2-$C_6H_4(COOH)_2$. Colourless crystals; m.p. 190–210°C. At m.p. it decomposes into water and phthalic anhydride. Prepared by oxidation of naphthalene or *o*-xylene, or by alkaline hydrolysis of phthalic anhydride.

It is a dibasic acid, and forms stable metallic salts. Distillation with soda lime gives benzene. Readily dehydrated to phthalic anhydride.* Its reactions are similar to phthalic anhydride in which form it is almost invariably used.

phthalic anhydride, $C_8H_4O_3$. Long silky

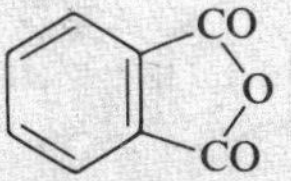

needles; m.p. 130°C, b.p. 284°C. Slowly converted into phthalic acid in water.

Prepared by the oxidation of naphthalene with sulphuric acid at 270–300°C in the presence of $HgSO_4$; under these conditions the phthalic anhydride sublimes out of the reaction mixture as formed. It is manufactured by the vapour phase oxidation of naphthalene or *o*-xylene over a V_2O_5 catalyst.

Treatment with PCl_5 gives phthalyl chloride; reduction with zinc and ethanoic acid or NaOH gives phthalide. Fusion with urea gives phthalimide.

It is a valuable dyestuffs intermediate. It can be condensed with benzene and $AlCl_3$ to give anthraquinone. This general reaction is much used for preparing substituted anthraquinones. With phenols and a condensing agent phthalic anhydride gives rise to phthalein dyestuffs, e.g. phenolphthalein, eosin. It also condenses with quinaldine derivatives to give 'quinoline' dyes. The chief commercial uses of phthalic anhydride are in the manufacture of dialkyl phthalate plasticizers and alkyd resins. U.S.production 1978 450 000 tonnes.

phthalic esters Because of their low vapour pressures and chemical stability, various esters of phthalic acid such as diethyl (b.p. 298°C), dibutyl (b.p. 340°C) and di-*n*-octyl (b.p. 248°C) are used as plasticizers. The dimethyl ester is used in the manufacture of polyethylene terephthalate. The dimethyl and dibutyl esters are also used as insect repellants. They are all manufactured by reacting phthalic anhydride with the alcohol in the presence of catalytic amounts of sulphuric acid. The dimethyl ester is used in alkyd resins, U.S. production 1981 2·2 megatonnes.

phthalide, $C_8H_6O_2$. M.p. 75°C. Formed by

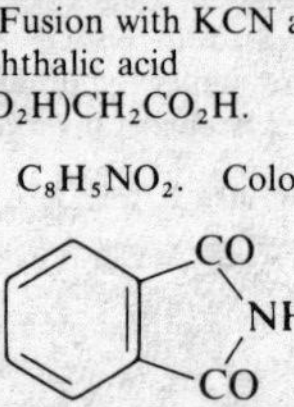

reduction of phthalic anhydride with Zn dust and NaOH. Fusion with KCN and hydrolysis gives homophthalic acid $1,2\text{-}C_6H_4(CO_2H)CH_2CO_2H$.

phthalimide, $C_8H_5NO_2$. Colourless plates,

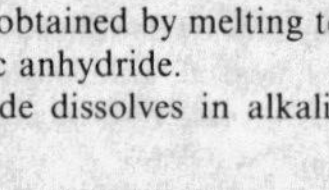

m.p. 230°C. It is prepared on the large scale by passing ammonia into molten phthalic anhydride. Also obtained by melting together urea and phthalic anhydride.

Phthalimide dissolves in alkalis to give at first N-metal derivatives, but on warming the solution the ring opens, and salts of phthalamic acid are obtained. Reduced to phthalide with zinc and NaOH solution; prolonged hydrolysis with acids or alkalis gives phthalic acid or its salts. Treatment with concentrated ammonia solution gives phthalamide.

An important laboratory use involves the Gabriel synthesis of α-amino-acids.

When treated with alkaline hypochlorite solution anthranilic acid is obtained. This reaction, the first stage in the indigo synthesis, is the most important commercial outlet for phthalimide.

phthalocyanines An important class of organic colouring matters capable of general application as pigments and of more limited use as dyestuffs. They are characterized by their great fastness and brilliance of shade. The shades are green to blue in colour.

They can be prepared by three general methods:

(1) Heating phthalic anhydride with urea (or ammonia) and a metallic salt.

(2) Heating *o*-cyanobenzamide with a metallic salt.

(3) Heating phthalonitrile with a metal or metallic salt.

Depending upon the metallic compound used, different metallic phthalocyanine derivatives are obtained, e.g. when copper chloride is used copper phthalocyanine (Monastral Fast Blue B) is obtained.

The phthalocyanines must be suitably dispersed to be used as pigments or they can be sulphonated to water-soluble forms for dyeing and for precipitation as lakes. The dispersion is carried out by solution in sulphuric acid, followed by precipitation in water.

The phthalocyanine molecule is remarkably stable to heat and chemical reagents. The metal-free and heavy metal compounds sublime practically unchanged at 550–580°C.

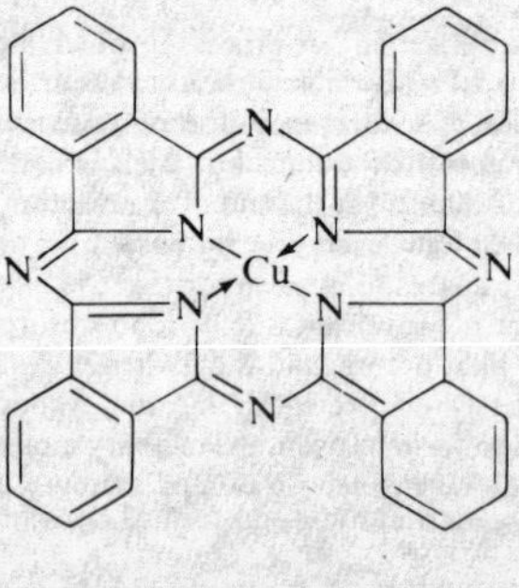

Structurally, the molecule is formed by the union of one atom of the metal with four molecules of phthalonitrile. Copper phthalocyanine is typical, and has the formula $C_{32}H_{16}CuN_8$.

The other phthalocyanines are derived from the above by replacement of the copper atom by two atoms of a monopositive or by one atom of a dipositive metal.

phthalonitrile, $C_8H_4N_2$, 1,2-$C_6H_4(CN)_2$. Colourless needles; m.p. 141°C, b.p. 290°C. Slowly hydrolysed to phthalic acid by dilute acids and alkalis.

It can be obtained by dehydration of *o*-cyanobenzamide or phthalamide with ethanoic acid. On a larger scale the dehydration is better carried out by means of phosgene in pyridine solution or to pass a mixture of ammonia and phthalimide over a catalyst at 350–450°C. Used to prepare phthalocyanines*.

phthienoic acids *See* phthioic acids.

phthioic acids, phthienoic acids A group of branched-chain fatty acids isolated from mycobacteria. The term is chiefly used for acids ranging from C_{22} to C_{28}, while the term 'mycolic acid' generally refers to hydroxy-acids of higher molecular weight, also found in mycobacteria, and derived by coupling two or more molecules of fatty acids.

phyllins *See* chlorophyll.

phytanic acid, 3,7,11,15-tetramethylhexadecanoic acid, $C_{20}H_{40}O_2$. A diterpenoid acid found in bovine plasma phospholipids.

phytic acid, $C_6H_6(OPO(OH)_2)_6$. *Meso*-inositol hexaphosphoric acid, a water-soluble syrup, sparingly soluble in organic solvents and strongly acidic. Occurs as its insoluble calcium magnesium salt, phytin, in seeds. It is also found in blood plasma phospholipids and in the erythrocytes of chicken blood.

phytin *See* phytic acid.

phytochemistry The chemistry of compounds associated with plants.

phytol, 3,7,11,15-tetramethyl-2-hexadecen-1-ol, $C_{20}H_{40}O$. A diterpenic alcohol obtained by the action of alkalis on chlorophyll. Colourless oil; b.p. 202–204°C/10 mm. On oxidation it yields a ketone $C_{18}H_{36}O$; b.p. 175°C/11 mm.

phytoxic Poisonous to plants.

pi bonding, π-bonding Bonding by interaction of orbitals of such shape that there are two regions of overlap. Thus ethene, C_2H_4, is considered to have a σ-bond formed by overlap of s-orbitals and a π-bond formed by overlap of p-orbitals. π-Bonding is of importance in co-ordination compounds where co-ordinate bonds may have π-character. The electrons can be donated from the donor or there may be electron transfer from the acceptor to the donor (back-bonding).

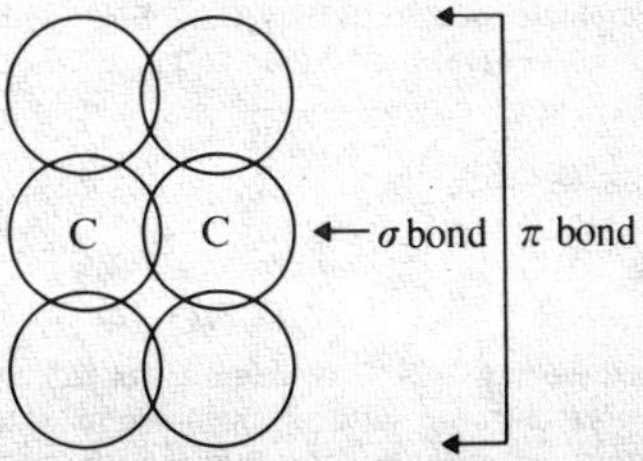

pickling The immersion of metal parts in dil. H_2SO_4 or HCl followed by washing to obtain a clean surface which can be subsequently treated, e.g. by galvanizing.

picolines, methylpyridines, C_6H_7N. Found with pyridine in bone oil and coal tar. α-Picoline (2-methylpyridine), b.p. 129°C; β-picoline (3-methyl), b.p. 144°C; γ-picoline (4-methyl), b.p. 143°C. α-Picoline is used in the manufacture of the monomer 2-vinylpyridine, β-picoline for nicotinic acid, and γ-picoline for isoniazid.

picramic acid, 2-amino-4,6-dinitrophenol, $C_6H_5N_3O_5$. Red needles; m.p. 168–169°C. Soluble in dilute acids and alkalis. Prepared by reduction of picric acid with sodium hydrogen sulphide. It is used for the preparation of azo-dyes, which can be 'after-chromed' by treatment with metallic salts owing to the presence of a hydroxyl group ortho to the amino-group.

picrates Derivatives (salts) of picric acid, i.e. $X^+[OC_6H_2(NO_2)_3]^-$. A convenient method of characterization of basic organic compounds. Some picrates, especially of metals, explode on heating. Aromatic hydrocarbons form charge transfer complexes with picric acid, also known as picrates.

picric acid, 2,4,6-trinitrophenol, $C_6H_3N_3O_7$. Bright yellow crystals; m.p. 122°C. Prepared by the nitration of phenolsulphonic acid. It is used in the dyeing industry and also as an explosive. Is not very sensitive but attacks metals and can form metallic picrates which are very sensitive to shock. Because of this and other disadvantages, it has largely been replaced by other materials as a high explosive. In the cast form it is known as lyddite.

picrolonic acid, $C_{10}H_8N_4O_5$. Fine yellow

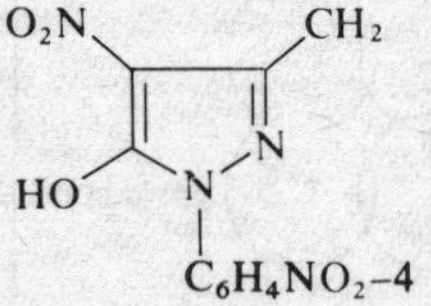

needles; m.p. 124°C (decomp.). Its calcium, copper and lead salts are insoluble in water. Forms crystalline compounds with many organic bases; these compounds are usually sparingly soluble in water or dilute alcohol and have sharply defined melting-points. Used for the isolation and identification of organic bases, and in tests for calcium, copper and lead.

piezoelectricity The electric charge developed in anisotropic crystals when subjected to stress (e.g. pressure). Used in electrical components and particularly in lighters.

pig iron The product from a blast furnace after cooling in sand moulds. Essentially the same as cast iron but the latter has generally been remelted and cast to a finished shape.

pigments Materials which are used to impart colour to surfaces, plastics, inks, etc., although the pigments may incidentally affect other properties of the substrate. Distinct from dyestuffs which operate on a molecular level, pigments tend to be particulate and insoluble. Many pigments are inorganic in nature. White (TiO_2, basic lead carbonate and sulphate, zinc oxide and sulphide), red and brown (iron oxides, red lead, cadmium reds), yellow (iron oxides, lead zinc and cadmium chromates), green (chromium derivatives), blue (cyanoferrates, ultramarines), black (C). Metal powders (Al, bronze, Zn) are also used in surface coatings. Many insoluble organic dyestuffs are also used as pigments.

pimelic acid, heptanedioic acid, $HOOC{\cdot}[CH_2]_5{\cdot}COOH$, $C_7H_{12}O_4$. Colourless prisms; m.p. 105°C. It is present in castor oil. It can be prepared from pentamethylene dibromide or cyclohexanone via the nitrile and by the reduction of salicylic acid followed by oxidation of the ring system. Also produced by oxidizing capric or oleic acids. Used in polymeric systems.

pinacol, pinacone, 2,3-dimethyl-2,3-butanediol, $C_6H_{14}O_2$. Colourless crystals; m.p. 38°C, b.p. 175°C. Crystallizes from water as a hexahydrate melting at 47°C. Prepared by treating a mixture of propanone and benzene with magnesium amalgam. When heated with sulphuric acid, it forms pinacolone. 2,3-Dimethylbutadiene is formed when the vapour of pinacol is passed over alumina at 400°C.

pinacolone, pinacolin, 3,3-dimethyl-2-butanone, $C_6H_{12}O$. Colourless liquid with a camphor-like odour; b.p. 103–106°C/746 mm. Prepared by heating pinacol hydrate with sulphuric acid and distilling the mixture. It is oxidized to trimethylacetic acid.

pinacol-pinacolone rearrangement When heated with mineral or organic acids, pinacols undergo a molecular rearrangement with loss of water to give the ketones, pinacolones. The change involves the loss of one hydroxyl group, as a neutral water molecule derived from the protonated pinacol, followed by an intramolecular migration of an alkyl group in the carbenium ion thus formed, and loss of a proton from the remaining hydroxyl group to give a ketone.

pinacols 1,2-Glycols where R_1, R_2, etc are

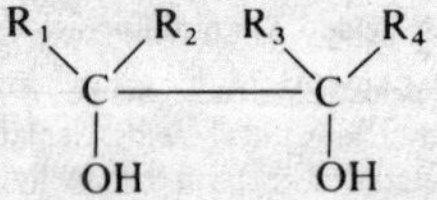

alkyl or aryl groups. They are formed by reducing aromatic ketones with zinc and dilute acid, or from aliphatic ketones by reduction with magnesium amalgam.

pinacone *See* pinacol.

pinane, $C_{10}H_{18}$. A saturated hydrocarbon produced by the catalytic hydrogenation of α- or β-pinene. It can exist in both *cis*- and *trans*-forms.

α-pinene, $C_{10}H_{16}$. A dicyclic terpene; m.p.

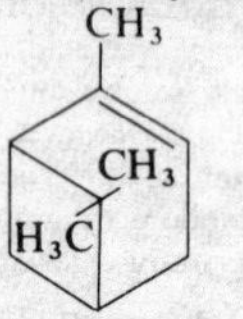

−50°C, b.p. 156°C. The most important of the terpene hydrocarbons. It is found in most essential oils derived from the *Coniferae*, and is the main constituent of turpentine oil. Contains two asymmetric carbon atoms. The (+)-form is easily obtained in a pure state by fractionation of Greek turpentine oil, of which it constitutes 95%. Pinene may be separated from turpentine oil in the form of its crystalline nitrosochloride, $C_{10}H_{16}ClNO$, from which the (±)-form may be recovered by boiling with aniline in alcoholic solution. When heated under pressure at 250–270°C, α-pinene is converted into dipentene. It can be reduced by hydrogen in the presence of a catalyst to form pinane. It is oxidized by air or oxygen, both wet and dry. With moist air it forms sobrerol, $C_{10}H_{18}O_2$, which forms the oxide pinol by the action of weak mineral acids.

By oxidation with permanganate it forms pinonic acid, $C_{10}H_{16}O_3$, a monobasic acid derived from cyclobutane. With strong sulphuric acid it forms a mixture of limonene, dipentene, terpinolene, terpinene, camphene and *p*-cymene. Hydrogen chloride reacts with turpentine oil to give $C_{10}H_{17}Cl$, bornyl chloride, 'artificial camphor'.

β-pinene, nopinene, $C_{10}H_{16}$. α-Pinene in

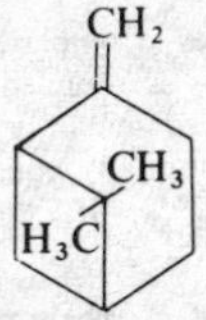

nature is usually accompanied by smaller but variable amounts of (−)-β-pinene. Natural (+)-β-pinene has only been found in one oil, extracted from the ripe fruits of *Ferula galbaniflua*. β-Pinene yields no crystalline derivatives. B.p. 162–163°C. On oxidation with potassium permanganate it forms nopinic acid, $C_{10}H_{16}O_3$; m.p. 126–127°C.

pink salt Ammonium hexachlorostannate.

pinocamphone, $C_{10}H_{16}O$. A dicyclic ketone,

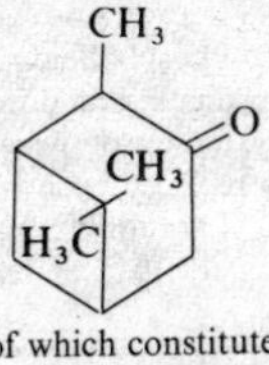

the (−)-form of which constitutes about 45% of oil of hyssop. The (±)-form is obtained by the reduction of nitrosopinene with zinc dust in ethanoic acid solution. It forms pinonic acid on oxidation. (−)-Pinocamphone has b.p. 211°C.

pinocarveol, $C_{10}H_{16}O$. A dicyclic terpene

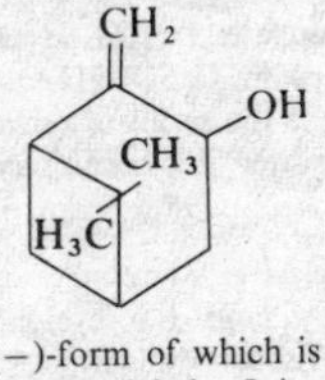

alcohol, the (−)-form of which is present in the oil of *Eucalyptus globulus*. It is also formed amongst other products by the auto-oxidation of β-pinene. (−)-Pinocarveol has m.p. 7°C, b.p. 208–209°C. The (±)-form has b.p. 215–218°C.

piperazine, hexahydropyrazine, $C_4H_{10}N_2$.

The hexahydrate forms colourless crystals, m.p. 44°C (104°C when anhydrous), b.p. 126°C. Made by the action of alcoholic ammonia on 1,2-dichloroethane, piperazine is used in human and veterinary medicine in the treatment of threadworm and roundworm infestations.

piperidine, hexahydropyridine, $C_5H_{11}N$.

Colourless liquid with a characteristic ammoniacal smell; m.p. −9°C, b.p. 106°C. Miscible with water. It is present in pepper as the alkaloid piperine from which it can be obtained by heating with alkali. It can also be prepared by the reduction of pyridine, either electrolytically or by other means. Piperidine is a strong base, behaving like the aliphatic amines.

piperine, $C_{17}H_{19}NO_3$. *See* piperidine.

piperitol, $C_{10}H_{18}O$. B.p. 165–170°C/200 mm.

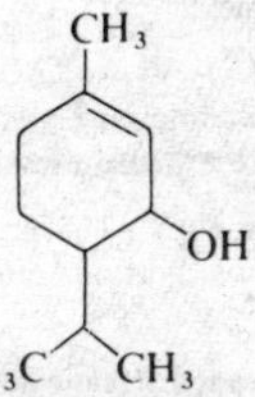

An optically active, secondary terpene alcohol. (−)-Piperitol is found in various eucalyptus oils and (+)-piperitol in the oil from a species of *Andropogon*. A somewhat viscous oil of pleasant smell. It yields piperitone on oxidation with chromic acid.

piperitone, $C_{10}H_{16}O$. An optically active terpene ketone, the oxidation product of piperitol, the (−)-form of which is found in a very large number of eucalyptus oils. The (+)-form is found in Japanese peppermint oil, and in the oil from Sudan Mahareb grass.

Piperitone is of considerable technical importance. It is a colourless oil of a pleasant peppermint-like smell. (−)-Piperitone has b.p. 109·5–110·5°C/15 mm. Piperitone yields thymol on oxidation with $FeCl_3$. On reduction with hydrogen in presence of a nickel catalyst it yields menthone. On reduction with sodium in alcoholic solution all forms of piperitone yield racemic menthols and *iso*menthols together with some racemic α-phellandrene.

piperonal, 3,4-methylenedioxybenzaldehyde,

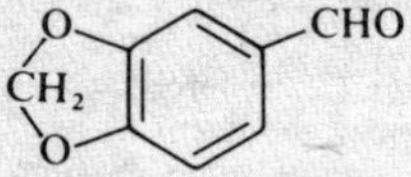

$C_8H_6O_3$. White crystals; m.p. 37°C, b.p. 263°C. Occurs associated with vanillin. Obtained on oxidation of various natural products such as piperine. Used extensively in soap perfumery.

piperonyl butoxide, $C_{19}H_{30}O_5$. An in-

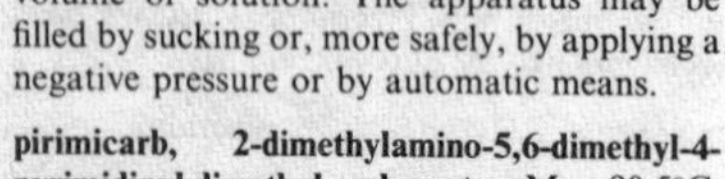

secticide synergist, which is often used in conjunction with pyrethrum.

pipette Apparatus for obtaining a given volume of solution. The apparatus may be filled by sucking or, more safely, by applying a negative pressure or by automatic means.

pirimicarb, 2-dimethylamino-5,6-dimethyl-4-pyrimidinyl dimethyl carbamate M.p. 90·5°C. A specific aphicide.

pitchblende, $UO_{2-2\cdot25}$. An important uranium ore also containing useful quantities of Ra. Th, Ce, Pb are also present.

pitch, coal tar pitch The product left when the more volatile portions of coal tar are distilled.

pivalic acid, trimethylacetic acid, 2,2-dimethylpropanoic acid, $(CH_3)_3C{\cdot}COOH$. Colourless solid; m.p. 35·5°C, b.p. 164°C. Prepared by the carboxylation of isobutene with carbon monoxide in sulphuric acid or by the oxidation of pinacone with NaOBr, and is an intermediate in the synthesis of various industrial products. It is esterified only with difficulty owing to the steric hindrance of the methyl groups.

pK For any acid HA which partially dissociates in solution, the equilibrium

$$HA(solv.) \rightleftharpoons H^+(solv.) + A^-(solv.)$$

is defined by an equilibrium constant K, where

$$K = \frac{[H^+][A^-]}{[HA]}$$

The negative logarithm of the equilibrium constant is termed the pK, i.e. pK = −log K. For acids, the pK is denoted by pK_a. Similarly for a base BOH, the pK_b is given by

$$pK_b = -\log K_b, \text{ where}$$

$$K_b = \frac{[B^+][OH^-]}{[BOH]}$$

planar complexes Complexes in which the acceptor atom is surrounded by four ligands in a plane. If predominantly covalent the hybridization is normally considered as dsp^2. E.g. $[Pt(NH_3)_4]^{2+}$, $[Ir(CO)Cl(PPh_3)_2]$ and $[PtCl_4]^{2-}$.

Planck's constant, *h* The constant relating the energy and frequency of radiation, $E = h\nu$. It has a value of $6{\cdot}6256 \times 10^{-27}$ erg.sec.

plane of symmetry *See* symmetry elements.

planetary electrons Extranuclear electrons. *See* electronic configuration.

plant hormones, plant growth substances Compounds other than nutrients which affect physiological processes in plants. E.g. auxins, gibberellins.

plasmalogens Phospholipids analogous to lecithins and cephalins but differing from these in that the acyl moiety of the α-hydroxyl group of glycerol is replaced by a *cis*-1-unsaturated long chain alkyl group. Plasmalogens account for approximately 10% of brain and muscle phospholipids.

plaster A material used to cover and produce a pleasing flat surface on a wall or other structure. The chief materials are (1) a mixture of lime and sand or of Portland cement and sand, and (2) materials consisting essentially of plaster of Paris.

plaster of Paris An incompletely hydrated calcium sulphate, $2CaSO_4,H_2O$, produced by heating gypsum, $CaSO_4,2H_2O$. The setting of the plaster is the result of rehydration due to the added water; it is accompanied by a slight expansion.

plastic explosives Explosives which can be moulded by hand. E.g. cyclonite mixed with an oily or rubbery binder.

plasticizers Materials incorporated into plastic resins to change workability, flexibility, flow and impact resistance. Thermoplastic resins (particularly PVC) undergo major changes in properties in the presence of plasticizers. Plasticizers are high mol.wt. liquids or low-melting solids generally carboxylic phosphate esters although hydrocarbons, halogenated hydrocarbons, ethers, polyglycols are also used.

plastics Artificial polymeric materials usually organic in nature which have generally been shaped by flow. Many artificial elastomers are plastics. Plastics can be classified on the basis of their thermal behaviour. Thermoplastics (e.g. polyethene, polyvinyl chloride) can be repeatedly softened by heating and rehardened on cooling. Thermoset or thermocured plastics (e.g. phenol-formaldehyde resins) are infusible and insoluble in their final moulded shape. Plastics are also characterized on their chemical derivation or final structure. Plastics are moulded and shaped whilst in their heat-softened stage and then cured by continued heating (thermosetting plastics) or cooling (thermoplastics). Extrusion can give pipe, rod, etc.

plastoquinones A family of trisubstituted benzoquinones with polyisoprenyl groups. They occur in chloroplasts and play an essential part in the photophosphorylation during photosynthesis.

plate, tray The vapour-liquid contacting units in an absorption or fractionating column*. The term 'plate' tends to be used in Britain, 'tray' in the U.S. *See* plate column, bubble-cap plate.

plate column, tray column The plate column, or tray column as it is known in the U.S., is the plant most widely used for vapour-liquid contacting, and as such is employed in gas absorption* and stripping* and for fractional distillation. It consists of a vertical cylindrical column containing a series of horizontal *plates* or *trays* spaced up to 1 m apart and set one above the other all the way up the column. Liquid flows down the column from plate to plate and vapour passes upwards through holes in them.

Since the transfer of material between phases takes place on the plates, the degree of gas absorption, or of separation in the case of a distillation column, depends directly on their number.

plate efficiency, tray efficiency If a plate in a distillation or absorption column were 100% efficient, perfect contacting would occur on it, and the liquid and vapour streams leaving the plate would be in equilibrium with each other. In practice the efficiency of plates is usually less than 100%, with the result that the number required to achieve a given degree of separation or absorption is greater than the theoretical number.

platforming *See* catalytic reforming.

platinum, Pt. At.no. 78, at.wt. 195·09, m.p. 1772°C, b.p. 3827°C, *d* 21·45. The most abundant of the platinum metals occurring naturally and in traces in heavy metal sulphide ores. Pt compounds are readily reduced to the metal which has the ccp structure. The metal is very malleable and ductile. It is attacked by F_2 and Cl_2 above 300°C, dissolved by aqua regia and fused alloys. The metal is used extensively in jewellery, laboratory ware, thermocouples (particularly Pt/Pt–Rh), electrical contacts, and catalysis (SO_3, NH_3, hydrocarbon cracking, methanal production). In catalysis the metal is generally impregnated on an inert support. World demand 1976 2 740 000 troy oz.

platinum ammines Ranges of Pt(II) (4-co-ordinate) and Pt(IV) (octahedral) complexes containing $Pt-NH_3$ (and amine) groups.

platinum black A finely divided precipitate formed from Pt(II) solutions by reducing agents. Used as a hydrogenation catalyst.

platinum chemistry Platinum is the heaviest element of the nickel group, electronic configuration $5d^96s^1$. It shows oxidation states of +6 and +5 (fluoride derivatives only), +4, +2, +1 and 0. Pt(II) and (IV) form extensive series of complexes generally with a square planar or octahedral co-ordination respectively. Halides, *N*-ligands (particularly amines and ammonia), phosphines form stable complexes with both oxidation states. Cyanide and carbonyl complexes are particularly stable with Pt(II). Some complexes are anti-tumour agents. Pt(0) complexes are stable with phosphine ligands; cluster derivatives are formed in many cases. Organic derivatives are known for Pt(II) and Pt(IV).

platinum halides Platinum fluorides known are volatile red PtF_6 (Pt plus F_2), m.p. 61°C, a very strong oxidizing agent (O_2 gives $O_2^+PtF_6^-$), red polymeric PtF_5 (Pt plus F_2), yellow brown PtF_4 (Pt plus BrF_3). Fluoroplatinates(V) and (IV) (e.g. K_2PtCl_6 plus BrF_3, gives K_2PtF_6) contain complex anions. All platinum fluoride species, except PtF_6^{2-}, are hydrolysed by water. Known chlorides are red–brown $PtCl_4$, green–black $PtCl_3$ (contains Pt^{II} and Pt^{IV}) and red–black $PtCl_2$ (all formed from Pt and Cl_2). Chloroplatinates containing, e.g. $PtCl_6^{2-}$ and $PtCl_4^{2-}$ ions are stable. Bromides and iodides are similar to chlorides; the stability of the complex halides increases with the mass of the halogen. Complexes are readily formed between Pt(II) and (IV) halides and donors, e.g. phosphines, thiols.

platinum metals The elements ruthenium and osmium (iron group), rhodium and iridium (cobalt group) and palladium and platinum (nickel group). They occur together as native alloys (native platinum, osmiridium), in mixed sulphide ores and as traces in Au, Ag and Cu sulphide ores. *See* the individual elements. World production 1976 183 tonnes.

platinum organic derivatives Stable alkyls and aryls, e.g. [*trans*-$PtBrMe(PEt_3)_2$] are formed by the action of Grignard reagents on the halides. Tetrameric species, e.g. $[Me_3PtCl]_4$, are formed by the reaction of, e.g. $PtCl_4$. Olefines and alkynes form π-bonded complexes, e.g. $K[PtCl_3(C_2H_4)]$, $[PtCl_2(C_2H_4)]_2$, $[(Ph_3P)_2Pt(PhC_2Ph)]$.

platinum oxides Brown PtO_2 is formed by dehydration of the hydrated oxide precipitated on hydrolysis of $PtCl_6^{2-}$. PtO ($PtCl_2$ plus KNO_3) and Pt_3O_4 (Pt electrode or Pt plus pressure of O_2) are more doubtful. Mixed oxides, e.g. $Tl_2^{III}Pt_2^{IV}O_7$, are known.

pleochroism The property of a crystal of having a different colour depending upon the direction of transmitted light through the crystal. Cyanoplatinates, e.g. $K_2Pt(CN)_4$, are strongly pleochroic.

plumbates *See* lead oxides.

plutonium, Pu. At.no. 94, m.p. 641°C, b.p. 3232°C, *d* 19·82. Traces of Pu occur naturally being formed by neutron capture by natural uranium. Pu is normally prepared by neutron irradiation of ^{238}U [^{239}Pu (24000 years)]. ^{238}Pu (86 years) is formed from Np. Pu is separated by selective oxidation and solvent extraction. The metal is formed by reduction of PuF_4 with calcium; there are six crystal forms. ^{239}Pu is used in nuclear weapons and reactors; ^{238}Pu is used as a nuclear power source (e.g. in space exploration). The ionizing radiation of plutonium can be a health hazard if the material is inhaled.

plutonium compounds Plutonium chemistry is similar to that of uranium with a greater stability of the lower oxidation states but also an unstable +7 state formed by electrolytic oxidation in alkaline solution

$$PuO_2^{2+}\ (\text{pink}) \xrightarrow{+0{\cdot}91} PuO_2^{+}\ (\text{red}) \xrightarrow{+1{\cdot}17} Pu^{4+}\ (\text{green}) \xrightarrow{+0{\cdot}98} Pu^{3+}\ (\text{blue}) \xrightarrow{-2{\cdot}05} Pu\ (\text{volts in 1 M acid})$$

The metal reacts slowly with water and rapidly with dilute acids. In aqueous solutions the +3, +4, +5 and +6 oxidation states co-exist. PuF_6 (red–brown, m.p. 52°C, b.p. 62°C) is formed from fluorine and PuF_4. PuF_4 is prepared from PuO_2, HF and oxygen, $PuCl_4$ is known in the gas phase. All PuX_3 are known. The stable oxide is PuO_2; Pu_2O_3 is also known. Complex fluorides are formed by Pu(V) and Pu(IV) and chlorides by Pu(IV). The PuO_2^{2+} ion and PuX_4 form complexes with oxygen donors. $Pu(C_5H_5)_3$ has been prepared.

plywood Layers of wood bonded by adhesives (generally phenol-formaldehyde, urea-formaldehyde, melamine-formaldehyde resins). A laminate.

Pm Promethium.

pnictogens The elements nitrogen, phosphorus, arsenic, antimony and bismuth of Group V. Pnictides contain X^{3-} species.

Po Polonium.

point group A set of symmetry elements passing through or referring to a single point. Refers to molecules – in lattices where symmetry elements may not all pass through one point the corresponding arrangement is referred to as a space-group.

polarimeter An instrument used to measure the amount by which the plane of polarization of plane-polarized light is rotated on passing through an optically active medium.

polarizability When a molecule is placed in an electric field the electrons tend to be drawn away from the nuclei with the result that a dipole is induced in the molecule. The strength of the dipole divided by the field strength is termed the polarizability of the molecule. Such *electronic* polarizability can be determined from measurements of refractive index at wavelengths in the visible spectral region. Similar measurements in the far i.r. give, additionally, the *atomic* polarizability. Polarizability has units of volume.

polarization When molecules possessing dipole moments are placed in an electric field they tend to align themselves in a certain direction within the field. This effect is called orientation polarization. Simultaneously the electrons within each molecule are displaced slightly in the direction towards the positive pole of the field. This effect is called electron polarization. In addition, the nuclei having a positive charge are slightly displaced relative to each other. This atomic polarization is of very small magnitude. The total molecular polarization is the sum of the orientation, electronic and atomic polarizations.

polar molecule As the two electrons in a covalent link are not shared equally (different electronegativities), there is a separation of charge over the bond and, if the effects are not cancelled out, over the whole molecule. Such a molecule is known as a polar molecule. Thus in covalent hydrogen chloride $\overset{\delta+}{H}-\overset{\delta-}{Cl}$ the chlorine has a greater share of the bonding pair than the hydrogen and the molecule has a resultant dipole moment. Lone pairs of electrons also make molecules strongly polar.

polarography A highly developed electrochemical method of analysis, particularly suitable for dilute solutions of substances which are susceptible to electrolytic reduction at a mercury cathode. A dropping-mercury electrode (DME) is used and the polarogram, the current-voltage curve, for the solution is recorded. This polarogram shows a step for each reducible species present in solution. This step is of characteristic potential and of height proportional to the concentration of the component. In favourable cases several components may be detected and determined quantitatively from the polarogram. The method is particularly useful for determining trace amounts of metal and it can be used for investigations of complexes in solution and for studies in non-aqueous solvents.

polar solvent A solvent, the molecules of which have good solvating power. Generally, the greater the dipole of the molecules of the solvent the better the liquid as a solvent for ionic substances.

pollucite, $Cs(AlSi_2O_6),xH_2O$. The major mineral source of Cs. Contains (Al, Si)O_4 tetrahedra linked into a three-dimensional network.

polonium, Po. At.no. 84, at.wt. 210, m.p. 254°C, b.p. 962°C. A decay product from heavier elements but prepared by neutron irradiation of ^{209}Bi. ^{210}Po has a half life of 138 days, it is an α-emitter and is a very hazardous material. Separated from Bi by electrochemical methods and sublimed. The metal is obtained by thermal decomposition of the sulphide. Used to remove static electricity and as a neutron source.

polonium chemistry Polonium is an element of Group VI, electronic configuration $5d^{10}6s^26p^4$ and shows many similarities to tellurium. However polonium hydroxide and dioxide are more basic than the corresponding tellurium compounds, and the dihalides are quite stable. The *halides* PoX_4 and PoX_2 are known. The tetrapositive halides are obtained by the action of the metal with the halogen and form complex halides, M_2PoX_6. Both $PoCl_2$, and $PoBr_2$ are obtained by thermal degradation of the tetrahalides. *Polonium dioxide*, PoO_2, is obtained from the elements at 250°C, it is basic in character; a monoxide, PoO, and a trioxide, PoO_3 are also known. Lead and mercuric polonides, containing the Po^{2-} ion, have been prepared by interaction of the elements. Most polonium salts are highly coloured, being red or yellow. All polonium compounds are readily hydrolysed and reduced.

poly- Polymeric derivative. *See* here or under particular heading. E.g. for polyacrolein *see* acrolein polymers.

polyacetaldehyde *See* aldehyde polymers.

polyacetals Polymers formed from polyols, e.g. $HO(CH_2)_nOH$, and carbonyl derivatives, e.g. methanal CH_2O, which react to give $HO[(CH_2)_nOCH_2O]_x(CH_2)_n-OH$. Ring structures are also formed. Pentaerythritol and glutonaldehyde are typical components. Used in coatings.

polyacrylamide
Polymeric CH_2:$CHCONH_2$, linear polymers through the vinyl groups. High mol.wt. material, gels, gums, plastics soluble in water. By reaction at the amide group can be converted into polyelectrolytes or thermosetting resins. Widely used polymer, particularly because of water solubility. Substituted acrylamides, e.g. methacrylamide, CH_2:$CMeCONH_2$, give similar polymers. World production in excess of 500 tonnes p.a.

polyacrylonitrile Polymeric acrylonitrile, polymeric propenenitrile*, $(CH_2{\cdot}CHCN)_n$. Important fibre, elastomer. Frequently used as a co-polymer with, e.g. butadiene, styrene or vinyl chloride. Polymerization initiated by free radicals or by light. U.S. production 1983 300 000 tonnes.

polyalkenes Polymeric olefines. *See*, e.g. butene polymers, butyl rubber, ethene polymers, propene polymers.

poly(alkylene oxides) *See* 1,2-epoxide polymers.

polyalkylidenes A series of polymers,

$(CHR)_n$, formed, e.g. from the reaction of diazomethane and alcohols or hydroxylamine derivatives in the presence of boron compounds or with metal compounds. Polymethylene is formally the same as polyethene and the properties of the various polymers depend upon the degree of polymerization and the stereochemistry.

polyamides *See* Nylon.

polyamines, poly (alkylene polyamines), $(HN-R-NH-R')_n$. Hydrophilic polar substances formed by reacting alkylene polyamines or simple amines with alkylene dihalides. Used as flocculants, for, e.g. cellulose fibre and mineral ore suspensions.

polybenzimidazoles Heat-resistant polymers

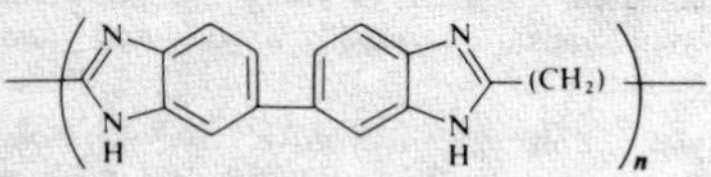

containing benzimidazole rings. The monomers are formed by condensation of aromatic diamine and dicarboxylic acid derivatives and the polymers are formed by condensation in a solution of P_2O_5 in phosphoric acid or a high boiling point solvent. Used in high strength adhesives and laminates.

poly blends Mixtures of structurally different polymers or co-polymers.

polycarbonates A group of thermoplastics

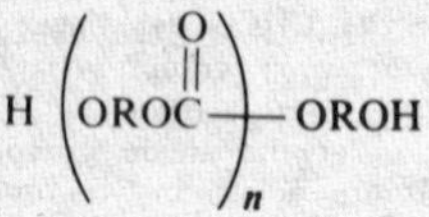

characterized by their toughness, high softening point and clarity. Polyesters of carbonic acid with aliphatic or aromatic dihydroxy compounds. A typical polycarbonate is formed from bisphenol A and phosgene.

polychloral *See* aldehyde polymers.

polychloroethylene *See* tetrachloroethene.

polychloroprene *See* neoprene.

polychlorotrifluoroethene *See* fluorine-containing polymers.

polydentate ligand *See* multidentate ligand.

polydispersion In most colloidal sols the dispersed particle sizes vary within very wide limits unless special means are adopted to exclude particles outside certain size limits. Such sols are said to be polydispersed.

polyelectrolyte A macromolecular compound containing many ionizable groups within the same molecule. A polyelectrolyte may be purely anionic or cationic, or may be amphoteric; the individual groups may be weak or strong acids. E.g. strongly acidic – polystyrene sulphonic acid; weakly acidic – polymethacrylic acid; strongly basic – polyvinylpyridinium bromide; weakly basic – polyvinylamine; amphoteric – polyglycine, proteins in general. Although the term is usually used for soluble substances, ion-exchange resins and fibrous proteins could be considered insoluble polyelectrolytes.

polyene antibiotics *See* macrolides.

polyenes Compounds with many carbon-to-carbon double bonds, e.g. carotenoids; *see* dienes.

polyesters Polymers formed by condensation of polyhydric alcohols (generally glycol or propylene glycol) and polybasic acids (generally maleic or terephthalic acid), i.e.

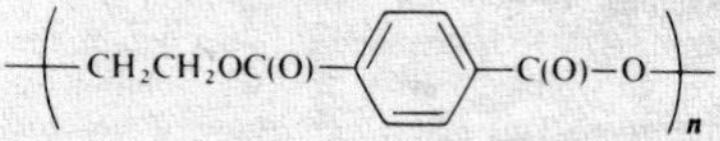

Used as fibres, particularly in textiles and film. Many other polyester polymers are of importance, e.g. unsaturated polyester resins from phthalic anhydride, propylene glycol and maleic anhydride used with reinforcement in boats, cars, etc. (alkyd resins). U.S. production 1983 1·7 megatonnes.

polyethers Polymers with $(C-O-C)_n$ units in the backbone. *See* e.g. aldehyde polymers, 1,2 epoxide polymers, poly(oxyphenylenes), phenoxy resins, polyacetals.

polyethylene glycols, $HO(CH_2CH_2O)_nH$. Polyether glycols having mol.wt. above 200. Prepared from ethylene oxide and water, dihydroxyethene or diethylene glycol plus base. Short chain members ($n = 1$, 2, etc.) are used as solvents. Polyethylene glycols are used in pharmaceuticals, and textiles.

polyethyleneimine *See* dihydroazirine.

polyformaldehyde *See* aldehyde polymers.

polyhalides Compounds containing complex anions formally formed by addition of halogens and interhalogens to halides. Anions of types $[AB_2]^-$, $[AB_4]^-$, $[AB_6]^-$ (B may be all the same or different (e.g. $[IClBr]^-$)). Polyiodides are specific polyhalides containing species such as I_3^-, I_7^-, I_9^-. Polyhalides often dissociate on heating the solid complexes.

polyhexafluoropropene *See* fluorine-containing polymers.

polyimides Film, plastics, wire enamels containing the imide group – C(O)NC(O) – derived from aromatic acids. These materials have good heat-resistant properties. Prepared from diamines and polycarboxylic acids, half esters, anhydrides (alkylene pyromellitimide from mellitic acid derivatives).

polyiodides *See* polyhalides.

polyketides Natural organic compounds regarded as arising from poly-β-ketonic intermediates derived by condensation of ethanoate units. Biosynthesis of fatty acids, many aromatic compounds, natural oxygen heterocyclics, tetracycline antibiotics and other fungal metabolites has been shown to occur via polyketide routes. Mevalonic acid is a special case, arising from only three ethanoate units.

polymeric reagents Solid, insoluble reagents for organic synthesis produced by introducing typical organic functional groups [e.g. Br, Li, SO_3H, CH_2Cl] on to a small proportion of benzene rings of cross-linked polystyrene. They allow selective reactivity, simplified reaction work-up (filtration) and high purity product isolation, low volatility of noxious reagents and polymer recycling. Catalyst systems in which active groups, e.g. PR_2, co-ordinated to metals and substituted on to a polymer are also available.

polymerization Any process in which two or more molecules of the same substance unite, to give a molecule (polymer) with the same composition as the original substance (monomer), but with a molecular weight which is an integral multiple of the original. Now the term includes any process that results in the formation of large molecules (macromolecules), consisting of repeated structural units. These structural units (described as 'mers') are contributed by the reacting monomers but, constitutionally, need not be identical with them.

Polymerization processes yielding polymers, whose mers are constitutionally identical to the reacting monomers are now classified as *addition polymerizations*. Thus styrene can be converted, by addition polymerization, to polystyrene:

$$PhCH=CH_2 \longrightarrow [PhCHCH_2-]_n$$

Such reactions can be initiated by free radicals, derived from compounds (initiators) such as benzoyl peroxide, ammonium persulphate or azobis-isobutyronitrile; or by ionic mechanisms by reactive ions, derived, for example, from BF_3 or $TiCl_4$. The process is also known in the plastics industry as 'vinyl polymerization' since the best-known examples are those involving vinyl or related monomers. Addition polymerizations are characterized by their extremely rapid rates, achieved at relatively low temperatures, especially in the case of ionic initiated polymerizations. The kinetics are typical of those expected of chain reactions. Industrially the process is important for the production of many polymers and copolymers used for the manufacture of plastics and synthetic rubbers.

In mass polymerization bulk monomer is converted to polymers. In solution polymerization the reaction is completed in the presence of a solvent. In suspension, dispersed mass, pearl or granular polymerization the monomer, containing dissolved initiator, is polymerized while dispersed in the form of fine droplets in a second non-reactive liquid (usually water). In emulsion polymerization an aqueous emulsion of the monomer in the presence of a water-soluble initiator is converted to a polymer latex (colloidal dispersion of polymer in water).

Condensation polymerization differs from addition polymerization in that the polymer is formed by reaction of monomers, each step in the process resulting in the elimination of some easily removed molecule (often water). E.g. the polyester polyethylene terephthalate (Terylene) is formed by the condensation polymerization (polycondensation) of ethylene glycol with terephthalic acid:

$$n\text{HO·}[CH_2]_2\text{·OH} + n\text{HO·CO}\text{—}\bigcirc\text{—}CO_2H$$

$$\downarrow$$

$$\left(\!\!\!-O\text{·}[CH_2]_2\text{·O·CO}\text{—}\bigcirc\text{—·CO}-\!\!\!\right)_n + n H_2O$$

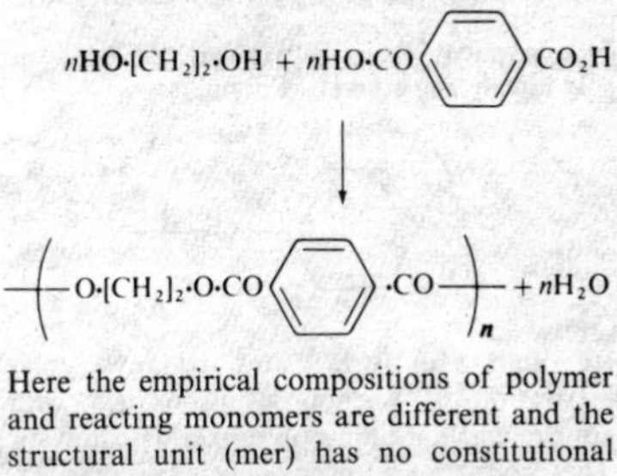

Here the empirical compositions of polymer and reacting monomers are different and the structural unit (mer) has no constitutional identity with the monomer components.

The kinetics of this type of polymerization are the same as for simple condensation; for this reason, the use of the term polycondensation is perhaps more appropriate. Unless kinetic evidence suggests otherwise, polymerizations involving the formation of chain polymers from cyclic compounds, following ring scission, are classed as condensation polymerizations. Some important condensation polymers are those derived from phenol, urea, or melamine and methanal; the polyesters; the polyamides (Nylons); the polysiloxanes (silicones).

polymethine dyes Dyestuffs consisting of two polar atoms (parts of heterocycles) joined by a methine chain $[Y(CH)_n=X]$ containing an odd number of atoms in the chain. Cyanine dyes belong to this class. The main use of polymethine dyes is as spectral sensitizers.

polymethylene glycols Compounds formed by the union of several molecules of methanal with a molecule of water. They are of the type

$$HO-CH_2OCH_2O\ .\ .\ .\ CH_2OH,$$

and are formed in aqueous solutions of methanal. The size of the molecule increases with increasing concentration of CH_2O. They are readily decomposed by heat to give CH_2O and show all its reactions.

polymorphism The existence of a substance in more than one crystal form, e.g. HgI_2 yellow-orthorhombic, red-tetragonal. The different forms are referred to as polymorphs. Substances which exist in two forms are referred to as dimorphic. Polymorphism of the elements is known as allotropy*. Thermodynamically unstable forms (e.g. the yellow form of HgI_2) are referred to as metastable*.

polymyxins A group of closely related antibiotics* produced by *Bacillus polymyxa*. Basic peptides, containing L-α,γ-diaminobutyric acid, L-threonine, two other amino-acids, and a fatty acid group.

poly(oxyphenylenes), poly(phenylene oxides), poly (phenylene ethers) Polymers,

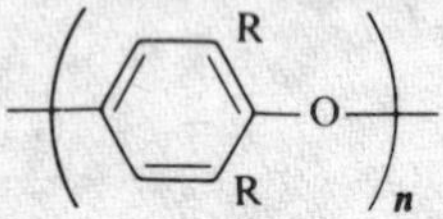

(R often Me) formed by oxidative polymerization of phenols using oxygen with copper and an amine (pyridine) as catalysts. The products are thermoplastics used in engineering applications and in electrical equipment.

poly(oxytetramethylene)glycols *See* tetrahydrofuran polymers.

poly(phenylene ethers) *See* poly (oxyphenylenes).

poly(phenylene oxides) *See* poly (oxyphenylenes).

poly (phenylene sulphides)

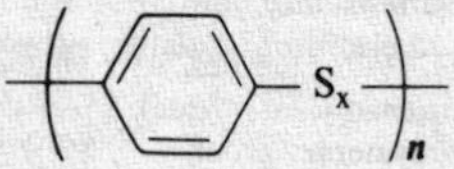

Formed commercially by the polymerization of S, Na_2CO_3 and PhX in a sealed container at 275–370°C, substituted derivatives are known but not fully evaluated. Used as high-temperature adhesives, for laminates and in coatings.

polyphosphates *See* phosphorus oxyacids.

polyprenols Primary monohydroxylic alcohols with a carbon skeleton composed of isoprene units linked head to tail. Isoprene residues are either all-*trans* or are a mixture of *cis* and *trans*. They are found in bacteria, yeasts, plants and mammals, and are grouped depending on their length and number of *cis* and *trans* residues. Solanesols are found in green leaves, dolichols in bacteria and mammals and betulaprenols in silver birch wood. Polyprenols play an important rôle as carriers (usually as their phosphates) in the transfer of sugars from nucleotide diphosphate sugars to a wide range of acceptors. The acceptors are usually membrane or wall-bound and include polysaccharides, glycopeptides, glycolipids, etc. Polyprenols are biosynthesized from mevalonic acid.

polysaccharides Carbohydrates derived from monosaccharides by the removal of $n-1$ molecules of water from n molecules of monosaccharides. All the higher carbohydrates are polysaccharides.

polysulphide polymers Polymers containing $-(SRSSRS)_n$ groups, formed from dihalides, e.g. bis-2-chloroethyl formal, $ClCH_2CH_2OCH_2OCH_2CH_2Cl$, and inorganic polysulphides (e.g. Na_2S_2) in aqueous suspension. They have good solvent resistance and elastomers are used in printing rolls, hose, diaphragms, etc. and paint spray.

polysulphides Species containing anions such as $[S_2]^{2-}$ (e.g. marcasite FeS_2), $[S_3]^{2-}$, $[S_4]^{2-}$, $[S_5]^{2-}$, $[S_6]^{2-}$ formed as aqueous solutions from sulphides and sulphur. Used in the production of polysulphide polymers*.

polysulphone Polyether derivative formed from bisphenol A and 44′-dichlorodiphenylsulphone.

polytetrafluoroethene, polytetrafluoroethylene, Teflon, Fluon, PTFE *See* fluorine-containing polymers. A very tough, translucent material of high softening point (320°C), excellent chemical resistance, low coefficient of friction and good electrical insulating properties.

poly(tetramethylene oxide) *See* tetrahydrofuran polymers.

polythionic acids, polythionates, $H_2S_nO_6$. A series of more or less unstable dibasic acids, obtained by the action of iodine solution on sodium sulphite, thiosulphate and mixtures of these, by the action of hydrogen sulphide on other members of the same series and of H_2S on H_2SO_3 at 0°C. The best known as alkali salts are: dithionic acid $H_2S_2O_6$, trithionic acid $H_2S_3O_6$, tetrathionic acid $H_2S_4O_6$, pentathionic acid $H_2S_5O_6$ and hexathionic acid $H_2S_6O_6$.

polyterephthalic acid *See* polyesters.

polyurethanes Condensation polymers formed from polyhydroxy compounds and polyisocyanates containing $[ROC(O)NHR^1-NHC(O)-O]_n$ repeating units. The main use is in foams. *See* isocyanates. U.S. use 1980 933 000 tonnes.

poly(vinyl acetals),

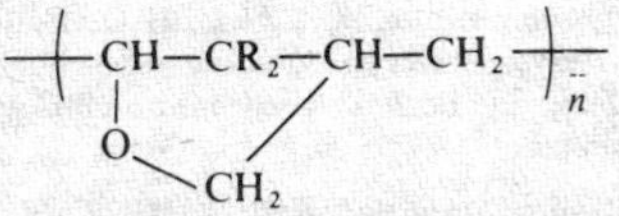

Resins formed from the reaction of poly(vinyl alcohol) with aldehydes. The formal derivative (from methanal) is used in wire coatings and adhesives and the butyral (from butanal) is used in metal paints, wood-sealers, adhesives and in safety glass interlayers.

poly(vinyl alcohol), $-(CHOH-CH_2)_n$. Prepared generally by ester interchange from polyvinylacetate (ethanoate) using methanol and base; also formed by hydrolysis of the acetate by NaOH and water. The properties of the poly(vinyl alcohol) depend upon the structure of the original polyvinyl acetate. Forms copolymers. Used as a size in the textile industry, in aqueous adhesives, in the production of polyvinyl acetates (e.g. butynal) for safety glasses. U.S. production 1980 64 000 tonnes.

porphobilinogen, $C_{10}H_{14}N_2O_4$. The im-

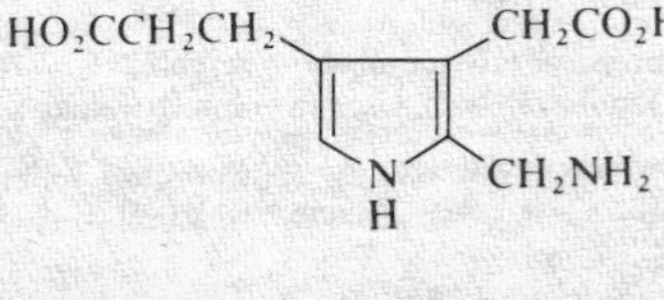

mediate biogenetic precursor of the porphyrins. It arises by the condensation of two molecules of δ-aminolaevulinic acid.

porphyrinogens Similar to porphyrins* except that they contain methylene bridges instead of the methine bridges between the pyrroles. The porphyrinogens are the actual biosynthetic intermediates and may be spontaneously oxidized to porphyrins.

porphyrins A group of naturally occurring pigments. Haemoglobin and other animal respiratory pigments and chlorophyll, the respiratory catalyst of plants, are compounds of porphyrins with metals. They are derivatives of the substance porphin which has the formula shown below, the rings and carbon atoms being numbered and lettered as indicated.

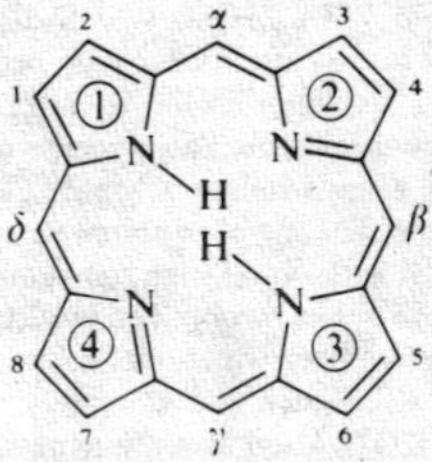

Important porphyrins include: *protoporphyrin,* $C_{34}H_{34}N_4O_4$, the porphyrin present in haem; *mesoporphyrin,* $C_{34}H_{38}N_4O_4$, obtained by treating haemin with hydriodic acid; *haematoporphyrin,* $C_{34}H_{38}N_4O_6$, obtained from haemin or haematin by the action of strong acids; *coproporphyrin,* $C_{36}H_{38}N_4O_8$, found in the faeces and in normal urine, in the serum of various animals, and in certain yeasts.

Portland cement A hydraulic cement* made from $CaCO_3$ and aluminium silicates.

positive ray analysis When a beam of positive particles (rays), formed by a high potential or electric discharge molecules, is passed through a magnetic and an electric field, the various ions are deflected to varying extents depending upon their velocities, masses and charges. Such analysis is the basis of a mass spectrometer.

positron, β^+. The positive counterpart of the electron. The ultimate indivisible unit of positive electricity.

positronium One electron and one positron; formally analogous to hydrogen.

post-actinide elements The elements following $_{103}Lw$ which should be part of a 6d transition series. Element 104 has been formed from

^{242}Pa with accelerated ^{22}Ne (^{260}104, $t_{\frac{1}{2}}$ ~ 0·3 sec and ^{259}104, $t_{\frac{1}{2}}$ 4·5 sec), from ^{249}Cf with ^{13}C and ^{12}C) (^{257}104 and ^{259}104), or from ^{248}Cm with ^{18}O (^{261}104, $t_{\frac{1}{2}} = 70$ sec). The element has been variously named Kurchatovium (Ku) and Rutherfordium (Rf). Element 105 has been formed similarly and named Hahnium (Ha). Element 106 has been claimed. All these elements have very short half-lives.

potash alum *See* alum.

potassamide, potassium amide, KNH_2. White solid, base on ammonia system (K in liquid NH_3 gives a blue solution which is a good reducing agent; this liberates H_2 and KNH_2 is formed on removal of excess NH_3).

potassium, K. At.no. 19, at.wt. 39·0983, m.p. 63·65°C, b.p. 774°C, *d* 0·86. Widely distributed in nature in silicate rocks, e.g. orthoclase, $KAlSi_3O_8$, in plants as the oxalate, tartrate and in blood, milk, etc., also occurs widely in salt beds, e.g. as carnallite, $KMgCl_3,6H_2O$ and in sea water. The metal is prepared by reduction of KCl with Na at high temperature and low pressure or can be obtained by electrolysis. It is a soft, silver-white metal, bcc, reacts violently with water and is readily oxidized (O_2 gives KO_2). Used as a reducing agent, an Na/K alloy may be used for heat transfer. Potassium compounds are used extensively as fertilizers (95%); KOH is used as an electrolyte in batteries, glass, ceramics, ^{40}K is radioactive. Total U.S. use of potassium salts (K_2O equivalent) 1982 2·4 megatonnes.

potassium acetate, KO_2CCH_3. *See* potassium ethanoate.

potassium antimonyl tartrate *See* antimony potassium tartrate.

potassium bromate, potassium bromate(V), $KBrO_3$. Formed from Br_2 and hot aqueous KOH. Decomposes to KBr and O_2 on heating. Used as a volumetric standard. *See* bromates(V).

potassium bromide, KBr. Obtained KOH or K_2CO_3 plus HBr or Br_2 plus KOH. M.p. 728°C, b.p. 1376°C. Used in photography and as a sedative.

potassium *t*-butoxide, $KOCMe_3$. M.p. 220°C. White solid (K in Me_3COH); very strong base. Hygroscopic.

potassium carbonate, pearl ash, K_2CO_3. Deliquescent powder made by modification of Solvay process using alcoholic solutions. Used in standard alkali solutions.

potassium chemistry Potassium is a typical alkali metal, electronic configuration $4s^1$. In its simple compounds it shows only the +1 oxidation state, generally 6-co-ordinate. Complexes are formed with water, NH_3 (relatively unstable) and macrocyclic ligands such as crown ethers.

potassium chlorate, potassium chlorate(V), $KClO_3$. Not very soluble in water, deposited from chlorate(V) solutions (Cl_2 plus $Ca(OH)_2$ or electrolysis of aqueous NaCl). On heating gives KCl and $KClO_4$ but decomposes to KCl at high temperatures.

potassium chloride, KCl. M.p. 776°C, sublimes 1500°C, *d* 1·984. Occurs naturally as sylvine and in sylvinite (KCl – NaCl), carnallite ($KMgCl_3,6H_2O$), kainite ($KCl{\cdot}MgSO_4,3H_2O$), 'hard salt' (KCl – NaCl – $MgSO_4$) and in many brines. Separated by fractional crystallization, soluble water and lower alcohols. Used in fertilizer production and to produce other potassium salts.

potassium chromate, K_2CrO_4. *See* chromates(VI).

potassium citrate, $K_3C_6H_5O_7,H_2O$. Colourless crystals; very soluble in water, slightly soluble in alcohol. Used medicinally as a diuretic.

potassium cyanate, KOCN. Obtained by PbO oxidation of aqueous KCN. Used to prepare cyanates.

potassium cyanide, KCN. M.p. 635°C. Prepared fused K_2CO_3 plus C plus NH_3 gas (Beilby's process). Hydrolysed in aqueous solution. Used to prepare cyanides. Very poisonous.

potassium dichromate, $K_2Cr_2O_7$. M.p. 396°C. Orange red crystals prepared from chromite, K_2CO_3 and CaO followed by acidification. Used in the preparation of chrome pigments, bleaching, as an oxidizing agent, in organic chemistry and in volumetric analysis.

potassium ethanoate, potassium acetate, KO_2CCH_3. Occurs in the sap of plants.

potassium ethoxide, KOC_2H_5. Base similar to potassium *t*-butoxide.

potassium ferricyanide, $K_3Fe(CN)_6$. *See* cyanoferrates(III).

potassium ferrocyanide, $K_4Fe(CN)_6$. *See* cyanoferrates(II). Used in volumetric analysis.

potassium fluoride, KF. M.p. 858°C, b.p. 1505°C, *d* 2·48. Obtained KOH or K_2CO_3

plus HF, has NaCl structure. Forms hydrates and adducts with HF. The latter acid salts are used as electrolytes in electrolysis to form fluorine. KF is poisonous in large quantities but is used in water treatment.

potassium hydrogen carbonate, $KHCO_3$. Less soluble in water than K_2CO_3, precipitated from an aqueous solution of K_2CO_3 and CO_2. Used in foods, medicine, fire extinguishers.

potassium hydrogen fluoride, KHF_2. *See* potassium fluoride.

potassium hydrogen sulphate, $KHSO_4$. Formed K_2SO_4 plus H_2SO_4. Other acid salts, e.g. $K_2SO_4{\cdot}3H_2SO_4$ are known.

potassium hydrogen tartrate, cream of tartar, $C_4H_5O_6K$. Colourless salt, soluble boiling water; occurs in grape juice, deposited as argol during fermentation. Used in baking powders (liberates CO_2 with $NaHCO_3$).

potassium hydroxide, KOH. M.p. 306°C, b.p. 1320°C. Prepared by electrolysis of KCl solution (also $Ba(OH)_2$ plus K_2SO_4 or H_2O/K amalgam). Forms hydrates. Strong base, used as electrolyte in batteries and as base.

potassium iodate, potassium iodate(V), KIO_3. Formed I_2 plus hot KOH solution. Forms acid salts, e.g. $KIO_3{\cdot}HIO_3$. Used in volumetric analysis.

potassium iodide, KI. Formed I_2 plus hot KOH solution. Soluble water and many organic solvents; dissolves iodine to solutions of polyiodides containing $[I_3]^-$, $[I_5]^-$, etc. Used medicinally.

potassium methoxide, $KOCH_3$. Similar to potassium *t*-butoxide*.

potassium nitrate, KNO_3. Prepared by fractional crystallization from a solution of $NaNO_3$ and KCl. Soluble in water, NH_3 and methanol. Used in gunpowder.

potassium nitrite, KNO_2. M.p. 440°C. Prepared KNO_3 plus Pb. Used in diazotization.

potassium, organic derivatives Potassium forms a limited range of almost ionic derivatives, e.g. $K^+[CPh_3]^-$. The alkyls are very reactive, and can be used in metallation reactions.

potassium oxalate, $C_2O_4K_2,H_2O$. The acid salt C_2O_4HK,H_2O is also known and decomposes to the quadroxalate or tetroxalate, $C_2O_4HK{\cdot}C_2O_4K_2,2H_2O$. Potassium oxalate (salts of sorrel, salts of lemon) is used for removing ink stains and iron mould. All formed from KOH or K_2CO_3 and oxalic acid.

potassium oxides Potassium burns in air to give mainly the orange superoxide KO_2 which decomposes to K_2O_2 and K_2O on heating. It gives H_2O_2, KOH and O_2 with water. Potassium peroxide, K_2O_2, formed from O_2 into a solution of K in liquid NH_3 at −60°C or by heating KO_2 gives H_2O_2 with acids. Potassium monoxide, K_2O, formed by heating KNO_3 with K or KN_3. Reacts violently with water to give KOH.

potassium perchlorate, potassium chlorate(VII), $KClO_4$. Formed by heating $KClO_3$ to 500°C or $HClO_4$ plus KOH. Relatively insoluble water (used for determination of K^+).

potassium periodate, potassium iodate(VII), KIO_4. Prepared KIO_3 solution plus Cl_2 and used as an oxidizing agent. Other iodates(VII), e.g. $K_4I_2O_9,9H_2O$, are known. *See* iodates.

potassium permanganate, potassium manganate(VII), $KMnO_4$. *See* permanganates. Used extensively as an oxidizing agent and as a volumetric reagent.

potassium peroxydisulphate, $K_2S_2O_8$. Obtained by double decomposition between $(NH_4)_2S_2O_8$ and K_2CO_3 or by electrolytic oxidation of K_2SO_4. The peroxymonosulphate, K_2SO_5, is also known.

potassium phosphates Many phosphates including K_3PO_4, $K_4P_2O_7$ and hydrogen phosphates are known.

potassium sulphate, K_2SO_4. Occurs naturally with other sulphates as glaserite, schönite* and syngenite, and purified by crystallization. Can be prepared from KOH and H_2SO_4. Used as fertilizer (U.S. production 500 000 tonnes p.a.) especially where Cl^- content must be kept down in irrigated soils. Forms acid salts, e.g. K_2SO_4. 1, 3, $\frac{3}{4}$, H_2SO_4, from solutions in aqueous H_2SO_4.

potassium sulphite, K_2SO_3. Formed KOH solution plus SO_2. Potassium pyrosulphite, $K_2S_2O_5$ is also formed from this solution.

potassium tetrafluoroborate, KBF_4. Precipitated from solution.

potassium thiocyanate, KSCN. Prepared KCN plus S (the reaction with a solution of S in, e.g., C_6H_6 or propanone is fast and quantitative). Used to prepare thiocyanates and in volumetric analysis.

potentiometric titration At the end-point in a titration there is a rapid change in the concentration of all of the reacting species. An inert electrode immersed in the solution thus shows a rapid change in e.m.f. as the end-point is approached and the end-point can be determined by observing the change in e.m.f. of the electrode during the titration. This method is especially useful when coloured solutions are used and the use of indicators is impractical.

potter's clay A term used for any clay or earth which can be used for making pottery, either alone or after admixture with non-plastic materials. The red colour of some pottery is due to iron compounds in the clay; these decompose on heating and liberate Fe_2O_3 or Fe(III) complexes which are red in colour. Other colours are produced by firing the kiln or oven under reducing conditions, or by adding coloured oxides to the clay.

powder metallurgy The techniques of sintering powders under pressure but below the melting point to produce solid objects. Used for metals, alloys and some plastics.

power kerosine A name sometimes given to tractor vaporizing oil (TVO)*.

Pr Praseodymium. Also propyl.

praseodymium, Pr. At.no. 59, at.wt. 140·91, m.p. 931°C, b.p. *c.* 3212°C, *d* 6·77. A typical lanthanide*. The metal is hcp (to 798°C) and bcc (to m.p.). Used in glasses and in thermoelectric materials.

praseodymium compounds
Praseodymium(III) compounds are typical lanthanide compounds* Pr^{3+} (f^2 pale green) $\longrightarrow$ Pr (−2·47 volts in acid). PrF_4 is formed by the action of HF on Na_2PrF_6 (F_2 on NaF/PrF_3). Pr(III) oxysalts are decomposed in air to Pr_6O_{11} and O_2 under pressure and give PrO_2 at 500°C – intermediate phases are known.

precipitation The formation of an insoluble compound from solution either by interaction of two salts, e.g.

$$NaCl + AgNO_3 \longrightarrow AgCl\downarrow + NaNO_3 \quad \text{(in water)}$$

or by temperature change effecting solubility. The formation of a precipitate is governed by the solubility product. In analysis precipitation is important in gravimetric and some forms of qualitative analysis. The form and purity of the precipitate depends upon the conditions used for the precipitation. Easily filtered precipitates are frequently formed by use of compounds which slowly release the precipitating reagent.

precipitation hardening, age hardening The hardening of certain alloys which takes place with time when a supersaturated solution tends to decompose with partial precipitation of the solute metal as an intermetallic compound.

precipitation indicator An indicator* which functions by giving a coloured precipitate at the end-point*. Thus K_2CrO_4 is used as indicator in the estimation of Cl^- by Ag^+; at the end-point there is a red precipitate of Ag_2CrO_4.

predissociation A term used in spectroscopy in connection with a certain type of molecular band spectrum, in which the bands are diffuse, showing no sharp convergence limit but merge gradually into a continuum. The behaviour is explained by supposing that the absorption of light excites the molecule initially to a metastable state, but after a few vibrations the molecule may dissociate.

prednisolone, 11*β*,17*α*,21-trihydroxypregna-1,4-diene-3,20-dione, $C_{21}H_{28}O_5$. Prepared from dihydrocortisone ethanoate, or by the action of certain micro-organisms on hydrocortisone. White crystalline powder. More potent than cortisone, it is used for the same purposes: as an immunosuppressive and for treating rheumatoid arthritis, rheumatic fever, etc. It is especially useful in the treatment of asthma.

pregnane, 17*β*-ethylaetiocholane, $C_{21}H_{36}$.

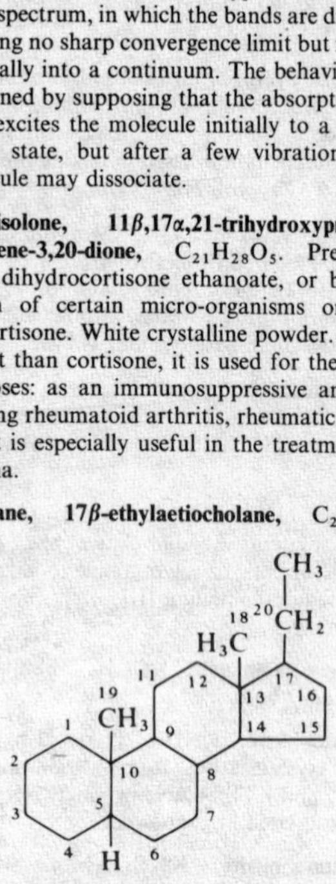

M.p. 83·5°C. Does not occur naturally, but has been obtained synthetically. It is the basic hydrocarbon skeleton of biologically and clinically important steroids.

pregnanediol, 5*β*-pregnane-3*α*,20*α*-diol, $C_{21}H_{36}O_2$. M.p. 238°C. There are four isomeric pregnane-3,20-diols differing only in the orientation of the hydroxyl groups at positions 3 and 20 and with the 5*β* configuration. Only the 3*α*,20*α* occurs naturally. It is formed by reduction of progesterone* in the liver and is the chief urinary metabolite of it, being excreted in the urine of women, as the glucosiduronate.

prehnitene 1,2,3,4-Tetramethylbenzene, $C_6H_2(CH_3)_4$.

premetallized dyes Dyestuffs are similar to metallizable dyes* except that the soluble metal complexes are formed prior to application to the fibre.

prephenic acid, $C_{10}H_{10}O_6$. An intermediate

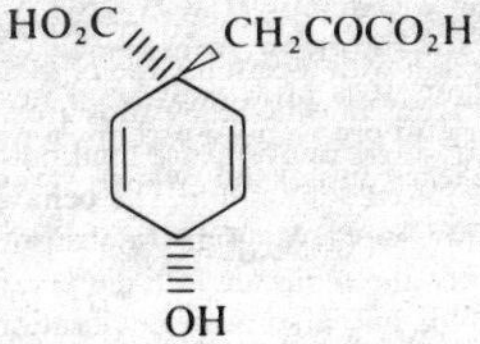

in the 'shikimic acid pathway' of aromatic biosynthesis, arising by rearrangement of chorismic acid*. It is the precursor of phenylpyruvic acid and *p*-hydroxyphenylpyruvic acid which yield phenylalanine and tyrosine respectively on transamination.

primeverose, $C_{11}H_{20}O_{10}$. Colourless crystals; m.p. 208°C. A disaccharide, glucose-6-*β*-D-xyloside, only found as a constituent of the glycosides gaultherin, rhamnicosin, primeverin, genticaulin, etc.

primitive lattice A lattice having only one equivalent point in the unit cell. Often a primitive lattice is expressed with the atom of molecule corresponding to the point at the corners of the unit cell.

principal quantum number *See* electronic configuration.

printing inks Normally mixtures of pigment (e.g. carbon black) dispersed in an oil of a suitable viscosity.

procaine, diethylaminoethyl-*p*-aminobenzoate, $NH_2{\cdot}C_6H_4{\cdot}COOC_2H_4{\cdot}N(C_2H_5)_2$, $C_{13}H_{20}N_2O_2$. M.p. ($2H_2O$) 51°C (61°C anhydrous). Its hydrochloride has m.p. 153–156°C and is a powerful local anaesthetic.

prochiral An atom C-linked to two identical ligands and to two other different ligands (C *aabc*) is said to be prochiral because if one of the ligands *a* is replaced by a ligand *d* different from *a*, *b* and *c*, a chiral centre is produced in (C *abcd*).

Procion dyes A group of azo dyestuffs which can form covalent bonds to cellulose by reactive groups.

producer gas A mixture of CO and N_2 produced by passing air through a bed of incandescent coke or coal in a *producer*. Carbon–air reactions are exothermic and in order to control the temperature in the producer the air is saturated with steam at about 50°C. This results in endothermic carbon–steam reactions – the water gas* reaction – which yields CO and H_2. The gas also contains some carbon dioxide and methane. The calorific value of producer gas from coke is of the order of 4600 kJ/m³ and from coal about 6400 kJ/m³.

progesterone, $C_{21}H_{30}O_2$. Crystallizes in two

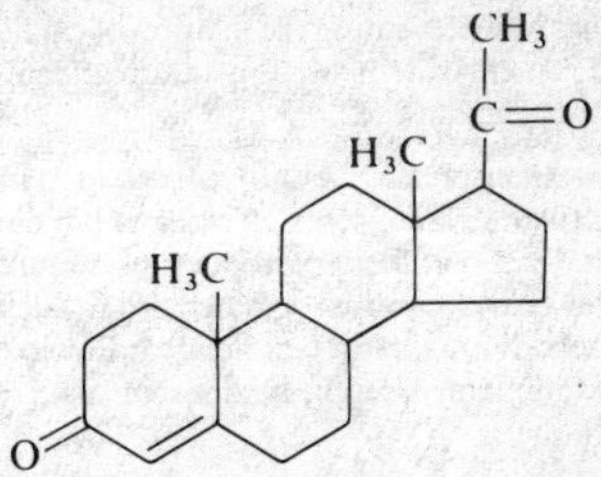

forms, i.e. prisms, m.p. 128·5°C; and needles, m.p. 121–122°C. Progesterone is the hormone of the corpus luteum and governs the growth and development of the uterus during pregnancy. It also exerts an antiovulatory effect when given during days 5 to 25 of the normal menstrual cycle: this is the basis for the use of some synthetic progestins as oral contraceptives.

Progesterone can be manufactured from cholesterol, from certain steroid sapogenins and from stigmasterol.

proguanil, $C_{11}H_{16}ClN_5$. Obtained as the

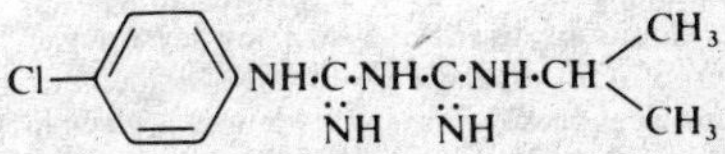

hydrochloride, m.p. 248°C, prepared by treating *p*-chlorophenyldicyandiamide with isopropylamine hydrochloride. It is used for the prevention and treatment of malaria.

prolactin One of the hormones obtained from the anterior lobe of the pituitary gland, the presence of which is necessary for the secretion of milk. It is a protein, with mol.wt. about 33 000.

prolamines *See* proteins.

proline, 2-pyrrolidinecarboxylic acid, $C_5H_9NO_2$ Colourless crystals; m.p. 220–222°C. The naturally occurring product is laevorotatory. While not strictly an amino-acid, it is commonly included in the list of those obtained by the hydrolysis of proteins.

promazine *See* chlorpromazine.

promethazine, 10-(2-dimethylaminopropyl)-phenothiazine, $C_{17}H_{20}N_2S$. Used as the hydrochloride, white powder, m.p. 223°C. Prepared by condensing phenothiazine with 2-chloro-1-dimethylaminopropane in the presence of sodamide and converting to the hydrochloride.

It is a powerful antagonist of histamine, antagonizing its effect on smooth muscle of the bronchioles, bladder and partially the intestines and preventing the dilation of capillaries. Promethazine is used in the treatment of allergic reactions.

promethium, Pm. At.no. 61, at.wt. (145), m.p. 1080°C, b.p. 2460°C. Traces of Pm occur naturally as a fission product of uranium, ^{147}Pm (2·64 years) is obtained from nuclear reactors. It is a typical lanthanide*. Previously called illinium. Used in luminescent paint for watch dials.

promethium compounds Promethium forms a single series of typical lanthanide compounds* in the +3 state

Pm^{3+} (f^4 rose pink) $\longrightarrow$ Pm
(−2·42 volts in acid)

proof spirit The legal standard used in assessing the strength of alcoholic liquors for excise purposes. Proof spirit is defined as a solution of ethanol in water which at a temperature of 51°F (283·5 K) weighs exactly 12/13 of an equal volume of distilled water. This corresponds to an alcoholic solution of relative density 0·9198 at 15·6°C (288·6 K) and containing 57·1% v/v or 49·2 w/w of ethanol.

Spirits are described as so many degrees over proof (OP) or under proof (UP). Sometimes, for alcoholic liquors, spirit strength is given in degrees proof, e.g. proof spirit is 100° proof and 70° proof 30° under proof (i.e. a solution containing 70% of proof spirit).

1,2-propadiene, allene, $CH_2=C=CH_2$, C_3H_4. Colourless gas prepared by the electrolysis of potassium itaconate, or by the action of zinc and alcohol on 1,3-dibromopropane. It is easily isomerized to propyne (methylacetylene), and is produced as a mixture with this substance from some reactions.

propanal, propionaldehyde, CH_3CH_2CHO. Colourless liquid; b.p. 48°C. Prepared by dehydrogenation of *n*-propanol over a catalyst, Fischer–Tropsch or oxo combination of C_2H_4, CO and H_2. Used for production of trimethylolethane by reaction with methanol (used in alkyd resins) and for hydrogenation to propanol or oxidation to propanoic acid.

propane, C_3H_8, $CH_3CH_2CH_3$. A colourless inflammable gas with a peculiar odour. M.p. −190°C, b.p. −44·5°C. It occurs in the natural gas from petroleum. Also obtained by reduction of propene. Has general properties of the paraffins. Used as a refrigerant and as a fuel.

propanedioic acid *See* malonic acid.

1,3-propane dithiol, $HS[CH_2]_3SH$. Liquid, b.p. 170–171°C, with an extremely disagreeable odour. It is principally used in organic chemistry for protecting a carbonyl group as a trimethylene dithioketal whilst transforma-

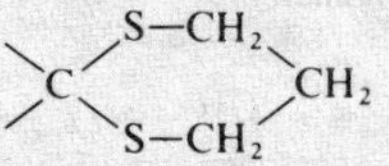

tions are effected at other centres of the molecule. The protecting group can be removed by treatment with dilute acid and $HgCl_2$.

propanoic acid, propionic acid, $C_3H_6O_2$, CH_3CH_2COOH. Colourless liquid with an odour resembling that of ethanoic acid; m.p. 24°C, b.p. 141°C. Occurs in the products of the distillation of wood. Prepared by the oxidation of propanol, propanal or by the reduction of propenoic acid. Used in production of esters and polymers.

1-propanol, *n*-propyl alcohol, C_3H_8O, $CH_3CH_2CH_2OH$. Colourless liquid with a pleasant odour; b.p. 97°C. It occurs in fusel oil and is separated by distillation. Also obtained by the hydrogenation of propene oxide. Oxidized to propanal and propanoic acid. Forms propene when heated over alumina. Used as a solvent and for the preparation of esters (propyl acetate).

2-propanol, isopropyl alcohol, $CH_3CH_2OHCH_3$. B.p. 82°C. Manufactured by hydrolysis of propene. Used in the production of acetone (propanone) by oxidation, for the preparation of esters (e.g. the ethanoate used as a solvent), amines (diisopropylamines, etc.), glycerol, hydrogen peroxide. The alcohol is used as an important solvent for many resins, aerosols, anti-freezes. U.S. production 1978 775 000 tonnes.

propanolamines *See* isopropanolamines.

propanone, acetone, dimethyl ketone, C_3H_6O, CH_3COCH_3. Colourless, volatile liquid with a pleasant, ethereal odour: highly inflammable, b.p. 56°C. It is largely manufactured by the dehydrogenation of 2-propanol over a copper catalyst at 500°C and 4 atm. By oxidation of cumene, phenol is also formed. Considerable quantities are also produced in the manufacture of other chemicals, notably glycerin, hydrogen peroxide and phenol. It occurs in significant amounts in blood and urine in certain pathological conditions. By oxidation of cumene, phenol is formed. Ammonia reacts with boiling propanone to give diacetonamine. Sodium hydroxide causes condensation to diacetone alcohol; under more vigorous conditions mesityl oxide and phorone are produced. These are also formed, together with mesitylene, by the action of small amounts of mineral acids. Reduction of propanone gives isopropanol and pinacol. Crystalline derivatives are formed with hydroxylamine, phenylhydrazine and semicarbazide. It is detected by the formation of iodoform when treated with iodine and sodium hydroxide. Propanone is used as a solvent and for the manufacture of methyl methacrylate (30%), methyl isobutyl ketone, methyl isobutyl carbinol and various other chemicals. U.S. production 1981 1·3 megatonnes.

propargyl alcohol *See* 2-propyn-1-ol.

propellants A general description for an explosive used to propel a rocket, bullet, shot or shell.

The word 'propellant' is also used for the liquefied gas in a pressurized aerosol container.

propenal, acrolein, acraldehyde, vinyl aldehyde, $CH_2{:}CH{\cdot}CHO$. A colourless, volatile liquid, with characteristic odour. The vapour is poisonous, and intensely irritating to eyes and nose; b.p. 53°C. It is prepared by the distillation of a mixture of glycerin, potassium sulphate and potassium hydrogen sulphate. It is manufactured by direct oxidation of propene or cross-condensation of ethanal with methanal.

When exposed to sunlight, it is converted to a white insoluble resin, disacryl. Oxidized by air to propenoic acid; small amounts of hydroquinone will inhibit this. Bromine forms a dibromide which is converted by barium hydroxide into DL-fructose. The acrid odour of burning fats is due to traces of propenal. It is used in the production of methionine and in controlled polymerization reactions to give acrolein polymers*.

propene, propylene, C_3H_6, $CH_3CH{=}CH_2$. Colourless gas; m.p. −185°C, b.p. −48°C. Formed by cracking petroleum. Prepared by passing the vapour of l-propanol over heated alumina. Reacts with sulphuric acid to give 2-propyl ether and 2-propanol, with halogens to give propene dihalides; with dilute chlorine water to give propene chlorohydrin, and with hydrogen in the presence of catalysts to give propane, with benzene to give cumene (route to phenol and propanone). Used in the manufacture of these compounds and others derived from them but principally to form polymers (70%) [polypropene (25%), polyacrylonitrile (15%)], cumene (10%) and solvents, Me_2CHOH (10%), propylene oxide (10%). U.S. production 1982 7 megatonnes. *See* propene polymers.

propenenitrile, acrylonitrile, vinyl cyanide, $CH_2{:}CH{\cdot}CN$. Volatile liquid; b.p. 78°C. Manufactured by the catalytic dehydration of ethylene cyanhydrin, by the addition of hydrogen cyanide to ethyne in the presence of CuCl or the reaction of propene, ammonia and air in the presence of a molybdenum-based catalyst.

Propenenitrile is very active, both in polymerization and in undergoing ready addition to compounds containing active hydrogen (e.g. the process of cyanoethylation). Polymers and copolymers of propenenitrile are industrially important as synthetic fibres, nitrile rubbers and as components in several thermoplastic compositions. It is an effective fumigant against stored grain insects. U.S. production 1978 800 000 tonnes.

propene polymers, polypropene, polypropylene An important group of polymers used as moulding resins and in extruded forms (e.g. film). Can be electroplated. Useful polymerization is by Ziegler catalysis and gives an isotactic material. U.S. production 1983 1·7 megatonnes.

propenoic acid, acrylic acid, vinylformic acid, $CH_2{:}CH{\cdot}COOH$. Colourless liquid having an odour resembling that of ethanoic acid; m.p. 13°C, b.p. 141°C. Prepared by oxidizing propenal with moist AgO or treating β-hydroxypropionitrile with sulphuric acid. Slowly converted to a resin at ordinary temperatures. Important glass-like resins are now manufactured from methyl acrylate, *see* acrylic resins. Propenoic acid itself can also be polymerized to important polymers – *see* acrylic acid polymers.

propenol, allyl alcohol, $CH_2{=}CHCH_2OH$. Colourless liquid with a pungent odour; b.p. 97°C. Prepared by heating glycerol with oxalic acid, or by reduction of propenal. Manufacture: (1) direct chlorination of propene at 500°C followed by hydrolysis of allyl chloride; (2) reaction of propenal and isopropanol (both derived from propene) at 400°C in the presence of a catalyst. It combines quantitatively with bromine, and may be estimated by this means. Forms insoluble compounds with mercury salts, and is oxidized to glycerol by potassium permanganate. Silver oxide produces propenal and propenoic acid.

propenyl isothiocyanate, allyl isothiocyanate, mustard oil, $C_4H_5N_5$, $CH_2{=}CHCH_2NCS$. Colourless liquid with an intensely pungent odour; b.p. 151°C. Forms the chief constituent of the oil derived from black mustard. Manufactured by treating a solution of propenyl iodide in alcohol with potassium thiocyanate. Reacts with ammonia to give propenyl thiourea. It is a strong vesicant and is used, suitably diluted, as a counter-irritant and rubefacient.

propenyl polymers *See* allyl polymers.

propenylthiourea, allylthiourea, thiosinamine, rhodallin, $C_4H_8N_2S$, $CH_2{:}CHCH_2NHCSNH_2$. Colourless crystalline solid with a faint garlic-like odour; m.p. 74°C. Manufactured by treating propenyl isothiocyanate with a solution of ammonia in alcohol. It has been given by injection in the treatment of conditions associated with the formation of excessive fibrous tissue. Toxic side reactions may occur. Propenyl thiourea is a chemical sensitizer for photographic silver halide emulsions.

β-propiolactone Colourless liquid; b.p.

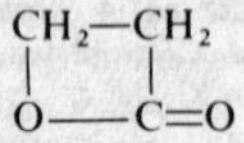

162°C (decomp.). Prepared by reacting ketene with methanol under carefully controlled conditions in the presence of anhydrous zinc chloride. This highly reactive compound has many synthetic uses, chiefly for adding the group $-CH_2CH_2-\overset{\overset{\displaystyle O}{\|}}{C}-$. It is a potent carcinogen.

propiolic alcohol *See* 2-propyn-1-ol.

propionaldehyde *See* propanal.

propionic acid *See* propanoic acid.

propranolol, 1-(isopropylamino)-3-(1-naphthyloxy)-2-propanol, $C_{16}H_{21}NO_2$. Used as the hydrochloride, $C_{16}H_{21}NO_2{\cdot}HCl$; m.p. 163°C. Propranolol is a competitive antagonist of the catecholamines* at various sites of action, most importantly in the heart. It is used to treat hypertension.

propyl The C_3H_7- group which occurs in two isomeric forms: normal propyl, or Pr^n, 1-propyl, $CH_3{\cdot}CH_2{\cdot}CH_2-$, and *iso* propyl, Pr^i, 2-propyl, $(CH_3)_2CH-$.

propyl alcohols *See* propanol.

propylene *See* propene.

propylene chlorohydrins, C_3H_7ClO. Both chlorohydrins are formed as colourless liquids by the action of dilute solutions of HOCl on propene, or by the action of HOCl on propylene oxide (1,2-epoxypropane). The product contains about 90% α-chlorohydrin and 10% β-chlorohydrin.

α-Propylene chlorohydrin, 1-chloro-2-hydroxypropane, $CH_2Cl{\cdot}CHOH{\cdot}CH_3$. B.p. 127°C. Used in organic syntheses particularly for propylene oxide.

β-Propylene chlorohydrin, 2-chloropropanol, $CH_3CHClCH_2OH$. B.p. 134°C.

Both of these chlorohydrins are converted to 1,2-dihydroxypropane by heating with solutions of sodium hydrogen carbonate; when heated with solid sodium hydroxide they give 1,2-epoxypropane. U.S. production 1976 4400 tonnes.

propylene dichloride, 1,2-dichloropropane, $CH_3CHClCH_2Cl$. Colourless liquid with a pleasant odour; b.p. 96°C. Manufactured by treating liquid chlorine with an excess of liquid propene. It is very similar in properties to ethylene dichloride, and is used for similar purposes.

propylene glycol *See* 1,2-dihydroxypropane.

propylene oxide, 1,2-epoxypropane Colour-

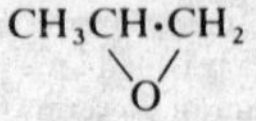

less liquid; b.p. 34°C. Manufactured by heating propylene chlorohydrin with solid NaOH or CaO. Reacts with water in the presence of sulphuric acid to give propylene glycol, and with alcohols and phenols to give ethers. It resembles ethylene oxide in properties, but is somewhat less reactive. Used in chemical syntheses particularly reactions with polyols to give polyglycols, and as a solvent for nitrocellulose. U.S. production 1979 930 000 tonnes.

propyne, allylene, methylethyne, $CH_3C{\equiv}CH$. B.p. −23°C. Prepared by the action of alcoholic potassium hydroxide on 1,2-dibromopropane, or by the reaction of dimethyl sulphate on sodium acetylide in liquid ammonia. The chemical properties of propyne resemble closely those of ethyne.

2-propyn-1-ol, propargyl alcohol, propiolic alcohol, $HC\vdots C{\cdot}CH_2OH$. B.p. 112°C. Easily polymerized by heat or alkali, and used widely in organic synthesis.

prostaglandins A group of related unsatur-

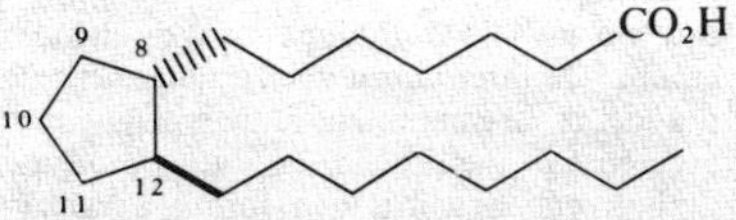

ated hydroxylated fatty acids occurring in mammalian organs, tissues and secretions and in some simple animals like the soft coral. The prostaglandins possess a broad spectrum of physiological activity including the ability to cause contraction of smooth muscle, regulate cell function and fertility, lower arterial blood pressure and inhibit aggregation of blood platelets. Unsaturated fatty acids like arachidonic acid* are biological precursors to prostaglandins which have now been chemically synthesized. *See also* thromboxanes.

prosthetic group The non-protein group of a conjugated protein. *See* enzymes*.

protactinium, Pa. At.no. 91, at.wt. 231·0359, m.p. 1565°C, *d* 15·4. Originally separated from pitchblende residues by co-precipitation with ZrO_2 followed by removal of Zr as $ZrOCl_2$ and of Ta as the soluble peroxide. Now separated from residues from uranium extraction plants by solvent extraction. The most stable isotope is ^{231}Pa (32 340 years). The metal has been obtained by reduction of PaF_4 with barium at 1400°C or PaC plus a deficiency of iodine. The silvery metal has a modified bcc structure. Protactinium compounds are very hazardous to health.

protactinium compounds In its compounds Pa shows oxidation states of +5 and +4

$Pa^{5+} \xrightarrow{-0{\cdot}1} Pa^{4+} \xrightarrow{-0{\cdot}9} Pa$ (volts in aq. solution)

the +5 state is most stable. All the halides PaX_5 and PaX_4 are known; brown PaF_4 is formed from Pa_2O_5, excess H_2 and HF; F_2 reacts with PaF_4 to give colourless PaF_5, $PaCl_5$ is yellow, volatile (sublimes 200°C) solid prepared from $SOCl_2$ and Pa_2O_5 at 350–500°C; it is reduced to $PaCl_4$ by H_2 at 400°C. All of the halides are hydrolysed to hydrates, then oxide halides $PaOX_2$ and $PaOX_3$, and eventually to hydrated oxides. Complex halides such as K_2PaF_7 are formed from solution and also complexes, e.g. between the halides and N-and O-donors. The hydrated pentoxide on heating gives Pa_2O_5, $PaO_{2\cdot3}$, then PaO_2 although there are very complex relationships in the Pa – O series. $Pa(C_5H_5)_4$ has been prepared from $PaCl_4$ and KC_5H_5. Pa metal gives PaH_3 with hydrogen.

protamines *See* proteins.

protecting group A group which is added to one functional group in a molecule before chemical treatment of a different functional group. A protecting group may prevent unwanted reaction at the site it is protecting or it may serve to direct a reagent to another site. Protecting groups are usually easily attachable without causing racemization at chiral centres and can be removed under conditions where the remainder of the molecule is unaffected. Protecting groups are used extensively in synthesizing peptides from amino-acids.

protective colloids Hydrophilic colloids e.g. gelatin, being themselves unaffected by small concentrations of electrolytes, due to hydration, are able, when added in very small amounts to protect hydrophobic sols from the coagulating influence of electrolytes. These are therefore termed protective colloids.

All hydrophilic colloids possess some degree of protective action and gelatin, starch and casein are used commercially for this purpose.

proteins The proteins are the chief nitrogenous constituents of living organisms. They contain about 50% carbon, about 25% oxygen, about 15% nitrogen, about 7% hydrogen and some sulphur. They are formed from a mixture of amino-acids* which they give on hydrolysis. Proteins are precipitated by alcohol, propanone and by strong solutions of salts. They give certain colour reactions which are caused by the presence of specific amino-acids in the molecule.

The proteins can be classified as follows but one particular protein may fall into more than one class.

Simple proteins:

Albumins. Soluble proteins both in water and in dilute aqueous salt solutions; found in all living tissue. Typical albumins are ovalbumin from eggs and lactalbumin from milk.

Globulins. Proteins insoluble in water, soluble in dilute salt solutions. They include such proteins as myosin from muscle, fibrinogen from blood and edestin from hemp.

Protamines. Strongly basic, low mol. wt. proteins which contain high levels of arginine, but no sulphur-containing amino-acids. They are soluble proteins, associated with nucleic acids and are obtained in large quantity from fish spermatozoa.

Histones. Like protamines, they are basic

proteins, but contain a greater variety of amino-acids. The histones are usually found in combination with the nucleic acid in the nuclei of the somatic cells of many organisms. They are not found in plants but can be extracted in large amounts from the thymus and pancreas. It is possible that their rôle is to act as specific inhibitors of nuclear genes.

Prolamines. Proteins insoluble in water, but dissolving in aqueous alcohol solutions. Found in the seeds of cereals.

Glutelins. Also insoluble in water and in 7% aqueous ethanol. Soluble in acids and alkalis. Also found in cereals.

Scleroproteins. Insoluble proteins obtained from the skeletal and connective tissues of animals. Typical classes are keratins* collagens* and elastin* classes.

Conjugated proteins Proteins with a prosthetic group*:

Phosphoproteins. Contain phosphates and include casein from milk and phosvitin from egg yolk.

Chromoproteins. Compounds of proteins with haem or some similar pigments.

Nucleoproteins. The prosthetic group of the nucleoproteins is nucleic acid, often linked through salt linkages with protamines or histones. The nucleoproteins are present in the nuclei of all cells. Chromasomes are largely nucleoproteins and some plant viruses and bacteriophages have been shown to be pure nucleoproteins. *See also* histones.

Lipoproteins. The lipid moiety of lipoproteins is quite variable both qualitatively and quantitatively. The α-lipoprotein of serum contains glyceride, phosphatide and cholesterol to about 30–40% of the total complex. The β-lipoprotein of serum contains some glyceride but the phosphatide and cholesterol account for nearly 75% of the total.

Glycoproteins or mucoproteins. Compounds of proteins with carbohydrates. All glycoproteins contain a hexosamine and usually sulphate, ethanoate and glucuronic acid. The carbohydrate–protein linkages are, in some cases covalent and in others of the salt type. Aqueous solutions of glycoproteins are extremely viscous.

The enzymes may be classified under some of the above headings.

Proteins consist of large numbers of amino-acids joined by the peptide link $-CO-NH-$ into chains, as shown in the diagram, where R′ and R″ are amino-acid residues. These chains are called peptides and may be broken into smaller chains by partial hydrolysis (*see* peptides). Proteins may contain more than one peptide chain: thus insulin consists of

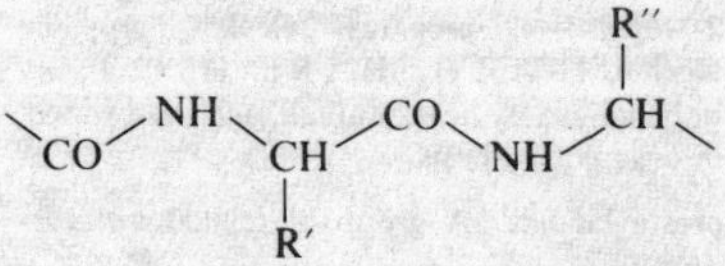

two chains joined through two disulphide groups. The three-dimensional arrangement of peptides is very important in determining the properties of the protein. In fibrous proteins the polypeptide chain is in the form of a regular helix, called the α-helix. Chains may be cross linked through hydrogen bonds; there may also be linkages to adjacent chains through amino-acid residues, particularly cysteine. Globular proteins consist of chains folded round themselves.

Modern methods of amino-acid and peptide analysis, have enabled the complete amino-acid sequence of a number of proteins to be worked out. The grosser structure can be determined by X-ray diffraction procedures. Proteins have molecular weights ranging from about 6 000 000 to 5 000 (although the dividing line between a protein and a peptide is ill defined). Edible proteins can be produced from petroleum and nutrients under fermentation.

prothrombin *See* thrombin.

protocatechuic acid, 3,4-dihydroxybenzoic acid, $C_7H_6O_4$. Crystallizes with $1H_2O$; m.p. 199°C. It occurs in the free state in the onion and other plants; is a constituent of one group of tannins, and is a product of the alkaline decomposition of resins.

protogenic *See* amphiprotic.

proton One of the units from which all forms of matter are built. The proton is identical to the nucleus of the hydrogen atom having a mass of one atomic unit and a unit positive charge equivalent to the negative charge on the electron. The proton is 1840 times the weight of the electron.

protonation Addition of a proton, e.g. H_2O is protonated by HCl to form $[H_3O]^+Cl^-$.

protonic acid Molecules which give H^+ by dissociation, e.g. HCl, H_2SO_4.

proton motive force *See* chemiosmotic hypothesis.

protophilic *See* amphiprotic.

protoporphyrin *See* porphyrins.

Prussian blue, $KFeFe(CN)_6$. *See* cyanoferrates.

prussiates An old name for cyanides*.

prussic acid, HCN. An old name for hydrogen cyanide*.

pseudobrookite structure A structure adopted by some A_2BO_5 materials. A has between 4- and 6-co-ordination, B is octahedrally co-ordinated. E.g. Fe_2TiO_5, Al_2TiO_5.

pseudocumene, 1,2,4-trimethylbenzene, $C_6H_3Me_3$. Liquid produced by catalytic cracking of methyl benzenes. Used to prepare trimellitic anhydride.

pseudohalogens Compounds which show a resemblance to the halogens in their reactions and in the properties of their compounds. Pseudohalogens X_2 form X^-, XO^- and XO_3^- anions; AgX compounds are insoluble in water. E.g. $(CN)_2$, $(SCN)_2$, $(SeCN)_2$, $(SCSN_3)_2$.

pseudoionone *See* citral.

pseudomorphic Having the same crystal form.

psychotomimetic drugs, psychopharmacological agents Drugs which produce changes in thought, perception and mood without seriously disturbing the autonomic nervous system. Examples are mescaline*, lysergic acid diethylamide (LSD-25*), psilocybin and the active constituents of marihuana, notably tetrahydrocannabinol*.

Pt Platinum.

PTA p-Terephthalic acid.

pterins Derivatives of pteridine which include

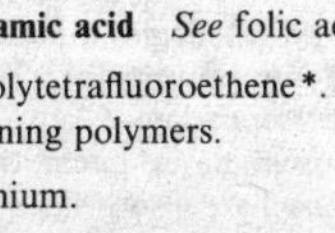

the butterfly pigments leucopterin and xanthopterin and the vitamin folic acid*.

pteroylglutamic acid *See* folic acid.

PTFE Polytetrafluoroethene*. *See also* fluorine-containing polymers.

Pu Plutonium.

pulp Fibrous cellulose material made from wood, cotton, linen, straw, etc., by mechanical disintegration with possibly some chemical treatment or by chemical treatment (sodium sulphide and sodium hydroxide or sodium hydrogen sulphite or bisulphite). Pulp is used in paper, rayon, cellulose production.

pump A device, usually mechanical, for moving a fluid against a pressure.

purine, $C_5H_4N_4$. A crystalline solid; m.p.

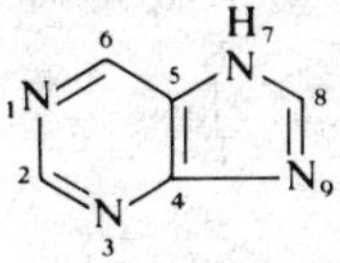

216–217°C. It can be prepared from uric acid, but is not obtained naturally, and is not physiologically important, although many of its derivatives are. It is the parent compound of a group of compounds of animal and vegetable origin, collectively called purines. The group includes adenine and guanine, which are constituents of the nucleic acid portion of nucleoproteins, their breakdown products, hypoxanthine, xanthine and uric acid, and the drugs caffeine, theobromine and theophylline.

Purple of cassius *See* gold.

putrescine, tetramethylenediamine, 1,4-diaminobutane, $H_2N{\cdot}[CH_2]_4{\cdot}NH_2$. M.p. 27–28°C, b.p. 158–159°C. Putrescine is found, associated with cadaverine, in putrefying tissue being formed by bacterial action on the amino-acid arginine. It is found in the urine in some cases of the congenital disease cystinuria. It is also present in ergot.

putty powder *See* tin oxides.

PVC Polyvinyl chloride. *See* vinyl chloride polymers.

pyknometer An instrument for measuring the density of a substance by determining the weight of a known volume.

pyranose The stable ring form of the sugars

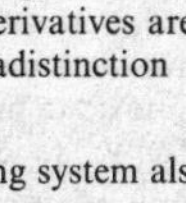

which contains a C_5O ring skeleton. Glucose is structurally a 1,5-glucopyranose with a primary alcohol group CH_2OH as a side chain. Such pyranose derivatives are stable and crystalline in contradistinction to the furanose sugars*.

pyrazine The ring system also known as 1,4-diazine.

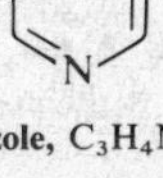

pyrazole, 1,2-diazole, $C_3H_4N_2$. A colourless

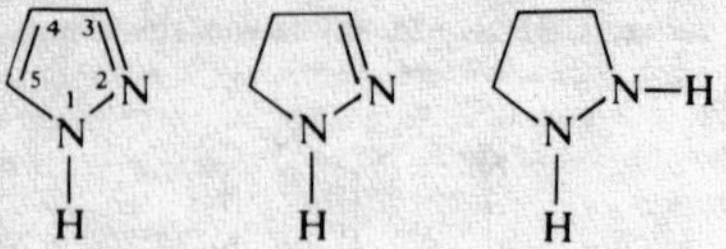

1 2 3

crystalline substance (1) m.p. 70°C. Obtained by passing ethyne into cold ethereal solutions of diazomethane. It is a weak base. Partial reduction gives pyrazoline (2) and complete reduction leads to pyrazolidine (3). Pyrazole is an aromatic compound, undergoing electrophilic substitution in the 4-position.

pyrazolidine *See* pyrazole.

pyrazoline *See* pyrazole.

pyrazolinones *See* pyrazolones.

pyrazolones, pyrazolinones, oxopyrazolines 5-oxo-1,3-substituted pyrazolines. Used extensively as dyes and pigments, in colour photography and in pharmaceuticals. Prepared from acetoacetic ester and phenylhydrazines (e.g. Fast Light yellow G).

pyrene Colourless crystals; m.p. 150–151°C.

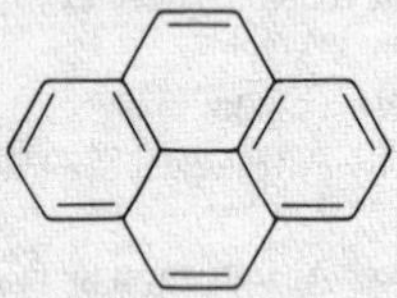

A tetracyclic benzenoid hydrocarbon suspected of being a carcinogen.

pyrethrins Constituents of pyrethrum.

pyrethrolone A ketonic alcohol, esters of which, with chrysanthemum carboxylic acids, are the pyrethrins A cinerolone with side chain $-CH_2CH=CH{\cdot}CH=CH_2$.

pyrethrum The mixture of substances known as pyrethrum, obtained by grinding or extracting the dried flowers of *Chrysanthemum cinerariaefolium*, is a powerful but non-persistent contact insecticide with rapid 'knockdown' effect. Being non-toxic to mammals it is used in the home and in the food industry, usually in conjunction with a synergist such as piperonyl butoxide*.

Pyrex A trade name for a heat-resisting borosilicate glass containing a high percentage of SiO_2 with some B, Al and alkalis. Pyrex has high mechanical strength, resists attacks by strong acids and alkalis, and withstands sudden changes of temperature without breakage; it softens at a higher temperature than ordinary soda glasses.

pyridine, C_5H_5N. Colourless refractive hy-

groscopic liquid; b.p. 115·3°C. Strong characteristic smell. Burns with a smoky flame. Strong base, forms stable pyridinium salts with mineral acids and quaternary compounds with alkyl halides, e.g. 1-methylpyridinium iodide $[C_5H_5NCH_3]I$. Forms a characteristic picrate and complexes with most metal salts. Reduced to piperidine, $C_5H_{11}N$, by sodium and alcohol, oxidized to pyridine-N-oxide by per-acids. Very poisonous.

Commercially, pyridine is manufactured from ethyne and ammonia. It is used as a solvent, particularly in the plastics industry, in the manufacture of nicotinic acid, various drugs and rubber chemicals.

pyridinium The ion formed by *N*-co-ordina-

tion of pyridine to a proton.

pyridoxal phosphate The coenzyme, or more

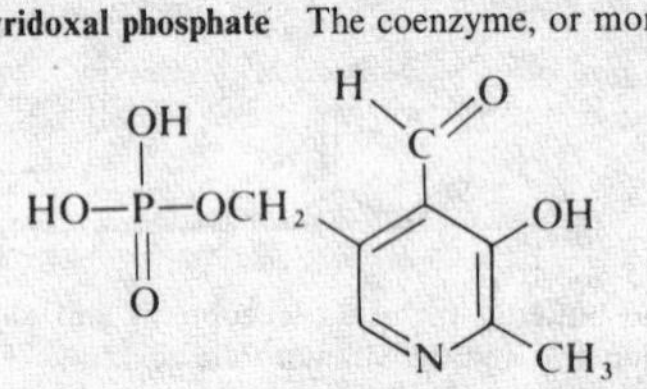

correctly the prosthetic group, of transaminations, i.e. the interconversions of amino-acids and their corresponding α-keto-acids, catalysed by transaminases (aminotransferases). Pyridoxal phosphate also plays an important rôle in the decarboxylation of amino-acids to 'biogenic amines'.

pyridoxine, vitamin-B₆, 2-methyl-3-hydroxy-4,5-hydroxy-methylpyridine, $C_8H_{11}NO_3$. Colourless needles; m.p. 160°C. Pyridoxine is needed by rats to cure dermatitis developed on a vitamin B-free diet, supplemented by thiamine and riboflavin. Its absence from the diet is also associated with anaemia. It is needed also by certain bacteria. It is present in rice husks, maize, wheat germ, yeast and other sources of vitamin B.

The related compounds pyridoxamine and pyridoxal, in which the CH_2OH group in the 4-position is replaced by CH_2NH_2 and CHO respectively, also possess vitamin B_6 activity and for certain bacteria are much more active than pyridoxine.

pyrimidine, $C_4H_4N_2$. Crystalline compound

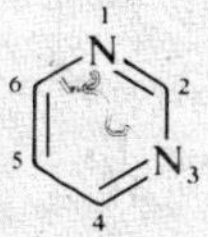

with a penetrating smell, m.p. 20–22°C, b.p. 124°C. It can be prepared from barbituric acid via trichloropyrimidine.

It is the parent substance of a group of compounds which includes cytosine, thymine and uracil, which are constituents of nucleic acids and barbituric acid and its derivatives, which are important medicinally.

pyrites, FeS_2. A common mineral, brass-yellow with a metallic lustre. The structure contains $S_2{}^{2-}$ species. Used for production of H_2SO_4.

pyro acids Acids or anions derived from hypothetical acids containing two oxygroups with a bridging (μ) oxygen, e.g. $[O_3SOSO_3]^{2-}$, pyrosulphate.

pyrocatechol *See* 1,2-dihydroxybenzene.

pyrochlore, pyrochlorite, $NaCaNb_2O_6F$. A niobium mineral sometimes containing some Ta.

pyrogallol, 1,2,3-trihydroxybenzene, $C_6H_6O_3$. White lustrous needles, m.p. 132°C, b.p. 210°C (decomp.). Alkaline solutions rapidly absorb oxygen from the air and become dark brown in colour. It is prepared by digesting gallic acid with water at 200°C. Extensively used as a photographic developer, and in gas analysis as an absorbent for oxygen.

2-pyrolidone, C_4H_7NO. M.p. 25°C, b.p.

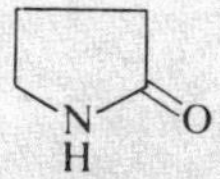

245°C. Manufactured from butyrolactone and ammonia. Easily hydrolysed to 4-aminobutanoic acid, its most important use is for the formation of N-vinylpyrrolidone* by reaction with ethyne.

pyroligneous acid The crude brown liquor obtained by the distillation of wood. Contains ethanoic acid. Methanol and propanone are also present.

pyrolusite, β-MnO_2. An important manganese mineral, iron-grey in colour with the rutile structure.

pyromellitic acid, 1,2,4,5-benzenetetracarboxylic acid, $C_6H_2(CO_2H)_4$. Prepared by oxidation of durene. M.p. 260°C (decomp.). Readily converted to pyromellitic dianhydride, m.p. 287°C, used as a cross-linking agent for epoxy resins and also with diamines giving polyimides with excellent high-temperature resistance.

pyromucic acid *See* furoic acid.

pyrones Compounds containing the ring

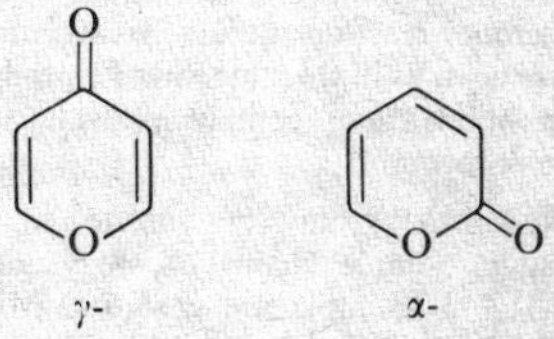

systems which occur in nature, e.g. as part of the anthocyans.

pyrophoric metals Many metals when produced in a porous condition or finely divided state by reduction at low temperatures are in a state of considerable activity and are often spontaneously inflammable. Thus pyrophoric iron can be obtained by reduction of iron(II) oxalate and many others by distillation of amalgams at low pressures.

Pyrophoric alloy is a Ce–Fe alloy used in lighters.

pyrophosphates Salts containing $[P_2O_7]^{4-}$ species. *See* phosphorus oxyacids.

pyrosols The electrolysis of fused salts results in cloudy solutions called pyrosols which are generally regarded as colloidal, the free metal being the dispersed material. Pyrosols are easily produced by dissolving the metal directly in the fused salt, e.g. zinc in fused zinc chloride.

pyrosulphuric acid, $H_2S_2O_7$. Present in fuming sulphuric acid. Pyrosulphates contain $[S_2O_7]^{2-}$ ions. *See* sulphur oxyacids.

pyrosulphurous acid, $H_2S_2O_5$. The free acid is unknown but pyrosulphites contain $[S_2O_5]^{2-}$ $[O_3SS(O)_2]^{2-}$ ions. *See* sulphur oxyacids. Formed from sulphites, $[SO_3]^{2-}$, and SO_2.

pyrotartaric acid, methylsuccinic acid,

$$\begin{array}{l} CH_3CHCOOH \\ \quad\;| \\ \quad CH_2COOH \end{array}$$

$C_5H_8O_4$. Colourless, crystalline solid; m.p. 115°C. Prepared by the dry distillation of tartaric acid or by reduction of itaconic or citraconic acids. Forms an anhydride when heated to 200°C.

pyrotechnics Combustible mixtures, including illuminating, incendiary, priming, warning and smoke-producing compositions. They contain an oxidant, such as nitrate or chlorate, and combustible substances such as charcoal, sulphur, antimony sulphide, etc. Metallic powders are incorporated to give illuminating effects and metallic salts provide coloured effects. In incendiary compositions a highly exothermic reaction occurs giving rise to intense heat, as in thermite. Smoke-producing compositions may contain phosphorus or hexachloroethane.

pyroxenes A group of silicate minerals containing simple chains of stoicheiometry $(SiO_3)_n^{2n-}$. Example are enstatite, $MgSiO_3$ and diopside $CaMg(SiO_3)_2$.

pyrrocoline ring system *See* indolizine ring system.

pyrrole, C_4H_5N. Colourless oil, b.p. 130°C.

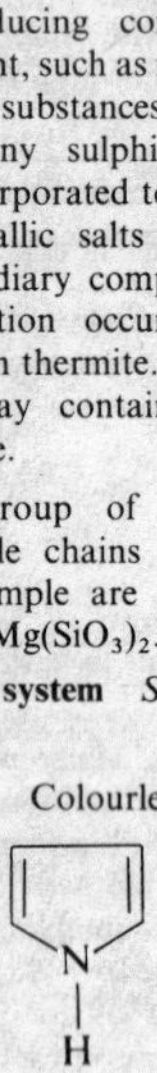

Prepared by heating ammonium mucate, or from butyne-1,4-diol and ammonia in the presence of an alumina catalyst. The pyrrole molecule is aromatic in character. It is not basic and the imino-hydrogen atom can be replaced by potassium. Many pyrrole derivatives occur naturally, e.g. proline, indican, haem and chlorophyll.

pyrrolidine, tetrahydropyrrole,

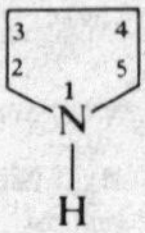

C_4H_9N. Almost colourless, ammonia-like liquid, b.p. 88–89°C, which fumes in air. Strong base. It occurs naturally in tobacco leaves, but is made industrially by hydrogenation of pyrrole.

2-pyrrolidone, 2-oxypyrrolidine, C_4H_7NO. B.p. 215°C. Prepared by electrolytic reduction of succinimide.

pyruvic acid, 2-oxopropanoic acid, $CH_3COCOOH$. A colourless liquid with an odour resembling that of ethanoic acid, m.p. 13°C, b.p. 65°C/10 mm. It is an intermediate in the breakdown of sugars to alcohol by yeast. Prepared by distilling tartaric acid with potassium hydrogen sulphate. Tends to polymerize to a solid (m.p. 92°C). Oxidized to oxalic acid or ethanoic acid. Reduced to (±)-lactic acid.

pyruvic alcohol *See* acetol.

Q

quadrupole moment The coupling of the nuclear quadrupole with the asymmetric electron fields in compounds. Measurement of quadrupole moments yields information about the electronic distribution over the atoms within a molecule or ion.

quadruple point The unique conditions of temperature and pressure at which four phases of a two-component system exist in equilibrium.

qualitative analysis Analysis for the identification of constituents.

quantitative analysis Analysis for the estimation of constituents.

quantum According to Planck's quantum theory the energy associated with electromagnetic radiation consists of discrete units, quanta; which, for light of given frequency, have definite energies. Thus the transfer of energy can only occur in definite parcels, or quanta, and not continuously as a stream of fluids. For radiation of frequency v, the unit of energy, the quantum, is equal to hv, where h is Planck's constant. When a quantum has particle properties it is known as a photon*.

quantum efficiency In radiation induced processes the actual number of species which are decomposed or reacted per quantum of energy absorbed.

quantum number The planetary electrons in an atom or the atoms in a molecule, possess energy due to type, rotation, vibration or spin. According to the quantum theory the energy possessed by such a particle is quantized, i.e. can only have certain definite values. Each such value for a particular form of energy is a multiple – which is invariably a small whole number, or sometimes one half – of a quantum, which is characteristic of the form of energy under consideration. This multiple which defines the energy of the particle, is called the quantum number.

quark Fundamental subatomic particle.

quartering *See* coning and quartering.

quartet A group of four closely spaced lines in a spectrum (n.m.r., etc.).

quaternary salts The products of reaction of, e.g., tertiary amines or phosphines with alkyl halides. Ionic salts. The quaternary ammonium salts are used extensively in fabric softening (cations react with fabrics), cationic emulsions, as germicides, antistatic agents.

quartz, SiO_2. The low temperature form of silica*. Quartz is used extensively for optical components, it is more transparent to u.v. radiation than is glass. World demand 1976 320 tonnes.

quartz glass *See* silica vitreous.

quercitrin, $C_{21}H_{20}O_4$. Pale yellow crystals,

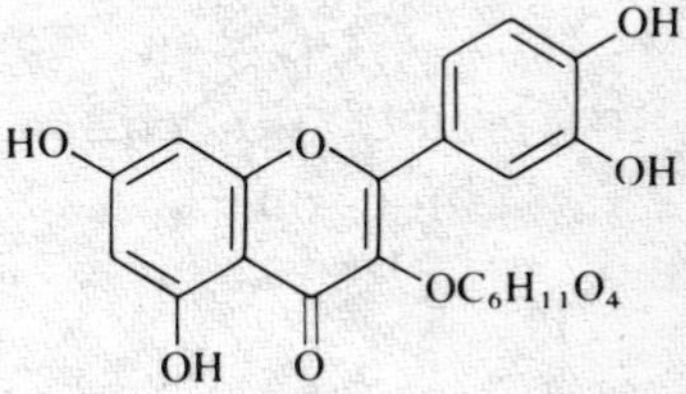

m.p. 169°C, a glycoside widely distributed in plants. Hydrolysed by acids to rhamnose and quercitin (3′,4′,5,7-tetrahydroxyflavanol). It is prepared commercially from oak bark, and is still used as a natural dyestuff (lemon flavin).

quicklime, CaO. Calcium oxide.

quicksilver An old name for mercury.

quinaldine, 2-methylquinoline, $C_{10}H_9N$. Colourless oily liquid, b.p. 246–247°C. Odour resembling quinoline. Volatile in steam. Forms stable salts and quaternary compounds. It is prepared by heating together aniline, ethanal and zinc chloride. It is used for the preparation of photosensitizing dyes.

quinhydrone, $C_{12}H_{10}O_4$. A molecular compound stable in the solid state consisting of alternate quinone and hydroquinone molecules linked together by hydrogen-bonds. It is prepared by mixing alcoholic solutions of quinone and hydroquinone and crystallizes in reddish-brown needles with a green lustre, m.p. 171°C. It is very sparingly soluble in cold water, and is split into its constituents by boiling water. It was used in the quinhydrone electrode. Other related compounds of quinone with phenols have been prepared and these are collectively called quinhydrones.

quinhydrone electrode An electrode used in pH measurement relying on the redox reaction hydroquinone ⇌ quinone. Can be used in non-aqueous solvents.

quinic acid, $C_7H_{12}O_6$. The carboxylic acid of a 1,3,4,5-tetrahydroxycyclohexane (*see* inositol). It is found in cinchona bark, coffee beans, bilberries, etc., often conjugated with caffeic acid. M.p. 162°C.

quinidine, $C_{20}H_{24}N_2O_2,2H_2O$. M.p. 172°C. Dextrorotatory stereoisomer of quinine. It occurs in cinchona bark and is obtained as a by-product in quinine* manufacture. It has similar medicinal properties to quinine, but has also a depressant action on the heart. It is chiefly used for the control of atrial fibrillation.

quinine, $C_{20}H_{24}N_2O_2,3H_2O$. White micro-

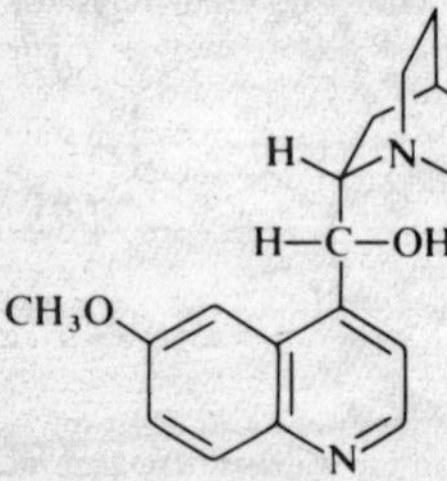

crystalline powder, m.p. 57°C (177°C when anhydrous). The principal alkaloid of cinchona bark synthesized from 7-hydroxyisoquinoline. Quinine and its salts are used for the treatment of malaria. It acts by destroying the malarial parasite. It is a general protoplasmic poison; it retards enzyme action and diminishes metabolism. It is an antipyretic and is a remedy for colds, influenza and cramp.

quinitol The trivial name for 1,4-cyclohexanediol.

quinizarin, 1,4-dihydroxy-9,10-anthraquinone, $C_{14}H_8O_4$. Crystallizes in yellowish-red leaflets, m.p. 195°C.

Prepared by condensing *p*-chlorophenol with phthalic anhydride in sulphuric acid solution in the presence of boric acid. The chlorine atom is replaced by hydroxyl during the condensation. It can also be prepared by oxidation of anthraquinone or 1-hydroxyanthraquinone by means of sulphuric acid in the presence of mercury(II) sulphate and boric acid.

Quinizarin is used for the preparation of various dyestuffs.

quinol An alternative name for hydroquinone*.

quinoline, C_9H_7N. Colourless oily refractive

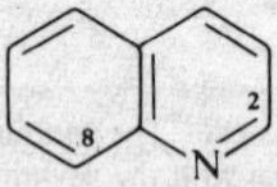

liquid; b.p. 238°C. Very hygroscopic, disagreeable odour. Oxidized by alkaline permanganate to quinolinic acid (pyridine-2-dicarboxylic acid). Reduced by tin and hydrochloric acid to the tetrahydro-compound. Basic, gives stable salts with mineral acids, and quaternary ammonium compounds with alkyl halides (quinolinium salts).

Occurs in the high-boiling fraction of coal tar. Most conveniently prepared by Skraup's reaction by heating a mixture of aniline, glycerol, sulphuric acid and nitrobenzene. Used in the manufacture of dyestuffs, and pharmaceutical products.

quinolinol 8-Hydroxyquinoline*.

quinone Another name for benzoquinone.

quinones Diketones formally derived by replacing the $>CH_2$ groups in a dihydroaromatic system by $>C{=}O$. They are highly coloured substances many of which occur in nature, e.g. vitamin K. They are prepared by oxidation of phenols. They give addition compounds with many types of organic compounds, especially quinols. They are readily reduced to dihydric phenols, have characteristic redox potentials and are used as dehydrogenating agents.

quintet A group of five closely spaced lines in a spectrum.

quinuclidine, $C_7H_{13}N$. A sublimable solid;

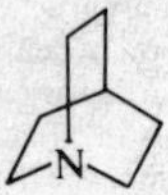

m.p. 158–159°C. A strong, hindered base (pK_b $=11$), it forms very soluble crystalline complexes with organometallic compounds of magnesium, lithium and zinc. These are useful for characterization and for conversion reactions like metalation*, e.g. this base promotes reaction of toluene with an alkyl-lithium to form benzyl-lithium.

R

Ra Radium.

Racah parameters The parameters used to express quantitatively the inter-electronic repulsion between the various energy levels of an atom. Generally expressed as B and C. The ratios between B in a compound and B in the free ion give a measure of the nephelauxetic effect*.

racemate, racemic compound An equimolecular compound or mixture of the dextro- and laevoratatory optically active forms of a given compound.

racemic acid *See* tartaric acid.

racemization The process in which an optically active stereoisomer is converted into a racemate. Epimerization* is a special case of racemization. Heat and the action of acids or bases can encourage racemization.

radical Term usually applied to species which have one or more free valencies. *See* free radicals. Formerly applied to a group of atoms, present in many compounds which can pass unchanged through a series of chemical reactions.

radii Covalent and ionic radii of selected elements.

Element	*Covalent radius* (Å)	*Ionic radius*† (Å)
H	0·29	2·08 (−1)
Li	1·23	0·60 (+1)
Be	0·89	0·31 (+2)
B	0·82	0·20 (+3)
C	0·77	0·15 (+4); 2·60 (−4)
N	0·70	0·11 (+5); 1·71 (−3)
O	0·66	0·09 (+6); 1·40 (−2)
F	0·64	0·07 (+7); 1·36 (−1)
Na	1·56	0·95 (+1)
Mg	1·36	0·65 (+2)
Al	1·25	0·50 (+3)
Si	1·17	0·41 (+4); 2·71 (−4)
P	1·10	0·34 (+5); 2·12 (−3)
S	1·04	0·29 (+6); 1·84 (−2)
Cl	0·99	0·26 (+7); 1·81 (−1)
K	2·03	1·33 (+1)
Ca	1·74	0·99 (+2)
Sc	1·44	0·81 (+3)
Ti	1·32	0·68 (+4); 0·76 (+3)
V	1·22	0·60 (+4); 0·74 (+3)
Cr	1·18	0·69 (+3); 0·84 (+2)
Mn	1·17	0·66 (+3); 0·80 (+2)
Fe	1·17	0·64 (+3); 0·76 (+2)
Co	1·16	0·63 (+3); 0·74 (+2)
Ni	1·15	0·62 (+3); 0·72 (+2)
Cu	1·17	0·96 (+1); 0·69 (+2)
Zn	1·25	0·74 (+2)
Ga	1·25	0·62 (+3); 1·13 (+1)
Ge	1·22	0·53 (+4); 0·93 (+2)
As	1·21	0·47 (+5); 2·22 (−3)
Se	1·17	0·42 (+6); 1·98 (−2)
Br	1·14	0·39 (+7); 1·95 (−1)

† Values in parentheses are the oxidation state of the element.

radioactive decay series The series of isotopes into which a radioactive nucleus is successively transformed.

radioactivity Becquerel (1896) found that uranium salts had a pronounced effect on a photographic plate even when they were separated by thin sheets of metal. This property was called radioactivity. Three main types of radiation are concerned, α-, β- and γ-rays, which arise from the spontaneous disintegration of the nucleus of the radioactive element. The emission of an α particle corresponds to a decrease in mass of the nucleus by four units; β particle emission corresponds to an increase in atomic number of the nucleus by one unit. The emission of γ-rays corresponds to loss in energy by the nucleus; they are often accompanied by X-rays.

There are other less common types of radioactive decay. Positron emission results in a decrease by one unit in the atomic number; K capture involves the incorporation of one of the extranuclear electrons into the nucleus, the atomic number is again decreased by one unit.

All elements of atomic number greater than 83 exhibit radioactive decay; K, Rb, Ir and a few other light elements emit β particles. The heavy elements decay through various isotopes until a stable nucleus is reached. Known half-lives range from 10^{-7} seconds to 10^9 years.

The rate of radioactive transformations cannot be altered by changing the conditions which are available in the laboratory. The process is a spontaneous one.

Radioactive elements may often be prepared artificially by bombarding the atoms of ordinary stable elements with, e.g. helium nuclei. *See* radioactivity, artificial.

radioactivity, artificial When an element undergoes a nuclear reaction, e.g. as a result of bombardment with α-particles, protons, deuterons or neutrons, the resulting atoms may be stable or metastable. In the latter case, they will be radioactive, and the phenomenon is known as artificial or induced radioactivity.

The reaction between the aluminium nuclei and the α-particles is expressed by the equation

$$^{27}_{13}Al + ^{4}_{2}He \longrightarrow ^{30}_{15}P + ^{1}_{0}n$$

(the symbol $^{1}_{0}n$ represents a neutron which is emitted in the process). The new phosphorus atoms disintegrate into atoms of a silicon isotope and positive electrons: $^{30}_{15}P \longrightarrow ^{30}_{14}Si + ^{0}_{1}e$; this process produces the observed artificial radioactivity. In this process, only a fraction of the α-particles causing disintegration actually give rise to active phosphorus, the majority following an (α, *p*) reaction giving $^{30}_{14}Si$ directly. The emission of positive electrons or other charged particles in such processes is detected experimentally by the ionization which they produce in gases; in counters this effect is used to count the number of particles emitted and hence to determine the rate of decay of the activity.

Normally, artificial radioactivity is only induced by bombardment with particles of appreciable mass – α-particles, protons, deuterons or neutrons, but γ-rays of very high energy are capable of disintegrating atoms with the ejection of a neutron from each nucleus and the formation of an isotope of the original element, which emits β-rays. Neutrons possess no electric charge and can penetrate the heaviest atoms without being subject to strong electrostatic forces. For this reason neutrons are particularly effective in producing artificial activity. The best sources of neutrons are nuclear reactors and the cyclotron, and by their use many radioactive isotopes have been prepared.

radiocarbon dating Carbon in living things contains a uniform amount of radioactive ^{14}C, which is produced constantly in the atmosphere by the action of cosmic radiation. Once the host dies, the exchange of nonradioactive carbon with the atmospheric ^{14}C ceases. Thus from the amount of ^{14}C in the dead sample the age of it may be determined. ^{14}C has a half-life of 5570 years and the method is useful for samples having ages up to 30 000 years.

radium, Ra. At.no. 88, at.wt. 226·0254, m.p. 700°C, b.p. 1140°C. A radioactive element, various isotopes are members of various radioactive decay series, ^{226}Ra is least active, half-life 1600 years. Ra is isolated from uranium ores; the element is obtained by electrolysis using a Hg cathode. Ra is a white metal which tarnishes in air and reacts with water (at least partly radiochemically). Used in self-luminous paints, as a neutron source and in radiotherapy (being supplemented by the artificial radioisotopes). World production 1968 200 g.

radium compounds Radium is an alkaline earth, electronic configuration $7s^2$. Its compounds are very similar to the corresponding barium derivatives although they are generally less soluble. Radium halides RaF_2, $RaCl_2$ and $RaBr_2$ are known. The carbonate, fluoroberyllate and iodate are insoluble in water.

radon, Rn. At.no. 86, at.wt. 222, m.p. −71°C, b.p. −61·8°C. Radon is a product of radioactive decay of the heavy elements. ^{222}Rn is obtained as a gas from $RaCl_2$ solution Rn is radioactive and has been used as a radiation source and as a gaseous tracer; it is a considerable hazard in uranium mines. Electronic configuration $6s^2 6p^6$. Rn forms compounds, particularly a fluoride and solid adducts between the fluoride and Lewis acid fluorides.

raffinate *See* liquid–liquid extraction.

raffinose, melitose, $C_{18}H_{32}O_{16}$. Large prisms as the pentahydrate; m.p. 80°C or m.p. 118–119°C when anhydrous. The best-known trisaccharide, it is composed of galactose, glucose and fructose. It is often found in considerable quantities in the sugar beet, but the best source is cotton-seed meal, which contains 8%. It has no reducing power.

Raman effect When light of frequency ν_0 is scattered by molecules of a substance, which have a vibrational frequency of ν_i, the scattered light when analysed spectroscopically has lines of frequency ν, where

$$\nu = \nu_0 \pm \nu_i$$

This spectrum is called a Raman spectrum and corresponds to the vibrational or rotational changes in the molecule. The selection rules for Raman activity are different from those for i.r. activity and the two types of spectroscopy are complementary in the study of molecular structure. Modern Raman spectrometers use lasers for excitation. In the resonance Raman effect excitation at a frequency corresponding to electronic absorption causes great enhancement of the Raman spectrum.

Ramsay–Shields equation An equation relating the molecular surface energy of a liquid with its temperature

$$\gamma\left[\frac{\mathrm{M}}{\mathrm{D}}\right]^{\frac{2}{3}}=k(T_{\mathrm{C}}-T-6)$$

where γ is the surface tension of the liquid, M its molecular weight, *D* its density, T_{C} is the critical temperature and *T* the temperature at which the measurement is made. *k* is known as the Ramsay–Shields constant and for many undissociated liquids, has an approximate value of 2·12.

Raney nickel A special form of nickel prepared by treating an Al–Ni alloy with NaOH solution. The nickel is left in a spongy mass which is pyrophoric when dry. This form of nickel is a most powerful catalyst, especially for hydrogenations.

Raoult's law When a solute is dissolved in a solvent, the vapour pressure of the latter is lowered proportionally to the mole fraction of solute present. Since the lowering of vapour pressure causes an elevation of the boiling point and a depression of the freezing point, Raoult's law also applies and leads to the conclusion that the elevation of boiling point or depression of freezing point is proportional to the weight of the solute and inversely proportional to its molecular weight. Raoult's law is strictly only applicable to ideal solutions since it assumes that there is no chemical interaction between the solute and solvent molecules.

rare earths Strictly the oxides of the lanthanide elements but often applied to the elements themselves. *See* lanthanides. World demand 1976 17 000 tonnes.

rare gases The noble gases*.

Raschig process *See* hydrazine.

rasorite *See* kernite, borax.

Rast's method A method of determining molecular weights by measuring the depression of freezing point of a solvent by a known weight of the solute whose molecular weight is required. Camphor is commonly used as the solvent because of its high molal depression constant. Molecular weights so determined are usually accurate to $\pm 10\%$; the method is useful for determining E_n when the empirical formula E is known.

rationality of indices, law of *See* rationality of intercepts, law of.

rationality of intercepts, law of The intercepts of the faces of a crystal upon the axes of the crystal bear a simple ratio to each other. This is sometimes referred to as the law of rationality of indices.

ratio of specific heats The ratio between the specific heat of a gas at constant pressure (C_{p}) and that at constant volume (C_{v}). This ratio ($C_{\mathrm{p}}/C_{\mathrm{v}}$) is characteristic of the number of atoms in each molecule of the gas; monatomic gases have a value of 1·67; diatomic, 1·40; and triatomic 1·33.

rayon Artificial fibres made of regenerated cellulose formerly known as viscose, artificial silk, fibre silk, etc. The most widely used process is the viscose process in which cellulose (generally wood fibre) is treated first with alkali and then carbon disulphide to give xanthated cellulose which is extruded into an acid bath. Copper ammonium hydroxide has been used for dissolving the cellulose. U.S. production 1978 270 000 tonnes.

Rb Rubidium.

Re Rhenium.

reaction velocity The rate of a chemical reaction expressed either as the rate of disappearance of reactant molecules or the rate of formation of product molecules.

reactive dyes Dyes which contain a functional group (chloro-substituted heterocycle, acrylamide, sulphonic acid derivative) capable of reacting directly with the fibre.

reactor Large-scale chemical processes are carried out in various environments, usually in tanks, tubes, towers or fluidized beds*. These are usually referred to as chemical reactors and form the heart of the overall chemical process. Reactor design involving scale-up from laboratory reactions is an important aspect of chemical technology or chemical engineering. The term reactor is also used for a nuclear reactor.

realgar The mineral form of As_4S_4, formerly used as an orange–red pigment. Formed artificially by distilling sulphur and excess of arsenic.

reboiler *See* rectification.

reciprocal proportions, law of A special case of the law of equivalent proportions* stating that 'the weights of two or more substances which separately react chemically with identical weights of another substance are also the weights which react with each other or are simple multiples of them'. Thus 23 g of Na react with 35·5 g of Cl_2 to give NaCl; 23 g of Na react with 1 g of H_2 to give NaH; 35·5 g of Cl_2 react with 1 g of H_2 to give HCl. The law is not of completely general application owing to the existence of variable oxidation states and non-stoicheiometric compounds.

recoil atom When an atomic nucleus disintegrates the fragments receive equal momentum. In the case of a heavy nucleus emitting a light fragment, the heavy remnant will recoil with a comparatively low velocity. This residue is termed a recoil atom. The recoil atom has a sufficiently high energy to rupture bonds which it may form with other atoms. The resulting group may not recombine (Szilard–Chalmers effect) and it may thus be possible to separate chemically the atoms which have undergone radioactive disintegration. The 'hot atom' may recombine with its original partner or may combine with another group to form a new chemical compound.

recrystallization The process of repeated crystallization carried out with the object of removing impurities or of obtaining more satisfactory crystals of a substance which is already pure.

rectification A method of separating a mixture into its components by fractional distillation. The terms rectification and fractional distillation are frequently used synonymously, despite the fact that two other methods are covered by the latter term.

A *rectifying* or *fractionating column* consists of a long vertical cylinder containing either a series of *plates*, or a *packing* material having a large specific surface, both these arrangements ensuring good liquid–vapour contacting over the length of the unit. At the top of the column there is a *condenser* and at the bottom a vaporizing unit called a *reboiler*.

Rectification is of the greatest industrial importance; in particular, it is the principal separation process employed in the petroleum industry.

See also packed column, plate column.

rectified spirit A solution of ethanol in water which contains 90% v/v ethyl alcohol. Made by dilution of 95% ethanol with water. Rectified spirit is 57·8° over proof (OP). *See* proof spirit.

rectifying column *See* rectification.

recuperators, regenerators Heat-exchangers used in furnace operations where combustion air or lean fuel gas may be preheated.

red lead, Pb_3O_4. *See* lead oxides.

redox An abbreviation for oxidation-reduction, e.g. redox potential – oxidation potential.

redox catalyst A combination of a free radical catalyst with a reducing ion or salt which increases the rate of free radical production. Used in polymerization reactions, which in consequence can either be carried out at lower temperatures or at faster rates. Many synthetic rubbers are produced by this method.

redox indicator A substance which undergoes a definite colour change during its reversible oxidation or reduction. The substance has a definite, and preferably narrow, potential range over which the colour change occurs. E.g. methylene blue is colourless when reduced, blue when oxidized.

reduced mass For any system of particles of masses m and M in harmonic motion about each other, the reduced mass μ is given by

$$\mu = \frac{mM}{(m+M)}$$

reducing agent A material which brings about reduction – and in the process is itself oxidized. E.g. Sn^{2+} reduces Fe^{3+} and is itself oxidized to Sn^{4+}.

reduction Chemical processes in which the proportion of electronegative substituent is decreased, e.g. $FeCl_3 \longrightarrow FeCl_2$, or the charge on an ion is made more negative, e.g. $Sn^{4+} \longrightarrow Sn^{2+}$ or $[Fe(CN)_6]^{3-} \longrightarrow [Fe(CN)_6]^{4-}$, or the oxidation number is lowered. In organic chemistry the most obvious effect is to increase the proportion of hydrogen and to reduce the number of multiple bonds, e.g. C_2H_4 (ethene)$\longrightarrow C_2H_6$ (ethane).

reductone, glucic acid, 1,2-dihydroxyacrolein, 3-hydroxy-2-oxo-propanone, $C_3H_4O_3$. Solid, $CH(OH){=}C(OH)CHO \rightleftharpoons CH_2OH{\cdot}C(O)CHO$.
m.p. 140°C (decomp.). Prepared by the action of alkalis on hexoses or on dihydroxyacetone It is a reducing agent, and a weak acid forming metallic salts. With diazomethane, methyl reductone, m.p. 67°C, is formed.

reductones Enediols stabilized by conjugation to a carbonyl or similar group.

reference electrode A standard electrode used in electrochemistry. Examples are the calomel electrode, $Ag-AgCl$, $Hg-Hg_2SO_4$ electrodes.

refining The removal of impurities from metals or alloys in metallurgy using, e.g. oxidation, electrolysis, chemical separation, distillation, zone refining.

refining, petroleum The general process of

converting crude oil into a wide range of petroleum products including fuels, lubricants, bitumens, waxes, etc. Specific refining operations include physical separation techniques such as distillation*, solvent extraction, absorption*, etc. Chemical refining operations involve many treatments including sulphuric acid treatment, sweetening*, desulphurization*, etc.

reflux Liquid from partial condensation of vapour which is returned to the top of a fractionating column and allowed to flow down the column counter-current to the ascending vapour.

Reformatski reaction Aldehydes and ketones react with α-bromo- fatty acid esters in the presence of zinc powder to give β-hydroxyesters which may be dehydrated to give α-, β-unsaturated esters. α-Chloroesters will react if copper powder is used in conjunction with the zinc.

reforming The process of subjecting straight-run gasoline from primary distillation to a thermal or catalytic treatment. Straight-chain alkanes are converted to isomers while cycloalkanes are dehydrogenated to form aromatics. Thermal reforming is similar to thermal cracking* but higher temperatures (500–600°C) and pressures (about 7000 kN/m^2) are used. The process results in gasoline of higher octane number* and wider boiling range. The reforming process is more usually carried out in the presence of a catalyst as lower temperatures and pressures may be used. Higher yields of better quality gasoline are obtained. *See* catalytic reforming.

refractory materials, refractories A general term for materials which are not damaged by heating to at least 1500°C in a clean (oxidizing) atmosphere.

Acid refractory materials include fireclays, flint clays, china clays (kaolins), silica, flint, chalcedony, ganister and titanium dioxide.

Neutral refractory materials include graphite, charcoal, coke, chromite and various carbides.

Basic refractory materials include lime, magnesia, various materials composed chiefly of alumina (bauxite, diaspore, laterite, gibbsite, etc.), dolomite and most of the rarer refractory oxides, particularly zirconia.

refrigerants Materials used for cooling, e.g. liquid nitrogen. *See* low-temperature baths.

refrigeration The most important method of achieving refrigeration, i.e. lowering the temperature of a material below that of its surroundings, is by vapour compression. The cycle consists of compressing a refrigerant vapour and condensing it, then passing the high-pressure liquid through a reducing valve. Reduction in pressure causes part of the liquid to vaporize resulting in a considerable lowering of temperature. The cold liquid passes to an evaporator where it is vaporized by heat exchange with the medium to be cooled, the vapour formed passing back to the compressor.

Many condensable vapours have been used as refrigerants, but the main ones are the chlorofluorocarbons (Freons*), ammonia and carbon dioxide.

regioselectivity The tendency for reactions which could proceed in two or more ways to proceed in one manner only, e.g. addition to olefine derivatives to give only 1,2-adducts.

Regnault's method Obsolete method for determining gas densities by direct weighing of a known volume of gas under known conditions of temperature and pressure.

regular system The cubic system*.

reinforced plastics Composites* between plastics, generally thermosetting resins, and strengthening materials which may run from carbon fibres and metal, glass fibres, fabrics, paper, etc. Laminates have the thermosetting resin built up by heat and pressure.

relative atomic mass, A_r The ratio of the average mass of an element to 1/12 of the mass of ^{12}C. Known as atomic weight*.

relaxation The process of loss of energy from an excited state to another excited state or to the ground state.

release agents *See* abherents.

remote handling facilities Equipment used to carry out chemical reactions with intensely radioactive elements, e.g. Am, or to manipulate the very active fuel elements from a nuclear reactor.

rennin The milk-clotting enzyme from the fourth stomach of ruminants. It liberates a glycopeptide from casein, the remaining portion reacts with calcium to yield the insoluble curd. Under the name of rennet it is used in cheesemaking and for making rennet casein and junket.

reserve acidity and alkalinity A buffer solution is sometimes referred to as a solution with reserve acidity or alkalinity.

resins High mol.wt. materials, often of variable but controlled mol.wt., which soften at high temperatures. Synthetic resins are formed by polymerization. Natural resins occur in vegetable products (rosin) or from insects (shellac).

resolution The separation of a racemate into its two enantiomorphic forms (i.e. (+) and (−) (R) and (S) forms).

resonance In the optical sense the term refers to the absorption of radiation (visible, i.r., microwave) by a system which it is capable of emitting. E.g. mercury vapour, when suitably excited in an electrical discharge, will emit light of wavelength λ 2536 Å. This same wavelength is very readily absorbed by Hg vapour, and the energy so absorbed may be utilized in producing photochemical reactions. The process is analogous to acoustic resonance.

The term resonance has also been applied in valency. The general idea of resonance in this sense is that if the valency electrons in a molecule are capable of several alternative arrangements which differ by only a small amount in energy and have no geometrical differences, then the actual arrangement will be a hybrid of these various alternatives. *See* mesomerism. The stabilization of such a system over the non-resonating forms is the resonance energy.

resonance ionization spectroscopy A technique allowing the selection, isolation and observation of as small an amount as a single distinct atom by using a laser to ionize that atom selectively.

resonance Raman spectroscopy *See* Raman effect.

resorcinol, 1,3-dihydroxybenzene, $C_6H_6O_2$. Colourless needles; m.p. 110°C, b.p. 276°C.

Obtained by KOH fusion of many resins. Prepared by fusion of *m*-benzenedisulphonic acid with caustic soda, also obtained to some extent in the NaOH fusion of *o*-and *p*-benzenedisulphonic acids.

It is extensively used in the preparation of dyestuffs. Combines with diazonium salts to form oxyazo-colouring matters. Gives rise to fluorescein dyes on fusion with phthalic anhydride. Used for production of plasticizers, resins, adhesives.

respiratory pigments A group of pigments, of which haemoglobin is the most important member, which act as oxygen carriers in living organisms.

retene, 1-methyl-7-isopropylphenanthrene, $C_{18}H_{18}$. Faintly yellow plate-like crystals; m.p. 101°C, b.p. 390°C. Occurs in the high boiling tar-oils from resinous pine wood and is the chief constituent of certain fossil resins of *Coniferae*. It is produced when abietic acid is dehydrogenated.

retinene, vitamin-A aldehyde, $C_{20}H_{28}O$. Orange crystals; m.p. 61–64°C. Isolated from retinas, obtained by oxidation of vitamin A and β-carotene, and by various syntheses. Retinene generally refers to the all-*trans* isomer which has essentially the same bioactivity as the corresponding vitamin-A alcohol. Retinene is the intermediate in the conversion in mammals of β-carotene to vitamin A by oxidative cleavage of its central double bond. *See also* vitamin A and rhodopsin.

reverse osmosis Separation of solute from a solution by causing the solvent to flow through a membrane at pressures higher than the normal osmotic pressure*.

Reverse osmosis is used for desalination of seawater, treatment of recycle water in chemical plants and separation of industrial wastes. More recently the technique has been applied to concentration and dehydrogenation of food products such as milk and fruit juices. *See* ultrafiltration.

reversible process For a system in equilibrium a change (process) is said to occur reversibly when the direction of change can be made to occur at will in either direction by an infinitely small change in one of the factors controlling the position of equilibrium (pressure, temperature, concentration). For example in a reaction such as

$$N_2O_4 \rightleftharpoons 2NO_2$$

which is at equilibrium, a slight increase in pressure will shift the position of equilibrium to the N_2O_4 side. However, if the pressure is restored to its original value the equilibrium will immediately return to its former position. An irreversible change is one in which a slight change in conditions results in a complete change in equilibrium which cannot be reversed when the conditions are restored to their original values.

R_f, R_s, R_r A measure of retention of solute in chromatography defined as the ratio of the distance travelled by the solute to the distance travelled by the mobile phase.

Rh Rhodium.

rH The log of the reciprocal of the partial pressure of hydrogen in equilibrium with a system; thus $rH = -\log p$. Used as a concept in redox reactions.

L-rhamnose, $C_6H_{12}O_5$. A constituent of

many glycosides, which is widely distributed in plants, particularly in combination with flavanol derivatives and in saponins. Crystallizes with $1H_2O$, m.p. 94°C; anhydrous, m.p. 122°C. As a methylpentose it has the formula

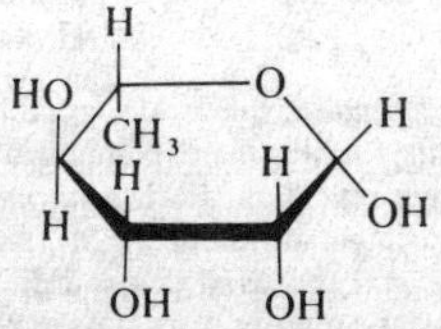

It shows the usual carbohydrate reactions and most resembles mannose in behaviour; it exists in α and β forms which exhibit mutarotation.

rhenates *See* rhenium oxides.

rhenium, Re. At.no. 75, at.wt. 186·2, m.p. 3180°C, b.p. 5627°C (est.), *d* 20·53. Re is a rare metal extracted from flue dusts after roasting Mo or Cu ores. The metal is obtained by reduction of Re compounds with H_2, it has a hcp structure; is silvery grey in colour, dissolves in H_2O_2 and reacts with hot Cl_2, S or O_2. Used as an additive in W- and Mo-based alloys, in Re–W thermocouples and in catalysts. ^{187}Re with a half-life of 7×10^{10} years can be used in determining the age of the universe. World demand 1976 5·5 tonnes.

rhenium chemistry Rhenium is an element of Group VII of the transition series, electronic configuration of the element $5d^5 6s^2$. It shows oxidation states +7 to −1, the higher oxidation states are much less strongly oxidizing than those of Mn. There is no chemistry of the hydrated cations, organometallic derivatives are similar to those of Mn. Lower oxidation state compounds contain Re–Re bonds. Extensive series of complexes are known, many containing Re=O or Re≡N groups. The $[ReH_9]^{2-}$ ion is formed by reduction of ReO_4^- with K in ethylenediamine and a series of phosphine hydride complexes, e.g. $ReH_7(PR_3)_2$, is known.

rhenium halides The only Re(VII) halide is ReF_7 ($Re + F_2$ 3 atm. at 400°C) an orange solid, b.p. 74°C. The hexahalides are pale yellow ReF_6 ($Re + F_2$) m.p. 19°C, b.p. 48°C and dark green $ReCl_6$ ($ReF_6 + BCl_3$) m.p. 29°C. Yellow–green ReF_5 ($ReF_6 + W(CO)_6$) (disproportionates to ReF_6 and ReF_4 on heating) and $ReCl_5$ (heat on $ReCl_6$) have halogen bridges. $ReCl_4$ ($ReCl_5 + C_2Cl_4$) has Cl bridges. Red $ReCl_3$ (heat on $ReCl_5$) has a triangle of three Re atoms. The bromides and iodides are similar: $ReBr_5$, $ReBr_4$, $ReBr_3$, ReI_4, ReI_3, ReI_2 and ReI are known. All of the higher halides are hydrolysed by water. Complex halides are known for Re(VI), $[ReF_7]^-$; (V), $[ReF_6]^-$ and $[ReOX_4]^-$, (IV), $[ReF_6]^{2-}$, $[ReCl_6]^{2-}$, (III), $[Re_3Cl_{12}]^{3-}$ and $[Re_2Cl_8]^{2-}$, the Re(III) species contain Re–Re bonds. Oxide halides, e.g. blue $ReOF_4$ and black $ReOF_3$, are known.

rhenium oxides Solid yellow Re_2O_7 ($Re + O_2$) contains linked ReO_6 and ReO_4 groups. It is a volatile material and the parent of rhenates(VII), ReO_4^-, salts. ReO_3 ($Re_2O_7 + CO$) has a classical ionic structure, rhenates(VI) are not known. ReO_2 is formed as a hydrate ($ReO_4^- + Zn + HCl$) which may be dehydrated. Colourless rhenate(VII) (perrhenate), ReO_4^-, is a mild oxidizing agent, $[ReO_6]^{5-}$ and $[ReO_5]^{3-}$ are formed in very basic solution or by fusion with peroxides.

rheopexy The phenomenon whereby the gelation of some thixotropic sols is accelerated by gentle mechanical agitation. An example is gypsum-water paste which, if allowed to stand, solidifies in about 10 minutes, but when gently agitated solidifies in seconds.

rhodallin *See* propenylthiourea.

rhodamines Dyestuffs obtained by Friedel–Crafts reactions involving phthalic anhydride and aminophenols.

rhodium, Rh. At.no. 45, at.wt. 102·9055, m.p. 1966°C, b.p. 3727°C, *d* 12·4. One of the platinum group metals. Found with other platinum metals and often in Ni–Cu deposits. After reduction to the mixed metals Rh is extracted from the aqua regia insoluble material with fused $KHSO_4$, extracted with water and precipitated as hydrated Rh_2O_3 which dissolves in HCl to give $(NH_4)_3RhCl_6$. The metal can be obtained by reduction with H_2, it is ccp and is very inert. It is mainly used as an alloying agent for Pt or Pd. World demand 1976 133000 troy oz.

rhodium chemistry Rhodium is a member of the cobalt group, electronic configuration $4d^8 5s^1$. It shows oxidation states from +6 (RhF_6) to −1 $[Rh(CO)_4]^-$. The +6, +5 and +4 states are strongly oxidizing and Rh(III) is the stablest state. The +2 state is uncommon but square planar complexes in the +1 state, often with phosphine ligands, are well established. The +1 complexes are active catalysts (Wilkinson's catalyst*) and are readily transformed to octahedral Rh(III) compounds. Many Rh(0) compounds are carbonyl derivatives, e.g. $Rh_2(CO)_8$, $Rh_6(CO)_{16}$. Rh(III) is stable in aqueous solution as the $[Rh(H_2O)_6]^{3+}$ ion, N-bonded complexes, particularly ammines, and O-complexes, e.g. sulphates, are stable. Rh(III) is generally available as $RhCl_3$ hydrates or halide complexes.

rhodium halides The known fluorides are black RhF_6 ($Rh + F_2$), m.p. 70°C, red $(RhF_5)_4$ ($Rh + F_2$), purple-red RhF_4 ($RhBr_3 + BrF_3$) and RhF_3 ($RhI_3 + F_2$). Fluororhodates containing $[RhF_6]^-$, $[RhF_6]^{2-}$ ions are known. $RhCl_3$ ($Rh + Cl_2$ at 300°C) is the only anhydrous chloride although hydrates are known from aqueous solution. $[RhCl_6]^{2-}$, $[RhCl_6]^{3-}$ and some aquo species are formed in solution. $RhBr_3$ and RhI_3 are stable.

rhodium oxide Grey Rh_2O_3 is formed by heating Rh in air to 600°C. Hydrated forms are precipitated from Na_3RhCl_6 with alkali. There is some evidence for a higher oxide. Na_2RhO_3 and other rhodates(IV) as well as rhodates(III), $M^{II}Rh_2O_4$ and $MRhO_2$ are formed by solid state reactions but rhodates are not stable in solution.

rhodopsin A major photosensitive pigment of the eye, composed of the protein opsin and neoretinene-b (11-*cis*-retinene). In the visual process, the latter is isomerized to all-*trans*-retinene (vitamin-A aldehyde) and dissociates from the opsin (bleaching of rhodopsin). Reconversion of retinene to neoretinene-b may then occur with regeneration of rhodopsin. *See* vitamin A and retinene.

rhombohedral system An alternative representation of the trigonal* or hexagonal* crystal systems.

riboflavin, $C_{17}H_{20}N_4O_6$. Crystallizes in

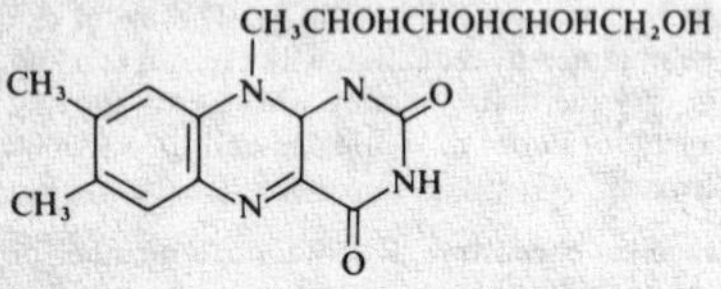

orange needles; m.p. 271°C (decomp.). Soluble in water. Riboflavin is that part of the original vitamin B_2 complex which stimulates the growth of rats. It occurs widely in nature, particularly in liver, milk and white of egg. It is the precursor of flavoproteins*. *See also* flavin-adenine dinucleotide and flavin mononucleotide.

D-ribose, $C_5H_{10}O_5$. M.p. 87°C. The sugar of ribonucleic acid; it is therefore present in all plant and animal cells. It has the furanose structure shown.

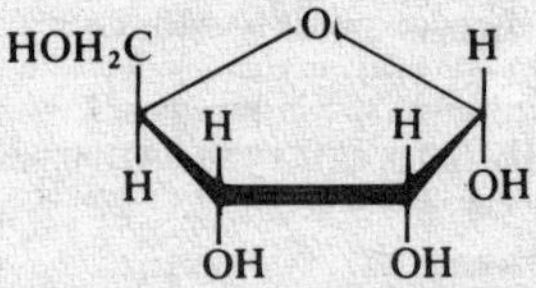

ribosomes Small, spherical, granular cell constituents, forming major components of the endoplasmic reticulum. Each ribosome has a molecular weight of about 2·5 M., and there may be several thousand in one cell, making up as much as a quarter of its mass. Ribosomes are the site of protein synthesis, each one forming one polypeptide chain at a time. They consist principally of protein (55%), ribonucleic acid (40%) and organic bases including spermine, cadaverine and putrescine. Ribosomal ribonucleic acids are of high molecular weight (0·5–1 M.). Ribosomes 'translate' the code in a messenger RNA molecule to form a polypeptide chain. More than one ribosome can be engaged in 'translation' of an *m*-RNA molecule at any particular time.

ribulose, $C_5H_{10}O_5$. Ribulose-1,5-diphosphate

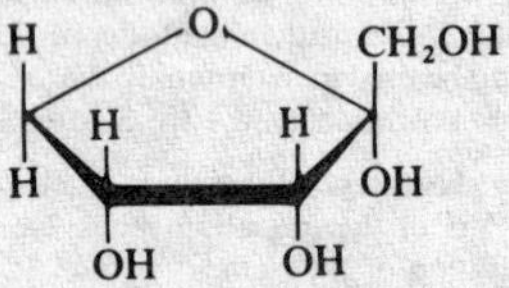

is a key intermediate in the photosynthetic fixation of atmospheric CO_2. It is formed from the 5-monophosphate by the action of ATP and a kinase. The combined action of CO_2, light, water and enzyme on the diphosphate results in the formation of two molecules of 3-phosphoglyceric acid. This may be in part reconverted to ribulose diphosphate, and in part transformed to give acetylcoenzyme A.

ricin A poisonous protein of the albumin class found in the seeds of the castor bean.

ricinoleic acid, $C_{18}H_{34}O_3$. *Cis*-12-hydroxy-9-octadecenoic acid. M.p. 4–5°C, b.p. 226–228°C/10 mm. It comprises about 85% of the acids of castor oil.

The name ricinoleic acid is used in commerce for the mixture of fatty acids obtained by hydrolysing castor oil, and its salts are called ricinoleates.

Rinmann's green, $ZnCo_2O_4$. A spinel formed when cobalt nitrate solution is placed on zinc oxide and the mixture heated to redness. The green colour forms a delicate test for Zn.

Rn Radon.

RNA *See* nucleic acids.

rochelle salt, $C_4H_4O_6KNa,4H_2O$. Sodium potassium tartrate*.

rock crystal, SiO_2. *See* quartz.

rock salt, NaCl. Natural sodium chloride*. Occurs in extensive deposits.

rodinol *See* aminophenols.

röntgen A measurement of a quantity of ionizing radiation. One röntgen of radiation produces 1 esu of charge/cm^3 of air at stp; this amount of radiation produces $2{\cdot}08 \times 10^9$ pairs of gaseous ions in air under standard conditions. A lethal whole-body single dose of radiation is about 500 röntgens for man.

Röntgen rays An alternative name for X-rays, which were discovered by Röntgen in 1895.

rosin A solid resin which occurs in the oil from pine trees and is also obtained from tall oil*. It is mainly composed of resin acids which are monocarboxylic acids related to abietic acids. Used in the preparation of various compounds (including esters) and in lacquers, plasticizers and flotation agents.

rotamers Isomers formed by restricted rotation.

rotational spectrum When a molecule absorbs energy it may increase its total rotational energy as whole (*see* band spectrum). Conversely, the transition from a state of higher to one of lower rotational energy corresponds to the emission of energy, which may be of light of frequency v; the energy E emitted being given by $E = hv$. The lines in the molecular spectrum corresponding to such transitions are termed rotational bands, which are characteristic of the molecule. Taken collectively the rotational bands comprise the rotational spectrum of the molecule. Rotational transitions are associated with lower energies than vibrational or electronic transitions, and hence the frequency of the emitted light is small. Rotational spectra are associated with the far i.r. and microwave regions of the electromagnetic spectrum.

rotatory power, specific The specific rotatory power of a pure liquid is given by

$$[\alpha]^t_D = (\alpha/ld)$$

where $[\alpha]^t_D$ is the specific rotatory power at temperature t for the sodium D-line; α is the rotation of the plane of polarization produced by a column of liquid of length l (decimetres) and density d. For solutions the corresponding equation is

$$[\alpha]^t_D = 100\alpha/lc$$

where c is the solute concentration (g/ml).

The molecular rotation is given by

$$[M]^t_D = M[\alpha]/100$$

where M is the molecular weight.

rouge A finely divided form of iron(III) oxide, Fe_2O_3, used for polishing purposes and in cosmetics on account of its smooth texture and freedom from gritty particles.

royal jelly The sole nutrient of queen bee larvae.

R_s. *See* R_f.

R,S convention *See* optical activity.

Ru Ruthenium.

rubber A high mol.wt. natural or synthetic polymer which exhibits elasticity at room temperature. Natural rubber is obtained almost exclusively from the tree *Hevea Braziliensis* as a latex containing 30–36% rubber exuded from grooves cut into the bark. The latex is strained, diluted with water coagulated with methanoic or ethanoic acids to a solid rubber. For use the rubber is vulcanized* which introduces a degree of cross-linking and wear and tensile strength is increased by adding a reinforcing filler (generally carbon black) in the process of compounding. Natural rubber is *cis*-polyisoprene. Synthetic rubbers are synthetic isoprene polymers*, butyl rubbers*, ethene–propene copolymers, various vinyl polymers, styrene–butadiene, butadiene polymers, neoprene. *See also* elastomers. World rubber use 1977, 12 megatonnes (natural 4 megatonnes, synthetic 8 megatonnes).

rubber conversion products Products made from rubber. *See* chlorinated rubber, cyclized rubber, oxidized rubber, hydrochlorinated rubber.

rubber hydrochloride *See* hydrochlorinated rubber.

rubber oxidation products *See* oxidized rubber.

rubber (synthetic) *See* isoprene polymers, rubber, buna rubbers, neoprene. Synthetic rubbers are generally more expensive than natural rubber but may be oil and chemical resistant.

rubbone *See* oxidized rubber.

rubidium, Rb. At.no. 37, at.wt. 85·468, m.p. 38·89°C, b.p. 688°C, *d* 1·53. An alkali metal which occurs in traces in salt deposits and also in the lithium mineral lepidolite. The metal, bcc, is formed from RbCl and Ca. It is silver-white and very reactive. ^{87}Rb is a weak β-emitter. The metal is used as a getter and in photocells. It may find use in ion-propulsion rocket motors. World demand 1976 1·5 tonnes.

rubidium chemistry Rubidium is a Group I element, electronic configuration 5s^1 It forms a single series of +1 compounds ($Rb^+ \longrightarrow Rb$ −2·92 V), mainly ionic although some complexes with, e.g, crown ethers, are known. A series of lower oxides containing metal cluster cations are formed by burning Rb in a deficiency of oxygen.

ruby Impure α-Al_2O_3 containing some chromium. Occurs as gemstones and made by fusion of Al_2O_3 and traces of Cr_2O_3. Used as a gemstone and in lasers.

Russell–Saunders coupling, LS coupling A system of arriving at the different energy levels associated with unfilled shells, in which electron spins are assumed to couple only with electron spins, and orbital momenta to couple only with orbital momenta. Used to describe the ions of the first row of the transition elements. The heavier elements require the use of jj-coupling in which spin-orbital interaction is assumed.

rusting The formation of a loose flaky layer of hydrated iron(III) oxide on the surface of iron or steel. The reaction proceeds in a moist atmosphere via Fe(II) ions; it is accelerated by atmospheric SO_2 or H_2SO_4 (cities) or Cl^- (coast). It is prevented by plating, painting or alloying (e.g. 12–18% Cr gives a protective oxide layer). Rusting is negligible in hot, dry areas as the critical humidity is too low.

ruthenates *See* ruthenium oxides.

ruthenium, Ru. At.no. 44, at.wt. 101·07, m.p. 2310°C, b.p. 3900°C, *d* 12·45. Ru is one of the platinum metals and occurs native as an alloy with the other platinum metals (osmiridium) and also in some sulphide ores. The elements are separated by electrolysis followed by chemical separation, ion-exchange, and solvent-extraction to give pure Ru. The metal is obtained from its compounds by H_2 reduction and is grey–white in colour, hcp. The metal is unattacked by most acids, O_2 and F_2 react at high temperatures and the metal dissolves in fused Na_2O_3, $KClO_3$, etc. Ru compounds are used in catalysis and the metal is used as a hardening agent for Pd and Pt. U.S. production 1970, 245 kg.

ruthenium chemistry Ruthenium is a platinum metal, electronic configuration 4d^{7}5s^1 which shows oxidation states from +8 to −2. The +8 state is more strongly oxidizing than for Os, the principal lower oxidation states are +3 and +2. There is some chemistry of the aquo-ions $[Ru(H_2O)_6]^{n+}$ ($n = 3,2$) although most anions complex with the metal. N-bonded complexes are particularly stable and many ammines (Ru(II) and Ru(III)), dinitrogen and nitrosyl complexes are known. Phosphine complexes are stable with lower oxidation states and chloride- and hydride-phosphine complexes are good hydrogenation catalysts. Many carbonyl derivatives, e.g. $[RuCl_2(CO)_2(PEt_3)_2]$, and $[Ru(CO)_5]$, are known and these and other low oxidation state compounds form organometallic derivatives similar to those of Fe.

ruthenium halides Dark-brown RuF_6 ($Ru + F_2$), m.p. 54°C, is octahedral molecular. Dark-green RuF_5 ($Ru + F_2$), m.p. 86°C, is tetrameric with fluorine bridges. Yellow RuF_4 ($RuF_5 + I_2$), and brown RuF_3 ($RuF_5 + I_2$ at high temperatures) are the known fluorides. $RuCl_3$ ($Ru + Cl_2$) is the only established chloride although many aquo-chloro complexes of Ru(IV), Ru(III) and Ru(II) are known. $RuBr_3$ and RuI_3 are also known. Complex fluorides of Ru(V) and Ru(IV), e.g. $KRuF_6$ and K_2RuF_6, are formed by fluorination in the absence of water. Some oxide fluorides, e.g. $RuOF_4$, are formed by hydrolysis.

ruthenium oxides Yellow RuO_4 ($RuO_2 + KIO_4$), m.p. 25°C, b.p. 40°C, is fairly unstable and a strong oxidizing agent cleanly converting secondary alcohols to ketones. It decomposes to blue RuO_2 ($Ru + O_2$ at 1000°C). Green ruthenates(VII), RuO_4^- (Ru fused in $KNO_3 + KOH$) decompose in water to orange ruthenates(VI), RuO_4^{2-} and oxygen.

rutherfordium Element 104. *See* post-actinide elements.

rutile Mineral TiO_2, a reddish brown material used as a source of Ti and in ceramics. The tetragonal structure has each Ti octahedrally co-ordinated by O and each O co-ordinated by three Ti in an approximate plane. This structure is adopted by many oxides and fluorides, e.g. PbO_2 and MnO_2. (*See diagram on next page*.)

rutin, $C_{27}H_{30}O_{16},3H_2O$. The 3-rhamnoglucoside of 5,7,3′,4′-tetrahydroxyflavanol. It

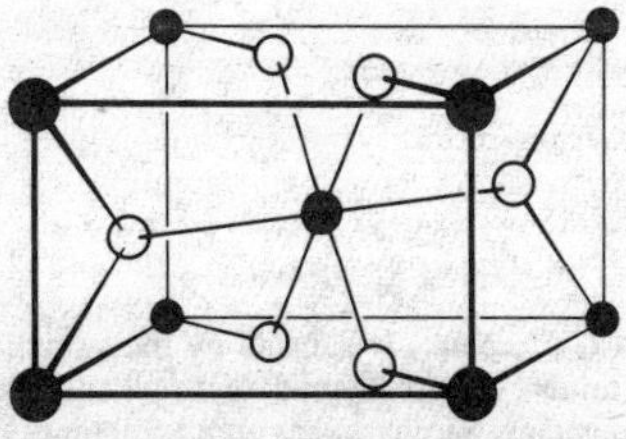

Crystal structure of rutile

can be extracted in crystalline form from buckwheat. It has vitamin P activity and is used in the treatment of hypertension associated with capillary fragility.

Rydberg's constant *See* Balmer series

S

S Sulphur.

sabinene, $C_{10}H_{16}$. Colourless oil; b.p. 165°C.

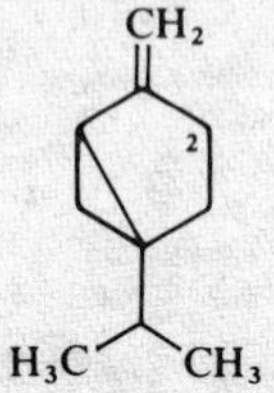

A dicyclic monoterpene, the (+)-form of which is found in oil of savin and many other essential oils. The (−)- and racemic forms occur occasionally in nature.

sabinol, $C_{10}H_{16}O$. The 2-hydroxy derivative of sabinene. Colourless oil; b.p. 208°C. (+)-Sabinol is found both free and as the ethanoate in oil of savin and also in the oil from *Juniperis phoenicea.* It can be reduced by hydrogen with Pt or Pd catalysts at ordinary temperatures to form (−)-dihydrosabinol which is identical with (−)-thujyl alcohol.

saccharates Normally the salts of saccharic acid, but the term is also used in the sugar industry for the alkali and alkaline earth compounds of sucrose: strictly these should be called sucrates.

saccharic acid The saccharic acids have the formula $HOOC{\cdot}[CHOH]_4{\cdot}COOH$, and are produced by the oxidation of the corresponding aldoses with nitric acid. Ten possible isomeric forms are known.

By saccharic acid is usually meant D-glucosaccharic acid, m.p. 125–126°C, obtained by the oxidation of glucose or starch. This exists in water solution in equilibrium with its two γ lactones, both of which can be obtained crystalline, though the acid itself does not crystallize readily.

saccharin, $C_7H_5NO_3S$. White crystals; m.p.

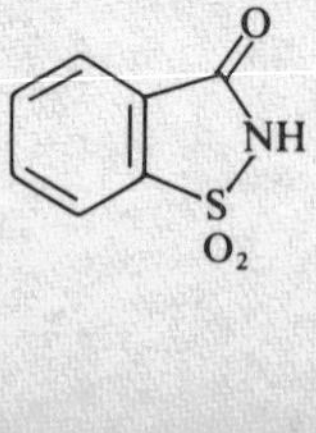

224°C (decomp.) It is made by the oxidation of toluene-*o*-sulphonamide with alkaline permanganate. Saccharin has about 550 times the sweetening power of sucrose, and is used extensively as a sweetening agent, usually in the form of the sodium salt. The use of saccharin is restricted in the U.S.

safflower oil An important oleic-linoleic acid oil from safflower seed used in drying oils* and in alkyd resins.

safrole, 3,4-methylenedioxyallylbenzene, shikimole, $C_{10}H_{10}O_2$. Colourless liquid; m.p. 11°C, b.p. 232°C. It is obtained from oil of camphor, and is the chief constituent of oil of sassafras and other essential oils.

sal ammoniac, NH_4Cl. *See* ammonium chloride.

salicin, $C_{13}H_{18}O_7$. The β-D-glucoside of *o*-hydroxybenzyl alcohol,
$C_6H_{11}O_5{\cdot}O{\cdot}C_6H_4{\cdot}CH_2OH$. Colourless, bitter crystals, m.p. 201°C; soluble in water and alcohol, insoluble in chloroform. It occurs in the leaves, bark and twigs of species of willow and poplar. On oxidation with dilute nitric acid it is converted into helicin, the glucoside of salicylaldehyde, which has been made the starting point of further syntheses. Gives populin with benzoyl chloride.

salicyl alcohol, saligenin, 2-hydroxybenzyl alcohol, $C_7H_8O_2$. White crystals; m.p. 87°C. It is obtained by the reduction of salicylaldehyde or by hydrolysis of the glucoside salicin.

salicylaldehyde, 2-hydroxybenzaldehyde,
$C_7H_6O_2$ Oily liquid of aromatic odour; b.p. 196°C. It is prepared by the action of chloroform and caustic potash on phenol (the Reimer–Tiemann reaction) or by the oxidation of the glucoside salicin. It is easily reduced to salicyl alcohol or oxidized to salicylic acid.

salicylamide, $C_7H_7NO_2$, 2-hydroxybenzamide. White crystalline powder, m.p. 140°C. It has antipyretic, analgesic and antirheumatic properties.

salicylic acid, 2-hydroxybenzoic acid, $C_7H_6O_3$. Colourless needles; m.p. 159°C. At 200°C it decomposes to phenol and carbon dioxide.

Occurs naturally as methyl salicylate*, oil

of wintergreen, from which it can be obtained by hydrolysis with alcoholic KOH. It is manufactured by heating NaOPh with carbon dioxide under pressure and acidifying the sodium salicylate formed. The production of various pharmaceuticals, principally aspirin, accounts for the bulk of salicylic acid consumption, but it is also used in the manufacture of dyes, liniments and rust-releasing fluids.

saligenin *See* salicyl alcohol.

salol, phenyl salicylate, $C_{13}H_{10}O_3$. Colourless crystals; m.p. 43°C. It is prepared by heating sodium salicylate and NaOPh in the presence of $POCl_3$. It is used as an enteric coating for pills, and as a paint for throat ailments. Its use as an intestinal antiseptic is doubtful as doses large enough to be effective may produce toxic effects due to liberation of phenol.

salt Commonly sodium chloride. More generally a salt is the substance produced by interaction of equivalent quantities of acid and base. To consider aquo-acids: if replacement is complete the salt is said to be normal (e.g. Na_2SO_4), and if only partial an acid salt is formed (e.g. $NaHSO_4$). In general, only the salts of strong acids with strong bases are stable in solution. If either or both constituents are weak, hydrolysis occurs and the solution becomes acidic or basic.

salt bridge In many electrochemical cells, electrical conductivity may be achieved whilst keeping the two solutions in the cell apart by means of a porous plug. The resistance of the solution in this case is very high. In practice the liquid-junction potential may be greatly reduced by means of a salt bridge, which consists of a tube containing a saturated solution of a salt (usually KCl) in an organic gel, e.g. agar. The ends of the salt bridge dip into the two half cells and the current is carried by the K^+ and Cl^- ions which have approximately the same ionic velocities.

salt hydrates Clathrates* formed by frameworks of hydrogen-bonded water and anions with cations occupying holes in the lattice. Examples have very high water content, such as $(Pr^n{}_3S)F,20H_2O$.

salting out A term used for both colloidal and non-colloidal solutions. In the context of colloids, salting out refers to the coagulation of hydrophilic colloids, e.g. gelatin or soaps, by the addition of large concentrations of electrolytes. This should not be confused with the coagulation of hydrophobic sols by small quantities of electrolytes. With non-colloidal solutions the term salting out is used to describe the reduction in solubility of a non-electrolyte by addition of a strong electrolyte.

sal volatile Ammonia, ammonium carbonate, oil of lemon and nutmeg in an alcoholic solution used as a first aid stimulant and also in dyspepsia medicines.

samarium, Sm. At.no. 62, at.wt. 150·36, m.p. 1072°C, b.p. 1778°C, *d* 7·52. A typical lanthanide*. $SmCo_5$ is used in permanent magnets. The oxide is used in optical glasses.

samarium compounds Samarium shows the +3 and +2 oxidation states. Sm(III) compounds are typical lanthanide compounds*,

$$Sm^{3+}\ (f^5 \text{ pale yellow}) \longrightarrow Sm\ (-2{\cdot}41 \text{ volts in acid})$$

SmX_2 (SmX_3 + Sm) are red–brown in colour, SmO is formed similarly.

$$Sm^{3+} \longrightarrow Sm^{2+}\ (-1{\cdot}55 \text{ volts})$$

and Sm^{2+} salts are rapidly oxidized by water.

sampling The process of obtaining a small quantity of material, e.g. for analysis, truly representative of the whole. Samples can be withdrawn by hand or using a mechanical sampler. Taking a sample of a sample is known as secondary sampling (*see* coning and quartering). Heterogeneous mixtures must be handled statistically, homogeneous samples (gases or liquids) are handled simply.

sand An accumulation of minerals, mainly SiO_2, resulting from rock disintegration.

Sandmeyer's reaction The replacement of a diazonium group by a halogen or pseudohalogen atom or group. It allows the conversion of an aromatic primary amine into the corresponding halogen compound. The amine is diazotized, and the diazonium salt solution is treated with CuCl, etc., causing the evolution of nitrogen and isolation of the halobenzene. Chloride, bromide and cyanide react easily; potassium iodide is sufficient for preparation of the iodo-compound. Aromatic fluorides are prepared after isolation and thermal decomposition of the diazonium fluoroborate salt. *See also* Gattermann's reaction.

sandwich compounds A trivial name, originally applied to the bis-π-cyclopentadienyl-metal compounds like ferrocene but since used variously to cover all the compounds where a metal atom is disposed between two hydrocarbon residues such as cyclopentadienyl, benzene, cyclohexadienyl, cyclobutadiene, tropylium, cyclo-octatetraene, etc. Those compounds having only one such ring are correspondingly called half sandwich compounds, e.g. cyclopentadienylmanganese tricarbonyl,

$C_5H_5Mn(CO)_3$. Triple-decker derivatives are also known. *See* organometallic compounds, metallocenes, ferrocene.

sandwich plastics *See* reinforced plastics.

santene, C_9H_{14}. Colourless oil; b.p. 140–

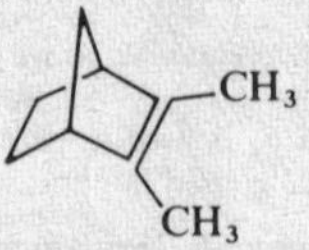

142°C. The only terpene-like natural hydrocarbon containing nine C atoms. It is found in East Indian sandalwood oil and also in Siberian, German and Swedish pine-needle oils. It has a rather unpleasant smell, resinifies easily and is difficult to obtain pure.

santenone, π-norcamphor, $C_9H_{14}O$. A crystalline solid, m.p. 56–59°C, of camphor-like smell. A saturated dicyclic ketone related to santene. (−)-Santenone is found in East Indian sandalwood oil. The optically inactive

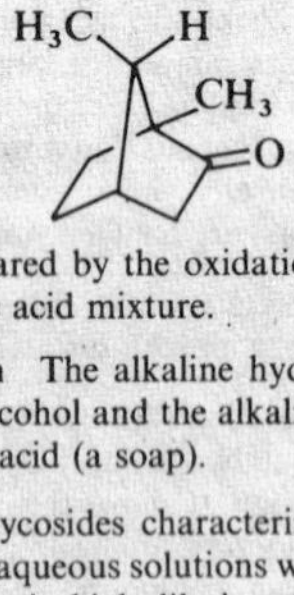

form is prepared by the oxidation of santene with chromic acid mixture.

saponification The alkaline hydrolysis of an ester to an alcohol and the alkali-metal salt of a carboxylic acid (a soap).

saponins Glycosides characterized by forming colloidal aqueous solutions which foam on shaking. Even in high dilution they effect hydrolysis of the red blood cells. The saponins themselves are hardly known in the pure state but their aglucones, termed sapogenins, have been characterized.

They are found in a great variety of plants and are strong fish poisons. On hydrolysis they yield a variety of sugars, frequently several molecules to each aglucone. Glucose, galactose, arabinose are the more common; pentoses, methylpentoses and glucuronic acid are also obtained.

sapphire As a gemstone a blue form of aluminium oxide* (corundum) containing some metal impurities. Now prepared artificially by fusing Al_2O_3. Used as a very inert material, e.g. for handling HF.

sarcosine, N-methylglycine, N-methylaminoethanoic acid, $C_3H_7NO_2$, $CH_3{\cdot}NH{\cdot}CH_2{\cdot}CO_2H$. Colourless crystals; m.p. 213°C (decomp.). It is obtained on hydrolysing creatine with alkalis.

satin white A mixture of $CaSO_4$ and $Al(OH)_3$ ($Al_2(SO_4)_3$ or alum plus $Ca(OH)_2$). Used as a pigment and for surfacing paper.

saturated compound A substance in which the atoms are linked by single bonds, and which has no double or triple bonds, e.g. ethane.

However, the term 'saturated' is often applied to compounds containing double or triple bonds which do not easily undergo addition reactions. Thus ethanoic acid is termed a saturated carboxylic acid and acetonitrile a saturated nitrile, whereas a Schiff base* is considered to be unsaturated.

sawhorse projections The Fischer projection* inadequately portrays the spatial relationship between ligands (atoms, groups) attached to adjacent atoms. The sawhorse projection attempts to clarify the relative location of the ligands. E.g. the Fischer projection of an enantiomer of the simple sugar threose is shown (i) together with a sawhorse projection (ii), or (iii) which is obtained by rotation of the rear group through 180°. The Newman projection

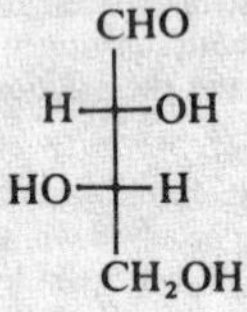

(i)

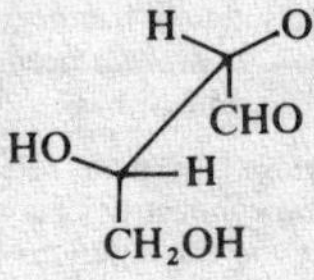

(ii)

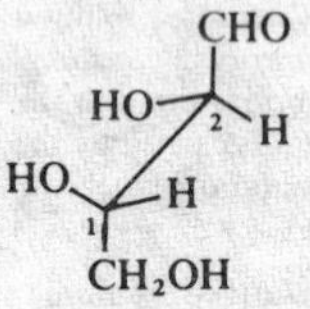

(iii)

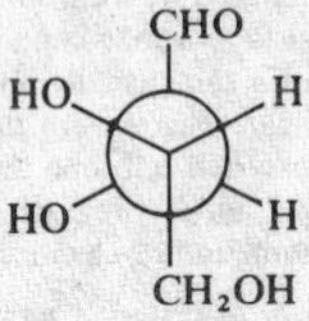

(iv)

(iv) is an alternative method of representing the structure when viewed along the C_1-C_2 axis in (iii), carbon 2 being represented as the full circle.

Sb Antimony.

SBP spirits Special boiling point spirits.

SBR Styrene butadiene rubber.

Sc Scandium.

scandium, Sc. At.no. 21, at.wt. 44·96, m.p.

1539°C, b.p. 2832°C, *d* 3·0. An element of Group III with electronic configuration $4s^2 3d^1$. The principal Sc ore is thortveitite ($Sc_2Si_2O_7$) and monazite and other lanthanide ores contain some Sc. Separated from the lanthanides by ion exchange. The metal ($ScF_3 + Ca$) is fairly reactive and is used to strengthen alloys. ScC hardens TiC to make it the second hardest substance known. Sc_2O_3 is used in electronic components, ScI_3 is added to high-intensity electric light bulbs. World demand 1976 10Kg.

scandium compounds Scandium shows only the +3 oxidation state and its chemistry is much more similar to that of aluminium* than that of the lanthanides*

$$Sc^{3+} \longrightarrow Sc\ (-1{\cdot}88 \text{ volts in acid})$$

The colourless Sc^{3+} is much smaller than the lanthanide^{3+} ions and the maximum co-ordination number is 6. Halides, ScX_3, are formed by interaction of the elements, by precipitation from aqueous solution (ScF_3) or by removal of water from, e.g., $ScCl_3,6H_2O$ with, e.g. $SOCl_2$. Compounds such as $ScBr_{2{\cdot}3}$, $ScI_{2{\cdot}17}$ ($ScX_3 + Sc$) probably contain metal–metal bonds. Complex fluorides of the cryolite type M_3ScF_6 are formed from solution or melts. $ScCl_3$ is hydrolysed by water to a hydrous oxide. ScO(OH), the complex ion $Sc(OH)_6^{3-}$ and mixed oxides, e.g. $LiScO_2$ ($Li_2O + Sc_2O_3$), are also known; Sc_2O_3 is very similar to Al_2O_3. Oxyacid salts include the sulphate and the nitrate.

scavengers When tetraethyl* or tetramethyl* lead are added to gasoline as anti-knock additives*, lead scavengers must be used to ensure that lead oxide is not deposited in combustion chambers. Mixtures of dibromoethane and dichloroethane are added to the lead alkyls in order to produce volatile lead halides which are removed with the exhaust gases.

Schäffer's acid, $C_{10}H_8O_4S$. 2-hydroxy-7-naphthalene sulphonic acid. Obtained by sulphonating 2-naphthol with a small amount of sulphuric acid at a higher temperature than is used for the preparation of crocein acid. A valuable dyestuff intermediate.

scheelite A tungsten mineral $CaWO_4$. An important structure type.

Schiff's bases *N*-Arylimides, $Ar{\cdot}N{=}CR_2$, prepared by reaction of aromatic amines with aliphatic or aromatic aldehydes and ketones. They are crystalline, weakly basic compounds which give hydrochlorides in non-aqueous solvents. With dilute aqueous acids the parent amine and carbonyl compounds are regenerated. Reduction with sodium and alcohol gives a secondary base. Secondary bases can also be obtained from the reactions with Grignard reagents and with alkyl halides.

Schiff's reagent A solution of rosaniline in water decolorized with sulphurous acid. Aliphatic aldehydes and aldose sugars give a magenta colour with this reagent; with aromatic aldehydes and aliphatic ketones the colour develops more slowly; aromatic ketones do not react.

Schomaker–Stevenson equation The equation $r_{A-B} = r_A + r_B - 0{\cdot}09\,(X_A - X_B)$ relating the bond length r_{A-B} to the individual radii r_A and r_B of the two atoms concerned and the electronegativities X_A and X_B of the two atoms concerned in the bond. This relation is only empirical and is not very accurate.

schönite, K_2SO_4, $MgSO_4,6H_2O$. A mineral obtained from the Stassfurt salt deposits and used as a source of potassium sulphate. A series of schönites, analogous to the alums, is formed containing dipositive ions such as $Zn^{2+}, Co^{2+}, Ni^{2+}, Fe^{2+}, Cu^{2+}, Mn^{2+}, V^{2+}$, in place of magnesium.

Schotten-Baumann reaction *See* benzoyl chloride.

Schottky defect *See* defect structures.

Schradan, octamethylpyrophosphoramide, $C_8H_{24}N_4O_3P_2$, $(Me_2N)_2P(O)OP(O)(NMe_2)_2$. It is not highly toxic to insects when used as a contact insecticide, but is readily absorbed by the roots and leaves of plants and translocated in the sap, so that the plant becomes toxic to species feeding on it.

Schrödinger wave equation The fundamental equation of wave mechanics which relates energy to field. The equation which gives the most probable positions of any particle, when it is behaving in a wave form, in terms of the field.

Schweinfürter green *See* emerald green.

Schweizer's reagent The dark blue solution obtained by dissolving $Cu(OH)_2$ in concentrated ammonia solution. Used as a solvent for cellulose, the cellulose is precipitated on acidification. Used in the cuprammonium process for the manufacture of rayon.

scintillation counting The detection and quantitative estimation of radioactivity by the scintillations produced when the ionizing radiation interacts with a scintillating medium. This may be a screen of material such as zinc sulphide (for α-particles), a lithium-doped sodium iodide crystal (for γ-rays) or liquid scintillators (β-particles).

scleroproteins *See* proteins.

scopolamine *See* hyoscine.

screened indicator *See* colour indicators.

screening Separation of solid material into fractions containing particles whose sizes lie within a certain range, by the use of sieves.

screens Used in screening.

scrubbers These remove impurity from a gas by washing it with a liquid.

Se Selenium.

sea water An aqueous system of pH 8·0–8·4 (buffered by carbonate, Ca^{2+}, etc.). The major constituents are (p.p.m.) Cl^- (18 980); Na^+ (10 556); SO_4^{2-} (2649); Mg^{2+} (1272); Ca^{2+} (400); K^+ (380); HCO_3^- (140); Br^- (65); H_3BO_3 (26); Sr^{2+} (8). Used for production of bromine and magnesium. Potential metal sources (Mn, etc.) are available on the sea bed.

seaweed colloids The polysaccharide materials agar, algin, carrageenan prepared by extraction of algae (particularly seaweeds). Important in foodstuffs and textiles.

sebacic acid, decanedioic acid, $HOOC{\cdot}[CH_2]_8{\cdot}COOH$, $C_{10}H_{18}O_4$. Colourless leaflets; m.p. 134°C. Manufactured by heating castor oil with alkalis or by distillation of oleic acid. Forms an anhydride, m.p. 78°C. The esters of sebacic acid are used as plasticizers, especially for vinyl resins.

secondary radiation Radiation produced by the absorption of some other, generally more energetic, radiation. Thus, when X-rays* strike a body it emits other X-rays, which are termed secondary rays, or secondary radiations, characteristic of the atom which acts as the emitter. This phenomenon is also observed with electrons and is then termed secondary emission.

sedimentation Gravitational settling of solid particles suspended in a liquid so that the original suspension is divided into a reasonably clear effluent and a sludge with an increased solid content. *See* clarification, thickening, segregation.

sedimentation potential The measurable potential difference which results from the motion of the disperse phase through the dispersion medium, when the particles of a suspension are allowed to settle under the influence of gravity.

seed crystals Crystals which are added to a supersaturated solution to facilitate crystallization of the whole.

segregation The tendency for solid particles, either dry or in suspension in a liquid, to arrange themselves in different layers according to size.

seignette salt, $C_4H_4O_6KNa,4H_2O$. Sodium potassium tartrate.

selection rules Rules derived from group theory which denote, e.g. whether a particular vibration is i.r. or Raman active or a particular transition between electronic energy levels is allowed or forbidden.

selecto A mixture of phenols and cresols formerly used for the solvent extraction of aromatics, etc., in the refining of petroleum products.

selenates Normally salts containing $[SeO_4]^{2-}$, $[HSeO_4]^-$, i.e. selenates(VI). Prepared from H_2SeO_4 or by electrolytic or HNO_3 oxidation of selenates(IV).

Selenates(VI) resemble sulphates but can be reduced more easily (E° to SeO_3^{2-} +1·15 volts). Alums* are formed. Pyroselenates, $[O_3SeOSeO_3]^{2-}$ are known.

Selenates(IV) contain the $[SeO_3]^{2-}$ and related species. They are hydrolysed in solution and give complexes.

Selenates are poisonous.

selenic acid, H_2SeO_4. *See* selenium oxyacids.

selenides Binary compounds of selenium with other elements. Generally prepared by reaction of the elements but some result from H_2Se on the element or a derivative. Similar to sulphides but less stable, more easily oxidized and more easily hydrolysed to H_2Se.

selenious acid, H_2SeO_3. *See* selenium oxyacids.

selenites Salts containing $[SeO_3]^{2-}$ ions. *See* selenates(IV).

selenium, Se. At.no. 34, at.wt. 78·96, m.p. 217°C, b.p. 684·9°C, *d* 4·79. Occurs as selenide impurities in sulphide ores and as clausthalite (PbSe) and crookesite ($(CuTlAg)_2Se$). Extracted from the flue dusts where sulphide ores are used, dissolved (KCN solution) and precipitated with HCl. Stable grey Se is metallic in appearance and contains Se_n chains with some cross-linking. It is a photoconductor (xerography, photo-electric cells). The red form is precipitated from solutions with SO_2 and contains Se_8 units – it is soluble in CS_2. Se reacts with many elements to give, e.g., oxides or halides. Additional uses are as a steel additive and in glass. Selenium is an essential trace element but is toxic in excess. World demand 1976 1300 tonnes.

selenium bromides Selenium monobromide, Se_2Br_2. Deep red liquid; b.p. 227°C (decomp.) ($SeBr_4$ or Br_2 on Se). Selenium tetrabromide, $SeBr_4$. Yellow solid readily loses Br_2 to give Se_2Br_2 (Se plus excess Br_2). $[SeBr_6]^{2-}$ formed with alkali bromides. Selenium dibromide, $SeBr_2$, is only stable in the vapour.

selenium chemistry Selenium is a member of the oxygen group, electronic configuration $4s^24p^4$. It is a non-metal and shows little true cation chemistry ($Se_8{}^{2+}$, etc. cations are formed with SbF_5). The normal oxidation states are +6 (octahedral in SeF_6, tetrahedral in $(SeO_4)^{2-}$), +4 (one lone-pair), +2 (two lone-pairs) and −2 (selenides, H_2Se, etc.). Selenium(IV) compounds particularly act as acceptors. Selenium(II) compounds are good donors.

selenium chlorides
Selenium monochloride, Se_2Cl_2; m.p. −85°C 130°C (decomp.). Reddish liquid, a good chlorinating agent (Cl_2 plus Se).

Selenium tetrachloride, $SeCl_4$; m.p. 305°C. Yellowish solid (Cl_2 plus Se_2Cl_2). Forms hexachloroselenates containing $(SeCl_6)^{2-}$ in conc. HCl solution.

Selenium dichloride, $SeCl_2$, is only stable in the gas phase.

selenium fluorides
Selenium tetrafluoride, SeF_4; m.p. −14°C, b.p. 106°C. Colourless liquid (F_2 on sublimed Se). Hydrolysed by water.

Selenium hexafluoride, SeF_6; m.p. −39°C, sublimes −47°C. Formed (with Se_2F_{10}) from Se and F_2. Chemically fairly inert.

SeO_2F_2 is also known ($BaSeO_4$ plus HSO_3F).

selenium, organic derivatives These are similar to the simpler sulphur derivatives. Much of the chemistry involves selenium(II) derivatives although some selenium(IV) organic derivatives are known.

selenium oxides
Selenium dioxide, SeO_2, sublimes 315°C. White solid (Se plus O_2) with a polymeric structure. Forms H_2SeO_3 with water. Used as an oxidizing agent in organic chemistry (α-hydrogens to C=O groups). $SSeO_3$ is formed from Se and molten SO_3.

Selenium trioxide, SeO_3. M.p. 120°C. Formed Se vapour plus O_2 in electric discharge. It is white and gives H_2SeO_4 with water.

selenium oxyacids The system is very simple by comparison with sulphur.

Selenic acid, H_2SeO_4. Formed by oxidation of selenates(IV) (Cl_2, $MnO_4{}^-$). A strong acid similar to H_2SO_4 but loses O_2 on heating.

Selenious acid, H_2SeO_3. Formed water on SeO_2. More stable than H_2SO_3, oxidized to H_2SeO_4, reduced to Se by SO_2. Gives selenates(IV) with oxides, hydroxides or carbonates.

Selenosulphuric acid, H_2SeSO_3. An analogue of thiosulphuric acid; the free acid does not exist (Se plus $(SO_3)^{2-}$). Selenosulphates readily decompose.

selenocyanates Derivatives containing the SeCN grouping (KSeCN from KCN plus Se).

selenonium ions Three co-ordinate selenium-containing cations, e.g. $[Me_3Se]^+$.

self-ionization The process of partial dissociation into cations and anions envisaged for self-ionizing solvents. E.g.

$$2H_2O \rightleftharpoons [H_3O]^+ + OH^-$$

$$2NH_3 \rightleftharpoons [NH_4]^+ + NH_2{}^-$$

Ionic species containing the characteristic cation – e.g. NH_4Cl – are acids; species containing the characteristic anion are bases, e.g. KNH_2.

semicarbazide, aminourea,
$NH_2{\cdot}CO{\cdot}NH{\cdot}NH_2$,$CH_5N_3O$. Colourless crystalline substance; m.p. 96°C. Prepared by the electrolytic reduction of nitrourea in 20% sulphuric acid at 10°C. Forms crystalline salts with acids. Reacts with aldehydes and ketones to give semicarbazones. Used for the isolation and identification of aldehydes and ketones.

semicarbazones Organic compounds containing the group $>C{:}N{\cdot}NH{\cdot}CONH_2$. They are formed by treating a solution of an aldehyde or ketone in alcohol, dilute ethanoic acid, or pyridine with a concentrated solution of semicarbazide hydrochloride. They have definite melting points and are used to isolate and identify aldehydes and ketones.

semi-conductors Imperfections or impurities in what are normally insulators may give rise to a temperature-dependent conductivity (metallic conductivity decreases with rise in temperature) arising because the highest occupied energy level is very close to an unoccupied level. Such materials are called semi-conductors. There are two main types: $MX_{1-x}(e)_x$ with an electron (e) in an anion vacancy, e.g. (NaCl) and $M_{1+x}X\text{-}(e)_x$ with an interstitial M and an electron in its vicinity (e.g. ZnO) which conduct by movements of the electrons and are called *n* type,

$(+)_xM_{1-x}X$ with a vacant metal site and a neighbouring +ve charge (M^{2+}) in the vicinity (e.g. Cu_2O) and $(+)_xMX_{1+x}$ with an interstitial anion and a neighbouring +ve charge (M^{2+}) (e.g. VO_2) conduct formally by movement of the +ve charge (actually by electron movement) and are called *p* type. Semi-conductors are of great importance in electronics and are used in, e.g. transistors, thermistors, solid rectifiers and microprocessors.

semi-permeable membrane A membrane which is permeable to some substances, e.g. water, but not to others, e.g. salt, ions, sugar molecules. Commonly used semi-permeable membranes are composed of parchment, pig's bladder, cellophane and copper(II) cyanoferrate(II), $Cu_2[Fe(CN)_6]$.

semi-polar bond A semi-obsolete name for a co-ordinate bond*.

sensitizing dyes *See* spectral sensitization.

Sephadex A trade name for an insoluble hydrophilic substance prepared by cross-linking dextran, and used in gel filtration*. It can also be linked to acidic or basic groups for ion exchange* or to alkanes for the chromatography of lipophilic compounds.

sequence rules In order to give an organic compound an unambiguous name (*see* isomerism) or three-dimensional configuration (*see* R,S convention) it becomes necessary to designate the order of groups or atoms attached to a particular atom. Several sequence rules have been introduced but those most widely accepted attach an agreed order of precedence to a ligand thus:

(1) Ligands are arranged in order of decreasing atomic number.

(2) When atoms are of similar atomic number (isotopes) they are arranged in decreasing mass number.

(3) When necessary a lone pair of electrons are considered a substituent and given a priority lower than ^1H.

(4) When similar atoms are attached to a particular atom, priority order is determined by the nature of the atoms further away.

(5) Multiple linkages are treated formalistically as multiple single bonds, e.g. $C{=}O$ becomes 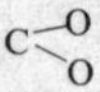

and $C{\equiv}N$ becomes 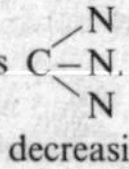

Thus in order of decreasing priority.

Atoms: I, Br, Cl, F, O, N, C, H, lone pair.

Isotopes: ^2H, ^1H.

Groups: $-C(CH_3)_3$, $-CH(CH_3)_2$, $-CH_2CH_3$, CH_3.

$-C(CH_3)_3$, $-CH(CH_3)CH_2CH_3$, $-CH_2CH(CH_3)_2$, $-(CH_2)_3CH_3$.

$-COOCH_3$, $-COOH$, $-CONH_2$, $-CHO$.

$-C{\equiv}N$, $-C_6H_5$, $-C{\equiv}CH$, $-CH{=}CH_2$.

sequestering agent, complexones Compounds which are very effective in forming complexes with metal cations (and thus with salts) and thus preventing them acting as simple hydrated cations. Ethylenediaminetetra-acetic acid (EDTA) and related derivatives form complexes with most M^{2+} and M^{3+} species. Gluconic acid and other hydroxy acids act similarly. Sequestering agents are used extensively, e.g. for preventing deleterious catalytic effects of traces of Fe or Cu and in analysis.

series, spectroscopic A group of lines the frequencies of which are simply related to each other and may be expressed by a simple formula (e.g. the Balmer series).

serine, 2-amino-3-hydroxypropanoic acid, $C_3H_7NO_3$, $CH_2OH{\cdot}CHNH_2{\cdot}CO_2H$. Colourless crystals; m.p. 228°C (decomp.). It is one of the amino-acids present in small quantities among the hydrolysis products of proteins.

sesqui[oxide] A semi-obsolete term to denote a ratio of 3 [oxygen] atoms to 2 of the cationic element (e.g. Fe_2O_3 – now denoted iron(III) oxide).

sesquiterpenes *See* terpenes.

sessile dislocation A dislocation in a crystal lattice which cannot glide since the Burger's vector is not in the plane of slip of the dislocation. *See* dislocation.

sewage treatment Waste water from industrial or domestic sources is initially filtered to remove solids and colloidal materials are then removed with micro-organisms which use the waste as food. Methane and other gases are evolved during anaerobic sludge digestion and can be used as fuels. High protein animal-feed supplements are also produced by this process.

sextet A series of six closely-spaced transitions in a spectrum.

shale oil Oil produced by heating oil shales in continuous retorts at temperatures up to 950 K.

Although estimated world reserves of shale are very high and shales contain other potentially useful materials, oil production has, in the past, been too expensive for large-scale exploitation.

shear structures Crystal structures in which it appears that a regular array of, e.g. corner-linked octahedra, is interrupted along a line by, e.g. edge-sharing octahedra. Found in heavy metal oxides, e.g. tungsten oxides.

shellac A resinous secretion of the insect *Tachardia lacca* used for its content of resinous matter. Shellac yields aleuritic acid, $C_{13}H_{26}O_4$, m.p. 101°C, on extraction with alkali. It is used for the preparation of French polish, as a resin binder for mica-based insulating materials and in sealing waxes.

sherardizing, vapour galvanizing Coating metal articles (iron) by heating with Zn dust in a closed vessel somewhat below the m.p. of Zn.

shift conversion, shift reaction An important reaction in which carbon monoxide reacts with steam to produce hydrogen and carbon dioxide

$$CO + H_2O \longrightarrow CO_2 + H_2$$

Carbon monoxide and excess steam are normally passed over a cobalt catalyst at about 250–300°C resulting in greater than 99% conversion of CO to CO_2. This conversion reaction is widely used in oil or solid fuel gasification processes for the production of town gas* or substitute natural gas*.

shift reagents Paramagnetic species which are magnetically anisotropic can cause large n.m.r. shifts in any species with which they form weak complexes. Stable paramagnetic species, particularly lanthanide diketonates, are added to specimens in order to simplify complex n.m.r. spectra.

shikimic acid, $C_7H_{10}O_5$. M.p. 190°C. A 3,4,5-trihydroxycyclohex-1-enecarboxylic

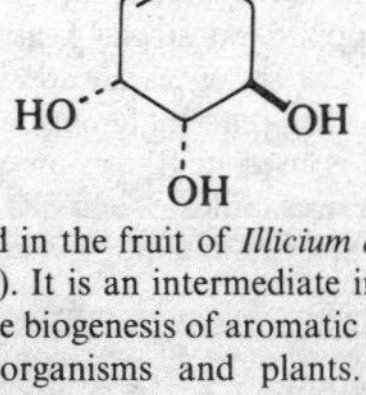

acid found in the fruit of *Illicium anisatum* (*I. religiosum*). It is an intermediate in one pathway for the biogenesis of aromatic compounds in micro-organisms and plants. Important compounds arising by the shikimic acid route include *p*-aminobenzoic acid, phenylalanine, tyrosine, tryptophan and *p*-hydroxybenzoic acid. *See* chorismic acid.

The initial step in the pathway is the condensation of erythrose-4-phosphate with phosphoenolpyruvate, yielding dehydroquinic acid, which by elimination of the elements of water affords dehydroshikimic acid; reduction of the 3-keto group to hydroxyl gives shikimic acid.

Si Silicon.

sialic acid A generic name for *N*- and *O*-acyl derivatives of neuraminic acid*, e.g. acetylneuraminic acid. They are ubiquitously distributed in tissues and bacterial cell wall structures.

sialons Si-Al-O-N systems used in machine tools.

side reactions Few reactions proceed quantitatively to one product. Generally a number of reactions occur. The predominant reaction is the main reaction; other reactions are termed side reactions.

Siemens's process A method of steel manufacture by addition of scrap metal, iron oxides and ferromanganese to molten pig (cast) iron.

sienna Earth containing hydrated Fe_2O_3 used as stain and coloured filler for wood and other materials.

sigma bond, σ-bonds Covalent bond con-

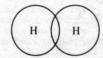

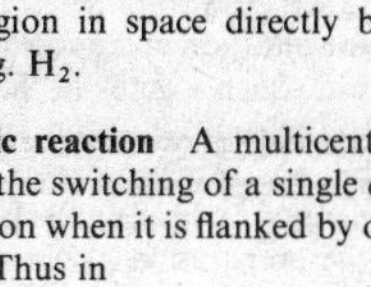

taining two electrons with the major overlap between the orbitals forming the bond being in one region in space directly between the nuclei. E.g. H_2.

sigmatropic reaction A multicentre reaction involving the switching of a single σ-bond to a new position when it is flanked by one or more π-bonds. Thus in

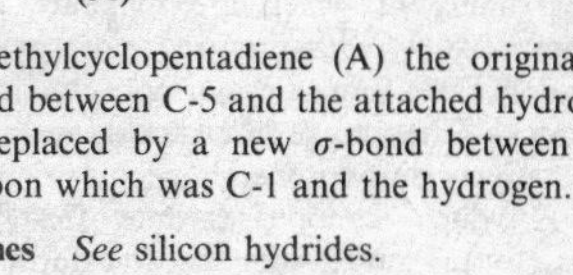

(A)

5-methylcyclopentadiene (A) the original σ-bond between C-5 and the attached hydrogen is replaced by a new σ-bond between the carbon which was C-1 and the hydrogen.

silanes *See* silicon hydrides.

silatranes Internal complexes of silicon. Some of these complexes have physiological activity. E.g.

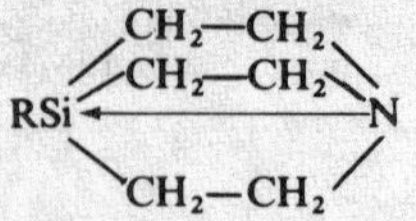

silazanes Compounds containing Si–N bonds. *See* silicon nitrides.

silica, silicon dioxide, SiO_2. One of the most important constituents of the earth's crust. Polymorphic, quartz (to 573°C), tridymite (to 1470°C), cristobalite (to 1710°C m.p.). Each form has two modifications and all consist of infinitely linked SiO_4 tetrahedra. The melt can solidify to a glass (naturally occurring form obsidian), quartz crystals are non-centro-symmetric and can be right- or left-handed. High pressure forms of SiO_2 are also known, coesite and keatite contain SiO_4 units whilst stishovite contains SiO_6 octahedra in the rutile structure. Naturally occurring silica is frequently coloured, e.g. by Fe; purple quartz is amethyst. Common flint is amorphous silica. Silica ware is used extensively in laboratories.

silica gel An amorphous form of hydrated silica produced by the precipitation, flocculation, or coagulation of a silica sol or the decomposition of some silicates. When freshly prepared the silica appears to be in a gelatinous state and, on standing, it will set and form a jelly. After being heated it cannot easily be reconverted into a sol.

Dehydrated silica gel is used commercially as an absorbent in the recovery of solvents, also for drying air, dehydrating gases, refining mineral oils, for filtration and as a support for various catalysts. Such a gel is in the form of hard granules, chemically and physically almost inert but highly hygroscopic. After use it can be regenerated by heat. In the laboratory it is used in the preparation of chromatographic columns.

Cobalt(II) compounds are often incorporated in the gel. When the gel has absorbed a fair amount of moisture it will turn pink but will revert to blue on heating.

silicates A very extensive group of substances, many of which are minerals, derivatives of SiO_2 acting as an acidic oxide. Natural silicates form a major constituent of most rocks. They range in composition from comparatively simple minerals, such as zircon, $ZrSiO_4$, which contain discrete $[SiO_4]^{4-}$ anions, to far more complex structures in which the silicate anion is an extended two- or three-dimensional structure in which oxygen atoms form bridges between the silicon atoms. All silicates are based on the $[SiO_4]$ tetrahedron with partial or complete sharing of oxygen atoms. Al, B, Be, may replace silicon in this anion and the cations are in the holes in the lattice. Typical silicates are feldspar, mica, zeolites, aluminosilicates.

The alkali silicates are soluble in water and are used industrially. *See* sodium silicates.

silica vitreous, quartz glass The amorphous vitreous product formed on cooling fused silica. Transparent to u.v. light down to about 2000 Å, used in optical work. Low coefficient of expansion so that apparatus constructed of vitreous silica may be subjected to irregular heating or to sudden changes of temperature without risk of fracture.

silicic acids Addition of dilute sodium silicate solution, to excess of dilute HCl followed by dialysis gives a colloidal solution of hydrated silica which is readily coagulated to a gel. None of the silicic acids can be isolated as such. However, volatile esters, such as $Si(OMe)_4$, b.p. 121°C, are well known.

silicides Binary compounds with silicon, e.g. Mg_2Si. Silicides are related structurally to carbides and are hard and intractable solids. They form a mixture of silicon hydrides and hydrogen with acids.

silicofluorides *See* fluorosilicates.

silicon, Si. At.no. 14, at.wt. 28·0855, m.p. 1410°C, b.p. 2355°C, *d* 2·33. Group IV element which does not occur native but is the second most abundant element in the earth's crust occurring as SiO_2 (*see* silica) and in the many silicates. Crude Si is obtained by reduction of SiO_2 with carbon in an electric furnace. For pure Si this is converted to $SiCl_4$ or other halide (Cl_2) which is purified by fractional distillation, reduced to Si on a hot wire and finally purified by zone-refining. Si has a diamond lattice, is oxidized on strong heating in air, reacts with F_2 or Cl_2, gives silicates with fused alkalis. The ultra-pure material, doped with, e.g. B or P, is used in transistors and other solid state devices. Silicates and SiO_2 are used in glass, refractories, building materials, etc.

silicon bromides Silicon forms ranges of compounds containing Si–Br bonds. *Silicon tetrabromide*, $SiBr_4$, m.p. 5°C, b.p. 155°C, is the lowest binary bromide (Si plus Br_2 at red heat).

silicon carbide, carborundum SiC. A hard

refractory material made by heating C and Si or SiO_2 to 2000°C. Used for refractory bricks, crucibles, abrasive tools and refractory cement.

silicon chemistry Silicon is a relatively electropositive element of Group IV, electronic configuration $3s^2 3p^2$. It is normally in the +4 oxidation state 4-co-ordinate (5- and 6-co-ordinate complexes are known). Catenation and bridging though, e.g. C, N, O, is possible so that there is an extensive range of silicon chemistry corresponding to aliphatic-carbon chemistry with cyclization, etc. Terminal Si–X bonds can be exchanged; Si–H are reducing. Cationic species, e.g. $[Ph_3Sibipyridyl]^+$ are known. Silicon(II) species are important reactive intermediates, *see* silicon fluorides. Normally multiply bonded Si – e.g. the C=Si group – is rare, but π-bonding involving the empty d orbitals of Si is common, e.g. linear SiOSi units.

silicon chlorides
Silicon tetrachloride, $SiCl_4$; m.p. −70°C, b.p. 57°C. Colourless liquid (Si plus Cl_2 followed by distillation). Hydrolyses in moist air or in water.

Higher chlorides, Si_2Cl_6 to Si_6Cl_{14} (highly branched – some cyclic) are formed from $SiCl_4$ plus Si or a silicide or by amine catalysed disproportionations of Si_2Cl_6, etc. Partial hydrolysis gives oxide chlorides, e.g. $Cl_3SiOSiCl_3$. $SiCl_4$ is used for preparation of silicones.

silicone rubbers, $\text{-(}Me_2SiO\text{)-}_n$ (also other alkyl derivatives). Silicone* derivatives. Retain elastic properties over very wide temperature range, inert to acids, alkalis, oils and oxidizing agents.

silicones Polymeric organosilicon derivatives containing Si–O–Si links. Organosilicon halides are prepared RX plus Cu–Si or Ag–Si alloy; RX plus $SiCl_4$ over Zn or Al; $SiCl_4$ or Si–H compounds plus olefine over catalyst (hydrosilylation*). (These derivatives also formed $SiCl_4$ plus, e.g., LiR, RMgX on laboratory scale). The halides are hydrolysed to give polymers with properties depending upon the degree of polymerization (extent of hydrolysis). Silicones are used in greases, hydraulic fluids, sealing compounds, enamels, varnishes, resins, synthetic rubbers*. Resins are used in electrical insulation and laminates. They are odourless, colourless and inert, immiscible with water and have high flash points.

silicon ester *See* ethyl silicate.

silicon fluorides *Silicon tetrafluoride,* SiF_4; m.p. −90°C, b.p. −86°C. Colourless gas hydrolysed (fumes) in air gives $F_3SiOSiF_3$ then SiO_2. Prepared $SiCl_4$ plus KF; SiO_2 plus HF or HSO_3F. Forms many complexes including $(SiF_5)^-$ and $(SiF_6)^{2-}$, fluorosilicate ions. Higher fluorides are formed SiF_4 plus Si. Si plus SiF_4 at 1100°C and low pressure gives silicon difluoride, SiF_2. Stable for minutes at low pressure otherwise polymerizes to higher fluorides. Very reactive material, e.g. H_2S gives $HSiF_2(SH)$. Many organosilicon fluorides are known (Si–Cl plus KF), etc.

silicon hydrides, silanes Monosilane, SiH_4; b.p. −112°C. Formed SiO_2 or $SiCl_4$ plus $LiAlH_4$. Higher hydrides Si_2H_6 to Si_8H_{18} are formed from Mg_2Si and dil. HCl or Mg_2Si plus NH_4Br in liquid NH_3 or by an electric discharge on SiH_4. All are spontaneously inflammable in oxygen or air. Alkalis give SiOR plus H_2. SiH compounds are good reducing agents.

silicon nitrides Si_3N_4 formed Si plus N_2 at 1500°C. Grey inert solid, resistant to acid, gives NH_3 and silicates with alkalis. Many compounds containing Si–N bonds are known, e.g. heptamethyl disilazane, $Me_3SiNMeSiMe_3$ used to introduce NMe groups. Highly polymeric materials are known but they are more reactive than silicones.

silicon, organic derivatives Silicon forms a wide range of organic derivatives containing C–Si bonds (for preparative methods *see* silicones). R_3SiX derivatives, particularly Me_3SiF are very stable leaving groups (e.g. Me_3SiNMe_2 plus PF_5 gives F_4PNMe_2 plus Me_3SiF).

silicon oxide chlorides Formed by controlled hydrolysis of silicon chlorides*.

silicon oxides
Silicon monoxide, SiO. Formed SiO_2 plus C in electric furnace. The impure brown powder is used as a pigment and abrasive (Monex). Stable in vapour phase (Si plus SiO_2).

Silicon dioxide, SiO_2. *See* silica.

silk A protein fibre from the cocoons of the silk moth.

siloxanes Molecular compounds containing SiOSi groups, e.g. hexachlorosiloxane, $Cl_3SiOSiCl_3$ (controlled hydrolysis of $SiCl_4$; $SiCl_4$ plus air at high temperature). Poly(organosiloxanes) are silicones*.

silver, Ag. At.no. 47, at.wt. 107·8682, m.p. 962°C, b.p. 2212°C, *d* 10·49. Occurs as the metal and also as sulphide ores (argentite or silver glance, Ag_2S; pyrargyrite or ruby silver,

Ag_3SbS_3; and silver–copper glance $(Cu,Ag)_2S$). Silver is frequently recovered after the work up of Cu and Pb ores and is generally extracted as a complex cyanide or thiosulphate complex and is recovered by reduction with Zn and purified electrolytically. The metal is a pure white ductile material with a ccp structure. It dissolves in HNO_3 and hot conc. H_2SO_4 but is inert to alkalis. The metal is used extensively in jewellery, electrical components particularly conductors, contacts and batteries, dental and surgical components. It was formerly used extensively in mirrors formed by depositing silver on glass by a reduction process and as a coinage metal. Silver compounds are extensively used in photography and catalysis. World production 1976 9420 tonnes.

silver bromide, AgBr. A pale yellow solid, m.p. 420°C, precipitated from aqueous solution. Similar to silver chloride.

silver carbonate, Ag_2CO_3. Prepared by precipitation from a silver nitrate solution. It is a light-sensitive white solid which becomes yellow. It decomposes to Ag_2O above 100°C.

silver chemistry Silver is an element of the copper group, electronic configuration $4d^{10}$ $5s^1$. It shows oxidation states +4 and +3, +2, +1. The three former are strongly oxidizing

$$(Ag(II) \longrightarrow Ag(I) + 2{\cdot}00\ V \text{ in } 4\text{M } HClO_4)$$
$$(Ag^+ \longrightarrow Ag\ +0{\cdot}8\ V)$$

but are stable in fluorides, complex fluorides and oxides, and as nitrogen complexes, e.g. $Ag(bipyridyl)_2(NO_3)_2$. Ag(I) forms a stable hydrated ion and many complexes particularly with N, P and S ligands and olefines.

silver chloride, AgCl. A white solid, m.p. 449°C, precipitated from aqueous solution (the precipitation is used as a test and for analysis of Ag^+ or Cl^- ions; the precipitate is soluble in aqueous ammonia or sodium thiosulphate). Forms ammines with ammonia. Used extensively in photography. Sheets of AgCl are transparent to i.r. radiation and are used as support materials in i.r. spectroscopy. AgCl darkens on exposure to light.

silver fluorides
Silver(I) fluoride, AgF, is prepared by evaporation of a solution of excess Ag_2O in HF after filtration or by heating anhydrous $AgBF_4$. The anhydrous salt is yellow; hydrates are known. It is very soluble in water and in many organic solvents. Used as a mild fluorinating agent. On treatment of a solution with Ag a subfluoride, Ag_2F, is formed.

Black *silver(II) fluoride*, AgF_2 $(Ag + F_2)$, is a strong fluorinating agent.

Complex *silver(III) fluorides* containing AgF_4^- or AgF_6^{3-} ions and also Cs_2AgF_6 containing Ag(IV) are formed with fluorine. They are also strong fluorinating agents.

silver fluoroborate, $AgBF_4$. Prepared by the action of bromine trifluoride on silver borate or by passing boron trifluoride through a solution or suspension of silver(I) fluoride in acetonitrile, nitromethane, sulphur dioxide or toluene. It is a white solid, very soluble in water and ether, moderately soluble in benzene.

silver halide grains The individual silver halide crystals in a photographic emulsion.

silver nitrate, $AgNO_3$. The most important compound of silver, obtained by dissolving the metal in dilute nitric acid, and crystallizing. On strong heating silver nitrate decomposes into silver, N_2O_4, and oxygen. It has m.p. 212°C.

silver oxides
Silver(I) oxide, Ag_2O, is obtained as a brown amorphous precipitate by the action of alkali on $AgNO_3$ solution. It cannot be obtained pure and is soluble in ammonia solution.

Silver(II) oxide, AgO, is a black solid, $Ag^IAg^{III}O_2$, obtained by anodic or persulphate oxidation of an $AgNO_3$ solution. Continued anodic oxidation gives impure Ag_2O_3. Argentates, e.g. KAgO, containing silver(I) are known.

silver perchlorate, $AgClO_4$. Prepared by the action of $HClO_4$ on Ag_2CO_3. $AgClO_4$ is extremely soluble in water and ether and is also soluble in benzene and toluene.

silver salt The commercial name for sodium anthraquinone-2-sulphonate. *See* anthraquinone sulphonic acids.

silver sulphate, Ag_2SO_4. Prepared by dissolving silver in concentrated H_2SO_4 or by addition of a sulphate to $AgNO_3$. Only sparingly soluble in water. Decomposes on heating to Ag, SO_2 and O_2.

silver sulphide, Ag_2S. Obtained as a black precipitate by the action of H_2S on a solution of any silver salt. It is very insoluble in water.

silylation Formation of R_3Si- derivatives (R=H or organic). *See* silicones.

silyl compounds Derivatives of the H_3Si- group.

simazine, 2-chloro-4,6-bis(ethylamino)-s-triazine, $C_7H_{12}ClN_5$. A pre-emergence herbicide.

Simmons–Smith reagent Named after the duPont chemists who discovered that diiodomethane would react with an active zinc–copper couple in ether to give a reagent with molecular formula ICH_2ZnI. The reagent adds stereospecifically *cis*- to alkenes to give cyclopropanes in high yields.

single bond A bond between two atoms with two electrons occupying the bonding orbital and no further electronic interaction between the atoms. Generally a σ-bond, e.g. as in the $C-H$ bonds in CH_4, but can be essentially a π-bond, as in $Ni(PF_3)_4$.

singlet A single transition in a spectrum with other levels at very different energies.

singlet state *See* spin multiplicity.

sinigrin, $C_9H_{16}KO_9S_2$. M.p. 126°C. The best known of the mustard glycosides; it is hydrolysed to allyl isothiocyanate, glucose and potassium hydrogen sulphate by the enzyme myrosin, which accompanies it in mustard seed.

sintering The process of bonding by atomic or molecular diffusion in which powders are heated under pressure but at a temperature below the melting point. Used for metals and high melting polymers, e.g. polytetrafluoroethene. The process depends upon such mobility of the atoms in the surface layers as to cause physical diffusion between the lattices of the separate particles and thus physical joining of the lattices.

sitosterol, 24-(24-*R*)ethylcholesterol, $C_{29}H_{50}O$. M.p. 137°C. The main plant sterol but is often accompanied by related sterols, from which it is difficult to separate. The glucoside, m.p. 250°C, is also very common in plants and has been described by a variety of names. Sitosterol is accumulated by many invertebrate marine animals.

size reduction equipment Plant for particle size reduction which generally falls under the description of *crushers*, *grinders*, *mills* or *disintegrators*.

sizing The process of filling the pores of paper and giving it some degree of water-repellency. This can be done either in the beaters by the addition of rosin and alum or a colloidal mixture, such as starch and water-glass, when the process is termed engine sizing, or by spraying the surface of the ready-made paper with glue or the like when it is known as surface sizing. *See* paper.

skatole, 3-methylindole, C_9H_8N. M.p. 95°C, b.p. 265–266°C. The chief volatile constituent of the faeces, being formed by the action of the intestinal bacteria on tryptophan. It occurs also in coal tar and in beetroot.

Skraup's reaction *See* anthranol.

slack wax An oily wax obtained from crude oil or oil fractions by chilling or by treatment with methyl ethyl ketone. Slack wax can be cracked to produce higher alkenes (C_8-C_{18}) for use in the production of detergents.

slag A liquid solution of oxides produced in the smelting and refining of metals. The primary function of slag in smelting is to take into solution the gangue of the ore and separate it from the liquid metal. To achieve this fluxes are added which reduce the melting point of the oxide mixture. Common fluxes are alumina, limestone or lime, silica and iron(III) oxide. In addition to removing gangue, the slag can also exert a refining function in smelting processes, e.g. the desulphurization of iron in the blast furnace. In smelting, slag composition depends on the metal being smelted.

Slags are also produced in metal refining processes where oxidation is used to separate the impurities, e.g. steelmaking, copper and lead refining. The aim in refining is to control the slag composition to effect the removal of impurities and this process invariably means that some of the parent metal is oxidized and passes into the slag. In steelmaking the aim is to remove C, S and P and this requires a slag high in CaO, i.e. a basic slag.

Slags are often sold for use in other industries. Thus blast furnace slag can be used for manufacturing cement, as road metal, as slag wool. The slag from the basic steelmaking process sometimes contains sufficient P_2O_5 to make it suitable as an agricultural fertilizer and it is sold as 'basic slag'; the high lime content of these slags is also useful. Converted to slagceram by crystallization of the glass from slag.

slaked lime, $Ca(OH)_2$. *See* calcium hydroxide.

slip planes The planes of weakness in crystals corresponding to crystalline boundaries and to planes of atoms with only weak forces between such planes (e.g. the layers in graphite).

slow combustion *See* cool flames.

slurry A liquid containing an appreciable quantity of suspended solid.

slush bath *See* low temperature baths.

Sm Samarium.

smekal cracks Crystal defects – dislocations.

smelting The melting of an ore with chemical reaction with other materials. A part of the process of winning an element from its ores.

smetic A phase of matter intermediate between solid and liquid. *See* liquid crystals.

smokeless fuel A fuel which, when burnt in an open domestic fireplace or boiler gives rise to minimal smoke. The fuel obtained from the low-temperature carbonization of coal is a smokeless fuel since the constituents which give rise to smoke have already been distilled off, but sufficient volatile matter (about 7–10%) has been left to allow the fuel to be easily ignitable and to burn freely. All smokeless fuels should be compact, strong and have a low ash content.

Sn Tin.

SNG Substitute natural gas.

soaps Sodium and potassium salts of fatty acids, particularly stearic, palmitic and oleic acids. Animal and vegetable oils and fats, from which soaps are prepared, consist essentially of the glyceryl esters of these acids. In soap manufacture the oil or fat is heated with dilute NaOH (less frequently KOH) solution in large vats. When hydrolysis is complete the soap is 'salted out', or precipitated from solution by addition of NaCl. The soap is then treated, as required, with perfumes, etc. and made into tablets.

Castile soap is manufactured from olive oil, transparent soap from decolorized fats and liquid green soap from KOH and vegetable oils. Soaps are sometimes 'superfatted' in that they contain some free fatty acid.

Soaps of other metals such as aluminium, calcium, cobalt, lithium, lead or zinc – *see* metallic soaps. European production 1976: toilet soap 307 000 tonnes, household soaps *c.* 140 000 tonnes, soap powders *c.* 35 000 tonnes.

soda ash, Na_2CO_3. Sodium carbonate*.

soda lime A granular material prepared from quicklime and NaOH solution followed by heating the product until dry. Used as an absorbent for CO_2 and as an alkali in testing for ammonium salts (NH_3 liberated on grinding).

sodamide, $NaNH_2$. *See* sodium amide.

sodium, Na. At.no. 11, at.wt. 22·98 977, m.p. 97·81°C, b.p. 882·9°C, *d* 0·97. An alkali metal which occurs widely, principally as NaCl in sea water and in salt deposits. Sodium metal (bcc) is obtained by electrolysis of fused NaCl or $NaCl-CaCl_2$. The metal is soft and silvery white. It reacts rapidly with water, oxygen, halogens. Dissolves in liquid NH_3 to give a blue reducing solution but ultimately $NaNH_2$. The metal is used in the preparation of some Na compounds and in $PbEt_4$ (anti-knock) preparation. Used as a heat-transfer medium, conductor, as a reducing agent, particularly as a dispersion, or as Na naphthenide. Sodium compounds are of great importance – *see* under separate headings. World production of sodium 1968 56 000 tonnes. Sodium ions are essential to life, both in the diet and for function.

sodium acetate, NaO_2CCH_3. *See* sodium ethanoate.

sodium acetylides Formed (e.g. HC_2Na, Na_2C_2) by the action of Na or NaH on alkynes. Used in synthesis.

sodium amide, $NaNH_2$. Formed Na plus NH_3, hydrolysed to ammonia. A strong base, particularly in liquid ammonia, used extensively in organic synthesis.

sodium ammonium hydrogen phosphate, microcosmic salt, $NaNH_4HPO_4,4H_2O$. Colourless salt (Na_2HPO_4 plus NH_4Cl or $(NH_4)_2HPO_4$ plus NaCl). The melt solidifies to glasses which give characteristic colours with certain metals.

sodium antimonyl tartrate, $NaSbOC_4H_4O_6$. Antimony sodium tartrate.

sodium arsenates Sodium arsenates(III), Na_3AsO_3, $NaAsO_2$; sodium arsenates(V), Na_3AsO_4, $NaAsO_3$, $Na_4As_2O_7$. *See* arsenates.

sodium azide, NaN_3. *See* azides.

sodium benzoate, $C_6H_5CO_2Na$. White powder, soluble in water, sparingly soluble in ethanol. Used as corrosion inhibitor and medicinally as an antiseptic and for gout and rheumatism.

sodium bicarbonate, $NaHCO_3$. *See* sodium hydrogen carbonate.

sodium bifluoride, $NaHF_2$. *See* sodium fluoride.

sodium bismuthate, $NaBiO_3$. Very strong oxidizing agent (Mn^{2+} gives $[MnO_4]^-$) prepared impure by heating Na_2O_2 and Bi_2O_3.

sodium bisulphate *See* sodium sulphates.

sodium borate, $Na_2B_4O_7,10H_2O$. *See* borax.

sodium borohydride, sodium tetrahydroborate, $NaBH_4$. *See* borohydrides. Prepared NaOMe plus B_2H_6. Used in organic reductions (e.g. ketones, acids to alcohols) preparatively and to eliminate impurities.

sodium bromate(V), $NaBrO_3$. Prepared by electrolysis of aqueous NaBr or Br_2 on hot NaOH solution.

sodium bromide, NaBr. M.p. 757°C, b.p. 1393°C. Formed Na_2CO_3 or NaOH plus HBr or (with $NaBrO_3$) Br_2 plus hot NaOH solution. Has NaCl structure.

sodium carbonate, Na_2CO_3. Manufactured in the ammonia-soda or Solvay process. Brine (NaCl solution in water) has added NH_3 then CO_2. $NaHCO_3$ is precipitated which decomposes to Na_2CO_3 at 175°C; NH_3 is recovered by treating the NH_4Cl solution with $Ca(OH)_2$ slurry. Forms hydrates. Anhydrous Na_2CO_3 is known as soda ash; $Na_2CO_3,10H_2O$ is washing soda. Na_2CO_3 occurs naturally as trona. Used in glass (55%), soap, chemicals, detergents and paper industries. U.S. production (natural and synthetic) 1982 7·9 megatonnes.

sodium chemistry Sodium is an alkali metal, electronic configuration $3s^1$. Except for the formation of the Na^- ion (in adducts of Na with cryptates) the chemistry is exclusively that of the +1 state ($E°$ $Na^+ \longrightarrow Na$ −2·71 V); predominantly ionic, 6-co-ordinate. Complexes are formed (e.g. $Na_2CO_3,10H_2O$ contains a $[(H_2O)_4Na(\mu\text{-}H_2O)_2Na(H_2O)_4]^+$ cation and cryptates are formed by the cyclic polydentate ligands).

sodium chlorate(V), $NaClO_3$. Formed Cl_2 on hot NaOH solution or by electrolysis of NaCl in water and allowing anode and cathode products to come into contact. Used in the pulp industry to generate ClO_2 for bleaching, in the manufacture of chlorates(V) and (VII) (perchlorates) and as a herbicide.

sodium chloride, NaCl. M.p. 801°C, b.p. 1439°C, *d* 2·17. Occurs naturally as rock salt

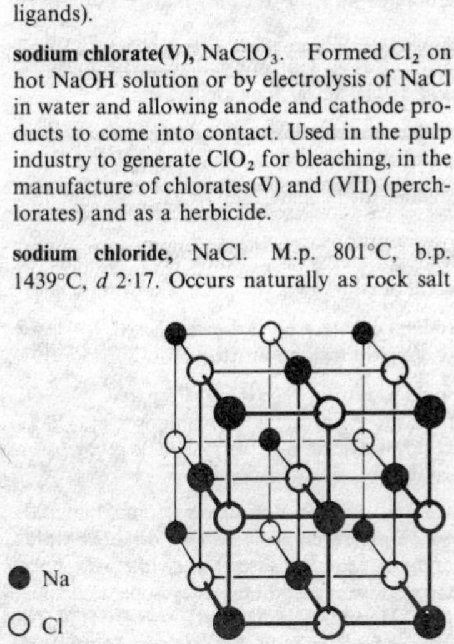

Sodium chloride structure

which is widely distributed and occurs extensively in thick beds and to about 3% in sea water. Obtained commercially by solar evaporation of sea water, by mining or usually as brine by passing water into underground deposits. The brine contains Ca^{2+}, Fe^{2+}, Mg^{2+} as impurities, these are removed by precipitation with Na_2CO_3 and NaOH. Pure NaCl is obtained by passing HCl gas into a saturated aqueous solution of NaCl. Sodium chloride is used for the formation of NaOH and Na_2CO_3 (for the alkali and glass industries) and chlorine (for the chlorocarbon industry) and is one of the chemical industries' most important raw materials. Solid NaCl is used for the removal of ice and snow and as a preservative. The sodium chloride structure is cubic with octahedral co-ordination about Na and Cl (both are ccp). Many other MX compounds crystallize with this structure. Total U.S. use 1978 44 megatonnes.

sodium chlorite, sodium chlorate(III), $NaClO_2$. Used for preparing chlorates(III).

sodium chromate(VI), $Na_2CrO_4,10H_2O$. *See* chromates.

sodium citrate, $C_6H_5O_7Na_3,2H_2O$ (also 5·5 H_2O). Salt of citric acid. Used medicinally as a blood anticoagulant and to reduce blood acidity and to prevent the formation of large curds in the stomach of infants.

sodium cyanate, NaOCN. Used to prepare cyanates.

sodium cyanide, NaCN. M.p. 564°C. Manufactured CH_4 and NH_3 to give HCN followed by neutralization with NaOH. Used to prepare cyanides and as a poison.

sodium cyanoborohydride, $NaBH_3CN$. Prepared by reacting HCN with sodium borohydride. A milder reducing agent than $NaBH_4$ especially valuable for its ability to reduce a given organic functional group in the presence of various other functional groups and for its stability in aqueous media over a broad range of pH, e.g. a $>C{=}\overset{+}{N}<$ group is reduced more rapidly than $>C{=}O$.

sodium cyclopentadienide, NaC_5H_5. Formed cyclopentadiene and Na, used in the formation of metal cyclopentadienyl derivatives.

sodium dichromate, $Na_2Cr_2O_7,2H_2O$. *See* chromates(VI).

sodium dithionite, $Na_2S_2O_4$. Formed by reduction of Na_2SO_3 plus excess SO_2 with Zn. Powerful reducing agent ($E°$ $S_2O_4{}^{2-}$ to $SO_3{}^{2-}$ in alkaline solution +1·12 volts; used in dyeing), removes O_2 from gases in presence of

2-anthraquinone sulphonate (Fieser's solution).

sodium ethanoate, sodium acetate, $NaO_2CCH_3,3H_2O$. Formed ethanoic acid and Na_2CO_3.

sodium ethoxide, sodium ethylate, $NaOCH_2CH_3$. White solid (Na in EtOH). Decomposed by water, gives ethers with alkyl halides; reacts with esters. Used in organic syntheses particularly as a base to remove protons adjacent to carbonyl or sulphonyl groups to give resonance-stabilized anions.

sodium fluoride, NaF. M.p. 902°C. Formed Na_2CO_3 or NaOH plus HF; gives HF with H_2SO_4. Forms $NaHF_2$ with excess HF. Used as a fluorinating agent.

sodium fluoroacetates (fluoroethanoates) *See* fluoroacetates.

sodium (mono)fluorophosphate, Na_2PO_3F. *See* fluorophosphoric acids.

sodium formate, NaO_2CH. Sodium methanoate.

sodium glutamate, monosodium glutamate, MSG, $C_8H_8NNaO_4,H_2O$. Flavouring agent extensively used as a food additive. Prepared from natural or synthetic L-glutamic acid.

sodium hexafluoroaluminate, Na_3AlF_6. Cryolite, used in aluminium production.

sodium hexametaphosphate *See* sodium phosphates.

sodium hydride, NaH. Prepared H_2 over Na at 350°C. Has NaCl structure. Used in organic reductions and condensations and in metallurgy as a descaling agent.

sodium hydridotrimethoxyborate, $Na[HB(OMe)_3]$. Reducing and hydroboration agent.

sodium hydrogen carbonate, sodium bicarbonate, $NaHCO_3$. Prepared CO_2 into Na_2CO_3 solution (also in Solvay process for sodium carbonate*); sparingly soluble in water. Decomposes to Na_2CO_3 on heating. Used in Na_2CO_3 manufacture, as an ingredient of baking powders, pharmaceutically, and in fire extinguishers.

sodium hydrogen sulphate, $NaHSO_4$. Formed Na_2SO_4 plus H_2SO_4; other acid salts known. $NaHSO_4$ gives the pyrosulphate, $Na_2S_2O_7$, on heating.

sodium hydrogen sulphite $NaHSO_3$.

sodium hydrosulphite, $Na_2S_2O_4$. Sodium dithionite.

sodium hydroxide, caustic soda, NaOH. M.p. 318°C, b.p. 1390°C, *d* 2·13. Manufactured by electrolysis of sodium chloride solution. In the mercury cell process (Castner–Kellner) a stream of Hg is used as cathode and the mercury amalgam formed is reacted with water. In the diaphragm cell process the electrolyte flows from anode to cathode and the diaphragm (asbestos or other membrane material) separates anode and cathode products. In both processes chlorine is the other product. NaOH is a white, translucent, hygroscopic solid; hydrates are known. Gives a strongly alkaline solution in water. A very important industrial chemical (U.S. production 1982 9·3 megatonnes) used in the manufacture of other chemicals (50%), rayon, pulp and paper (15%), aluminium (10%), petrochemicals, textile (5%), soap (5%) and detergents.

sodium hypochlorite, sodium chlorate(I), NaOCl. Formed (with NaCl) from Cl_2 plus cold NaOH solution. Forms hydrates. Aqueous solutions are used as bleaches and antiseptics.

sodium hyposulphite, $Na_2S_2O_4$. *See* sodium dithionite.

sodium iodate, sodium iodate(V), $NaIO_3$. Formed HIO_3 on NaOH or I_2 plus $NaClO_3$. Forms hydrates and acid salts. Used in volumetric analysis.

sodium iodide, NaI. M.p. 660°C, b.p. 1300°C. Obtained aqueous HI plus Na_2CO_3.

sodium lactate, $C_3H_5O_3Na$. Used in calico printing as glycerin alternative, textile finishing and as plasticizer for casein.

sodium metabisulphite, $Na_2S_2O_5$. *See* sodium pyrosulphite.

sodium methanoate, NaO_2CH. Sodium formate.

sodium methoxide, $NaOCH_3$. Similar to sodium ethoxide*.

sodium nitrate, Chile saltpetre, $NaNO_3$. Used to prepare sodium nitrite.

sodium nitrite, $NaNO_2$. Prepared $NaNO_3$ plus Pb. M.p. 271°C, decomposes at 320°C. Used in dyestuffs industry and as a corrosion inhibitor.

sodium, organic derivatives Ionic materials which are rarely isolated (Na dispersion plus organic halide) used for olefine polymerization. Sodium cyclopentadienide, NaC_5H_5 and sodium acetylides, e.g. CH_3C_2Na are used in synthesis. Aromatic hydrocarbons, ketones, etc., give radical anions, e.g. $(C_6H_6)^-$

with Na in tetrahydrofuran at low temperatures. Used as reducing agents.

sodium oxalate, $Na_2C_2O_4$. Colourless crystals. Sodium hydrogen oxalate, $NaHC_2O_4$.

sodium oxides Sodium normally reacts with O_2 to give Na_2O_2. Sodium monoxide, Na_2O formed Na plus deficiency of O_2 or by reduction of Na_2O_2 (C, Ag, Na) or NaN_3 plus $NaNO_3$. White or yellowish solid, reacts violently with water. Sodium peroxide, Na_2O_2. Formed Na plus O_2. The impure material contains some superoxide, O_2^- ions. Strong oxidizing agent. Forms octahydrate from cold water but generally reacts with water to give NaOH plus H_2O_2.

sodium perborate $NaBO_3,4H_2O$. Used in textile bleaching, dye oxidation, cleaners.

sodium percarbonate *See* percarbonates.

sodium perchlorate, sodium chlorate(VII), $NaClO_4$. Formed by heating $NaClO_3$ (*see* chlorates). Used to form perchlorates.

sodium periodates, sodium iodates(VII) Various sodium periodates are known and are used as oxidizing agents and to form periodates. $Na_2H_3IO_6$, $Na_3H_2IO_6$, Na_5IO_6 have octahedrally co-ordinated iodine, $NaIO_4$ contains a tetrahedral $(IO_4)^-$ species.

sodium permanganate, $NaMnO_4$. Impure material used as disinfectant and oxidizing agent.

sodium peroxide, Na_2O_2. *See* sodium oxides.

sodium peroxydisulphate, $Na_2S_2O_8$. Used as oxidizing agent and in preparation of peroxydisulphates.

sodium phosphates Various salts are known, e.g. Na_2HPO_4 (Na_2CO_3 plus H_3PO_4) (used to form NaH_2PO_4(H_3PO_4) and Na_3PO_4(NaOH)). Heat on Na_2HPO_4 gives the pyrophosphates $Na_2H_2P_2O_7$ and $Na_4P_2O_7$. Heat on NaH_2PO_4 followed by quenching gives a polyphosphate $(NaPO_3)_x$. The 'tripolyphosphate' is formed by heating $1NaH_2PO_4$:$2Na_2HPO_4$ (U.S. production 1978 660 000 tonnes). The various sodium phosphates are used principally as detergents but also in water conditioning, foodstuffs and medicine. Sodium pyrophosphates and sodium ammonium phosphates are used in baking powders.

sodium potassium tartrate, rochelle salt, seignette salt, $C_4H_4O_6NaK,4H_2O$. Used as reducing agent (silver mirrors) and as a saline aperient.

sodium pyrosulphite, sodium metabisulphite, $Na_2S_2O_5$. Obtained NaOH solution saturated SO_2 at 100°C. Forms 7 and 6 hydrates. Used in photography.

sodium saccharine, $C_7H_4NNaO_3S,2H_2O$. Sweetening agent.

sodium salicylate, o-$C_6H_4(OH)CO_2Na$. White powder, very soluble in water. Used in treatment of gout and rheumatism and has antipyretic properties.

sodium selenate Na_2SeO_4.

sodium silicates Many different sodium silicates are known; some occur as minerals. Silicates prepared from aqueous solution and salts of orthosilicic acid, H_4SiO_4. Fusion of Na_2CO_3 and SiO_2 in a furnace gives a range of products with the Na:Si ratio varying from 4:1 to 1:4. Na_2SiO_3 is sodium metasilicate; those with ratios Na:Si 3·2 to 4 are the water-glasses and are soluble in water. Used in the manufacture of silica gel (treatment with acid) as adhesives, sizes, in metal-cleaning and in detergent manufacture. U.S. production 1978 750 000 tonnes.

sodium stannate, $Na_2SnO_3,3H_2O$. Prepared by fusing SnO_2 with NaOH. Used as a mordant.

sodium sulphate, Glauber's salt, Na_2SO_4. M.p. 884°C, b.p. 1429°C. Manufactured from NaCl and H_2SO_4 at high temperature or NaCl, SO_2, air, water vapour; recovered from various chemical processes and natural brines. Forms hydrates and double salts with other sulphates. Forms hydrogen sulphates with excess H_2SO_4. Used in wood pulp production (66%), glass, detergent and chemical manufacture. Used as a mild aperient. U.S. production 1978 1·1 megatonnes.

sodium sulphides
Sodium sulphide, Na_2S, formed by reduction Na_2SO_4 with CO or H_2. Aqueous solutions are oxidized to sodium thiosulphate.

Sodium polysulphides, Na_2S_x ($x = 2, 4, 5$) are formed from Na in liquid NH_3 plus S.

sodium sulphite, Na_2SO_3. Formed SO_2 and NaOH solution, readily oxidized to Na_2SO_4 by air. Forms hydrates. Used to remove chlorine after bleaching and in the paper industry to dissolve lignin.

sodium tellurate Na_2TeO_4.

sodium tetraborate $Na_2B_4O_7$. *See* borax.

sodium tetrahydroborate, $NaBH_4$. *See* sodium borohydride.

sodium tetraphenylborate, $NaB(C_6H_5)_4$. Used in the direct estimation of K^+ which gives insoluble $KB(C_6H_5)_4$.

sodium thiocyanate, NaNCS. Formed fused NaCN plus sulphur. Used to prepare thiocyanates and as a volumetric reagent.

sodium thiosulphate, $Na_2S_2O_3$. Manufactured by treating a solution containing Na_2CO_3 and Na_2S with SO_2 or from Na_2SO_3 plus sulphur. Forms many hydrates. Used in photography ('hypo') because it dissolves silver halides. Also used in tanning, preparation of mordants, as a fermentation preventative in dyeing and in chemical manufacture.

sodium tungstate $Na_2WO_4,2H_2O$.

soft acids and bases *See* hard and soft acids and bases.

soft detergents Detergents which are biodegradable. They consist largely of alkylbenzene sulphonates with linear alkyl groups produced by reacting linear 1-alkenes with benzene. *See* detergents.

softeners Materials added to the rubber mix to improve the mix.

sol A colloidal solution; term generally used for colloidal dispersions of inorganic solids in liquids.

solder Solders are fusible alloys used for bonding metals. Hard solder or brazing metal is a low-melting brass with m.p. 800°C. Soft solders are normally Pb – Sn entectic alloys.

soldering Causing a melted metal to adhere to the surface of a solid metal by a process of diffusion brought about by chemical solution. Used to bond metals.

solid foams Solids formed, often in a liquid system, with gas evolution so that a porous network enclosing the gas results. Charcoal is an example and expanded foam plastics are of great importance.

solid-phase synthesis A term applied to the technique of carrying out chemical transformations on a substrate chemically bonded to an insoluble, solid polymer molecule. This technique is used for the synthesis of polypeptides from amino-acids using another amino-acid bonded to a polystyrene matrix. The advantage is that the products at each stage of a multi-step synthesis need not be isolated and purified. Instead the polymer to which the peptide is attached is washed free of excess reagents and impurities and virtually no loss of product occurs. The method has been adapted to automatic control and addition of reagents. At the completion of the desired synthesis the product is liberated from the polymer surface by a suitable reagent.

solid solution A solid solution is formed when two or more elements or compounds share a common lattice. The composition may vary within very wide limits. A substitutional alloy results when atoms of solute replace atoms of solvent in its crystal lattice, e.g. nickel–copper alloys. An interstitial solid solution occurs when a small atom, such as carbon, enters the interstices of the host lattice of larger atoms such as metals.

solid state electrode *See* ion-selective electrodes.

solid state reactions Reactions occurring between two or more solid compounds or decompositions occurring from the solid state. Gases may be evolved in such reactions. Most ternary oxides are formed by solid state reactions.

solids, structures The structures of solids depend upon the bonding (ionic, covalent, coordinate, dipole–dipole (hydrogen bonding)) between the atoms and groups. In compounds which are substantially ionic the structures depend on the relative sizes (radius ratios) of the ions. Many structures approximate to close-packed arrangements of the largest species present.

solidus curve *See* liquidus curve.

solubility The maximum quantity of one phase dissolved by another under specified conditions. In the case of solutions of solids or liquids in liquids, the solubility is usually expressed as the weight dissolved in a given weight or volume of the solvent at a specified temperature. *See* Henry's law for solubility of gases.

Substances are generally soluble in like solvents. Organic molecules in molecular solvents such as CCl_4, C_2H_5OH, ether, propanone. Inorganic salts are often soluble in water and less soluble in organic solvents.

solubility curve A graphical representation of the variation of the solubility with temperature.

solubility product When a saturated solution of an electrolyte is in contact with some undissolved electrolyte the following equilibria may be considered

$$AB_{solid} \rightleftharpoons AB_{dissolved} \rightleftharpoons A^+ + B^-$$

the equilibrium constant for this process may be written as

$$K = \frac{[A^+][B^-]}{[AB]_{dissolved}}$$

However because of the equilibrium between the solid and dissolved undissociated AB, the concentration of the latter may be considered to remain constant and hence

$$[A^+][B^-] = K_{sp}$$

where K_{sp} = solubility product of the electrolyte. It can be seen that if an excess of A^+ or B^- ions are added to the solution, the K_{sp} is exceeded and the salt will be precipitated to restore the equilibrium.

soluble oil Normally a mineral oil containing emulsifiers enabling stable emulsions to be obtained when the oil is added to water. Soluble oils are used in metal cutting and grinding operations for lubrication and cooling purposes. Certain other additives may be used, such as extreme-pressure, anti-freeze or anti-foam agents. *See also* emulsion.

solute The substance dissolved in the solvent to yield a solution.

solution A liquid or solid may be dispersed through a liquid either by being in suspension, or by being dissolved in it, in the form of a solution. The qualitative characteristics of a solution are homogeneity, and the absence of any tendency for the dissolved substance to settle out again. Unfortunately from the point of view of these criteria, a suspension of extremely minute particles (colloidal suspension) of a pure chemical compound behaves in a similar manner. A solution is better defined as a homogeneous mixture of substances which is separable into its components by altering the state of one of them (freezing or boiling out one component), and whose properties vary continuously with the proportions of the components between certain limits. Solutions are not confined to liquids and solids; solutions of gas in liquids, and of gases, liquids and solids in solids are also formed. The amount of a substance dissolved in a given amount of another substance is described as the concentration of the solution, and may be expressed in grams per litre, or as a molar fraction (*see* molarity). A solution containing only a small proportion of the dissolved substance is called dilute; one containing a high proportion is said to be concentrated. There is a limit to the solubility of most substances, and a solution containing the limiting amount of the dissolved substance is said to be saturated with respect to the substance, and is termed a saturated

solution. Such solutions are prepared, e.g. by shaking the liquid with excess of the powdered solid until no more is taken up by the solution.

The maximum concentration attainable under such conditions is termed the solubility of the substance at the specific temperature used in the experiment, since solubility generally increases with rising temperature. Solubility is usually expressed in grams per 100 g of solvent, or grams per 100 g of solution. Sometimes, for practical convenience, it may be expressed in grams per 100 ml of solvent or solution.

solvation The process of interaction of a solvent with a solute. The general case corresponding to hydration *.

Solvay process *See* sodium carbonate.

solvent In a solution, the substance which makes up the bulk of the solution is usually termed the *solvent*. The substance dissolved in the solvent is called the *solute*. These terms are somewhat arbitrary except in special cases when, e.g. a chemical reaction occurs between certain constituents of a solution. In this case the inert medium used to disperse the reactants is invariably termed the inert solvent.

solvent extraction *See* extraction.

solvolysis The reaction when a compound reacts with a solvent. In the case of water the special term hydrolysis is used, e.g. the reaction of a salt with water giving the free acid and base.

$$BA + H_2O \longrightarrow HA + BOH$$

somatropin, growth hormone A protein hormone isolated from the anterior lobe of the pituitary gland, which stimulates many of the processes of growth.

sorbic acid, 2,4-hexadienoic acid, $C_6H_8O_2$, $CH_3 \cdot CH:CH \cdot CH:CH \cdot COOH$. White needles; m.p. 134°C, b.p. 228°C (decomp.). Sorbic acid can be obtained from rowan berries, and is prepared by condensing crotonaldehyde and keten. It is a selective growth inhibitor for yeasts, moulds and certain bacteria, and is used as a food preservative. Its esters can be used in the preparation of quick-drying oils.

D-sorbitol, D-glucitol, $C_6H_{14}O_6$. M.p. 110°C. The alcohol corresponding to glucose.

$$CH_2OH \cdot \overset{\displaystyle HO}{\underset{\displaystyle H}{C}} \cdot \overset{\displaystyle H}{\underset{\displaystyle OH}{C}} \cdot \overset{\displaystyle OH}{\underset{\displaystyle H}{C}} \cdot \overset{\displaystyle OH}{\underset{\displaystyle H}{C}} \cdot CH_2OH$$

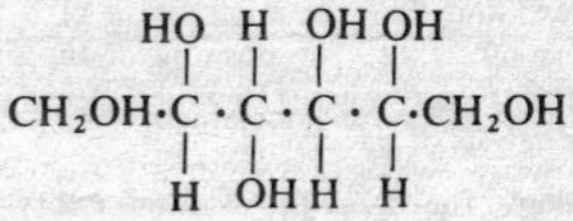

It is present in rowan berries and is characteristic of the fruits of the *Rosaceae*.

Sorbitol is manufactured by the reduction of glucose in aqueous solution using hydrogen with a nickel catalyst. It is used in the manufacture of ascorbic acid (vitamin C), various surface active agents, foodstuffs, pharmaceuticals, cosmetics, dentifrices, adhesives, polyurethane foams, etc.

L-sorbose, $C_6H_{12}O_6$. Colourless crystals,

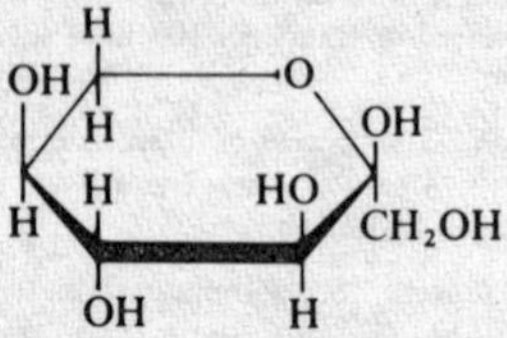

m.p. 160°C. A hexose obtained by the fermentation of sorbitol with certain *Acetobacter*. It is an intermediate in the manufacture of ascorbic acid.

'sour' products Petroleum fractions, generally gasoline or kerosine, containing mercaptans with an unpleasant odour. 'Sour' fractions are 'sweetened' to convert the mercaptans to disulphides. *See* sweetening.

soxhlet Equipment for the continuous extraction of a solid by a solvent. The material to be extracted is usually placed in a porous paper thimble, and continually condensing solvent allowed to percolate through it, and return to the boiling vessel, either continuously or intermittently. Such extraction procedures are frequently used in the first stages of isolating natural products from raw materials.

space group The whole array of symmetry-elements in a crystal framework or lattice. The total number of possible space groups is 230. The atoms in any crystal must be arranged so that they fit into one of these groups. The framework composing a space group is termed a *space lattice**.

space lattice A term applied to the array of points in any crystalline structure for which the pattern repeats. By joining these points the crystals can be divided into a series of parallel *unit cells*, each of which contains a complete unit of pattern. The atoms composing the crystal occupy the points in the lattice so that a space lattice affords a convenient means of representing the atomic structure of a crystalline substance. The position of the points relative to each other is found by means of X-ray analysis of the crystal.

spalling The break-up of a refractory with ultimate mechanical failure. Spalling is likely to occur: (1) when there is a large high temperature gradient through the brick, especially when the brick contains substances of high coefficients of expansion such as magnesite; (2) when bricks are laid badly, giving stresses; (3) when there is reaction between the furnace gas or material in the furnace with the constituents of the refractory.

spark spectrum *See* arc spectrum.

special boiling point spirits (SBP) Refined solvents of specially selected boiling ranges distilled from gasoline fractions. SBP are usually classified by means of their boiling ranges, e.g. SBP 62/82°C. They are used for many industrial purposes such as solvents for rubber and for the extraction of oil from seeds. *See* white spirits.

specific conductance *See* conductivity.

specific refractivity The quantity (r) defined by

$$r = \frac{n^2 - 1}{n^2 + 2} \cdot \frac{1}{d}$$

where n is the refractive index of the substance of density d. The molar refractivity is equal to the product of the specific refractivity and the molecular weight of the substance.

specific rotatory power *See* rotatory power, specific.

spectral sensitization The process by which the natural sensitivity of a photographic silver halide emulsion is extended to light of longer wavelengths by means of sensitizing dyes (spectral sensitizers). Using appropriate sensitizers, orthochromatic and i.r. sensitive emulsions are produced. For colour photography, tri-pack materials incorporate three sensitive layers sensitized to red, green and blue light.

spectral sensitizers *See* spectral sensitization. Structurally, the majority of sensitizers are polymethine dyes, e.g. cyanines.

spectrochemical series Ligands arranged in order of their tendency to cause splitting of the d orbitals in complexes (*see* crystal field theory). The order is obtained from changes in magnetism and the frequencies of d – d transitions in the u.v.–visible–near i.r. spectra. The series is $I^- < Br^- < Cl^- < F^- < OH^- < H_2O <$ pyridine $< NH_3 <$ en $< NO_2^- < CN^-$.

spectrophotometer An instrument that measures (often automatically) the relation between absorption of electromagnetic radiation and frequency of that radiation (or wavelength). Spectrophotometers operate in various regions of the electromagnetic spectrum, e.g. u.v., visible, i.r., microwave.

spectrophotometric methods of analysis Methods which use a spectrophotometer.

sphalerite, (Zn,Fe)S. Zinc blende* containing some iron. An important zinc ore.

sphingomyelins Phosphatides with the struc-

$$\left[\begin{array}{c} O\cdot C_{18}H_{33}(OH)NH\cdot R \\ O=P-OH \\ O\cdot CH_2\cdot CH_2\cdot N \begin{array}{c} CH_3 \\ CH_3 \\ CH_3 \end{array} \end{array} \right]^+ \quad OH^-$$

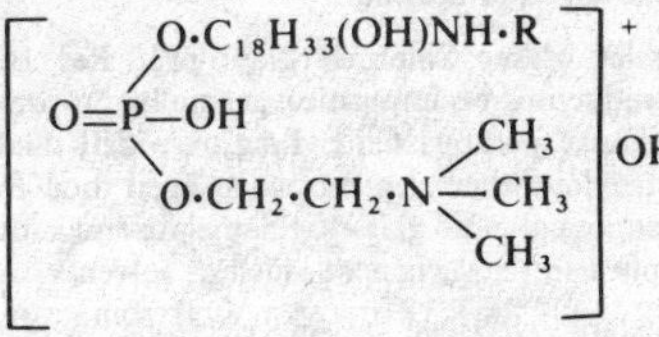

ture where R is a fatty acid radical, usually lignoceric acid, though others are found. Sphingomyelins occur abundantly in the brain and to a lesser extent in other animal tissues. They are always found in association with the cerebrosides, which they resemble in many properties, and consequently are not easily separated from them. On hydrolysis they split up into the bases choline and sphingosine, phosphoric acid and a fatty acid.

sphingosine,
$C_{18}H_{37}NO_2, CH_3\cdot[CH_2]_{12}\cdot CH:CH\cdot CH(OH)\cdot CH(NH_2)\cdot CH_2OH$. A base, forming part of the molecules of sphingomyelins and the cerebrosides, from which it splits off on hydrolysis. It crystallizes in needles from ether.

spindle oil Any low-viscosity mineral lubricating oil.

spinel Mineral $MgAl_2O_4$.

spinels A group of mixed metal oxides, generally $M^{2+}M_2^{3+}O_4$ (other combinations of cations are possible) (M^{2+} being usually Mg, Fe, Co, Ni, Zn, Mn; M^{3+} being Al, Fe, Mn, Cr, Rh). The crystals are cubic with close-packed oxygen atoms. In a normal spinel each M^{2+} is in tetrahedral co-ordination, in an inverse spinel half of the M^{3+} are in tetrahedral co-ordination, half in octahedral co-ordination. Intermediate arrangements are also found.

spin label A free radical or other paramagnetic probe (e.g. Cu^{2+}, Mn^{2+}) bound to a biological macromolecule in an effort to determine molecular structure or biological function by using electron spin resonance spectroscopic techniques. Spin labels are very sensitive to the local environment, allow the measurement of very rapid molecular motion in either

optically transparent or opaque solution and do not produce problems of interfering signals from the environment. Substituted organic nitroxides are the commonest class of free radicals used as spin labels. The relatively unreactive NO group contains an unpaired electron necessary to produce an e.s.r. signal whilst the functional group R R = OH or NH_2 or CO_2H provides the chemical means of attachment to the macromolecule.

spin moment An electron spinning on its own axis produces a magnetic moment, termed the spin moment. For a multielectron system the magnitude of the spin moment, μ_S, is given by

$$\mu_S = g\sqrt{S(S+1)}$$

where g is the gyromagnetic ratio and S is the sum of the spin quantum numbers (m_S) of the individual electrons.

spin multiplicity For an electronic state containing n unpaired electrons, the total spin quantum number, S, is equal to ($n/2$). The spin multiplicity of the state is given by $(2S+1)$. When this is equal to unity, it is called a singlet state; when it is equal to three it is a triplet state, etc.

spin-orbit coupling The interaction of spin and angular momentum.

spin quantum number *See* electronic configuration.

spin–spin coupling The process of interaction between nuclear spins giving rise to fine structure in n.m.r. spectra.

spirans, spiro-compounds Bicyclic compounds with one, and only one, atom common to both rings. E.g. spiro[4,5]decane as shown.

$$\begin{array}{c} H_2C-CH_2 \\ | \quad\quad C \\ H_2C-CH_2 \end{array} \begin{array}{c} CH_2\cdots CH_2 \\ CH_2-CH_2 \end{array} CH_2$$

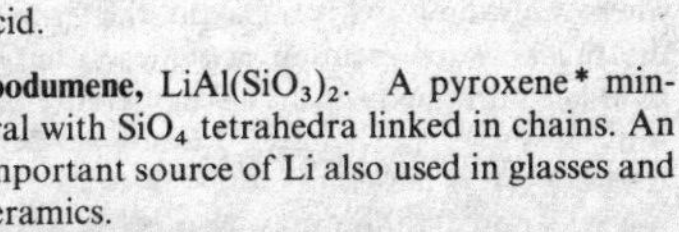

The atom common to both rings is called the spiro-atom; this may also be nitrogen, phosphorus, etc.

spirit of salt An old name for hydrochloric acid.

spodumene, $LiAl(SiO_3)_2$. A pyroxene* mineral with SiO_4 tetrahedra linked in chains. An important source of Li also used in glasses and ceramics.

spray ponds A method of cooling water by atmospheric evaporation. The water is sprayed through fine nozzles and collected in open ponds.

spray tower The simplest form of scrubber*.

The gas to be purified passes up the tower and is met by a stream of liquid introduced by a series of sprays. The space inside the tower is empty without packing.

squalene, $C_{30}H_{50}$. An acrylic triterpene

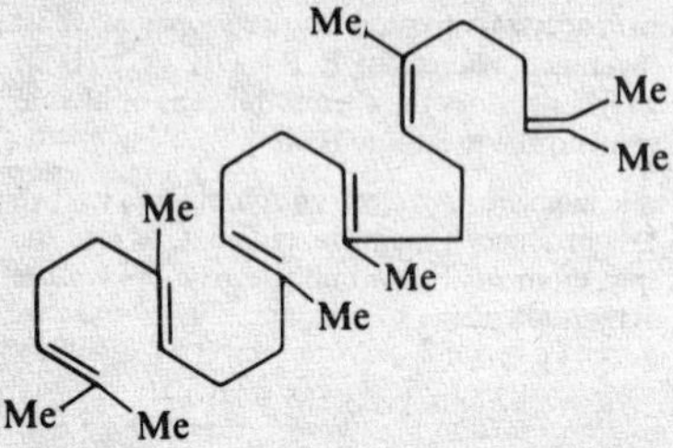

originally found in the livers of elasmobranch fish, and shown to be of wide occurrence and significance as the natural precursor of cholesterol and of other sterols and triterpenoids of plants and animals. It is the major hydrocarbon of human skin surface lipid, forming about 10% of total lipid. Natural squalene is the all-*trans*-isomer of 2,6,10,15,19,23-hexamethyltetracosa-2,6,10,14,18,22-hexaene and is a colourless oil, b.p. 261°C/9 mm, m.p. −5°C. Squalene, and the more stable saturated hydrocarbon squalane, are used as non-toxic vehicles for cosmetics, and for promoting absorption of drugs applied to the skin.

square antiprismatic co-ordination Co-ordination by eight neighbours in a square antiprismatic arrangement (i.e. the two square faces at 45°). Found in $[Mo(CN)_8]^{4-}$. A square antiprism has 2 square and 8 triangular faces.

Sr Strontium.

stability constant When a complex is formed in solution between a metal ion and a ligand the equilibrium may be expressed by a constant which is related to the free energy change for the process:

$$M + A \rightleftharpoons MA; \quad K_1 = \frac{[MA]}{[M][A]}$$

where K_1 is the stability constant, the terms in the brackets representing activities. Charges have been dropped for the sake of clarity and

$$-\Delta G = RT \ln K_1$$

Step-wise equilibria may also be expressed in this way, e.g.

$$M + A \rightleftharpoons MA; \quad K_1 = \frac{[MA]}{[M][A]}$$

$$MA + A \rightleftharpoons MA_2; \quad K_2 = \frac{[MA_2]}{[MA][A]}$$

$$MA_{(n-1)} + A \rightleftharpoons MA_n; \quad K_n = \frac{[MA_n]}{[MA_{(n-1)}][A]}$$

K_1, K_2 . . . K_n being called the step-wise stability constants. Overall stability constants, designated β, may be evaluated by multiplication of the individual stability constants, e.g.

$$M + 2A \rightleftharpoons MA_2; \quad \beta_2 = \frac{[MA_2]}{[M][A]^2} = K_1 K_2$$

stachydrine, A betaine*.

stained glass Coloured glass produced by introduction of impurities generally by ion-exchange methods. Cu and Ag are widely used but many other elements have been used to produce coloured glasses. Enamels can also be applied to the surface of glasses.

standard electrode An electrode which produces a standard e.m.f. Electrode potentials are normally quoted relative to the hydrogen electrode, although in practice reference is usually made to the calomel electrode*.

standard state Using thermodynamics it is possible to predict whether or not reaction will occur when substances are placed together in a closed system. However, to do this it is necessary to define a standard state of the system as one which contains all the reactants and products: (1) in the pure state if they are crystalline substances or liquids; (2) at one atmosphere pressure if they are gases; (3) at a concentration of 1F (one formula weight in grams per litre of solution) if they are dissolved in a solvent; and (4) at a specified temperature, usually, but not necessarily, 298 K.

Thermodynamic quantities which refer to the standard state are denoted by superscript zeros (^{0}), e.g. $\Delta G_T{}^0$, $\Delta H_T{}^0$, $\Delta S_T{}^0$, the subscript denoting the temperature T of the system.

standard temperature and pressure (stp) Arbitrary reference conditions of 1 standard atmosphere at 0°C. This is represented by $1{\cdot}013 \times 10^5$ N/m² and 273·15 K.

stannane, tin hydride, SnH_4. M.p. −150°C, b.p. −52°C. Colourless gas ($SnCl_4$ plus $LiAlH_4$ in ether at −30°C). Decomposes to tin and H_2 even at 0°C. Acts as a reducing agent. Sn_2H_6 is also known.

stannates *See* tin oxides.

stannic compounds Tin(IV) compounds.

stannite, Cu_2FeSnS_4. A mineral with a structure derived from that of zinc blende. Stannites were formerly mixed oxides containing Sn(II).

stannous compounds Tin(II) compounds.

starch, $(C_6H_{10}O_5)_X$. The carbohydrate

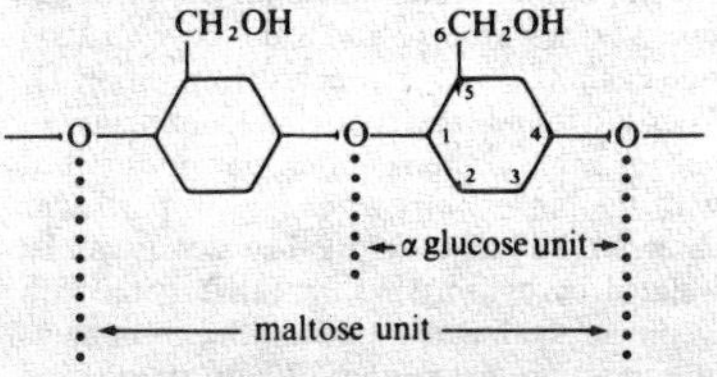

which is being continuously formed and broken down in the living cell and which also serves as a reserve material. It is, like cellulose, made up of a long chain of glucopyranose units joined together through oxygen by α-glucosidic bonds. It yields glucose alone on complete hydrolysis, maltose when broken down by enzymes and dextrin under other conditions.

Starch consists of amylose, which is water-soluble and retrogrades on concentration forming an insoluble precipitate, and amylopectin, a mucilaginous substance with the characteristic paste-forming properties. Amylose is composed of long straight chains containing 200–1000 glucose units linked by α-1,4-glycoside links; amylopectin consists of comparatively short chains (about 20 glucose units) cross-linked by α-1,6-glycoside links. Both amylose and amylopectin have been synthesized from glucose-1-phosphate by the action of the enzyme phosphorylase.

Starch is insoluble in cold water, but in hot water the granules gelatinize to form an opalescent dispersion. It is made from corn, wheat, potatoes, rice and other cereals by various physical processes such as steeping, milling and sedimentation. It is used as an adhesive, for sizing paper and cloth, as an inert diluent in foods and drugs, and for many other purposes.

Starch can be split into amylose and amylopectin by a commercial process based on selective solubilities. Amylose is used for making edible films, and amylopectin for textile sizing and finishing, and as a thickener in foods.

Stassfurt deposits Deposits of various salts mined at Stassfurt chiefly for potassium (carnallite). Kainite, kieserite and gypsum are also present, Br_2 is obtained as a by-product.

state symbols Symbols which designate the spin multiplicity and electronic state of a species. E.g. the ground state of a d^2 species is ^3F, the spin multiplicity symbol (3) designates 2 unpaired electrons, the symbol F designates an angular momentum quantum number of 3.

stationary phase *See* chromatography.

stationary state In a reversible chemical reaction, a condition is reached at which the rate of the forward reaction is exactly equal to the rate of the backward reaction. At this point the reaction system has reached a state of equilibrium often referred to as a stationary state. In the stationary state the concentration of the reacting species remain constant.

steam Water vapour generally at temperatures above 100°C. Used for power generation, heat distribution, as a chemical reactant (e.g. water-gas), in fire fighting and for cleaning.

steam distillation Distillation carried out in a current of steam with the object of distilling out a particular component of a mixture, or of avoiding too high a temperature during distillation. As normally carried out, the operation consists of distilling a mixture of two immiscible liquids in a current of steam. The ratio of the weights of the two substances distilling is (m_1p_1/m_2p_2), where m_1 and m_2 are the molecular weights of the two substances and p_1 and p_2 are their respective vapour pressures at the distillation temperature. Substances with a relatively high molecular weight may be steam distilled over in relatively large amounts in spite of the fact that they may have relatively low vapour pressures at the temperature of distillation.

steam reforming The reaction of a hydrocarbon (natural gas) plus steam over a catalyst at 800–1000°C to give CO plus H_2. After the water–gas shift reaction * the process is used for hydrogen * production.

stearic acid, *n*-octadecanoic acid, $C_{18}H_{36}O_2$, $CH_3 \cdot [CH_2]_{16} \cdot COOH$. Crystallizes in leaflets; m.p. 70°C, b.p. 376°C (decomp.). Soluble in ether and hot alcohol, insoluble in water. Stearic acid is one of the most common fatty acids, and occurs as glycerides in most animal and vegetable fats, particularly in the harder fats with high melting-points. A solid mixture of stearic and palmitic acids, 'stearine', is used for making candles. The soaps are the sodium and potassium salts of stearic and palmitic acids.

stearine *See* stearic acid, palmitic acid.

stearyl alcohol, 1-octadecanol, $C_{18}H_{38}O$. M.p. 58°C, b.p. 210°C. Important fatty alcohol obtained from naturally occurring glycerides. Used in the pharmaceutical and cosmetics industries and as a gel stabilizer for greases.

steel The name given to alloys of iron and carbon containing 0·05–1·5% C. *See* iron.

Alloy steels are steels containing deliberate

additions of alloying elements over and above the amounts which are residual from the steel-making process.

Austenitic steels retain the ccp structure right down to room temperature. For this reason these steels cannot be hardened by quenching.

Stainless steel. A group of Cr or Cr – Ni steels showing an unusually high resistance to corrosion by the atmosphere and many chemical reagents. In order to achieve this resistance to corrosion > 12% Cr is necessary.

stereochemistry The study of the spatial arrangements of atoms in molecules and complexes.

stereoisomerism *See* isomerism and conformation.

stereospecific reactions Reactions in which bonds are broken and made at a single asymmetric atom (usually, but not necessarily carbon), and which lead largely to a single stereoisomer, are said to be stereospecific. If the configuration is altered in the process, the reaction is said to involve *inversion of configuration*. If the configuration remains the same, the transformation occurs with *retention of configuration*.

steric hindrance or geometry A term intended to denote the influence exerted on a reacting group by the spatial arrangement of neighbouring atoms. The optical activity of the di-ortho substituted derivatives of diphenyl is probably due entirely to the size of the substituting groups, which obstruct the free rotation of the benzene rings about the single bond joining them and so produce an asymmetric molecule. This is a case of true steric hindrance.

steroids A generic term for substances having the nuclear carbon skeleton of the sterols, or a very similar structure. The natural steroids, which arise from the oxidative cyclization of squalene, include sterols, bile acids, sex hormones, adrenocortical hormones, cardiac-active glycosides, sapogenins, alkaloids and insect hormones such as ecdysone. Many synthetic steroids are known, and some of these are important drugs.

sterols Alcohols which possess the following

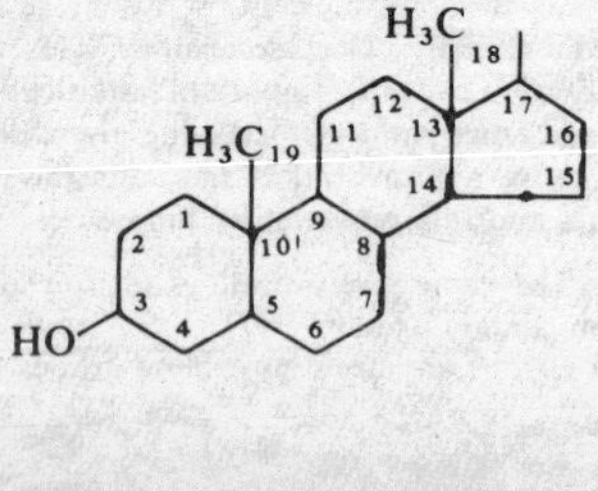

ring structure numbered as shown. They may contain one or more double bonds, often between carbon atoms 5 and 6 and have an alkyl side chain ($C_7 - C_{11}$) attached to position 17. They are found in every animal and plant cell and in some bacteria, often partly esterified with the higher fatty acids (*see* in particular cholesterol and sitosterol). The sterols can be isolated by hydrolysing the fat fraction with alcoholic alkali and extracting the unsaponifiable portion with ether. Plant sterols are known as phytosterols.

stibine *See* antimony hydride.

stibnite, Sb_2S_3. The only important antimony ore.

stigmasterol, (24*S*)-24-ethyl-5,22-cholestadien-3β-ol, $C_{29}H_{48}O$. First obtained from *Physostigma venenosum* (Calabar bean) and is readily isolated from many plant sources where it often co-occurs with sitosterol*. It has served as a starting substance for the synthesis of steroid hormones.

stilbene, *trans*-1,2-diphenylethene, $C_{14}H_{12}$. Colourless crystals; m.p. 124°C, b.p. 306–307°C. It can be prepared by the action of benzyl magnesium bromide on benzaldehyde, and by many other methods. Derivatives of stilbene are important in the dye industry and some have oestrogenic activity (*see* stilboestrol). U.v. light converts stilbene into its *cis* isomer, isostilbene, a yellow oil.

stilbite A zeolite, $Na_2Ca[(Al_2Si_6)O_{16}],6H_2O$, used for softening hard water.

stilboestrol, 4,4′-dihydroxy-αβ-diethylstilbene, $C_{18}H_{20}O_2$. White crystals, m.p. 168–171°C. Prepared from deoxyanisoin by ethylation, conversion to the alcohol, dehydration and demethylation. It is an oestrogenic substance which is highly active when administered orally. It is used for treating menopausal symptoms, for the suppression of lactation and for treatment of cancer of the prostate.

still Any apparatus or plant in which distillation is carried out.

stoicheiometric A stoicheiometric compound is one in which the ratio of the number of atoms to each other, as determined from the atomic weights, is a ratio of small whole numbers. In nonstoicheiometric compounds there are defects in the crystal lattice, or part replacement of atoms of one element by another.

stoicheiometry Generally applied to reacting

species. The relative numbers of reacting atoms or molecules and of the products.

stopped flow spectrophotometry A technique used to investigate the rates of fast chemical reactions, with half-lives in the range 10 seconds to 5×10^{-3} seconds. The technique involves measuring the optical density of a segment of the resulting solution one or two milliseconds after the reactants have been mixed.

stp Standard temperature and pressure.

straight-run A term used to denote fractions produced from crude petroleum by distillation but without further processing.

strain hardening Work hardening*.

stream double refraction Some colloidal sols such as aged vanadium pentoxide, whose dispersed particles are rod-shaped, are non-polarizing – as can be shown by placing the sol between crossed nicols, there being total extinction. When the liquid is stirred, however, or made to flow, polarization takes place, the stream lines being shown in a brightly illuminated image. This is due to the orientation of the rod-shaped molecules on stirring.

streaming potential When a liquid is forced by pressure through a diaphragm a potential difference, termed the streaming potential, is set up. The streaming potential, which is the converse of electro-osmotic flow, may be measured by forcing water through a capillary of the material under investigation and determining the potential difference between two electrodes, one placed at each end of the capillary.

strengths of acids and bases The strength of an aqueous acid is its ability to give hydroxonium ions (H_3O^+) and that of an aqueous base its ability to give hydroxyl ions or to accept protons. For Lewis acids and bases, acidic and basic character are measured by the tendency to react with standard bases and acids respectively. Since the degree of dissociation varies with the dilution, it is necessary to compare the relative strengths of acids or bases in solutions of equivalent concentrations. This can be done by measuring the concentrations of hydroxonium ion or of hydroxyl ion. Those acids like hydrochloric and nitric which are largely dissociated into their ions are strong acids; those like ethanoic and tartaric which are only slightly dissociated are weak acids. Potassium, sodium and barium hydroxides which are largely dissociated are strong bases, whilst the slightly dissociated ammonium hydroxide is a weak base.

streptomycin, $C_{21}H_{39}N_7O_{12}$. Antibiotic*

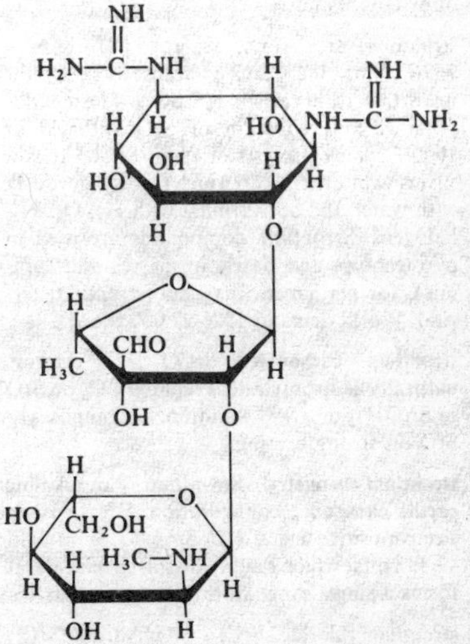

and complex organic base, produced by *Streptomyces griseus*. It is used in the form of its hydrochloride or sulphate, both white solids. It is active against both Gram-negative and Gram-positive bacteria. It was formerly used, in combination with other drugs, in the treatment of tuberculosis, and still finds application in certain infections due to organisms not susceptible to other antibiotics. Disadvantages are that bacteria may become resistant to the drug during treatment, and that it has some toxic action, especially on the auditory nerves.

Streptomycin B (mannosidostreptomycin) has a mannose molecule attached to the methylglucosamine group, and is the first antibacterial product made, but is enzymatically converted to streptomycin later in the fermentation.

Dihydrostreptomycin, in which the CHO group in the middle ring is replaced by CH_2OH, is made by the catalytic reduction of streptomycin, and has similar antibacterial properties.

stripping The separation of the more volatile component(s) of a liquid mixture from the less volatile component(s) in such a way that the latter is obtained in a pure state, but the former not necessarily so. Stripping may be carried out either by a process of fractional distillation* or by bringing the mixture into contact with an inert gas into which the light material will diffuse and be carried away. *See* rectification.

strontianite, $SrCO_3$. Mineral used as a source of Sr compounds.

strontium, Sr. At.no. 38, at.wt. 87·62, m.p. 769°C, b.p. 1384°C, *d* 2·6. Occurs as strontianite ($SrCO_3$) and celestine ($SrSO_4$). The metal is obtained by reduction of SrO with Al at 1000°C or electrolysis of fused $SrCl_2$. It is a silvery-white metal ccp which reacts vigorously with water and on warming with H_2, O_2, N_2, halogens. Strontium compounds are used in pyrotechnics and flares, in glasses and ceramics, for permanent magnets (strontium ferrite). World demand 1976 22 000 tonnes.

strontium carbonate, $SrCO_3$. Occurs naturally as strontianite. Prepared CO_2 on SrO or $Sr(OH)_2$ or a Sr^{2+} solution. Decomposes to SrO on strong heating.

strontium chemistry Strontium is an alkaline earth, electronic configuration $5s^2$. It is an electropositive metal (E° Sr^{2+} to Sr in acid −2·89 volts) which readily dissolves in acids. It forms a single range of Sr(II) compounds.

strontium chloride, $SrCl_2$. Anhydrous $SrCl_2$ formed from Sr plus Cl_2 or HCl over SrO or $SrCO_3$. Hydrates are formed.

strontium fluoride, SrF_2. Almost insoluble in water.

strontium hydroxide, $Sr(OH)_2$. Strong base, particularly in aqueous solution. Formed SrO plus water. Hydrates are known.

strontium nitrate, $Sr(NO_3)_2$. Formed $SrCO_3$ plus HNO_3. Forms 4 and 0 hydrates.

strontium oxide, SrO. M.p. 2430°C. Formed by ignition $SrCO_3$, $Sr(NO_3)_2$ or $Sr(OH)_2$. Forms $Sr(OH)_2$ in water.

Strontium peroxide, SrO_2, is formed from SrO and O_2 at dull red heat under pressure. $SrO_2,8H_2O$ is formed from $Sr(OH)_2$ plus H_2O_2 in cold aqueous solution.

strontium sulphate, $SrSO_4$. Occurs naturally as celestine. Prepared SrO, $Sr(OH)_2$ or $SrCO_3$ plus H_2SO_4. Sparingly soluble in water.

strychnine, $C_{21}H_{22}N_2O_2$. Strychnine is the

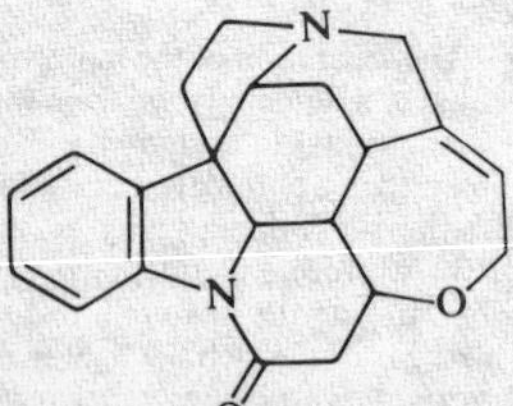

principal alkaloid of *Nux vomica* and other plants. Colourless prisms, m.p. 270–280°C.

Base, it stimulates all parts of the nervous system and in large doses produces convulsions. It is used for killing vermin.

styphnic acid, 2,4-dihydroxy-1,3,5-trinitrobenzene, $C_6H_3N_3O_8$. Yellow crystals, m.p. 178°C, obtained by nitrating 1,3-dihydroxybenzene; sparingly soluble in water, soluble in alcohol and ether. Has properties similar to those of picric acid. Its salts are more violent explosives than the picrates, and lead styphnate is an initiator.

styrene, ethenylbenzene, vinylbenzene, C_8H_8. Colourless aromatic liquid; b.p. 146°C. Polymerizes on heating to a glassy resinous substance, metastyrene, which regenerates styrene on heating. Occurs in liquid storax. Most conveniently prepared by treating cinnamic acid with hydriodic acid and then heating the product with aqueous potash.

Styrene is manufactured by alkylating benzene with ethene followed by dehydrogenation, or from petroleum reformate co-production with propylene oxide. Styrene is used almost exclusively for the manufacture of polymers, of which the most important are polystyrene, ABS plastics* and styrene-butadiene rubber. U.S. production 1980 3 megatonnes.

styrene-butadiene rubber, SBR The most important synthetic rubber copolymer generally containing about 25% styrene, 75% butadiene. U.S. production 1982 1 megatonne.

styrene oxide, 1,2-epoxyethylbenzene, phenyloxinane, $PhCHCH_2O$, C_8H_8O. B.p. 194°C.

Prepared by epoxidation of styrene with peroxyethanoic acid. Reactions are similar to those of aliphatic epoxides (*see*, e.g. ethylene oxide). Reacts with alcohols to give monoethers, e.g. $PhCH(OMe)CH_2OH$. Phenols give resins.

styrene polymers Styrene, $PhCH{=}CH_2$, undergoes polymerization on heating or with free radical initiators. Radiation and Ziegler–Natta catalysts also give useful forms of polystyrene. Polystyrene is used in many copolymers (e.g. acrylontrile-butadiene-styrene, ABS). Polystyrene is a thermoplastic used for moulding (good electrical properties), film, as latexes for paints and paper, in coatings, in laminates, foams, beads. U.S. production 1983 1·5 megatonnes.

sub-boric acid, $(HO)_2BB(OH)_2$. *See* boric acid.

suberane The trivial name for cycloheptane, C_7H_{14}, a colourless liquid, b.p. 118°C. The derived ketone is called suberone and monosubstituted derivatives called suberyl, e.g. suberyl alcohol is cycloheptanol.

suberic acid, 1,6-hexanedicarboxylic acid, $HO_2C[CH_2]_6CO_2H$, $C_8H_{14}O_4$. Important dicarboxylic acid obtained by oxidizing ricinoleic acid (from castor oil); also obtained by oxidation of cyclo-octene or cyclo-octadiene; formerly obtained from cork. Used in the formation of alkyd resins and polyamides. Esters are used as plasticizers and heavy duty lubricants and oils.

sublimation The volatilization of a solid substance into the vapour phase without passing through the liquid phase. Also used to describe the process of purification in which the vapour is condensed directly from the vapour phase to a solid (on a 'cold-finger' often cooled by refrigerant). In the latter case this substance may melt during the initial vaporization. Used for purification.

sublimation point (temperature). The temperature at which the vapour pressure above a solid is equal to the external pressure.

submicron A term used to describe particles, in the size range $10 \, m\mu - 1 \, \mu$, which are invisible in an ordinary microscope, but sufficiently large to appear in an ultra-microscope.

substitute natural gas A manufactured fuel produced from coal or oil and used as a complete or peak demand substitute for natural gas.

substitution reactions Reactions in which there is replacement of one atom or group in a molecule by another atom or group. E.g. the chlorination of benzene to give chlorobenzene in which one of the hydrogen atoms is substituted by a chlorine atom. This is an example of electrophilic substitution since the chlorine species is Cl^+. Nucleophilic reactions are named as S_N1 (substitution, nucleophilic, monomolecular) and S_N2 (bimolecular) reactions. S_N1 reactions proceed via an intermediate carbenium ion and are favoured by t-alkyl halides. If the starting material is chiral a racemic product is formed. S_N2 reactions characterize the reactions of primary alkyl halides (e.g. with cyanide ion) and involve a transition state.

Octahedral substitution reactions (e.g. those involving cobalt(III) complexes) may proceed by both S_N1 or S_N2 reactions. In the S_N1 case a slow dissociative mechanism (bond breaking) may take place. Reaction with the substituting ligand then yields the product. In S_N2 reactions direct participation of the incoming group causes the co-ordination number of the transition state intermediate to increase. Here bond making and bond breaking are equally important. Substitution reactions of square planar complex, e.g. those involving platinum-(II) complexes, have also been shown to proceed via S_N1 and S_N2 reactions.

substrate The substrate in an enzymic reaction is the molecule on which the enzyme acts. The three-dimensional enzyme-substrate interaction is highly favourable and necessary for the progression of the reaction.

succinic acid, butanedioic acid, $C_4H_6O_4$, $HO_2C(CH_2)_2CO_2H$. Colourless prisms; m.p. 182°C, b.p. 235°C. Occurs in amber, algae, lichens, sugar cane, beets and other plants, and is formed during the fermentation of sugar, tartrates, malates and other substances by a variety of yeasts, moulds and bacteria. Manufactured by the catalytic reduction of maleic acid or by heating 1,2-dicyanoethane with acids or alkalis. Forms an anhydride when heated at 235°C. Forms both acid and neutral salts and esters. Used in the manufacture of succinic anhydride and of polyesters with polyols.

succinic anhydride, $C_4H_4O_3$. White crystalline

$$\begin{array}{c} CH_2CO \\ | \diagdown \\ O \\ | \diagup \\ CH_2CO \end{array}$$

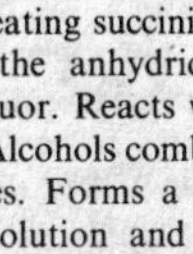

talline solid; m.p. 120°C, b.p. 261°C. Decomposed by water. Manufactured by heating succinic acid at 235°C. Conveniently prepared by heating succinic acid with ethanoyl chloride; the anhydride crystallizes on cooling the liquor. Reacts with water to give succinic acid. Alcohols combine to give monoalkyl succinates. Forms a mono-amide with ammonia in solution and this, on heating, changes to succinimide. Combines with primary amines in a similar manner. Reduced to butyrolactone by sodium amalgam. Used in the manufacture of dyestuffs and polymers.

succinimide, $C_4H_5NO_2$,

$\overline{|(CH_2)_2C(O) \cdot NH \cdot C(O)}$. Colourless plates; m.p. 126°C, b.p. 287°C (decomp.). Soluble in propanone; sparingly soluble in water. Prepared by heating ammonium succinate. Decomposed by hot water to give ammonium succinate. Reacts as an acid and forms salts with metals; these replace the hydrogen atom of the $>NH$ groups. Distillation with zinc dust gives pyrrole. Forms a chloroimide used for disinfecting water supplies.

sucrase *See* invertase.

sucrates *See* saccharates and sucrose.

sucrose, $C_{12}H_{22}O_{11}$. Cane or beet sugar, 1-

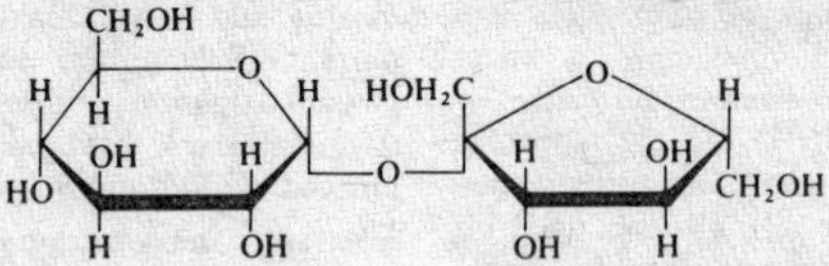

sucrose Glucopyranose Fructofuranose

α-glucosido-2-β-fructofuranose. M.p. 185–186°C. It has no reducing properties. It is readily hydrolysed by dilute acids to glucose and fructose and even more rapidly by invertase; the hydrolysed product which has a negative rotatory power is termed invert sugar. Sucrose has not been synthesized by purely chemical means, but has been prepared by the condensation of the potassium salt of D-glucose-1-phosphate by means of a bacterial phosphorylase. Forms sucrates with the alkali and alkaline earth metals. It contains 8 hydroxyl groups but gives rise to one series of derivatives only.

sugar of lead, $Pb(O_2CCH_3)_2,3H_2O$. *See* lead ethanoates.

sugars The sugars are carbohydrates, the majority of the natural sugars containing six or twelve carbon atoms in the molecule. They are crystalline, very soluble in water and generally have a sweet taste. The sugar of commerce, called cane sugar or beet sugar according to its origin, is the disaccharide sucrose*.

sulfur North American spelling of sulphur*

sulphacetamide, *p*-aminobenzene sulphonacetamide, $C_8H_{10}N_2O_3S$, p-$H_2NC_6H_4SO_2$-$NHCOCH_3$. White crystals, m.p. 181–183°C. Soluble sulphacetamide is the sodium salt which is soluble in water. It is prepared by acetylating sulphanilamide and hydrolysing one acetyl group. Being more soluble than most of the sulphonamides* it is used in treating infections of the urinary tract and of the conjunctiva.

sulphadiazine, 2-(*p*-aminobenzenesulphonamido)-pyrimidine, $C_{10}H_{10}N_4O_2S$. White powder, which darkens on exposure to light; m.p. 255–256°C. Prepared by condensing *p*-acetamidobenzenesulphonyl chloride with 2-aminopyrimidine and subsequent hydrolysis. Soluble sulphadiazine is the sodium salt. Sulphadiazine is the least toxic of the more potent sulphonamides*.

It is effective against streptococci, pneumococci, staphylococci and other bacteria.

sulphadimidine, 2-(*p*-aminobenzenesulphonamido)-4,6-dimethylpyrimidine, $C_{12}H_{14}N_4O_2S$. M.p. 198°C. It is effective against the same bacteria as sulphadiazine.

sulphaguanidine, *p*-aminobenzenesulphonylguanidine, $C_7H_{10}N_4O_2S$. White needle-like powder which darkens in air. Anhydrous; m.p. 189–190°C. It is prepared by fusing cyanoguanidine with sulphanilamide. It is used for treating intestinal bacterial infections since it is poorly absorbed by the gut.

sulphamates Salts of sulphamic acid.

sulphamerazine, 2-(*p*-aminobenzenesulphonamido)-4-methylpyrimidine, $C_{11}H_{12}N_4O_2S$. M.p. 235–236°C. Used to produce a high and persistent concentration in the blood. Its therapeutic properties resemble those of sulphanilamide*.

sulphamethoxazole, N′-(5-methyl-3-isoxazolyl)sulphanilamide, $C_{10}H_{11}N_3O_3S$. Colourless crystals; m.p. 167°C. A sulphonamide* antimicrobial drug, particularly effective when used in combination (called cotrimoxazole) with trimethoprim.

sulphamethoxypyridazine, 3-(4-aminobenzenesulphonamido)-6-methoxypyridazine, $C_{11}H_{12}N_4O_3S$. White or yellowish white, crystalline powder; m.p. 180–183°C. It is a long-acting sulphonamide*.

sulphamic acid, $H_3N^+{\cdot}SO_3^-$. A colourless crystalline solid normally existing as a zwitterion; prepared by the reaction of sulphur dioxide with acetoxime, or, commercially, by the treatment of urea with fuming H_2SO_4. Sulphamic acid is very soluble in water and is a strong acid, forming a series of salts, the sulphamates. It is an excellent acidimetric primary standard. It is used in weed-killers, flame-proofing agents, electrolytes in metal deposition processes and sweeteners.

sulphamide, $SO_2(NH_2)_2$ and **sulphimide,** $(SO_2NH)_n$. Colourless crystalline solids formed by the action of ammonia on a solution of sulphuryl chloride in benzene; free sulphimide exists only in the polymerized form.

Both compounds are soluble in water and are readily hydrolysed to sulphamic acid, $H_3N^+ \cdot SO_3^-$ and ammonia; the hydrogen atoms are in each case replaceable by metals to form salts. Many derivatives of sulphamide and cyclic derivatives of sulphimide are known.

sulphanes The hydrogen sulphides H_2S_2 to H_2S_6 and beyond (for H_2S *see* hydrogen sulphide). Formed S in aqueous Na_2S and addition to HCl or S_2Cl_2 and H_2S. Decompose into H_2S and S.

sulphanilamide, *p*-aminobenzenesulphonamide, $C_6H_8N_2O_2S$. Colourless crystals, m.p. 164·5–166·5°C. It is usually prepared by treating *p*-acetamidobenzenesulphonyl chloride with ammonia, and hydrolysing the acetyl derivative to the base. Used for the treatment of streptococcal infections, gonorrhoea, meningococcal meningitis and urinary infections. Liable to cause unpleasant reactions, such as nausea, cyanosis and skin rashes.

sulphanilic acid, 4-aminobenzenesulphonic acid, $C_6H_7NO_3S$. Colourless crystals.

It is prepared by heating aniline sulphate for 8 hours at 190°C. It readily diazotizes and is used as first component in a large variety of azo dyes.

sulphatases Enzymes which catalyse the hydrolysis of esters of sulphuric acid. One type splits the sulphuric esters of phenols and is widely distributed in plants and animal tissues. Another type from bacteria, splits the ethereal sulphates of aliphatic compounds.

sulphates Formally any compound containing an anionic oxy-sulphur species. Practically derivatives of sulphuric acid; i.e. sulphates(VI) containing $(SO_4)^{2-}$ groups which may be ionic or co-ordinated.

sulphathiazole, 2-(*p*-aminobenzenesulphonamido)-thiazole, $C_9H_9N_3O_2S_2$. White powder, m.p. 200–203°C. Sulphathiazole has the typical actions of sulphonamides*, being effective against *β*-haemolytic *Streptococci*, *Pneumococci*, *Gonococci* and *Staphylococci*.

sulphides Compounds of other elements with sulphur. Generally prepared by direct interaction of the elements. Metal sulphides are often very insoluble, occur as minerals and can be used for separation of the elements. Sulphides of the most electropositive elements give H_2S with water. Covalent derivatives formed with the less electropositive elements, e.g. liquid CS_2, S_2Cl_2.

sulphinyl Containing the >SO group.

sulphites Salts of sulphurous acid, i.e. sulphates(IV) containing $[SO_3]^{2-}$ species.

sulpholane, tetrahydrothiophen-1,1-dioxide, tetramethylenesulphone, $C_4H_8O_2S$. A viscous liquid, b.p. 285°C, which readily solidifies

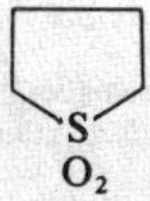

to a colourless solid, m.p. 28°C when freed from water, with which it is miscible. Used as a selective solvent for liquid-vapour extractions especially by gas-liquid chromatography. It will also extract benzene from alkanes and is used as a non-aqueous polar solvent. Used as a stationary phase the material is very suitable for the analysis of complex C_2-C_5 hydrocarbon mixtures. Most easily prepared by reduction of 2,5-dihydrothiophene-1,1,-dioxide, which is obtained from butadiene and sulphur dioxide.

sulphonamides Compounds containing the grouping $-SO_2N<$. They are obtained by the action of ammonia and primary and secondary amines on sulphonyl halides. The term is also used for a group of drugs related to sulphanilamide*. The use of the sulphonamides as drugs dates from 1934 when the dye prontosil was shown to be an effective curative agent for certain infections caused by bacteria. Later it was found that sulphanilamide was the active portion of the prontosil molecule. Subsequently other and more effective derivatives of sulphanilamide have been introduced. They are prepared in general by condensing *p*-acetamidobenzenesulphonyl chloride with the appropriate amine and removing the acetyl group by hydrolysis.

The various sulphonamides differ in their specificity to various bacteria and in their ease of absorption and excretion. They are bacteriostatic (inhibiting growth) and not bactericidal, acting by allowing the natural body mechanisms to destroy the bacteria.

The action of the sulphonamides is antagonized by *p*-aminobenzoic acid and they act by inhibiting the uptake and utilization of 1-aminobenzoic acid by bacterial cells, which require this as a precursor of folic acid.

See sulphacetamide, sulphadiazine, sulphadimidine, sulphaguanidine, sulphamerazine, sulphamethoxypyridazine, sulphanilamide, sulphathiazole.

sulphonation Introduction of sulphonic groups. *See* sulphonic acids.

sulphones Organic compounds containing an $>SO_2$ group united directly to two carbon atoms, e.g. sulpholane. They are produced by oxidation of organic sulphides with nitric acid or potassium permanganate. They are colourless solids and are very stable.

sulphonic acids (aromatic), sulpho-compounds Organic compounds containing the $-SO_3H$ group.

The sulphonic acids are usually prepared by the action of sulphuric acid upon a compound. The concentration of the acid and the temperature of reaction are varied according to the reactivity of the compound. Often oleum is used or even chlorosulphonic acid. Alternatively sulphur trioxide complexed to pyridine or dioxan can be used with reactive substrates. Aminosulphonic acids such as sulphanilic and naphthionic acids are most conveniently prepared by heating the sulphate of the amine at 180°C.

Prolonged sulphonation of benzene gives 1,3-benzenedisulphonic acid. In the case of naphthalene, sulphonation gives 1-substitution at low temperatures and 2-substitution at higher temperatures.

The sulphonic acids are strongly acidic compounds, very soluble in water and readily give water-soluble metallic salts.

The most important reaction of the sulphonic acids is their conversion into phenols by fusion with caustic alkalis. When they are fused with potassium cyanide, nitriles are obtained, e.g. benzonitrile from benzenesulphonic acid.

Sulphonation is widely used for the solubilization of insoluble azo dyestuffs.

sulphonphthaleins A group of dyes, many of which are used as indicators, made by heating *o*-sulphobenzoic anhydride with phenols or substituted phenols. *See*, e.g. phenol red.

sulphonyl halides Organic compounds $RS(O)_2X$, where R is an alkyl or aryl group and X is a halogen atom. A particularly important example is toluene-*p*-sulphonyl chloride ($R = p\text{-}CH_3{\cdot}C_6H_4$, $X = Cl$). This is prepared by the action of chlorosulphonic acid on toluene, the *ortho* and *para* isomers being separated by centrifugation. Toluene-*p*-sulphonyl chloride is a white crystalline solid, m.p. 69°C, which forms characteristic crystalline derivatives with many alcohols, phenols, sugars and amines. The properties of sulphonyl halides thus resemble those of carboxylic acid chlorides. *See* acetyl chloride, toluene-4-sulphonyl chloride.

sulphoxides Organic compounds containing an $>SO$ group linked to two C atoms. They are obtained by the oxidation of organic sulphides with hydrogen peroxide. They are usually hygroscopic liquids and form addition compounds with acids. On oxidation, sulphones are produced. Dimethyl sulphoxide* is a useful solvent and chemical intermediate.

sulphoxylic acid, H_2SO_2. Two forms, symmetric, $S(OH)_2$, by hydrolysis of $S(NR_2)_2$ or $S(OR)_2$ (thiosulphite esters plus NaOEt) and asymmetric HS(O)OH (reaction intermediate and some esters). Cobalt(II) sulphoxylate is formed $NaHSO_3$ plus cobalt ethanoate plus NH_3.

sulphur, S. At.no. 16, at.wt. 32·06, m.p. 112·8°C, b.p. 444·674°C, *d* 2·07. Yellow nonmetallic element of the oxygen group. Occurs native (extracted from subterranean sources with superheated steam – Frasch process). Also occurs as sulphates, e.g. gypsum ($CaSO_4{,}2H_2O$), anhydrite ($CaSO_4$) and sulphides, e.g. iron and copper pyrites (used as SO_2). Occurs as H_2S in some natural gas and oil (recovered by reaction with SO_2 over a catalyst). Sulphur exists in various allotropic forms, α-S, rhombic sulphur, contains puckered S_8 cyclic units. β-S, monoclinic sulphur also contains S_8 rings (the conversion temperature $\alpha \rightleftharpoons \beta$ is 95·5°C). γ-S (also monoclinic) is deposited from some solvents and also contains S_8 rings. ρ-sulphur, rhombohedral, is formed from $Na_2S_2O_3$ and conc. HCl at 0°C followed by extraction with toluene, it contains S_6 chair rings. Melting of sulphur gives S_λ still containing S_8 rings but above 160°C spiral chains of μ-S are formed. π-S (probably S_6) is also present. Plastic sulphur (molten S into water) contains μ-S chains. Sulphur vapour contains S_2 (predominates, paramagnetic), S_4, S_6, S_8 species. Sulphur is a reactive element, combining with most other elements. The principal use is in H_2SO_4 production (90%) but also used in the wood pulp industry, photography, leather, rubber, dyestuffs industries, in detergents, fertilizers, insecticides and fungicides. U.S. production 1981 27·6 megatonnes (recovered 18, Frasch 9·6).

sulphur bromide, sulphur monobromide, S_2Br_2. The only known sulphur bromide.

sulphur chemistry Sulphur is an element of the oxygen group. It shows oxidation states +6, +4, +2, 0, −2. It behaves principally as a non-metal, forming covalent species (S_5^{+}, S_4^{2-}, etc. are also formed). It shows a great tendency to catenation. Onium cations are readily formed (e.g. Ph_3S^+).

sulphur chlorides Sulphur and chlorine react in all proportions.

Sulphur monochloride, S_2Cl_2, m.p. −80°C, b.p. 138°C. Yellow liquid hydrolysed by water to SO_2, HCl and S. Lower halides S_xCl_2 (*x* up to 5) are formed S_2Cl_2 plus H_2 at a hot surface with freezing of products. S_2Cl_2 is used in the rubber industry as a solvent for S.

Sulphur dichloride, SCl_2, b.p. 59°C. Red liquid.

Sulphur tetrachloride, SCl_4, m.p. −30°C. Stable only in the solid. The oxide chloride of sulphur(IV) is $SOCl_2$, thionyl chloride*.

Sulphur(VI) forms the oxide chloride SO_2Cl_2, sulphuryl chloride (*see* sulphuryl halides).

sulphur dyes Dyestuffs which are applied in a reduced state from solutions containing S^{2-}, SH^- or polysulphide. The dyestuff is reoxidized on the fibre. The actual dyestuff may or may not contain sulphur.

sulphuretted hydrogen, H_2S. *See* hydrogen sulphide.

sulphur fluorides

Sulphur hexafluoride, SF_6, m.p. −51°C. Formed S plus F_2. Very inert material used as an inert dielectric. S_2F_{10} (toxic) is also formed from S plus F_2 and there is an extensive chemistry of SF_5 derivatives (e.g. SF_5Cl, ClF plus SF_4).

The oxide fluorides SO_2F_2 and SOF_4 and $N{\equiv}SF_3$ are known.

Sulphur tetrafluoride, SF_4, m.p. −121°C, b.p. −40°C. Reactive gas (SCl_2 plus NaF in MeCN). Used as fluorinating agent (e.g. $>C{=}O$ gives $>CF_2$).

The oxide fluoride SOF_2 ($SOCl_2$ plus NaF) and $(NSF)_4$ are known.

Disulphur difluoride, S_2F_2. Two isomers, FSSF and $S{=}SF_2$ formed AgF plus S. Reactive material.

sulphuric acid, H_2SO_4, $S(O)_2(OH)_2$. M.p. 10°C, b.p. 340°C, *d* 1·83. Very important industrial acid, top tonnage chemical. Manufactured from SO_2 plus air over a Pt or V_2O_5 catalyst (500–600°C then 400–450°C). Emergent gases, SO_3, (100°C), are absorbed in 97% acid (contact process). In the virtually obsolete lead chamber process SO_2, NO_2 air and steam are sprayed into lead chambers to produce 62–68% acid. SO_2 is obtained from combustion of sulphide ores or H_2S or by heating $CaSO_4$ with C and clay. H_2SO_4 is miscible in all proportions with water and forms an azeotrope b.p. 339°C containing 98·3% H_2SO_4. A solution of SO_3 in H_2SO_4 is fuming sulphuric acid (*see* sulphuric acid, fuming). It decomposes to SO_3 and H_2O above the b.p. It is a very powerful dehydrating agent (the reaction with H_2O is very violent and exothermic). The hot concentrated acid is a strong oxidizing agent (e.g. C gives CO_2). It is a strong acid although dissociation of $[HSO_4]^-$ is not always complete. Low volatility and therefore displaces most other acids (or their anhydrides) from their salts. Can be handled in glass, cast iron or steel. Used in the production of phosphate (70% used for fertilizers), ammonium sulphate, aluminium sulphate, explosives, pigments, petrochemicals, pickling iron and steel. U.S. production 1982 43 megatonnes. The salts of sulphuric acid are the sulphates*.

sulphuric acid, fuming; Nordhausen sulphuric acid; oleum A solution of SO_3 in H_2SO_4 containing some $H_2S_2O_7$, etc. SO_3 distils from the acid on heating. Used in organic synthesis.

sulphur nitrogen derivatives There is an extensive chemistry of sulphur nitrogen derivatives, many of them being cyclic derivatives. S_4N_4 (S_2Cl_2 plus NH_4Cl, S plus NH_3 in CCl_4) gives an extensive series of derivatives, e.g. $SOCl_2$ gives $(S_4N_3)Cl$. $(SN)_x$ is a polymer which is a good electrical conductor. S_7NH, $S_5(NH)_3$ and $S_4(NH)_4$ (S_2Cl_2 plus NH_3) all have structures related to S_8. *See also* sulphonamides.

sulphurous acid, H_2SO_3, $OS(OH)_2$. Sulphur dioxide in water gives $SO_2{,}7H_2O$, a gas hydrate, and free H_2SO_3 does not exist. The solutions react with bases to give sulphites, sulphates(IV), containing the $[SO_3]^{2-}$ or $[HSO_3]^-$ ions. Sulphites react with acids to give SO_2. Sulphites are reducing agents.

sulphur oxide halides *See* under the halides. Thionyl and sulphuryl halides, e.g. SO_2F_2, SOF_4, SOF_2, SO_2Cl_2, $SOCl_2$ are known.

sulphur oxides

Disulphur monoxide, S_2O, SSO. Unstable, glow discharge on SO_2.

Disulphur trioxide, S_2O_3. Blue–green solid S plus SO_3.

Sulphur dioxide, SO_2, m.p. −72·7°C, b.p. −10°C. Colourless gas with characteristic smell. Formed by burning S, metal sulphides, H_2S in air or acid on a sulphite or hydrogen sulphite. Powerful reducing agent, particularly in water. Dissolves in water to give a gas hydrate; the solution behaves as an acid – *see* sulphurous acid. Used in the production of SO_3 for sulphuric acid.

Sulphur trioxide, SO_3, m.p. 17°C, b.p. 49°C. Formed SO_2 plus O_2 over a catalyst (contact process – *see* sulphuric acid). The solid exists in several forms. Readily absorbed by water (to give sulphuric acid) or H_2SO_4 (to give fuming sulphuric acid).

sulphur oxyacids Sulphur forms an extensive range of oxyacids which are dealt with under separate headings. The acids are:

$S(OH)_2$, $HS(O)OH$	sulphoxylic acid
$OS(OH)_2$	sulphurous acid
$O_2S(OH)_2$	sulphuric acid
$S(O)S(OH)_2$	thiosulphuric acid
$O(HO)SS(OH)O$	dithionous acid
$O(HO)SS_nS(OH)O$	polythionic acids
$O_2(HO)SS(OH)O$	pyrosulphurous acid, disulphurous acid
$O_2(HO)SS(OH)O_2$	dithionic acid
$O_2(HO)SOS(OH)O_2$	pyrosulphuric acid, disulphuric acid
$H_2S_3O_{10}$	trisulphuric acid
$H_2S_4O_{13}$	tetrasulphuric acid
$O_2S(OH)(OOH)$	peroxymonosulphuric acid
$O_2(HO)SOOS(OH)O_2$	peroxydisulphuric acid

sulphuryl Containing the $>SO_2$ group.

sulphuryl halides, SO_2X_2. Sulphuryl chloride, SO_2Cl_2, b.p. 69°C (SO_2 plus Cl_2 plus light) is used as a chlorinating agent. SO_2F_2 and SO_2Br_2 are also known.

sulphydryl The $-SH$ group, also called thiol.

super acids Mixtures, e.g. $HF-SbF_5$, HSO_3F-SbF_5, which act as very strong acids and, e.g. induce carbenium ion formation, protonation, hydride abstraction.

superconductivity At low temperatures certain metals and a large number of intermetallic compounds possess superconductivity, have zero sensitivity and undetectable magnetic permeability. Although the origin of these properties is not fully understood, they are of considerable engineering interest.

supercooling The phenomenon of cooling a liquid to a temperature below that at which crystallization would normally occur without the separation of solid. E.g. pure water may be cooled to several degrees below its freezing point provided crystallization nuclei, such as suspended dust particles, are absent. Glass is an extreme example of supercooling in which solidification has occurred without crystallization. Solutions which have been cooled below the saturation temperature for the quantity of solute present, without crystallization occurring, are said to be supersaturated.

superheavy elements *See* post-actinide elements.

superoxide dismutase, SOD This enzyme catalyses the reaction

$$2O_2^- + 2H^+ \longrightarrow H_2O_2 + O_2$$

It has been isolated from many plant, animal and bacterial erythrocytes and heart. Organisms that utilize oxygen for respiration produce some superoxide radicals and superoxide dismutase is thought to protect cells against the deleterious action of these radicals. *See also* xanthine oxidase.

superoxides Compounds containing the paramagnetic O_2^- ion, e.g. KO_2.

superphosphates Insoluble calcium phosphate, $Ca_3(PO_4)_2$, treated with concentrated sulphuric acid gives calcium hydrogen phosphate, $Ca(H_2PO_4)_2$, which is soluble in water and is a very effective fertilizer. The dry product consists of the soluble calcium salt together with calcium sulphate and sulphates of other metals present in the raw material, and is known as 'superphosphate'. The grades of superphosphate most widely used have about 13·7% of phosphoric acid (30% tricalcium phosphate) and 16–17% of phosphoric acid.

supersaturation *See* supercooling.

surface active agents Water, possessing powerful intermolecular attractive forces, has a high surface tension (72·8 dynes/cm at 20°C). Many soluble substances (mainly organic) when dissolved in water reduce the surface tension even in very low concentrations.

These surface active agents have weaker intermolecular attractive forces than the solvent, and therefore tend to concentrate in the surface at the expense of the water molecules. The accumulation of adsorbed surface active agent is related to the change in surface tension according to the Gibbs adsorption equation

$$\Gamma = -\frac{c}{RT}\frac{d\gamma}{dc}$$

where Γ is the surface excess, c is the bulk concentration of the surface active agent and γ is the surface tension. Surface active agents, which include certain dyestuffs, wetting agents, detergents, antibiotics, bactericides, etc., may be electrically neutral, non-dissociating substances (non-ionic), or they may ionize to give surface active anions or cations with small oppositely charged counterions (usually metallic ions and halogen ions respectively).

surface activity A phenomenon which arises from the non-symmetrical distribution of attractive forces between molecules (and ions) residing within about 5–10 Å of any surface (i.e. within the surface layer). This results in their being attracted away from the surface towards the interior, and with liquids leads to the contraction of the surface to the smallest possible area. Work has therefore to be performed to cause the surface to expand, the amount, measured in ergs/cm^2, being numerically equal to the surface tension (in dynes/cm). The surface activity of solids, particularly when finely divided, often leads to gas adsorption, with which catalysis is frequently associated.

surface combustion The burning of gaseous fuel and air in such a way that catalytic oxidation occurs at the surface of a porous refractory. Combustion is rapid and complete within a very thin flame section and very high rates of heat release are obtained. Many industrial furnaces and kilns employ this method.

surface compounds The surfaces of many solids become coated with a unimolecular layer of chemisorbed gas in which the gas molecules form ordinary chemical bonds with the surface atoms leading to the formation of surface compounds. Such layers often give the surface its characteristic properties. Thus many metals become coated with an oxide layer which render them stable to further attack.

surface energy The surface energy of a system is defined as the work necessary to increase the surface by unit area against the force of surface tension.

surface orientation When a film of an organic material is formed on water, the molecules in the surface layers usually orientate themselves according to their structure. With large molecules there may be, e.g. groups such as carboxylate or sulphonate present; these will tend to confer solubility upon the molecule. Such groups are said to be hydrophilic, i.e. attracted to the water, and the molecule will be anchored to the surface by these groups which will tend to penetrate below the surface of the water. Conversely, hydrocarbon groups are hydrophobic and will tend to remove themselves from the water. Thus, e.g. long chain fatty acids will tend to orientate themselves so that the hydrocarbon groups are perpendicular to the water surface into which the acid groups are dipping.

surface potential The change in work function, relative to a clean surface, when an adsorbed layer is present on the surface.

surface pressure Thin films of spreading oils, e.g. oleic acid, lower the surface tension of the liquid (generally water) on which they are spread. Consequently, they exert a force of repulsion on floating particles. The force exerted by a spreading film per cm of boundary is numerically equal to the lowering of surface tension; it is called the surface pressure of the film by analogy (in two dimensions) with the pressure of a gas. The surface pressure exerted by a solid or liquid present in excess is termed its spreading pressure.

surface tension The molecules in the body of a liquid are equally attracted in all directions by the other molecules. Those at the surface, on the other hand, will have a residual inward attraction. This is associated with a *surface tension* which acts in a direction parallel to the boundary surface and tends to reduce that surface to a minimum. Work has to be done to increase the area of the surface against this force. This explains why an emulsified system always tends to be unstable and breaks down unless there is present some substance to lower the surface or interfacial tension. *See* surface energy.

surface viscosity Unimolecular films on liquid surfaces may be either readily mobile (e.g. a 'gaseous' film) or slow to flow under the action of a two-dimensional stress (*see* surface pressure). They therefore possess surface viscosity, the two-dimensional analogue of ordinary viscosity. Some surface films show surface plasticity, i.e. they behave as solids until a critical shearing stress is applied.

surfactants Surface-active agents which must be soluble, contain groups of opposite polarity and solubilizing tendencies, form oriented monolayers at phase interfaces, form micelles, have detergency, foaming, wetting, emulsifying and dispersing properties. Long chain sulphonic acids are typical surfactants; fluorocarbon derivatives are even more active.

surgical spirit Industrial methylated spirits* further denatured by the addition of castor oil (2·5%), diethyl phthalate (2·0%) and methyl salicylate (0·5%).

suspended transformation A system in a metastable state*.

suspending agents For many purposes the suspension of a small amount of material in a large volume of liquid is required. To prepare this the material to be suspended should be ground as finely as possible and given an electrical double layer by adsorption of an ionic detergent. A further method is to obstruct sedimentation by the use of a bulky hydrated precipitate which will hold up the solid to be suspended. Bentonite and related clays may be used for this purpose.

suxamethonium chloride, succinyl dicholine,

$$\left[\begin{array}{l} CH_2\cdot CO\cdot O\cdot CH_2\cdot CH_2\cdot N(CH_3)_3 \\ | \\ CH_2\cdot CO\cdot O\cdot CH_2\cdot CH_2\cdot N(CH_3)_3 \end{array}\right]^{2+} \begin{array}{l} 2Cl^-, \\ 2H_2O \end{array}$$

$C_{14}H_{30}Cl_2N_2O_4$. White powder prepared from dimethylaminoethanol and succinyl chloride, followed by methylation. Neuromuscular blocking agent used to relax skeletal muscles during certain types of surgical operation.

sweating The process of removing entrapped oil and low-melting waxes from the filter cake obtained in dewaxing processes.

sweetening The process of treating 'sour'* gasoline and kerosine fractions for the purpose of converting or extracting mercaptans to remove objectionable odours. All involve oxidation of mercaptans to dialkyl disulphides.

sweetening agents The sugars, particularly sucrose, fructose, lactose and maltose are the classical sweetening agents. Sorbitol and mannitol are sweet-tasting polyols. Synthetic sweeteners have no food value. Cyclamates are possibly carcinogenic. Saccharin appears safer.

swelling of colloids When a piece of dry gelatin is exposed to moist air or placed in water, it becomes hydrated and swells considerably. This is an effect common to all hydrophilic colloids – most natural organic matter, e.g. cellulose, wood, carbohydrates. The swelling produces considerable pressure in the colloid and, in consequence, still occurs even when opposed by an external pressure of many atmospheres.

The swelling of gels is markedly affected by the presence of electrolytes, this effect being a minimum at the isoelectric point of the material. In general, sulphates, tartrates, etc. inhibit swelling, while iodides and thiocyanates promote the swelling. Thus gelatine disperses completely in iodide solution even at low temperatures.

sydnones Heterocyclic compounds for which satisfactory formulae using normal covalent bonds cannot be written. Often termed 'mesoionic' and are pseudoaromatic. Sydnone itself can be represented as:

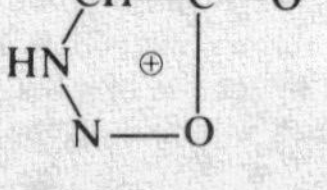

Sydnones are neutral, highly crystalline, stable compounds, soluble in most organic solvents. N-Arylsydnones typically are obtained by treating N-nitroso-N-aryl-glycines with ethanoic anhydride. The parent glycine is regenerated when the sydnone is heated with dilute alkali.

sylvestrene, $C_{10}H_{16}$. An optically active monocyclic terpene, the dihydrochloride of which is obtained by the action of HCl on various pine-needle oils, and on Finnish and Indian oils of turpentine. Sylvestrene is not present in these oils, but is formed by the isomerization of Δ^3- and Δ^4-carene. Sylvestrene

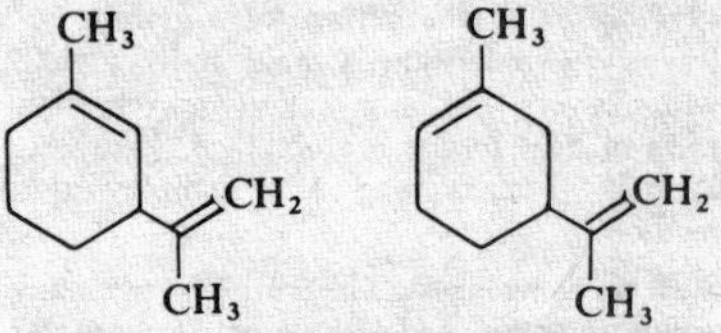

is liberated from the dihydrochloride by treatment with aniline. Both (+)-and (−)-forms are known, as is also the racemic form, carvestrene. These hydrocarbons are inseparable mixtures of compounds of the two constitutional formulae given above. Carvestrene can be obtained by various synthetic reactions from carvone. Sylvestrene gives a crystalline dihydrochloride of m.p. 72–73°C. Carvestrene dihydrochloride melts at 52–53°C.

sylvine, sylvite, KCl. Occurs in salt deposits – an important source of potassium salts.

sylvinite, KCl – NaCl. A source of potassium salts.

symmetry elements The operators which give repetition and pattern in molecules and lattices. Symmetry elements comprise:

(1) Axes of symmetry. An axis about which rotation of the body through an angle of $2\pi/n$ (where n is an integer) gives an identical pattern; 2-fold, 3-fold, 4-fold and 6-fold axes are known in crystals; 5-fold axes are known in molecules. In a lattice the rotation may be accompanied by a lateral movement parallel to the axis (screw axis).

(2) Planes of symmetry. Planes through which there is reflection to an identical point in the pattern. In a lattice there may be a lateral movement parallel to one or more axes (glide plane).

(3) Identity operator. Present in all lattices and molecules.

(4) Centre of symmetry. A point through which there is reflection to an identical point in the pattern.

symproportionation, conproportionation The opposite of disproportionation.

syndiotactic *See* syntactic polymers.

syneresis The separation of liquid from a gel on standing. All gels show syneresis to some extent and the phenomenon may be regarded as a continuation of the process of gelation, the gel material adhering at more and more places and, in consequence, shrinking and squeezing out a portion of the dispersion medium. Viscose gels synerize to a large extent. The phenomenon is commonly met in the cooking of foods and is thought to play an important part in gland secretion.

synergism The phenomenon by which the additive properties of two components is much greater than the sum of the effects of the two separately.

synergist A compound which, whilst formally inactive or weakly active, in a particular context can give enhanced activity to another compound present in the system.

***syn* isomer** *See* isomerism and anti-isomer.

syntactic polymers Polymers with a stereoregular structure in which the substituent groups lie alternately above and below the plane of the carbon chain are said to be syntactic. Such structures are also referred to as syndiotactic configurations.

synthesis In a chemical sense the building up of complex molecules from smaller ones.

synthesis gas A mixture of hydrogen and carbon monoxide, normally produced from the reaction of methane and steam under pressure or, less usually, by the partial oxidation of methane. Synthesis gas is used in the manufacture of methanol, oxo* alcohols and hydrocarbons. *See* Fischer–Tropsch reaction.

synthetic fibres *See* fibres

synthetic rubbers *See* rubber, elastomers, etc.

systemic insecticides Compounds which when applied to the leaves, stems and sometimes the roots of plants, are absorbed and translocated in the plant in the course of normal nutrition in concentrations safe for the plant but lethal to insects feeding on the plants. The most important systemic insecticides are certain organo-phosphorus compounds and these are thought to act as inhibitors of esterase activity in the organism. For examples of systemic insecticides in practical use, *see* dimefox, Metasystox, Phosdrin and Schradan.

Szilard–Chalmers effect *See* recoil atom.

T

T Tritium.

2,4,5-T, 2,4,5-trichlorophenoxyacetic acid, $C_8H_5Cl_3O_3$. M.p. 155°C. Used as a selective herbicide. It is made from 2,4,5-trichlorophenol and sodium chloroacetate. Ester sprays and combined ester sprays with 2,4-D are available. 2,4,5-T products are of particular value in that they control many woody species, and eradicate perennial weeds such as nettles in pastures.

Ta Tantalum.

tactosols Colloidal sols which contain non-spherical particles, which are capable of orientating themselves. E.g. aged vanadium pentoxide sol separates into rod-shaped particles in a concentrated tactosol and a dilute isotropic atactosol. In a magnetic field the particles of such a sol arrange themselves along the lines of force.

talc, $Mg_3(OH)_2Si_4O_{10}$. A silicate with layers of linked SiO_4 tetrahedra, the layers are electrically neutral so that there is only weak interaction between layers. The crystals are soft and used as a lubricant (French chalk). Also used as an extender and filler in paint, rubber, insecticides, and tar and asphalt roofing materials, as a constituent of toilet powders and in some types of porcelain. Steatite is a pure compact variety, soap stone is a massive impure form. *See* clays. World use 1976 5·9 megatonnes.

tall oil A major by-product of the kraft or sulphite pulping process of wood. Mainly fatty acids but contains some unsaponifiable material. Used in protective coatings (30%), soaps and detergents (15%), esters and plasticizers (30%). World production greater than 1 megatonnes p.a.

talose *See* hexose.

Tanabe–Sugano diagram Graphs showing the relation between the splitting of electronic energy levels (E/B) and crystal field stabilization energy (Δ/B) where B is a Racah parameter. Used in assigning electronic transitions in complexes and obtaining values of Δ and B.

tanacetyl alcohol *See* thujyl alcohol.

tannase A number of moulds, e.g. *Aspergillus niger*, contain an enzyme which splits tannins of the ester type, of which methylgallate is the simplest equivalent.

tannic acid, gallotannic acid, tannin

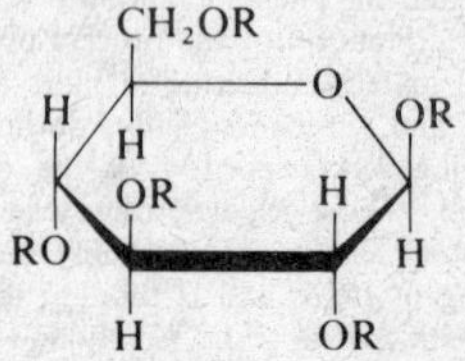

R:galloyl; *m*-digalloyl; *m*-trigalloyl.

A yellowish-white powder, m.p. 210–215°C (decomp.), with an astringent taste. It is obtained from oak-galls by fermenting them and extracting with ether. It gives precipitates with most metallic salts, with proteins and with alkaloids. On hydrolysis it gives glucose and gallic acid.

tanning The process of changing skins and hides into useful leather. Skins are cleaned and soaked in water, treated with lime, and hair and flesh removed. Neutralized with buffering salts [$(NH_4)_2SO_4$,NH_4Cl] and proteolytic enzyme and finally tanned with either tannin, obtained from vegetable sources which combine with collagen and other proteins in the hides to form leather, or with a basic chromium(III) sulphate which forms complexes with the collagen.

tanning development The use of the products of photographic development, dihydroxybenzenes, to react with the photographic gelatine to produce cross-linking and hardness in the developed areas which remain stable whilst the rest of the gelatine may be washed away. Used for copying engineering drawings as positives and for the preparation of printing plates.

tannins A large class of amorphous substances present in plants. They have an astringent taste, give a blue or green colour with iron salts and are precipitated from water solution by proteins and by alkaloids. They are condensation products of various phenols, of which the most important are pyrogallol and catechol. The tannins have similar structures to tannic acid*. The tannins can be obtained by extracting the raw materials with water or other solvents and precipitating with lead ethanoate. It is their property of precipitating gelatine which is responsible for the use of tannins in the treatment of hides to make leather. They are also used as mordants in the textile industry.

tantalates *See* tantalum oxides.

tantalite $(Fe,Mn)(Nb,Ta)_2O_6$ with excess Ta. The principle ore of tantalum*.

tantalum, Ta. At.no. 73, at.wt. 180·9479, m.p. 2996°C, b.p. 5425°C, *d* 16·6. An element of Group V, electronic configuration $6s^25d^3$. Occurs with Nb in tantalite $(Fe,Mn)(Nb,Ta)_2O_6$. The ores are treated by NaOH fusion followed by acid washing to give mixed pentoxides. Ta is separated by liquid extraction from HF with isobutylmethylketone. The metal is prepared from Ta_2O_5 by reduction with alkali metals or by electrolysis of fused fluorides, it is bcc, remaining lustrous in air but reacts with oxygen or steam at high temperatures and dissolves in HNO_3-HF. Used in resistant alloys particularly construction for the chemical industry and in surgical appliances. World demand 1976 1000 tonnes.

tantalum chemistry Largely that of the +5 oxidation state and very similar to niobium chemistry*. The major difference is the absence of stoicheiometric lower oxides for Ta and minor differences in complex species, e.g. $TaCl_6^-$ and $NbOCl_5^{2-}$ in HCl solutions.

tantalum halides Known for all TaX_5 and TaX_4 (except TaF_4). TaX_5 ($Ta+X_2$; $Ta_2O_5+C+Cl_2$ or CCl_4 give $TaCl_5$) contain bridged polymeric species ($TaCl_5$ dimer). TaF_5, m.p. 95°C, b.p. 229°C; $TaCl_5$, m.p. 211°C, b.p. 241°C. Hydrolysis gives oxide halides, e.g. $TaOF_3$, TaO_2F. Complexes are formed with many donors and complex fluorides, e.g. TaF_6^-, TaF_7^{2-}, TaF_8^{3-}, $TaCl_6^-$ are formed in acid solution. TaX_4 (TaX_5+ Ta) are dark coloured solids with some Ta–Ta interaction. Adducts are formed with various ligands. Lower halides, e.g. $[Ta_6Cl_{12}]Cl_{4/2}$, $[Ta_6Cl_{12}]Cl_{6/2}$ (not fluorides) are formed by further reduction – they contain metal-cluster species and there is an extensive chemistry based on, e.g. $[Ta_6Cl_{12}]^{n+}$ ($n=2$, 3 or 4) species. TaF_3 (Ta+HF at 225°C) has an essentially ionic structure.

tantalum oxides Ta_2O_5 is the most important oxide and is prepared by dehydration of the hydrated oxides formed by hydrolysis of halides. It is a dense white inert material, reduced to $Ta_{2\cdot5}O_5$. The pentoxide forms tantalates, e.g. K_3TaO_4, as mixed metal oxides by fusion with oxides or carbonates, reduction gives bronzes. Isopolytantalates, e.g. $[H_xTa_6O_{19}]^{(8-x)-}$ ($x=0$, 1 or 2) are formed by fusing Ta_2O_5 with alkali and dissolving the melt in water.

tapiolite, $FeTa_2O_6$. A tantalum mineral.

tar The non-aqueous liquid condensate resulting from the destructive distillation of carbonaceous materials. Types are:

(1) High-temperature tar from coke ovens.

(2) Low-temperature tar from the low-temperature carbonization of coal.

(3) High-temperature gasworks tar.

(4) Oil tar from water gas processes.

(5) Wood tar or Stockholm tar from wood carbonization.

tar acids These consist largely of phenols, and are obtained from tar by reaction with sodium carbonate to obtain the sodium phenolate and treatment of this compound with CO_2 to regain the phenol. *See* cresylic acids.

tar oils Heavy tar oils, mainly aromatic hydrocarbons, applied to trees in the dormant season to kill eggs of certain insects. They are very phytotoxic.

tar sands *See* Athabasca tar sands.

tartar emetic *See* antimony potassium tartrate.

tartaric acid, 2,3-dihydroxybutanedioic acid, $C_4H_6O_6$,$HO_2CCH(OH)CH(OH)CO_2H$.

Occurs in two optically active and two inactive forms. (+)-Tartaric acid, m.p. 170°C, very soluble in water. It occurs in the grape and a few other fruits. Manufactured from tartar argol and wine lees by precipitation as the calcium salt and decomposition of this with sulphuric acid. Potassium hydrogen tartrate is sparingly soluble in water. Many complex tartrates are known. Its chief use is in the manufacture of effervescent drinks but also used in textiles, dyeing, metal cleaning. (−)-Tartaric acid is obtained from racemic acid by fractional crystallization of the cinchonine salt.

Racemic acid, (±)-tartaric acid, is a compound of the two active forms. M.p. 273°C (with $1H_2O$), m.p. 205°C (anhydrous). Less soluble in water than (+)-tartaric acid. Formed, together with mesotartaric acid, by boiling (+)-tartaric acid with 30% NaOH solution, or by oxidation of fumaric acid. Potassium hydrogen racemate is very insoluble.

Mesotartaric acid crystallizes in plates ($1H_2O$), m.p. 140°C (anhydrous). Very soluble in water. Obtained from the mother-liquors in the preparation of racemic acid; or by oxidation of maleic acid. Potassium hydrogen mesotartrate is soluble in water.

tartrazine A pyrazolone dyestuff.

tartronic acid, hydroxymalonic acid, $C_3H_4O_5$, $HO_2CCH(OH)CO_2H$. Colourless crystals with $1H_2O$ lost at 60°C. M.p. 160°C (decomp.). Prepared by heating dinitrotartaric acid in aqueous alcohol.

taurine, aminoethylsulphonic acid, $C_2H_7NO_3S$, $NH_2{\cdot}CH_2{\cdot}CH_2{\cdot}SO_3H$. Crystallizes in columns, decomposing at 317°C. In combination with cholic acid it forms one of the bile acids. It is formed in the liver from cysteine.

taurocholic acid, cholyltaurine, $C_{26}H_{45}NO_7S$. The ester of taurine and cholic acid. It occurs in the bile as the sodium salt. *See* bile salts.

tautomerism, dynamic isomerism *See* isomerism.

tau value, τ A measure of shift in nuclear magnetic resonance*.

Tb Terbium.

Tc Technetium.

TCNE Tetracyanoethylene.

Te Tellurium.

technetium, Tc. At.no. 43, at.wt. 98·9062, m.p. 2172°C, b.p. 4877°C, *d* 11·5. ^{99}Tc (half-life 2×10^5 years) is obtained from fission products. All Tc isotopes are radioactive. The metal is obtained by hydrogen reduction of Tc compounds. It is silvery grey in colour and has an hcp structure. It dissolves in H_2O_2 and is attacked by hot Cl_2, S and O_2. Pertechnetates, TcO_4^-, are effective corrosion inhibitors. ^{99m}Tc has been used in clinical medicine.

technetium chemistry Technetium is an element of Group VII of the transition series, electronic configuration $4d^65s^1$. Its chemistry is very similar to that of rhenium* although the higher oxidation states are slightly more strongly oxidizing.

The halides known are yellow TcF_6 and TcF_5 (both Tc and F_2). Red–brown $TcCl_4$ ($Tc + Cl_2$) is the only known chloride. Oxides Tc_2O_7, TcO_3 and TcO_2 are known. The complexes are similar to those of Re; K_2TcH_9 is known.

Teflon *See* polytetrafluoroethene and fluorine-containing polymers.

teichoic acids Polymers of sugar alcohols linked by phosphate groups and found in the cell walls and cell contents of Gram-positive bacteria. They also bear glucosyl or *N*-acetylglucosaminyl residues and each sugar alcohol (usually ribitol or glycerol) is linked to D-alanine. Teichoic acids may be involved in the immunological specificity of bacteria.

TEL, Et_4Pb. Lead tetraethyl*.

tellurates Formally salts containing oxy-anions of tellurium. Tellurates(VI) contain octahedral $[TeO_6]^{6-}$ species and derivatives of these. Tellurates(IV), tellurites, contain $[TeO_3]^{2-}$ species.

telluric acid, $Te(OH)_6$. *See* tellurium oxy-acids.

tellurides Binary derivatives of Te with other elements. Similar structurally to sulphides.

tellurites Tellurates(IV)*.

tellurium, Te. At.no. 52, at.wt. 127·60, m.p. 449·5°C, b.p. 989·9°C, *d* 6·24. Sulphur-group element occurring naturally as tellurides in sulphide ores. Recovered from the flue dust after combustion of sulphide ores (particularly Au, Ag, Cu). It is taken into solution with H_2SO_4 and reduced to Te with zinc. The grey 'metallic' form has a chain structure with low electrical conductivity. The amorphous form is grey-black in colour. Te combines readily with O_2, halogens and metals, and dissolves in oxidizing acids. Used in alloys (particularly with Pb and also in Cu and stainless steel). Te compounds are poisonous. World demand 1976 100 tonnes.

tellurium bromides
Tellurium dibromide, $TeBr_2$, m.p. 210°C, b.p. 339°C. Unstable compound formed Te plus $TeBr_4$ in Et_2O.

Tellurium tetrabromide, $TeBr_4$, m.p. 380°C, b.p. 420°C. Red solid, Te plus excess Br_2. The vapour is dissociated to $TeBr_2$ and Br_2. Forms many complexes, including $[TeBr_6]^{2-}$ ions.

tellurium chemistry Tellurium is a sulphur-group element, electronic configuration $5s^25p^4$. It is a non-metal showing oxidation states -2, $+2$, $+4$ and $+6$ but is much more electropositive than sulphur giving Te_4^{2+} (and other cluster) compounds and almost cationic character in TeO_2. Te(IV) derivatives form complexes, e.g. $[TeCl_6]^{2-}$.

tellurium chlorides
Tellurium dichloride, $TeCl_2$. Unstable solid $TeCl_4$ plus Te.

Tellurium tetrachloride, $TeCl_4$. Colourless solid, m.p. 225°C, b.p. 390°, obtained Te plus excess Cl_2. Hydrolysed by water, forms complexes including $[TeCl_6]^{2-}$ ions.

tellurium fluorides
Tellurium tetrafluoride, TeF_4; m.p. 130°C. Colourless deliquescent solid prepared SeF_4 plus TeO_2. Hydrolysed by water.

Tellurium hexafluoride, TeF_6, m.p. $-35{\cdot}5$°C, sublimes -39°C. Stable gas prepared Te plus F_2. Reacts to give many derivatives, e.g. TeF_5OR, TeF_5OH (forms $[TeF_5O]^-$ derivatives).

Te_2F_{10}, and oxide fluorides, e.g. TeF_5OTeF_5, are also formed during the fluorination of tellurium oxides.

tellurium, organic derivatives Tellurium forms organic derivatives in the +2 and +4 states. The +2 compounds are similar to divalent sulphur derivatives although less stable. Tellurium(IV) derivatives are comparatively unstable.

tellurium oxides
Tellurium monoxide, TeO. Black solid resulting from thermal decomposition of $TeSO_3$ (Te plus SO_3).

Tellurium dioxide, TeO_2. Obtained by burning Te in air or heating tellurates(IV). Has rutile or brookite structures. Almost insoluble in water but gives tellurates(IV) with bases and, e.g., $TeCl_4$ with acids.

Tellurium trioxide, TeO_3. Yellow or grey solid formed by heating telluric acid, H_6TeO_6. Oxidizing agent being reduced to TeO_2. Gives tellurates(VI) with hot concentrated alkalis.

tellurium oxyacids
Tellurous acid, H_2TeO_3, does not exist as TeO_2 is insoluble in water. *Tellurates(IV)*, tellurites containing $(TeO_3)^{2-}$ species can be formed from TeO_2 and bases.

Telluric acid, $Te(OH)_6$, is formed from Te in aqua regia and a chlorate(V). Forms tellurates, e.g. $Na[TeO(OH)_5]$, $Na_2[TeO_2(OH)_4]$.

Fluorotellurates containing $(-OTeF_5)^-$ groups are also known – *see* tellurium fluorides.

tellurous acid *See* tellurium oxyacids.

telomerization Polymerization to a very low mol.wt. polymer, or condensation involving a few molecules of olefine. E.g. $[C_2F_4]_4$ is a telomer; $[C_2F_4]_n$ (*n* large) is a polymer.

tempering The process of reheating a quenched steel to relieve internal stresses, reduce hardness and tensile strength, and restore ductility and resistance to impact loading. Tempering is carried out within the temperature range 200–550°C for plain carbon steels. For alloy steels the temperatures may be somewhat higher. The higher the tempering temperature the greater the loss in hardness.

template reaction A reaction, often a condensation reaction or a polymerization reaction, carried out in the presence of a metal species which, by co-ordination induces a particular configuration and a particular product in the reaction.

teratogen A chemical which produces malfunction, generally in the form of mutations or tumours.

terbium, Tb. At.no. 65, at.wt. 158·93, m.p. 1360°C, b.p. 3041°C, *d* 8·28. A typical lanthanide*. The metal is hcp. The oxide has potential as a green phosphor for colour television. Sodium terbium borate has been used as a laser material.

terbium compounds Terbium shows +3 and +4 oxidation states. Tb(III) compounds are typical lanthanide compounds.*

$$Tb^{3+}\ (f^8\ \text{pale pink}) \longrightarrow Tb\ (-2{\cdot}39\ \text{volts in acid})$$

TbF_4 ($TbF_3 + F_2$ at 300°C) and Cs_3TbF_7 are known and Tb oxides intermediate between Tb_2O_3 and TbO_2 are formed by heating Tb compounds in air. These higher oxides liberate oxygen from water. Tb_2Cl_3 and TbCl ($TbCl_3$ plus Tb) have metal–metal bonding.

terebene A mixture of terpene hydrocarbons obtained by the action of various acids on pinene, which contains *p*-cymene, camphene and terpinolene. It has b.p. 160–190°C. It has a pleasant characteristic odour and is used medicinally as an inhalant, an expectorant, a deodorant and a mild antiseptic.

terephthalic acid, 1,4-benzenedicarboxylic acid, $C_8H_6O_4$. Crystallizes in colourless needles; m.p. 300°C (sublimes). Manufactured by the oxidation of *p*-xylene and used in the production of Terylene (*see also* polyesters). U.S. production 1980 2·05 megatonnes.

term symbols The spectroscopic notation used to denote an electronic state.

ternary compound A compound containing three elements, e.g. $KNbO_3$.

terpenes In the strict sense terpenes are volatile, aromatic hydrocarbons of the empirical formula $C_{10}H_{16}$. In a wider sense the term includes sesquiterpenes, $C_{15}H_{24}$, diterpenes, $C_{20}H_{32}$, and higher polymers. In a still looser sense the term includes various oxygen-containing compounds derived from the terpene hydrocarbons, such as alcohols, ketones and camphors. The terpenes are of great scientific and industrial importance, being characteristic products of many varieties of vegetable life and important constituents of most odorants, natural and synthetic, employed in perfumery. Many of them, e.g. the constituents of many eucalyptus oils, menthol and camphor, are of pharmaceutical importance. Practically without exception the terpene hydrocarbons may be considered as polymers of isoprene, C_5H_8, and may be either open-chain compounds or may contain one or more C_6 and other rings. They are chemically unsaturated, very reactive and in many cases form characteristic derivatives. Most of them have highly characteristic and usually pleasant odours. They are usually obtained from vegetable products by steam distillation, or better, by chromatography.

terphenyl, *p*-diphenylbenzene. Used in heat-transfer fluids.

1,8-terpin, terpine, $C_{10}H_{20}O_2$. Monocyclic

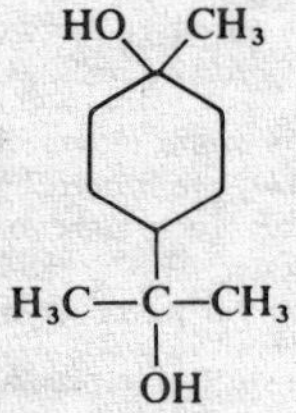

dialcohol obtained together with α-terpineol and other substances by the action of dilute alkali on *trans*-dipentene dihydrochloride, by the cyclization of linalool, geraniol and nerol, and by various other reactions. It has *cis*- and *trans*-isomers. *cis*-Terpin forms a hydrate, m.p. 116–117°C, with a loss of water, forming *cis*-terpin, m.p. 104–105°C. *trans*-Terpin, which forms no hydrates, has m.p. 158–159°C.

terpinenes, $C_{10}H_{16}$. The three isomers, α-, β- and γ-terpinene, are monocyclic terpenes, $C_{10}H_{16}$, which all yield the same dihydrochloride (1).

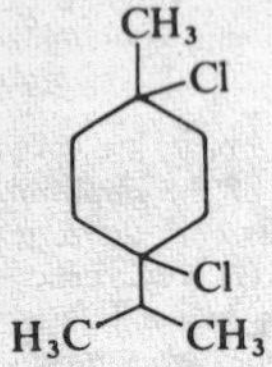

(1)

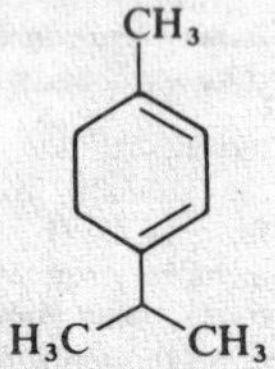

(2)

α-Terpinene, (2) *p*-menthadiene, b.p. 180–182°C. This is found in cardamom, marjoram and coriander oils. It is also obtained as the main constituent of the product obtained by treating linalool and geraniol with concentrated methanoic acid, also by treating α-pinene with strong sulphuric acid or α-terpineol with oxalic acid. No method of complete purification is known. It smells of lemons.

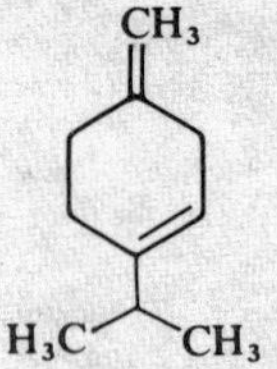

(3)

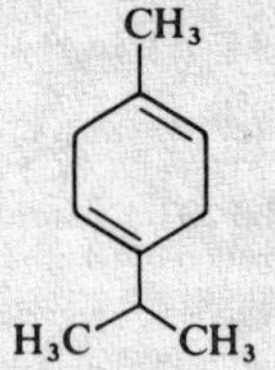

(4)

β-Terpinene (3) is always present in α-terpinene but has not been obtained pure from natural sources. It has been prepared synthetically from sabinene. B.p. 173–174°C.

γ-Terpinene (4). B.p. 182°C. A monocyclic terpene found in coriander, lemon, cumin, ajowan and samphire oils. It also occurs, together with α-terpinene, in the mixture of hydrocarbons obtained by acting on terpin hydrate with weak sulphuric acid or by treatment of pinene with an alcoholic solution of sulphuric acid.

terpinenols, $C_{10}H_{18}O$. Monocyclic alcohols obtained from terpinenes by the addition of one molecule of water.

terpineols, $C_{10}H_{18}O$. Monocyclic terpene alcohols obtained by the dehydration of terpin.

α-Terpineol (unsaturated ring). The (+)-form is found in petitgrain, neroli and other oils, the (−)-form in camphor oils, and the racemic form in cajuput oil. α-Terpineol is widely used in the perfumery industry, and is most conveniently prepared by the dehydration of terpin. It can also be prepared directly by the action of sulphuric acid in ethanoic acid or alcoholic solution on turpentine oil. By all these technical processes a mixture of isomers is obtained from which α- and β-terpineols can be separated by fractional distillation. (+)-α-Terpineol has m.p. 37°C, b.p. 104°C/15 mm. Other terpineols have unsaturation outside the ring.

terpinolene, $C_{10}H_{16}$. B.p. 185°C. A mono-

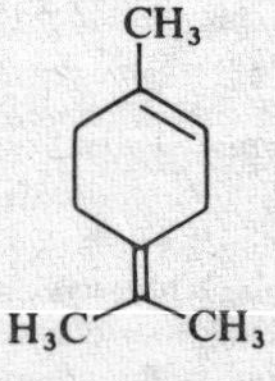

cyclic terpene which is found as a constituent of terebene which is in turn obtained by the action of acids on pinene. It is also formed by the action of dilute acids on terpin, α-terpineol and other similar compounds. It forms a crystalline tetrabromide of m.p. 116°C. It is doubtful whether it has been obtained in a pure state. Mineral acids convert it into its isomer terpinene.

testosterone, 17β-hydroxy-4-androsten-3-one, $C_{19}H_{28}O_2$. M.p. 154°C. Testosterone is the androgenic hormone formed in the testes: it controls the development and maintenance of the male sex organs and secondary sex characteristics. It is used, together with synthetic analogues, in treating disorders due to impaired secretion of the natural hormone. Testosterone can be prepared commercially from dehydroepiandrosterone.

tetraborane (10), B_4H_{10}. *See* boron hydrides.

tetracene, 1-(5-tetrazolyl)-4-guanyl tetrazene hydrate, $C_2H_8N_{10}O$. Pale yellow solid

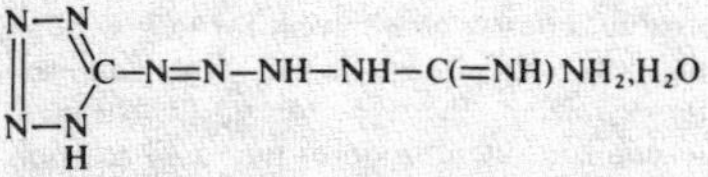

formed from nitrous acid and amino guanidine. Insoluble water and most solvents. Used as a detonator.

2,3,7,8-tetrachlorodibenzo-*p*-dioxin, dioxin, $C_{12}H_4Cl_4O_2$. A by-product in the preparation of 2,4,5-trichlorophenol from 1,2,4,5-tetrachlorobenzene, sodium hydroxide and ethylene glycol. Causes chloracne in humans.

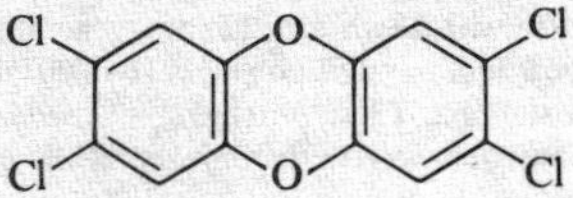

***sym*-tetrachloroethane, acetylene tetrachloride,** $HCCl_2CHCl_2$. Colourless toxic liquid with a chloroform-like odour, b.p. 146°C. Manufactured by passing chlorine and ethyne separately into a solution of $SbCl_5$ in tetrachloroethane. Reacts with dilute alkalis to give trichloroethene;.

tetrachloroethene, tetrachloroethylene, polychloroethylene, perchloroethylene, $CCl_2{:}CCl_2$. M.p. −22°C, b.p. 121°C. Manufactured by reacting pentachloroethane, (Cl_2 and $CHCl{:}CCl_2$), with $Ca(OH)_2$; by the direct chlorination of light hydrocarbons at 500–650°C; by the direct chlorination of ethene. It is noninflammable and is used very largely as a dry-cleaning solvent; another use is for metal degreasing.

tetrachloromethane *See* carbon tetrachloride.

tetrachlorophthalic anhydride, $C_8O_3Cl_4$.

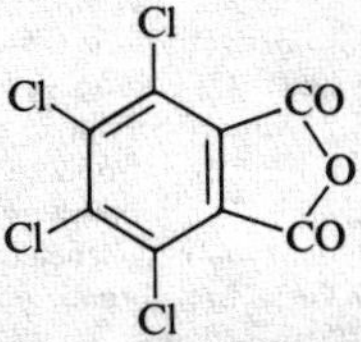

M.p. 255–257°C. Prepared by the direct chlorination of phthalic anhydride in 50–60% oleum in the presence of iodine, the temperature being raised from 50–200°C as the reaction proceeds.

It is used instead of phthalic anhydride in the preparation of certain eosin dyes (phloxines, etc.) which are bluer and brighter than those from unchlorinated intermediates.

tetracyanoethylene, TCNE, $(NC)_2C{:}C(CN)_2$. A sublimable white crystalline solid, smelling of hydrogen cyanide, m.p. 200°C, which has high thermal and oxidative stability. Reacts with most compounds containing an active hydrogen and with dienes in typical Diels–Alder additions. It is probably the strongest π-acid known; it forms a series of coloured complexes with aromatic hydrocarbons and gives salts of the radical anion with many metals, e.g. K^+TCNE^-.

7,7,8,8-tetracyanoquinonedimethane, TCNQ

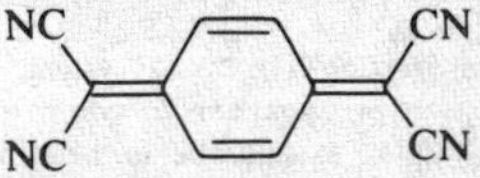

M.p. 296°C. Accepts an electron from suitable donors forming a radical anion. Used for colorimetric determination of free radical precursors, replacement of MnO_2 in aluminium solid electrolytic capacitors, construction of heat-sensitive resistors and ion-specific electrodes and for inducing radical polymerizations. The charge transfer complexes it forms with certain donors behave electrically like metals with anisotropic conductivity. Like tetracyanoethylene* it belongs to a class of compounds called π-acids.

tetracyclines An important group of antibiotics* isolated from *Streptomyces* spp., having structures based on a naphthacene skeleton. Tetracycline, the parent compound, has the structure:

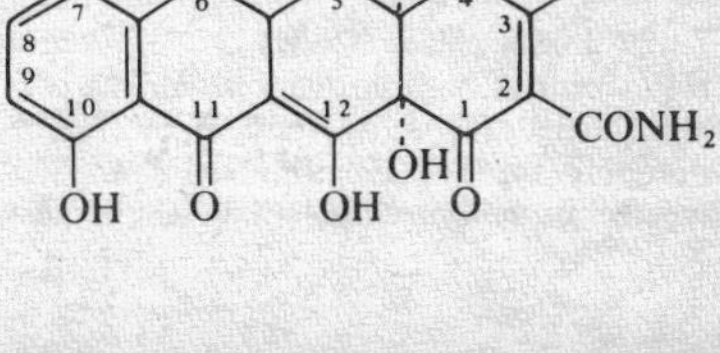

The 7-chloro-derivative, the first of the group to be isolated (1948) is known as chlortetracycline. The 5-hydroxy-derivative is oxytetracycline. More recently introduced tetracyclines are 6-demethyl-7-chlorotetracycline and 5-hydroxy-6-deoxy-6-methylenetetracycline.

Tetracyclines are broad-spectrum antibiotics, effective against Gram-positive and Gram-negative bacteria, also against Rickettsiae (typhus fever) and certain other organisms.

tetracyclone *See* tetraphenylcyclopentadienone.

tetraethyl lead, TEL *See* lead tetraethyl.

tetraethyl pyrophosphate, TEPP, $C_8H_{20}O_7P_2$, $(EtO)_2P(O)OP(O)(OEt)_2$. An insecticidal substance which is prepared by reaction of triethyl phosphate with P_2O_5 or $POCl_3$. It is a colourless mobile hygroscopic liquid, b.p. 135–138°C/1 mm. Used as a direct aphicide or acaricide. It is acutely toxic to mammals.

tetrafluoroberyllates Containing the $[BeF_4]^{2-}$ ion.

tetrafluoroborates Containing the $[BF_4]^-$ ion.

tetrafluorohydrazine, N_2F_4. *See* nitrogen fluorides.

tetragonal system The crystal system * with a 4-fold axis as principal axis of symmetry. The unit cell has its *z*-axis parallel to the 4-fold axis and the other two equal axes mutually perpendicular at right angles to this axis. E.g. rutile, TiO_2.

tetrahedral co-ordination Regular co-ordination by four ligands situated at the corners of a tetrahedron. Methane, CH_4, has tetrahedral co-ordination about the carbon. A tetrahedron has four triangular faces.

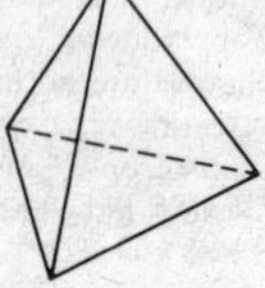

Tetrahedron

tetrahedron *See* tetrahedral co-ordination.

tetrahydrocannabinol, $C_{21}H_{29}O_2$. Active component of hashish (marihuana), the psychotomimetically active resin of the female flowering tops of *Cannabis sativa* L. The member of the active fraction present in most hemp plants is the Δ^1-3,4-*trans* isomer, but the isomeric Δ^6-3,4-*trans* isomer (with the double bond between positions 1 and 6) also occurs naturally and is similar to the Δ^1 isomer in pharmacological activity. The Δ^1-3,4-*trans* isomer is a liquid with b.p.$_{0\cdot05}$ 157°C, which is soluble in ethanol and chloroform.

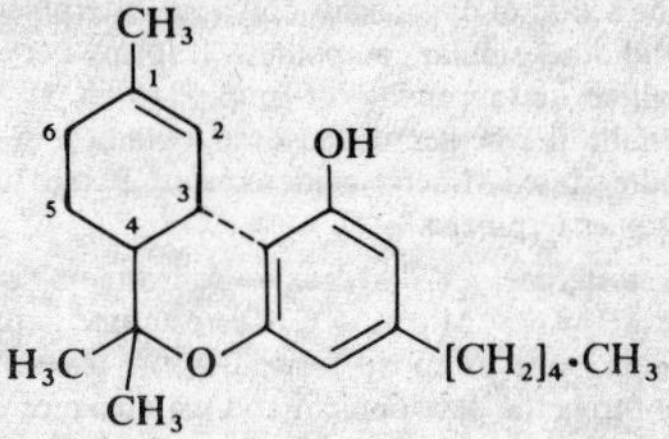

tetrahydrofuran, THF, C_4H_8O. Colourless liquid, b.p. 66°C, m.p. −108·5°C It is widely used in industry as a solvent for resins, many plastics (especially of the PVC type) and elastomers; as an ether type solvent for chemical reactions it often produces increased reaction rates and yields. Also used in tetrahydrofuran polymers. Prepared via furfural, furoic acid and furan by acid hydrolysis of the polysaccharides in oat husks.

tetrahydrofuran polymers, poly(oxytetramethylene) glycols, poly(tetramethylene oxide), polytetrahydrofuran Tetrahydrofuran undergoes ring opening and polymerization with carbenium, diazonium and trialkyloxonium ions, and also with Lewis acid halides, generally to give low mol.wt. diol polymers (1–3000) used as plasticizers. These polymers are used with polyurethanes to give thermoplastics and elastomers.

tetrahydrothiophen, THT, tetramethylene sulphide Colourless mobile liquid, with a penetrating odour; b.p. 118–119°C. Can be obtained by the catalytic reduction of thiophen or tetrahydrothiophen-1,1-dioxide (sulpholane). Used to give an odour to domestic gas supplies.

tetralin, 1,2,3,4-tetrahydronaphthalene, $C_{10}H_{12}$. Colourless liquid; b.p. 207°C. Obtained by the catalytic hydrogenation of naphthalene. Owing to the presence of one aromatic ring it can be nitrated and sulphonated. It is non-toxic and is used as a solvent for fats, oils and resins.

1,2,3,4-tetramethylbenzene Prehnitene.

1,2,3,5-tetramethylbenzene Isodurene.

1,2,4,5-tetramethylbenzene *See* durene.

3,3′,5,5′-tetramethylbenzidine, $C_{16}H_{20}N_2$.

M.p. 168–169°C. This reagent has superseded benzidine* for blood detection, showing same sensitivity and specificity but it is non-carcinogenic in doses at which benzidine produces a high incidence of neoplasms.

N,N,N′,N′-tetramethyl-1,2-diaminoethane, N,N,N′,N′-tetramethylethylene diamine, TMED, $(CH_3)_2NCH_2CH_2N(CH_3)_2$. B.p. 122°C; a hygroscopic base which forms a hydrocarbon-soluble stable chelate with lithium ions and promotes enhanced reactivity of compounds of lithium, e.g. $LiAlH_4$, LiC_4H_9, due to enhanced kinetic basicity of the chelate. Used in polymerization catalysts.

tetramethyl lead, TML *See* lead tetramethyl.

tetramethylsilane, TMS, $(CH_3)_4Si$. Used as an internal reference standard for proton magnetic resonance spectroscopy. On the δ-scale TMS protons absorb at 0·00 δ and most protons bonded to carbon then appear downfield (positive δ-value) from TMS.

The older τ scale places TMS at 10τ with most protons bonded to carbon having smaller τ values.

tetramethylthiuram disulphide *See* thiuram disulphides.

tetranitromethane, $C(NO_2)_4$. Colourless liquid; m.p. 129°C, b.p. 126°C. It is made by nitration of ethanoic anhydride with anhydrous nitric acid. Decomposed by alcoholic solutions of KOH to nitroform. Reacts with alcohols in the presence of alkali to give the nitrate of the alcohol. Gives a yellow colour with derivatives of cyclopropane and unsaturated compounds. Used as a test for unsaturation.

tetraphenylcyclopentadienone, tetracyclone An intense purple crystalline solid; m.p. 219–220°C. One of the few monomeric cyclopentadienone derivatives, most of which spontaneously undergo self Diels–Alder type dimerization. Used as a diene in many studies of various aspects of the Diels–Alder reaction*.

tetrathionic acid, $HO(O)_2SS_2S(O)_2OH$. *See* polythionic acids.

tetrazolium salts Quaternary salts derived from tetrazoles. They are colourless or yellow substances, usually soluble in alcohol and water which in the presence of a trace of alkali oxidize aldoses, ketoses and other α-ketols and are themselves reduced to insoluble formazan pigments, e.g. 'blue tetrazolium' gives an intense blue pigment. This sensitive test distinguishes between reducing sugars and simple aldehydes. Also used for detection of dehydrogenase in tissues, cells and bacteria and for determination of cortical steroids. Tetrazolium salts are used in colour photography.

tetritol *See* erythritol.

tetrolic acid The trivial name for 1-carboxyprop-1-yne $CH_3-C\equiv C-CO_2H$.

tetrose A carbohydrate with four carbon atoms. There are four aldo-tetroses, the two stereoisomers of erythrose and threose, and two possible keto-tetroses.

tetryl, N-2,4,6-trinitrophenyl-N-methylnitramine, $C_7H_5N_5O_8$. A pale yellow powder, m.p. about 129°C, which decomposes rapidly above this temperature. It is obtained by nitrating a solution of dimethylaniline in concentrated sulphuric acid. Tetryl is a very stable explosive, more powerful than TNT and picric acid, but more sensitive to shock, and is much used as a primer or booster. It is also known as C.E. (composition exploding).

textiles Apart from the natural fibres (wool, cotton) or synthetics (acrylics, polyesters, Nylons, etc.), many chemicals are used in textiles. Dyebath additives are principally dyestuffs but include solvents (C_2Cl_4, aromatics), dispersing agents (e.g. sulphuric acid, quaternary ammonium salts), anti foams (e.g. silicones), acids (e.g. ethanoic), wetting agents (ethoxylated alcohols). Finishing agents include non-fibrous polymers, stain repellants (e.g. fluorochemicals), durable press agents (glyoxal), softeners and lubricants (silicones), flame retardants. Printing chemicals include thickeners (e.g. starch, alginates), binders (acrylates), acids, wetting agents, bacteriocides, solvents (hydrocarbons, urea), softeners and dispensing agents.

Th Thallium.

thallates Mixed oxides of thallium(III).

thallation *See* thallium(III) trifluoroacetate.

thallic Thallium(III) compounds.

thallium, Tl. At.no. 81, at.wt. 204·383, m.p. 303·5°C, b.p. 1457°C, *d* 11·85. Element of Group III, occurs in some sulphide and selenide ores – e.g. crookesite (Ag,Cu,Tl)–Se, recovered from the flue dusts after oxidation of sulphide ores and reduced to the metal electrolytically. The soft metal is similar to Pb. Fairly reactive metal – combines halogens, S, only slowly soluble in dil. acids as Tl(I) salts insoluble, dissolves rapidly in HNO_3. The metal itself is not used; Tl compounds are used in poisons, glasses and some electronic applications. World production 1976 3 tonnes. Thallium(I) compounds are very poisonous.

thallium bromides
Thallium(I) bromide, TlBr. Insoluble in water.
Thallium(III) bromide, $TlBr_3$. Formed Tl plus Br_2. Unstable, decomposes to $Tl^I [TlBr_4]^-$.

thallium(I) carbonate, Tl_2CO_3. Formed by CO_2 on solution of TlOH.

thallium chemistry Thallium is a Group (III) element, electronic configuration $6s^26p^1$. The group oxidation state is strongly oxidizing (E° Tl(III) to Tl +1·25 volts). Thallium(I) is the stable state (E° Tl(I) to Tl −0·33 volts in acid solution) and is very similar in its chemistry to Ag(I). Compounds which have stoicheiometries corresponding to other than Tl(III) or Tl(I) contain mixed oxidation states, e.g. $TlBr_2$ is $Tl^ITl^{III}Br_4$. Thallium(I) forms few complexes; Tl(III) is readily complexed.

thallium chloride
Thallium(I) chloride, TlCl, m.p. 430°C, b.p. 806°C. Precipitated from aqueous solution of a Tl(I) salt by HCl. White solid.
Thallium(III) chloride, $TlCl_3,4H_2O$. Formed by passing Cl_2 through a suspension of TlCl in water. Hygroscopic, loses Cl_2 at 100°C. The $[TlCl_2]^+$ ion is stable; chloro complexes up to $[TlCl_6]^{3-}$ are formed.

thallium chromate, Tl_2CrO_4. Yellow powder precipitated from aqueous solution of Tl(I) salt by Na_2CrO_4. Used to estimate Tl(I).

thallium fluorides
Thallium(I) fluoride, TlF. Colourless solid soluble in water (TlOH in HF). Forms $TlHF_2$.
Thallium(III) fluoride, TlF_3 white solid (Tl_2O_3 plus F_2), immediately hydrolysed by water.

thallium hydroxide, $TlOH,H_2O$. Formed Tl_2SO_4 plus $Ba(OH)_2$. Crystallizes from aqueous solution as yellow needles. Soluble in water to give strongly basic solution (used to prepare other Tl(I) salts). The solution absorbs CO_2 to give Tl_2CO_3.

thallium iodides
Thallium(I) iodide, TlI yellow or red solid precipitated from aqueous solution.
Thallium(I) triiodide, $Tl[I_3]$. Formed from Tl(III) solution plus KI, readily loses I_2.

thallium nitrates
Thallium(I), $TlNO_3$. Formed by dissolving Tl, Tl_2CO_3 or TlOH in HNO_3; soluble in water. Decomposes at 300°C.
Thallium(III) nitrate, $Tl(NO_3)_3,3H_2O$. Formed Tl plus conc. HNO_3. Immediately hydrolysed by water but is soluble in dilute mineral acids, alcohols and diglyme. Used in oxythallation*.

thallium organic derivatives Formed TlX_3 plus Grignard reagents; $ArTlX_2$ derivatives can be formed by direct thallation (*see* thallium trifluoroacetate). Simple derivatives, e.g. Me_3Tl, are spontaneously inflammable in air. The $[R_2Tl]$ groups (linear) are stable. TlC_5H_5 (polymeric) is formed from TlOH and C_5H_6; it is used in the synthesis of metal cyclopentadienides. TlOR (e.g. TlOMe) are polymeric.

thallium oxides
Thallium(I) oxide, Tl_2O. Black powder formed by heating TlOH, gives Tl(I) salts with acids.
Thallium(III) oxide, Tl_2O_3. Dark solid obtained by Tl plus O_2 or Tl(III) salt plus base. Insoluble in water and alkalis but dissolves in acid (often gives Tl(I) salts).

thallium sulphate
Thallium(I) sulphate, Tl_2SO_4. Formed Tl plus hot conc. H_2SO_4 or TlOH plus H_2SO_4. Moderately soluble in water; forms alums and double sulphates.
Thallium(III) sulphate, $Tl_2(SO_4)_3,7H_2O$. Probably formed Tl_2O_3 in conc. H_2SO_4. Very readily hydrolysed by water.

thallium trifluoroacetate, $Tl(O_2CCF_3)_3$. Solid formed Tl_2O_3 plus CF_3CO_2H. Used in thallation of aromatics; Ar−H plus $Tl(O_2CCF_3)_3$ gives $ArTl(O_2CCF_3)_2$. The reaction proceeds specifically depending upon the substituents present and the thallium can then be replaced specifically, e.g. KI gives ArI.

thenyl *See* thienyl ring.

theobromine, 3,7-dimethylxanthine, $C_7H_8N_4O_2$. M.p. 337°C, an alkaloid obtained from cacao seeds or prepared synthetically. Constitutionally it is similar to caffeine, and is also a weak base. It is usually administered as the sodium compound combined with either sodium ethanoate or sodium salicylate, and is employed almost entirely as a diuretic. Physiologically theobromine resembles caffeine, but its effect on the central nervous system is less, while its action on the kidneys, is more pronounced.

theophylline, 1,3-dimethylxanthine, $C_7H_8N_4O_2$. Occurs to a small extent in tea, but is chiefly prepared synthetically. Like caffeine*, it is a very weak base which forms water-soluble compounds with alkalis. It has a similar pharmacological mechanism to that of caffeine and is used, in combination with ethylenediamine, as a diuretic and a bronchodilator.

theoretical plate A theoretical plate in a distillation or absorption column* is one on which perfect liquid-vapour contacting occurs, so that the two streams leaving it are in equilibrium with each other.

In the case of a plate column* the performance of a real plate is related to the performance of a theoretical one by the *plate efficiency**. In the case of a packed column* the *height equivalent to a theoretical plate* (*HETP*) gives a measure of the contacting efficiency of the packing.

thermal analysis Analytical techniques used for identification and also for investigation of the products of the action of heat on a compound or complex. In differential thermal analysis (DTA) the sample is heated, frequently in an inert atmosphere, and weight is measured as a function of temperature. In differential scanning calorimetry (DSC) electrical heat is added or removed from the sample as the temperature is raised; enthalpy changes due to thermal decomposition are thus followed.

thermal cracking Formerly, the thermal cracking of gas oil and similar petroleum fractions was used to increase the gasoline yield from a crude oil. This involved subjecting the petroleum fraction to temperatures of about 500°C at pressures up to 25 bar to yield gases, hydrocarbons in the gasoline and gas oil range and heavier fractions. While catalytic cracking has replaced thermal cracking for the production of gasoline the latter process is still used for certain purposes such as coke production, or the production of feedstocks for special purposes, e.g. the cracking of wax to produce long-chain alkenes. *See* catalytic cracking.

thermal reforming *See* reforming.

thermite reaction A reduction process involving Al powder and a metal oxide (generally Fe_2O_3) used to produce heat for welding or incendiary devices. Generally ignited with Mg ribbon; the reaction gives, e.g. Al_2O_3 and Fe.

thermobalance A balance in which the sample can be heated to allow thermal analysis*.

thermochemistry Most chemical reactions occur with the evolution or absorption of heat. Thermochemistry is that branch of chemistry concerned with the heat changes accompanying chemical reactions.

thermochroism Change in colour (and spectrum) with temperature.

thermodynamic data Standard enthalpies of formation (ΔH_f°), standard free energies of formation (ΔG_f°) and absolute entropies of selected substances at 298 K.

By definition all pure elements have $\Delta H_f^\circ = 0$ and $\Delta G_f^\circ = 0$.

Substance	ΔH_f° (*kJ mol*$^{-1}$)	ΔG_f° (*kJ mol*$^{-1}$)	S° (*J*$^-$ *mol*$^{-1}$)
$CaCO_3$ (s)	−1207	−1129	93
CH_4 (g)	−75	−51	186
C_2H_6 (g)	−85	−33	229
C_3H_8 (g)	−104	−24	270
C_6H_6 (g)	83	130	269
C_6H_6 (l)	49	125	173
C_2H_4 (g)	52	68	219
C_2H_2 (g)	227	209	201
CO (g)	−111	−137	198
CO_2 (g)	−394	−395	214
Cl_2 (g)	0	0	223
H_2 (g)	0	0	131
HCl (g)	−92	−95	187
H_2O (g)	−242	−229	189
H_2O (l)	−286	−237	70
Mg (s)	0	0	33
MgO (s)	−602	−570	27
N_2(g)	0	0	192
O_2 (g)	0	0	205
NH_3 (g)	−46	−17	35

thermodynamics, first law of This law is a consequence of the law of conservation of energy and states that mechanical energy and work are quantitatively interconvertible. Alternatively, 'In any process energy can be changed from one form to another, but it is neither created nor destroyed.'

thermodynamics, second law of This states that 'Heat cannot of itself pass from a colder to a warmer body.' Thus, if heat is to be transferred from the cold to the hot body, work must be provided by an external agency. An alternative and useful expression of the second law is; 'Any system of its own accord will always undergo change in such a way as to increase the entropy.'

thermodynamics, third law of This states that 'for a perfect crystal at absolute zero on the Kelvin scale the entropy is zero'. This follows the predictions of Einstein (1907) that the specific heats of all substances would approach zero at 0 K, and the conclusions of Planck (1912) that the entropy of all pure solids and liquids approach zero at this temperature.

thermogram A plot of weight against temperature obtained in thermal analysis*.

thermonuclear reaction *See* atomic energy.

thermoplastic resins Resins (plastics) which can be repeatedly softened by heating and harden again on cooling. The most important types are the various forms of polyethene, polyvinyl chloride, polypropene and polystyrene. U.S. production 1978 12 megatonnes.

thermosetting resins Resins (plastics) which once formed in their final moulded state are infusible and insoluble. They are said to be thermocured. The most important types are phenolic resins, polyesters, urea-resins, epoxide resins and melamine resins. U.S. production 1978, 2 megatonnes.

thexyl Trivial name for the 1,2,3-trimethylpropyl group.

THF Tetrahydrofuran.

thiamine, aneurine, vitamin B_1 The anti-neur-

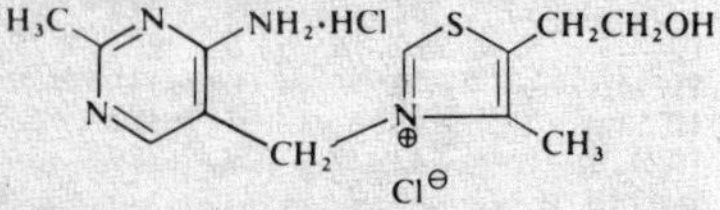

itic factor, the absence of which from the diet of man leads to the disease beri-beri, and from that of mammals and birds to polyneuritis, the most fundamental symptom of which is a general nervous atrophy. It is isolated as the chloride hydrochloride, $C_{12}H_{18}Cl_2N_4OS,H_2O$. It is destroyed by heating above 100°C in water solution. The richest natural sources of the vitamin are yeast, eggs and the germ of cereals. It is not present in polished rice and other highly purified cereal products. The minimum daily dose is believed to be about 2 mg.

thiamine pyrophosphate The biochemically active derivative of thiamine, the pyrophosphate ester of thiamine. It is a coenzyme which is concerned in a number of important metabolic processes. These include the decarboxylation of α-oxoglutaric acid in the citric acid cycle*, and the conversion of alanine, via pyruvic acid, to acetyl coenzyme A. The actions of thiamine pyrophosphate are all similar in that they involve the intermediacy of 'active aldehyde' attached (as carbinol substituents) to the thiazole ring at position 2.

thiazole dyes Dyestuffs containing the thiazole ring. Often show good dyeing properties to cellulosic substrates.

thiazole ring The ring numbered as shown.

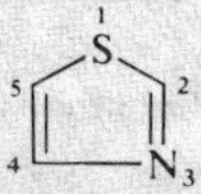

thiazyl ions Species, e.g. $[S_5N_5]^+$; cationic derivatives of sulphur nitrogen compounds.

thickening Treatment of a slurry or suspension to recover the solids in the form of a more concentrated slurry. A thickener is equipment for such an operation.

thienyl ring The ring system numbered as shown.

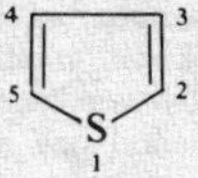

thiirane The 3-membered sulphur hetero-

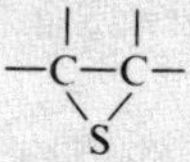

cycle. The sulphur analogue of an epoxide (oxirane).

thin layer chromatography *See* chromatography.

thio- Sulphur-containing, e.g. thioantimonates (antimony sulphides), thioarsenates (arsenic sulphides).

thiocarbanilide, *sym*-diphenylthiourea, $C_{13}H_{12}N_2S$, PhNHC(S)NHPh. Colourless flakes; m.p. 151°C. Prepared by boiling aniline with carbon disulphide. It is used commercially as a rubber accelerator.

thiocarbonates Salts of thiocarbonic acid*.

thiocarbonic acid, H_2CS_3. Formed, mixed with H_2CS_4, by acidifying thiocarbonates. Na_2CS_3 formed from solution of Na_2S plus CS_2. $(NH_4)_2CS_3$ formed CS_2 plus concentrated aqueous NH_3. $BaCS_3$ (relatively stable) formed $Ba(HS)_2$ plus CS_2. Mixed oxide thiocarbonates containing, e.g. $[COS_2]^{2-}$ ions are also known. Thiocarbonates are slowly hydrolysed and react with CO_2 to CS_2 and sulphides. Used in destroying the fungus phylloxera which occurs on vines.

thiocarbonyl complexes Complexes containing CS ligands analogous to carbonyls. As CS does not exist formed from CS_2, e.g. $(C_5H_5)Mn(CO)_2(CS)$.

thiochrome, $C_{12}H_{14}N_4OS$. Yellow prisms,

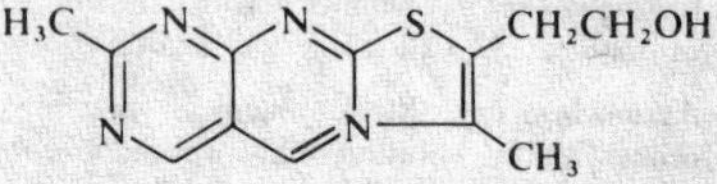

m.p. 227–228°C, soluble in methanol, fairly soluble in water. It gives an intense blue fluorescence in neutral or alkaline solution. The colouring matter of yeast. It can be extracted from yeast, prepared by oxidizing thiamine or synthesized.

thiocyanates Salts of thiocyanic acid, HSCN. Complexes and esters can be N or S bonded. Formed cyanides plus S. Used in estimation of

Ag^+ (precipitate of AgSCN) and Fe^{3+} (red colouration). Photographic sulphur sensitizers. With oxidizing agents (MnO_2, Br_2) give thiocyanogen, $(SCN)_2$, which polymerizes above 0°C.

thiocyanic acid, HSCN. Prepared $KHSO_4$ plus KSCN. Gas at room temperature but rapidly polymerized. Forms thiocyanic esters RSCN and isothiocyanic esters or mustard oils, RNCS.

thiocyanogen, $(SCN)_2$. *See* thiocyanates.

thioglycollic acid, mercaptoethanoic acid, $HS{\cdot}CH_2{\cdot}COOH$, $C_2H_4O_2S$. When pure it is a colourless liquid with a faint odour; the odour of the impure material is penetrating and unpleasant; b.p. 123°C/29 mm. Obtained by electrolytic reduction of the dithioglycollic acid formed by the action of sodium hydrosulphide on chloroethanoic acid. It oxidizes readily in the air and is very reactive. It adds across double bonds of unsaturated compounds and has been used as a test for unsaturation. The hydrogen of the –SH group is replaceable by metals. It is an important constituent of the cold process for waving hair. It is also used as a test for iron, giving a purple colour with ammoniacal solutions of iron salts. The calcium salt is a depilatory.

thiokols Polysulphide polymers*.

thiols The systematic name for mercaptans.

thiomersalate, thimerosal, $C_9H_9HgNaO_2S$, *o*-$EtHgSC_6H_4CO_2Na$. Sodium salt of ethyl mercurithiosalicylic acid. Made from ethylmercury chloride and thiosalicylic acid. It is a cream-coloured powder. It is used for treating fungal infections of the skin, and for other sterilizing purposes, including the preservation of pharmaceutical products against microbial attack.

thiomolybdates Species such as $[MoS_4]^{2-}$, $[MoO_2S_2]^{2-}$. Can act as bidentate ligands. Thiotungstates are similar.

thionic acids, $H_2S_nO_6$. *See* polythionic acids.

thionyl Containing $>$SO groups.

thionyl chloride, $SOCl_2$. M.p. –105°C, b.p. 79°C. Formed SO_2 plus PCl_5. Hydrolysed by water to HCl, SO_2 and H_2O. Used to convert COH groups to CCl and for the dehydration of metal chloride hydrates.

thionyl halides, SOF_2 ($SOCl_2$ plus SbF_3 or anhydrous HF); SOClF (as for SOF_2); $SOBr_2$ ($SOCl_2$ plus HBr); $SOCl_2$ (*see* thionyl chloride).

thiopentone sodium, $C_{11}H_{17}N_2O_2SNa$. Monosodium salt of 5-ethyl-5-(1-methylbutyl)-2-thiobarbituric acid. It is a yellowish white powder, unstable in air. It is administered by intravenous injection to produce short duration general anaesthesia.

thiophene, C_4H_4S. A colourless liquid with a

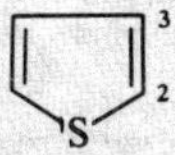

faint odour resembling that of benzene, m.p. –38°C, b.p. 84°C. It occurs to the extent of about 0·5% in commercial benzene and can be prepared by heating sodium succinate with P_2S_5. It is manufactured from butane and sulphur. It can be nitrated, sulphonated and brominated, and gives rise to two series of monosubstituted derivatives (2 and 3). In derivatives the ring system shown is called the thienyl ring.

thiophenol, phenylthiol, phenylmercaptan, mercaptobenzene, C_6H_5SH, PhSH. The sulphur analogue of phenol. A colourless, foul-smelling liquid; b.p. 168°C. Manufactured by the reduction of benzene sulphonyl chloride and forms a series of complexes (mercaptides) with metal salts.

thiosinamine *See* propenylthiourea.

thiosulphates Salts of thiosulphuric acid*.

thiosulphuric acid, $H_2S_2O_3$. The free acid is only known (SO_3 plus H_2S) at –78°C as an etherate. Thiosulphates, containing the $[S_2O_3]^{2-}$ ion are formed from sulphites (sulphates(IV)) plus S. Form complexes – particularly with Ag^+. Thiosulphates are used in photography. React quantitatively with I_2; reaction used in volumetric analysis. Acids on $[S_2O_3]^{2-}$ give SO_2 and S.

thiotungstates *See* thiomolybdates.

thiouracil, $C_4H_4N_2OS$. White powder, prepared by condensing sodium ethylformyl ethanoate with thiourea. Thiouracil decreases the activity of the thyroid gland by interfering with the iodination of tyrosine and so preventing thyroxine production.

thiourea, CH_4N_2S, $S{:}C(NH_2)_2$. Colourless crystals, m.p. 172°C. Decomposed on heating with water to give ammonium thiocyanate, but in many of its properties it resembles urea. It forms complexes with many metal salts. It is manufactured by the action of hydrogen sulphide on cyanamide. It is used as a photographic sensitizer.

thiram *See* thiuram disulphides.

thiuram disulphides, $R_2NC(S)SSC(S)NR_2$. Obtained by oxidation of dithiocarbamates (by, e.g. H_2O_2, Cl_2) and used as polymerization initiators and vulcanization accelerators. The tetramethyl derivative is Antabuse, the agent for treatment of alcoholism, and in agricultural application to prevent the 'damping off' of seedlings. TMTD, TMTDS is tetramethylthiuram disulphide, m.p. 155°C.

thixotropy The isothermal gel-sol transformation brought about by shaking or other mechanical means. It may be regarded as a packing phenomenon. E.g. $Fe(OH)_3$ gels on shaking form sols which rapidly set again when allowed to stand. Thixotropy is commonly found with suspensions of clay or oil paints and emulsions.

thorium, Th. At.no. 90, at.wt. 232·04, m.p. 1750°C, b.p. 4790°C, *d* 11·7. Thorium is a plentiful element 12 p.p.m. (comparable to Pb). The most important ores being thorite and thorogummite (silicates), thorianite (ThO_2) and monazite (a mixed phosphate, the principle lanthanide ore). The ores are decomposed with alkali or acid and Th is extracted from acid solution using solvents such as tributyl phosphate. The metal has been prepared by Ca reduction of ThF_4; it is tarnished in air and slowly attacked by hot water and dilute acids; low temperature ccp, higher temperature bcc. ThO_2 was used in the production of incandescent gas mantles; the fissile isotope ^{233}U is obtained from ^{232}Th and thermal neutrons; ThO_2 is an important catalyst (Fischer–Tropsch) and is used in strengthening nickel. The metal is used as an oxygen remover (getter) in the electronics industry. World demand 1976 100 tonnnes.

thorium compounds Thorium chemistry is almost entirely that of the +4 state $Th^{4+} \xrightarrow{-1·9V} Th$. Compounds which are formally in lower oxidation states ThI_3 and ThI_2 (prepared from ThI_4 and Th in a sealed container) contain Th^{4+} ions, I^- ions and electrons, Th(IV) compounds are partially hydrolysed in water. Salts crystallized from water are heavily hydrated; O- and N-donors also form complexes. High co-ordination numbers (e.g. 11 in $[Th(NO_3)_4,5H_2O]$ and 10 in $[Th(NO_3)_4(OPPh_3)_2]$) are common. Thorium diketonates (e.g. acetylacetonates) are particularly stable. Some organometallic derivatives of thorium are known e.g. $Th(C_5H_5)_4$ is formed from $ThCl_4$ and KC_5H_5.

thorium halides All ThX_4 are known

$$ThO_2 + CCl_2F_2 \text{ or } HF \xrightarrow{400°C} ThF_4,$$

$$ThCl_4 \text{ aq.} \xrightarrow{HF} ThF_4,$$

$$Th \xrightarrow{Br_2 700°C} ThBr_4,$$

$$Th \xrightarrow{I_2 400°C} ThI_4$$

ThF_4 is insoluble in water (but $ThF_4,4H_2O$ can be formed); the other halides are soluble in water. Halide oxides, $ThOX_2$, are formed from ThO_2 and ThX_4. Complex halides, e.g. K_5ThF_9, Na_4ThF_8, Rb_2ThCl_6 are formed from acid solution and have high co-ordination numbers about the metal. For lower halides *see* thorium compounds.

thorium oxide, ThO_2. Thorium salts are hydrolysed in aqueous solution and the hydrated oxide is precipitated by alkalis. Ignition of this hydrated oxide, not other oxyacid salts, gives white ThO_2.

thorium oxyacid salts Commercially available thorium salts include the nitrate ($4H_2O$ and H_2O), the oxalate ($6H_2O$) and sulphate (0, 4, 6, 8 and $9H_2O$). The anion is generally co-ordinated to the metal; the salts are eventually hydrolysed in aqueous solution.

thortveitite, $Sc_2Si_2O_7$. An important source of Sc. Contains $Si_2O_7^{6-}$ ions.

three-centre bond A bonding orbital formed by two electrons between three atoms. *See* multi-centre bonds.

threitols *See* erythritol.

threo- *See* erythro-.

threonine, 2-amino-3-hydroxybutanoic acid, $C_4H_9NO_3$, $CH_3CHOHCHNH_2COOH$. An amino-acid; m.p. 251–252°C. Occurs widely in proteins, and has been shown to be an essential component of foods.

threose, $C_4H_8O_4$. A tetrose sugar. The D-

CHO
|
HOCH
|
HCOH
|
CH_2OH

form has been obtained crystalline, m.p. 126–132°C. It is very hygroscopic and soluble in water and alcohol. L-Threose has been obtained only in solution.

threshold limit values (TLV) A list of values for chemical substances in the form of time-weighted average concentrations, representing conditions which can be tolerated in working environments. For some materials the designation 'C' is added to indicate that the value shown is a ceiling value which should not be exceeded. E.g. benzene – 25 C (i.e. 25 p.p.m.); xylene – 100; pentane – 500.

thrombin The enzyme which catalyses the conversion of fibrinogen to fibrin in clotting blood. Thrombin itself is first formed from the precursor protein prothrombin in the presence of ionized calcium and an activation factor thromboplastin.

thromboxanes Thromboxanes are cellular regulatory compounds similar to prostaglandins*, both of them being derived biosynthetically from prostaglandin endoperoxides.

thujane, $C_{10}H_{18}$. A dicyclic hydrocarbon obtained by the catalytic hydrogenation of α-and β-thujene and sabinene, and by other reactions. (+)-Thujane only has been described, and is a colourless mobile oil with a faint smell, b.p. 157°C/758 mm.

α-thujene, $C_{10}H_6$. A mobile oil with a rather

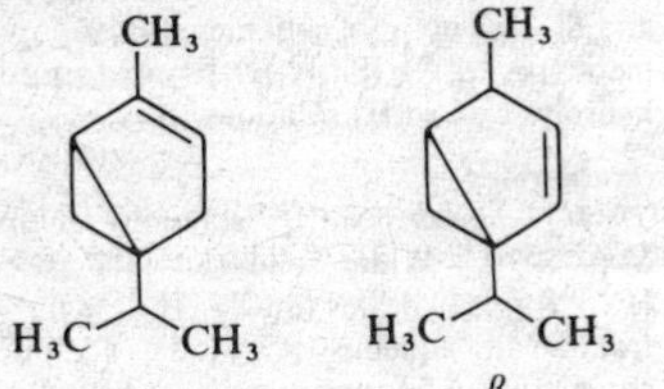

α β

penetrating smell. A dicyclic monoterpene, the (+)-form of which is present in the oil from the gum-oleo-resin of *Boswellia serrata.*

β-Thujene, $C_{10}H_{16}$. B.p. 147°C/739 mm. A dicyclic monoterpene.

thujone, $C_{10}H_{16}O$. Colourless oil the smell

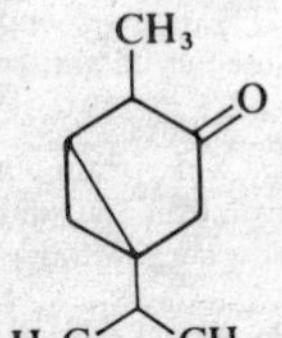

of which resembles that of menthol. α-Thujone has b.p. 74·5°C/9 mm, β-thujone has b.p. 76°C/10 mm. This dicyclic ketone occurs in thuja, tansy, wormwood and many other oils.

thujyl alcohol, $C_{10}H_{18}O$. A secondary dicyclic alcohol, also known as tanacetyl alcohol, found both free and combined with various acids in wormwood oil. It can also be prepared by reducing thujone with sodium in alcohol.

thulium, Tm. At.no. 69, at.wt. 168·93, m.p. 1545°C, b.p. 1727°C, *d* 9·332. A typical lanthanide*. The metal is hcp. The metal or its oxides are widely used as a portable source of X-radiation after neutron irradiation.

thulium compounds Thulium forms a series of +3 compounds which are typical lanthanide compounds*.

Tm^{3+} (f^{12} pale green) ⟶ Tm (−2·28 volts in acid)

$TmBr_2$ ($TmBr_3$ + Tm) is known.

thymidine The nucleoside of thymine and deoxyribose, i.e. the β-N-deoxyribofuranoside of thymine*. *See* nucleosides.

thymine, 5-methyluracil, 5-methyl-2,6-dioxytetrahydropyrimidine, $C_5H_6N_2O_2$. Crystal-

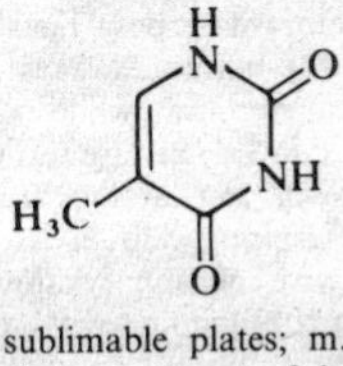

lizes in sublimable plates; m.p. 321–325°C. Thymine is a constituent of deoxyribose nucleic acid.

thymol, 2-hydroxy-1-isopropyl-4-methylbenzene, $C_{10}H_{14}O$. Colourless plates; m.p. 51·5°C, b.p. 233·5°C. It has the pungent odour of thyme. It is a constituent of oil of thyme and numerous other essential oils. It is manufactured from piperitone and used to a limited extent in antiseptic mouthwashes and gargles.

thymol blue, thymolsulphonphthalein One of the sulphonphthalein group of indicators. It has two useful pH ranges, 1·2 (red) to 2·8 (yellow) and 8·0 (yellow) to 9·6 (blue).

thymolphthalein, $C_{28}H_{30}O_4$. An indicator, prepared from phthalic anhydride and thymol. It has a pH range of 9·3 (colourless) to 10·5 (blue).

thyroid stimulating hormone, thyrotropin, thyrotropic hormone A glycoprotein hormone, mol.wt. about 10 000, secreted by the anterior lobe of the pituitary gland. Its action is to stimulate the thyroid gland to produce thyroxine; the level of circulating thyroxine or tri-iodothyronine* in turn controls the secretion of thyrotropin, i.e. negative feedback. It probably stimulates the activation of enzymes which split thyroxine from the proteins to which it is bound for storage.

thyroxine, $C_{15}H_{11}I_4NO_4$. M.p. 231–233°C

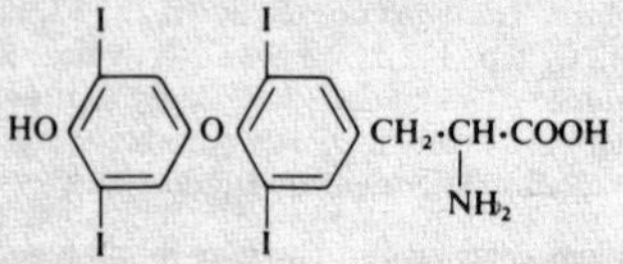

(decomp.). Hormone from the thyroid gland. It is formed from protein-bound tyrosine and iodine (derived from the blood) and occurs as a protein-bound form, thyroglobulin. Natural thyroxine is the L-isomer; the D-isomer has very little activity.

Thyroxine or tri-iodothyronine* (one less iodine) are secreted by the thyroid into the blood, especially when the gland is stimulated by thyrotropin, and circulate in combination with a specific protein, thyroxine-binding globulin (TBG). The thyroid hormones are essential regulators of the basal metabolic rate. Deficiency of the hormones leads to goitre, myxoedema and cretinism, while excessive secretion causes Graves's disease (exophthalmic goitre). Thyroxine also acts as a metabolic stimulant on respiring cells *in vitro*: it 'uncouples' oxidative phosphorylation, i.e. increases oxygen consumption without generating more ATP.

Ti Titanium.

tiglic acid, *cis*-1,2-dimethylacrylic acid, *trans*-(2-methyl-2-butenoic acid), $C_5H_8O_2$. M.p.

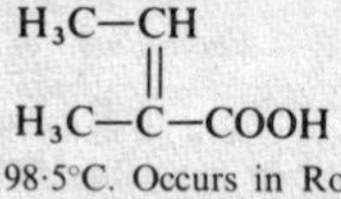

64°C, b.p. 198·5°C. Occurs in Roman oil of cumin and in croton oil. *See* angelic acid.

tin, Sn. At.no. 50, at.wt. 118·69, m.p. 232°C, b.p. 2270°C, *d* 7·28–7·31. Group IV metal easily obtained from its ores and hence known since prehistoric times. Occurs as cassiterite or tinstone, SnO_2, which is reduced with C (also recovered electrolytically). The element has three allotropic forms, metallic tin is silvery in colour, soft and malleable, tetragonal (distorted cp lattice). Below 13·5°C the stable form is grey tin (α) diamond lattice, brittle powder (the phase change is slow in the absence of grey tin). γ-Sn, another brittle modification is stable above 161°C. The metal is attacked by halogens (Cl_2 used in Sn recovery) but because of high over voltage is not attacked by cold dilute acids. Concentrated acids will dissolve Sn. Used in coating steel and in alloys* (solder*, bronze*) (Sn – Nb alloys are superconductors). Compounds are used in glass coatings and as fungicides. World demand 240 000 tonnes p.a.

tin alloys Tin is a component of many alloys including solders*, Britannia metal, fusible metal, anti-friction metal, Babbitt metal, bell metal, bronze, gun-metal and pewter.

tin bromides
Tin(II) bromide, $SnBr_2$. M.p. 215°C, b.p. 619°C. Very similar to tin(II) chloride.

Tin(IV) bromide, $SnBr_4$. M.p. 33°C, b.p. 203°C, prepared from the elements. Forms many complexes, including $[SnBr_6]^{2-}$ ions.

tincal Natural borax*, $Na_2B_4O_7,10H_2O$.

tin chemistry Tin is a Group IV element, electronic configuration $5s^25p^2$. It shows the group oxidation state of +4 (largely covalent except, e.g., SnO_2) and also the +2 state which has some reducing properties; SnX_2 compounds can co-ordinate through the lone-pair. Both oxidation states form complexes (Sn(IV) compounds 5- or 6-co-ordinate). Sn – Sn bonds are formed but are much less stable than for C and Si. Has no pure cationic chemistry, polymeric species, e.g. $[Sn_3(OH)_4]^{2+}$ are formed by hydrolysis of Sn(II) solutions.

tin chlorides
Tin(II) chloride, $SnCl_2$, stannous chloride. M.p. 247°C. White solid (Sn plus gaseous HCl), forms hydrates ($SnCl_2,2H_2O$ is tin salt) from Sn and aqueous HCl. Acts as acceptor in complexes and forms complexes with transition metals. Used as a mordant.

Tin(IV) chloride, $SnCl_4$, stannic chloride. M.p. −33°C, b.p. 114°C. Colourless fuming liquid (Sn plus Cl_2) hydrolysed in water but forms $SnCl_4,5H_2O$ and $[SnCl_6]^{2-}$ from acid solutions, soluble in organic solvents. Used as a mordant.

tincture of iodine Alcoholic solution of iodine containing 2½% iodine and 2½% potassium iodide, used as an antiseptic.

tin fluorides
Tin(II) fluoride, SnF_2, stannous fluoride. M.p. 213°C. Formed from Sn and aqueous HF. Forms complexes, e.g. $MSnF_3$. Used in toothpastes.

Tin(IV) fluoride, SnF_4, stannic fluoride. Polymeric solid formed Sn plus F_2 or $SnCl_4$ plus HF. Very hygroscopic, forms fluorostannates(IV) containing $[SnF_6]^{2-}$ ions.

tin hydrides SnH_4, b.p. −52°C, Sn_2H_6 – *see*

stannane. Compounds containing SnH bonds, e.g. Me_3SnH, are good reducing agents.

tin iodides *Tin(II) iodide*, SnI_2; *Tin(IV) iodide*, SnI_4. Very similar to tin chlorides.

tinning The coating of iron with tin, chiefly used for the manufacture of tin-cans. The sheet iron is pickled and passed into a bath of molten tin; the surface of the bath is protected from oxidation by a $ZnCl_2-NH_4Cl$ melt. Sheet iron was formerly dipped in a tallow bath before tinning. A cheap form of tinplate, 'terne plate' is coated from a Pb – Sn bath.

tin, organic derivatives Tin(IV) derivatives are formed by use of Grignard reagents on $SnCl_4$; catenated derivatives, e.g. $Ph_3SnSnPh_3$ are known (Ph_3SnCl plus Na). Used as fungicides, wood preservatives, as stabilizers in plastics (e.g. PVC) and as catalysts (particularly for the formation of polyurethanes). There is an extensive organotin chemistry, e.g. hydrides R_3SnH are good reducing agents.

tin oxides
Tin(II) oxides. Lower tin oxides: SnO (white, NH_4OH to $SnCl_2$ solution; black, heat on white SnO; red), form a complex system.

Tin(IV) oxide, SnO_2. Occurs naturally as cassiterite or tinstone. Precipitated as hydrate from Sn(IV) solution; can be dehydrated to SnO_2. Forms mixed metal oxides and species containing $[Sn(OH)_6]^{2-}$ ions (stannates). Used, under the name putty powder, for polishing glass and metal.

tin salt, $SnCl_2,2H_2O$. *See* tin(II) chloride.

tinstone, SnO_2. Cassiterite. *See* tin oxides.

tin sulphate, $SnSO_4$. Tin(II) sulphate.

tin sulphides
Tin(II) sulphides, SnS. Grey solid, Sn plus S at 900°C.

Tin(IV) sulphide, SnS_2. Precipitated from Sn(IV) solution with H_2S or Sn plus S under pressure. NH_4Cl, Sn, S heated gives a yellow solid (mosaic gold). Used as a pigment.

titanates *See* titanium oxides. Organic titanates, $Ti(OR)_4$, are used in catalysis, for cross-linking resins etc.

titanium, Ti At. no. 20, at. wt. 47. 88, m.p. 1660°C, b.p. 3287°C, *d* 4.5. Common element, the most important minerals are ilmenite, $FeTiO_3$, and rutile, TiO_2. The metal is obtained by reducing $TiCl_4$ with Mg. It is used as a light weight construction material; TiO_2 is used extensively as a white pigment and in catalysts. The massive metal has a protective layer of oxide in air. World demand 1976 metal 54 000 tonnes, compounds 1·7 megatonnes.

titanium alloys Used in the aircraft industry, engines, chemical plant. Additives include Al, Ti, Mn, Cr, Fe.

titanium carbide, TiC. Steel grey solid of high m.p., formed TiO_2 plus charcoal in an electric furnace. Very resistant to chemical attack. Used in tool tips and as a deoxidizer in steel manufacture.

titanium chemistry Titanium is Group IV transition element, electronic configuration $3d^2 4s^2$. The most stable oxidation state is +4 (e.g. TiO_2) largely covalent co-ordination numbers 4 in halides, 6 (in $[TiF_6]^{2-}$) and higher ($E°$ $[TiO]^{2+}$ to Ti(III) in acid solution −0·1 volt). Titanium(III) is more ionic, the $[Ti(H_2O)_6]^{3+}$ ion is violet; titanium(III) compounds are reducing agents. Titanium(II) is strongly reducing and has no aqueous chemistry. Ti(0) and Ti(−1), ($[Tibipy_3]^-$) are generally octahedral.

titanium chlorides
Titanium(II) chloride, $TiCl_2$. Black powder ($TiCl_4$ plus Ti or heat on $TiCl_3$). Strong reducing agent, immediately reduces water. Forms some complexes.

Titanium(III) chloride, $TiCl_3$. Violet or brown solid ($TiCl_4$ plus H_2 at 700°C; $TiCl_4$ plus AlR_3 (brown form)). Forms violet 6 hydrate. Used as a reducing agent. The fibrous brown form is an active agent in the Ziegler–Natta stereoregular polymerization of olefines.

Titanium(IV) chloride, $TiCl_4$, m.p. −25°C, b.p. 136°C, *d* 1·76. Colourless liquid, important intermediate in Ti production (TiO_2 plus C plus Cl_2). Hydrolysed via oxide chlorides to TiO_2. Forms many complexes including $[TiCl_6]^{2-}$.

titanium dioxide, TiO_2. Occurs naturally in three modifications: rutile, brookite and anatase. A most important white pigment formed by hydrolysis of purified $TiOSO_4$ or $TiCl_4$ or $TiCl_4$ plus O_2 through flame. Used as a pigment (66%) and in paper, rubber, fabrics, plastics, leather, printing inks. U.S. production 1981 775 000 tonnes.

titanium fluorides
Titanium(IV) fluoride, TiF_4. White solid ($TiCl_4$ plus anhydrous HF). Forms $[TiF_6]^{2-}$ ion.

Titanium(III) fluoride, TiF_3. Formed Ti or $TiH_{1\cdot7}$ plus anhydrous HF at 700°C. Blue stable solid. Complexes containing $[TiF_6]^{3-}$ ions are known.

titanium halides In addition to the chlorides* and fluorides*, $TiBr_4$, $TiBr_3$, TiI_4, TiI_3, TiI_2 are known.

titanium oxides
Titanium(IV) oxide, TiO_2. *See* titanium dioxide. Dissolves in concentrated alkali hydroxides to give titanates. Mixed metal oxides, many of commercial importance, are formed by TiO_2. $CaTiO_3$ is perovskite. $BaTiO_3$, perovskite related structure, is piezoelectric and is used in transducers in ultrasonic apparatus and gramophone pickups and also as a polishing compound. Other mixed oxides have the ilmenite structure (e.g. $FeTiO_3$) and the spinel structure (e.g. Mg_2TiO_4).

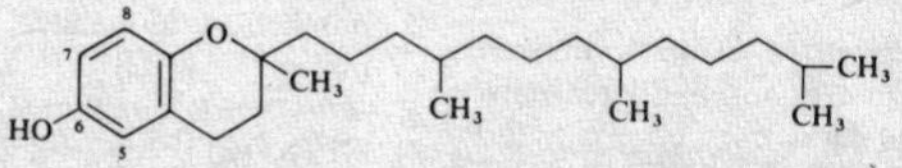

tocopherols

Titanium(III) oxide, Ti_2O_3. Violet, formed by reducing TiO_2 with H_2 at high temperatures.

Titanium(II) oxide, TiO. Has the NaCl structure but is non-stoicheiometric (TiO_2 plus Ti).

titanium sulphates
Titanium(IV) sulphate, $Ti(SO_4)_2$ also $TiO(SO_4)$. Readily hydrolysed salts formed $TiCl_4$ plus conc. H_2SO_4.

Titanium(III) sulphate, $Ti_2(SO_4)_3,8H_2O$. Formed as violet salt $TiCl_3$ plus dil. H_2SO_4 with exclusion of O_2.

titration A process for determining the volume of one solution required to react quantitatively with a given volume of another in which one solution is added to the other, a small amount at a time until just sufficient has been added to complete the reaction (equivalence point*). The equivalence point is determined as an end-point* by, e.g. cessation of precipitation (Ag^+ plus Cl^-), colour change (oxalate plus MnO_4^-) use of an indicator*, or by use of electrical methods (electrometric*, conductiometric*, titrations). Titrations may be carried out by hand from a burette or automatically.

Tl Thallium.

TLV Threshold limit values.

Tm Thulium.

TMA *See* trimellitic acid.

TMED Tetramethylethylene diamine.

TML Tetramethyl lead.

TMS Tetramethylsilane.

TMTD, TMTDS *See* thiuram disulphides.

TNT Trinitrotoluene.

toad venoms The poisonous substances of the toad are secreted by the skin glands and have an action on the heart of animals similar to that of digitalis and related plant glycosides, but without the persistence of action associated with digitalis. They are conjugated compounds consisting of genins of a sterol-like structure combined as esters with suberylarginine. The free genins are also present. The best-known toad venom is bufotoxin.

tocopherols The methylated derivatives of tocol (formula shown)

They are widely distributed in vegetable lipids, and in the body fat of animals, though animals cannot synthesize them. They have vitamin E activity and can protect unsaturated lipids against oxidation. Four are found naturally:

α-tocopherol, $C_{29}H_{50}O_2$, is 5,7,8-trimethyltocol,
β-tocopherol, $C_{28}H_{48}O_2$, is 5,8-dimethyltocol,
γ-tocopherol, $C_{28}H_{48}O_2$, is 7,8-dimethyltocol,
δ-tocopherol, $C_{27}H_{46}O_2$, is 8-methyltocol.

α-tocopherol has the strongest vitamin E activity.

tolan The trivial name given to diphenylethyne (diphenyl acetylene), $C_6H_5{\cdot}C{\equiv}C{\cdot}C_6H_5$. Obtained as volatile colourless crystals, m.p. 61°C, by HgO oxidation of benzil bis-hydrazone.

tolidines, $C_{14}H_{16}N_2$. The commonest isomer is *o*-tolidine, 3,3′-dimethylbenzidine, 3,3′-dimethyl-4,4′-diaminobiphenyl. Made by zinc reduction of *o*-nitrotoluene and subsequent acid catalysed rearrangement of the intermediate *o*-hydrazotoluene. A light-sensitive white solid, m.p. 130–131°C, soluble in polar solvents. Used in dyes. Because it is less carcinogenic than benzidine it is becoming more popular as a colour testing reagent for blood.

Tollens reagent An ammoniacal solution of silver oxide which is used as a test for aldehydes, which, unlike ketones, cause the deposition of a silver mirror.

toluene, C_7H_8, $C_6H_5CH_3$. B.p. 111°C, m.p. −95°C. Colourless refractive liquid with characteristic smell. Burns with a smoky flame. Very volatile in steam.

It is produced from petroleum fractions rich in naphthenes by catalytic reforming in the presence of hydrogen (hydroforming); in this process dehydrogenation and dealkylation take place simultaneously and a mixture of aromatics is produced. Also obtained from the carbonization of coal.

Chromic or nitric acids cause oxidation to benzoic acid; milder oxidation gives benzaldehyde and benzoic acid. The action of chlorine on boiling toluene gives mainly side chain substitution to benzyl chloride, benzal chloride and benzotrichloride, all used for the preparation of benzaldehyde and benzoic acid. Cold chlorination in the presence of iron causes nuclear substitution to a mixture of *o*- and *p*-chlorotoluenes. Easily nitrated in the cold to mixture of *o*- and *p*-nitrotoluenes; more vigorous conditions produce ultimately 2,4,6-trinitrotoluene, which is widely used as an explosive (TNT). Sulphonates to a mixture of *o*- and *p*-isomers, the latter preponderating; the sulphonic acids are used for the preparation of saccharine and chloramine-T respectively. Toluene is used as a constituent of high-octane aviation and motor gasolines, as a solvent and as a raw material in the manufacture of benzene, caprolactam, phenol, many dyestuffs and various other chemicals. U.S. production 1978 4·2 megatonnes.

toluene-2,4-diisocyanate, TDI, $C_9H_6N_2O_2$, $C_6H_4(NCO)_2$. B.p. 120°C. Manufactured by nitration of toluene, reduction to the diamine and condensation with phosgene. Used in condensation reactions with glycols to give polyurethanes. U.S. production 1983 290 000 tonnes.

toluene-4-sulphonyl chloride, tosyl chloride, $C_7H_7ClO_2S$, *p*-$CH_3C_6H_4SO_2Cl$. Colourless crystals, m.p. 71°C, formed by the action of chlorosulphonic acid on toluene. Esters of toluenesulphonic acid are frequently called tosylates and their formation tosylation. Many tosylates are easily obtained crystalline, and the reaction is thus of considerable importance.

***o*-toluidine, 2-aminotoluene,** C_7H_9N. Colourless liquid; b.p. 198°C. It is prepared by the reduction of 2-nitrotoluene. It is basic, forms a stable hydrochloride and sulphate, both moderately soluble in water. The sulphate is converted at 200°C into *o*-toluidinesulphonic acid. Readily diazotizes and in this form used as first component in many dyestuffs. Used for condensation with benzaldehyde in preparation of magenta.

***p*-toluidine, 4-aminotoluene,** C_7H_9N. Colourless leaflets; m.p. 45°C, b.p. 200°C. It is basic and gives well-defined salts with mineral acids. Prepared by reduction of 4-nitrotoluene with iron and hydrochloric acid. Easily sulphonated. It is used as a first component in azo dyes and in the manufacture of magenta.

toluyl- The group $CH_3C_6H_4CO-$.

tolyl The group $CH_3C_6H_4CH_2-$.

topaz, $Al_2SiO_4(F,OH)_2$. Occurs in igneous rocks and pegmatites, contains discrete SiO_4 tetrahedra. Used in glasses and glazes and in slags in the steel industry and after heating as a refractory. Coloured transparent crystals are used as gemstones.

topotactic A reaction in which the product has an ordered geometric relation to the reactant.

tor A pressure unit. One tor is 1 mm Hg pressure.

tosyl *See* toluene-4-sulphonyl chloride.

tosylation A chemical transformation which substitutes the toluene-4-sulphonyl* (tosyl-) group into a molecule.

total pressure In a mixture of gas and vapours the total pressure exerted by the mixture is equal to the sum of the partial pressures of the constituents. *See* partial pressure.

total reflux A distillation column is said to be operating under total reflux when all the vapour leaving the column is condensed and returned to it. In this case no products are withdrawn from the system and the reflux ratio is infinity. Total reflux is frequently employed when starting up a column. *See* rectification.

town gas The general name for manufactured gas of calorific value* about 18 MJ/m³. Various processes have been used, e.g. coal carbonization at about 1250°C; the complete gasification of coal, as in the Lurgi process*; and oil-based processes, such as the catalytic steam-reforming of naphtha.

All these fuel gases contain more than 50% hydrogen and 10–30% methane, the other main components being CO, higher hydrocarbons, CO_2 and N_2. In many parts of the world natural gas* of calorific value of approximately 38 MJ/m³ has become the widely-used gaseous fuel.

toxaphene A chlorinated camphene insecticide of approximate formula $C_{10}H_{10}Cl_8$. It is a light yellow waxy solid, melting range 65–90°C. The technical product is prepared by chlorination of camphene and is a mixture of compounds.

tracer A substance added to a system to enable the experimenter to follow the course of a process, e.g. reaction or self-diffusion, without seriously altering the conditions. More particularly, the term is chiefly used for isotopic species added as tracers, especially radioactive isotopes or rare isotopes of the lighter elements (e.g. ^{18}O). The former are detected and determined with the aid of a counter, the latter with a mass spectrometer. As these detectors are very sensitive, only small proportions of tracer are needed.

tractor vaporizing oil (TVO) Kerosine used for spark-ignition engines of low compression ratio.

tragacanth *See* gums.

tranexamic acid, trans-4-(aminomethyl)cyclohexanecarboxylic acid, $C_8H_{15}NO_2$. White crystals, soften at 270°C. An antifibrinolytic agent, used to treat thrombosis in the deep veins.

trans Term used in designating isomers to indicate isomer with non-adjacent (opposite) like-ligands. *Compare cis.*

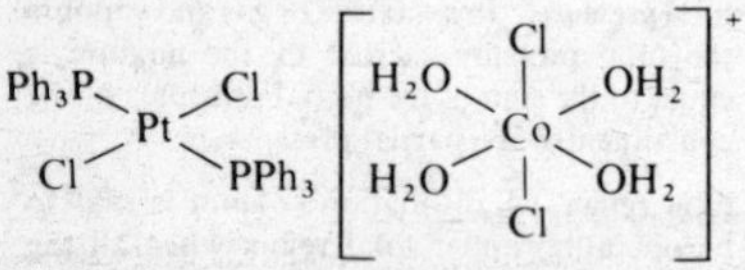

trans-planar *trans*-octahedral

trans-actinides *See* post-actinide elements.

transaminases Enzymes catalysing the transfer of amino groups, and found in all organisms. Many transfer NH_2 to and from glutamine and asparagine. Glutamic acid is thus of considerable importance in the formation of amino-acids from keto-acids and ammonia.

trans effect In substitution in complexes the

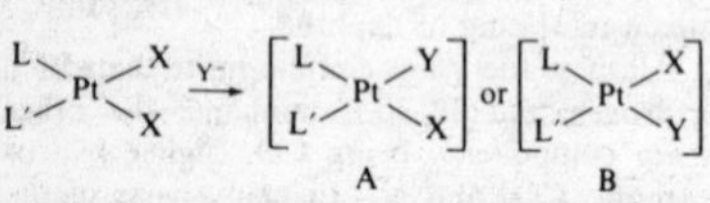

proportions of *cis* and *trans* isomers formed depends markedly upon the other ligands present. Thus in square-planar platinum complexes the proportion of the two products A and B depends upon the relative *trans*-directing effects of L and L′. Ligands can be placed in a series of ability to direct *trans* substitution: some common ligands in order are $H_2O < NH_3 < Cl^- < CO$.

transferrin A protein which is the main mode of transport of iron.

transformer oils Highly refined oil of low viscosity, high resistance to oxidation used for cooling and insulation in electrical installations, e.g. transformers and switch gear. Often replaced by synthetic materials.

transition elements The elements of the periodic table in which there is a partially filled d shell. The elements from Sc to Zn, Y to Cd and La to Hg are classed as transition elements, the first named element in each series having one d electron and the last ten d electrons. Transition elements are characterized by variable oxidation state, coloured ions and a tendency to form complexes. Many of the elements and their salts are paramagnetic.

transition interval The visible colour change of an indicator is limited to a definite range of hydrogen-ion concentrations or of potential. The region between the limiting values, expressed in terms of pH or rH, is the transition interval, or region of change of the indicator, e.g. litmus changes from red to blue in the interval pH 5·0–8·0. *See* hydrogen-ion concentration; indicators, oxidation-reduction.

transition point Transition temperature.

transition state theory Since most rate processes (e.g. diffusion or electrical conduction through solid and liquids, viscous flow in gases and liquids and chemical reactions) obey the Arrhenius equation*, it is implied that they proceed by thermal activation. The transition state (or activated state) is the high energy configuration through which the reactants must pass before becoming products.

A quantitative theory of rate processes has been developed on the assumption that the activated state has a characteristic enthalpy, entropy and free energy; the concentration of activated molecules may thus be calculated using statistical mechanical methods. Whilst the theory gives a very plausible treatment of very many rate processes, it suffers from the difficulty of calculating the thermodynamic properties of the transition state.

transition temperature The temperature at which one (allotropic) form of a substance is converted into another allotropic form. E.g. rhombic sulphur is converted into monoclinic sulphur at the transition temperature of 95·6°C. Below this temperature the rhombic form is the stable allotrope; the monoclinic allotrope is the stable form above this temperature. Transition temperatures vary with the pressure exerted on the system.

transmethylases Enzymes which catalyse the transfer of methyl groups in biological systems. The methyl groups are transferred from the amino-acid methionine which is converted to homocysteine.

transmutation The process of transforming one element into another. It has occupied the attention of scientists since the earliest times. The search for the Philosopher's Stone, a substance by means of which base metals like iron could be converted into gold, was one of the chief preoccupations of the alchemists, but it is only since 1919 that successful transmutation of any kind has been achieved using particle accelerators and reactors. A proton fired with a high velocity at an atom of lithium, may yield two atoms of helium.

$$^{7}_{3}\mathrm{Li} + ^{1}_{1}\mathrm{p} \longrightarrow 2\ ^{4}_{2}\mathrm{He}$$

Numerous nuclear transformations have been induced by processes in which atoms have been bombarded with neutrons, protons, deuterium, carbon atoms and ions.

The new elements neptunium and plutonium have been produced in quantity by neutron bombardment of uranium. Subsequently many isotopes have been obtained by transmutation and synthetic isotopes of elements such as Ac and Pa are more easily obtained than the naturally occurring species. Synthetic species of lighter elements, e.g. Tc and Pm are also prepared.

The transformations of the radioactive elements, whereby, e.g. uranium ultimately becomes lead, are not usually regarded as instances of transmutation because the processes are spontaneous, and cannot be controlled by the experimenter.

transport number The fraction of the current carried by a particular ion during electrolysis.

tranylcypromine, 2-phenylcyclopropylamine, $C_9H_{11}N$. Liquid, b.p. 79°C. Used as the hydrochloride, $C_9H_{11}N{\cdot}HCl$, with m.p. 164°C. Tranylcypromine is an inhibitor of monamine oxidase and is administered orally in the treatment of depressive illness.

traumatic acid, trans-2-dodecenedioic acid,

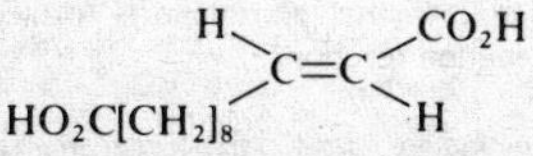

$C_{12}H_{20}O_4$. M.p. 165°C. A plant growth hormone, which is produced in damaged plant tissue, and on diffusing into adjacent intact tissue cells stimulates them to divide. Traumatic acid has been isolated from the pods of green beans.

tray *See* plate.

tray column *See* plate column.

trehalose, (α-D-glucosido)-α-D-glucoside,

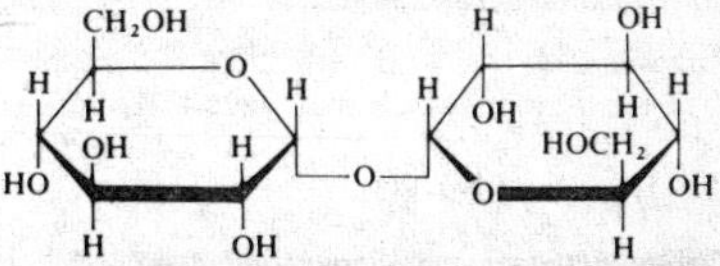

$C_{12}H_{22}O_{11},2H_2O$. M.p. 97°C. A non-reducing disaccharide, which forms the principal carbohydrate of insect haemolymph. It comprises about 25% of trehala manna, the cocoons of a parasitic beetle. Trehalose also occurs in fungi, e.g. *Amanita muscaria*, generally replacing sucrose in plants lacking chlorophyll and starch.

tremolite, $(OH)_2Ca_2Mg_5(Si_4O_{11})_2$. An amphibole, an asbestos mineral. Used in acid-resistant filters.

triacetin *See* acetins.

triacetoneamine, 2,2,6,6-tetramethyl-γ-piperidone, $C_9H_{17}NO$. Colourless needles, m.p.

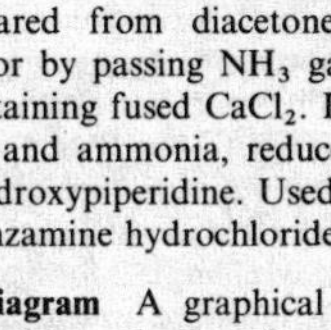

35°C, b.p. 205°C, forms monohydrate m.p. 58°C. Prepared from diacetoneamine and propanone or by passing NH_3 gas into propanone containing fused $CaCl_2$. Decomposes to phorone and ammonia, reduced to tetramethyl-3-hydroxypiperidine. Used in the synthesis of benzamine hydrochloride.

triangular diagram A graphical representation of the phase-rule data for a system of three components.

triarylmethane dyes *See* triphenylmethane dyes.

triazine herbicides Derivatives of s-triazine

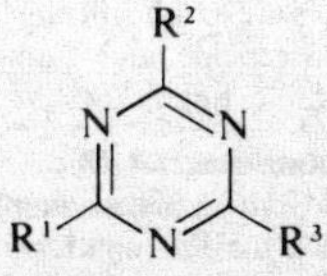

(from cyanuric chloride). Various derivatives are used as selective herbicides, e.g. simazine (R^1, R^2 = EtNH; R^3 = Cl), atrazine (R^1 = EtNH, R^2 = Pr^iNH, R^3 = Cl)

triazines Compounds containing sym- C_3N_3

rings. *See* e.g. melamine, diallylmelamine. Used in dyestuffs, herbicides, etc.

tribasic acid An acid with three replaceable hydrogen atoms which may yield three series of salts, e.g. H_3PO_4 gives NaH_2PO_4, Na_2HPO_4 and Na_3PO_4.

tribromoethanal, tribromoacetaldehyde, bromal, CBr_3CHO. An oily liquid; b.p. 174°C. Prepared by action of bromine on ethanol and decomposed by alkali to bromoform $CHBr_3$. Forms a dihydrate $CBr_3CH(OH)_2,H_2O$, m.p. 53·5°C.

tribromomethane *See* bromoform.

tributyl phosphate *See* phosphate esters.

tricarboxylic acid cycle *See* citric acid cycle.

trichloroacetic acid *See* chloroethanoic acids.

1,1,1-trichloroethane, $CCl_3{\cdot}CH_3$. Prepared by chlorination of ethane or HCl on vinyl chloride. Used in metal cleaning and as a solvent. U.S. production 1979 275 000 tonnes.

trichloroethanoic acid *See* chloroethanoic acids.

trichloroethene, trichloroethylene, $CHCl{:}CCl_2$. Colourless liquid with a chloroform-like odour, b.p. 87°C. It is non-inflammable but toxic to the liver.

Trichloroethylene is manufactured by the dehydrochlorination of tetrachloroethane derived from the chlorination of ethyne with lime or by vapour phase cracking.

Trichloroethylene is not attacked by dilute acids or alkalis, but when heated with sodium hydroxide under pressure it yields sodium glycollate. In the presence of light and oxygen dichloroethanoyl chloride is formed, which can react with any moisture present to give small amounts of highly corrosive HCl. Numerous stabilizers have been patented.

Most of the trichloroethylene produced is used for metal degreasing. Other important uses are in the scouring of wool and as an extractive solvent, e.g. for olive and soya bean oils. Minor uses are as a heat transfer medium, anaesthetic, insecticide and fumigant, paint remover and fire extinguisher.

trichlorofluoromethane, Cl_3CF. A colourless liquid; b.p. 24°C. Used as an internal standard with $\delta = 0$ for ^{19}F n.m.r. spectroscopy. Prepared by fluorination of CCl_4 with $SbCl_5$ and HF.

2,4,6-trichlorophenol, $C_6H_3Cl_3O$. Colourless crystalline substance, m.p. 68°C. Used medicinally as an antiseptic and disinfectant.

triclinic system The crystal system* without symmetry planes or axes. A unit cell has three unequal axes at mutually oblique angles. Example $CuSO_4,5H_2O$.

tricresyl phosphate *See* phosphate esters.

tridymite, SiO_2. The form of silica* stable between 573°C and 1470°C. It is a desirable constituent of silica bricks because of its low coefficient of expansion.

triethanolamine *See* ethanolamines.

triethoxymethane *See* triethyl orthoformate.

triethylaluminium *See* aluminium organic derivatives.

triethylamine *See* ethylamines.

triethylene glycol, triglycol, $HOCH_2CH_2OCH_2CH_2OCH_2CH_2OH$. Similar to glycol (*see* dihydroxyethane). Used as desiccant and in humidity control; esters are used as plasticizers.

triethyl orthoformate, orthoformic ester, triethoxy-methane, $HC(OC_2H_5)_3$. B.p. 146°C; prepared by reaction between chloroform and sodium ethoxide in ethanol. A valuable reagent for the introduction of the formyl group, $-CHO$ (as its diethylacetal $-CH(OC_2H_5)_2$) into aliphatic and aromatic substrates. Other uses include acetal formation, ethyl ester formation and heterocyclic synthesis.

triethyl phosphate, $(C_2H_5O)_3P=O$. Colourless liquid, b.p. 215–216°C, soluble in water and used as a polar reaction medium. Prepared from $POCl_3$ and ethanol.

triethyl phosphite, $(C_2H_5O)_3P$. Colourless liquid, b.p. 156°C, insoluble in water. Prepared from PCl_3 and ethanol. Reacts with alkyl halides to form quaternary salts $[(EtO)_3P-R]^+X^-$ which readily lose C_2H_5Cl to form phosphonates $(EtO)_2P(O)R$ which may be used in modified Wittig reactions. Used as a deoxygenating desulphurizing agent. The more stable triethyl phosphate is formed in deoxygenation reactions.

trifluoroethanoic acid, trifluoroacetic acid, CF_3CO_2H. Colourless liquid, b.p. 72·5°C, fumes in air. Trifluoroacetic acid is the most important halogen-substituted acetic acid. It is a very strong acid ($pK_a = 0{\cdot}3$) and used extensively for acid catalysed reactions, especially ester cleavage in peptide synthesis.

It forms an anhydride $(CF_3CO)_2O$, b.p. 39·5°C, with P_2O_5. The anhydride is used for trifluoroacetylation of $-OH$ and $-NH_2$

groups. As a catalyst in esterification reactions and acylation reactions of activated aromatic substrates it is frequently preferable to conc. sulphuric acid.

trifluoromethyl derivatives Compounds containing the electronegative CF_3- group. This confers important pharmaceutical properties on compounds and may radically alter properties, e.g. CF_3SO_3H is a very strong acid.

triglyme, $C_8H_{18}O_4$. The dimethyl ether of triethylene glycol.

$$CH_3OCH_2CH_2OCH_2CH_2OCH_2CH_2OCH_3$$

trigol The trivial name for triethylene glycol, $HO(CH_2)_2O(CH_2)_2O(CH_2)_2OH$, a colourless hygroscopic liquid, b.p. 285°C. Used as a solvent.

trigonal bipyramidal co-ordination Co-ordi-

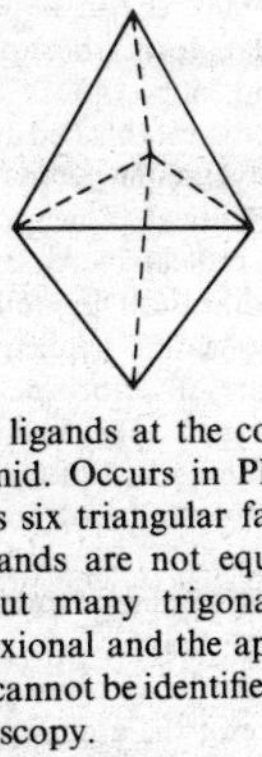

nation of five ligands at the corners of a trigonal bipyramid. Occurs in PF_5. A trigonal bipyramid has six triangular faces. The three equatorial ligands are not equivalent to the two apical but many trigonal bipyramidal species are fluxional and the apical and equatorial ligands cannot be identified separately by n.m.r. spectroscopy.

trigonal prismatic co-ordination Co-ordina-

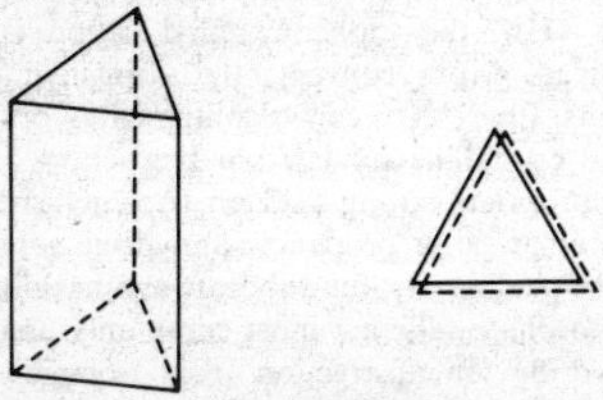

trigonal prism

tion of six ligands at the corners of a trigonal prism. Occurs for Mo in MoS_2. A trigonal prism has two triangular and three rectangular faces.

trigonal system The crystal system* with a 3-fold axis as principal axis. The unit cell is generally taken with the *z*-axis parallel to the 3-fold axis and the other two axes mutually inclined at 60° and perpendicular to the 3-fold axis. Rhombohedral unit cells (three equal axes equally at an angle [not 90°]) can be expressed in terms of trigonal cells. Example trigonal $CaCO_3$.

trigonelline A betaine.

trihydroxypropane *See* glycerol.

tri-iodomethane *See* iodoform.

3,5,3′-tri-iodothyronine, $C_{15}H_{12}I_3NO_4$. The principal circulating thyroid hormone, present in blood plasma and in the thyroid gland. It is about five times as active as thyroxine in preventing developing goitre in rats which have undergone thyroidectomy.

Trilene A trade name for trichloroethylene* when used as an anaesthetic. It is given by inhalation for short operative procedures.

trimellitic acid, 1,2,4-benzene tricarboxylic acid M.p. 218°C. Obtained by oxidizing

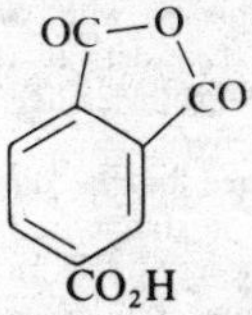

pseudocumene. Converted to trimellitic anhydride, TMA, m.p. 168°C, used in plasticizers, alkyd resins, polyesters.

trimethylamine *See* methylamines.

trimethylamine oxide, C_3H_9NO, Me_3NO. Crystallizes from water in colourless needles containing $2H_2O$, m.p. 255–257°C (hydrate); 96°C (anhydrous). Occurs widely distributed in fish and animal tissues. Prepared by treating an aqueous solution of trimethylamine with hydrogen peroxide. Forms alkaline solutions in water and combines with acids to give crystalline salts. Decomposes when heated at 180°C into trimethylamine and methanal.

1,2,3-trimethylbenzene Hemimelletene.

1,2,4-trimethylbenzene *See* pseudocumene.

1,3,5-trimethylbenzene *See* mesitylene.

trimethylene *See* cyclopropane.

trimethylolethane, 2-hydroxymethyl-2-methyl-1,3-propenediol, $CH_3C(CH_2OH)_3$. Colourless solid, m.p. 202°C, b.p. 283°C, prepared by condensation of methanal with CH_3CH_2CHO in the presence of NaOH. Used for alkyd resins particularly for coatings.

trimethylolpropane, 2-ethyl-2-hydroxymethyl-

1,3-propanediol, $EtC(CH_2OH)_3$. Hygroscopic solid; m.p. 57–59°C, b.p. 160°C/15 mm. Prepared by condensation of methanal with butyraldehyde. Used (as esters) in the production of polyurethane foams.

2,4,6-trimethylpyridine, s-collidine, $C_7H_{11}N$. B.p. 171°C; a base, generally superior to pyridine for elimination reactions.

trimethylsilyl derivatives, Me_3SiX. Used for introduction of X groups.

1,3,5-trinitrobenzene, $C_6H_3N_3O_6$. A yellow crystalline solid; m.p. 122°C. Best prepared by oxidizing trinitrotoluene to 2,4,6-trinitrobenzoic acid with sodium dichromate, and then boiling a suspension of the acid in water.

Its charge transfer complexes with aromatic hydrocarbons have characteristic melting points and may be used for the identification and purification of the hydrocarbons.

2,4,6-trinitrotoluene, TNT, $C_7H_5N_3O_6$. Yellow crystals; m.p. 81°C. It forms charge transfer complexes with aromatic hydrocarbons and their derivatives. It condenses with 4-nitrosodimethylaniline and 4-dimethylaminobenzaldehyde.

It is prepared by the direct nitration of toluene with a mixture of nitric and sulphuric acids. TNT is a very stable, violent and powerful high explosive, but less sensitive to shock and friction than picric acid. It is widely used as a filling for shells, bombs, etc. often mixed with ammonium nitrate and other high explosives. The lower grades of TNT may contain isomers which under hot storage conditions may give rise to exudation.

triose A carbohydrate with three carbon atoms. Three trioses are known, dihydroxyacetone and two stereoisomers of glyceric aldehyde.

trioxymethylene *See* paraformaldehyde.

tripentaerythritol, $(HOCH_2)_3CCH_2OCH_2C(CH_2OH)_2CH_2OCH_2C(CH_2OH)_3$. White compound; m.p. 250°C. Prepared during the preparation of pentaerythritol. Used in the preparation of surface-coating materials.

triphenylene, 9,10-benzphenanthrene, $C_{18}H_{12}$.

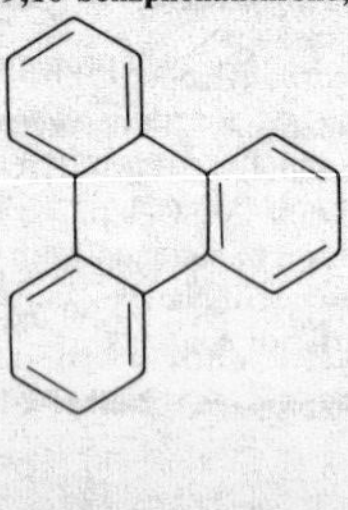

Colourless needles, m.p. 199°C, sublimes easily. Solutions exhibit blue fluorescence. Occurs in coal tar but may also be obtained from the spontaneous trimerization of benzyne.

triphenylmethane dyes A major class of dyestuffs, formally part of the triarylmethane series, the most useful dyestuffs are derivatives of triphenylmethane, Ph_3CH, and diphenylnaphthylmethane. The dyestuffs have brilliant hue but poor fastness. The dyestuffs are carbenium ions $[Ar_3C]^+X^-$ and are reduced to the 'leuco-dyes', the triarylmethanes, Ar_3CH. Widely used in copying papers and as phosphotungstates or molybdates. Many of the dyestuffs are also used as antiseptics.

triphenylmethyl, Ph_3C. The anion, cation and radical. The discovery of the radical, the first known free radical, was announced by Gomberg in 1900. It can be isolated only in the bimolecular form, hexaphenylethane, a colourless solid, m.p. 145–147°C. In solution an equilibrium is established between colourless hexaphenylethane and the yellow triphenylmethyl radical. The proportion of the paramagnetic radical increases with dilution or rise in temperature. The equilibrium mixture is obtained by heating triphenylmethyl chloride with mercury, silver or zinc in the complete absence of oxygen. The free radical is very reactive, combining with atmospheric oxygen to give a peroxide and with iodine to give triphenylmethyl iodide. It will also react with nitrogen dioxide, quinone and alkenes.

If triphenylmethyl chloride in ether is treated with sodium, a yellow colour is produced due to the presence of the anionic species Ph_3C^-. Alternatively, if triphenylmethyl chloride is treated with silver perchlorate in a solvent such as THF, the triphenylmethyl cation is obtained. More conveniently, triphenylmethyl salts, $Ph_3C^+X^-$, can be obtained as orange–red crystalline solids from the action of the appropriate strong acid on triphenylcarbinol in ethanoic or propanoic anhydride solution. The perchlorate, fluoroborate and hexafluorophosphate salts are most commonly used for hydride ion abstraction from organic compounds (e.g. cycloheptatriene gives tropylium salts). The salts are rather easily hydrolysed to triphenylcarbinol.

The radical and ions are exceptionally stable due to resonance: the free electron or charge is not localized on the methyl carbon atom but is distributed over the benzene rings.

triphenylphosphine, $(C_6H_5)_3P$. Colourless crystalline solid; m.p. 82°C, sublimes easily. It is used widely as a stabilizing ligand for low oxidation states of many transition metal derivatives (many of the complexes are used as catalysts), and, *via* the appropriate phosphonium salt, for conversion to a phosphorane used in the Wittig alkene synthesis.

triple bond A bond using three bonding pairs of electrons between two atoms.

triple point In a one component system three phases can only co-exist in equilibrium at a certain temperature and pressure. Thus, e.g. in the system ice, water and water vapour, the three phases are in equilibrium at a pressure of 4 torr and a temperature of 0·0075°C. This point is termed the triple point for water.

triplet Three closely spaced transitions in a spectrum.

triplet state *See* spin multiplicity.

tripolite A variety of diatomite*.

triptycene, $C_{20}H_{14}$. Tribenzobicyclo-[2,2,2]-

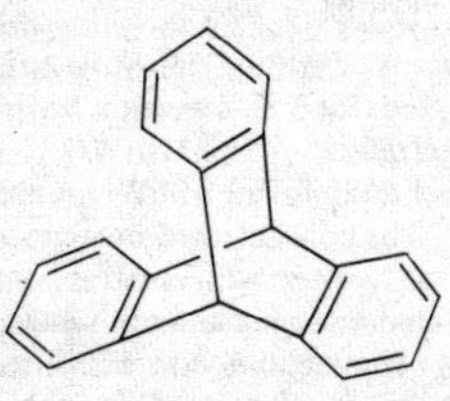

octatriene. Colourless solid, m.p. 255–256·5°C, which is best obtained by a Diels–Alder reaction between benzyne and anthracene.

triterpenes Triterpenoid compounds are widely distributed in nature. They contain 30 C atoms and are derived from the hydrocarbons $C_{30}H_{50}$. They include squalene and tetracyclic compounds such as lanosterol.

trithiocarbonates, M_2CS_3. Prepared MSH plus CS_2. Contain planar $[CS_3]^{2-}$ ions.

trithionic acid, $H_2S_3O_6$. *See* polythionic acids.

tritium, ${}^3_1H$. The heaviest isotope of hydrogen, occurring in a proportion of less than 1 in 10^{17} in natural hydrogen. It is β-active and is used as a radioactive tracer and in thermonuclear reactions. It may be prepared by the deuterium bombardment of various light atoms or by the thermal neutron reaction ${}^6Li(n\alpha){}^3H$.

tritolyl phosphate Tricresyl phosphate. *See* phosphate esters.

Triton B Trade name for benzyltrimethylammonium hydroxide usually as a 40% solution in methanol. A strong base, soluble in many solvents used as a catalyst. *See* phase transfer chemistry.

trityl The triphenylmethyl group*. Triphenylmethyl chloride reacts with pyridine solutions of certain hydroxy compounds and sugar derivatives to form trityl ethers.

trona A naturally occurring sodium carbonate, Na_2CO_3, $NaHCO_3,2H_2O$, formed by evaporation of soda lakes. Known also as urao. An important natural source of sodium carbonate.

tropilidine, 1,3,5-cycloheptatriene, C_7H_8. *See* 1,3,5-cycloheptatriene.

tropine, 3-tropanol, $C_8H_{15}NO$. M.p. 63°C,

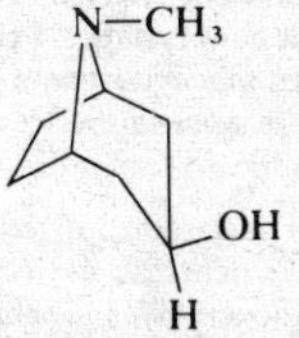

b.p. 229°C. It is formed by hydrolysis of its esters, which are called tropeines, and are important medically. *See* atropine, hyoscine, hyoscyamine.

tropolones 2-Hydroxy-derivatives of cyclo-

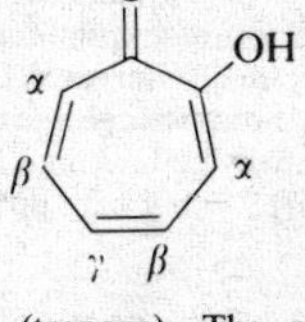

heptatrienone (tropone). The parent compound, *tropolone*, forms colourless needles, m.p. 49–50°C. Forms metal derivatives. It may be prepared from cyclohepta-1,2-dione by bromination followed by dehydrobromination. A number of tropolones occur naturally including colchicine, stipitatic acid – a β-hydroxy-β-carboxytropolone from *Penicillium stipitatum* – and the thujaplicins, which are isopropyltropolones present in the essential oils of the *Cupressaceae*.

tropylium The trivial name given to the symmetrical cation $(C_7H_7)^+$ of 1,3,5-cycloheptatriene, obtained when a hydride ion is abstracted from 1,3,5-cycloheptatriene by, e.g., trityl salts. It is an aromatic system having 6 π electrons distributed equally over seven carbon atoms, and is an unusually stable carbonium ion. Alkaline hydrolysis produces tropone. Complexes of tropylium with many transition metal carbonyls, etc. are known.

Trouton's rule The latent heat of vaporization (ΔH_{vap}) of a liquid of molecular weight (M) and boiling point (T K) are related by the expression

$$(M\Delta H_{vap}/T) = \text{constant}$$

For most normal liquids the constant has a value of approx. 88 J mol K^{-1}. Associated liquids show marked variations from this value. The Trouton constant is the molar entropy of vaporization.

tryparsamide, tryparsone, $C_8H_{10}AsN_2NaO_4$. Sodium salt of N-phenylglycineamide-*p*-arsonate. Colourless crystals. Prepared from sodium-*p*-aminophenylarsonate and chloroacetamide. It is used in the treatment of trypanosomiasis.

trypsin An important digestive enzyme, formed in the pancreas as its inactive precursor, trypsinogen. This proenzyme is activated in the intestine by enterokinase or, autocatalytically, by trypsin itself. In the process, the hexapeptide Val•Asp•Asp•Asp•Asp•Lys is split off. Both trypsin and trypsinogen have been obtained crystalline. Trypsin has a molecular weight of 24 000; it acts optimally at pH 7–9, and is specific for cleavage of peptide linkages adjacent to arginine or lysine residues. The reactive site of trypsin contains a serine residue. Trypsin also serves as the activator to convert chymotrypsinogen into chymotrypsins* accompanying intestinal enzymes which also act at pH 7–9 but have different specificities.

tryptophan, α-amino-β-indolylpropionic acid, $C_{11}H_{12}N_2O_2$. M.p. 289°C. L-Tryptophan is an essential amino-acid, its presence in the food of animals is necessary for proper growth. It is present in small quantities in the hydrolysis products of most proteins, although absent in certain vegetable proteins.

tung oil, china-wood oil A rapidly drying oil* obtained from the nuts of *Aleurites cordata* and *Aleurites fordii*, trees of the euphorbia order that are native in China and Japan and now cultivated in America and elsewhere. Tung oil is used on a considerable scale with natural resins in the preparation of oil varnishes; when it is heated with certain metallic oxides driers* are formed; these are known as tungates.

tungstates Formed by dissolving WO_3 in NaOH solution and subsequently by cation exchange. The normal tungstates, e.g. $Na_2WO_4,2H_2O$, contain tetrahedral WO_4^{2-} ions. Most metal salts other than alkali metal salts are insoluble. Acidification of a solution of WO_4^{2-} ions gives polymeric anions – isopolytungstates – containing WO_6 octahedra joined corner to corner or edge to edge; $[HW_6O_{21}]^{5-}$, $[H_2W_{12}O_{42}]^{10-}$ and $[W_{12}O_{39}]^{6-}$ are well established. If the acidification is carried out in the presence of another oxyanion, heteropolytungstates in which MO_6 or MO_4 groups (M = heteroatom) are incorporated into the polyanion are formed. E.g. $[FeW_{12}O_{40}]^{5-}$ and $[PW_{12}O_{40}]^{3-}$. It seems unlikely that free tungstic acids of any of these forms are stable. Further acidification of isopolytungstates gives $WO_3,2H_2O$. Heteropolytungstates are used to form W-containing catalysts by heating.

tungsten, W. At.no. 74, at.wt. 183·8, m.p. 3410°C, b.p. 5660°C. The main ore of tungsten is wolframite $(Fe,Mn)WO_4$; scheelite ($CaWO_4$) and stolzite ($PbWO_4$) are also important. The concentrated ores are fused with NaOH and, after water extraction, WO_3 is precipitated with acid and reduced to the metal with H_2. The metal is bcc. The fused metal is lustrous silver–white and is only attacked slowly even by HNO_3-HF. Dissolves in KNO_3-NaOH or Na_2O_2, attacked by O_2 at red-heat. The metal is used extensively in steel alloys, in electric lamp and heating filaments and in electric contacts. Tungsten carbides are very hard and are used in cutting tools. World production WO_3 1981 52 000 tonnes.

tungsten alloys The most important alloys are the tungsten steels which contain up to 18% of tungsten. Stellite, an extremely hard alloy of W with Cr and Cu, is used for high-speed cutting tools. Electric contacts for switch-gear are made from Cu and Ag–W alloys.

tungsten blue Partially reduced WO_3 or tungstates.

tungsten bronzes Coloured compounds M_nWO_3 (M = unipositive metal $0<n<1$) prepared by reducing tungstates with, e.g., Na. The compounds contain W^{VI} and W^{V}.

tungsten carbides W_2C and WC are obtained by heating W powder with C. Both compounds are extremely hard and are used for making cutting tools and dies. Ternary carbides are also used in cutting tools and other metallic carbides, borides, silicides or nitrides are added to improve the mechanical properties of the mixture.

tungsten carbonyl, $W(CO)_6$. Very similar to molybdenum carbonyl* but less readily substituted.

tungsten chemistry Tungsten is a typical element of transition element Group VI and shows oxidation states from +6 to −2 and, particularly in its oxides, forms many non-stoichbeiometric compounds. There is little aqueous chemistry except that of complex oxy-anions and some complex halides. The hexahalides are molecular but lower halides are polymeric and the lowest halides show extensive W−W bonding (more than Mo). Carbonyl and phosphine derivatives are typical low oxidation state compounds. Complexes are formed, particularly by O- and S-ligands in higher oxidation states and by P-ligands in low oxidation states. Complex cyanides are well established.

tungsten halides The known fluorides are colourless WF_6 ($W+F_2$), b.p. 17°C, mixed tungsten(VI) chloride fluorides (e.g. WF_5Cl) and derivatives of WF_6 (e.g. WF_5OMe), tetrameric WF_5 ($W+WF_6$) and WF_4 (heat on WF_5). Complex fluorides of W(VI) and W(V) contain ions such as WF_7^- and WF_6^-. The known chlorides are WCl_6 ($W+Cl_2$) an olefine metathesis reagent, green WCl_5 ($W+Cl_2$), involatile WCl_4 (WCl_6+Al), WCl_3 and WCl_2 (both WCl_6+H_2). WCl_3 and WCl_2 both contain metal-clusters $(W_6Cl_{12})Cl_6$ and $(W_6Cl_8)Cl_4$ respectively. Complex chlorides containing WCl_6^-, WCl_6^{2-}, $(W_2Cl_9)^{3-}$ are known ($(W_2Cl_9)^{3-}$ contains three bridging Cl and a W−W bond). All of the halides form extensive ranges of complexes with various ligands. Hydrolysis of the higher oxidation state fluorides and chlorides gives oxide halides and these compounds may also be prepared directly from oxides and halogenating agents. WOF_4, WOF_5^-, WOF_4^{2-}, $WO_2F_3^{3-}$, $WOCl_4$, WO_2Cl_2, $WOCl_3$ and $WOCl_5^{2-}$ are known and their chemistry is similar to that of the halides. Tungsten bromides, WBr_6, WBr_4, WBr_3, WBr_2 and iodides WI_4, WI_3 and WI_2 are similar to the chlorides.

tungsten, organic derivatives Generally similar to the molybdenum derivatives but higher oxidation state alkyls, e.g. WMe_6 (WCl_6+LiMe) are known.

tungsten oxides Yellow WO_3 is the final product from heating tungsten or other tungsten oxides in excess oxygen. It is used in yellow glazes for ceramics. Brown WO_2 (WO_3+H_2) and many intermediate phases containing shear structures* (e.g. $W_{18}O_{49}$) are formed by reduction of WO_3. Mixed metal oxides are formed by fusing WO_3 with other metal oxides.

tungstic acid *See* tungstates.

turbidimetry A method of quantitative analysis which involves the spectrophotometric estimation of absorption by a colloidal dispersion of a precipitate. *See* nephelometry.

turbidity indicators The use of, e.g. a weak organic acid which flocculates in the presence of a slight excess of hydrogen ions as an indicator for pH titrations. The applications are restricted but are used in titrations of, e.g. glycine where the change in pH near the equivalence-point is slow.

turbidity point The point in a titration at which a turbidity indicator* shows flocculation.

turkey-red oil A sulphonated castor oil, consisting chiefly of ricinoleosulphuric acid, ricinoleic acid and anhydrides of ricinoleic acid. It is used in the preparation of the cotton fibre for dyeing with turkey red and as an anti-foaming agent.

Turnbull's blue Various blue cyanoferrates*.

turpentine Various types of turpentine are produced; all are light volatile essential oils obtained as exudates or by distillation from coniferous trees. Turpentine is a mixture of cyclic terpene hydrocarbons, chiefly α-pinene; it begins to boil over at 150°C, has *d* about 0·85–0·875. It is used as a thinner for paints and varnishes and as a solvent.

turquoise A blue basic hydrated aluminium phosphate, $Al_2(OH)_3PO_4,H_2O$ containing some Cu. Usually microcrystalline but amorphous in appearance with a waxy lustre; used as a gemstone.

Tutton salts The isomorphous salts $M_2{}^{I}SO_4$, $M^{II}SO_4,6H_2O$ where M^I is an alkali metal and M^{II} is a dipositive transition metal.

twinning A term used in crystallography to denote a set of crystals that have one or more planes or faces in common but whose main developments away from that plane are not those of a single crystal.

two-component system A system consisting of two components; most commonly met in problems involving the distillation of two liquids.

Tyndall effect When a beam of light is passed

through a disperse system, the path of light is visible because of the scattering caused by the particles of the disperse phase. This is the Tyndall effect, which is shown both by colloidal sols as well as coarse suspensions. A common example is the illumination of dust particles by sunlight. The light produced by Tyndall scattering is polarized. The effect is greatest for light of short wavelengths and the Tyndall cone is thus bluish in colour.

tyramine, 4(β-aminoethyl)phenol, $C_8H_{11}NO$. Colourless crystals, m.p. 164°C. A base found in ergot, and in putrefying animal and vegetable material and certain cheeses where it is formed by bacterial action on tyrosine. It is usually made synthetically and has a weak and prolonged pressor action, caused by its ability to release noradrenaline from sympathetic nerve endings and from the adrenal medulla.

tyrian purple, $C_{16}H_8Br_2N_2O_2$. A purple vat dye of great antiquity. Occurs in the shell fish *Murex brandaris* from which it was once extracted for making royal purple.

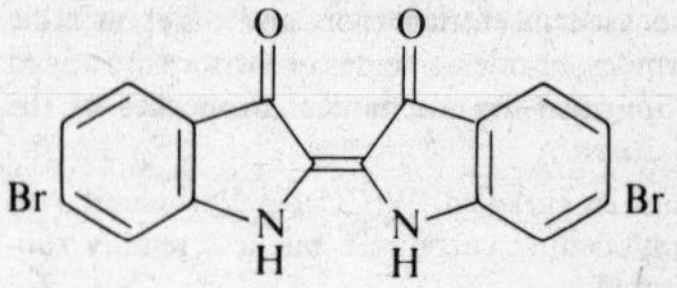

tyrosinase An enzyme of the oxidase class: it is responsible for the darkening of the cut surface of potatoes and other plant tissues. Tyrosine is first converted to a red compound and then to black melanin.

tyrosine, p-hydroxyphenylalanine, 2-amino-3(4-hydroxyphenyl)-propanoic acid, $C_9H_{11}NO_3$. Colourless crystals; m.p. 314–318°C. The naturally occurring substance is laevorotatory. Tyrosine is the least soluble amino-acid obtained on the hydrolysis of proteins, and it is thus the easiest to prepare from protein hydrolysates.

U

U Uranium.

ubiquinones, coenzymes Q A family of

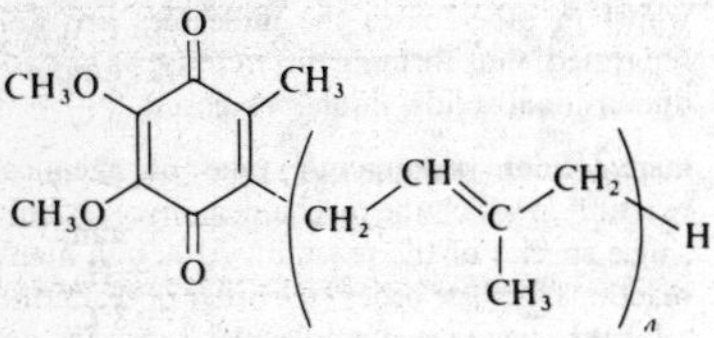

naturally occurring tetrasubstituted 1,4-benzoquinones, the members differing only in the length of the 3-polyisoprenoid chain. In the typical ubiquinone of animal tissues there are 10 units (C_{50}: 'Q-10'); yeasts produce Q-6 and Q-7 and some bacteria produce Q-8. Ubiquinones play an important rôle in the electron transport chain*.

ulexite, $NaCaB_5O_9,8H_2O$. An important source of borates*.

Ullman reaction The synthesis of diaryls by the condensation of aromatic halides with themselves or other aromatic halides, with the concomitant removal of halogens by a metal, e.g. copper powder: thus bromobenzene gives diphenyl. The reaction may be extended to the preparation of diaryl ethers and diaryl thioethers by coupling a metal phenolate with an aryl halide.

ultimate analysis A determination of the proportion of every element in a compound without regard to its molecular structure.

ultracentrifuge A device used to bring about the sedimentation of colloidal sols by subjecting them to large forces. An ultracentrifuge may be used to determine the molecular weights and size distribution of sol particles.

ultrafiltration A method of filtration in which particles of colloidal dimensions are separated from molecular and ionic substances by drawing the sol liquid through a membrane whose capillaries are very small. Membranes of collodion supported on filter paper are often used and these may be formed with different porosities so that particles of different maximum sizes may be retained. The process may also be used to cause the sedimentation of large organic molecules in aqueous solution and of heavy inorganic ions such as caesium.

The mechanism of ultrafiltration is not simply a sieve effect, but depends also upon the electrical conditions of both the membrane and the colloid.

ultramarine An aluminosilicate with an $(Al,Si)O_2$ framework containing sulphur some of which is present as S_2 groups. It has a blue colour and is manufactured as a pigment from china clay, sulphur, Na_2CO_3, SiO_2 and some minor constituents.

ultramicroscope A device used to examine particles having colloidal dimensions (5–10 mμ). It consists of a microscope which is used to observe the Tyndall effect produced when a beam of light is brought to a focus in a colloidal sol. The particles themselves are not visible, only the halo effect produced by the light scattered by the particles. This effect gives some idea of the shape and the movement of the sol particles.

ultrasonics Ultrasonic vibrations have frequencies of the order of 2×10^5 cycles per second; they may be set up by applying an alternating current to quartz, tourmaline or rochelle salt, or by the effect of oscillating magnetism on a rod of magnetic material immersed in a liquid. They can be used to bring about the liquefaction of gels, the depolymerization of macromolecules and to bleach solutions of coloured dyestuffs. With their help it is possible to make an emulsion of mercury and water, or to sterilize milk. They are also used in descaling metals, during soldering and for crack detection.

Ultrasonic absorption is used in the investigation of fast reactions in solution. If a system is at equilibrium and the equilibrium is disturbed in a very short time (of the order of 10^{-7} seconds) then it takes a finite time for the system to recover its equilibrium condition. This is called a relaxation process. When a system in solution is caused to relax using ultrasonics, the relaxation time of the equilibrium can be related to the attenuation of the sound wave. Relaxation times of 10^{-4} to 10^{-9} seconds have been measured using this method and the rates of formation of many mono-, di- and tripositive metal complexes with a range of anions have been determined.

ultra-violet absorbers, light stabilizers Compounds which absorb u.v. light and dissipate energy harmlessly. Used to protect polymers against photolytic degradation and also in sun-tan lotions. Most light stabilizers are aromatic compounds with the aromatic ring conjugated with a C=O or N system.

ultra-violet light, u.v. Radiation which occurs beyond the visible violet light in the spectrum, i.e. light of wavelength less than about 3600 Å, is termed ultra-violet light. Radiation of yet shorter wavelength is called X-rays. Ultra-violet light possesses much greater energy than visible radiation, and is generally much more effective in inducing photochemical reactions, but possesses much less penetrating power. Ordinary glass is opaque to light of wavelength less than about 3600 Å. Quartz is transparent down to about 1800 Å, and is therefore used for making prisms and lenses in ultra-violet optical apparatus.

umbellic acid, 2,4-dihydroxycinnamic acid, $C_9H_8O_4$. A yellow powder, darkening at 240°C and decomposing at 260°C. It occurs in asafoetida as an ester with its anhydride, umbelliferone.

umbelliferone, 7-hydroxycoumarin, $C_9H_6O_3$. Crystals; m.p. 223–224°C. It occurs in many plants.

umbellulone, $C_{10}H_{14}O$. B.p. 93°C/10 mm.

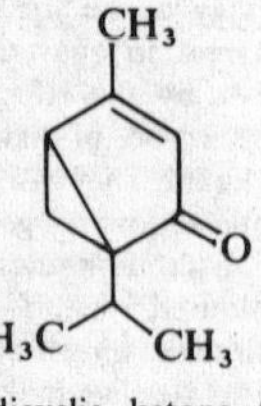

Unsaturated dicyclic ketone found in the essential oil from the leaves of *Umbellularia californica*, with pungent mint-like smell.

umber Natural $Fe(OH)_3$ often containing MnO_2. A brown powder used as a pigment.

uncertainty principle *See* Heisenburg uncertainty principle.

undecane *See* hendecane.

undecenoic acid, hendecenoic acid, Δ^{10}-undecylenic acid, $CH_2{:}CH[CH_2]_8COOH$. Pale yellow liquid, m.p. 20–24°C. Prepared by the vacuum distillation of castor oil. It is a fungicide used in treating tinea pedis.

uniaxial Applied to crystals which have one principal axis, the other two axes being equivalent. Hexagonal, tetragonal and rhombohedral crystals are uniaxial.

unimolecular films When certain insoluble oils and fats (e.g. stearic acid) are dispersed on the surface of water it is possible to prepare these as unimolecular monolayers. In these unimolecular films the hydrophobic group is orientated away from the water; the hydrophilic group is orientated towards it. In some cases the molecules form close-packed arrays, whilst in other cases the molecules are well separated, thus forming different types of two-dimensional solids, liquids or gases.

unimolecular reaction A type of chemical reaction in which the mechanism involves only single species of the reactant. Although many reactions are first order (*see* order of reaction), relatively few true unimolecular reactions are known. The radioactive decay process is an example as is the thermal decomposition of certain alkyl halides.

unit cell The characteristic portion of the crystal which, if repeated indefinitely in the directions of the crystal axes, reproduces the crystal. The portion of the crystal is contained within a parallelopiped called the unit cell, whose sides are parallel to the crystal axes.

universal indicator *See* colour indicators.

unsaturated compound *See* saturated compound.

unsaturated polyesters, polyester alkyds Low mol.wt. polymers containing carboxylic ester groupings and double bonds derived from unsaturated diols and saturated acids. Used as copolymers with styrene and other vinyl monomers in free radical polymerization to give very strong cross-linked structures.

uracil, 2,6-dioxytetrahydropyrimidine,

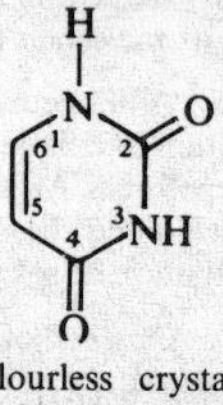

$C_4H_4N_2O_2$. Colourless crystalline powder, turning brown at 280°C and melting at 338°C (decomp.). Uracil is a constituent of ribose nucleic acid. Used as a diuretic and derivatives have pharmaceutical importance. 5-Fluorouracil is used in cancer treatment.

uramil *See* violuric acid.

uranates *See* uranium oxides.

uranium, U. At.no. 92, at.wt. 238·029, m.p. 1132·3°C, b.p. 3818°C, *d* 19·07. Uranium is widely spread in nature (4 p.p.m. in earth's crust) and important ores are pitchblende (U_3O_8), uraninite ($UO_2 + X$), carnotite ($KUO_2VO_4,1{\cdot}5H_2O$), coffinite (a silicate) and autunite and torbernite (hydrated double phosphates). Extraction is by flotation, acid extraction under oxidizing conditions, ion exchange, precipitation as a hydrated oxide and purification by solvent extraction. The metal is best prepared by reduction of UF_4 with Mg at 700°C; low temperatures form orthorhombic, 665–770°C to m.p. bcc. The metal is rapidly tarnished in air and attacked by hot water or acids. Both the fissile isotope ^{235}U and the more common isotope ^{238}U (half-life $4{\cdot}51 \times 10^9$ years) are of importance in nuclear reactions and in the formation of the actinide elements. Fissile ^{235}U is used in nuclear weapons. The isotopes are separated commercially by gaseous diffusion of UF_6 although other processes are now under development. Uranium is used as an additive in steels; uranium carbide is a catalyst in the formation of NH_3. World production U_3O_8 1976 22 300 tonnes.

uranium chemistry Uranium compounds occur in the oxidation states +3 to +6

$$UO_2^{2+}\text{ (yellow)} \xrightarrow{+0{\cdot}063} UO_2^{+}\text{ (blue)} \xrightarrow{+0{\cdot}62}$$

$$U^{4+}\text{ (green)} \xrightarrow{-0{\cdot}631} U^{3+}\text{ (red–brown)} \xrightarrow{-1{\cdot}80} U$$

(volts in 1M $HClO_4$)

The ultimate product of oxidation in moist air is the linear UO_2^{2+} species which is present in many uranium(VI) compounds. The UO_2^{+} ion disproportionates in aqueous solution although U(V) can be stabilized by, e.g., fluoride and alkoxide groups (UF_5 and $U(OR)_5$). U(III) is rapidly oxidized by air or water. Many uranium compounds show high co-ordination numbers, e.g. UX_4(8), UO_2Cl_2(7). Organometallic derivatives include $U(C_5H_5)_4$ and $U(C_5H_5)_3X$ (X = halides, OR, alkyls and aryls) ($UCl_4 + KC_5H_5$) and uranocene, $U(C_8H_8)_2$, ($K_2C_8H_8 + UCl_4$) which has uranium sandwiched between two planar, parallel, cyclo-octatetraenyl rings.

uranium halides The +6 halides are UF_6 ($U + F_2$, $UF_4 + O_2$, m.p. 64°C, sublimes 57°C) and UCl_6 ($UF_6 + AlCl_3$) both of which are immediately hydrolysed by water to, e.g. UOF_4 and UO_2F_2. Other uranyl halides, UO_2X_2, are formed directly (e.g. $UCl_4 + O_2$/350°C gives UO_2Cl_2) and O-bonded complexes, including hydrates are readily formed and complex fluorides are also known. The +5 halides are UF_5, UCl_5 and UBr_5 (e.g. $UF_6 + HBr$ gives UF_5) and the UF_6^- ion is stable in aqueous HF. All of the tetrahalides, UX_4, are known ($UO_2 + HF$/500°C gives UF_4; $U_3O_8 + C_3Cl_6$) and all form O- and N-bonded complexes and a wide range of complex fluorides is known. The +3 halides UF_3, UCl_3 and UBr_3 are formed by H_2 reduction of UX_4; hydrates, ammines and other N-bonded complexes are known.

uranium oxides Uranium forms an extensive series of oxides including UO (semi-metallic), UO_2, U_3O_8 (the stable oxide in air), U_4O_9, $U_{2{\cdot}25}$–$U_{2{\cdot}40}$ and UO_3. The hydrated peroxide $UO_4,2H_2O$ is precipitated from an aqueous solution of a U(VI) salt by H_2O and on heating gives UO_3 which may be reduced with hydrogen. Complex insoluble uranates(VI), generally polymeric mixed oxides, are precipitated from solutions of uranium(VI) salts or are formed by solid state reactions, many uranates(VI) contain linear UO_2 groupings. U(V) and U(IV) mixed metal oxides are formed by solid state reactions (e.g. $BaUO_3 + UO_3$/550°C gives $Ba(UO_3)_2$). So-called sodium di- or pyro-uranate, uranium yellow, $Na_2U_2O_7,6H_2O$, is an insoluble sodium salt used in glass.

uranocene *See* uranium chemistry.

uranyl derivatives Uranium(VI) species containing the (linear) UO_2 grouping. Amongst commercially available compounds are the ethanoate, hydrogen phosphate, magnesium ethanoate, nitrate, perchlorate, sulphate and zinc ethanoate. Many compounds are hydrated, show high co-ordination numbers with co-ordinated anions, and water may be replaced with other N-and O-donors. Zinc uranyl ethanoate forms a precipitate from water with Na^+ ions. Most uranyl derivatives have a strong yellow colour and show a bright green fluorescence when irradiated.

urao *See* trona.

urea, carbamide, CH_4N_2O, $(H_2N)_2C = O$. Colourless crystals, m.p. 132°C. It is a weak base, and forms salts with strong acids. Urea was first obtained synthetically by Wöhler in 1828 by evaporating a solution of ammonium cyanate. It can be made by the addition of water to cyanamide, by passing water into fused phenylcyanamide, and by numerous other methods. Commercially it is made by reacting carbon dioxide with ammonia at

200°C and pressures up to 400 atm., the ammonium carbamate formed initially largely dehydrating to urea.

Urea occurs in the urine of all mammals and in small quantities in the blood of mammals and fish (*see* urea cycle).

Urea is largely used as a fertilizer (75%), and as a non-protein feed supplement for sheep and cattle. The most important chemical use, which however accounts for only a small part of urea production, is in the manufacture of urea-formaldehyde resins. It is also used in the manufacture of adhesives, pharmaceuticals, dyes and various other materials. U.S. production 1981 7·0 megatonnes; urea resins 1983 6 megatonnes.

urea adduction The basis of processes for the separation of normal alkanes, especially those from petroleum distillates in the C_{25}–C_{30} range. In the presence of *n*-alkanes urea crystallizes in a form which allows only the normal alkanes to be clathrated. The clathrate formed is decomposed into two phases by heating with water at 80°C.

urea cycle The cyclic sequence of reactions occurring in the liver, which results in the excretion of excess nitrogen as urea. Carbamyl phosphate is first formed from CO_2, ammonia and ATP. This is also the first step in the biosynthesis of pyrimidines. Carbamyl phosphate then condenses with ornithine to yield citrulline which is converted to arginine via arginosuccinate. Arginase* then cleaves arginine to give back ornithine, which can condense again with another molecule of carbamyl phosphate, and also gives a molecule of urea which is excreted in the urine.

urea-formaldehyde resins Typical aminoresins* used in adhesives, foams and moulding. *See* urea.

urease The enzyme which catalyses the hydrolysis of urea to ammonia and CO_2. It is present in many plants but not in vertebrates.

ureides Compounds formed between urea and organic acids. Interaction can be with either or both nitrogens and dicarboxylic acids or α-hydroxy acids give cyclic ureides, e.g. barbituric acid, hydantoin. Colourless crystalline solids, sparingly soluble in water, soluble in ethanol, decomposed by alkalis, form salts.

urethane, ethyl carbamate, $NH_2{\cdot}COOC_2H_5$. Colourless prisms; m.p. 49–50°C, b.p. 184°C. Prepared by the action of ammonia on ethyl chloroformate. Used as a long-acting anaesthetic for laboratory animals.

urethanes Esters of carbamic acid, $NH_2{\cdot}COOH$. Colourless crystalline solids; they distil without decomposition. They are prepared by the action of ammonia on carbonic or chlorocarbonic esters, or by treating alcohols with urea nitrate or isocyanates. When heated with ammonia they form urea and an alcohol. Phenyl- and α-naphthylurethanes are used for characterizing alcohols and phenols. The term is also used for polyurethanes, major plastics used in foams (85%) and elastomers. World production 1979 3·0 megatonnes.

uric acid, 2,6,8-trihydroxypurine, $C_5H_4N_4O_3$. A colourless microcrystalline powder, odourless and tasteless, which decomposes on heating above 250°C. It is dibasic and forms two series of salts. It occurs in small quantities in human urine, being excreted as the end product of nucleic acid metabolism. In the excrement of birds and reptiles it is present as ammonium urate and is the chief nitrogenous substance. It is prepared commercially from guano by extracting with alkali and reprecipitating with acid. It can be synthesized by fusing glycine with urea, and by other processes.

uricase An enzyme in liver and kidney which oxidizes uric acid to allantoin and CO_2.

uridine *See* nucleosides.

uronic acids Formed by oxidation of the primary alcohol groups of sugars, and named after the sugar from which they are formed, thus glucuronic acid from glucose. Only D-glucuronic, D-mannuronic and D-galacturonic acids are important. Polyuronides are constituents of the gums and mucilages; alginic acid is a polymer of mannuronic acid and pectic acid of galacturonic acid.

u.v. Ultra-violet.

V

V Vanadium.

vaccenic acid, trans-11-octadecenoic acid, $CH_3{\cdot}[CH_2]_5{\cdot}CH{:}CH{\cdot}[CH_2]_9{\cdot}CO_2H$. A white solid, m.p. 43–44°C, which is present in small quantities in animal fats and in milk. It is the only naturally occurring fatty acid with the *trans* configuration.

vacuum crystallizer *See* crystallizers.

vacuum pump A pump designed to extract gas from a space where the pressure is considerably below atmospheric and discharge it to atmosphere. Mechanical pumps used for this purpose are normally of the rotary, positive displacement type and can provide pressures down to about $1{\cdot}3\ N/m^2$. In the chemical industry steam jet ejectors* are widely used, especially with distillation columns and evaporators, pressures of 0·13 bar to $6{\cdot}5 \times 10^{-4}$ bar being attained, depending on the number of stages. For very high vacua, as in molecular distillation, diffusion pumps are employed.

valence isomerization The isomerization of

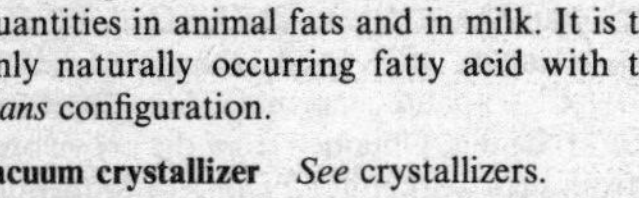

molecules which involve structural changes resulting only from relocation of single and double bonds. If a dynamic equilibrium is established between the two isomers it is also referred to as valence tautomerism, e.g. for the cyclo-octa-1,3,5-triene system.

valence shell electron pair repulsion theory, VSEPR theory Theory for the prediction of molecular shape – *see* valency, theory of.

valency A term formerly used to denote oxidation state*. Frequently used to designate the number of bonded neighbour atoms, *see* co-ordination number.

valency electrons The outer electrons of an atom which determine the chemical reactivity of an element, the compounds formed by that element and the geometry of those compounds. Generally the electrons which are additional to those of the preceding noble gas.

valency, theory of The oxidation states, co-ordination number and stereochemistry of any atom are determined by the outermost or valency electrons of the atom. The elements in which the outer s or p shells are full are relatively inert chemically and simple valency theory is based on the concept that elements tend to gain, lose or share electrons in order to complete their outer electron shells. The elements having these shells filled are the noble gases and other elements tend to approach the noble gas structures.

For the transition metals it is often impossible to reach a noble gas structure except in covalent compounds (*see* effective atomic number rule) and it is found that relative stability is given by having the sub-shells (d or f) filled, half-filled or empty.

The *electrovalent bond* is formed by electrostatic attraction between oppositely charged ions. Thus Na, with one outer electron, loses this electron to achieve the noble gas Ne structure, while Cl with seven outer electrons, gains one electron to achieve the Ar structure.

$$Na(1s^2 2s^2 2p^6 3s^1) \longrightarrow Na^+(1s^2 2s^2 2p^6) + e$$
$$Cl(1s^2 2s^2 2p^6 3s^2 3p^5) + e \longrightarrow Cl^-(1s^2 2s^2 2p^6 3s^2 3p^6)$$

Crystalline sodium chloride consists of Na^+ and Cl^- ions.

Covalent bonds are formed by the sharing of electrons. Thus the carbon atom, with four equivalent electrons shares with the electrons from four hydrogen atoms.

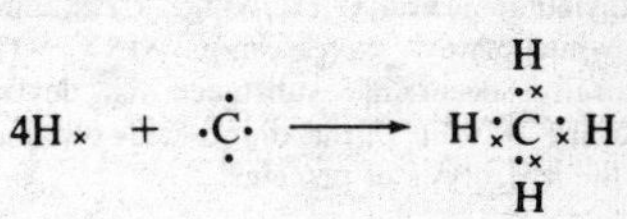

The carbon atom has a share in eight electrons (Ne structure) whilst each hydrogen atom has a share in two electrons (He structure). This is a gross simplification of covalent bonding, since the actual electrons are present in molecular orbitals which occupy the whole space around the five atoms of the molecule.

Co-ordinate bonds are formed by the sharing of electrons, both electrons being donated by the same atom. Thus the hydrogen ion, H^+, has no outer electrons whilst ammonia has eight, six shared with hydrogen atoms and one lone-pair. This lone-pair is donated to the hydrogen ion and the ammonium ion is formed:

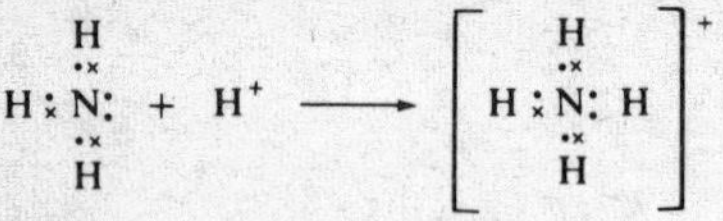

The N retains its share in eight electrons (Ne structure) and each hydrogen atom has a share in two electrons (He structure). Each hydrogen atom is equivalent.

The shapes of covalent compounds are determined by the tendency for bonding pairs to be as far apart as possible whilst lone pairs have a greater effect than bonding pairs (VSEPR theory).

The co-ordination number in ionic compounds is determined by the radius ratio – a measure of the necessity to minimize cationic contacts. More subtle effects are the Jahn–Teller effect* (distortions due to incomplete occupancy of degenerate orbitals) and metal–metal bonding.

valeric acids, valerianic acids, pentanoic acids, $C_5H_{10}O_2$. *n-Valeric acid,* $CH_3{\cdot}CH_2{\cdot}CH_2{\cdot}CH_2{\cdot}COOH$, is a colourless liquid with an unpleasant odour, b.p. 186°C. Prepared from *n*-butyl-iodide by a Grignard reaction or by oxidation of *n*-amyl alcohol.

Isovaleric acid, Me_2CHCH_2COOH, is a colourless liquid with the unpleasant odour of valerian, b.p. 177°C. Occurs in the roots of valerian and angelica together with an optically active form of methylethylethanoic acid. Prepared by oxidation of isoamyl alcohol. A mixture of acids similar to that obtained from valerian roots is prepared by oxidation of fusel oil.

valine, α-aminoisovaleric acid, 2-amino-3-methylbutanoic acid, $C_5H_{11}NO_2$. Crystallizes in white shiny leaves; m.p. 315°C. The naturally occurring substance is dextrorotatory. It is one of the amino-acids obtained on the hydrolysis of proteins.

valium Diazepam*, used as a sedative.

valve tray *See* plate, tray.

vanadates Anionic derivatives of vanadium. Practically vanadates(V) although vanadates-(IV), mixed metal oxides, are also known. Free vanadic acids are not known. *Vanadates(V)* are oxy-salts of vanadium(V). Vanadium pentoxide, V_2O_5, dissolves in strong bases to give hydroxy species – e.g. $[VO_3(OH)]^{2-}$, $[VO_2(OH)_2]^-$ – which are colourless. On acidification these give orange and then yellow species, e.g. $[HV_{10}O_{28}]^{5-}$. Orthovanadates, e.g. $Na_3VO_4{,}12H_2O$, $K_3VO_4{,}6H_2O$ can be crystallized from alkaline solution and contain tetrahedral $[VO_4]$ units. The polymeric species of KVO_3 contain chains of tetrahedra. The $[V_{10}O_{28}]^{6-}$ unit from acid solution (decavanadate) is an isopolyanion with VO_6 units sharing edges. Solid state mixed metal oxides derived from V_2O_5 are also known.

vanadic acids *See* vanadates.

vanadium, V. At.no. 23, at.wt. 50·9415, m.p. 1890°C, b.p. 3380°C, *d* 5·96. Group V transition element occurring naturally as patronite (V_2S_5 plus S), carnotite ($KUO_2VO_4{,}1{\cdot}5H_2O$ – also uranium ore), and vanadinite ($Pb_5(VO_4)_3Cl$ – apatite structure). Also occurs in some crude oils. Obtained from the above ores or from ores worked for Fe or Ti production. The ore is converted to vanadate(V) by roasting with NaCl, leached and acidified to V_2O_5 which is purified *via* NH_4VO_3. V_2O_5 is reduced with Ca in a $CaCl_2$ flux and refined by electrolysis of fused $NaCl-LiCl-VCl_2$ or by a van Arkel process. Often used as ferrovanadium (Al reduction in the presence of Fe). The metal is bcc; soft grey or silver. It burns on heating in O_2 or Cl_2 and dissolves in HNO_3 and slowly in hot conc. H_2SO_4 and fused alkalis. Used mainly as a steel additive (80%) and in V–Al alloys (10%). Used in catalysts. World demand 1976 31 000 tonnes.

vanadium bromides
Vanadium(II) bromide, VBr_2. Reddish brown crystals formed from VBr_3 and H_2.

Vanadium(III) bromide, VBr_3. Dark green or black (V plus Br_2) gives green solution in water and green crystalline $VBr_3{,}6H_2O$. Forms many complexes.

Vanadium oxide tribromide, $VOBr_3$. Dark red deliquescent liquid formed by heating V_2O_3 plus Br_2.

Vanadium oxide dibromide, $VOBr_2$. Yellow powder obtained by heating $VOBr_3$ or Br_2 plus S_2Br_2 over heated V_2O_5.

vanadium carbides VC (NaCl structure) and V_4C_3 are formed from the elements and are important in the constitution of vanadium steels.

vanadium chemistry Vanadium is Group V transition element electronic configuration $3d^34s^2$. It shows oxidation states -1 to $+5$. The $+5$ state is largely covalent or in complexes; it is colourless or yellow–red in oxides, etc. Vanadium(IV) is blue in aqueous solution; its chemistry is part ionic and there is a very stable $[VO]^{2+}$ group particularly in square pyramidal co-ordination. Vanadium(III) is green in aqueous solution, predominantly ionic, octahedral or tetrahedral. Vanadium(II) is violet in aqueous solution, ionic but strongly reducing. Vanadium (1), (0) and (−1), e.g. $[V\,dipy_3]^+$, $[V(CO)_6]$, $[V(CO)_6]^-$ are largely covalent. All states readily form complexes. Redox potentials are $[VO_3]^-$ to $[VO]^{2+}$ +1·0 volts; $[VO]^{2+}$ to $[V(H_2O)_6]^{3+}$ +0·3 volts; $[V(H_2O)_6]^{3+}$ to $[V(H_2O)_6]^{2+}$ −0·25 volts.

vanadium chlorides
Vanadium(IV) chloride, VCl_4, b.p. 154°C. Reddish brown liquid formed V plus Cl_2. Decomposes slowly to VCl_3 and Cl_2; hydrolysed by water.

Vanadium(III) chloride, VCl_3. Violet solid formed V plus HCl gas. $VCl_3,6H_2O$ formed from solution. Forms many complexes.

Vanadium(II) chloride, VCl_2. Green solid formed VCl_4 plus H_2.

Vanadium oxide trichloride, $VOCl_3$, vanadyl chloride. Readily prepared yellow liquid, b.p. 127°C, formed Cl_2 plus heated V_2O_5 plus C. Readily hydrolysed by water.

Vanadium oxide dichloride, $VOCl_2$. Green crystals $VOCl_3$ plus H_2.

vanadium fluorides
Vanadium pentafluoride, VF_5, m.p. 19·5°C, b.p. 48°C. White solid immediately hydrolysed by water via VOF_3 and VO_2F. VF_5 prepared V plus F_2. Forms hexafluorovanadates(V), MVF_6, most easily in BrF_3.

Vanadium tetrafluoride, VF_4. Green solid prepared HF on VCl_4. Forms hexafluorovanadate(IV) ion $(VF_6)^{2-}$. VOF_2 formed by hydrolysis.

Vanadium trifluoride, VF_3. Yellow green (VCl_3 plus HF at 600°C). Forms $(VF_6)^{3-}$ ion and $VF_3,6H_2O$ from aqueous HF.

Vanadium difluoride, VF_2. Blue solid formed H_2 plus HF on VF_3 or HF on VCl_2 at 600°C.

vanadium oxides
Vanadium pentoxide, V_2O_5. Orange–yellow oxide formed by heating NH_4VO_3. Only sparingly soluble in water; dissolves alkalis to give vanadates(V)*. V_2O_5 dissolves in acids to give vanadyl salts. V_2O_5 is used in catalyst systems for the contact process (sulphuric acid) and some other reactions. Reduction of V_2O_5 gives lower oxides.

Vanadium dioxide, VO_2 is dark blue (V_2O_5 plus SO_2) but is readily reduced further to $V_{0\cdot186}$–$V_{1\cdot66}$. VO_2 gives the $(VO)^{2+}$ ion with acids and vanadates(IV) with alkalis and as mixed metal oxides.

Vanadium trioxide, V_2O_3. Black powder (V_2O_5 plus H_2 under heat). Readily reoxidized to V_2O_5. Stable down to $VO_{1\cdot35}$.

Vanadium monoxide, VO. Formed V plus V_2O_3, stable over composition range $VO_{0\cdot85}$ to $VO_{1\cdot15}$.

vanadium sulphates The only important oxyacid salts.
Vanadium(IV) forms blue $VOSO_4$, (0, 3 and $5H_2O$), vanadyl sulphate, which forms ranges of double salts. Prepared by SO_2 reduction of V_2O_5 in H_2SO_4.

Vanadium(III) forms green $V_2(SO_4)_3$ (electrolytic or Mg reduction of V_2O_5 in H_2SO_4). Alums and other double salts are formed.

Vanadium(II) forms violet $VSO_4,7H_2O$ (electrolytic or Na/Hg reduction). Forms double salts.

vanadyl species Compounds containing the blue VO^{2+} group, e.g. $VOSO_4$, $VOCl_2$.

van Arkel–de Boer process Process for purification of metals, e.g. Ti, where TiI_4 is vaporized and decomposed on a hot wire.

van der Waals' adsorption An alternative name for the physical adsorption of species upon an adsorbent in which the adsorbate–adsorbent bonds are van der Waals' forces. *See* adsorption, physical.

van der Waals' equation Because the molecules in a gas have a definite volume and exert forces of attraction on each other, real gases do not obey the ideal gas laws to a high degree of accuracy. The effects of real gases were taken into account by van der Waals in his equation of state:

$$\left(P+n^2\frac{a}{V^2}\right)(V-nb)=nRT$$

where a and b are constants characteristic of the gas under consideration. The term (a/V^2) allows for the reduction of pressure from the ideal value due to intermolecular attractive forces; b is the 'co-volume' and allows for the volume occupied by the molecules, it has a value of about four times the actual volume of the molecules.

van der Waals' forces The weak forces between molecules arising from weak electronic coupling. Act only over a very short distance.

vanillin, 4-hydroxy-3-methoxybenzaldehyde, $C_8H_8O_3$. Fine white needles, m.p. 82°C, b.p. 285°C, strong vanilla odour, characteristic taste. It occurs extensively in nature, and is the odoriferous principle of the vanilla pod; it can be obtained from the glucoside coniferin. Vanillin is made commercially from the lignosulphonic acid obtained as a by-product in the manufacture of wood pulp. It is one of the most important flavouring and perfuming materials, and large quantities are used in the manufacture of foodstuffs, toilet goods and pharmaceuticals.

van't Hoff isochore The name given to the relation

$$\frac{d \ln K}{dT} = \frac{\Delta H}{RT^2}$$

where K is the equilibrium constant of a reversible reaction, ΔH the enthalpy of reaction, T the absolute temperature and R the gas constant. This equation permits the calculation of enthalpies of reaction by measuring the variation of K with temperature.

van't Hoff isotherm In a reversible chemical reaction of the type

$$aA + bB \rightleftharpoons cC + dD$$

the free energy change ΔG is given by the equation

$$\Delta G = -RT \ln K + RT \ln \left[\frac{[C]^c[D]^d}{[A]^a[B]^b}\right]$$

where R is the gas constant, T the absolute temperature, K the equilibrium constant, [A], [B], etc., the activities of A, B, etc., under the reaction conditions.

vapour density For a given substance the ratio of the weight of a given volume of its vapour to that of the same volume of hydrogen, measured under identical conditions of temperature and pressure, is called the vapour density of the substance.

vapour galvanizing *See* sherardizing.

vapour lock The restriction of supply of volatile fluids being pumped through pipe systems due to the formation of bubbles of vapour at critical points in the system. This can be caused by high temperature or low pressure conditions.

vapour pressure Over every liquid or solid there is a certain pressure of its vapour. In a closed vessel after sufficient time, equilibrium is established, so that as many molecules leave the liquid surface to form vapour as return from the vapour to the liquid. At equilibrium at the specified temperature the pressure exerted by the liquid or solid is termed the vapour pressure at that particular temperature.

Vaseline A trade name for soft paraffin. Yellow and white semi-solid, partly translucent mixtures of hydrocarbons of the paraffin series ranging from $C_{15}H_{32}$ to $C_{20}H_{42}$. Obtained from the high-boiling fractions of petroleum or shale oil. They are chemically very inert, and are used as bases for ointments and as dressings for wounds. *See* petrolatum.

Vaska's compound, *trans*-$[Ir(CO)(Ph_3P)_2Cl]$. Undergoes ready oxidative addition to give Ir(III) complexes.

vasopressin A cyclic peptide hormone secreted by the posterior lobe of the pituitary gland. Injection of vasopressin leads to a prolonged rise in blood pressure. Physiologically it is extremely important as an antidiuretic.

vat dyes Water-insoluble dyestuffs that become water-soluble, and in a reactive state towards fibres, on reduction in alkaline solution. On reoxidation, generally with air, the insoluble dyestuff is precipitated within the fibre. Sodium dithionite, $Na_2S_2O_4$, is generally used as reducing agent and C=O groups are reduced to colourless C–OH or C–ONa leuco compounds. Vat dyes include anthraquinone and indigoid types.

Végard's law In phases in which there is a range of composition of solid solution then the cell dimensions vary linearly with the molar proportions of the constituents. The law is rarely followed exactly.

velocity of reaction The velocity of a chemical reaction is the amount of reactants transformed, or of product produced, per unit time. Thus velocity is usually expressed in gram moles, or gram atoms, per second.

veratraldehyde, 3,4-dimethoxybenzaldehyde, $C_9H_{10}O_3$. A compound of largely historical interest for its rôle in establishing the structure of many natural products. Methylation of vanillin gives veratraldehyde which may be oxidized to veratric acid. Veratric acid was identified as a degradation product of the alkaloid papaverine.

veratric acid *See* veratraldehyde.

veratrole The trivial name for 1,2-dimethoxybenzene. Veratryl compounds are 3,4-dimethoxybenzene derivatives, e.g. veratric acid is 3,4-dimethoxybenzoic acid. *See also* veratraldehyde.

verbenone, $C_{10}H_{14}O$. An unsaturated ketone

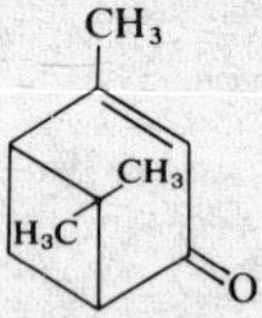

found in Spanish verbena oil in association with citrals -*a* and -*b*. Both (+)- and (−)-verbenones are produced by the auto-oxidation of (+)- and (−)-pinenes respectively. It is a colourless oil and smells both of camphor and peppermint. The (+)-form has m.p. 10°C, b.p. 228°C. The (−)-form has b.p. 253°C.

Verdet's constant *See* magnetic polarization of light.

verdigris The green compound to which copper or bronze is converted on exposure to the atmosphere. It is usually a basic copper carbonate, but near the sea will be a basic chloride. The term is also used for basic copper ethanoate which is used as a pigment.

veridian, Cr_2O_3. *See* chromium(III) oxide.

vermiculite A hydrated silicate clay*. Used as a thermal insulator and in agriculture as a rooting medium and soil additive. World demand 1976 475 000 tonnes.

vermilion red, HgS. Occurs naturally as cinnabar. The red pigment in Chinese lacquer painting, red ink and temple paint are due to vermilion. Eventually Pb_3O_4 replaced HgS as red pigment.

vetivone, $C_{15}H_{22}O$. Two stereoisomeric

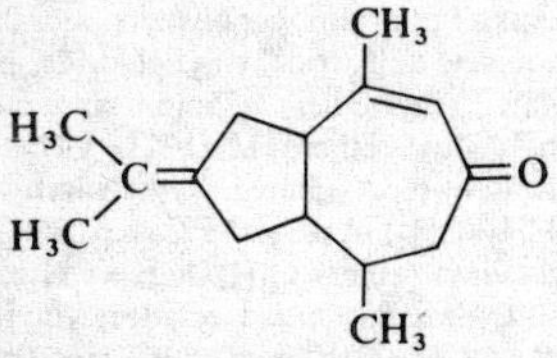

ketones have been isolated from oil of vetiver. The α-isomer has m.p. 51°C. The β-isomer has m.p. 44°C. Both isomers on heating with Se give vetivazulene, $C_{15}H_{18}$.

Vetrocoke process *See* Catacarb process.

vibrational spectrum When a molecule absorbs energy the vibrational energy of the constituent atoms relative to each other may be increased. Conversely, when a transition occurs from a higher vibrational level to a lower one, it corresponds to the emission of energy, generally as radiation, of frequency ν, where the energy loss E is related to ν by the equation

$$E = h\nu$$

Vibrational transitions, usually associated with simultaneous rotational transitions, occur in the i.r. and near i.r., and give rise to bands characteristic of the vibrational and rotational changes. These bands constitute the vibrational–rotational spectrum.

vibrational–rotational spectrum *See* vibrational spectrum.

vic- The prefix *vic-* (short for vicinal) in the name of an organic compound indicates that substituent atoms or groups are bonded to adjacent carbon atoms. Thus 1,2,3-trichlorobenzene can be called *vic*-trichlorobenzene.

Victor Meyer method for vapour densities Method using the volume occupied by a known weight of vapour (generally measured as expelled air).

Vilsmeier reagent The reagent obtained from $POCl_3$ mixed with either N,N-dimethylformamide or N-methylformanilide. Used for introducing the methanoyl (formyl) (−CHO) group into activated aromatic substrates.

vinegar A dilute (4–10%) solution of ethanoic acid prepared by the oxidation of alcoholic liquors by various species of *Acetobacter* – usually *Acetobacter aceti*. In the U.K. vinegar is obtained from malt, the sugars of which are converted to alcohol by yeast. The alcoholic liquor is then treated with *Acetobacter*. Synthetic ethanoic acid is also used for flavouring. In Europe much vinegar is made from poor quality wine and is known as wine vinegar.

vinyl, ethenyl The $CH_2{=}CH-$ group.

vinyl acetate, ethenyl ethanoate, vinyl ethanoate, $CH_2{:}CHOOC{\cdot}CH_3$. Colourless liquid, b.p. 73°C, with an ethereal odour; its vapour is lachrymatory. Manufactured by the vapour-phase reaction of ethyne and ethanoic acid in the presence of a zinc ethanoate catalyst at 200°C or by a solution phase reaction of ethene and ethanoic acid in the presence of a palladium(II) chloride catalyst. Purified vinyl acetate absorbs free oxygen, reacting to give ethanal and the free acid. In the absence of catalysts, vinyl acetate shows little tendency to polymerize, but conversion to polyvinyl acetate occurs with the free radical catalysts. Used in the manufacture of adhesives (30%), paints (20%), coatings (10%), paper and textile finishes (15%), moulding compounds, etc., polyvinyl acetate is of great importance as an intermediate in the manufacture of polyvinyl alcohol and polyvinyl acetals. The copolymers of vinyl acetate, particularly those with vinyl chloride, are also of commercial importance. U.S. production 1981 900 000 tonnes. *See* vinyl ester polymers.

vinyl acetylene *See* monovinylacetylene.

vinyl alcohol, $CH_2{:}CHOH$. The enol form of ethanal. It cannot be isolated in the pure state but vinyl esters and ethers are known and hydrolysis of these gives either ethanal or a polymer of vinyl alcohol. The palladium(II) catalysed hydration of ethyne to ethanal must proceed via vinyl alcohol which tautomerizes. *See* polyvinyl alcohol.

vinylation A term used to describe the reaction between ethyne and compounds such as alcohols, carboxylic acids and amines which contain active hydrogen atoms. Under the influence of catalysts, addition occurs across the triple bond to give vinyl (ethenyl) compounds, e.g.

$$H-C\equiv C-H+HX\longrightarrow CH_2=CHX$$

vinyl chloride, monochloroethylene, chloroethene, $CH_2{:}CHCl$. Colourless gas, b.p. −14°C, with a pleasant ethereal odour, but now known to be carcinogenic. It is manufactured by reacting ethyne with HCl over a catalyst, by the pyrolysis of ethene dichloride or by the hydrolysis of dichloroethene using dilute NaOH.

Vinyl chloride is used almost exclusively for the manufacture of polymers and copolymers. U.S. production 1983 2·6 megatonnes. *See* vinyl chloride polymers.

vinyl chloride polymers Polymers containing $(CH_2-CHCl)_n$ units including polyvinyl chloride, PVC and many copolymers. Vinyl chloride was generally polymerized in an aqueous system containing an emulsifying agent or suspension stabilizer but solution and bulk processes are now in use. PVC resin is compounded with various plasticizers and other additives and pigments are added. It is used in rigid materials (e.g. for construction), flooring, electrical coverings, clothing, furnishing, packaging, toys, luggage. Chlorination gives a material with greater solubility in organic solvents used in lacquers and fibres.

vinyl ester polymers The most important such ester is polyvinylacetate (ethanoate). Vinyl ethanoate* monomer $CH_3CO_2CH=CH_2$ is prepared from ethyne and ethanoic acid or ethene, ethanoic acid and $PdCl_2$, b.p. 73°C. It is polymerized by free radical processes. Polyvinyl acetate is used in adhesives, thickeners, solvents, plasticizers, in textiles, in concrete additives and particularly for the production of polyvinyl alcohol by hydrolysis. Other vinyl esters are produced by transvinylation with the appropriate acid. Polymers derived from vinylproprionate (b.p. 95°C), vinyl caproate (b.p. 166°C), vinyl laurate (b.p. 142°C/10 mm), vinyl stearate (b.p. 187°C/4·3 mm), or vinyl benzoate (b.p. 203°C) have or have had some importance.

vinyl esters *See* vinyl ester polymers, vinyl acetate.

vinyl ethanoate *See* vinyl ester polymers and vinyl acetate.

vinyl ether, diethenyl ether, C_4H_6O, $CH_2{:}CH{\cdot}O{\cdot}CH{:}CH_2$. Colourless inflammable liquid, with a characteristic ethereal odour, b.p. 28–31°C. Prepared by the action of KOH on 2,2′-dichlorodiethyl ether. It is unstable, breaking down to methanal and methanoic acid.

vinyl ether polymers *See* vinyl ethers.

vinyl ethers, $ROCH=CH_2$. Prepared commercially by reaction between alkynes (generally C_2H_2) and alcohols at about 150°C under pressure and in the presence of base; also prepared by thermal cracking of acetals (formed aldehydes plus alcohols plus acid catalysts or alkynes plus alcohols). Hydrolysed in presence of acid to alcohol plus aldehyde. Polymerized to homopolymers by acidic catalysts (e.g. BF_3, $AlCl_3$). The vinyl ether polymers are used as adhesives, in processing, lubricants, fibres, film and in moulding components. Important copolymers are formed with maleic anhydride, vinyl chloride, methyl acrylate. Vinyl ethers of importance include methyl vinyl ether, $MeOCH=CH_2$, m.p. −222°C, b.p. 5·5°C; ethylvinyl ether $EtOCH=CH_2$, m.p. −115°C, b.p. 36°C; isobutyl vinyl ether $C_4H_9OCH=CH_2$, m.p. −132°C, b.p. 83°C; and octadecylvinyl ether $C_{18}H_{37}OCH=CH_2$, m.p. 30°C, b.p. 182°C.

vinyl fluoride, $CH_2=CHF$. M.p. −160°C, b.p. −72°C. Prepared HF plus ethyne or, in the laboratory, $CF_2H{\cdot}CBrH_2$ plus zinc or RMgX or pyrolysis of $CF_2H{\cdot}CH_3$. It is relatively stable to polymerization but Ziegler–Natta catalysts give homopolymers. Used in coatings and laminates particularly where there is out-door exposure. *See* fluorine-containing polymers.

vinylidene chloride, 1,1-dichloroethene, $CH_2{:}CCl_2$. Colourless liquid, b.p. 32°C, manufactured by the dehydrochlorination of trichloroethane. In the presence of light and air, it decomposes with the evolution of HCl, phosgene, and methanal and deposition of some polyvinylidene chloride. Consequently it must be stored away from light and in the presence of dissolved inhibitors (such as phenols and amines). Under the influence of free radical and ionic polymerization catalysts, the monomer can be readily converted to give the industrially important vinylidene chloride polymers*.

vinylidene chloride polymers, Saran polymers Polymeric vinylidene chloride generally produced by free radical polymerization of $CH_2=CCl_2$. Homopolymers and copolymers are used. A thermoplastic used in moulding, coatings and fibres. The polymers have high thermal stability and low permeability to gases, and are self extinguishing.

vinylidene fluoride, $CH_2=CF_2$. M.p. −144°C, b.p. −84°C. Prepared by dehydrohalogenation or dehalogenation of, e.g., CH_3CClF_2 (prepared C_2H_2 plus HF then Cl_2 or $CH_2=CCl_2$ or CH_3CCl_3 plus HF). Vinylidene fluoride is normally stable but can be polymerized with free-radical initiators. The polymer is very stable and is used in electrical applications, as exterior coatings and as elastomeric copolymers with hexafluoropropene, $CClF=CF_2$, etc.

vinylogs Compounds related by the introduction or removal of $-CH=CH-$ units in a chain. Vinylogous functional groups have two parts of a functional group separated by a $-CH=CH-$ unit.

vinyl polymers and resins Polymers derived by polymerization of $CH_2=CHX$. *See* various entries under vinyl and polyvinyl.

vinyl pyridines The important vinyl pyridines are 2-vinylpyridine, b.p. 158°C; 4-vinylpyridine, b.p. 65°C at 15 mm; and 2-methyl-5-vinylpyridine, b.p. 75°C at 15 mm. Prepared by catalytic dehydrogenation or dehydration of ethyl or hydroxyethyl derivatives. Polymerized similarly to styrene by free-radical and anionic mechanisms. Used as copolymers for synthetic rubbers and fibres, and as chemical reagents.

N-vinylpyrrolidone, C_6H_9NO. Prepared

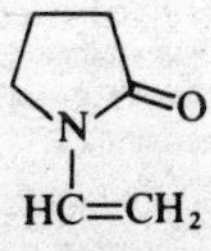

from pyrrolidone and ethyne in the presence of a base, b.p. 96°C at 14 mm, m.p. 13·5°C. Polymerizes in the presence of free radical initiators so that the monomer is generally stored with 0·1% NaOH to prevent hydrolysis and polymerization. Poly(vinylpyrrolidone) and other N-vinyl amides are water soluble. Poly(vinylpyrrolidone) is used in adhesives, detergents, pharmaceuticals, toiletries and in the textile industry for fibre treatment. It has been used in blood plasma.

violaxanthin, $C_{40}H_{56}O_4$. A carotenoid pigment, esters of which are present in viola and other blossoms. Reddish-brown spears, m.p. 20°C.

violuric acid, 5-isonitrosobarbituric acid, alloxan-5-oxine, $C_4H_3N_3O_4$. Pale yellow crystalline substance; m.p. 203–204°C (decomp.). Prepared by treating barbituric acid with a solution of sodium nitrite or by the action of hydroxylamine on alloxan. It dissolves in water to give a violet solution; forms intensely coloured salts with metals. Reduced to uramil which has NH_2 for NOH.

virial equation For an ideal gas the compressibility factor, PV/RT, is unity. However, for real gases the compressibility factor is a function of temperature and volume which may be expressed in terms of a virial equation of the form

$$\frac{PV}{RT}=1+\frac{B}{V}+\frac{C}{V^2}+\cdots$$

where B and C are functions of temperature, referred to as the second and third virial coefficients respectively.

visco-elasticity Many plastics and other long-chain macro-molecular compounds show rheological properties intermediate between those of an elastic solid and a Newtonian liquid. In rubber-like plastics, the application of a stress immediately produces a large strain ('high elasticity') which is recoverable only if its duration is very short. If the stress is maintained, the material acquires a permanent strain. It is as though the chain molecules, at first simply extended, slowly flow past one another as in a very viscous liquid. This is visco-elasticity. The higher the temperature, the more marked the flow.

At the other extreme are materials which flow like liquids but show a tendency to recoil when the stress is removed. This has been called 'flow-elasticity' or 'elastico-viscosity'. Such liquids (e.g. molten Nylon) can be spun to form threads.

viscose *See* rayon.

viscosity All fluids show a resistance to flow, called viscosity. Mobile liquids, e.g. water, have a low viscosity whereas liquids such as oil or treacle have a high viscosity and flow only with difficulty. Viscosity measurements are often used to determine the molecular weights of polymeric substances.

viscosity, coefficient of The force required per unit area of fluid to maintain unit difference of velocity between two parallel planes in the fluid at a fixed distance apart. For plates one centimetre apart, the unity of viscosity is the poise.

vitamins Vitamins are substances other than proteins, carbohydrates, fats and mineral salts, that are essential constituents of the food of animals. In their absence the animal develops certain deficiency diseases or other abnormal conditions. Vitamins might also be defined as substances that play an essential part in animal metabolic processes, but which the animal cannot synthesize, although certain animals can synthesize certain vitamins and all animals needing vitamin D can manufacture it from ergosterol in the presence of u.v. light. The precise mechanism of action of many vitamins is still poorly understood.

vitamin A Vitamin A is the original fat-soluble vitamin. Its absence from the diet leads to a loss in weight and failure of growth in young animals, to the eye diseases xerophthalmia and night blindness, and to a general susceptibility to infections. The most fundamental effect of its deficiency is a keratinization of epithelial tissues. Vitamin A is present in animal fats, butter, yolk of egg, and in particularly large quantities in fish-liver oils, especially halibut liver oil. Carotene is converted into vitamin A in the liver, hence good sources of carotene, such as green vegetables, are good potential sources of vitamin A. Vitamin A is structurally related to carotene. It has the empirical formula $C_{20}H_{30}O$ and the structural formula

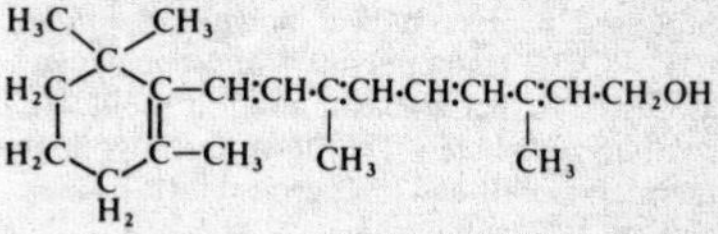

Two molecules of vitamin A are formed from one molecule of *β*-carotene. Vitamin A crystallizes in pale yellow needles; m.p. 64°C. It is optically inactive. It is unstable in solution when heated in air, but comparatively stable without aeration. Vitamin A is manufactured by extraction from fish-liver oils and by synthesis from *β*-ionone. The rôle of vitamin A in vision seems to be different from its systemic function. *See also* retinene and rhodopsin.

vitamin B The original vitamin B has been shown to consist of a number of different substances. The B vitamins include vitamin B_1 (thiamine), vitamin B_{12} (cyanocobalamine), vitamin B_c folic acid (pteroylglutamic acid), vitamin B_6 (pyridoxine), pantothenic acid and biotin.

vitamin B_{12}, cyanocobalamine,

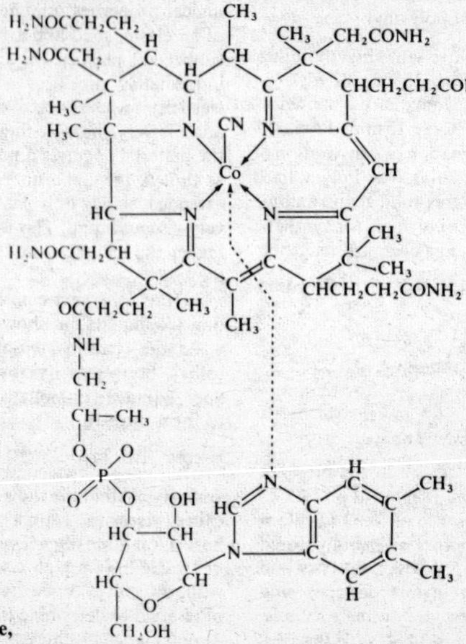

vitamin B_{12}, cyanocobalamine,

$C_{63}H_{90}CoN_{14}O_{14}P$. Dark red crystals. Vitamin B_{12} has been prepared synthetically.

Vitamin B_{12} is produced by the growth of certain micro-organisms, and occurs also in liver, being the extrinsic anti-pernicious anaemia factor the isolation of which was sought for many years.

The vitamin deficiency is often due to failure to absorb B_{12} from the stomach and can be alleviated by giving mg doses with extracts of hog's stomach which contains the intrinsic anti-pernicious anaemia factor (a mucoprotein), which promotes the absorption.

Vitamin B_{12} is concerned in the biosynthesis of methyl groups of choline and methionine.

vitamin C *See* ascorbic acid.

vitamin D The anti-rachitic vitamin. The

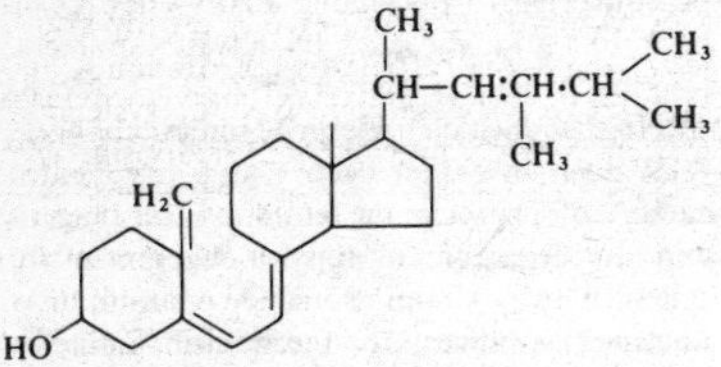

absence of vitamin D in the food of young animals leads to the development of rickets unless the animal is exposed to sunlight or u.v. irradiation. It is soluble in fats and fat solvents, and is present in animal fats, milk, butter and eggs. The richest sources of vitamin D are fish-liver oils, particularly those of the halibut and the cod. The first vitamin D to be discovered was a crude mixture called vitamin D_1. Irradiation of ergosterol with u.v. light gives calciferol or vitamin D_2. Irradiation of 7-dehydrocholesterol gives the natural vitamin D or vitamin D_3, which differs in structure only in the side chain, which is

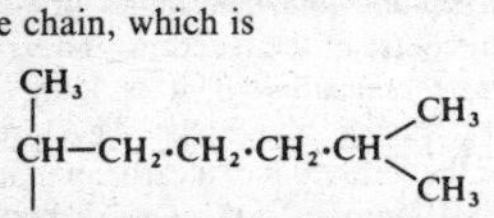

Vitamin D_2 has m.p. 115–117°C and D_3 m.p. 82–83°C. Both vitamins, which have almost identical actions, are used for the prevention and cure of infantile rickets; they are essential for the normal development of teeth, and are used for treating osteomalacia and dental caries. They are necessary for the absorption of Ca and P from the gut.

vitamin E Essential for fertility and reproduction. Deficiency in rats leads to abortion in the female and loss of fertility in the male. The structure is that of the tocopherols*. The richest sources are seed embryos and their oils, and green leaves. The vitamin is supposed to work as an antioxidant.

vitamin K An accessory factor needed by chickens, ducks and geese, the absence of which is characterized by haemorrhages due to a failure of the blood to clot properly. The factor is associated in some way with prothrombin, and may be part of the prothrombin molecule. It is fat-soluble, and found in liver fats, vegetables and to a lesser extent in cereals. It is stable to heat and light and destroyed by alkalis.

Several substances having vitamin K activity have been isolated from natural sources. Vitamin K_1 from alfalfa oil, is 2-methyl-3-phytyl-1,4-naphthoquinone.

The term vitamin K_2 was applied to 2-methyl-3-difarnesyl-1,4-naphthoquinone, m.p. 54°C, isolated from putrefied fish meal. It now includes a group of related natural compounds ('menaquinones'), differing in the number of isoprene units in the side chain and in their degree of unsaturation. These quinones also appear to be involved in the electron transport chain and oxidative phosphorylation.

vitellin The chief protein of egg yolk. It is a phosphoprotein and can be obtained as a yellow powder, insoluble in water, neutral salts and dilute acids.

vitreous state The glassy state. *See* glass.

Volhard method Titration of Ag^+ with NCS^- in the presence of Fe^{3+}. A deep red colour is formed at the end-point.

VSEPR theory *See* valency, theory of.

vulcanite *See* ebonite.

vulcanization The process of conferring more cross-linking upon a rubber and so altering its structure that it becomes less plastic and sticky, more resistant to swelling by organic liquids and has enhanced elasticity. The rubber is heated with a vulcanizing agent, generally sulphur although other compounds, e.g. S_2Cl_2, tetramethyl thiuram disulphide, Se, Te, organic peroxides are used with additives to affect the rate of vulcanization.

W

W Tungsten.

Wacker process The oxidation of ethene to ethanal by air and a $PdCl_2$ catalyst in aqueous solution. The Pd^{2+} is reduced to Pd^0 in the process but is reoxidized to Pd^{2+} by oxygen and Cu^{2+}.

$$C_2H_4 + O_2 \xrightarrow[H_2O]{Cu^{2+}/Pd^{2+}} CH_3CHO$$

Using ethene in ethanoic acid at 70–80°C it is possible to prepare vinyl acetate (ethenyl ethanoate) in good yields.

Wagner–Meerwein rearrangement A rearrangement of the carbon skeleton of a compound occurring in the course of an addition reaction to an alkene, an alkene-forming elimination or a substitution reaction. The rearrangement may involve a migration of an alkyl group or a change in ring structure, and it is frequently encountered in the chemistry of the bicyclic terpenes. The mechanism is essentially the same as that of the pinacol-pinacolone rearrangement*. An example of the change is

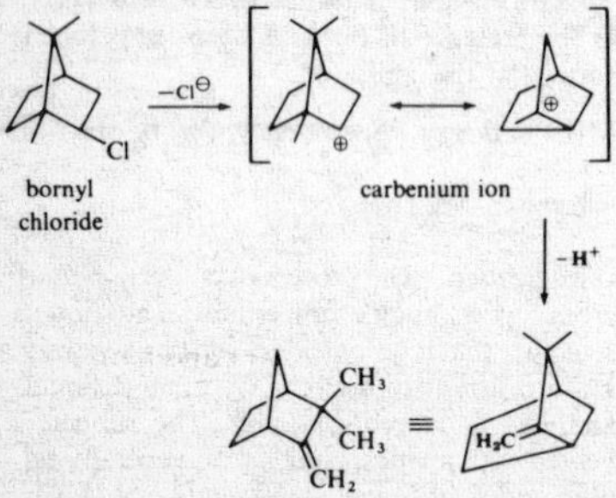

bornyl chloride

carbenium ion

camphene

This involves the formation of a carbenium ion which is best described as a hybrid of the two structures shown. This then rearranges by migration of a bond, and in so doing forms a more stable tertiary carbenium ion. Elimination of a proton yields camphene.

Walden inversion A phenomenon discovered in 1895 by Walden. When one of the atoms or groups attached to the asymmetric carbon atom in an optically active compound is replaced by a different atom, the product is sometimes a derivative of the optical isomer of the original compound. It is thus possible to pass from one isomer to the other without the formation and separation of a racemic compound. (+)-Malic acid, when treated with PCl_5 gives (−)-chlorosuccinic acid, which may be converted to (−)-malic acid by Ag_2O or back to (+)-malic acid by KOH. Similarly, (−)-malic acid is converted to (+)-chlorosuccinic acid which undergoes similar changes. A Walden inversion occurs at a tetrahedral carbon atom when the entry of the reagent and the departure of the leaving group are synchronous – the so-called bimolecular nucleophilic substitution mechanism. Since the reagent must approach from the side of the molecule opposite to that of the leaving group an *inversion* of optical configuration results.

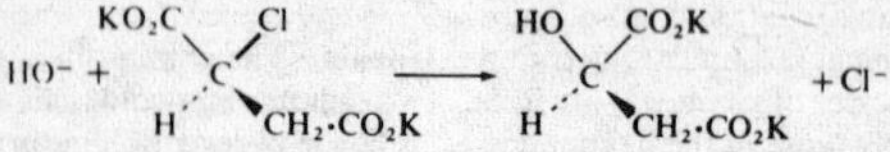

Walden inversion D- L-

An alternative mechanism of substitution is unimolecular and involves ionization of the leaving group to give a carbenium ion which reacts rapidly with the reagent. The lifetime of this carbenium ion determines the stereochemical course of the reaction – *inversion* if it is short, *racemization* if it is long. In the examples quoted above the Ag_2O reaction involves the formation of a carbenium ion, but the carboxyl group forms a weak bond with the developing centre of positive charge so that the approach of the reagent from this side is blocked and must occur from the side of the leaving group. The original optical configuration is thus *retained* because of neighbouring group participation in the reaction.

warfarin, 3-(α-acetonylbenzyl)-4-hydroxycoumarin, $C_{19}H_{16}O_4$. Anti-coagulant rodenticide prepared by condensing benzylidene acetone with 4-hydroxycoumarin. The product is a racemic mixture; it is a colourless, practically odourless and tasteless solid, m.p. 159–

161°C. Warfarin baits need contain only 0·025% active principle, and rats are killed after ingesting about 5 doses; the bait can be left down and the risk of acute toxicity to man or domestic animals is not serious. In common with other coumarin derivatives, warfarin reduces the clotting power of blood and death is caused by haemorrhages initiated by any slight injury. Warfarin is a vitamin K antagonist, and large oral doses of the vitamin can be given as an antidote.

The sodium derivative is used therapeutically as an anticoagulant.

washing soda Sodium carbonate decahydrated, $Na_2CO_3,10H_2O$.

waste heat boiler A boiler which produces steam by utilizing the heat in the gases or liquid from a chemical process, e.g. a calcining operation.

water, H_2O. M.p. 0°C, b.p. 100°C, *d* at 0°C 0·99987. The simplest oxygen hydride, a colourless or very pale blue–green liquid. Maximum *d* at 4°C. Solid water (ice) has an essentially tetrahedral arrangement of oxygens about each other with hydrogen bridges (there are different forms depending upon the detailed geometry). Hydrogen bonding and residual order is carried through into solution. Forms hydrates with many salts (e.g. $CuSO_4,5H_2O$); bonding to the cation is through the lone-pairs; bonding to anions is by hydrogen bonding. In many solids the essential structure is determined by the hydrogen bonding. Behaves as a neutral oxide; dissociation to $[H_3O]^+$ and $[OH]^-$ is small. Reacts with electropositive metals (e.g. Na, Ca) to give hydrated oxides or hydroxides and hydrogen; reacts with non-metal oxides (e.g. SO_3, P_2O_5) to form acids and with oxides of the most electropositive metals to form hydroxides (e.g. NaOH). Reacts with many halides, particularly non-metal halides and metal halides in high oxidation states with hydrolysis – e.g. $SOCl_2$ gives SO_2 and HCl. Of major industrial importance as a reagent for cooling and as a solvent. Domestically important for drinking, cleaning and sanitation. Sea water can give fresh water (desalination) by evaporation, freezing, extraction, reverse osmosis and ionic processes but these are all expensive. Industrial waste water is purified for re-use and domestic water is also purified, *see* sewage treatment.

water gas *See* blue water gas.

water-glass *See* sodium silicates.

water-proofing, water repellancy Waterproofing renders a fabric impermeable to water. Impregnation with oils, varnishes, rubber, etc., can confer waterproof properties. Water repellant coatings repel liquid water but allow passage of air and water vapour. Silicones and fluorocarbon materials are widely used to confer water repellancy.

wavelength, λ The distance between corres-

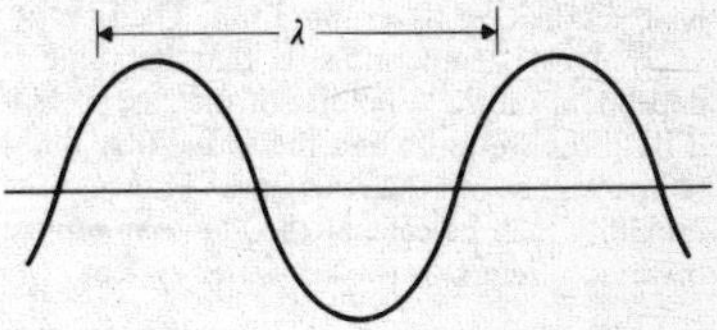

ponding points in the profile of wave motion.

wave mechanics The electronic theory of matter is confronted by certain difficulties which have only been overcome by assumptions of a very arbitrary nature. To overcome these difficulties by a general theorem, de Broglie and Schrödinger independently devised systems of wave mechanics which are essentially mathematical, and may only be very approximately described in terms of physical pictures. Wave mechanics gives allowed energy levels (orbitals). Transitions between these states gives rise to various forms of electromagnetic radiation. Wave mechanics has led to a considerable clarification of many difficulties inherent in the older quantum theory and in the electronic theory of matter, but it is still necessary to use these concepts because of their more concrete nature.

wave number The reciprocal of wavelength, i.e. the number of waves in a given length (generally 1 cm). Often referred to as frequencies.

waxes Formally fatty acid esters with monohydric fatty alcohols having wax-like properties but now arbitrarily extended to any organic material having such properties. Waxes are water-repellant and have plasticity. They are used in paper coating, polishes, electrical insulation, textiles, leather, pharmaceuticals. Typical examples are cetyl palmitate in spermaceti and melissyl palmitate in bees' wax. Synthetic waxes include some forms of polyethers.

weedkillers Herbicides used to control selectively unwanted vegetation on agricultural land. They may act either selectively (2,4-D)* or may be applied before the crop emerges.

Weston cell A widely used standard cell $Cd(Hg)/CdSO_4,8/3H_2O(s),//Hg_2SO_4(s)/Hg$. The electrolyte is a saturated solution of $CdSO_4$. At 20°C $E=1{\cdot}01485$ volts.

wet-bulb temperature The dynamic equilibrium temperature attained by a liquid surface when exposed to a stream of gas under adiabatic conditions*. The rate of heat transfer to the liquid is then equal to the rate at which heat is taken up by evaporation. The importance of this temperature is that its value is dependent on the humidity of the gas, so that if both the wet bulb and the actual (dry bulb) temperature of the gas are known, the humidity may be calculated. The term almost invariably refers to the air–water system.

wet-bulb thermometer A thermometer for measuring wet-bulb temperature in which the surface of the bulb is kept damp by means of a wick surrounding it, liquid being allowed to evaporate adiabatically into the gas stream.

wetting agents Water, because of its powerful intermolecular attractive forces, does not readily spread over many surfaces. The addition of surface active agents, which decrease the contact angle between the water and the other surface, thereby lowering the surface tension, permits the wetting of the surface. Such surface active agents, e.g. detergents, usually consist of molecules which have an oil-attracting hydrophobic group (e.g. an alkyl chain) and a water-attracting (hydrophilic) group (e.g. a negatively charged carboxylic or sulphonic acid group).

white arsenic Crude arsenic(III) oxide*, As_2O_3.

white lead, $Pb_3(OH)_2(CO_3)_2$. *See* lead carbonate.

white oils Oils used in medicinal and toilet preparations obtained by drastic refining of lubricating oil stocks. *See* liquid paraffin.

white spirits, mineral solvents Essentially a mixture of hydrocarbons, usually from a petroleum source, b.p. 130–220°C. White spirits are used as paint thinners and as solvents for a variety of purposes.

whiting A form of $CaCO_3$ made by grinding chalk and collecting the finer sediments from water. It is extensively used in the chemical and many other industries.

Wilkinson's catalyst $[Rh(Ph_3P)_3Cl]$. An important catalyst for homogeneous hydrogenation which also catalyses oxidation by O_2 and CO abstraction from organic derivatives.

Williamson ether synthesis Alkyl halides react with sodium or potassium alkoxides or phenoxides to give ethers.

$$RX+NaOR' \longrightarrow R-O-R'+NaX$$

Although the yields are often poor, especially for halides other than primary alkyl halides, it remains a valuable method for synthesizing unsymmetrical ethers.

Wiswesser line notation The Wiswesser line-formula notation (WLN) is a method for expressing the more usual graphical structure of a chemical compound as a linear string of symbols. The resulting alternative notation is unambiguous, short and particularly suitable for computer processing and retrieval but can also be understood easily by chemists after minimal training in its use.

The basic characters from which the notations are constructed comprise the upper-case letters A–Z of the alphabet, the numerals zero (symbolized ø) to nine (ø–9), three punctuation marks hyphen (-), ampersand (&) and oblique (/) and a blank space. Many of the normal atomic symbols such as B, F, P, I, etc., are also employed unchanged but frequently occurring important elements and groups are assigned a single letter notation (e.g. chlorine ≡ G; $-NH_2 \equiv Z$; carbonyl = C(O) ≡ V; hydroxyl $-OH \equiv Q$; unfused benzene ring ≡ R). The presence of hydrogen atoms in an organic structure is usually implied. Numerals are assigned to unbranched internally saturated alkyl chains of the indicated number of carbon atoms (e.g. $CH_3 \equiv 1$; $CH_3CH_2 \equiv 2$). The notation for a given formula is obtained by linking the appropriate characters for the fragments in an established order.

Simple examples of WLN are: C_2H_5OH is Q2 $CH_3CO{\cdot}OCH_3$ is 1VO1 For branch chain and fused ring structures rules determine the order of notation. It is claimed that over 50% of all organic structures can be represented by less than 25 characters.

witherite, $BaCO_3$. The white mineral form of barium carbonate. Used as a source of Ba compounds and in the brick and ceramic industries.

Wittig reaction The reaction between an alkylidene phosphorane, $\begin{matrix}R\\R'\end{matrix}\!\!>C=PR''_3$ and an aldehyde or ketone, $\begin{matrix}R^{\circ}\\R^{\times}\end{matrix}\!\!>C=O$ to produce an alkene, $\begin{matrix}R\\R'\end{matrix}\!\!>C=C\!\!<\begin{matrix}R^{\circ}\\R^{\times}\end{matrix}$ Although the position of the double bond in the alkene is known, the reaction is not completely stereospecific and can produce *cis-trans* mixtures.

WLN Wiswesser line notation.

wolfram An alternative name for tungsten.

wolframite, (Fe,Mn)WO_4. The principal ore of tungsten*.

wood Wood is a complex substance with cell walls containing carbohydrates (cellulose, hemicellulose, etc.) and lignin together with some inorganics and many organic materials (hydrocarbons, phenols, aldehydes, ketones, terpenes, etc.). Wood is used in construction (may be treated to lower flammability), and in the pulp and paper industries, for charcoal, rayon and for various chemicals – *see* tall oil, rosin, terpenes, tannic acid.

wood flour Finely ground timber used as filler in plastics, linoleum and also in sheet metal work.

wood naphtha, wood spirit *See* methanol.

Woodward-Hoffmann rules Orbital symmetry rules (*see* frontier orbital symmetry) relating to ring formation and ring opening.

wool A natural protein fibre (keratin) mainly from the fleece of sheep.

work function The energy required to remove an electron from the highest occupied level inside a solid to a point *in vacuo* outside the surface.

work hardening Increase in strength and hardness, decrease in ductility and toughness of a metal alloy resulting from plastic deformation at a temperature below its recrystallization temperature (normally room temperature). Beneficial in production of high tensile wire by cold drawing and spring strip by cold rolling. Results from mutual interference of dislocations moving on intersecting slip planes and also of grain boundary obstruction of dislocations.

Wulff–Bock crystallizer *See* crystallizers.

Wurster's salts Stable radical cations formed

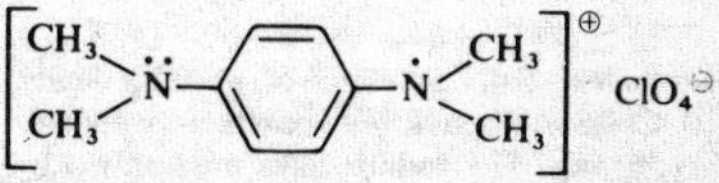

by one electron oxidation of N,N,N′,N′-tetramethyl-*p*-phenylendiamine can be isolated as salts, e.g. the brilliant blue perchlorate salt is often called 'Wurster's Blue'.

wurtzite A form of ZnS which gives its name

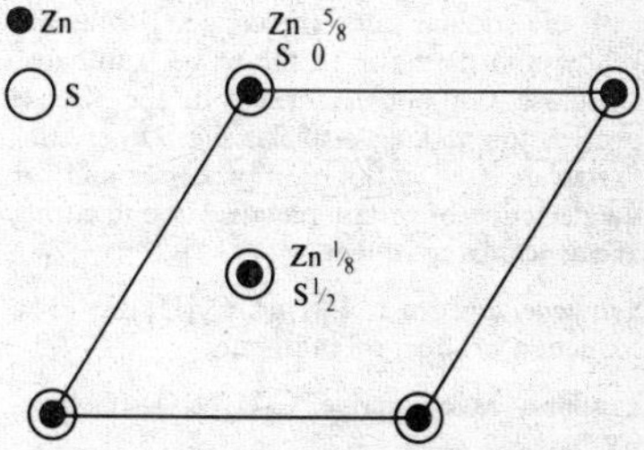

to an important structural type. Hexagonal with both Zn and S tetrahedrally co-ordinated. Has hcp S with Zn in tetrahedral holes. Other compounds with this structure include ZnO, BeO, AlN, GaN.

Wurtz synthesis Alkyl halides react with sodium in dry ethereal solution to give hydrocarbons. If equimolecular amounts of two different halides are used, then a mixture of three hydrocarbons of the types R–R, R–R′ and R′–R′, where R and R′ represent the original radicals, will be formed. The yields are often poor owing to subsidiary reactions taking place.

X

xanthates Salts or esters of unstable acids, ROC(S)SH, where R may be either an alkyl or aryl group. The sodium salts are formed by treating an alcohol or some hydroxy compounds with carbon disulphide and sodium hydroxide or by treating a sodium alkoxide with carbon disulphide. The esters are formed by treating the sodium salts with alkyl halides. The free acids are very unstable; other metallic xanthates are formed by double decomposition with the sodium salts; the copper(I) salts are yellow and gave rise to the name 'xanthate'. Cellulose xanthate is formed in the viscose process for making artificial silk. Other xanthates are used in flotation processes and for the detection of certain metals. Used in curing and vulcanizing rubber.

xanthene, dibenzo-1, 4-pyran, $C_{13}H_{10}O$. The reduction product of xanthone.

xanthine, 2,6-oxypurine, $C_5H_4N_4O_2$. Colour-

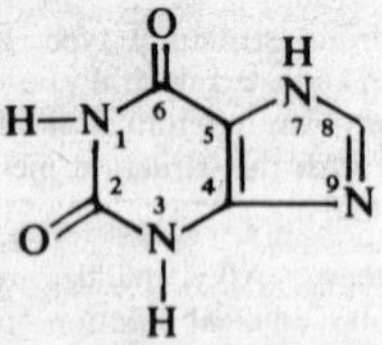

less crystalline powder with $1H_2O$ which it loses at 125–130°C. Decomposes without melting. It is a breakdown product of nucleic acid metabolism, being formed from guanine by the action of the enzyme guanase and from hypoxanthine by oxidation. It is itself oxidized in the body to uric acid.

xanthine oxidase An enzyme of the oxidoreductase class, present in liver and other organs and also in milk. Xanthine oxidase produces superoxide radicals ($O_2{}^-$) during the course of its oxidation of substrates and is thus inhibited by superoxide dismutase*.

xanthone, dibenzo-4-pyrone, $C_{13}H_8O_2$.

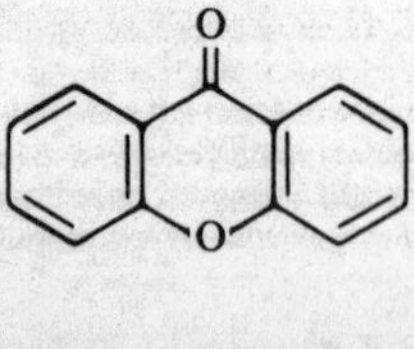

Colourless crystals; m.p. 174°C. It is obtained by the action of heat on phenyl salicylate. It may be reduced to xanthene. It is the parent substance of the xanthone group of dyestuffs.

xanthophyll A substance of formula $C_{40}H_{56}O_2$ found with carotene* in most green plant material. It consists of two isomers, lutein and zeaxanthin, which are oxidation products of α- and β-carotene respectively. The term is now applied to encompass most hydroxylated carotenoids. Xanthophylls are yellow pigments.

xanthosine, xanthine riboside Crystallizes but

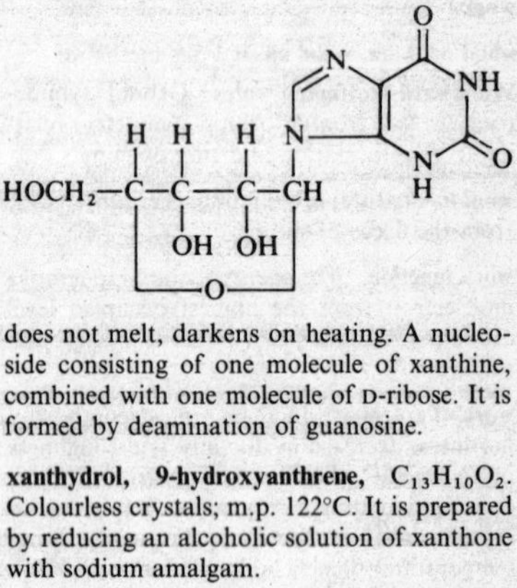

does not melt, darkens on heating. A nucleoside consisting of one molecule of xanthine, combined with one molecule of D-ribose. It is formed by deamination of guanosine.

xanthydrol, 9-hydroxyanthrene, $C_{13}H_{10}O_2$. Colourless crystals; m.p. 122°C. It is prepared by reducing an alcoholic solution of xanthone with sodium amalgam.

Xe Xenon.

xenon, Xe. At.no. 54, at.wt. 131·30, m.p. −111·9°C, b.p. −107·1°C. One of the noble gases (% abundance $8{\cdot}7 \times 10^{-6}$ in air) separated by fractional distillation of liquid air. Used in lamps and discharge tubes and in bubble chambers. Moderately soluble in water.

xenon chemistry Xenon has electronic configuration $5s^25p^6$ and shows oxidation states +2 [XeF_2 (Xe plus F_2 plus light), $XeCl_2$ (Xe plus Cl_2 plus electric discharge) both linear], +4 [XeF_4 (Xe plus F_2) square planar and its adducts with acid fluorides], +6 [XeF_6 (Xe plus F_2), explosive XeO_3 (XeF_6 plus H_2O)] and +8 [$(XeO_6)^{4-}$ ($(HXeO_4)^-$ plus base)]. Other derivatives including oxyfluoro anions, e.g. $[XeO_3F]^-$, and oxide fluorides, e.g. $XeOF_2$ and $XeOF_4$, are also known.

xenylamine, 4-aminobiphenyl, $C_{12}H_{11}N$. Colourless crystals, m.p. 53–54°C. Obtained by the iron, or tin, and aqueous acid reduction of 4-nitrobiphenyl.

Substituted derivatives are called xenylamines. All xenylamines are suspected carcinogens.

xerogels A method of classification of gels. Xerogels are relatively free of the dispersion medium; lyogels are rich in the dispersant.

xerography A reprographic process in which a photoconductor (generally Se) on a surface is charged, exposed to the image which discharges all except the image, removal of Se from the non-image areas, developing the image with an oppositely charged pigment, transferring the image to paper electrostatically and fixing the image on paper by baking.

X-ray Electromagnetic radiation of wave length *c.* 1 Å. X-rays are generated in various ways, including the bombarding of solids with electrons, when they are emitted as a result of electron transitions in the inner orbits of the atoms bombarded. Each element has a characteristic X-ray spectrum.

X-rays may be detected either photographically or with an ionization counter. They have great penetrating power which increases with their frequency, and owing to this are used to photograph the interior of many solid objects, notably the human body and in monitoring for faults in construction.

X-rays find wide applications in X-ray photography and in crystallography. Prolonged exposure of the human body to the rays induces a dangerous form of dermatitis, and even sterility, but controlled exposures are applied to alleviate cancer.

X-ray diffraction A powerful technique for the study of crystal structure, discovered by von Laue (1912) and developed for crystal analysis by W. H. and W. L. Bragg (1912–1913). The atomic nuclei in a crystal lattice act as diffraction gratings; the rows of atoms have spacings of a few Ångstrom units, which are comparable with the wavelength of X-rays. Strong scattering of the rays by the crystal therefore occurs in certain directions, according to Bragg's equation*. Various techniques are available for applying X-ray diffraction to the study of single crystals, powders, fibres, etc., and by computation it is possible to work out 3-dimensional electron-density maps of the solid lattice from the recorded X-ray patterns.

X-ray fluorescence A method of analysis used to identify and measure heavy elements in the presence of each other in any matrix. The sample is irradiated with a beam of primary X-rays of greater energy than the characteristic X-radiation of the elements in the sample. This results in the excitation of the heavy elements present and the emission of characteristic X-ray energies, which can be separated into individual wavelengths and measured. The technique is not suitable for use with elements of lower atomic number than calcium.

X-ray spectrometer An apparatus used in the X-ray study of crystals in which a fine beam of monochromatic X-rays impinges at a measured angle on the face of a crystal mounted in its path, and in which the intensity of the X-rays diffracted in various directions by the crystal is measured with an ionization chamber mounted on an arm of the spectrometer table, or is recorded photographically.

X-ray spectroscopy Analytical method by which a sample is irradiated with X-rays, characteristic radiation being emitted after scattering from the specimen. The detection limits for various elements are of the ordering cm^{-2}.

X-ray tube The apparatus employed for producing X-rays.

xylan Occurs in association with cellulose in lignified cell walls, in the wood of deciduous trees, in the bran and straw of cereals and in similar plant substances. Xylan from esparto grass has been obtained in a relatively pure state; it is made up of 18 or 19 xylopyranose units, connected by 1,4-β-linkages, with a terminal arabofuranose residue. It is strongly laevorotatory.

xylene, dimethylbenzene, C_8H_{10}. As usually obtained xylene is a colourless refractive liquid of characteristic smell, burns with a smoky flame and is a mixture of the three possible isomers, *o*-xylene, b.p. 144°C, *m*-xylene b.p. 139°C and *p*-xylene b.p. 138°C. Oxidation with chromic acid or permanganate gives the corresponding dicarboxylic acids.

Commercially, xylene is obtained by the catalytic reforming of naphthenes in the presence of hydrogen (*see* toluene) or was formerly obtained from coal tar. The material so-produced is suitable for use as a solvent or gasoline ingredient, these uses accounting for a large part of xylene consumption. If xylene is required as a chemical, separation into the isomers is usually necessary, and although the *o*-compound can be readily removed by fractional distillation, the separation of the *m*- and *p*-isomers requires techniques such as fractional crystallization, solvent extraction or clathration. *o*-Xylene is used in the manufacture of phthalic anhydride, and the *m*-and *p*-isomers in the manufacture of isophthalic and terephthalic acid respectively. Total U.S. production of xylene 1978 2·8 megatonnes; *p*-xylene 1·6 megatonnes; *o*-xylene 470 000 tonnes.

xylenols, $C_8H_{10}O$, $C_6H_3(CH_3)_2OH$. Hydroxydimethylbenzenes of which six isomers are possible. The pure substances are low-melting solids having the general properties of phenols. Xylenol is the name given to a mixture of the isomers separated from the phenolic fraction of coal tar and used as a solvent. The chlorinated derivative known as *p*-chloro-*m*-xylenol is used as a disinfectant.

xylic acid The trivial name given to 2,4-dimethylbenzoic acid.

xylidenes, $C_8H_{11}N$, $C_6H_3(CH_3)_2NH_2$. The mixture of isomeric xylenes obtained from coal-tar is usually nitrated without separation of the isomers. Reduction of the nitroxylenes with iron and hydrochloric acid gives a mixture of five aminoxylenes or xylidines.

The mixture of xylidines has been used as a first component of azo-dyes. The chief constituent of the mixture is '*m*-xylidine' (4-amino-1,3-xylene). It can be separated by crystallization from glacial ethanoic acid. It is also used for the preparation of azo-dyes.

D-xylose, wood sugar, $C_5H_{10}O_5$. M.p.

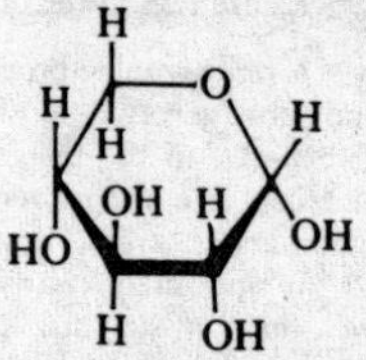

144°C. The pentose sugar of straw, cotton-seed hulls and various hemicelluloses, and of some glycosides, including the primeverosides. It is not fermentable and behaves chemically as other sugars.

Y

Y Yttrium.

Y alloy *See* aluminium alloys.

Yb Ytterbium.

yellow ammonium sulphide *See* ammonium sulphides.

ylides Internal salts, e.g.

$$R_2C{=}PPh_3 \longleftrightarrow R_2C^{-}{-}P^{+}Ph_3$$

with the heteroatom formally in a cationic state (N, P, S are amongst elements forming ylides). Intermediates in the Wittig reaction.

ytterbium, Yb. At.no. 70, at.wt. 173·04, m.p. 824°C, b.p. 1193°C, *d* 6·98. A typical lanthanide*. The metal is ccp (to 798°C) and bcc (to m.p.). Some Yb based garnets are used as synthetic gemstones.

ytterbium compounds Ytterbium shows +3 and +2 oxidation states. Yb(III) species are typical lanthanide compounds*.
Yb^{3+} (f^{13} colourless) $\longrightarrow$ Yb
(−2·27 volts in acid)

The dihalides YbX_2 are yellow (YbX_3 + Yb) and are genuine Yb^{2+} salts.

$$Y^{3+} \longrightarrow Yb^{2+} \ (-1{\cdot}15 \text{ volts})$$

yttrium, Y. At.no. 39, at.wt. 88·91, m.p. 1523°C, b.p. 3337°C, *d* 4·472. An element of Group III with electronic configuration $5s^24d^1$. The principal ore is gadolinite (a silicate also containing lanthanides). Y_2O_3 containing Eu is used as a red phosphor in colour television. Yttrium iron garnets are used as microwave filters.

yttrium compounds Yttrium only forms compounds in the +3 oxidation state
Y^{3+} (colourless) $\longrightarrow$ Y (−2·37 volts in acid)
The chemistry is similar to that of the lanthanides*.

Z

zeaxanthin, zeazanthol, $C_{40}H_{56}O_2$. Yellow crystals; m.p. 215°C.

One of the xanthophyll pigments present in various leaves, seeds and fruits, and in yolk of egg. It is often present in company with lutein, of which it is an isomer. It bears the same relation to β-carotene as lutein does to α-carotene.

zeaxanthin, zeazanthol,

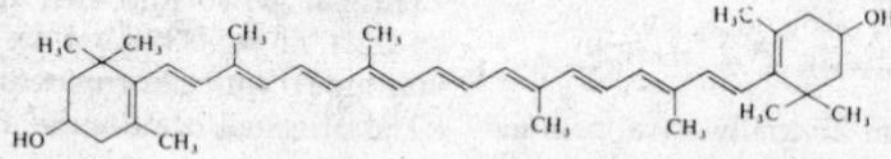

zein A protein of the prolamine class present in maize. It contains no lysine or tryptophan. Zein has been used in the plastics, paints, paper and other industries, in a variety of ways.

Zeise's salt, $K[PtCl_3(C_2H_4)]$. A π-ethylene complex, the first such complex isolated.

zeolites Aluminosilicates containing $(Si,Al)_nO_{2n}$ frameworks with a negative charge which is balanced by cations present in the cavities. The cations are easily exchanged and water and gases can be selectively absorbed into the cavities. Various types of zeolites are known including analcite, chabazite, faujasite, the synthetic zeolite A, natrolite and thomsite. Sodium-containing zeolites are used to soften water, replacing Ca^{2+} by Na^+, the zeolite (permutit) is regenerated with concentrated NaCl solution. Zeolites are used to remove molecules of specific sizes by absorption into the pores of the zeolite (molecular sieves); for drying solvents and absorbing gases.

zeta (ζ) potential An alternative name for the electrokinetic potential*.

Ziegler catalysts Complex catalysts prepared by interaction between an organometallic derivative and a transition metal derivative. A typical catalyst is the product of the interaction of $TiCl_4$ and $AlBu_3$. These catalysts polymerize olefins, particularly ethylene, to polyolefins, the polymerization generally being in a stereoregular manner.

Ziegler–Natta polymerization Stereospecific polymerization of olefines using a Ziegler catalyst*. *See* titanium(III) chloride.

zinc, Zn. At.no. 30, at.wt. 65·38, m.p. 419·58°C, b.p. 907°C, *d* 7·13. Transition element occurring as zinc blende, sphalerite (Zn,Fe)S calamine or smithsonite ($ZnCO_3$), willemite (Zn_2SiO_4), franklinite ($ZnFe_2O_4$). Extracted by roasting to ZnO and reduction with carbon. The metal is bluish-white (deformed hcp) fairly hard and brittle. Burns in air and combines with halogens and sulphur. Reacts with steam at red heat. Dissolves in dilute acids and hot solutions of alkalies. Used in alloys (e.g. brass, solder) for die-casting and in protecting (galvanizing) steel by coating. Used in anodes for protection against corrosion. ZnO is used in paints, rubber, plastics, textiles, etc., and as an inert filler and in electrical components. ZnS is used in luminous paint, X-ray and TV screens, fluorescent lights. Production 1976 6·4 megatonnes. The Zn^{2+} ion is the active site in many enzymes including carbonic anhydrases and dehydrogenases.

zinc amalgam A solid Zn – Hg solution, frequently used as a reducing agent.

zincates Ions such as $[Zn(OH)_4]^{2-}$, $[Zn(OH)_3H_2O]^-$ formed by the action of excess base on Zn^{2+} solutions.

zinc blende This cubic form of zinc sulphide

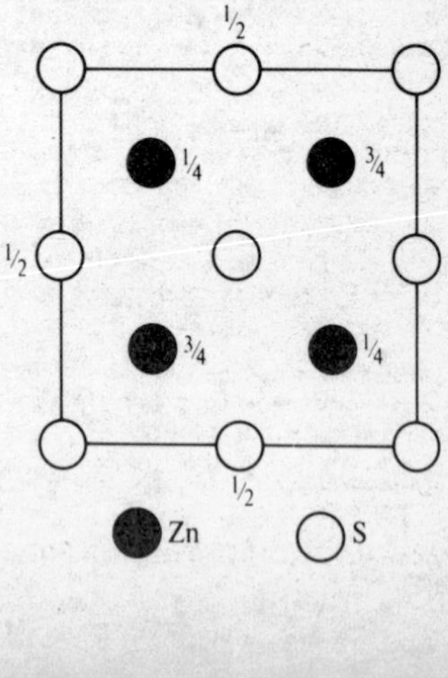

gives its name to an important crystal structure type. The Zn and S are tetrahedrally co-ordinated. Has ccp S with Zn in tetrahedral holes. Other compounds with this structure include BN, SiC, AlP, CuCl.

zinc borates Compounds of rather indefinite composition prepared by treating a solution of borax with zinc sulphate. Widely used for fireproofing textiles and in the ceramic and pharmaceutical industries.

zinc carbonate A basic carbonate is precipitated by addition of an alkali carbonate to a solution containing Zn^{2+}. $ZnCO_3$ may result from the use of, e.g., $NaHCO_3$. Hydrolysed to ZnO on boiling with Na_2CO_3 solution. The basic carbonate is used medicinally for external application. *See* calamine.

zinc chemistry Zinc is an electropositive Group II transition element, electronic configuration $3d^{10}4s^2$
(E° $Zn^{2+} \longrightarrow Zn$ in acid solution $-0{\cdot}76$ volts)
Apart from possible Zn^+ and ${Zn_2}^{2+}$ in some melts all zinc compounds are in the $+2$ state, generally in octahedral or tetrahedral co-ordination. Readily forms complexes, particularly with O and N ligands.

zinc chloride, $ZnCl_2$. White mass formed Zn plus HCl gas or in solution Zn or ZnO plus aqueous HCl. Forms hydrates. Used as wood preservative, in batteries, as a dehydrating agent, in dental stoppings.

zinc chromate Yellow pigment precipitated from $ZnSO_4$ solution by K_2CrO_4. Mixed with Prussian blue gives zinc greens which are particularly stable and light fast. Zinc chromates are used in rust-inhibiting paints.

zinc dithionite, zinc hydrosulphite, ZnS_2O_4. Made SO_2 on aqueous suspension of Zn dust. Used in bleaches and the vat dyeing process.

zinc ethanoate, zinc acetate, $Zn(O_2CCH_3)_2,2H_2O$. Also forms basic salt $Zn_4O(O_2CCH_3)_6$.

zinc fluoride, $ZnF_2,4H_2O$. Forms ZnF_2 at 100°C, formed ZnO plus aqueous HF or Zn plus HF or F_2. Used as catalyst and wood preservative.

zinc fluorosilicate, $ZnSiF_6,6H_2O$. Water-soluble salt used as wood preservative and fungicide and in plastics industry.

zinc hydrosulphite, ZnS_2O_4. *See* zinc dithionite.

zinc hydroxide, $Zn(OH)_2$. Precipitated from Zn^{2+} solutions by, e.g., NaOH. Occurs hydrated and loses water on heating. Dissolves in acids and in excess alkali (to give zincates). Used as rubber filler and as an absorbent in surgical dressings.

zinc nitrate $Zn(NO_3)_2,xH_2O$.

zinc, organic derivatives Zinc dialkyls and diaryls result from Zn or a Zn–Cu couple with RI in an autoclave at 150°C, also prepared using LiR, RMgX and HgR_2. ZnR_2 are immediately oxidized by air. Used, with $TiCl_4$, as polymerization catalysts for olefines.

zinc oxide, ZnO. Soft powder. White cold, yellow hot. Obtained by burning Zn or from ZnS ores. Dissolves in acids and bases. Used as filler in rubber and plastics for assisting curing of plastics, in ceramics, and medicinally for external application and as a white pigment (Chinese white). *Zinc peroxide,* ZnO_2. Formed Zn^{2+} solution plus Na_2O_2 at low temperatures. Used as a disinfectant in pharmacy.

zinc perchlorate $Zn(ClO_4)_2,6H_2O$.

zinc sulphate, white vitriol, $ZnSO_4$. Forms 7, 6, 1 and 0 hydrates (ZnO plus H_2SO_4). Used in the textile industry and in arsenical sprays in agriculture.

zinc sulphide, ZnS. Occurs naturally as zinc blende (sphalerite) and wurtzite (*see* separate headings for structures). Formed Zn plus S or precipitated from a Zn^{2+} solution by ammonium sulphide. Impure ZnS is phosphorescent. Used as a pigment and in luminous paints.

zingiberene, $C_{15}H_{24}$. The main constituent of

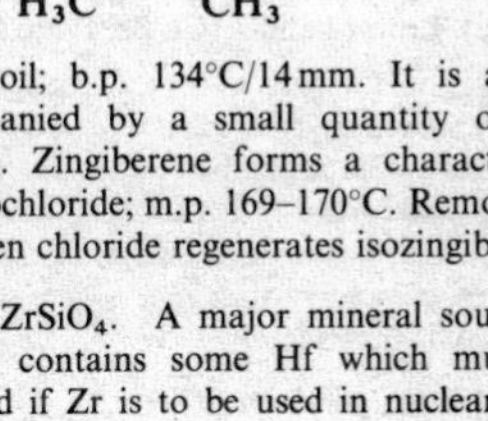

ginger oil; b.p. 134°C/14 mm. It is always accompanied by a small quantity of bisabolene. Zingiberene forms a characteristic dihydrochloride; m.p. 169–170°C. Removal of hydrogen chloride regenerates isozingiberene.

zircon, $ZrSiO_4$. A major mineral source of Zr but contains some Hf which must be removed if Zr is to be used in nuclear technology.

zirconium, Zr. At.no. 40, at.wt. 91·22, m.p. 1852°C, b.p. 4377°C, *d* 6·506. A transition element of Group IV, electronic configuration $5s^2 4d^2$. Mineral sources are baddeleyite (ZrO_2) and zircon ($ZrSiO_4$), generally occurs with Hf which is separated by ion-exchange. The metal ($ZrCl_4 + Mg$ at 1150°C) is hcp (to 862°C) and bcc (to m.p.). The metal is steel-like in appearance, very resistant to corrosion but is dissolved by aqua regia or HF and burns in oxygen at high temperatures. The element has low neutron absorption and alloys are used in reactor construction, the metal is also used for laboratory ware. Nb−Zr alloys are used in super-conducting magnets. World production 1976: metal 9 000 tonnes; compounds 315 000 tonnes.

zirconium compounds Most zirconium compounds are in the +4 oxidation state and the aqueous chemistry is confined to this state. The Zr^{4+} ion is extensively hydrolysed in aqueous solution and polymeric species, e.g. $[Zr_3(OH)_4]^{8+}$, are present. Zr(IV) compounds have high co-ordination numbers and show a wide variety of stereochemistries. Lower oxidation state halides ZrX_3(Cl,Br,I), ZrX_2(Cl,Br,I) and ZrCl($ZrCl_4$ + Zr) are only stable in the solid state. Complexes formed by the tetrahalides, e.g. $ZrCl_4(OPCl_3)_2$, and O-bonded complexes, e.g. zirconium diketonates, are stable. Organometallics, e.g. $Zr(CH_2C_6H_5)_4$, and relatively stable cyclopentadienyls, e.g. $(C_5H_5)_2ZrCl_2$ and $(C_5H_5)_2Zr(CO)_2$, are known. Alkoxides and alkylamides are easily formed as polymers.

zirconium halides Formed by interaction of the elements or ZrF_4 (gaseous HF on ZrO_2) $ZrCl_4$ (sublimes 331°C, $ZrO_2 + CCl_4$ or $C + Cl_2$). ZrF_4 has m.p. 932°C and forms 1 and 3 hydrates. The halides other than ZrF_4 are readily hydrolysed in water and, e.g. $ZrOCl_2$ and polymeric hydroxo cations are formed. Complex halides are known for F^-, Cl^- and Br^-. $ZrF_8{}^{2-}$ is 8-co-ordinate square antiprism, $ZrF_7{}^{3-}$ has pentagonal bipyramidal ions and polymeric species are known, e.g. K_2ZrF_6 (8-co-ordinate) $(NH_4)_2ZrF_6$ (7-co-ordinate). Lower halides (Cl, Br, I) are formed in the solid state (ZrX_4 + Zr); there is extensive metal–metal bonding in ZrCl.

zirconium oxides The dioxide is the stable species and occurs naturally as baddeleyite. It is very inert and has been used as a refractory and insulator. ZrO_2 exists in one form with 7-co-ordinate Zr and also one form with the calcium fluoride* structure – the latter form is stabilized by addition of CaO and this is used in refractories. A lower oxide, ZrO, is known.

zirconium oxyacid salts The nitrate ($5H_2O$) is obtained from conc. HNO_3 solution, the anhydrous compound ($ZrCl_4 + N_2O_5$) is volatile and an oxide nitrate is available commercially. Oxide sulphates, perchlorates and ethanoates are available as hydrates – all have high co-ordination numbers for Zr.

Zn Zinc.

zone electrophoresis *See* chromatography.

zone refining A method of refining metals and certain inorganic and organic compounds depending on the difference in solubility of impurities in the liquid and solid states. A narrow molten zone is caused to move along a bar or column of the substance; those impurities which lower the melting point will tend to remain in solution and will be moved in the direction of zone travel. Solutes which raise the melting point will preferentially freeze out and will move in a direction opposite to that of zone travel. By successive traverses the substance can be obtained in a very pure state. The method is applied to the production of pure germanium and other metals required very pure but in small quantities.

Zr Zirconium.

zwitterion A zwitterion is an electrically neutral ion with both a positive and a negative charge. Thus the amino-acid glycine exists in solution at the isoelectric point as the zwitterion $^+H_3N{\cdot}CH_2{\cdot}COO^-$.

zymase The enzyme present in yeast which converts glucose to ethanol but can also induce other reactions.

Published by Penguins

THE SOUL OF A NEW MACHINE

Tracy Kidder

The making of the 32-bit mini-computer, Eagle, was comparable to flying upside down.

It was a colossus of an enterprise, demanding the waking hours and sleeping thoughts of a brilliant and aggressive team of young computer wizards who worked for the multi-million-dollar American company, Data General. They were given one year: and day-by-day, wire-by-wire, the 'microkids' traced and retraced Eagle's labyrinthine arteries, fought back at its tantrums and wooed its obedience. It forced them to the limit of their endurance.

Tracy Kidder's electrifying account of a computer revolution is as thrilling as Le Carré, Forsyth and Ludlum rolled into one.

'All the incredible complexity and chaos and exploitation and loneliness and strange, half-mad beauty of this field ... I didn't think a book like this could be done' – Robert Pirsig, author of *Zen and the Art of Motorcycle Maintenance*

Published by Penguins

THE MIND'S I: FANTASIES AND REFLECTIONS ON SELF AND SOUL

Composed and arranged by Douglas R. Hofstader and Daniel C. Dennett

'What is the mind?'
'Who am I?'
'Can machines think?'

'Anyone who confronts these questions runs headlong into perplexities. We conceived this book as an attempt to reveal those perplexities and make them vivid . . . This book, then, is designed to provoke, disturb and befuddle its readers, to make the obvious strange and, perhaps, to make the strange obvious' – The Authors

Here is their selection of strange thought experiments, mind-stretching fantasies and humorous dialogues all designed to entice and tease the imagination into new and unexpected places. Like *Gödel, Escher, Bach* (published in Penguin) it is an extraordinary combination of science and art that will instruct and delight for years to come.

'A remarkable read' – *New Society*

Published in Pelicans

THE CHEMISTRY OF LIFE

SECOND EDITION

Steven Rose

Biochemistry has been at the centre of a vast expansion of biological understanding. Its methods and findings unify our knowledge of the mechanisms of life: biochemical theories approach questions of the origins of life on earth; biochemical techniques help to explain the interactions of viruses with their hosts, and to tailor-make the vital drugs in the pharmacologist's armoury.

And the pace of biochemical research shows no sign of slackening, from the determination of the structures of the complex molecules of life to the myriad of chemical reactions that drive the workings of the cell. Today, biochemists are beginning to discover where individual molecules are located within the cell, how cells of different organs and organisms resemble and differ, how a specialized biochemistry underlies the workings of brain and muscles, and how a coordinated series of reactions results in the development of the cell and the organism.

Written with his customary authority and clarity, this is a new and expanded edition of Professor Rose's well-known work which has been an Open University set text since 1971.

Published in Pelicans

TIME, SPACE AND THINGS

B. K. Ridley

'This is an attempt to survey, in simple terms, what physics has to say about the fundamental structure of the universe . . .

'It aims to extract, from a whole range of specialized activities, the basic essential concepts and to present them in plain non-mathematical language. There are some splendidly bizarre ideas in physics, and it seems a pity to keep them locked up in narrow boxes, available only to a small esoteric crowd of key holders.'

Dr Ridley deals here with all the basic concepts which physicists use to understand the universe – electro-magnetism, gravity, mass, energy, time, motion, space – and explains the techniques with which they manipulate those concepts.

'A clear presentation of difficult modern ideas in physics and [Dr Ridley's] own wonder and excitement in physics are well conveyed' – *School Science Review*

'Excellent' – *Physics Education*

'Clear and witty' – *Bookseller*

Published in Pelicans

THE NEW SCIENCE OF STRONG MATERIALS
or
WHY YOU DON'T FALL THROUGH THE FLOOR

J. E. Gordon

Why isn't wood weaker than it is? Why isn't steel stronger? Why does glass sometimes shatter and sometimes bend like a spring? Why do ships break in half? What is a liquid . . . and is treacle one?

All these are questions about the nature of materials. All of them are vital to engineers, but also intrinsically fascinating as scientific problems. During the two hundred and fifty years up to the 1920s and 1930s they had largely been answered by seeing how materials behaved in practice. But materials continued to do things which they 'ought' not to have done. Only in the last forty years have these questions begun to be answered by a new approach.

Now materials scientists, of whom Professor Gordon is one, have started to look more deeply into the make-up of materials – at the atoms and molecules on which their mechanical properties depend. They have found many surprises: above all, perhaps, that how a material behaves depends on how perfectly – or imperfectly – its atoms are arranged. If we could make a perfect crystal whisker of Epsom Salts, it would be far stronger than steel piano wire.

Now revised and updated – using both SI and imperial units – Professor Gordon's account of this fast developing science is a perfect demonstration of the sometimes curious and entertaining ways in which scientists isolate and solve problems.

Published in Pelicans

STRUCTURES

J. E. Gordon

For those who find difficulty in communicating with engineers, Professor Gordon is a godsend. His *The New Science of Strong Materials* made plain the secrets of materials science, and now with this volume he explains the importance and properties of different structures in a way which will appeal to everyone.

Engineers will of course understand why the Greeks took the wheels off their chariots at night, or why we get lumbago, why birds have feathers and how much science is involved in dressmaking as well as the strength of bridges, boats and aeroplanes. Professor Gordon explains all these things, showing how the need to be strong and to support various loads has influenced the development of all sorts of creatures and devices – including man. For this book is about modern views on the structural element in nature, technology and everyday life. It is up-to-date, enthralling and entertaining into the bargain.

'A splendid book ... written in entertaining and thought-provoking style ... [it] should be required reading for all of us whether aged 16 or 60' – A. G. Atkins

Published in Pelicans

THE LAW MACHINE

Marcel Berlins and Clare Dyer

Based on the LWT documentary series, *The Law Machine* aims to introduce the reader to and demystify the procedures, customs and controversial areas of the legal system. Written by two highly experienced journalists with legal backgrounds, it discusses how the system evolved, what happens if you find yourself entangled in its machinery, how lawyers work and – taking its lead from the television series which stages the most realistic trial ever seen on British television – vividly describes the trial process.

At every stage the authors have asked the question: is it as good as we have a right to expect? This critical approach adds zest to this readable – and unique – introduction to an enormous and fascinating subject.

Published in Pelicans

SUPERPOWERS IN COLLISION
THE NEW COLD WAR

Noam Chomsky, Jonathan Steele and John Gittings

As US–Soviet relations continue to deteriorate and as the arms race escalates, our outlook on world politics becomes easily confused and vulnerable to the superpowers' propaganda. Here three leading commentators cut through the myths to tackle the most basic questions, analysing the roles of the United States, the Soviet Union and China in the deepening crisis.

'If we hope to recover into our own hands the future of our own lives,' the authors stress, 'then the Dangerous Decade must become the Decade of Debate.' Informed and accessible, *Superpowers in Collision* is a vital starting-point for that debate.

Penguin Reference Books

THE PENGUIN DICTIONARY OF COMPUTERS

SECOND EDITION

Anthony Chandor with John Graham and Robin Williamson

The Penguin Dictionary of Computers has been designed to assist both technical readers and the increasing number of non-specialists whose work is to some extent affected by a computer. Exactly what computer men mean when they talk about 'graceful degradation' or 'garbage', about 'truth tables' or 'crippled leap-frog tests', about the 'flip-flop' or the 'output bus driver' is helpfully explained here in an alphabetical list of definitions with cross-references.

The entries are interspersed, too, with seventy general articles which cover, more fully, the major computer topics and such business processes as 'budgetary control' and 'systems analysis' which are more and more being handled by computers.

Penguin Reference Books

THE PENGUIN DICTIONARY OF PHYSICS

Edited by Valerie H. Pitt

The Penguin Dictionary of Physics is an abridgement, specially designed for a Penguin readership, of the latest thoroughly revised version of Gray & Isaacs' *A New Dictionary of Physics*. It includes information gleaned from the latest research, and it is intended to provide a concise accurate guide to the terminology of Physics and related disciplines. Longer entries on words of major importance go beyond mere definition to discuss their significance in relation to other words and ideas within the wider context of science. SI units are used throughout.

Penguin Reference Books

THE PENGUIN DICTIONARY OF SCIENCE

FIFTH EDITION: COMPLETELY REVISED AND ENLARGED

E. B. Uvarov, D. R. Chapman and Alan Isaacs

This latest edition of the Penguin *Dictionary of Science* has been completely revised and now gives all numerical information in SI UNITS, which are fully defined.

Many new words, like fluidics, holography, mascons, polywater, pulsars, and the key words in computer terminology are explained simply and accurately. Some 600 additional chemical compounds have been added, together with many new tables.

Apart from this new material, the reader will find reliable definitions and clear explanations of the basic terms used in astronomy, chemistry, mathematics, and physics, with a smattering of the words used in biochemistry, biophysics, and molecular biology. Notes are also included on all the chemical elements and most of their important compounds.

As a student's handbook and for the specialist outside his own field the Penguin *Dictionary of Science* remains an indispensable aid, and for the general reader it provides an up-to-date guide to the numerous scientific and technical words which are increasingly coming into daily life.

Penguin Reference Books

THE PENGUIN DICTIONARY OF BIOLOGY

SEVENTH REVISED EDITION

M. Abercrombie, C. J. Hickman and M. L. Johnson

In this dictionary the authors' aim is to explain biological terms which a layman may meet when reading scientific literature; to define the terms which a student of biology has to master at the beginning of his career – the thousand or so words which so grimly guard the approaches to the science; and to provide a reminder for the professional biologist reading outside his own narrow field. The entries are not restricted to a bare definition: some information about most of the things named is given, so as to convey something of their significance in biological discussion.

The authors have tried, as it were, to interpret a foreign language as it is actually used. It would be wrong to rely on etymology as a guide to correct usage. The meaning of a Greek root may be unequivocal, but biologists are not talking Greek: they are using a living language, and the proof of the meaning is in the speaking.